- **H. Andrew Michener**
  University of Wisconsin, Madison

- **John D. DeLamater**
  University of Wisconsin, Madison

- **Shalom H. Schwartz**
  Hebrew University of Jerusalem

- Under the general editorship of
  **Robert K. Merton**
  Columbia University

# Social Psychology
## Second Edition

HBJ

**Harcourt Brace Jovanovich, Publishers**
San Diego   New York   Chicago   Austin   Washington, D.C.
London   Sydney   Tokyo   Toronto

# Preface

**ABOUT THIS BOOK.** This new edition emerged in an evolutionary fashion and retains many of the distinctive characteristics of the first edition. Most important, the book still covers the whole range of phenomena that are of interest to social psychologists. While treating intrapsychic processes in detail, it provides strong coverage of social interaction and group processes and of large-scale phenomena such as intergroup conflict and social movements.

Our goal in writing this book has been to describe contemporary social psychology and to present the theoretical concepts and research findings that make up this broad field. We have drawn on the work of all kinds of social psychologists—those with sociological, psychological, and even anthropological perspectives. This book stresses the impact of social structure and group membership on the social behavior of individuals, but at the same time it covers the intrapsychic processes of cognition, attribution, and learning that underlie social behavior. Throughout the book we have used the results of empirical research—surveys, experiments, and observational studies—to illustrate these processes.

**NEW FEATURES IN THIS EDITION.** In developing this edition, we sought not only to keep the text abreast of changes within the field of social psychology but also to improve and strengthen the presentation of various topics. Thus, a number of new features are included in this edition. (1) This book has a new chapter on research methods (Chapter 2) that provides expanded and improved coverage of this important topic. (2) This edition offers expanded coverage of sex roles. A new chapter (Chapter 16) discusses sex roles and the influence of gender on behavior throughout life. (3) The sequence of chapters in the early part of the book has been reorganized to provide greater flexibility of use. (4) Coverage of major theories has been expanded and the text reworked to highlight important theoretical concepts. (5) Throughout the book, coverage of research studies has been updated to reflect advances in the field, and important new findings are presented throughout.

**CONTENT AND ORGANIZATION.** This edition opens with a chapter on theoretical perspectives in social psychology (Chapter 1) and a chapter on research methods (Chapter 2). These provide the groundwork for all that follows. The remainder of the book is divided into four substantive sections.

Section one focuses on individual social behavior and includes chapters on socialization (Chapter 3), self and identity (Chapter 4), social perception and attribution (Chapter 5), and attitudes (Chapter 6).

Section two is concerned with social interaction, the core of social psychology. Each of these chapters discusses human interaction. They cover such topics as communication (Chapter 7), social influence and persuasion (Chapter 8), self-presentation and impression management (Chapter 9), altruism and aggression (Chapter 10), and interpersonal attraction (Chapter 11).

Section three provides extensive coverage of groups. It includes chapters on group cohesiveness and conformity (Chapter 12), status processes in interaction (Chapter 13), group performance and leadership effectiveness (Chapter 14), and intergroup relations (Chapter 15).

Section four considers the relations between individuals and the wider society. These chapters treat the influence of life course and gender roles (Chapter 16), the impact of social structure on the individual (Chapter 17), deviant behavior (Chapter 18), and collective behavior and social movements (Chapter 19).

**EASE OF USE.** Because there are many different ways in which an instructor can organize an introductory course in social psychology, each chapter in this book has been written as a self-contained unit. Later chapters do not presume that the student has read earlier ones. This will enable instructors to assign chapters in whatever sequence they wish.

Chapters share a similar format. To make the material interesting and accessible to students, each chapter's introductory section poses four to six thought-provoking questions. These questions provide the issues to be discussed in the chapter. The remainder of the chapter consists of four to six major sections, each addressing one of these issues. A summary at the end of each chapter is similarly organized.

The text discussion is supplemented by a number of pedagogic aids. Tables are used to emphasize the results of important studies. Figures are used to illustrate important social psychological processes. Photographs dramatize essential ideas from the text. Boxes in each chapter highlight interesting or controversial issues and studies and discuss the applications of social psychological concepts to daily life. Key terms appear in boldface type and are listed alphabetically at the end of each chapter. A glossary of key terms is included at the end of the book.

**ACKNOWLEDGEMENTS.** Many of our colleagues reviewed one or more chapters of the book and provided useful comments and criticisms. We extend thanks to: Robert F. Bales, Harvard University; Philip W. Blumstein, University of Washington; Marilyn B. Brewer, University of California at Los Angeles; Bella DePaulo, University of Virginia; Glen Elder, Jr., University of North Carolina at Chapel Hill; Viktor Gecas, Washington State University; Christine Grella, University of California at Los Angeles; Allen Grimshaw, Indiana University; Elaine Hatfield, University of Hawaii —Manoa; George Homans, Harvard University; Michael Inbar, Hebrew University of Jerusalem; Dale Jaffe, University of Wisconsin–Milwaukee; Edward Jones, Princeton University; Lewis Killian, University of Massachusetts; Melvin Kohn, National Institute of Mental Health and Johns Hopkins University; Robert Krauss, Columbia University; Marianne LaFrance, Boston College; Steven Lybrand, University of Wisconsin–Madison; Patricia MacCorquodale, University of Arizona; Armand Mauss, Washington State University; Douglas Maynard, University of Wisconsin–Madison; William McBroom, University of Montana; John McCarthy, Catholic University of America; Kathleen McKinney, Illinois State University; Howard Nixon II, University of Vermont; Pamela Oliver, University of Wisconsin– Madison; James Orcutt, Florida State University; Daniel Perlman, University of Manitoba; Jane Allyn Piliavin, University of Wisconsin–Madison; Michael Ross, University of Waterloo, Ontario; Melvin Seeman, University of California at Los Angeles; Roberta Simmons, University of Minnesota; Sheldon Stryker, Indiana University; Robert Suchner, Northern Illinois University; James Tedeschi, State University of New York–Albany; Elizabeth Thomson, University of Wisconsin– Madison; Mark P. Zanna, University of Waterloo, Ontario; Morris Zelditch, Jr., Stanford University; Louis Zurcher, University of Texas.

Although this book has benefitted greatly from feedback and criticisms from colleagues, the authors accept responsibility for any mistakes that may remain.

We express thanks to the many students who used the first edition and who provided us with feedback about the book. We have used this feedback to improve the presentation, pace, and style of the new edition.

We extend thanks to Mary Rasmussen for her secretarial assistance in developing the manuscript for this edition.

We also express thanks to the many professionals at Harcourt Brace Jovanovich who contributed to the process of turning the manuscript into a book. Marcus Boggs, executive editor, has been a continuing source of support. Rick Roehrich, associate editor for the social sciences, worked directly with us in preparing the second edition. Martha Berlin, through her diligent work as manuscript editor, significantly improved the book's clarity and conciseness. Linda Miller and Stacy Simpson developed the design and artwork. Sarah Randall and Joan Harlan kept everything moving in the right direction. Our appreciation to them all.

Last, we express our gratitude to those who are close to us. They endured our absence when we were working long hours, listened to our complaints, provided helpful advice when asked, and shared our joy as we made progress on the book.

*H. Andrew Michener*
*John D. DeLamater*
*Shalom H. Schwartz*

# Contents
## in Brief

# Contents

# 11

# Interpersonal Attraction and Relationships

# 12

# Group Cohesiveness and Conformity

# 18
# Deviant Behavior and Social Reaction    520

# 19
# Collective Behavior and Social Movements    552

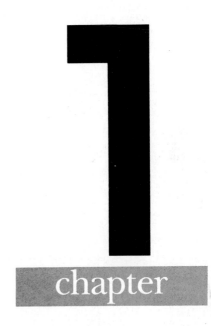

chapter

# Introduction to Social Psychology

## ■ Introduction

—Why are some persons effective leaders and others not?

—What makes people fall in love? What makes them fall out of love?

—Why is it so easy to achieve cooperation between persons in some instances but so difficult in others?

—What effects do major life events like getting married, having a child, or losing one's job have on physical health, mental health, and self-esteem?

—What causes conflict between groups? Why do some conflicts persist far beyond the point where participants can expect to achieve any real gains?

—Why do some people conform to norms and laws, whereas others violate them?

—Why do people present different images of themselves in various situations? What determines the particular images they present?

—What causes harmful or aggressive behavior? What causes helpful or altruistic behavior?

—Why are some groups so much better at doing their work than others?

—What causes people to develop unique conceptions of themselves? How do these self-concepts change?

—Why are some persons more persuasive and influential than others? What techniques do they use?

—Why do stereotypes persist even in the face of information that obviously contradicts them?

Questions such as these have probably puzzled you, just as they have perplexed others down through the ages. You might think about these questions merely because you want to understand better the social world around you. Or you might want the answers for more practical reasons, such as increasing your own effectiveness and influence in day-to-day social situations.

Answers to questions such as these come from various sources. Personal experience is one such source. Answers obtained by this means are often insightful, but they are usually limited in scope and generality. Occasionally they are even misleading. Another source is informal knowledge or advice from others. Again, answers obtained by this means are sometimes reliable, sometimes not. A third source is thinkers of various orientations—philosophers, novelists, poets, and men and women of practical affairs—who, over the centuries, have written about these issues. To a remarkable degree, their answers have filtered down and are available today in the form of sayings or aphorisms that comprise commonsense knowledge. Common sense covers a wide range of topics. We are told, for instance, that punishment is essential to successful child rearing ("Spare the rod and spoil the child") and that joint effort is an effective way to accomplish large jobs ("Many hands make light work"). Principles such as these may reflect certain truths, and they appear to provide guidelines for action.

Although commonsense knowledge has certain merits, it also presents some difficulties, not the least of which is that it often contradicts itself. For example, we are told that persons who are similar will like one another ("Birds of a feather flock together") but also that persons who are dissimilar will like each other ("Opposites attract"). We learn that groups are wiser and smarter than individuals ("Two heads are better than one") but also that problem solving by groups entails many compromises and inevitably produces mediocre results ("A camel is a racehorse designed by a committee"). Each of these seemingly contradictory statements may hold true under particular conditions. But without a clear statement of when they apply and when they do not, aphorisms ultimately provide little insight regarding relations among people. They provide even less guidance in situations where we need to make decisions. For example, when facing a choice that entails risk, what should we believe—"Nothing ventured, nothing gained" or "Better safe than sorry"?

If sources such as personal experience and commonsense knowledge have limited value, how are we to achieve an understanding of social interaction and relations among persons? Are we forever restricted to intuition and speculation, or is there a better alternative?

One answer to this problem is that offered by *social psychology*. Social psychologists propose that accurate and comprehensive knowledge about social behavior can be obtained by applying the methods of science to these issues. That is, by taking systematic observations of behavior and formulating theories that are subject to test and disconfirmation, we can attain a valid understanding of human social relations.

One objective of this book is to summarize many of the facts discovered through systematic social psychological research. To set the stage for this presentation, this chapter will first describe the core concerns of social psychology and provide a formal definition of the field. This definition will indicate in general terms what social psychology does and does not include. Second, we will summarize some of the broad theories (called theoretical perspectives) within social psychology today. These theories recur throughout this book and provide useful frameworks for understanding relations among people. Third, we will consider the question "Is social psychology a science?" To answer it, we will review the properties that characterize any science and then examine social psychology in light of them.

## ■ What Is Social Psychology?

### Core Concerns of Social Psychology

There are several ways to characterize social psychology. Perhaps the most direct approach is to describe the things social psychologists actually study. Social psychologists investigate human behavior, of course, but their primary concern is human behavior in a social context. There are four *core concerns* or major themes within social psychology. These are (1) the impact that one individual has on another individual; (2) the impact that a group has on its individual members; (3) the impact that individual members have on the group; and (4) the impact that one group has on another group. The four core concerns are shown schematically in Figure 1-1.

The first concern, the effect of one individual on another, is perhaps the most researched topic within social psychology. When the behavior of one person produces an immediate change in the behavior of a second person, we speak of *social influence.* For example, a bank robber brandishing his pistol is obviously exercising influence over the teller's behavior when he issues the threat "Put the money in this sack. Don't hit the alarm button. Do as I say and you won't get hurt." In other situations, the effects of one person on another's behavior can be more subtle. Merely staring at a stranger across a room, for example, may make that person uncomfortable enough to change seats or leave the room.

Beyond influencing others' overt behavior, one person may also affect another's psychological state. By this we mean that one person may affect another's *perceptions* and *cognitions* about the world, as well as his or her *attitudes.* These attitudes, of course, may pertain to objects or people. Interpersonal attitudes such as liking, disliking, loving, and hating hold the interest of social psychologists, as do the behaviors they spawn (helping, hurting, and so on). Under certain circumstances, one person may try intentionally to change another's beliefs and attitudes, a process referred to as *persuasion.* For example, Carol might try to persuade Debbie that nuclear power plants are dangerous, undesirable, and therefore should be closed. Carol's persuasion attempt, if successful, would likely be reflected in Debbie's future actions (such as picketing nuclear power plants) or in her verbal behavior (advocating nonnuclear power). Social psychologists are interested in the conditions under which such changes in personal beliefs and attitudes occur.

The second concern of social psychology is the impact of a group on the behavior of its individual members. Individuals typically belong to many different groups—families, work groups, seminars, and clubs. Consequently, they

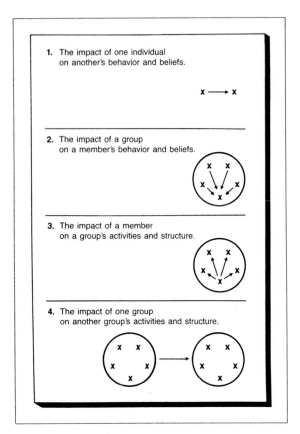

■ FIGURE 1-1

THE CORE CONCERNS OF SOCIAL PSYCHOLOGY

spend many hours interacting with others in group contexts. To regulate the behavior of their members, groups usually establish norms or rules. One consequence of social interaction in groups is *conformity,* the process by which a group member adjusts his or her behavior to bring it into line with group norms. For example, college fraternities and sororities have norms— some formal and some informal—that stipulate how members should dress, what meetings they should attend, whom they can date and whom they should avoid, how they should behave at parties, and so on.

Groups also exert substantial, long-term influence on their members through a process called *socialization.* Simply stated, this means that groups regulate what their members learn. They do this so that the members will be adequately trained to enact the roles they play in society. Socialization shapes the knowledge, values, and skills of group members. One product of socialization is language skills. Italian children are taught Italian, not Greek or Japanese, to prepare them for life in Italy. Likewise, students in medical school are taught thousands of technical terms—virtually an entire new language—to prepare them for professional responsibilities. Another product of socialization is a person's religious and political beliefs and attitudes. Socialization affects not only members' values and attitudes on a wide variety of issues, but also molds their *self-concept*. This in turn allows members to regulate their own behavior and to participate effectively in group activities.

The third concern of social psychology is the impact of individual members on the activities and structure of the groups to which they belong. Just as any group will influence the behavior of its members, these persons in turn may influence or redirect the group itself. One way of doing this is by *innovation,* wherein an individual or an active minority within a group introduces changes in norms or culture that are accepted by the group's majority. Another way of doing this is by *leadership,* which involves the enactment of several functions (planning, organizing, controlling) necessary for successful group performance. Without effective leadership, coordination among members will be lacking, and the group will drift or fail. Both leadership and innovation, of course, depend on the initiative, insight, and risk-taking propensity of individuals. Through leadership and innovation, individual members can exert a substantial impact on the group.

The fourth concern of social psychology is the impact of one group on the activities and structure of another group. Relations between two groups may be friendly or hostile, cooperative or competitive. This relationship can affect the structure and activities of each. Of special interest is *intergroup conflict,* with its accompanying tension and hostility. Violence may flare up, for instance, between two teenage street gangs disputing territorial rights or between racial groups competing for scarce jobs. Conflicts of this type affect the interpersonal relations between groups as well as within each group. Social psychologists have long been interested in the emergence, persistence, and resolution of intergroup conflict.

## A Formal Definition of Social Psychology

As indicated in the above discussion, social psychology is a broad field covering many topics. In essence, **social psychology** may be defined as the systematic study of the nature and causes of human social behavior.

Note certain features of this definition. First, it indicates that the main concern of social psychology is human social behavior. This includes many things—the activities of individuals in the context of others, the processes of social interaction between two or more persons, and even the relationships between individuals and the groups of which they are part.

Social psychology is concerned not only with the nature of social behavior, but also with the causes of such behavior. That is, it seeks to discover the conditions that produce any given social behavior. At its simplest, a causal statement takes the form "Whenever X occurs, then Y occurs." In social psychology, however, few causal statements are this elementary or unconditional. More typical are qualified statements like "Whenever conditions A, B, C, and D are present and X occurs, then Y occurs." Such statements obviously are more restricted in generality and explanatory power. Causal statements are the important building blocks of theory, and theory in turn is crucial for the prediction and control of social behavior.

Finally, the definition indicates that social psychologists approach the study of social behavior in a systematic fashion. They rely explicitly on research methodology, which is scientific in character and includes formal procedures such as experimentation, structured observation, and sample surveys to test theories of social behavior. A more detailed description of the research methodology used by social psychologists appears in Chapter 2.

## Relation to Other Fields

Social psychology bears a close relationship to several other fields, especially sociology and psychology. To understand this relationship, first consider these other fields.

*Sociology* is the scientific study of human society. It includes topics such as social institutions (family, religion, and politics), stratification within society (class structure, race and ethnicity, and sex roles), basic social processes (socialization, interaction, and social control), and the structure of social units (groups, networks, formal organizations, and bureaucracies).

In contrast, *psychology* is the scientific study of the individual and individual behavior. This behavior may be social in character, but it need not be. Psychology includes such topics as human learning, perception, memory, intelligence, emotion, motivation, and personality.

*Social psychology* bridges the gap between sociology and psychology. In fact, it is sometimes viewed as an interdisciplinary field. Both sociologists and psychologists have contributed to social psychological knowledge. Social psychologists working in the sociological tradition rely primarily on sample surveys and other systematic observational techniques to gather data. The subject that interests them most is the relationships between individuals and the groups to which they belong. This leads to an emphasis on such processes as socialization, conformity and deviation, social interaction, leadership, recruitment to membership, cooperation and competition, and the like. Social psychologists working in the psychological tradition rely heavily on laboratory experimental methodology. Their primary concern is how an individual's behavior and internal states are affected by social stimuli (often other persons). This includes such topics as person perception and attribution, attitudes and atti-

tude change, personality differences in social behavior, social learning and modeling, altruism and aggression, and so on.

Thus, sociologically oriented and psychologically oriented social psychologists differ in their outlooks and concerns. As might be expected, this leads them to formulate different theories and to conduct different research programs. Yet in the end, these differences are best viewed as complementary rather than as conflicting. Social psychology as a field is richer for them.

# ■ Theoretical Perspectives in Social Psychology

Yesterday was a bad day for Warren. First thing in the morning, he went to work and told his boss that he would not be able to complete his project on schedule. The boss got furious, screamed at Warren, and told him to complete the work by Monday—or else! Warren was not entirely sure what to make of the older man's behavior, but he decided to take the threat seriously. That evening, talking with his girlfriend Alice, Warren announced that he could not take her out to a party on Friday evening as originally planned because he had to work overtime at the office to complete the project. Alice immediately got mad at Warren and threw a frying pan at him. By now, Warren was very upset.

Reflecting on these events, Warren noticed that they had some features in common. To explain the behavior of his boss and his girlfriend, he formed a general proposition: "If you fail to deliver on promises and block someone's goals, he or she will get mad at you." He was happy with this theory until he read a strange newspaper story: "MAN IS FIRED FROM JOB, THEN SHOOTS HIS DOG IN ANGER." Warren wondered about this event and then concluded that his own theory needed revision. The new version included several propositions: "If someone's goals are blocked, he or she will become frustrated. If someone is frustrated, he or she will become aggressive. If someone is aggressive, he or she will attack either the source of the frustration or a convenient surrogate."

On an informal basis, Warren is starting to do the same thing that social psychologists do more systematically. Working from some observed facts regarding social behavior, Warren is attempting to formulate a theory to explain what he observed.

Defined in general terms, a **theory** is a set of interrelated propositions that organizes and explains a set of observed phenomena. Theories are not meant to be about particular events but about whole classes of events. Moreover, as the example of Warren indicates, a theory goes beyond mere observable facts, because it postulates (causal) relations between concepts. If a theory is valid, it enables its user to explain the phenomena under consideration and to make predictions regarding events not yet observed.

Theory is a crucial element in social psychology. The field includes not just one theory, but a large number of them. Some social psychological theories are scientific-causal in nature; that is, they are formulated in terms of cause and effect. Typically, these causal theories are narrow and tightly focused, and they seek to explain the conditions that give rise to some specific form of social behavior. For example, one such theory attempts to explain how people will react if they are treated unfairly or inequitably. Another theory specifies the conditions under which contact between members of different racial groups will cause stereotypes to fade or disappear. Theories of this type are presented in various places throughout this book.

Social psychology also includes another class of theories—usually termed **theoretical perspectives**—that are much broader in scope. Theories of this type are not always causal in character, but they make sweeping assumptions about human nature and offer general explanations of diverse social behaviors in a wide variety of situations.

Theoretical perspectives have a particular value. By adopting specific assumptions regarding human nature, a theoretical perspective provides a point of view from which we can examine a wide range of social behaviors. Any theoretical perspective provides a sharp focus on certain features of social reality. By highlighting certain features and downplaying others, a theoretical perspective enables us to more clearly "see" characteristics of a given social situation. The fundamental value of any theoretical perspective lies in its applicability across many situations; it provides a frame of reference in terms of which we can observe and interpret a wide range of social situations and behaviors.

Like many fields of study, social psychology includes several distinct theoretical perspectives. Four of the more important ones are (1) role theory, (2) reinforcement theory, (3) cognitive theory, and (4) symbolic interactionist theory. Each of these perspectives will be discussed here.

### Role Theory

Several months ago, Barbara was asked to participate in a stage production of Molière's comedy *The Learned Woman*. She was offered the role of Martine, a kitchen servant who is dismissed from her job for using poor grammar. Barbara enthusiastically accepted the offer and learned her part well. The theater group presented the play six times over a period of three weeks. Barbara played the role of Martine in the first four shows, but then she got sick. Fortunately, another woman (Barbara's understudy) was able to substitute as Martine during the final two shows. Barbara's performance was very good, but so was the understudy's. In fact, one critic wrote that it was difficult to tell them apart.

Barbara's friend, Craig, is more interested in football than in theater. A member of the college football team, Craig plays the position of fullback. Although very large and strong, he is a third-string player. The coach has sent him in to play in several games, but he still makes mistakes. He has the unfortunate habit of fumbling the ball, usually at the worst possible moment. Craig believes, however, that with another year's experience and some improvements in his technique, he will be able to perform better than the other fullbacks and win a place in the team's starting lineup.

Although active in different arenas, Craig and Barbara have something in common; they are both performing roles. When Barbara appears on stage, she performs the role of kitchen servant. When Craig appears on the football field, he performs the role of fullback. In both cases, their behavior is guided by a set of culturally specified plans or blueprints. Barbara's role is very specific. Her part calls for her to say certain things and perform certain actions at specified points in the plot. There is little room for her to improvise or deviate from her lines. Craig's role is also quite specific. He has to carry out certain assignments on each of the plays run by his team. Although he still makes mistakes, he tries hard to perform his role.

In everyday life, we all perform roles. Anyone who holds a job can be viewed as performing a role. For example, an advertising executive's job description does not dictate exactly what lines are to be spoken, but it will certainly specify what goals must be accomplished and what performances are required to attain these goals.

The perspective of **role theory** (Turner, 1978; Biddle, 1979; Heiss, 1981; Winnubst & Ter Heine, 1985; Biddle, 1986) is based on a theatrical metaphor. It holds that a substantial proportion of observable, day-to-day social behavior is simply persons carrying out their roles, much as actors carry out their roles on the stage or ball players theirs on the field. The following propositions are central to the role theory perspective.

1. People do not usually live in social isolation; instead, they spend much of their lives participating as members of groups and organizations.

2. Within these groups people occupy distinct positions (kitchen servant, fullback, advertising executive, and so on).

3. Each of these positions entails a **role,** which is a set of functions performed by the person on behalf of the group. A person's role is defined by how other group members expect him or her to perform.

4. These expectations are formalized as **norms,** which are rules specifying how a person should and should not behave.

5. In most cases, individuals perform in accordance with prevailing norms. In other words, people are conformists; they try to meet the expectations held by others.

6. Group members evaluate an individual's performance to determine whether it is aligned with the norms, and they respond accordingly. If the individual meets the role expectations held by others, then he or she will be rewarded. If, on the other hand, he or she fails to perform as expected, then group members will embarrass, punish, or even expel that person from the group. The anticipation that sanctions will be applied by others helps to ensure that persons perform as expected.

Role theory implies that when we are given information about the role expectations for a specified position, we should be able to predict a significant portion of the behavior of the person occupying that position. Moreover, role theory implies that to change a person's behavior, it is necessary either to modify the expectations that define the person's role or to shift the person into an entirely different role. For example, if the football coach shifted Craig from fullback to tackle, Craig's behavior would quickly change to match the role expectations of his new position.

Role theory maintains that a person's role determines not only behavior but also beliefs and attitudes. In other words, individuals bring their attitudes into congruence with the expecta-

On the assembly line, these men perform tasks and enact roles specified by their work group.

tions that define their roles. A change in role would lead to a change in attitude. One illustration of this appears in a classic study of factory workers by Lieberman (1965). In the initial stage of this study, researchers measured the attitudes of workers in a midwestern home-appliance factory toward union and management policies. During the following year, a number of these workers changed roles. Some were promoted to the position of foreman, a managerial role. Others were elected to the position of shop steward, a union role.

About a year after the initial measurement, workers' attitudes were reassessed. The attitudes of persons who had become foremen or shop stewards were compared to those of workers who had not changed roles. The recently promoted foremen expressed more positive attitudes than the nonchangers toward the company's management and the company's incentive system (which paid workers in proportion to what they produced). Similarly, recently elected shop stewards expressed more positive attitudes than the nonchangers toward the union and favored an incentive system based on seniority, not productivity. In effect, the workers' attitudes shifted to fit their new roles, as predicted by role theory.

Thus, in general, the roles that people occupy not only channel their behavior but also shape their attitudes. Roles can influence the values that people hold and even affect the direction of their personal growth and development. These topics are discussed in more depth in Chapters 3, 13, and 17.

Despite its usefulness, role theory runs into problems with respect to certain kinds of social behavior. Foremost among these is *deviant behavior,* which is any behavior that violates or contravenes the norms defining a given role. Virtually all forms of deviant behavior, whether simply a refusal to perform as expected or something more serious like commission of a crime, are disruptive to ongoing interpersonal relations. The occurrence of deviant behavior poses a challenge to role theory, because it flies in the face of the assumption that people conform to norms. Of course, a certain amount of deviant behavior can be explained by the fact that people are sometimes ignorant of the norms or may face conflicting and incompatible expectations from several other people (Miles, 1977). In general, however, deviant behavior is an unexplained and problematic exception from the standpoint of role theory. Chapters 12 and 18 discuss the conditions that give rise to deviant behavior, as well as the reactions of others to such behavior.

Even harsh critics of role theory acknowledge that a substantial portion of all social behavior can be explained in terms of conformity to established role expectations. But role theory does not and cannot explain how role expectations come to be what they are in the first place. Nor does it explain when and how role expectations change. Without accomplishing these tasks, role theory can provide no more than what is ultimately an incomplete explanation of social behavior.

## Reinforcement Theory

**Reinforcement theory,** a second major perspective on social behavior, is based on the premise that social behavior is governed by external events, especially rewards and punishments. The central insight is that people will tend to perform a particular behavior if it is followed by something pleasurable and need-satisfying, and they will tend to avoid behavior that is followed by something unpleasant. The pleasant (unpleasant) stimulus that follows the behavior and that makes the response more (less) likely is termed a positive (negative) **reinforcer.**

One illustration of the use of reinforcement comes from a classic study by Verplanck (1955). The point of this study was to show that the course of conversation between persons can be altered by the use of social approval (a reinforcer). Students conducting this study sought out situations in which each could be alone with another person and conduct a conversation. During the first 10 minutes, the student engaged the other in polite but neutral chitchat; the student was careful neither to reject nor to support opinions expressed by the other. During this period the student privately noted the number of opinions expressed by the other and recorded this information unobtrusively by doodling on a piece of paper.

After this initial period, the student shifted behavior and indicated approval whenever the other ventured an opinion. The student expressed approval by saying such things as "I agree," "That's so," or "You're right" or by smiling and nodding in agreement. The student continued this for 10 minutes, all the while noting the number of opinions mentioned by the other.

Next, the student shifted behavior again and suspended positive reinforcement. Any opinions expressed by the other were met with noncommittal remarks or subtle disagreement. As before, the number of opinions was recorded.

The results of the study are clear-cut. First, during the "reward period" (when the student expressed approval), all persons expressed opinions at a higher rate than they had during the initial period. Then, during the "extinction period" (when the student suspended approval), about 90 percent reduced the rate at which they expressed opinions. Overall, their behavior was substantially influenced by social approval.

Reinforcement theory has a long tradition, beginning at the turn of the century with research by Pavlov and by Thorndike and evolving through the work of Allport (1924), Hull (1943), and Skinner (1953, 1971). The reinforcement perspective holds that social behavior is determined primarily by external events, not by internal psychological states. Thus, the central concepts of reinforcement theory refer to events that are directly observable. Any event that leads to an alteration or change in behavior is called a *stimulus*. For example, a traffic light that changes to red is a stimulus, as is a wailing tornado siren. The change in behavior induced by a stimulus is called a *response*. Drivers respond to red lights by stopping; families respond to tornado sirens by rushing to their basements for shelter. Anything that strengthens the response is a reinforcer. In Verplanck's study, the students' social approval was a positive reinforcer that strengthened the response of expressing opinions.

Reinforcement theory has been used to explain many processes of interest in social psychology. Reinforcement theory has made especially important contributions in two areas—social learning and social exchange.

**Social Learning Theory**   The basic proposition of **social learning theory** is that a person learns (that is, acquires new responses) through the application of reinforcement. The name given to this process is **conditioning.** Specifically, if a person performs a particular response, and if this response is then reinforced, the response is strengthened. The person will be more likely to emit the same response again. For example, if a young child, Karl, helps his father rake the leaves, and if his father reinforces this behavior with approval or money, then Karl will be increasingly inclined to help with yard work in the future.

A closely related process, *discrimination learning,* occurs when a person comes to identify the exact conditions under which a response will be reinforced. For example, Karl has learned that if his mother rings the dinner bell (a stimulus), he should respond by coming indoors, washing his hands, and sitting in the appropriate place at the table. His mother will then put food

on his plate (a reinforcer). He has also learned, however, that if he performs the response (washing his hands and sitting down) without first hearing the stimulus (dinner bell), his mother will merely tell him that he's too early and cannot have food until later. He has further discovered that a partial response (sitting down at the table without first washing) is met not with food but with reprimands to wash up. Thus, through discrimination learning, Karl understands that reinforcement (food) is obtained only by making the full response in the presence of the bell stimulus.

Although conditioning is the most fundamental mechanism of learning, people frequently acquire new responses through the process of **imitation** (Bandura, 1977). Imitation differs from conditioning in that the learner neither performs a response nor receives any reinforcement. Instead, the learner watches another person's response and observes whether that person receives any reinforcement.

For example, a young girl might observe that her older sister puts on makeup before going out with friends and that her sister is rewarded (with social approval) for doing this. Noting the connection between the response (applying makeup) and the subsequent reinforcement from others, she might infer that if she behaves like her sister, she will receive reinforcement herself. Thus, she may mimic or copy her sister's behavior and apply cosmetics.

In sum, social learning theory holds that persons acquire new responses both through conditioning and through imitation. In this way, the theory can explain how persons acquire complex responses such as those learned during socialization. Chapter 3 discusses this in more detail.

**Social Exchange Theory**   Another theory based on the principle of reinforcement is the **social exchange theory** (Homans, 1961, 1974; Kelley & Thibaut, 1978; Emerson, 1981; Cook, 1987). Social exchange theory uses the concept of

A form of exchange takes place as a hustler scalps football tickets to a waiting father. In transactions of this type, the price is often determined through negotiation between buyer and seller.

reinforcement to explain stability and change in interaction between individuals. It assumes that individuals have freedom of choice and often face social situations in which they must choose among alternative actions. Any action provides some rewards and entails some costs. Rewards can assume many forms; they include not only money, goods, and services, but also prestige, status, and approval by others. The theory posits that individuals are hedonistic—they try to maximize rewards and minimize costs. Consequently, they will choose whatever actions produce the best outcomes (outcomes = rewards − costs).

Social exchange theory maintains that people will establish stable relationships only if they find it profitable to continue their exchange of goods and/or services. An individual judges the attractiveness of a relationship by comparing the outcomes it provides against the outcomes available in other, alternative relationships. The outcome available in the best alternative relationship is termed the individual's *comparison level for alternatives*. To illustrate this concept, suppose that an engineer employed by an aircraft manufacturer is offered a highly attractive, profitable job by a competing firm. The new job entails some additional responsibilities, but it also pays a much higher salary and provides more benefits.

This job offer has the effect of substantially increasing the engineer's comparison level for alternatives. In this case, exchange theory predicts that the engineer either will leave her job for the new one or will stay with her current employer if she is promoted to a new position with higher rewards (more salary, benefits, and so on).

These ideas apply not only to work relations but also to personal relations. For instance, a recent study of heterosexual couples in long-term dating relationships shows that rewards and costs can explain whether persons stay in or leave such relationships (Rusbult, 1983; Rusbult, Johnson, & Morrow, 1986). Results of this study indicate that persons are more likely to stay when their partner is physically and personally attractive, when the relationship does not entail undue hassle (high monetary costs, broken promises, arguments), and when romantic involvements with attractive outsiders are not readily available. In other words, they are more likely to stay when the rewards are high, the costs are low, and the comparison level for alternatives is low. Findings of this type are predicted by social exchange theory.

Exchange theory also predicts the conditions under which persons will try to change or restructure their relationships. Central to this is the concept of **equity** (Adams, 1963; Walster, Walster, & Berscheid, 1978). Equity prevails in a relationship when people feel that the rewards they each receive are proportional to the costs they each bear. For example, a supervisor may earn more money than a line worker and receive better benefits on the job. But the worker may still feel the relationship is equitable because the supervisor bears more responsibility and has a higher level of education.

When equity is absent—that is, when people feel their relationship is inequitable—the relationship is potentially unstable. Persons find inequity difficult to tolerate—they feel cheated or exploited and become angry. Social exchange theory predicts that people will try to modify an inequitable relationship. Most likely, they will attempt reallocation of the costs and rewards so that equity is established.

Despite its usefulness in illuminating why relationships change and how people learn, reinforcement theory has been criticized on various grounds. One criticism is that reinforcement theory portrays persons primarily as reacting to environmental stimuli, rather than as initiating behavior based on imaginative or creative thought. Thus, the theory does not account for creativity, innovation, or invention. A second criticism is that reinforcement theory largely ignores or downplays other motivations. It characterizes social behavior as hedonistic, with persons striving to obtain rewards or maximize outcomes. Thus, it cannot easily explain selfless behavior such as altruism and martyrdom. Despite its limitations, reinforcement theory has enjoyed substantial success in explaining how persons learn new behaviors and influence the behavior of others. Ideas based on reinforcement and exchange theory are discussed throughout this book, especially in Chapters 3, 8, 10, 11, and 13.

## Cognitive Theory

The basic premise of **cognitive theory** is that the mental activities of the individual are important determinants of social behavior. These mental activities, called **cognitive processes,** include perception, memory, and judgment, as well as problem solving and decision making. Cognitive theory does not deny the importance of external stimuli, but it maintains that the link between stimulus and response is not automatic or mechanical. Instead, the individual's cognitive processes intervene between external stimuli and behavioral responses. According to cognitive theory, the mind not only interprets the meaning of stimuli but also selects the actions to be made in response to stimuli.

The cognitive approach to social psychology has been influenced by the ideas of Koffka, Kohler, and other theorists in the *gestalt* movement within psychology. Central to gestalt psychology is the principle that people respond to configurations of stimuli rather than to a single, discrete stimulus. In other words, people understand the meaning of a stimulus only by viewing it in the context of an entire system of elements in which it is embedded. A chess master, for example, would never assess the importance of a chess piece on the board without considering its location and strategic capabilities vis-à-vis all the other pieces on the board. To comprehend any element, one must look at the whole of which it is a part.

Cognitive theorists (Cantor & Kihlstrom, 1981; Hastorf & Isen, 1982; Fiske & Taylor, 1984; Wyer & Srull, 1984; Markus & Zajonc, 1985) depict humans as active in selecting and interpreting stimuli. According to this view, people do more than react to their environment; they actively structure their world. First, because they cannot possibly attend to all the complex stimuli that surround them, they select only those stimuli that are important or useful to them. Other stimuli are ignored. Second, they actively control the categories or concepts they will use to interpret the stimuli in the environment. An implication of this, of course, is that several individuals can form dramatically different impressions of a complex stimulus in the environment.

Consider, for example, what happens when several people view a vacant house displaying a bright "for rent" sign. When a building contractor passes the house, he scrutinizes the construction closely. He sees lumber, bricks, shingles, glass—and some repairs that should be made. Another person, a potential renter, sees the house very differently. She notes that it is located close to her job and wonders whether the neighborhood is safe and whether the house is expensive to heat in winter. The realtor who is trying to rent the house construes it in still different terms—cash flow, occupancy rate, depreciation, mortgage, and amortization. One of the young kids living in the neighborhood has yet another view; observing that no person has lived in the house for several months, he is convinced the house is haunted.

Individuals view an object such as a diamond in different ways. A jeweler sees it as a marketable piece of hard carbon with various attributes—color, clarity, cut, and carat. A recently engaged woman sees it as a symbol of enduring love.

**Cognitive Structure and Schemas** Central to this perspective is the concept of **cognitive structure,** which refers broadly to any form of organization among cognitions. Because a person's cognitions (that is, concepts and beliefs) are interrelated, cognitive theory gives special emphasis to exactly how they are structured and organized in memory, as well as to how they affect a person's judgments.

Social psychologists have proposed that individuals use specific cognitive structures called **schemas** to make sense of complex information about other persons, groups, and situations. The term "schema" is derived from the Greek word for "form," and it refers to the form or basic sketch of what we know about people and things. For example, our schema for "law student" might be a set of traits thought to be characteristic of such persons: intelligent, analytic and logical, argumentative (perhaps even combative), thorough and workmanlike with an eagle eye for details, able to learn quickly, skillful in interpersonal relations, and (sometimes) committed to seeing justice done. Our schema, no doubt,

reflects our own experience with lawyers and law students, as well as our conception of what traits are necessary for success in the legal profession. That we hold this schema does not mean we believe that everyone with this set of characteristics is a law student or that every law student will have all of these characteristics. We might be surprised, however, if we met someone who impressed us as unmethodical, illogical, withdrawn, inarticulate, inattentive, sloppy, and not very intelligent and then later discovered that she was a law student.

Schemas are important in social interaction because they help us to interpret the environment efficiently. Whenever we encounter a new person, we do not try to interpret him or her entirely from scratch. Instead, we rely on our knowledge of similar persons we have met from the past—that is, we use our schemas for persons of a similar type. Schemas help us process information by enabling us to recognize which personal characteristics are important in the interaction and which are not. They structure and organize our information about the person, and they help us remember information better and process it more quickly. Sometimes they fill gaps in knowledge and enable us to make necessary inferences and judgments regarding others.

For example, a law school admissions officer (who has the job of deciding which of many candidates to admit as students) may use a schema in processing applications. His schema for "law student" will likely be formulated in terms of traits thought to be typical of persons who succeed in law school and beyond. The admissions officer doubtless will pay close attention to information regarding candidates that is relevant to his schema for law students, and he will most likely ignore other information. Eye color doesn't matter, LSAT scores do. Ability to throw a football doesn't matter, undergraduate GPA does.

Schemas are rarely perfect as predictive devices, and the admissions officer will probably make some mistakes (that is, admit some candidates who will fail to complete law school as well as turn down some candidates who would have succeeded). Then, too, another admissions officer with a different schema might admit a different set of students to law school. Despite their drawbacks, schemas are often superior to using no systematic framework at all. They are discussed in more detail in Chapter 5.

**Cognitive Consistency** One useful way to study cognitive structure is to investigate the changes that occur in cognitions when a person's beliefs are under attack or strain. The changes that result will reveal facts about their underlying structure or organization. An important idea emerging from this approach is the **principle of consistency** (Heider, 1958; Newcomb, 1968). This principle maintains that if a person holds several ideas that are incongruous or inconsistent with each other, an internal conflict will be experienced and one or more of the ideas subsequently will be changed to render them consistent. Only by changing ideas and making them consistent will the conflict be resolved.

As an illustration, suppose you hold the following cognitions about your friend Jeff: (1) Jeff has been a good friend for six years; (2) you dislike hard drugs and the people who use them; and (3) Jeff has recently started using hard drugs. These cognitions are obviously interrelated, and they are also incongruous with one another. The principle of consistency predicts that some change in cognitions will likely occur. That is, you will change either your negative attitude toward drugs or your positive attitude toward Jeff, or possibly you will intervene and try to change Jeff's behavior. Social psychologists have developed several useful theories based on the general notion of consistency. Among these are *balance theory* and the *theory of cognitive dissonance,* which are discussed in Chapter 6.

Cognitive theory has made many important contributions to social psychology. It treats such diverse phenomena as self-concept (Chapter 4), perception of persons and attribution of causes

(Chapter 5), attitude change (Chapter 6), impression management (Chapter 9), and group stereotypes (Chapters 5 and 15). In these contexts, this perspective has produced many insights and striking predictions regarding individual and social behavior.

One drawback of cognitive theory is that, while dealing with phenomena that are highly complex, it necessarily simplifies—and sometimes oversimplifies—the ways in which people process information. Another drawback is that cognitive phenomena are not directly observable (that is, they are located "in the head" of the individual and must be inferred from overt behavior); this means that compelling and definitive tests of theoretic predictions are often difficult to conduct. Overall, however, the cognitive perspective is among the more popular and productive approaches within social psychology.

### Symbolic Interactionist Theory

A fourth perspective in social psychology is **symbolic interactionist theory** (Mead, 1934; Blumer, 1969; Morrione, 1975; Stryker, 1980, 1987). Like the cognitive perspective, symbolic interactionism stresses cognitive process (thinking, reasoning, planning), but places more emphasis on interaction between persons.

The basic premise of symbolic interactionism is that human nature and social order are products of communication among people. In this perspective, a person's behavior is constructed through give and take during interaction with others. Thus, behavior is not merely a response to stimuli; nor is it an expression of inner biological drives, profit maximization, or conformity to norms. Rather, a person's behavior emerges through communication and interaction with others.

People succeed in communicating with one another only to the extent that they ascribe similar **meanings** to objects. The meanings of an object to a person depends not so much on the properties of the object itself but on what the person might do with the object. In other words, an object takes on meaning only in relation to a person's plans. A wine merchant, for example, might see a glass bottle as a container for his product; an interior decorator might see it as an attractive vase for some paper flowers; a man in a drunken brawl might see it as a weapon with which to hit his opponent.

Symbolic interactionist theory views humans as goal seeking and proactive. Persons formulate plans of action to achieve their goals. Many plans, of course, can be brought to realization only through cooperation with other people. To establish cooperation with others, meanings of things must be shared and consensual. If the meaning of something relevant is unclear or involves differences, an agreement must be developed through give and take before cooperative action is possible. For example, if a man and woman have just begun to date one another, and he invites her up to his apartment, exactly what meaning does this proposed visit have? One way or another, they will have to achieve some agreement regarding the purpose of the visit before joint action is possible. A symbolic interactionist would describe this as developing a consensual *definition of the situation.* The couple might achieve this through explicit negotiation or even through nonverbal communication. But without agreement on the definition of the situation, the woman will have difficulty in deciding whether to accept the invitation, and cooperative action will be impossible.

Symbolic interactionism portrays social interaction as having a tentative, developing quality. To fit their actions together and achieve consensus, persons interacting with one another must continually negotiate new meanings or reaffirm old meanings. Each person plans some actions, tries them out, and then adjusts his or her plans and behaviors in response to the other. Thus, social interaction always has some degree of unpredictability and indeterminacy.

Central to social interaction is the process of *role taking,* in which one individual imagines how he or she looks from the other person's standpoint. For example, if an employee is seeking an increase in salary, he might first imagine how his boss would react to one type of request or another. To do this, he might draw on knowledge of his past interactions with her, as well as what he has heard from others about her reactions to such salary requests. By viewing his own action from her standpoint, he may be able to anticipate what type of request would produce the desired effect. If he then actually makes a request of this type and if she reacts as expected, his role taking has succeeded. Through the role-taking process, cooperative interaction is established.

Symbolic interactionists maintain that for interaction to proceed smoothly, a person must achieve consensus with respect to his or her **self** and the identity of others. In other words, for each person there must be an answer to the question "Who am I in this situation, and who are the other people?" Only when this question is answered consensually can each person understand the implications (meanings) that others have for his or her own plan of action.

Sometimes a person's self is very unusual, and in consequence interaction becomes awkward, difficult, or even impossible. One example is the old tale by Cervantes of a man, temporarily deranged, who thought he was made of glass (Shibutani, 1961). This man's conception of himself created problems both for him and for others. Whenever people came near, he screamed and implored them to keep away for fear they would shatter him. He refused to eat anything hard and insisted on sleeping in beds of soft straw. Concerned that loose tiles might fall on him from the rooftops, he walked only in the middle of the street. Once, when a wasp stung him in the neck, he did not swat it away because he was afraid of smashing himself. Since glass is transparent while skin is not, he claimed that his body's unusual construction enabled his soul to perceive things more clearly than others, and he offered to assist persons perplexed by difficult problems. Gradually, he developed a reputation for astonishing insight, and many persons came to him seeking advice. In the end, a wealthy

Easy, fluid communication between two persons requires a high degree of consensus regarding personal identities and the meaning of symbols.

patron hired a bodyguard to protect him from outlaws and mischievous boys who threw stones at him.

These days, of course, it is not likely that we would meet someone who believes that he or she is made of glass. But we might encounter persons who believe they are unusually fragile or remarkably strong or superhumanly intelligent or in contact with the supernatural. Persons with unusual self-concepts can create problems in social interaction, because they make it difficult to achieve consensus regarding identities. Cooperative action is virtually impossible without such consensus, for people will simply not know what

to expect from individuals who insist they are Superman, Napoleon, Goldilocks, or Jesus Christ.

The self occupies a central place in symbolic interactionist theory, because social order is hypothesized to rest largely upon self-control. Coordination between individuals is possible because each person controls himself from within. The individual strives to maintain self-respect in his own eyes, but because he is continually engaging in role taking, he sees himself from the standpoint of the other persons with whom he interacts. Thus, he can maintain self-respect only by meeting the standards of others.

Of course, the individual cares more about the opinions and standards of some persons than about those of others. The persons whose opinions he cares most about are called **significant others.** Typically, these are persons who control important rewards or who occupy central positions in groups to which the individual belongs. Because their standards are valued, significant others will have relatively more influence over the individual's behavior.

In sum, the symbolic interactionist perspective has several strong points. It recognizes the importance of the self in social interaction. It stresses the central role of symbolic communication and language in personality and society. It explicitly treats the processes involved in achieving consensus and cooperation. And it illuminates why people try to maintain face and avoid embarrassment. Many of these topics are discussed in detail in later chapters. The self is discussed in Chapter 4, symbolic communication and language are taken up in Chapter 7, and self-presentation and impression management are treated in Chapter 9.

Critics of symbolic interactionism have pointed to various shortcomings. One criticism concerns the balance between rationality and emotion. Some critics argue that this perspective overemphasizes rational, self-conscious thought and deemphasizes unconscious or emotional states. A second criticism concerns the implicit model of the individual. In symbolic interactionist theory, the individual appears to be a particular personality type (the "other-directed" person) who is concerned primarily with maintaining self-respect by meeting others' standards. A third criticism of symbolic interactionism is that it places too much emphasis on consensus and cooperation and neglects conflict. The perspective does recognize, however, that interacting persons may fail to reach consensus, despite their best efforts to achieve it. The symbolic interactionist perspective is at its best when analyzing fluid, developing encounters with significant others; it is less useful when

analyzing self-interested behavior or principled action.

## A Comparison of Perspectives

The four theoretical perspectives discussed above differ in the issues they address and the issues they choose to ignore. They also differ with respect to the variables they each consider important and those they treat as irrelevant or incidental. In effect, each perspective makes different assumptions about social behavior.

In this section, we will compare the various perspectives in terms of four dimensions: (1) the theory's central concept or focus; (2) the primary social behaviors explained by the theory; (3) the theory's basic assumptions regarding human nature; and (4) the factors that, according to the theory, must change in order to produce a change in a person's behavior. Table 1-1 summarizes this comparison by showing the position of each perspective on each of these dimensions.

**Central Concepts**  Each of the four theoretical perspectives places primary emphasis on different concepts. Role theory places emphasis on roles, which are defined by group members' expectations regarding performance. Reinforcement theory explains observable social behavior in terms of the relationship between stimulus and response and the application of reinforcement. Cognitive theory stresses the importance of cognitions and cognitive structure in determining behavior. Symbolic interactionist theory emphasizes the self and role taking as crucial to the processes of social interaction.

**Behaviors Explained**  Although overlapping to some degree, the four theoretical perspectives differ with respect to the behaviors or outcomes they attempt to explain. Role theory places the greatest emphasis on role behavior and on the attitude change that results from occupying roles. Reinforcement theory focuses on the acquisition of new response patterns (that is, learn-

■ **TABLE 1-1**

COMPARISON OF THEORETICAL PERSPECTIVES IN SOCIAL PSYCHOLOGY

| | Theoretical Perspective | | | |
| --- | --- | --- | --- | --- |
| Dimension | Role Theory | Reinforcement Theory | Cognitive Theory | Symbolic Interactionist Theory |
| Central Concepts | Role | Stimulus-Response, Reinforcement | Cognitions, Cognitive Structure | Self, Role Taking |
| Primary Behaviors Explained | Behavior in role | Learning of new responses; exchange processes | Formation and change of beliefs and attitudes | Sequences of acts occurring in interaction |
| Assumptions about Human Nature | People are conformist and behave in accordance with role expectations | People are hedonistic; their acts are determined by patterns of reinforcement | People are cognitive beings who act on the basis of their cognitions | People are self-monitoring actors who use role taking in interaction |
| Factors Producing Change in Behavior | Shift in role expectations | Change in amount, type, or frequency of reinforcement | State of cognitive inconsistency | Shift in others' standards, in terms of which self-respect is established |

ing) and on the impact that rewards and punishments have on social interaction. Cognitive theory centers primarily on the mediating effect the cognitions have on a person's overt response to social stimuli, but it also treats changes in beliefs and attitudes. Symbolic interactionism is concerned primarily with sequences of behaviors occurring in interaction among people.

**Assumptions about Human Nature**   The four theoretical perspectives differ also with respect to their fundamental assumptions regarding human nature. Role theory, for instance, assumes that people are largely conformist. It views people as characteristically acting in accord with role expectations held by group members. In con-

trast, reinforcement theory views people's acts —what they learn and how they perform—as determined primarily by patterns of reinforcement. Cognitive theory stresses that people perceive, interpret, and make decisions about the world. They formulate concepts and develop beliefs, and they act on the basis of these structured cognitions. Symbolic interactionist theory assumes that the person is a conscious, self-monitoring being who uses role taking to achieve goals in interactions with others.

**Change in Behavior**   The four theoretical perspectives differ in their conception of what produces change in behavior. Role theory maintains that to change someone's behavior, it is necessary to change the role he or she occupies. Different behavior will result when the person shifts roles, because the new role will entail different demands and expectations. Reinforcement theory, in contrast, argues that change in behavior results from changes in the type, amount, and frequency of reinforcement received. Cognitive theory maintains that change in behavior results from changes in beliefs and attitudes; it further postulates that changes in beliefs and attitudes usually result from efforts to resolve inconsistency or internal conflict among cognitions. Symbolic interactionism holds that people try to maintain self-respect by meeting the standards of others; it also stresses that the relevancy of standards is usually resolved through negotiation. To produce change in behavior, the standards held by others and accepted as relevant must shift first. A person would detect this shift in standards (by role taking) and consequently change his or her behavior.

## ■ Is Social Psychology a Science?

Social psychology is the field that systematically studies the nature and causes of human social behavior. The field includes not only theoretical perspectives but also predictive theories and a large body of accumulated facts obtained through empirical investigation. Despite this, one can still ask whether social psychology is truly a science. That is, can we consider social psychology to be a scientific field in the same sense that we might consider physics or biology to be scientific?

### Characteristics of Science

Any scientific field rests on three basic assumptions. First, scientists assume there is a real, external world that exists independently of ourselves. This world is subject to investigation by observers. Second, scientists assume that relations in this world are organized in terms of

cause and effect. This assumption—which in practice is a working hypothesis—is the **principle of determinism.** In its starkest form, this principle holds that there are discoverable causes for all events in a science's domain of interest. Third, scientists assume that knowledge concerning this external world is objective; facts can be discovered by one scientist and then verified by others.

In addition to these assumptions, any field that is a science has certain characteristics or hallmarks. These include those listed below.

1. Any science is based on *observation of facts.* No field without observation can be a science. This means that "armchair" disciplines lacking practitioners who take observations cannot be scientific fields.

2. Any science uses explicit, formal *methodology.* This methodology is a set of procedures that must be followed by scientists when establishing facts as known. Because any investigator in the field can use this methodology, one scientist's findings can, in principle, be verified by others.

3. Any science involves the *cumulation of facts and generalizations.* That is, once relationships are observed to exist in the world, this knowledge is never lost. Of course, facts sometimes undergo reinterpretation of meaning, but the essential information is still available.

4. Any science includes a body of *theory.* This consists of at least one (but often many) theories that serve to systematize and organize empirical observations. Theory also serves to guide new, empirical investigation.

5. After it attains a reasonable level of development, any science provides at least some degree of *prediction and control* over selected aspects of the environment.

If you hold up a well-developed natural science such as physics or biology against this list, you will see immediately that it has all these hallmarks. These sciences are based on observation of the world, have a formal methodology to guide research, have an accumulation of many facts, possess a body of well-developed theory, and provide at least a moderate degree of prediction and/or control regarding selected aspects of the world.

## Social Psychology as a Science

Can social psychology be considered a science? That is, if we hold it up against the criteria listed above, will it measure up? Social psychology certainly has some of the hallmarks of science. For instance, consider the first of them—reliance on empirical observation. Social psychology clearly is based on empirical observation and classification of facts. The field consists of many thousands of empirical studies. Social psychology also meets the second hallmark, for it relies on widely shared methodological procedures for conducting empirical investigation. Among the most widely employed methods within social psychology are experimentation and systematic sample surveys. (Chapter 2 discusses methodology in social psychology.)

It is also fair to say that social psychology meets the third hallmark, the cumulation of observed facts. Social psychologists continue to gather facts regarding the conditions under which specific behaviors occur. Of course, more is known about some types of behavior than others, but increasingly sophisticated studies have continued to expand the frontiers of knowledge regarding social behavior.

Social psychology also tests moderately well against the fourth criterion, the reliance on theory. While social psychology has no single, unified theory covering all phenomena in the field, it does include several theoretical perspectives (such as those reviewed above). It also has numerous limited, or "middle-range," theories that make predictions regarding specific types of social behavior under restricted conditions. We will encounter many of these middle-range theories in subsequent chapters of this book.

If any problem arises for social psychology as a science, it is primarily with respect to the fifth hallmark. This hallmark holds that after attaining a reasonable level of development, any science should be able to provide some useful degree of prediction and/or control with respect to the phenomena investigated by the field. There is some question whether social psychology can accomplish this feat.

Although social psychology does fairly well in "explaining" social behavior (that is, identifying the conditions under which various forms of social behavior occur), it does less well in predicting future events or in providing a basis for control of behavior. Of course, it is possible to point to some successes in prediction and control. For example, political election forecasts based on sample surveys have been fairly accurate in recent years, and programs for modification of interpersonal behavior based on reinforcement principles have proved effective.

Nevertheless, social psychology does not excel in prediction and control. A large percentage of practicing social psychologists believe that, in the coming years, the field's capacity to predict interpersonal behavior will improve to some degree (Lewicki, 1982). At the present time, however, social psychology is no match for the mature physical sciences in this respect.

Part of the problem stems from the nature of social psychological theory. Few of the existing theories that make explicit predictions have much generality. They cover only limited ranges of phenomena or apply only under very restrictive (and sometimes artificial) conditions. Often, the theories fail to predict accurately when attempts are made to apply them to new settings.

If the problem ran no deeper than this, we might be very optimistic that social psychology would someday be able to predict social behavior with great accuracy. We might conclude that social psychology merely needs some better, more refined theories. To some degree this is true, but unfortunately the problem is more complex. As noted above, any science is based on the principle of determinism; it assumes the

## MILESTONES IN THE HISTORY OF SOCIAL PSYCHOLOGY

Although social psychology can trace its historical roots back to the philosophers of ancient Greece, and even more firmly to the psychological and social theorists of nineteenth-century Europe, the discipline as we know it today is largely a product of efforts by twentieth-century researchers and theorists, many of them American (Allport, 1985; Jones, 1985).

The following list summarizes what have come to be regarded as some of the important milestones in social psychology's development from the beginning of this century up to 1960. This list starts with the first social psychological experiment, which was conducted in 1897–98.

**1898** Norman Triplett publishes the first social psychological experiment ("The dynamogenic factors in pacemaking and competition"). This is an investigation of social facilitation, a process whereby a person's performance on a familiar task improves in the presence of others performing the same task.

**1902** Charles Horton Cooley publishes an influential book, *Human Nature and the Social Order,* which presents the idea that the self and society are ultimately the same thing, although viewed at different levels of abstraction.

**1908** W. McDougall and E. H. Ross independently publish the first textbooks in the field. Although both books are titled *Social Psychology,* they differ greatly in content. The book by McDougall (a psychologist) stresses the importance of instincts and innate drives in determining behavior, while that by Ross (a sociologist) discusses groups, crowds, and crazes and emphasizes interpersonal processes such as suggestion and imitation.

**1918** W. I. Thomas and F. Znaniecki begin their field study of attitudes within immigrant populations (Polish peasants) in Chicago.

**1922** Morton Prince establishes the first major social psychology journal, the *Journal of Abnormal and Social Psychology* (which, in 1965, becomes the *Journal of Personality and Social Psychology.*)

**1924** Floyd Allport, writing from a stimulus-response (behaviorist) perspective, publishes a social psychology textbook that is one of the first systematic treatments of the field. This book advocates the use of the experimental method in social psychology and sets forth a research agenda for the next decade.

**1928** L. L. Thurstone publishes a pathbreaking paper showing how attitudes can be measured.

**1934** George Herbert Mead, a symbolic interactionist, publishes his seminal work on the self.

**1934** J. L. Moreno develops sociometry, a system for measuring patterns of social interaction based on individuals' choices regarding with whom they would prefer to associate.

**1934** Richard T. LaPiere, investigates inconsistencies between attitudes (racial prejudice) and related behaviors (discrimination) in a field setting.

Box 1-1

**1936** Muzafer Sherif, by creating social norms in a controlled setting, demonstrates that complex and realistic social situations can be studied experimentally in a laboratory.

**1936** George Gallup develops methods for conducting public opinion polls and surveys.

**1937** J. L. Moreno founds *Sociometry,* a journal devoted to research on group structure and process. (The journal is later renamed *Social Psychology Quarterly.*)

**1939** Kurt Lewin, Ronald Lippitt, and Ralph White, using Lewin's field theory, study group members' reactions to various styles of leadership (autocratic, democratic, and laissez-faire).

**1943** Theodore Newcomb investigates the effects of social pressures on attitudes among students at Bennington College.

**1943** W. Foote White uses the technique of participant observation to study the activities of teenage street gangs.

**1946** Solomon Asch demonstrates that cognitive set can influence the impressions people form of others.

**1950** Robert Freed Bales develops a categoric framework for systematically observing communication and role differentiation in task groups.

**1950** George Homans publishes *The Human Group,* a seminal, theoretical treatise on group structure and process.

**1951** Solomon Asch demonstrates that most individuals in a group will conform to the position of a majority when their beliefs are questioned.

**1953** Carl Hovland and coworkers at Yale University publish the results of a programmatic study of persuasion and attitude change.

**1954** Gordon Allport publishes *The Nature of Prejudice,* an important analysis of intergroup prejudice and stereotyping.

**1957** Leon Festinger proposes the theory of cognitive dissonance, an approach to attitude change based on the idea that people strive for consistency between behavior and attitudes.

**1958** Fritz Heider publishes *The Psychology of Interpersonal Relations,* which lays the foundation for attribution theory and research.

**1959** John Thibaut and Harold Kelley publish *The Social Psychology of Groups,* a general theory of social exchange and interpersonal relations.

Since 1960, social psychology has continued to expand rapidly and develop as a field. The 1960s saw a large number of laboratory studies of cognitive dissonance, as well as an increased concern with such phenomena as altruism and aggression and interpersonal attraction. The 1970s and 1980s witnessed the growth of attribution theory, a renewed concern with the social self, and an expanded interest in interpersonal relations and the cognitive aspects of social behavior. In a fundamental sense, the chapters of the book you are now reading constitute a summary and distillation of the central concerns of social psychology as the field stands today.

world is organized in terms of cause and effect. Science tries to develop laws based on the notion that if X causes Y on one occasion, then X will again cause Y on some similar occasion in the future. Although the assumption of determinism works well for physical phenomena, it may not work as well for human or social behavior. Human beings are conscious and self-aware, and they are capable of exercising some control over their own behavior. Unlike atoms or rocks, they are capable of making decisions and suddenly changing their behavior if they want to. In other words, they have at least some degree of free will. For this reason, some theorists have argued that there will never be anything approaching true "universal laws" describing social behavior. Certainly, it is difficult to reconcile the scientific assumption of determinism with the concept of human free will. Of course, this problem is not unique to social psychology. It besets all the social sciences.

As we have shown, social psychology displays many of the characteristics of the more mature physical sciences. It is based on observation of the social world and relies on a formal methodology to guide research. It has accumulated many descriptive facts regarding human social behavior, and it possesses bodies of formal theory. Nevertheless, it has not yet achieved the same degree of accuracy in prediction as the mature physical sciences. Although it does offer compelling explanations for many types of observed social behavior, it has provided only a modest degree of predictability and control of social behavior.

# ■ Summary

This chapter has considered the fundamental characteristics of social psychology as well as important theoretical perspectives in the field.

**What Is Social Psychology?** There are several ways to characterize social psychology. (1) One way is to list the core concerns of social psychology. These concerns are the impact of one individual on another's behavior and beliefs, the impact of a group on a member's behavior and beliefs, the impact of a member on the group's activities and structure, and the impact of one group on another group's activities and structure. (2) By definition, social psychology is the field that systematically studies the nature and causes of human social behavior. (3) Social psychology has a close relationship with other social sciences, especially sociology and psychology. Although they emphasize different issues, both psychologists and sociologists have contributed to social psychology.

**Theoretical Perspectives in Social Psychology** A theoretical perspective is a broad theory based on particular assumptions about human nature that offers explanations for a wide range of social behaviors. In this chapter, four theoretical perspectives were discussed: role theory, reinforcement theory, cognitive theory, and symbolic interactionist theory. (1) Role theory is based on the premise that people conform to norms defined by the expectations of others. It is most useful in explaining the regular and recurring patterns apparent in day-to-day activity. (2) Reinforcement theory assumes that social behavior is governed by external events, especially rewards and punishments. Reinforcement is useful in explaining not only how people learn but also when social relationships will change. (3) Cognitive theory holds that such processes as perception, memory, and judgment are significant determinants of social behavior. The theory treats cognitions as organized into structures (schemas) and relies on various principles (such as the principle of consistency) to explain attitude change. Differences in cognitions help to illuminate why persons may behave differently in a given situation. (4) Symbolic interactionist theory holds that human nature and social order are products of communication among people. It stresses the importance of the self, of role taking, and of consensus in social interaction. It is most useful in explaining fluid, contingent encounters among people.

**Is Social Psychology a Science?** (1) To answer the question whether social psychology is a science, we must first identify the hallmarks that characterize any science. There are five such hallmarks: scientists engage in empirical observation of the world, use a formal research methodology, cumulate knowledge of facts, develop formal theories to explain facts, and employ these theories to provide some degree of prediction and control. (2) Social psychology meets the first four of these hallmarks, but falls somewhat short of meeting the fifth. To date, social psychology has provided only a modest degree of predictability and control over human social behavior.

## Key Terms

| | |
|---|---|
| COGNITIVE PROCESSES | 13 |
| COGNITIVE STRUCTURE | 14 |
| COGNITIVE THEORY | 13 |
| CONDITIONING | 11 |
| EQUITY | 12 |
| IMITATION | 11 |
| MEANING | 15 |
| NORM | 8 |
| PRINCIPLE OF CONSISTENCY | 15 |
| PRINCIPLE OF DETERMINISM | 20 |
| REINFORCEMENT THEORY | 10 |
| REINFORCER | 10 |
| ROLE | 8 |
| ROLE THEORY | 8 |
| SCHEMA | 14 |
| SELF | 6 |
| SIGNIFICANT OTHER | 18 |
| SOCIAL EXCHANGE THEORY | 11 |
| SOCIAL LEARNING THEORY | 11 |
| SOCIAL PSYCHOLOGY | 5 |
| SYMBOLIC INTERACTIONIST THEORY | 15 |
| THEORETICAL PERSPECTIVE | 7 |
| THEORY | 7 |

2

chapter

# Research Methods in Social Psychology

# ■ Introduction

To a considerable extent, the field of social psychology is based on *empirical research,* the systematic investigation of observable phenomena (behavior, events) in the real world. Research is intended to gather information about external events in an accurate and unbiased form. This information may be either quantitative or qualitative, and it is used by researchers both to describe reality in close detail and to evaluate theories about social behavior.

Empirical research is usually conducted in accordance with a **methodology,** a set of systematic procedures. Social psychologists plan in advance how the information will be collected, monitor the process of gathering it, and systematically analyze the data in order to arrive at certain conclusions. Throughout this process, investigators follow specific procedures designed to maximize the validity of the findings.

When investigators report their research to the wider community of social psychologists, they describe not only their results but also the exact methodological procedures that were followed in generating the results. By reporting their methods, they create the possibility that other investigators in the field may independently verify their findings.

The independent verification of research findings is one of the hallmarks of any science. Suppose, for instance, that an investigator were to report some unanticipated empirical findings that ran contrary to established theoretical viewpoints. Other investigators might wish to replicate the study in order to see whether the same findings could be obtained in other settings with different human subjects. To the extent that investigators with differing perspectives evaluate the work, biases will be identified and eliminated. If the same results occur again and again in replications, they stand a better chance of being accepted by other social psychologists.

## Questions about Research Methods

The purpose of this chapter is to acquaint you with the research process in contemporary social psychology. This information can be used to better understand and evaluate the empirical studies discussed throughout this book. The following are the major questions addressed in this chapter.

1. What are the basic goals that underlie social psychological research? Where do research hypotheses come from? What steps can be taken by researchers to ensure the validity of their findings?
2. What are the defining characteristics of major research methods such as social surveys, laboratory and field experiments, naturalistic observation, and archival research? What are the strengths and weaknesses of each?
3. What are some important ethical issues in the conduct of social psychological research? What safeguards are available to protect the rights of human subjects?

## ■ Characteristics of Empirical Research

Later in this chapter we will discuss some of the major research techniques used by social psychologists—surveys, experiments, field observation, and archival studies. Before looking at those techniques in detail, however, it is useful to review some considerations common to all forms of empirical research. Thus, in this section we will consider the objectives that underlie empirical research, the nature of hypotheses or questions that guide research, and the factors that strengthen (or weaken) the validity of research findings.

### Objectives of Research

In this section we discuss briefly four major objectives of social psychological research: describing reality, identifying correlations among variables, testing causal hypotheses, and constructing and testing theories.

**Describing Reality**   Frequently, the social psychologist wants simply to learn more about some behavior or social process or to describe its characteristics. This goal may be paramount when a researcher is investigating a previously unexplored phenomenon (where little or nothing is known). It may also prevail when a researcher wants to know the frequency with which some attitude or behavior occurs in a specified group or population. For instance, during election years researchers routinely conduct public opinion polls to learn how Americans feel about political candidates and issues. Their goal is to describe public sentiment with great accuracy.

**Identifying Correlations among Variables**   A second goal of research is to ascertain whether there is a relationship between two or more behaviors or characteristics. Researchers are often interested, for example, in whether one specific attribute is associated with other attributes or behaviors. We might be interested, for example, in whether sexual intercourse before marriage is more common among Protestants than among Catholics. It turns out that Catholics are less likely to have intercourse before marriage than are Protestants (DeLamater & MacCorquodale, 1979). To say that two variables are correlated, however, is not to say that one causes the other or vice versa.

**Testing Causal Hypotheses**   A third goal of research is to discover the causes of some behavior or event. One might be interested in discovering, for instance, what factors cause a person to identify with the Republican party or to engage in premarital intercourse. Before actually collecting data, the researcher first develops a **hypothesis,** which is a statement that a specific behavior or event is caused by some other event or social process. The data collected can then be used to test or evaluate this hypothesis. Frequently, the best way to test a causal hypothesis is to conduct an experiment, a topic discussed in greater detail below.

**Testing Theories**  As this term is used here, a theory is a set of interrelated hypotheses that explains some observable phenomena and that can be used as a basis for prediction. For example, several social psychologists have developed theories stating that differences in social class cause differences in attitudes and behavior. One such theory asserts that conditions related to one's work determine whether a person will place greater value on autonomy and self-direction or on conformity with the expectations of others (Kohn, 1969). Within this framework, one hypothesis states that it is the prestige associated with the job (that is, the rewards of money and social status one receives) that determines one's values. An alternative, competing hypothesis states that values are determined by the opportunities the work provides for autonomy and independence, for working on one's own. Kohn and Schooler (1973) have tested these two hypotheses, and the results support the second. In addition, this theory hypothesizes that occupational conditions determine (that is, cause) one to hold particular values rather than one's values determining the choice of job to hold. Data collected over time and analyzed by complex statistical techniques support this causal hypothesis (Kohn & Schooler, 1982).

## Research Hypotheses

Social psychological research begins with a question. Some studies are purely exploratory and are driven more by wide-ranging curiosity than by explicit hypotheses or formal theory. Many other studies, however, involve one or more hypotheses at the start. Before actually collecting data, investigators obviously must have an explicit idea regarding which variables they will measure and/or manipulate, and hypotheses provide important guidance in this respect.

Research hypotheses assume various forms. All state a relationship between two or more variables. Many hypotheses in social psychological research are causal in nature. A causal hypothesis relating two variables (X and Y) might take the form "X causes Y" or perhaps "An increase in X will cause an increase in Y." Sometimes, of course, causal hypotheses are more explicit and qualified regarding their scope: for example, "When conditions A and B are present, an increase of 1 unit in X will cause an increase of (some specified number of) units in Y."

Causal hypotheses always include at least two variables (here, X and Y). The term **independent variable** refers to any variable that is considered to cause some other variable(s). In contrast, a **dependent variable** is any variable caused by some other variable. In the above example, X is the independent variable (the causal source), whereas Y is the dependent variable (the outcome or effect that is caused).

Social psychologists often use the term **extraneous variable** to designate any variable that is not explicitly included in the hypothesis but has a causal impact on the dependent variable. Extraneous variables are important in social psychology, because most dependent variables of interest have more than one cause—and many have a very large number of causes. In some studies, researchers have an explicit idea regarding which variables (besides the independent variable X) may cause Y; typically, however, they know only a few of the extraneous variables relevant to Y.

**Origins of Research Hypotheses**  Where do research hypotheses come from? We noted above that research begins with a question. The questions or hypotheses that investigators assess may be based on observations of one sort or another or on an examination of existing knowledge (McGuire, 1973).

With regard to observation, one specific source of hypotheses is close observation of daily events. For instance, numerous hypotheses about the influence of child rearing on personality were drawn from Sigmund Freud's analyses of individual patients whom he interviewed. A second source of hypotheses is the attempt to account for a paradoxical incident. For example, several studies of the influence of a cult on its members grew out of the incident at Jonestown, where 900 persons committed murder or suicide at the exhortation of their leader, Jim Jones. A third source is the attempt to account for an observed pattern, for example, that middle-class persons are less frequently arrested than lower-class persons. A fourth source of hypotheses is the use of analogy, such as drawing on phenomena or principles from some other field. For instance, studies of animal communities indicate that if their populations grow too large, events such as migration and starvation will occur with increasing frequency and reduce their numbers. Investigators have tried to study whether analogous effects of density occur in human populations.

Hypotheses, especially causal ones, are frequently drawn from an examination of existing knowledge. First, in some cases, an investigator examines generalizations established by prior research and, through logical deduction, arrives at a hypothesis. Second, scrutiny of the conflicting results of prior research may suggest a hypothesis that explains why the findings conflict. Third, any exception to an otherwise general finding may suggest a hypothesis. Finally, an investigator's theory may generate hypotheses to be empirically tested.

## Validity of Findings

Suppose an investigator is studying deviant behavior. In particular, she is investigating the extent to which cheating occurs on exams by college students. She reasons that it is more difficult for persons monitoring an exam to keep all students under surveillance in large classes than in smaller ones. Hence, as a working hypothesis, she expects that a higher rate of cheating will occur on exams in large classes than in small ones. To test this hypothesis, she collects data on cheating in both large classes and small ones and then analyzes the data. Results show that more cheating per student occurs in the larger classes. Thus, it appears that the data support the investigator's research hypothesis.

A few days later, however, someone points out that the large classes in her study all used multiple-choice exams, whereas the small classes all used short-answer and essay exams. The investigator realizes immediately that an extraneous variable may be operating as a cause in her data and that the apparent support for her research hypothesis (more cheating in large classes) may be nothing more than an artifact. Perhaps the true effect is that more cheating occurs on multiple-choice exams than on essay exams, irrespective of class size.

The findings of a study are said to have **internal validity** to the extent that they are free from contamination by extraneous variables. Obviously, the investigator's findings regarding the effect of class size on cheating have low internal validity. Internal validity is very important, for it is the basic ingredient without which a study's results are uninterpretable.

Extraneous variables, should they be present, can reduce internal validity in either of two basic ways. First, an extraneous variable can mask the true effects of the independent variable (X) on the dependent variable (Y). This can happen in cases where the hypothesis (that is, X causes Y) is true but where the extraneous variable influences the dependent variable in a way that is opposite to the effect of the independent variable. Thus, even though X actually causes Y, the data may fail to reveal this fact. Second, an extraneous variable can create an apparent causal effect of X on Y even when none truly exists. This is what happened in the study of class size and cheating. The results initially appeared to show an effect of class size on cheating, but it later became clear that this effect may have been due to the exam format, an extraneous variable (Campbell & Stanley, 1963; Cherulnik, 1983).

To achieve results with higher internal validity, an investigator may repeat the study with an improved design. For instance, an investigator might repeat the study on cheating with only one exam format (say multiple choice) in both large and small classes and then test whether class size affects the rate of cheating. By holding constant the extraneous variable (exam format), her new study design will have greater internal validity. Better still, she might use a more complex design that includes all logical possibilities (small class-multiple choice, small class-essay, large class-multiple choice, large class-essay) and then analyze the data to estimate separately the impact of each of these variables on cheating. This design, in effect, converts an extraneous variable (exam format) into a second independent variable. Even so, it is by no means a perfect design, because other, unknown extraneous variables could still be operating as causes of cheating—and they may be confounded with class size and exam format.

As important as internal validity is, it is not the only concern of the investigator. Another concern is **external validity,** which refers to the extent to which the results of a study can be generalized to other populations, settings, or times. Even if an investigator's results are internally valid, they may hold only for the specific group and setting studied. For instance, if the investigator looking at the relation between cheating and class size conducted her study in a two-year college, there is no assurance that the findings (whatever they turned out to be) would apply to students in other settings, such as high schools or four-year colleges or universities. In general, external validity is important, because the results of a study usually will assume practical importance only if they generalize beyond the particular setting in which they were obtained.

# ■ Research Methods

There are many ways of collecting data about social behavior. Social psychologists rely heavily on four methods: surveys, experiments, naturalistic observation, and archival research based on content analysis. We will discuss each of these methods in turn.

## Surveys

A **survey** is a procedure for collecting information by asking members of some group a number of questions and systematically recording their responses. Many research projects in social psychology are based on surveys. For example, to test the hypothesis that occupational conditions influence values, a group of employed men were asked questions about their work and their values (Kohn, 1969). The researcher compared the attitudes of men whose work was less closely supervised, nonroutine, and complex with those whose work was closely supervised, routine, and simple. As predicted, men in the first group valued self-direction, while men in the second group valued conformity.

**The Purpose of a Survey**   Surveys are generally conducted to obtain information from individuals about their own attributes—that is, their attitudes, behavior, and experiences. Surveys are commonly used to determine the frequency (or distribution) of an attribute in the population of interest. They are also used to determine whether a relationship exists between two or more attributes of interest. When analyzed via sophisticated statistical techniques, data from surveys can be used to test causal theories.

One form of survey—the public opinion poll—has become very common in the United States. Several organizations specialize in conducting surveys that are designed to measure the frequency and strength of favorable or unfavorable attitudes toward public issues, political figures, and candidates for office. These polls play a major role in American politics, as public policy and the positions taken by political figures are increasingly influenced, at least in part, by the poll results (Halberstam, 1979). Reportedly, many of the decisions made by the Democratic and Republican presidential candidates during the 1988 campaign were based on such polls.

Surveys are also widely used to obtain data relevant to various social problems. Government agencies and individual researchers have conducted surveys on the effects of exposure to pornography, on frequency and patterns of alcohol and other drug use, and on teenage pregnancy and contraceptive use among teenagers (Zelnik & Kantner, 1981). Gathering information about the extent of such activities and the characteristics of persons involved in them is prerequisite to developing effective social policies.

Finally, many surveys have been conducted with the primary objective of making basic theoretical contributions to social psychology. For instance, many studies of socialization processes and outcomes, attitude structure and attitude-behavior relationships, psychological well-being, discrimination and prejudice, and collective behavior have been based on survey methodology.

**Types of Surveys** There are two basic types of surveys: the interview survey and the questionnaire survey. In the **interview,** a person serves as an interviewer and records each of the respondent's answers. The advantage of using an interviewer is that he or she can adjust the questioning to the individual; the interviewer can look for both verbal and nonverbal signs that the respondent does not understand a question and repeat or clarify as necessary. To ensure that each respondent is asked the same questions, the interviewer usually works from an *interview schedule,* which specifies the order and exact wording of questions in advance. In some studies, however, the interviewer is instructed to make sure that certain topics are covered but is given flexibility in determining the exact order and wording of questions.

In a **questionnaire,** the questions are printed and the respondents read and answer them at their own pace. No interviewer is used. Questionnaires have a clear advantage over interviews in terms of cost. The cost of a national survey using trained personnel to conduct face-to-face interviews is about $210 per completed interview. The same survey using mailed questionnaires would cost approximately $50 per completed form. The major disadvantage of questionnaires is the **response rate**—the percentage of those contacted who complete the

survey. Whereas an interview survey can obtain response rates as high as 75 or 80 percent or better, mailed questionnaires rarely attain more than a 50 percent response rate.

Some have argued that an advantage of the questionnaire over the interview is that a respondent may be more likely to answer questionnaire items truthfully, particularly when the questions deal with embarrassing or threatening topics (Sudman & Bradburn, 1974). However, data from two studies suggest that questionnaires do not really have this advantage over interviews (Johnson & DeLamater, 1976). Using well-trained age peers as interviewers, investigators found no significant differences between those who were interviewed and those who completed a questionnaire containing identical questions regarding sexual behavior.

A compromise between interviews and questionnaires is the *telephone interview.* This is now the standard method used by public opinion polling organizations, such as Gallup and Roper, and is increasingly used in basic research as well. It retains the use of a trained interviewer to ask the questions, but it sacrifices the visual feedback available in a face-to-face interview. It is much cheaper (about $75 per completed interview), although it typically has a somewhat lower response rate (about 65 percent). Many surveys now use Computer Assisted Telephone Interviewing (CATI). With CATI, the computer randomly selects and dials telephone numbers; the interviewer takes over when a call is answered. Once a potential respondent is on the line, the computer selects the questions and alters later questions in light of earlier answers as appropriate. The interviewer reads each question and enters the answer directly into the computer as the respondent gives it.

**Measurement Reliability and Validity** In surveys, as in any form of research, the quality of measurement is an important concern. There are two basic considerations with respect to measurement: the reliability and the validity of the instruments. **Reliability** is the extent to which a given measuring instrument produces the same results each time it is employed. A reliable instrument produces consistent results across independent measurements of the same phenomenon. If an instrument is reliable, it should yield the same results when two independent researchers use it. Reliability is a matter of degree; some instruments are highly reliable, others less so. Obviously, reliability is a very desirable property of a measuring instrument.

There are several ways to assess the reliability of an instrument. One of these, the *test-retest* method, uses the same measuring instrument on the same population at two different times. That is, a given instrument is used to measure a set of respondents on two occasions, and then their first responses are compared against their second responses. If the test-retest correlation is high, the instrument is taken to be reliable; if the correlation is low, the instrument has only low reliability.

A second technique is the *split-half* method. To illustrate this, suppose you have twenty questions that you believe measure psychological well-being. The respondent is asked, for example, how often he or she is sad, nervous, depressed, tense, or irritable and how often he or she has trouble concentrating, working, or sleeping. You give your scale to 300 male students. To use the split-half technique, you would randomly divide the questions into two groups of ten, calculate a score for each respondent on each half of the scale, and then compute a correlation between the two scores. Again, a high correlation is taken as evidence that your scale is reliable.

Assuming that a measure is reliable, the next concern is its **validity**—that is, does the measuring instrument actually measure the concept we intend to measure? There are several types of validity, including face validity, construct validity, and predictive validity. In some instances, it is easy to develop a measure with *face validity*—one that is manifestly similar to the behavior or process of interest. If a researcher is interested in measuring the frequency of sexual intercourse, for example, the question "How often do you engage in sexual intercourse?" has face validity. *Construct validity* refers to the extent to which a construct or concept accounts for performance on the measure. Truthful reports of frequency of sexual intercourse are a valid measure of the concept sexual experience. At times, construct validity is more problematic, for example, when one is attempting to measure an abstract concept such as intellectual development. In such cases, there is no readily observable referent, no behavior or physical object we can immediately point to. What the researcher measures depends on his or her definition of the construct. Finally, a measure of one variable that accurately predicts standing on some other variable of theoretical interest is said to have *predictive validity*.

**The Questions**   A major influence on the reliability and validity of a survey is the exact nature of the questions asked. Creating good survey questions is as much art as science, but there are certain guidelines that help in their formulation. First, the more precise and focussed a question, the more reliable and valid it will be. A question expressed in vague, ambiguous, abstract, or global terms often will be interpreted by different respondents to mean different things and, hence, will produce uncontrolled variation in responses. A second consideration in formulating survey questions is the exact choice of words used. In general, it is desirable to avoid jargon or specialized terminology unless one is interviewing a group of specialists. The words and grammar employed should be appropriate to the educational level of the respondents. Another important consideration is the length of questions. There is evidence that questions of moderate length elicit more complete answers than very short ones (Sudman & Bradburn, 1974; Anderson & Silver, 1987).

**The Sample**   Suppose a survey researcher is interested in learning the extent of prejudice toward blacks among white adults in the United

States. These white adults constitute the **population** of interest—that is, the set of all people whose attitudes are of interest to the researcher. Because it would be virtually impossible (and enormously expensive) to interview all people in the population of white adults, the researcher instead selects a **sample,** or subset of that population, to interview. Selection of the sample is one of the most important aspects of any survey. Typically, the researcher will take care to select a sample that is representative of the population. Representativeness is important, because it affects the extent to which the results of the survey (for example, information regarding racial prejudice obtained from the persons in the sample) can be generalized to the entire population. The nature of the sample, therefore, has a major influence on the external validity of the survey.

Three types of samples are common in social psychological surveys. One is the **convenience sample,** which is selected solely on the basis of availability; the researcher questions persons who happen to be handy. For example, collecting data from a class of students or students who happen to be in a dormitory or a student union yields a convenience sample. Although convenience samples have the appeal of low cost and easy access, their disadvantage is that the researcher cannot know how representative such a sample is of some larger population.

To ensure that a sample is representative, researchers often employ a technique called random sampling. In drawing a **simple random sample,** the researcher selects units, usually individuals, from the population in such a way that every unit is equally likely to be selected. In order to use this technique, the researcher needs a complete list of members of the population. At a university, for example, a list of all students can frequently be obtained from the registrar. At the city or county level, voter registration lists might be employed. A frequent problem, especially when the population being sampled is large, is the absence of a complete list. Under these circumstances, researchers usually fall back on

some substitute, such as a telephone directory. This may introduce bias into the sample, however, because persons who are poor or who move frequently may not have telephones, and some people choose not to list their numbers in the directory.

Once a complete list of the population is obtained, the researcher selects a sample. A common way to do this is to number the people on the list consecutively and then use a table of random numbers to determine which persons will be in the sample. Once a random sample has been chosen, the researcher must take steps to ensure that all the members of the sample are interviewed. In other words, the researcher must maximize the response rate; otherwise, bias may be introduced by differences between people who participate in the study and those who refuse to do so.

In a very large population, there may be too many units to list them all. Under these conditions, researchers frequently employ a **stratified sample.** That is, they subdivide the population into groups according to characteristics known or thought to be important, select a random sample of groups, and then draw a sample of units within each selected group. For example, public opinion polls designed to represent the entire adult population of the United States often involve stratified samples. The population is first stratified on the basis of region (South, Northeast, Midwest, Southwest, and West). Next, the population within each region is further stratified into urban versus rural; within urban areas, there may be still further stratification by size of urban area. The result is a large number of *sampling units*—population subgroups of known regional and residential type. Some units are then sampled in proportion to their frequency in the population as a whole. Thus, one would sample more urban units from the Northeast than from the South or Midwest; conversely, one would select more rural units in the latter regions. Finally, within each unit, persons are selected at random to serve as respondents. Using this technique, one can represent the adult population of the United States with a sample of 1,500 persons and obtain responses accurate within plus or minus 3 percent.

**Causal Analysis of Survey Data** In recent years, some social psychologists have begun to use computer-based techniques that assist in the causal interpretation of survey data. Techniques of this type (such as LISREL and path analysis) generally require the data analyst to postulate a structure of cause-and-effect relations among a set of variables (Joreskog & Sorbom, 1979). The computer then estimates coefficients of effect from the data. These coefficients provide a test of whether the causal linkages postulated on the basis of theory are indeed present in the data. Using this approach, an analyst can test many alternative theories. Most of the time, some theories will turn out to be inconsistent with the data; these can be rejected in favor of alternative theories that survive the test. The difficulty with this approach is that for problems involving many variables (say, 10 or 15), there exist a large number of plausible alternative theories. Even though many will be eliminated, more than one may survive as a tenable model.

**Advantages of Surveys** Surveys are very useful in providing a valid description of the attitudes and social characteristics of populations or of subgroups within populations. When a survey is conducted competently—that is, when the researcher uses reliable and valid measures, employs a sampling design that guarantees representativeness, and takes steps to ensure a high response rate from respondents—it can be a very powerful means to obtain a clear portrait of a population with respect to the attitudes and characteristics of interest.

Beyond population attitudes and characteristics, surveys can also be used to study frequency of social behavior. A survey asking people to report their behavior is usually more efficient and cost effective than observational studies of actual behavior, particularly if the behavior occurs only infrequently or in private settings.

Finally, survey data can be used to develop and test theories about phenomena such as attitude structure and change, the impact of mass media, and the effects of social class on individual values and behavior. As noted above, under certain circumstances, survey data can be used to test causal theories. The research by Kohn and his colleagues on the effects of social class on individual values described earlier in this chapter is a good example of this use of surveys.

**Disadvantages of Surveys** As with any methodology, there are certain drawbacks to the survey technique. Both questionnaires and interviews are based on reports by respondents about themselves. Self-reports, however, can be invalid sources of information under some conditions. First, some people may not respond truthfully to questions about themselves. This is not generally a major problem, but it can become troublesome if the survey deals with activities that are illegal or otherwise embarrassing to reveal. Second, even when respondents are motivated to tell the truth, they may give wrong reports due to imperfect recall or poor memory. This can be a nettlesome problem, especially in surveys investigating the past (for example, historical events, childhood, and so on). As an illustration, consider the question "When were you last vaccinated?" This may look simple and straightforward, but it often produces incorrect responses because many people cannot remember dates. Third, some respondents answering self-report questions have a tendency to fall into a *response set;* that is, they tend to answer all questions the same way (for example, agree or disagree), and/or they give extreme answers too frequently. If a large number of respondents adopt a response set, this will introduce bias into the survey's results.

### Experiments

The **experiment** is the most highly controlled of the research methodologies available to social psychologists. For a study to be considered a true experiment, it must have two specific characteristics. First, the researcher must be able to manipulate one or more independent variables in order to observe their causal impact on various outcomes. The experimenter creates at least two levels of each independent variable (for example, high versus low, presence versus absence, and so on), although in some cases he or she will create more than two levels. Second, the researcher must be able to randomly assign the subjects to the various *treatments*—that is, to the different levels of each of the independent variables. The term **random assignment** means the placement of subjects in experimental conditions on the basis of chance, as by flipping a coin or using random numbers. Random assignment serves as a means of controlling the effects of extraneous variables. By using random assignment, the researcher creates groups of subjects that are equivalent except for their exposure to different levels of the independent variables. This removes the possibility that these groups will differ systematically on extraneous variables, such as intelligence, personality, or motivation. With random assignment, the investigator can infer that any observed differences between groups in the dependent variable are due only to the differences in levels of the independent variables, not to extraneous variables (Kirk, 1982).

Experimental researchers manipulate independent variables in order to observe their impact on a dependent variable, the outcome of theoretical or practical interest. Dependent variables are measured (not manipulated) by experimenters. Dependent variables can be measured in many ways—for example, by monitoring subjects' physiological arousal, by administering short questionnaires that assess subjects' attitudes, by recording the interactions that occur between subjects, or by scoring the performance of subjects on tasks that are part of the experimental setting. The exact type of measurement used by researchers will depend heavily on the specific nature of the dependent variable of interest.

**Laboratory and Field Experiments** It is useful to distinguish between *laboratory experiments* and *field experiments*. Laboratory experiments are those conducted in a laboratory setting, where the investigator has essentially complete control over the subjects' physical surroundings. In the laboratory, the investigator can determine what stimuli, tasks, information, or situations will be presented to subjects. This control enables the experimenter to manipulate independent variables and to measure dependent variables. It also enables the experimenter to hold constant extraneous variables. For instance, in some laboratory experiments of group processes, investigators may wish to restrict or limit the forms of interaction among subjects. Communication may be limited to written notes or messages transmitted by electronic equipment. This practice not only eliminates the (possibly contaminating) influence of nonverbal communication, but it also permits the contents of any messages sent to be later analyzed by the experimenter.

Field experiments, in contrast to laboratory experiments, are studies where investigators manipulate variables within natural (nonlaboratory) situations. Usually, these situations are already familiar to the subjects. Field experiments can investigate just about anything imaginable, ranging from pay inequity in large bureaucratic organizations to helping behavior on street corners or in subway cars. Field experiments are less common in social psychology than are laboratory experiments. Field experiments usually have high external validity, their great advantage. Because they are conducted in natural, uncontrived settings, they usually have much greater realism than laboratory experiments. Moreover, participants in field experiments may not be particularly self-conscious of their status as experimental subjects, a fact that minimizes subjects' reactivity or their desire to be seen in a positive light. The major weakness of field experiments, of course, is that experimenters often have relatively little control over extraneous variables in natural settings. This means that the internal validity of field experiments is lower—

Laboratory research allows the investigator to manipulate independent variables and measure behavior in various ways. In this study, trained persons observe the subject through a one-way mirror to collect data on aggressive behavior in children.

sometimes much lower—than that of comparable laboratory experiments.

**Conduct of Experiments**  To illustrate how experiments are conducted, consider the following laboratory experiment, which sought to determine the influence of certain independent variables on whether one person will help another in an emergency (Darley & Latane, 1968). The study was conducted at a university in New York City. Men and women students were recruited to come to the laboratory to participate in a discussion of problems encountered in adjusting to the university. Each subject was placed in a separate room and asked to communicate with other

subjects via intercom. The rationale given was that this procedure would permit them to remain anonymous while discussing personal problems.

The independent variable was the number of other persons who the subject believed were participating in the discussion (and who would, therefore, later be witness to an emergency). Depending on experimental treatment, subjects were told there were one, two, or five other participants; assignment to the various levels of this independent variable was done randomly.

The discussion proceeded with each participant speaking in turn for two minutes over the intercom. Thus, depending on experimental treatment, the subject heard the voices of one, two, or five others. In reality, the subject was hearing a tape recording of other people, not the voices of actual research subjects. (This was the real reason for putting subjects in separate rooms and having them communicate via intercom.) One of these recorded voices admitted that he was subject to nervous seizures. In his second turn, he started to speak normally, but his speech suddenly became disorganized, then lapsed into gibberish and choking sounds, and then silence. Clearly an emergency was occurring. The subject believed that all participants could hear it, although the intercom prevented them from talking to one another.

The dependent variables were whether the subject would leave the room to offer help and how quickly he or she did so. Those subjects electing to help the victim typically came out of their room looking for the victim. The experimenter timed the speed of the subject's response from the beginning of the victim's speech. The results verified the research hypothesis that the greater the number of witnesses, the less likely a subject was to offer help.

This carefully controlled experiment allowed a reasonably straightforward test of the hypothesis. The manipulated independent variable (number of witnesses) and the measured dependent variable (speed of helping response) were relatively unambiguous. Confounds from extraneous variables can be ruled out due to the random assignment of subjects to treatments. One reservation, however, is that the causal effect was shown to hold only under conditions prevailing in the laboratory, some of which were fairly unusual. (When was the last time you interacted over an intercom with five strangers in other rooms?) Thus, from this study alone, it is not clear to what extent we can generalize the cause-and-effect findings to everyday, face-to-face situations. The relationship between the number of others present and a person's reaction to an emergency might be different in other situations.

While this experiment answers some questions, it opens up others. Why, for instance, should the number of witnesses present affect a person's willingness to help in an emergency? The researchers were sensitive to this question. Based on data from a brief questionnaire administered after the experiment, they proposed that subjects in larger groups were slower to help because the responsibility for helping was diffuse and less focussed than in smaller groups. Although this "diffusion of responsibility" explanation is interesting, note that this particular experiment did not demonstrate it to be either true or false. The experiment showed only that, under the conditions in the lab, the number of witnesses present affected helping behavior.

**Strengths of Experiments** Probably the greatest strength of experimental studies lies in their high level of internal validity. The results of a carefully done experiment admit direct, clear interpretation. This makes the experiment especially well-suited for testing causal hypotheses and theories. Experiments excel over other methods (surveys, field observation, and so on) in this respect.

Experiments have high internal validity precisely because they are designed to control or offset all factors other than the independent variable that might affect the dependent variable. Techniques to accomplish this include (1) randomly assigning subjects to treatments, (2) holding constant known extraneous variables, (3) incorporating extraneous variables as factors in the research design—that is, manipulating them as independent variables—so that they are not confounded with the main independent variables of interest, and (4) measuring extraneous variables and including them in the statistical analysis as covariates with the independent variables.

Both laboratory experiments and field experiments can be designed to have high internal validity. In practice, however, it often happens that laboratory experiments have higher internal validity than comparable field experiments. This happens, quite simply, because researchers often have more control over extraneous variables in the laboratory than in the field. With respect to external validity, however, field experiments often outstrip laboratory experiments.

**Weaknesses of Experiments** There are many weaknesses of experiments. One weakness is that many social phenomena simply cannot be investigated experimentally because the researchers are not able to manipulate the independent variables of interest. In daily life, there are substantial moral, financial, and practical restrictions on what can be manipulated experimentally.

Even when manipulation of the independent variable can be accomplished, experiments face several threats to internal validity. First, there is the possibility that the experimental manipulation may fail. This might occur, for example, if the subjects perceived or interpreted the manipulation as meaning something other than what the researcher intended. The usual remedy for this problem is to use *manipulation checks*—measures taken after the manipulation that check whether the subjects perceived the manipulation as intended. Use of manipulation checks is widespread in social psychological experiments.

Another threat to internal validity of experiments is the existence of *demand characteristics.* This term refers to cues in the experimental setting that practically demand certain kinds of responses (Orne, 1969). For instance, an experimenter studying cooperation and competition may inadvertently stress the word "competition" when giving instructions to subjects. If they detect this emphasis, the subjects may decide that they are supposed to compete in the situation. To reduce demand characteristics, many studies use standardized instructions (for example, tape-recorded messages). Then, too, some designs disguise the nature of the research, providing a plausible but false description of its purpose.

Another threat to internal validity is *experimenter expectancies.* This refers to the possibility that if an experimenter knows the research hypotheses, he or she may unwittingly telegraph that expectation to the subjects (Rosenthal, 1966). Then the participants—trying to be "good" subjects—may respond in such a fashion as to make the hypotheses come true! This is especially troublesome if the expectancies conveyed by the experimenter change systematically as a function of experimental treatment. Expectancy effects can be eliminated by keeping the research personnel "blind" to the experimental hypotheses and to the treatment condition to which each subject belongs; if they do not know the hypotheses, they cannot communicate them to subjects. The influence of experimenter expectancies can be prevented by minimizing contact between experimenters and subjects and by strictly standardizing the behaviors of the research staff in the experimental setting.

Beyond internal validity, experiments also face problems with external validity. Some experiments take place in settings that seem artificial to subjects and have low apparent realism. This is especially true of laboratory experiments, although less true of field experiments. Researchers often distinguish between *mundane realism* and *experimental realism* (Aronson, Brewer, & Carlsmith, 1985). Mundane realism refers to the extent to which the experimental setting appears similar to real-world situations. Experimental realism, in contrast, refers to the amount of impact the experimental situation creates—that is, the degree to which the subjects are involved in the situation.

Many laboratory experiments have been criticized as having low mundane realism, and this is often a correct characterization. It is quite possible, however, for a study to have low mundane realism but high experimental realism. Subjects were highly involved, for example, in the study discussed above where the experimenters staged an emergency in the laboratory. Many subjects were quite nervous and expressed concern when they came out of their room looking for the victim; and most expressed great surprise when they subsequently learned that the seizure was simulated, not genuine.

There is no single resolution to the problem of experimental realism. Some investigators use a mix of laboratory and field experiments when investigating a phenomenon. This approach is often successful, for it addresses both internal and external validity. Other investigators simply say they are more concerned with experimental realism than with mundane realism. If the situation seems real and involving to the subjects, they maintain, then the behavior by the subjects is equally real and worthy of study.

## Naturalistic Observation

Observational research involves collecting data about ongoing or naturally occurring events. Typically, it involves an observer or witness who makes observations of social behavior and records these systematically. Whereas a survey often intrudes on or interrupts ongoing activity and an experiment involves manipulation of the independent variable(s) and random assignment to treatment, observational research attempts to minimize or eliminate such intrusion and to study the events of interest *in situ* as they occur naturally. We will discuss two types of observational research: the case study and the field study.

**Case Studies** Some years ago on a Friday afternoon at 5:30 P.M.., a tornado roared through White County, Arkansas, completely destroying a town of 1,100 people. Altogether, there were 46 deaths, more than 600 injuries, and 3.5 million dollars in property damage. Such tragedies are of great interest to some social scientists who, by studying these events, hope to learn how communities can effectively cope with disaster. In this instance, a team of four researchers arrived in the area three days after the tornado. They interviewed county and other governmental officials. They also conducted a survey of residents of the affected areas and of a control group of people in nonaffected areas. In addition, they talked to representatives of various organizations about how their operations were affected by the disaster. The resulting data from this case study constitutes one of the most

comprehensive descriptions available of a community disaster (Barton, 1969).

A **case study** is an intensive investigation of a single incident or event. Case studies typically entail the use of multiple sources of information about the event in question. Interviews may be conducted with key participants or with some or all of the affected persons, as in the tornado study. Documents such as newspaper and media articles and official records may also be studied. When the event is ongoing or contemporary, trained observers may videotape it or record information about it as it occurs. The resulting data are usually rich in descriptive detail and reflect a variety of perspectives regarding what has taken place.

Case studies offer substantial advantages in certain circumstances. First, they are very useful for investigating social phenomena that are unpredictable and of relatively short duration. Case studies have frequently been employed to study such ephemeral phenomena as natural disasters (Drabeck & Stephenson, 1978) and collective behavior (Kerckhoff, Back, & Miller, 1965). Second, case studies are very useful in the beginning stages of research, where little is known about the phenomenon of interest. Because they involve information from diverse sources, case studies often yield excellent descriptive information that can help a thoughtful investigator formulate hypotheses about the phenomenon of concern. For this reason, a case study is sometimes used prior to a larger, more systematic investigation. Third, case studies can often be conducted with relatively little preparation. This means they can be used in circumstances where timeliness or speed is paramount, as in the study of current events.

There are two major disadvantages to the case study. First, the investigator has little or no control over the type of data available and is often forced to rely on information collected by others—newspaper and media personnel, public or organizational officials, and the diaries or informal records of participants. The information available may be highly selective and reflect particular biases. Other research methods such as surveys and experiments give the investigator much greater control over what information is collected and in what form.

The second disadvantage of case studies is that they usually treat only a single unit. In most case studies, such as the investigation of the tornado disaster, there is no way to determine whether a particular case is representative, so the external validity of any conclusions is essentially unknown. To overcome this limitation, a detailed case study may be combined with a survey of a sample of similar events. The case study provides detailed information, and the survey helps to assess the external validity of the findings.

**Field Studies**   A **field study** involves collecting observational data about ongoing activity in everyday settings—that is, in the "field." Most of the data come from one or more observers, members of the research team who have been trained to observe the activity of people and record information about it. For example, researchers have observed and recorded data about social interaction between couples in informal settings (Zimmerman & West, 1975), between judges and attorneys in the courtroom (Maynard, 1983), and between police and juveniles on city streets (Piliavin & Briar, 1964). One study (Suttles, 1968) observed the use of stores, churches, and parks by members of different ethnic groups in one neighborhood; this study was able to demonstrate that ethnicity was a major determinant of utilization patterns.

Field studies vary in how the information is collected and recorded by observers. In some instances, observers watch carefully while the phenomenon of interest is occurring and record the information later. The advantage here is that the observer does not arouse the curiosity or suspicion of participants; the disadvantage is that selective perception and memory may influence what is later recorded. In other studies, information is recorded concurrently with the observed behavior. For instance, in research on police-citizen encounters (Black & Reiss, 1970), trained observers coded the interaction as it occurred; they classified each behavior into one of a number of predetermined categories. In still other studies, researchers have made audio or video recordings of interactions, and subsequently coded and analyzed them (Whalen & Zimmerman, 1987). Such recordings maximize the information obtained, although interactions may be influenced if participants are aware that they are being taped.

When the behavior of interest occurs in public settings (such as restaurants, courtrooms, or retail stores), researchers can simply go to the setting and observe the action without participating directly in it. However, when the behavior of interest occurs only in private or restricted settings (such as the use of illegal drugs, intimate sexual activity, or recruiting new members for a cult), observation is usually difficult or impossible. To investigate activities of this type, researchers sometimes use the method of *participant observation* in which members of the research team not only make systematic observations but also play an active role in the ongoing events. In participant observation, researchers do not engage in coding or any other activity that would intrude on or disrupt the interaction.

Years ago, a researcher wanted to investigate sexual activity in "tearooms," public restrooms where men meet and engage in oral sex (Humphreys, 1970). By questioning police officers, he learned the location of these restrooms in the city and began to observe what occurred there. Sex in this setting typically involves three men, the two engaging in sex and a lookout. The researcher always adopted the role of lookout, which enabled him to observe without engaging in sexual activity. He was especially struck by the impersonality of the sexual interaction; the men did not speak and instead relied on gestures for communication. Had he tried to study this phenomenon by interviewing these men in their homes, it is unlikely that they would have reported this activity and very unlikely that the impersonality of it would have been as evident. One

Archival research, which involves the analysis of previously collected data, is facilitated by computer technology. Computers allow easy retrieval of large quantities of material.

drawback of this method, of course, is that the researchers' own behavior may influence the activity, producing outcomes that would not otherwise occur.

Field studies are sometimes able to utilize *unobtrusive measures,* techniques that do not intrude on the activity being studied and do not run the risk of eliciting reactions that otherwise might not occur (Webb et al., 1966). For example, some unobtrusive measures may be based on physical evidence left behind by people. This was illustrated in an unpublished work by Duncan, who discovered that the rate at which vinyl floor tiles needed replacement in the Chicago Museum of Science and Industry was a good indicator of the popularity of exhibits.

Overall, field observational techniques allow researchers to study ongoing activity in real-world settings. Their advantage is that they can provide a wealth of information about behavior as it actually occurs in natural settings. Disadvantages of field observation include its sensitivity to the specific recording methods used. Observations recorded after the fact are generally less reliable and valid than those recorded on the spot and those based on audio or video taping. In addition, the external validity of field observational studies can be problematic, since research of this type frequently focusses on only one group or organization or on a sample of interactions selected on the basis of convenience.

## Archival Research and Content Analysis

Although researchers prefer to collect original data, it is sometimes possible to test hypotheses and theories by analyzing data that already exist. The term **archival research** designates the acquisition and analysis (or reanalysis) of existing information that was collected by others. One significant advantage of archival research is its relatively low cost. When archival data of suitable quality exist, a researcher may find that to analyze them is an attractive alternative to the task of collecting and analyzing new data.

There are numerous sources of archival data. In the U.S., one important source is government agencies. The Census Bureau makes available much of the data it has collected over the years. Census data are a rich source of information about the U.S. population, and they often include repeated measures taken at different points in time that allow one to assess historical trends. The Bureau of Labor Statistics, the Federal Bureau of Investigation, and other agencies also make data available to investigators. Another source of archival data in the U.S. is the data banks maintained at various large universities. These archives serve as locations where researchers can deposit data they have collected so that others can use them. They include, among others, the Interuniversity Consortium for Political and Social Research at the University of Michigan and the Data Archive on Adolescent Pregnancy and Pregnancy Prevention. A third source of archival data—less used by social psychologists but still quite important —is individual organizations, such as insurance companies.

Yet another source of archival information for research is newspapers. Newspaper files are frequently used as a source of information about past events. For instance, an investigator wishing to study the student protests against racism that occurred on several college campuses in 1987 and 1988 might use local and campus newspapers as her data source. Beyond newspapers, other types of printed material and documents (for example, corporate annual reports or government documents and reports) often provide archival data amenable for use in research.

In some cases, an investigator relying on newspaper articles, government documents, or annual reports as archival sources can use information directly as it appears. All the investigator has to do is extract the information and analyze it (usually via computer). In other cases, however, the investigator faces the problem of how to interpret and code the information from the source. This problem can be especially severe, for example, if newspapers serve as the source. The verbal content of newspaper articles has to be coded into a form suitable for systematic analysis. Under these circumstances, an investigator typically will use the technique of **content analysis,** which involves a systematic scrutiny of documents or messages to identify specific characteristics and then making inferences based on their occurrence (Holsti, 1968). In doing a content analysis, the first step is to identify the information unit to be studied; is it the word, the sentence, the paragraph, or the theme? The next step is to develop and define the categories into which the units will be sorted. Finally, the researcher codes the units in each document into the categories.

As an example of content analysis, consider a study of the relationship between rhetorical forms and applause during speeches. Heritage and Greatbatch (1986) hypothesized that certain rhetorical forms—for example, a three-element list—are used by speakers to signal the audience to applaud. The raw data in this study were the texts of 476 speeches delivered by political leaders at party meetings. The researchers carefully defined each rhetorical device and identified its use in the speeches. They then counted the number of times each device was used and also noted whether each use was followed immediately by applause. As they hypothesized, applause was more likely to follow a three-element list.

Two advantages of archival research were noted earlier. These are the lower cost of using existing data compared to collecting new data and the possibility that a study can be done more quickly. A third advantage is that hypotheses can be tested with respect to phenomena that occur over time, since many records have been kept for decades or even centuries (such as marriage licenses). A fourth advantage is that archival methods and content analysis involve measures that are unobtrusive and do not in any way affect the behavior of interest. The major disadvantage of archival research is the lack of control over the type and quality of information; investigators are forced to work with whatever was collected by others. Another disadvantage is that creation of a reliable and valid content analytic scheme for use with records can be difficult. Finally, some sets of records contain substantial amounts of inconsistent and missing information.

## Comparison of Research Methods

The discussion of strengths and weaknesses of each method is summarized in Table 2-1. As you can see, there is no one best method of empirical investigation. Which method is most appropriate depends on the phenomenon being studied and which characteristic of research is most important to the researcher. In order to obtain descriptive data about some group or population, the investigator could conduct a survey, a field study, or look for relevant archival data. All three are characterized by a moderate degree of external validity. They differ substantially in how difficult they are to conduct. If lack of resources is a serious problem, the investigator may choose to do archival research.

Alternatively, consider a researcher who wants to test a causal hypothesis about attitude change. She could carry out a longitudinal survey (questioning the same people two or more times), conduct a laboratory experiment, or do a field experiment. One disadvantage of the survey is its limited degree of validity, which may prompt her to opt for an experiment. Both laboratory and field experiments are moderately difficult to conduct, and both have some associ-

■ **TABLE 2-1**
STRENGTHS AND WEAKNESSES OF RESEARCH METHODS

| | | | Method | | | |
|---|---|---|---|---|---|---|
| | Survey | Laboratory Experiment | Field Experiment | Case Study | Field Study | Archival Research |
| Internal Validity | Moderate | High | Moderate | Moderate | Low | Low |
| External Validity | Moderate | Moderate | High | Low | Moderate | Moderate |
| Investigator Control | Moderate | High | Moderate | Low | Moderate | Low |
| Intrusiveness of Measures | Moderate | Moderate | Low | Low | Moderate | Low |
| Difficulty of Conducting Study | Moderate | Moderate | High | Moderate | Moderate | Low |
| Ethical Problems | Few | Some | Some | Few | Many | Few |

ated ethical problems. However, the laboratory experiment is high in internal validity and only moderate in external validity, whereas the field experiment is the opposite. If this investigator is primarily concerned about the internal validity of her results, she should do a laboratory experiment; otherwise, perhaps a field experiment.

# ■ Ethical Issues in Research

In the past 20 years, there has been growing concern about ethical issues in research on humans. Specifically, a consensus has developed among investigators and other persons and agencies affiliated with the scientific community that subjects who participate in research have certain rights that must be protected. In some cases, protecting those rights requires investigators to limit or modify research practices.

In the following discussion of ethical issues, we focus first on several sources of potential harm to subjects. Then we discuss various safeguards, such as risk-benefit analysis and informed consent, to protect subjects' rights.

## Potential Sources of Harm

Harm to subjects from research can assume a variety of forms. These include physical harm, psychological harm, harm from breach of confidentiality, and social loss to a group of people. We will discuss each of these potential sources of harm.

**Physical Harm** These days, exposure to physical harm in social psychological research is relatively uncommon. Studies designed to measure the effects of stress sometimes utilize an exercise treadmill or the cold pressor test, where subjects immerse one hand in ice water. As precautions, prospective subjects can be thoroughly screened for relevant medical problems and any risks made known to them so that they can decide whether they might be harmed by participating. Then too, in studies involving physical stress, investigators can closely monitor participants for adverse effects throughout the research.

**Psychological Harm** A more common risk in social psychological research is psychological harm to subjects. This is most likely to occur as a result of subjects receiving negative information about themselves in the course of the study. For example, a common experimental manipulation is to give subjects false feedback about their physical attractiveness, others' reactions to them, or their performance on various tests or tasks. Such feedback can be used to raise or lower self-esteem, induce feelings of acceptance or rejection by others, or create perceptions of success or failure on important tasks. These manipulations are effective precisely because they do influence the subject's self-perception. Negative feedback may cause psychological stress or harm, at least temporarily. For this reason, some investigators believe that such techniques should not be employed in research. Others, however, believe that they are acceptable, and may be used if alternative, less potentially harmful manipulations are not available. When false feedback is used, it is important to minimize the time between presenting it and giving the subject a thorough *debriefing*, providing him or her with a full description of the study and emphasizing the falsity of the feedback.

Another potential source of psychological harm lies in inducing a subject to engage in behavior that, upon reflection, may have a negative impact on his or her self-image. A commonly cited example is Milgram's research (1965), in which subjects were asked to follow the experimenter's instructions and administer what they believed to be increasingly strong electric shocks to another person. Many subjects persisted even after the "victim" indicated that he had a heart condition and demanded to be released from the apparatus. Even though subjects were subsequently debriefed and informed that no shocks were actually received by the "victim," they were faced with the recognition that they did comply with the experimenter's orders to administer shock. Some people have suggested that it may have caused these subjects considerable anguish to be made aware of their capacity to behave in this way (Baumrind, 1964). Again, a postexperimental debriefing that emphasizes the ways in which the subject was induced to comply, as well as checkups on subjects following the experiment, are recommended safeguards.

**Breach of Confidentiality** Confidentiality is a special issue in survey and observational research. Interviewers and observers are frequently in a position to identify participants, and they may recall details of participants' answers to questions or of their behavior. Were the confidentiality of these data to be broken, the effects could be damaging to the participants. This concern arises especially in surveys inquiring about sexual behaviors or sensitive personal matters; it also arises in observational studies of deviant or criminal activities. For instance, in his observational study of sex in the tearoom, Humphreys (1970) saw the faces of dozens of the men; he did not disclose to anyone information about whom he observed. An important precaution is to avoid using research team members who are apt to have social contacts with respondents in other settings. In addition, some investigators avoid attaching any identifying information such as names and addresses to data after it has been collected. Identifying information can be kept separate from questionnaires or behavioral records to prevent later breaches of confidentiality.

Observational and case study research often deals with a specific group or organization. In the course of their investigation, researchers may gather a great deal of information both about the organization itself and about various members. When the findings are reported, the organization is typically assigned a pseudonym, and members are referred to by role only. This practice usually suffices to prevent outsiders from identifying the unit and its members, although it may not prevent members from identifying each other. There are obvious risks to members' positions, reputations, or jobs within the organization if certain kinds of information become known to other members.

**Social Harm**  Still another area of concern is the potential of research to harm entire groups or segments of society. There has been a great deal of research in recent years on ethnic minorities in the U.S. Research on differences between blacks and whites in intelligence, family structure, and academic success is potentially useful as a basis for social policy. But there is no guarantee that data on blacks or other minorities will be used to design better educational and job training programs. These data could as readily be used to restrict the access of blacks to better educational institutions and jobs. Although no one has yet advocated curtailing research of this type, there is heated debate over researchers' responsibilities with respect to the utilization of their findings.

## Institutional Safeguards

In the discussion of sources of harm we noted steps that researchers can take to prevent harm to participants. Many people feel that voluntary self-regulation is enough. Others feel that voluntary regulation, by itself, is insufficient and that some person or group other than the researcher should review research designs to protect the rights and interests of the participants. Accordingly, some institutional safeguards have been developed and put into place. Most importantly, these involve two practices, risk-benefit analysis and obtaining informed consent from all participants.

**Risk-Benefit Analysis**  Because it is a major provider of funds for research in the social and biomedical sciences, the U.S. Department of Health and Human Services has been instrumental in developing criteria for the review of research involving human subjects. Under these regulations, it is the responsibility of both investigators and institutions to minimize the risk, of

whatever type, to participants in research. The rules encourage researchers to develop designs that expose subjects to no more than *minimal risk,* defined as risk no greater than that "ordinarily encountered in daily life or during the performance of routine physical or psychological examinations or tests" (*Code of Federal Regulations, 46,* 1983, p. 6).

In addition, each institution receiving funds from various federal agencies is required to establish an institutional review board, which is responsible for reviewing proposed research involving human subjects. The review board (sometimes called a human subjects or research ethics committee) assesses the extent to which participants in each proposed study will be placed at risk. As noted earlier, many social psychological studies involve no foreseeable risks to participants. If the members of the board believe that participants might be harmed—physically, psychologically, or due to breach of confidentiality—it must make a further assessment. That is, it must conduct a **risk-benefit analysis,** weighing potential risks to the subjects against anticipated benefits to subjects and the importance of the knowledge that may result from the research. The review board will not approve research involving risk to participants unless it concludes that the risks "are reasonable in relation to anticipated benefits."

**Informed Consent**  The other major safeguard against risk is the requirement that informed consent be obtained from all individuals, groups, or organizations who participate in research studies. **Informed consent** means that potential participants have been told what their participation will entail and that researchers have obtained their voluntary consent to participate.

Specifically, six elements are essential to informed consent. First, potential subjects are entitled to a brief description of the procedures to be employed; they need not be told the purpose or hypothesis of the research. Second, they should be given information about foreseeable risks that participation may entail as well as any benefits they may receive. Third, participants should be told to what extent the information they provide will be kept confidential; if possible, they should be given a guarantee that confidentiality or anonymity will be maintained. Fourth, investigators should include information about what medical or psychological resources, if any, are available to subjects who are adversely affected by participation. Fifth, researchers should offer to answer questions about the study, whenever possible. Finally, potential subjects should be informed that they have the right to terminate their participation at any time.

In many survey and case-study settings, informed consent is implemented by orally providing subjects with this information. In experiments, especially those involving some risk to subjects, investigators often obtain written consent from each participant. Table 2-2 lists nine ethical questions that apply to many studies.

There is increasing acceptance in the social sciences of the principle that all potential participants in research should be given the opportunity to exercise judgement about whether to participate. In general, the mechanisms of documenting informed consent have served to make investigators more aware that their relationships with participants are interactive rather than unilateral and that participants have rights that cannot be ignored in the interest of pure science.

# ■ Summary

This chapter considers the methods used by social psychologists to systematically investigate social phenomena.

**Characteristics of Research** (1) The objectives of research include describing reality, identifying correlations among variables, testing causal hypotheses, and testing theory. (2) Research is usually guided by a hypothesis, which specifies a relationship between two or more variables. Hypotheses are drawn from observation of the world or from the examination of existing knowl-

---

■ **TABLE 2-2**
ETHICAL CONSIDERATIONS IN RESEARCH DESIGN

The following list provides some ethical questions commonly asked by investigators and review boards regarding the design of proposed social psychological studies.

1. Is it possible that participants might be harmed physically, for example, by strenuous exercise?
2. Does the study involve giving participants false information about themselves or any other form of deception?
3. Does the study involve inducing participants to engage in behavior that might threaten their self-respect?
4. If audio or video tapes of the participants are made, will their permission be obtained to use the tapes as a source of data?
5. Have appropriate steps been taken to preserve the confidentiality of information obtained about the participants?
6. Will people be told in advance about foreseeable risks which their participation may entail?
7. Will people be given an opportunity to ask questions before they consent to participate?
8. Will people be informed that they have the right to terminate their participation at any time?
9. At the end of the study, will participants be fully debriefed, informed about the real nature of the study and its procedures?

edge. (3) Ideally, the findings of empirical research should be high in both internal validity and external validity.

**Research Methods** Social psychologists rely heavily on four methods. (1) The survey involves asking questions and systematically recording the answers. Surveys are used to gather various types of information. The quality of the data obtained depend on the reliability and validity of the measures used. (2) The experiment involves the manipulation of independent variables, the measurement of the effect(s) on one (or more) dependent variable(s), and the random assignment of subjects to experimental conditions. Some experiments are conducted in a laboratory, where the investigator has a high degree of control, while others are conducted in natural settings. (3) Naturalistic observation involves collecting data about naturally occurring events. The case study is the intensive analysis of a single incident; this incident may have occurred in the past. The field study is the collection of observational data about an incident or event as it occurs. (4) Archival research involves the analysis of existing information collected by others. Sources of archival data include the Census Bureau and other federal agencies, data archives, and newspapers. Content analysis is used to systematically study textual material, such as speeches. (5) Each method has particular strengths and particular weaknesses.

**Ethical Issues in Research** (1) There are four potential sources of harm to participants in research. These are physical harm, psychological harm, breach of confidentiality, and social harm to a group. There are various steps that individual investigators can take to prevent such harm. (2) In addition, there are institutional safeguards against harm. These include the requirement that investigators minimize risks to participants and that informed consent be obtained from all participants. Research designs are monitored by review boards to make sure these conditions are met by investigators.

## Key Terms

| | |
|---|---|
| ARCHIVAL RESEARCH | 43 |
| CASE STUDY | 41 |
| CONTENT ANALYSIS | 44 |
| CONVENIENCE SAMPLE | 35 |
| DEPENDENT VARIABLE | 30 |
| EXPERIMENT | 36 |
| EXTERNAL VALIDITY | 32 |
| EXTRANEOUS VARIABLE | 30 |
| FIELD STUDY | 42 |
| HYPOTHESIS | 29 |
| INDEPENDENT VARIABLE | 30 |
| INFORMED CONSENT | 47 |
| INTERNAL VALIDITY | 31 |
| INTERVIEW | 33 |
| METHODOLOGY | 28 |
| POPULATION | 35 |
| QUESTIONNAIRE | 33 |
| RANDOM ASSIGNMENT | 37 |
| RELIABILITY | 33 |
| RESPONSE RATE | 33 |
| RISK-BENEFIT ANALYSIS | 47 |
| SAMPLE | 35 |
| SIMPLE RANDOM SAMPLE | 35 |
| STRATIFIED SAMPLE | 35 |
| SURVEY | 32 |
| VALIDITY | 34 |

# 3

chapter

# Socialization

# ■ Introduction

A change has come over me. I do things differently from the way I used to. The change has been subtle, slow, but other people have noticed it as well. My wife complains, but it's not my fault. I am suffering from creeping Harry Cohenism.

Harry Cohen is my father. For years, this was nothing more than a statement of fact. He is my father and I am his son and other than that, we had very little in common. He could spell and I could not. He could do math in his head and I could not. He could not sleep well or at all past 6 A.M. and I could sleep soundly and forever and he was always, but always, way older than I was.

Now, however, we are both the same age—middle-aged. He is upper middle-aged, and I am younger middle-aged, but we are essentially the same age.

I sleep like him now. I used to sleep like me—deeply and endlessly. When I was a kid, this would drive my father mad. He would look at me sound asleep in my bed at, say, 1 o'clock in the afternoon, and explode with frustration: "Get up, get up!" He could not understand how anyone could sleep so late. I could not understand how anyone could not. Sleeping, in fact, was one of the very few things I did really well. I excelled at sleeping.

But no more. I wake up at 6 in the morning. My eyes open and that is it. I try to go back to sleep, but I cannot. I lie and look at the clock and wait for the sun to come up, but there is no going back to sleep. For a long time, I worried that something was wrong with me. I could not figure out what was happening, and then it occurred to me. I was becoming my father.

I know that by all logic I should be becoming my mother, too. After all, half my genes are hers and, knowing her, they are the dominant ones. But for some reason, I feel that her genes declared themselves early and that from here on out it is my father who is taking over.

After all, it is my father and not my mother who falls asleep in front of the television set. It is my father and not my mother who gets bogged down in detail in the middle of telling a story. And it is my father who will tell you the same story twice—only each time a little bit differently. I am even beginning to like opera.

For a time, I thought that I was my own man, that I could be exactly what I wanted to be. I recognize now that there is such a thing as a genetic manifest destiny—a case of the future really being the past. I don't have any problems with that, though. My father is a grand man—sweet and sensitive and smart. I didn't realize it until recently, but in many ways we're a lot alike—not like my son. Boy can that kid sleep!

I wonder where he gets it from. (Cohen, 1982)

Richard Cohen expresses well one of the striking features of social life. There is a great deal of continuity from one generation to the next, continuity in both physical characteristics and in behavior. Genetic inheritance is one source of continuity. But a major contributor to intergenerational similarity is **socialization,** the ways in which individuals learn skills, knowledge, values, motives, and roles appropriate to their position in a group or society (Bush & Simmons, 1981).

How does an infant become "human"—that is, an effective participant in society? The answer is through socialization. As we grew from infancy, we interacted continually with others. Like Richard Cohen, we learned to speak English, a prerequisite for participation in our society. We learned basic interaction rituals, such as greeting a stranger with a handshake and a loved one with a kiss. We also learned the socially accepted ways to achieve various goals, both material (food, clothing, and shelter) and social (the respect, love, and help of others). It is obvious that socialization makes us like most other members of society in important ways. It is not so obvious that socialization also produces our individuality. The self and the capacity to engage in self-oriented acts (discussed in Chapter 4), are a result of socialization.

This chapter will first examine childhood socialization. By "childhood" we mean the period from birth to adolescence. Childhood is a social concept, shaped by historical, cultural, and political influences (Denzin, 1977). In con-temporary American society, we define children as immature, in need of training at home and of a formal education. At the end of this chapter, we will consider the continuing socialization of adults.

The discussion focuses on the following five questions.

1.  What are the basic perspectives in the study of socialization?

2.  What are the socializing agents in contemporary American society?

3.  What are the processes through which socialization occurs?

4.  What are the outcomes of socialization in childhood?

5.  What is the nature of socialization in adulthood?

# ■ Perspectives on Socialization

As noted at the beginning of this chapter, Richard Cohen attributed the similarity between himself and his father to "genetic manifest destiny." What is the more important influence on behavior—nature or nurture, heredity or environment? This question has been especially important to those who study the child. Although both are influential, one view emphasizes biological development (heredity), whereas another emphasizes social learning (environment).

## The Developmental Perspective

The human child obviously undergoes a process of maturation. He or she grows physically, develops motor skills in a sequence that appears to be relatively uniform, and begins to engage in various social behaviors at about the same age as most other children.

Some theorists view socialization as largely dependent on processes of physical and psychological maturation, which are biologically determined. Gesell and Ilg (1943) have documented the sequence in which motor and social skills develop and the ages at which each new ability appears in the average child. They view the development of many social behaviors as primarily due to physical and neurological maturation, not social factors. For example, toilet training requires voluntary control over sphincter muscles and the ability to recognize cues of pressure on the bladder or lower intestine. According to developmental theory, when children around age 2½ develop these skills, they learn by themselves without environmental influences.

Table 3-1 indicates sequences of development of various abilities that have been identified by observational research. The ages indicated are approximate; some children will exhibit the behavior at younger ages, while others will do so later.

As an example, consider the development of responsiveness to other persons. As early as 4 weeks, many infants respond to close physical contact by relaxing. At 16 weeks, babies can discriminate the human face and usually smile in response. They also show signs of recognizing the voice of their usual caregiver. By 28 weeks, the infant clearly differentiates faces and responds to variations in facial expression. At 1 year, the child shows a variety of emotions in response to behavior by others. He or she will seek out interaction with adults or with siblings by crawling or walking to them and tugging on clothing. Thus, recognition of, responsiveness to, and orientation toward adults seem to follow a uniform developmental pattern. The ability to interact with others depends in part on the development of visual and auditory discrimination.

## The Social Learning Perspective

While the developmental perspective focuses on the unfolding of the child's own abilities, the social learning perspective emphasizes the child's acquisition of cognitive and behavioral skills from the environment. Successful socialization requires that the child acquire considerable information about the world. First, many physical or natural realities must be learned (Baumrind, 1980), such as what animals are dangerous and which things are edible. Children must also learn about the social environment. They must learn the language used by people in their environment in order to communicate their needs to others. They also need to learn the meanings their caregivers associate with various actions. Children need to learn to identify the kinds of persons encountered in the immediate environment. They need to learn what behaviors they can expect of people, as well as the expectations for their own behavior.

According to the social learning perspective, socialization is primarily a process of children learning the meanings that are shared in the groups in which they are reared (Shibutani, 1961). Such variation in meanings gives groups and subcultures (and societies) their distinctiveness. While the content—what is learned—varies from group to group, the processes by which it takes place appear to be universal. This viewpoint emphasizes the adaptive nature of socialization. The infant learns the verbal and interpersonal skills necessary to interact successfully with others. Having acquired these skills, children can perpetuate the meanings that distinguish their social groups and even add to or modify these meanings by introducing innovations of their own.

Recent research on socialization takes into account both the importance of developmental processes and the influence of social learning. The developmental age of the child obviously determines which acts the child can perform. Infants less than 6 months old cannot walk. All cultures have adapted to these developmental limitations by coordinating the performance expectations placed on children with the maturation of their abilities. At the same time, developmental processes are not necessarily sufficient for the emergence of complex social behavior. In addition to developmental readiness, social interaction—learning—is necessary for the development of language. This is illustrated by the case of Isabelle, who lived alone with her deaf-mute mother until the age of 6½. When she was discovered, she was unable to make any sound other than a croak. Yet within two years after she

■ **TABLE 3-1**
THE PROCESS OF DEVELOPMENT

|  | 16 Weeks | 28 Weeks | 1 Year | 2 Years | 3 Years |
|---|---|---|---|---|---|
| **Visual Activity** | Eyes follow objects<br>Eyes adjust to objects at varying distances | Watches activity intently<br>Hand–eye coordination | Watches moving objects (TV picture) | Responds to stimuli in peripheral vision<br>Looks intently for long periods | |
| **Interpersonal** | Smiles at human face<br>Responds to caregiver's voice<br>Demands social attention | Responds to variation in tone of voice<br>Differentiates people (fears strangers) | Engages in responsive play<br>Shows emotions, anxiety<br>Shows definite preferences for some persons | Prefers solitary play<br>Rudimentary concept of ownership | Plays cooperatively with older child<br>Strong desire to please<br>Sex differences in choice of toys, materials |
| **Vocal Activity** | Vocalizes pleasure (coos, gurgles, laughs)<br>Babbles (strings of syllable-like sounds) | Vocalizes vowels and consonants<br>Tries to imitate sounds | Vocalizes syllables<br>Practices 2 to 8 words | Vocalizes constantly<br>Names actions<br>Repeats words | Uses three-word sentences<br>Likes novel words |
| **Bodily Movement** | Holds head up<br>Rolls over | Sits up | Stands<br>Climbs up and down stairs | Runs<br>Likes large-scale motor activity—push, pull, roll | Motion fluid, smooth<br>Good coordination |
| **Manual Dexterity** | Touches objects | Grasps with one hand<br>Manipulates objects | Manipulates objects serially | Good control of hand and arm | Good fine-motor control—uses fingers, thumb, wrist well |

*Sources: Adapted from Gesell and Ilg, 1943; Caplan, 1973.*

Responsiveness to another person develops early in life. By 16 weeks of age, a child smiles in response to a human face. By 28 weeks, a child can distinguish caregivers from strangers.

entered a systematic educational program, her vocabulary numbered more than 1,500 words, and she had the linguistic skills of a 6-year-old (Davis, 1947).

Thus, behavior is influenced by both nature and nurture. Developmental processes produce a readiness to perform certain behaviors. The content of these behaviors is determined primarily by social learning—by cultural influences.

## The Impact of Social Structure

A third perspective emphasizes the influence of social structure. Socialization is not a random process. Teaching new members the rules of the game is too important to be left to chance. Socialization is organized according to the sequence of roles that newcomers to the society ordinarily pass through. In American society, these include familial roles, such as son

or daughter, and roles in educational institutions, such as preschooler, elementary school student, and high school student. These are age-linked roles; we expect transitions from one to another to occur at certain ages. Distinctive socialization outcomes are sought among those who occupy each of these roles. Thus, young children are expected to learn language, and basic norms governing such diverse activities as eating, dressing, and bowel and bladder control. Most preschool programs will not enroll a child who has not learned the latter.

Furthermore, social structure designates the persons or organizations responsible for producing desired outcomes. In a complex society such as ours, there is a sequence of roles and a corresponding sequence of agents. From birth until age 5, the family is primarily responsible for socializing the child. From ages 6 to 12, a child is an elementary school student; we expect elementary school instructors to teach the basics to their students. Next, the adolescent becomes a high school student, with yet another group of agents to further develop their knowledge and abilities.

This view is sociological; it considers socialization as a product of group life. It calls our attention to the changing content of and responsibility for socialization throughout the individual's life.

# ■ Agents of Childhood Socialization

Socialization has four components. It always involves (1) someone who serves as a source of what is being learned; (2) a learning process; (3) a person who is being socialized; and (4) something that is being learned. We refer to these four components as *agent, process, target,* and *outcome.* This section will consider the three primary agents of childhood socialization—family, peers, and school. Subsequent sections will focus on the processes and outcomes of childhood socialization.

## Family

At birth, infants are primarily aware of their own bodies. Hunger, thirst, or pain create unpleasant and perhaps overwhelming bodily tension. The infant's primary concern is to remove these tensions and satisfy bodily needs. In order to meet the infant's needs, adult caregivers must learn to read the infant's signals accurately (Ainsworth, 1979). At the same time, infants begin to perceive their principal caregivers as the source of need satisfaction. These early experiences are truly interactive (Bell, 1979). The adult learns how to care effectively for this infant, and the infant forms a strong emotional attachment to the caregiver.

**Is a Mother Necessary?** Does it matter who responds to and establishes a caring relationship with the infant? Must there be a single principal caregiver in infancy and childhood in order for effective socialization to occur?

Psychoanalytic theory (Freud) asserts that an intimate, emotional relationship between infant and caregiver (almost always the mother at the time Freud wrote) is essential to healthy personality development. This was one of the first hypotheses to be studied empirically. To examine the effects of the absence of a single, close caregiver on children, researchers have studied infants who are institutionalized. In the earliest reported work, Spitz (1945, 1946) studied an institution in which 6 nurses cared for 45 infants less than 18 months old. Although the infants' basic biological needs were met, they had little contact with the nurses, and there was little evidence of emotional ties between the nurses and infants. Within one year, the infants' scores on developmental tests fell dramatically from an average of 124 to an average of 72. Within two years, one-third had died, 9 had left, and the 21 who remained in the institution were severely retarded. These findings dramatically support the hypothesis that an emotionally responsive caregiver is essential.

The findings of research on institutionalized children led to the conclusion that infants need a secure **attachment**—a warm, close relationship with an adult that produces a sense of security and provides stimulation—in order to develop the interpersonal and cognitive skills needed for proper growth (Ainsworth, 1979).

Could the argument be taken one step further? Can these studies be interpreted as indicating that infants need not merely an attachment, but a warm, intimate and continuous relationship with a *mother* (Bowlby, 1965)? Perhaps only in the mother-infant relationship can there occur the sense of security and emotional warmth and responsiveness necessary for development. Other potential caregivers may have less emotional interest in the infant and may not provide adequate substitutes. This position—if sustainable—clearly implies that (1) mothers should remain in the home and care for their children instead of working outside the home and (2) day-care centers will have detrimental effects on child development. Yet, in American society more and more women with children are employed full- or part-time outside the home. According to the Bureau of Labor Statistics, 57 percent of the women in two-parent families with preschool children are working. These women are vitally concerned with the effects of maternal employment and child care on children.

**Effects of Maternal Employment**  What are the effects of maternal employment on children of various ages? Studies that compare children whose mothers are employed with similar children whose mothers are not employed report few differences (Lamb, 1982). There is some evidence that children whose mothers return to work within a few weeks or months of the child's birth have an insecure maternal attachment at 12 months of age. This effect may be transient. Preschool-age children of employed mothers do not differ in attachment from children of non-employed mothers. Among grade-school children from lower-class families, sons of employed mothers show poorer psychological adjustment than do sons of nonemployed mothers. Finally, among adolescents, boys and girls whose mothers are employed have less stereotyped sex-role attitudes and expectations.

What about the effects of day care? It appears to have no effect on intellectual or emotional development (Belsky, Steinberg, & Walker, 1982). On the other hand, children in all-day day care for an extended time interact more with peers, both positively (for example, cooperatively) and negatively (for example, aggressively), than children cared for by their mothers. They are also less compliant with adult requests. Reviews of the effects of day care on preschoolers stress that the effects depend on the quality of the care: training of the staff, staff involvement with the children, and continuity in the caregiver-child relationship.

**Family Composition**  Does the makeup of the family affect the way in which parents socialize children? Research suggests that the answer is

yes. Both the number of children and the number of parents influence the process and the outcomes.

A study compared 52 sets of male twins and 44 "singleton" boys age 2½ (Lytton, Conway, & Sauve, 1977). Data collection involved interviews with parents and observation of parent-child interaction in the home and in the laboratory. Twins received less verbal interaction (per child) than the singletons. Parents gave twins fewer directions and were less likely to follow up to see whether the boy(s) had completed the directions. Twins and singletons also differed in their behavior. The quantity and quality of speech of the twins was lower than that of the singletons. Twins also showed a stronger need for contact or affiliation with their parents. Parents of twins obviously have less time per child, and the reduced rates of interaction appear to influence both verbal and emotional development.

The size of the family also influences patterns of authority. As more children are born, authority becomes centralized in the parents, and the rules applied to children's behavior become more formal (Clausen, 1966). A study of high school students found that those from larger families reported more often that their parents were autocratic and that they did not explain the rules to which the children were subject (Elder & Bowerman, 1963). At the same time, children from larger families are more likely to experience independence, as parental attention is spread over more children.

Because of divorce, many children are being raised by one parent, usually the mother. Research comparing these children with their counterparts in intact families has identified a number of differences. The period following the parents' separation is often characterized by stress and change, including changes in residence, school, and the custodial parent's work schedule. These changes have many effects on the children's intellectual, emotional, and social development (Hetherington, Cox, & Cox, 1982). These effects peak about one year after the separation, and many of the differences between children living with a single parent and children living with both parents lessen considerably in the second year.

A study of white preschool-age children found that children in divorced families have more negative interactions with both mothers and peers, compared to children in intact families. The differences were especially evident in boys. In addition, there was a loss of control over the child by the mother and more oppositional behavior by the children in single-parent families. A study of older children whose parents had been divorced for several years found that the children perceived less closeness and sharing in their families and reported less maternal control over their behavior than children in intact families. Adolescents in mother-only families are less likely to complete high school (Garfinkel & McLanahan, 1986). Daughters in such families are more likely to have children and marry as teenagers.

**Social Class and Socialization Techniques** An important influence on socialization is social class. In the U.S., the lower and middle classes differ from one another in their discipline techniques. These two classes differ in their level of education. Educated persons are more likely to read books and to rely on the opinion of experts (such as Dr. Spock) about child rearing. A survey of recent parents found that virtually all couples read at least one such "primer" on how to parent (Clarke-Stewart, 1978), but, as parental education increased, so did the number of books read.

Since 1945, child-rearing books have advised parents to rely on reasoning rather than physical punishment when children violate norms. If the more-educated parents read more of these books, we would expect them to follow this advice more closely than less-educated parents. Several studies support this hypothesis; there is a negative relationship between social class and the use of physical punishment (Gecas, 1979). Middle- and upper-class parents are more likely than lower-class parents to rely on reasoning and to use love withdrawal (shame and guilt)

in disciplining their children. Such a parent responds to a rule violation by saying "I'm ashamed of you" or "Good children don't do things like that." These techniques are obviously based on the child's emotional attachment to the parent.

Kohn (1969) argues that the goal of discipline varies with social class. Middle-class parents are likely to pay attention to intentions and punish intentional rule violations but not accidental ones. In working-class families, children are more likely to be punished for acts with harmful consequences, regardless of intentions. Research generally supports these hypotheses (Gecas, 1979). In a study of parents of third graders (Gecas & Nye, 1974), both husbands and wives were asked the following questions: "If your child is playing and accidentally breaks something of value, what would you do?" and "If your child intentionally disobeys after you have told him to do something, what would you do?" The indicator for social class was based on the husband's occupation, which was either white collar or blue collar. White-collar parents frequently reported they would respond differently to the two situations, whereas blue-collar parents were more likely to report similar responses.

Thus, socioeconomic status is a major influence on discipline techniques; higher-status parents are more likely to use psychological influence and to base rewards and punishments on the child's intentions.

## Peers

As the child grows, peers become increasingly important as socializing agents. The peer group differs from the family on several dimensions. These differences influence the type of interaction and, thus, the kinds of socialization that occur. The family consists of persons who differ in status or power, while the peer group is composed of status equals. From an early age, the child is taught to treat parents with respect and deference. Failure to do so will probably result in discipline, and the adult will use the incident as an opportunity to instruct the child about the importance of deference (Denzin, 1977; Cahill, 1987). Interaction with peers is more open and spontaneous; the child does not need to be deferential or tactful. Thus, children at the age of 4 bluntly refuse to let children they dislike join their games. With peers, they may say things that adults consider insulting, such as "You're ugly," to another child. The development of social skills is facilitated by daily exposure to other children (Roopnarine, 1985).

Membership in a particular family is ascribed, whereas peer interactions occur on a voluntary basis (Gecas, 1981). Thus, peer groups offer children their first experiences in exercising choice over whom they relate to. The opportunity to make such choices contributes to the child's sense of social competence and allows interaction with other children, who complement the developing identity.

Unlike the child's family, peer groups in early and especially middle childhood (ages 6 to 10) are usually homogeneous in regard to sex and age. A survey of 2,299 children in third through twelfth grade measured the extent to which they belonged to tightly knit peer groups, the size of such groups, and whether they were homogeneous by race and gender (Shrum & Cheek, 1987). The results are displayed in Table 3-2. The proportion belonging to a group peaked in sixth grade and then declined. The size of peer groups declined steadily from third through twelfth grade, as did heterogeneity by race and gender.

Peer associations make a major contribution to the development of the child's identity. In peer interaction, the role of friend is learned, contributing to greater differentiation of the self (Denzin, 1977). Peer and other relationships found outside the family provide a basis for establishing independence; the child ceases to be exclusively involved in the role of offspring, sibling, grandchild, and cousin. These alternate, nonfamilial identities may provide a basis for actively resisting parental socialization efforts (Stryker & Serpe, 1982). For example, a parent's

■ **TABLE 3-2**

PEER GROUP CHARACTERISTICS, BY GRADE IN SCHOOL

| Grade | Children Belonging to a Group (%) | Heterogeneous Groups (%) | Average Group Size |
|-------|-----------------------------------|--------------------------|--------------------|
| 3     | 34.9                              | 46                       | 8.0                |
| 4–6   | 46.3                              | 27                       | 7.9                |
| 7–8   | 32.8                              | 10                       | 6.5                |
| 9–12  | 23.2                              | 19                       | 6.3                |

*Source: Adapted from Shrum and Cheek, 1987, Tables 1 and 2.*

attempt to enforce certain rules may be resisted by the child whose friends make fun of children who behave that way.

## School

Unlike the peer group, school is expressly designed to socialize children. In the classroom, there is typically one adult and a group of children of similar age. There is a sharp status distinction between teacher and student. The teacher determines the skills to be acquired and relies heavily on instrumental learning techniques (with such reinforcers as praise, blame, and privileges) to shape student behavior (Gecas, 1981). School is the child's first experience with formal and public evaluation of performance. Every child's behavior and work is evaluated in terms of the same standards, and the judgments are made publicly to others in the class, as well as to parents.

Schools are expected to teach reading, writing, and arithmetic. But they do much more than that. Teachers use the rewards at their disposal to reinforce certain personality traits, such as punctuality, perseverance, and tact. Schools teach children which selves are desirable and which are not. Thus, children learn a vocabulary that they are expected to use in evaluating themselves and others (Denzin, 1977). The traits chosen are those thought to facilitate social interaction throughout life in a particular society. In this sense, schools civilize children.

Social comparison has an important influence on the behavior of school children. Because teachers make public evaluations of the children's work, each child can judge his or her performance relative to others. These comparisons are especially important to the child because of the relative homogeneity of the classroom group. Even if the teacher deemphasizes a child's low score on the spelling test, the child interprets the performance as a poor one relative to those of classmates. A consistent performance will affect a child's image of self as a student.

An observational study of children in kindergarten, first, second, and fourth grades documented the development of social comparison in the classroom (Frey & Ruble, 1985). In kindergarten, comparisons were to personal characteristics—for example, liking ice cream. Comparisons of performance increased sharply in first grade; at first these were blatant but became increasingly subtle in second and fourth grades.

# ■ Processes of Socialization

How does socialization occur? Frequently, it involves learning. We will examine three processes that are especially relevant: instrumental conditioning, observational learning, and internalization.

## Instrumental Conditioning

When you dressed this morning, chances are you put on a shirt or blouse, pants, a dress, or a skirt that had buttons, hooks, and/or zippers. When you were younger, learning how to master buttons, hooks, zippers, and shoelaces undoubtedly took considerable time, trial and error, and slow progress accompanied by praise from adults. These skills were probably acquired through **instrumental conditioning,** a process wherein a person learns what response to make in a situation in order to obtain a positive

reinforcement or avoid a negative reinforcement. The person's behavior is "instrumental" in the sense that it determines whether he or she is rewarded or punished.

The most important process in the acquisition of many skills is a type of instrumental learning called shaping (Skinner, 1953, 1957). **Shaping** refers to learning in which an agent initially reinforces any behavior that remotely resembles the desired response and subsequently requires increasing correspondence between the learner's behavior and the desired response before providing reinforcement. Shaping thus involves a series of successive approximations, where the learner's behavior comes closer and closer to resembling the specific response desired by the reinforcing agent.

In socialization, the degree of similarity between desired and observed responses required by the agent depends in part on the learner's past performance. In this sense, shaping is interactive in character. In teaching children to clean their rooms, parents initially reward them for picking up their toys. When children show they can do this consistently, parents may require that the toys be placed on certain shelves as the condition for a reward. Shaping is more likely to succeed if the level of performance required is consistent with the child's abilities. Thus, a 2-year-old may be praised for drawing lines with crayons, whereas a 5-year-old may be expected to draw recognizable objects or figures.

The first statement of the influence of rewards on behavior is called the *law of effect:* An act that is followed by satisfaction is more likely to recur than is an act that is followed by discomfort (Thorndike, 1907). The reinforcement may be either positive or negative.

**Reinforcement Schedules** When shaping behavior, a socialization agent can use either positive reinforcement or negative reinforcement. Positive reinforcers are stimuli whose presentation strengthens the learner's response; positive

Shaping is the process through which many complex behaviors are learned. Initially the socializer (teacher or parent) rewards behavior that resembles the desired response. As learning progresses, greater correspondence between the behavior and the desired response is required to earn the reward.

reinforcers are typically such things as food, candy, money, or high grades. Negative reinforcers are stimuli whose withdrawal strengthens the response, such as the removal of pain.

In everyday practice, it is rare for a learner to be reinforced each time the behavior desired by the agent is performed. Instead, reinforce-ment is given only some of the time. In fact, it is possible to structure when reinforcements are presented to the learner, using a reinforcement schedule.

There are a variety of possible reinforcement schedules. One of these—the *fixed-interval* schedule—involves reinforcing the first correct response after a specified time period has elapsed. This schedule produces the fewest correct responses per unit time; if the person is aware of the length of the interval, he or she will respond only at the beginning of the interval. It is interesting to note that many schools give examinations at fixed intervals, such as the middle and end of the semester; perhaps that is why many students study only just before an exam. The *variable-interval* schedule involves reinforcing the first correct response after a variable time period. In this case, the individual cannot predict when reinforcement will occur, so he responds at a regular rate. Grading a course on the basis of several surprise or "pop" quizzes utilizes this schedule.

The *fixed-ratio* schedule provides a reinforcement following a specified number of nonreinforced responses. Paying a worker on a piece rate, such as five dollars for every three items produced, utilizes this pattern. If the reward is sufficient, the rate of behavior may be high. Finally, the *variable-ratio* schedule provides reinforcement after several nonrewarded responses, with the number of responses between reinforcements varying. This schedule typically produces the highest and most stable rates of response. An excellent illustration is the gambler, who will insert quarters in a slot machine for hours, receiving only occasional, random payoffs.

**Punishment** By definition, **punishment** is the presentation of a painful or discomforting stimulus (by a socializing agent) that decreases the probability that the preceding behavior (by the learner) will occur. Punishment is one of the major child-rearing practices used by parents. Up to 90 percent of parents questioned in surveys reported that they punished their chil-

Box 3-1

## THE HIDDEN COSTS OF REWARD

Do rewards always increase the likelihood that a child will repeat the behavior?

We usually assume that the effects of reinforcement on behavior are positive, that rewarding someone for a behavior will make that person more likely to repeat it. Usually that assumption is correct, but there are exceptions to the rule. Depending on the nature of the reward and of the behavior, reinforcement can produce a decrease in the likelihood that the behavior will be repeated.

Behavior may result from two kinds of motivation (Deci, 1975). **Extrinsically motivated behavior** is engaged in in order to obtain an external reward, such as food, money, or praise. **Intrinsically motivated behaviors** are those performed in order to achieve an internal state that the individual finds rewarding. Extrinsic behavior is a means to an end; intrinsic behavior is enacted for its own sake. Many people work in order to earn money, whereas painting a picture, completing a crossword puzzle, or solving a riddle may be rewarding in and of themselves.

Research suggests that providing an extrinsic reward for intrinsically motivated behaviors may reduce the frequency or quality of the activity. One study employed three groups of preschool children who had demonstrated high levels of interest in drawing (Lepper, Greene, & Nesbit, 1973). The first group was told they would get a certificate (symbolic reward) if they completed the task. The second group was simply given the certificate after they finished. Those in the control group neither expected nor received the reward for drawing. Children who expected the reward spent less time drawing during the free-play period. In subsequent research using a similar design, investigators offered children either money, a symbolic reward, or verbal praise (Anderson, Manoogian, & Reznick, 1976). Whereas both monetary and symbolic reinforcement reduced interest in the task, verbal praise produced an increase in time spent in the task during the free period. A recent study of preschool children found that the decline in performance produced by rewards may be temporary (McCullers, Fabes, & Moran, 1987). If the child subsequently performs the same task and is not rewarded, performance improves dramatically.

Why do symbolic rewards have a different effect than verbal praise? It has been suggested that extrinsic rewards serve two functions: either a controlling function that maintains or changes behavior or an information function that indicates how well the person is doing (Deci, 1975). Tangible rewards such as money may serve a controlling function, while verbal praise may provide information that motivates further behavior. These hypotheses were tested by rewarding preschool and elementary school children with pretzels or stars or verbal praise for a maze performance task (Dollinger & Thelen, 1978). Consistent with prior research, pretzels and stars earned upon completion of the task reduced intrinsic motivation, while verbal praise during the task—which conveyed information about the quality of task performance—did not affect interest.

A review of research in this area (Condry, 1977) suggests that at age 5, children become aware that sometimes they do things because they want to and other times they do them because people promise them rewards. When the child is intrinsically motivated, the self is centrally involved. The focus and pace of activity are determined by the child. The child assesses performance in terms of internal standards. When someone else introduces extrinsic rewards, the child focuses effort in the direction desired by the adult. Such activity involves the self to a much smaller degree, and is evaluated in terms of the standards of the person providing the reward. This interpretation implies that games should elicit intrinsic motivation, while tests should stimulate extrinsic motivation. A study of college students found higher levels of intrinsic motivation when a task was presented as a game than when the same task was presented as a test (Koestner, Zuckerman, & Koestner, 1987).

dren occasionally (Sears, Maccoby, & Levin, 1957). Punishment is frequently physical; in a survey of 1,146 families, 71 percent of the parents reported slapping or spanking children (Gelles, 1980). Because punishment is so widely used, it is important to inquire whether it actually works as a socialization technique.

Research indicates that punishment is effective in some circumstances but not others. One influence on its effectiveness is timing. In a study of fourth- and fifth-grade boys, each was presented with a pair of toys and told to select one of the pair. The punishment consisted of a verbal reprimand: "No. That's for older boys." The time between the act and the occurrence of punishment was systematically varied. In one group, the boys were reprimanded just as they touched the toys. In another, the reprimand was given after the boys picked up the toy. In a control group, the boys were warned prior to their choice not to touch one of the pair. The researchers then measured the boys' behavior when they were subsequently left alone with the two toys. Boys who had been reprimanded just as they picked up the toy were least likely to play with the toy. The results suggest that the longer the delay between act and punishment, the less effective the punishment will be (Aronfreed & Reber, 1965).

The effectiveness of punishment may be limited to the situation in which it is given. Since punishment is usually administered by a particular person, it may be effective only when that person is present. This probably accounts for the fact that when their parents are absent, children may engage in activities that their parents earlier had punished (Parke, 1969, 1970).

Another factor is whether punishments are accompanied by a reason (Parke, 1969). Providing a reason allows the child to generalize the prohibition to a class of acts and situations. Yelling "No!" as a child reaches out to touch the stove may suppress that behavior. Telling the child not to touch it because it is "hot" enables him to learn to avoid hot objects generally.

Finally, consistency between the reprimands given by parents and their own behavior makes punishment more effective than if parents do not "practice what they preach" (Mischel & Liebert, 1966).

**Self-Reinforcement and Self-Efficacy** Children learn hundreds if not thousands of behaviors via instrumental learning. The performance of some of these behaviors will remain extrinsically motivated—that is, they are dependent on whether someone else will reward appropriate behaviors or punish inappropriate ones. However, the performance of other activities becomes intrinsically motivated—that is, independent of extrinsic rewards and punishments. As children are socialized, they learn not only specific behaviors but also performance standards. Children learn not only to write but to write neatly. These standards become part of the self; having learned them, the child uses them to judge his own behavior and, thus, becomes capable of **self-reinforcement** (Bandura, 1978). The child who has drawn a house and comes running up to her father with a big smile saying "Look what I drew," has already judged the drawing as a good one. If her father agrees, standards and self-evaluation are confirmed.

Successful experiences with an activity over time create self-perceptions of efficacy at that activity (Bandura, 1982). If the individual executes an activity successfully, that experience contributes to a sense of competence, or *self-efficacy*. That, in turn, makes the individual more likely to seek out opportunities to engage in that behavior. The greater one's sense of efficacy, the more effort one will expend at a task and the greater one's persistence in the face of difficulty. For instance, a young girl who perceives herself as a good basketball player is more likely to try out for a team. Conversely, experiences of failure to perform a task properly, or of the failure of the task to produce the expected results, creates the perception that one is not efficacious. Perceived lack of efficacy is likely to lead to avoidance of the task. A boy who perceives himself as poor at spelling will probably not enter the school spelling bee.

## Observational Learning

Children love to play "dress-up." Girls put on skirts, step into high-heeled shoes, and totter around the room; boys put on sport coats and drape ties around their necks. Through observing adults, children have learned the patterns of appropriate dress in their society. In similar fashion, children often learn interactive rituals —such as shaking hands or waving goodbye—by watching others perform the behavior and then doing it on their own.

**Observational learning,** or modeling, refers to the acquisition of behavior based on observation of another person's behavior and of its consequences for that person (Shaw & Costanzo, 1982). Many behaviors and skills are learned this way. By watching another person (the model) perform skilled actions, a child can increase his or her own skills. The major advantage of modeling is its greater efficiency compared to trial-and-error learning.

Does observational learning lead directly to performance of the learned behavior? No; research has shown that there is a difference between learning a behavior and performing it. People can learn how to perform a behavior by observing another person, but they may not perform the act until the appropriate opportunity arises. A great deal of time may elapse before the observer is in a situation where the eliciting stimulus is present. A parent in the habit of muttering "damn" when he spills something may, much to his chagrin, hear his 3-year-old say "damn" the first time she spills milk. Children may learn through observation many associations between situational characteristics and adult behavior, but they may not perform these behaviors until they occupy adult roles and find themselves in such situations.

Even if the appropriate stimulus occurs,

Modeling is an important process through which children learn the behaviors considered appropriate by their social groups.

people may not perform behaviors learned through observation. An important influence is the consequences experienced by the model following the model's performance of the behavior. For instance, in one study (Bandura, 1965), nursery school children watched a film in which an adult (model) punched, kicked, and threw balls at a large, inflated rubber Bobo doll. Three versions of the film were shown to three groups of children. In the first, the model was rewarded for his acts: a second adult appeared and gave the model soft drinks and candy. In the second version he was punished: the other adult spanked the model with a magazine. In the third version, there were no rewards or punishments. Subsequently, each child was left alone with various toys, including a Bobo doll. The child's behavior was observed through a one-way mirror. Children who observed the model who was punished were much less likely to punch and kick the doll than the other children.

Did these children not learn the aggressive behaviors, or did they learn them by observation but not perform them? In order to answer this question, the experimenter returned to the room and offered a reward for each act of the model that the child could reproduce. Following this offer, children in all three groups were equally able to reproduce the acts performed by the model. Thus, a child is less likely to perform an act learned by observation if the model experienced negative consequences.

Whether children learn from observing a model also depends on the characteristics of the model. Children are more likely to imitate high-status and nurturant models than models who are low in status and nurturance (Bandura, 1969). Preschool children given dolls representing peers, older children, and adults consistently chose adult dolls as people they would go to for help and older children as people they would go to for teaching (Lewis & Brooks-Gunn, 1979). Children, also, are more likely to model themselves after nurturant persons than cold and impersonal ones. Thus, socialization is much more likely to be effective when the child has a nurturant, loving primary caregiver.

### Internalization

Often we feel a sense of moral obligation to perform some behavior. At other times, we experience a strong internal feeling that a particular behavior is wrong. Usually we experience guilt if these moral prescriptions or prohibitions are violated.

**Internalization** is the process by which initially external behavioral standards (for example, those held by parents) become internal and subsequently guide the person's behavior. An action is based on internalized standards when the person engages in it without considering possible rewards or punishments. Various explanations have been offered of the process by which internalization occurs, but all of them agree that children are most likely to internalize the standards held by more powerful or nurturant adult caregivers.

Internalization is an important socializing process. It results in the exercise of self-control; people conform to behavioral standards even when there is no surveillance of their behavior by others and, therefore, no rewards for that conformity. People who are admired for taking political or religious actions that are unpopular, for standing up for their beliefs, often do so because those beliefs are internalized.

## ■ Outcomes of Socialization

Persons undergoing socialization acquire new skills, knowledge, and behavior. In this section, we will discuss some specific outcomes of the process, including gender roles, linguistic and cognitive competence, moral development, and orientation toward achievement and work.

### Gender Role

"Congratulations, you have a girl!" Such a pronouncement by a birth attendant may be the single most important event in a new person's life. In large part, the gender assigned to the infant—male or female—has a major influence on the socialization of that child.

Every society has differential expectations regarding the characteristics and behavior of males and females. In our society, men have traditionally been expected to be competent—competitive, logical, able to make decisions easily, ambitious; women have been expected to be high in warmth and expressiveness—gentle, sensitive, tactful (Broverman et al., 1972). Parents employ these as guidelines in socializing their children, and differential treatment begins at the moment of birth. Male infants are handled more vigorously and roughly, whereas females are given more cuddling (Lamb, 1979). Boys and girls are dressed differently from infancy and may be given different kinds of toys to play with.

In addition, mothers and fathers differ in the way they interact with infants. Mothers engage in behavior oriented toward fulfilling the child's physical and emotional needs (Baumrind, 1980), whereas fathers engage the child in rough-and-tumble, physically stimulating activity (Walters & Walters, 1980). Thus, almost from birth, infants are exposed to models of masculine and feminine behavior. At about age 1, fathers begin to pay special attention to their sons and reduce the interaction with their daughters (Lamb, 1979). At the same time, the child increasingly prefers the same-sex parent. More frequent interaction with and a preference for the same-sex parent account for Richard Cohen's observation that he is much more like his father than his mother.

By age 2, the child's gender identity—his or her conception of self as male or female—is firmly established (Money & Ehrhardt, 1972). Boys and girls show distinct preferences for different types of play materials and toys by this age. Between the ages of 2 and 3, differences in aggressiveness become evident, with boys displaying more physical and verbal aggression and behaving more aggressively in play (Maccoby & Jacklin, 1974). By age 3, children more frequently choose same-gender peers as playmates; this increases opportunities to learn gender-appropriate behavior via modeling (Lewis & Brooks-Gunn, 1979). By age 4, the games typically played by boys and girls differ; groups of girls play house, enacting familial roles, while groups of boys play cowboys.

Parents are an important influence on the learning of **gender role,** the behavioral expectations associated with one's gender. Children learn gender-appropriate behaviors by observing their parents' interaction. Children also learn by interacting with parents, who reward behavior consistent with gender roles and punish behavior inconsistent with these standards. The child's earliest experiences in relating to members of the other gender occur in interaction with the opposite-gender parent. A female may be more likely to develop the ability to have warm, psychologically intimate relationships with males if her relationship with her father was of this type (Appleton, 1981).

Obviously, boys are not all alike in our society, and neither are girls. The specific behaviors and characteristics that the child is taught depend partly on the gender-role expectations held by the parents. These, in turn, depend on the network of extended family—grandparents, aunts and uncles, and other relatives—and friends of the family. The expectations held by these people are influenced by the institutions to which they belong, such as churches and work organizations (Stryker & Serpe, 1982). With regard to religion, research suggests that the differences between denominations in socialization techniques and in outcomes such as gender-role attitudes have declined in the past 30 years (Alwin, 1986). The data suggest that church attendance is more influential than the denomination to which one belongs.

Schools also teach gender roles. Teachers tend to reward appropriate gender-role behavior; they often reinforce aggressive behavior in boys and dependency in girls (Serbin & O'Leary, 1975). A more subtle influence on socialization is the content of the stories that are read and told in preschool and first-grade classes. Many of these stories portray males and females as different. Typically, men are depicted as rulers, adventurers, and explorers; women are wives (Weitzman et al., 1972). As children learn to read, the books they are given elaborate their notion of gender roles. Stories such as Snow White, for instance, convey the expectation that men work outside the home and women do the housework.

## Linguistic and Cognitive Competence

Another important outcome of socialization is the ability to interact effectively with others. We shall discuss two specific competencies: language and the ability to cognitively represent the world.

**Language**  Learning language is a prerequisite for full and effective participation in social groups (Shibutani, 1961). Three main components of language are the sound system (phonology), the words and their associated meanings (lexicon), and rules for combining words into meaningful utterances (grammar).

Chronologically, infants master sounds first. In the first year of life, children babble and emit a vast range of sounds. Soon, however, their vocalizations are restricted to those which are meaningful. Three socialization processes discussed earlier contribute to learning phonology. Children imitate the sounds they hear from family members. Instrumental learning is also important. When children experience discomfort, they make sounds in order to attract a caregiver to remove the discomfort. The removal of discomfort reinforces those vocalizations. Shaping also occurs; as the child vocalizes, parents respond positively to some sounds and ignore others.

Next, children begin to master the second component of language—lexicon. They verbalize meaningful single words. Such utterances have much broader and diverse meaning for the child than for adults. For example, when a 15-month-old says "car," the child may mean "I see the car," "I want to go for a ride in the car," "Here comes a car," or "Give me my toy car." An adult caregiver is faced with the task of determining the child's meaning. The adult often engages in a trial-and-error process, offering several responses in succession until the child seems satisfied. Children's discovery that others do not initially respond "correctly" motivates them to improve the precision of their speech (Miller & McNeill, 1969). Thus, language acquisition proceeds via children's desire to satisfy their needs and adults' demands for greater articulateness before granting such satisfaction.

Around age 2, children begin to speak two- or three-word sentences. Examples of such sentences include "See truck, Mommy" and "There go one." Such speech is *telegraphic*—that is, the number of words is greatly reduced relative to adult speech (Brown & Fraser, 1963). At the same time, such utterances are clearly more precise than the single-word utterances of the 1-year-old child.

To utter full sentences, children must first acquire the third component of language—grammar. How do children learn the rules for combining words into meaningful sequences? In an early study, important insights were obtained by studying the speech of a 27-month-old boy named Adam (Brown & Bellugi, 1964). Adam used three classes of words: nouns, verbs, and "pivots"—or modifiers such as "a," "my," or "more" (Braine, 1963). Theoretically, these three classes of words could be combined into 8 types of sentences containing two words (2 cubed) and 27 types containing three words (3 cubed). But of these 35 possible types, only 4 types of the two-word sentences and 8 types of three-word utterances, using Adam's vocabulary, were grammatically permissible in the English language. Amazingly, all of Adam's sentences recorded by the researchers were of the permissible types. He did not utter even one combination that is not allowed in English grammar.

How does a young child learn to make grammatically correct sentences? An important process is speech expansion. That is, adults often respond to children's speech by repeating it in expanded form. In response to "Eve lunch," the mother might say "Eve is eating lunch." One study showed that mothers expanded 30 percent of the utterances of their 2-year-old children (Brown, 1964). Adults probably expand on the child's speech in order to determine the child's specific meaning. The fact that adults do this indicates that they assume children can comprehend more complex utterances than the children spontaneously emit (Fraser, Bellugi, & Brown, 1963). Speech expansion contributes to language acquisition by providing children with a model of how to convey more effectively the meanings they intend.

The next stage of language development is highlighted by the occurrence of *private speech,* in which children talk loudly to themselves, often for extended periods of time. Private speech begins at about age 3, increases in frequency until age 5, and disappears by about age 7. Such talk serves two functions. First, it contributes to the child's developing sense of self. Private speech is addressed to the self as object, and it often includes the application of meanings to the self such as "I'm a girl." Second, private speech helps the child develop an awareness of the environment. It often consists of naming aspects of the physical and social environment. The repeated use of these names solidifies the child's understanding of the environment. Children also often engage in appropriate actions as they speak, reflecting their developing awareness of the social meanings of objects and persons. Thus, a child may label a doll a "baby" and dress it and feed it.

Gradually, the child begins to engage in dialogues, either with others or with himself or herself. These conversations reflect the ability to adopt a second perspective. Thus, at age 6, when one child wants a toy that another child is using, the first child frequently offers to trade. She knows that the second child will be upset if she merely takes the toy. This movement away from a self-centered view may also reflect maturational changes. Dialogues require that the child's own speech meshes with that of another.

**Cognitive Competence**   Children must develop the ability to represent in their own minds the features of the world around them. This capacity to represent reality mentally is closely related to the development of language.

The child's basic tasks are to learn the regularities of the physical and social environment and to store past experience in a form that can be used in current situations. We employ three modes of storing information about the environment (Bruner, 1964). The first mode is *enactive,* the representation of past events in our

musculature. Examples include the behaviors necessary to maintain one's balance while riding a bicycle and the representation of routes that we drive frequently as a series of turns and directions. The second mode is *iconic,* in which we retain a concrete image of an object or event. Our ability to remember faces is an example of iconic representation. The third mode is *symbolic,* in which an arbitrary symbol represents an event, object, or person. Language is a particularly useful symbol system because we can use it to represent abstract as well as concrete phenomena.

In a complex society, there are so many physical objects, animals, and persons that it is not possible to remember each as a distinct entity. Things must be categorized into more general groupings, such as birds, dogs, houses, and girls. In doing this, we employ a superordinate dimension that ignores differences on more specific dimensions. Researchers can study the ability to utilize general categories by asking children to sort objects, pictures, or words into groups. Young children (ages 6 to 8) rely on visual features, such as color or word length, and sort objects into numerous categories. Older children (ages 10 to 12) increasingly use functional or superordinate categories, such as foods, and sort objects into fewer groups (Olver, 1961; Rigney, 1962). With age, children become increasingly adept at classifying diverse objects and treating them as equivalent.

These skills are very important in social interaction. Only by having the ability to group objects, persons, and situations can one determine how to behave toward them. Categories of persons and their associated meanings are especially important to smooth interaction. Even very young children differentiate persons by age (Lewis & Brooks-Gunn, 1979). By about 2 years of age, children correctly differentiate babies and adults when shown photographs. By about 5, children employ four categories: little children, big children, parents (age 13 to 40), and grandparents (age 40 and older). Another commonly used set of categories involves religion. Children's conceptions of members of various denominations have been studied by asking a child "What is a Catholic?" or "What is a Jew?" (Elkind, 1961, 1962, 1963). At age 6, the typical response is "A person." At 8, children refer to behavior ("A person who goes to temple"). At age 10, children focus on more abstract differences in beliefs ("A person who believes in one God"). The increasing abstractness of these responses with age parallels the increasing use of superordinate principles in categorization.

As children learn to group objects into meaningful categories, they learn not only the categories but also how others feel about such persons. Children learn not only that Catholics are people who believe in the Trinity of Father, Son, and Holy Spirit but also that their parents like or dislike Catholics. Thus, children acquire positive and negative attitudes toward the wide range of social objects they come to recognize. The particular categories and evaluations children learn are influenced by the social class, religious, ethnic, and other subcultural groupings to which those who socialize them belong.

## Moral Development

In this section we discuss moral development in children and adults. Specifically, we focus on acquisition of knowledge of social rules and on the process through which children become capable of making moral judgments.

**Knowledge of Social Rules**    In order to interact effectively with others, people must also learn the social rules that govern interaction and adhere to them. Beliefs about which behaviors are acceptable and which are unacceptable for specific persons in specific situations are termed **norms.** Without norms, coordinated activity would be very difficult, and we would find it hard or impossible to achieve our goals. Therefore, each group, organization, and society develops rules governing behavior.

Early in life, an American child learns to say "please"; a French child, "s'il vous plaît"; and a Serbian child, "molim te." In every case, the child is learning the value of conforming to arbitrary norms governing requests. Learning language trains the child to conform to linguistic norms and serves as a model for the learning of other norms. Gradually, through instrumental as well as observational learning, the child learns the generality of the relationship between conformity to norms and the ability to interact smoothly with others and achieve one's own goals.

What influences which norms children will learn? The general culture is one influence. All American children learn to cover most parts of the body with clothing. The position of the family within the society is another influence. Parental expectations reflect social class, religion, and ethnicity. Thus, the norms taught vary from one family to another. Interestingly, parents often hold norms that they apply distinctively to their own children. Mothers and fathers expect certain behaviors of their own sons or daughters but may have different expectations for other people's children (Elkin & Handel, 1978). For instance, they may expect their children to be more polite than other children in interaction with adults. Parental expectations are not constant over time; they change as the child grows older. Parents expect greater politeness from a 10-year-old than from a 5-year-old. Finally, parents adjust their expectations to the particular child. They take into account level of ability and past experiences relative to other children; they expect better performance in school from a child who has done well in the past than from one who has had problems in school. In all of these ways, each child is being socialized to a somewhat different set of norms.

When children begin to engage in cooperative play, at about 4 years of age, they begin to experience normative pressure from peers. The expectations of age-mates differ in two important ways from those of parents. First, children

At school, children get their first exposure to universal norms—behavioral expectations that are the same for everyone. Although parents and friends treat the child as an individual, teachers are less likely to do so.

bring different norms from their separate families and, therefore, introduce new expectations. Thus, through peers, children first become aware that there are other ways of behaving. In some cases, peers' expectations conflict with those of parents. For example, many parents do not allow their children to play with toy guns, knives, or swords. Through involvement with their peers, children may become aware that other children routinely play with such toys. As a result, some children will experience normative conflict and discover the need to develop strategies for resolving such conflicts.

Another way that peer-group norms differ from parental norms is that the former reflect a child's perspective (Elkin & Handel, 1978). Many parental expectations are oriented toward socializing the child for adult roles. Children react to each other as children and are not concerned with long-term outcomes. Thus, peers encourage impulsive, spontaneous behavior rather than behavior directed toward long-term goals. Peer-group norms emphasize participation in group activities, whereas parental norms may emphasize homework and other educational activities that may contribute to academic achievement.

When children enter school, they are exposed to a third major socializing agent, the teacher. In school, children are exposed to *universalistic* rules—norms that apply equally to all children. The teacher is much less likely than the parents to make allowances for the unique characteristics of the individual; children must learn to wait their turn, to control impulsive and spontaneous behavior, and to work without a great deal of supervision and support. In this regard, the school is the first of many settings where the individual is treated primarily as a member of the group rather than as a unique individual.

Thus, school is the setting in which children are first exposed to universalistic norms and the regular use of symbolic rewards such as grades.

Such settings become increasingly common in adolescence and adulthood, in contrast to the individualized character of familial settings.

**Moral Judgment**  We not only learn the norms of our social groups, but we also develop the ability to evaluate behavior in specific situations by applying certain standards. The process through which children become capable of making moral judgments is termed **moral development.** It involves two components: (1) the reasons why one adheres to social rules and (2) the bases used to evaluate actions by self or others as good or bad.

How do children evaluate acts as good or bad? One of the first persons to study this question in detail was Jean Piaget, the famous Swiss developmental psychologist. Piaget's methodology was to read to a young child a set of stories in which the central characters performed various acts that violated social rules. In one story, for example, the central character was a young girl who, contrary to rules, was playing with scissors and made a hole in her dress. Piaget asked the children to evaluate the behaviors of the various characters in the stories (that is, to indicate which characters were naughtier) and then to explain their reasons for these judgments. Based on this work, Piaget concluded that there were essentially three bases for moral judgments: amount of harm/benefit, actor's intentions, and the application of agreed-upon rules or norms (Piaget, 1965).

More recently, Kohlberg has extended Piaget's work by analyzing in greater detail the reasoning by which people reach moral judgments. Like Piaget, he uses stories involving conflict between human needs and social norms or laws. For example,

In Europe, a woman was near death from cancer. One drug might save her, a form of radium that a druggist in the same town had recently discovered. The druggist was charging $2,000, ten times what the drug cost him to make. The sick woman's husband, Heinz, went

to everyone he knew to borrow money, but he could only get together about half of what it cost. He told the druggist that his wife was dying and asked him to sell it cheaper or let him pay later. But the druggist said, "No." The husband got desperate and broke into the man's store to steal the drug for his wife. (Kohlberg, 1969)

Respondents are then asked: Should Heinz have done that? Was Heinz right or wrong? What obligations did Heinz and the druggist have? Should Heinz be punished?

Kohlberg proposes a developmental model with three levels of moral reasoning, each level involving two stages. This model is summarized in Table 3-3.

Kohlberg argues that the progression from stage 1 to stage 6 is a standard or universal one and that all children begin at stage 1 and progress through the stages in order. Practically no one consistently reasons at stage 6, and relatively few regularly use stage 5 considerations. Most people usually reason at stages 3 or 4. Several studies have shown that such a progression does occur (Kuhn, Langer, Kohlberg, & Haan, 1977). If the progression is universal, then children from different cultures should pass through the same stages in the same order. Again, data suggest that they do (White, Bushnell, & Regnemer, 1978). On the basis of such evidence, Kohlberg claims that this progression is the natural human pattern of moral development. He also believes that attaining higher levels is better or more desirable.

While moral development is an interesting topic of study in its own right, some investigators have explored the relationship between moral judgment and behavior. Studies of cheating in schoolwork, one involving sixth graders (Krebs, 1967) and two involving college students (Schwartz, Feldman, Brown, & Heingartner, 1969; Malinowski & Smith, 1985), found that those whose moral development had reached higher stages were less likely to cheat. Participants in political activities such as demonstrations and sit-ins often claim their behavior is

■ **TABLE 3-3**
KOHLBERG'S MODEL OF MORAL DEVELOPMENT

**Preconventional Morality**
*Moral judgment based on external, physical consequences of acts.*

*Stage 1:* Obedience and punishment orientation. Rules are obeyed in order to avoid punishment, trouble.
*Stage 2:* Hedonistic orientation. Rules are obeyed in order to obtain rewards for the self.

**Conventional Morality**
*Moral judgment based on social consequences of acts.*

*Stage 3:* "Good boy/nice girl" orientation. Rules are obeyed to please others, avoid disapproval.
*Stage 4:* Authority and social-order maintaining orientation. Rules are obeyed to show respect for authorities and maintain social order.

**Postconventional Morality**
*Moral judgments based on universal moral and ethical principles.*

*Stage 5:* Social-contract orientation. Rules are obeyed because they represent the will of the majority, to avoid violation of rights of others.
*Stage 6:* Universal ethical principles. Rules are obeyed in order to adhere to one's principles.

*Source: Adapted from Kohlberg, 1969, Table 6.2.*

based on moral principles. But is it? A study by Haan, Smith, and Block (1968) explored the relationship between political activities and moral development in young adults. The results are depicted in Table 3-4 (none of the subjects were classified as being in stage 1, which is consistent with their age). In general, the highest rates of political activity were found among men and women who reasoned at stages 5 or 6, while the

lowest rates were found for those who reasoned at stages 3 or 4. In most instances, there was a consistent increase from those classified in stage 3 to those classified in stage 6 in percentage reporting an activity.

Many of the subjects in this study reported involvement in a sit-in at the University of California at Berkeley. As Kohlberg might predict, the reasons people reported for participating varied by level of moral thought. Preconventional subjects (stages 1 and 2) said they did so because the protest was an opportunity to improve their status in their peer groups. Postconventional subjects (stages 5 and 6) viewed their participation as growing out of their commitment to individual civil liberties. In general, the conventional subjects (stages 3 and 4) avoided the sit-in because it constituted a confrontation with legitimate authorities. These differences in the reasons given by subjects at different levels are consistent with the theory; they also demonstrate that moral judgment is the outcome of an interaction between the characteristics of a situation and the level of moral thought achieved by those present.

Kohlberg's model is an impressive attempt to specify a universal model of moral development. However, there are limitations to it. First, like Piaget, he locates the determinants of moral judgment within the individual and does not recognize the influence of the situation. Studies of judgments of aggressive behavior (Berkowitz et al., 1986) and of decisions about reward allocation (Kurtines, 1986) found that both moral stage and type of situation influenced moral judgment.

Second, Kohlberg's model has been criticized as sexist, not applicable to the processes women use in moral reasoning. Gilligan (1982) identifies two conceptions of morality: a morality of justice and a morality of caring. She argues that the former is characteristic of men and is the basis of Kohlberg's model, while the latter is more characteristic of women. A study of the considerations that men and women had used in

■ **TABLE 3-4**

RELATIONSHIP BETWEEN MORAL DEVELOPMENT AND SOCIOPOLITICAL ACTIVITY

| Activity | Stage | Percentage of Subjects Reporting Activity | | | | | | | | | |
| | | Men | | | | | Women | | | | |
| | | 2 | 3 | 4 | 5 | 6 | 2 | 3 | 4 | 5 | 6 |
|---|---|---|---|---|---|---|---|---|---|---|---|
| Picket | | 56 | 12 | 9 | 44 | 75 | 0 | 8 | 13 | 44 | 55 |
| Sit-in | | 25 | 5 | 5 | 27 | 31 | 14 | 7 | 4 | 26 | 27 |
| Petition work | | 25 | 14 | 18 | 35 | 38 | 0 | 4 | 10 | 35 | 18 |
| Attend meetings | | 62 | 46 | 41 | 60 | 69 | 40 | 33 | 48 | 56 | 63 |
| Distribute literature | | 44 | 11 | 20 | 42 | 50 | 20 | 10 | 21 | 44 | 55 |
| Demonstrate | | 56 | 14 | 8 | 48 | 75 | 20 | 11 | 10 | 47 | 55 |
| Peace march | | 56 | 12 | 5 | 40 | 62 | 0 | 11 | 10 | 44 | 64 |

Source: Adapted from Haan, Smith, and Block, 1968, Table 7.

resolving personal moral dilemmas reports results consistent with Gilligan's thesis (Ford & Lowrey, 1986).

Third, Kohlberg shows little interest in the influence of social interaction on moral reasoning. In response to this limitation, Haan (1978) has proposed a model of interpersonal morality. Moral decisions and actions often result from negotiations among people in which the goal is a "moral balance." Participants attempt to balance situational characteristics, such as the options available, with their individual interests to arrive at a decision that allows them to preserve their sense of themselves as moral persons. Haan (1978, 1986) presented moral dilemmas to groups of friends and asked them to arrive at a decision. In some cases, the decisions were more influenced by individual moral principles; in others, by the group interaction.

## Achievement and Work Orientations

One of the persistent questions about human behavior is what guides its direction and explains its intensity. All of us make choices among alternatives, appear fairly consistent in our behavior across situations, and often persist in attempting to achieve a goal even in the face of adversity. Social psychologists employ the concept of motivation to account for these phenomena. A *motive* is a disposition within the person that produces behavior directed toward goals; social motives are those developed in interaction with other persons.

**The Achievement Motive** One very important social motive is the **achievement motive,** a conscious or unconscious desire to reach high standards of excellence (McClelland, 1961). The strength of this motive has been found to be positively related to academic performance, even among people who are equal in intellectual ability. It is also positively associated with the tendency to engage in entrepreneurial activity, including taking risks when the outcome is under one's control. People with high levels of this motive are more likely to engage in innovative activity and to try to anticipate future events.

The achievement motive develops early in the person's socialization. Differences among children in the strength of the motive are evident at 5 years of age (McClelland, 1958). Moreover, achievement scores of 6-year-old children are positively related to the scores of the same persons as adults (Moss & Kagan, 1961).

**■ FIGURE 3-1**
What Do You Think Is Happening in This Picture?

Various techniques for measuring achievement motivation have been developed. The most common involves analyzing the stories people write to describe what is happening in pictures of people they are shown. For example, look at the drawing in Figure 3-1. What is happening here? What is the boy thinking? What will happen? Two stories follow that illustrate how some individuals interpreted this picture.

1. A boy in a classroom is daydreaming. He is recalling an incident that was more fun than being in school. He will probably get called on to recite and be embarrassed.

2. The boy is a student taking an exam. The exam is two-thirds over and he is thinking about his answers. He can't remember some of the material. He will miss most of the items he can't remember and be disgusted with himself for not learning them. (Adapted from McClelland, 1961)

Once stories of this type are written by people, they are analyzed to determine the extent to which they portray characters as striving to achieve some standard. In the first story, for example, there is no mention of standards of excellence or achievement. The second story, by contrast, is rich in achievement imagery. How much concern with achievement and meeting standards of excellence is evident in the story you created? When analyzed by researchers, each story is given a score based on the total number of achievement-related ideas contained in it; this measure is termed a need achievement score, or *n ach.*

The strength of this motive is a result of two factors. One is the emphasis on excellence and success in childhood socialization (Rosen & D'Andrade, 1959). Parents and others who reward children for successfully competing against some standard are contributing to the development of this motive. The second factor is the extent to which parents train their children to be independent (Winterbottom, 1958). Children need autonomy in order to display achievement-oriented behavior.

Research indicates that high levels of the achievement motive do not necessarily result in high performance. For performance to occur, one must not only have high achievement motivation but also hold the expectation that high performance will in fact be rewarded (McClelland & Winter, 1969).

Recent research suggests that whether mothers are employed influences the achievement motivation of their children (Lamb, 1982). Interestingly, the effect depends on the child's gender. Adolescent daughters of working mothers have higher achievement motivation than daughters whose mothers are at home, whereas adolescent sons of working mothers are less motivated to achieve.

**Orientations toward Work** Work is of central importance in social life. In recognition of this, occupation is a major influence on the distribution of economic and other resources. We identify others by their work; its importance is evidenced by the fact that one of the first questions we ask a new acquaintance is "What do you do?"

Most adults are motivated to work at jobs that provide economic and perhaps other rewards. Therefore, it is not surprising that a major part of socialization is the learning of orientations toward work. By age 2, the child is aware that adults "go work" and asks why. Adults respond with a variety of concrete answers, such as "Mommy goes to work to earn money." The child in turn learns that money is needed to obtain food, clothing, and toys. The child of a physician or nurse might be told "Mommy goes to work to help people who are ill." Thus, from an early age the child is taught the social meaning of work.

Occupations vary tremendously in character. One dimension on which jobs differ is closeness of supervision; a self-employed auto mechanic has considerable freedom, while an assembly-line worker may be closely supervised. The nature of the work varies; mechanics deal with things, salespersons deal with people, attorneys deal with ideas. Finally, occupations such as attorney require self-reliance and independent judgment, whereas an assembly-line job does not. So the meaning of work depends on the type of job the individual has.

Adults in different occupations should have different orientations toward work, and these orientations should influence how they socialize their children. Based on this hypothesis, extensive research has been conducted on the differences between social classes in the values transmitted through socialization (Pearlin & Kohn, 1966; Kohn, 1969). Fathers are given a list of traits, including good manners, success, self-control, obedience, and responsibility, and asked to indicate how much they value each one for their children. Underlying these specific characteristics a general dimension, "self-direction versus conformity" is usually found. Data from fathers of 3- to 15-year-old children indicate that the emphasis on self-direction and reliance on internal standards increases as social class increases.

These differences in the evaluations of particular traits reflect differences in the conditions of work. In general, middle-class occupations involve the manipulation of people or symbols, and the work is not closely supervised. Thus, these occupational roles require people who are self-directing and who can make judgments based on knowledge and internal standards. Working-class occupations are more routinized and more closely supervised. Thus, they require workers with a conformist orientation. Kohn argues that fathers value those traits in their children that the father associates with success in his occupation.

Do differences in the value parents place on self-direction influence the kinds of activities they encourage their children to participate in? A study of 460 adolescents and their mothers (Morgan, Alwin, & Griffin, 1979) examined how maternal emphasis on self-direction affects the young person's grades in school, choice of curriculum, and participation in extracurricular activities. The researchers reasoned that parents who value self-direction will encourage their children to take college-preparatory courses, because a college education is a prerequisite to jobs that provide high levels of autonomy. Similarly, they expected mothers who value self-direction to encourage extracurricular activities, because such activities provide opportunities to develop interpersonal skills. The researchers did not expect differences in grades, however. The results confirmed all three predictions. Thus, parents who value particular traits in their children do encourage activities that they believe are likely to produce those traits.

**Adolescent Employment**  It is often suggested that part-time employment during adolescence has positive effects on the youth's development of work orientations. Many people believe that working for pay will cause the adolescent to become more responsible and mature. Many teenagers are employed; in 1980, from 36 to 45 percent of high school sophomores and from 55 to 65 percent of seniors worked.

Does work teach adolescents to be more responsible? In order to answer this question, two investigators surveyed 211 employed teenagers and 319 youth who had never held a steady job (Greenberger & Steinberg, 1986). Work does seem to increase self-reliance, the sense of being independent and in control. Young people who worked more hours also attained higher scores on a measure of work orientation—dependability, persistence, and motivation to perform well. On the other hand, there is little evidence that employed adolescents gain new information or skills from their jobs.

# ■ Adult Socialization

The process of socialization occurs not only in childhood but continues throughout life (Bush & Simmons, 1981). Its focus changes, however. In childhood, socializing efforts are directed at such basic outcomes as gender role, the acquisition of language, and the learning of social norms. In adolescence, socialization is focussed on the acquisition of traits, such as independence, responsibility, and the ability to relate to others. In adulthood, it is concerned with equipping the individual to function effectively in adult roles. In this section, we will discuss three processes important in adult socialization—role acquisition, anticipatory socialization, and role discontinuity.

## Role Acquisition

Throughout our lives, we move out of some roles and into new roles. The major roles we acquire as adults include intimate partner or spouse, parent, new work roles and, later, the roles of grandparent and retiree. Each of these changes involves *role acquisition*—learning the expectations and skills associated with the new role and entry into the role.

In recognition of the need to train people for new roles, many groups and organizations provide socialization opportunities. Certain

agents are given the responsibility for teaching potential or new role occupants the necessary information and skills. Often this training occurs in service, or on the job. The novice checker in a supermarket will work with an experienced clerk at first. Initially, the experienced clerk will perform the work, perhaps instructing the novice as she does so. Gradually, the roles will be reversed. As the novice becomes more skilled at both the mechanics of using the price scanner and interacting with shoppers, she will do more and more of the work. Eventually, if she demonstrates the requisite competence, she will be on her own. This process can be observed in hundreds of occupational settings. In some cases, there is a formal role designation for those undergoing socialization, such as "trainee."

Alternatively, there may be a separate period of formal training outside the organization before the person occupies the role. Many educational programs exist to prepare people for roles they will play in the future. Training programs or schools for beauticians, flight attendants, dental technicians, and truck drivers are only a few examples of this type of socialization.

## Anticipatory Socialization

In addition to intentional training before and after a role is acquired, there may be **anticipatory socialization**—activities that provide people with knowledge, skills, and values of a role they have not yet assumed. The teenager learning about sexual activity from the boasts of an older friend or from an X-rated movie is undergoing anticipatory socialization. So is the aspiring diplomat or politician who attends closely to the behavior of the president of the United States in a television interview. Anticipatory socialization is different from explicit training because it is not intentionally designed as role preparation by socialization agents (Clausen, 1968; Heiss, 1981).

Anticipatory socialization can ease the transition into new roles, but it is more effective for some roles than for others (Thornton & Nardi,

Through anticipatory socialization, individuals acquire skills, knowledge, and values of roles they hope to assume in the future. Highly visible aspects of roles (makeup) are much easier to acquire than hidden aspects (attitudes toward husbands).

1975; Bush & Simmons, 1981). First, anticipatory socialization usually works best with respect to future roles that are highly visible. Socialization of this type usually prepares children more effectively for the parent than for the spouse role, for example. Children see the parent role directly when interacting with their own parents, whereas important aspects of interaction among spouses occur when children are away or asleep.

Second, anticipatory socialization eases the transition if future roles are presented accurately. The interactions between spouses that children do observe are often intentionally laundered to hide negative feelings and conflict, resulting in poor anticipatory socialization and incorrect expectations about role demands.

Third, anticipatory socialization works better if there is certainty or agreement regarding role demands and expectations. Anticipatory socialization for retirement is difficult, for example, because we lack clear, consensual norms for how the elderly should behave after they leave the work force (Rosow, 1974; Matthews, 1977). Should they "ease up and accept dependence gracefully" or "stay active and insist on their independence?"

Successful anticipatory socialization entails goal setting, planning, and preparation for future roles. Only by setting at least tentative occupational and family goals during our teenage years, for example, can we effectively plan our educational and social lives. Preparation occurs through part-time jobs, special courses, reading, talking with informed individuals, and so on. People also prepare for transitions by trying out elements of their anticipated roles. This is what couples planning marriage are doing when they take joint vacations, live together for a trial period, and share purchases.

## Role Discontinuity

The acquisition of a new role does not always proceed smoothly, even if there has been anticipatory socialization. Changing roles can be stressful. A new role can involve meeting new expectations, performing new tasks, and interacting with new (types of) people. It may necessitate moving to a new location. Furthermore, acquiring a role may involve losses as well as gains. The person may have to leave a prior role in order to move into the new one.

Entering a new role can be especially difficult when there is **role discontinuity**—that is, when the values and identities associated with a new role contradict those of earlier roles (Benedict, 1938). Upon entering a discontinuous role, we must revise our former expectations and aspirations. Retirement, for example, often creates role discontinuity. During their working years, career-oriented adults are expected to strive for autonomy and productivity and to build their identities around their work. Retirement introduces contradictory expectations for such people. They must now assume more dependent and less productive roles and rebuild their identities around these discontinuous expectations (Mortimer & Simmons, 1978).

Role transitions of particular social importance are sometimes marked by *rites of passage*—public ceremonies or rituals in which the individual's new status is affirmed (Van Gennep, 1960; Glaser & Strauss, 1971). Christenings, bar mitzvahs, graduations, marriage ceremonies, and retirement parties are common rites of passage in American society. They signify both to the individual and to others that the person now has a new identity and that new behaviors, rights, and duties are appropriate. These ceremonies serve as social occasions for giving emotional support, advice, and material aid to those adopting new roles.

## ■ Summary

Socialization is the process through which infants become effective participants in society. It makes us like all other members of society in certain ways (shared language), but distinctive in other ways.

**Perspectives on Socialization** (1) One approach to the study of socialization emphasizes biological development; it views the emergence of interpersonal responsiveness and the development of speech and of cognitive structure as influenced by maturation. (2) The other approach emphasizes learning and the acquisition of skills from other persons. Society organizes this process by making certain agents responsible for particular types of socialization of specific persons.

**Agents of Childhood Socialization** There are three major socializing agents in childhood. (1) The family provides the infant with a strong attachment to one or more caregivers. This bond seems to be necessary in order for the infant to develop interpersonal and cognitive skills. Family composition, size, and social class all influence socialization by influencing the amount and kind of interaction between parent and child. (2) Peers provide the child with equal-status relationships and are an important influence on the development of self. (3) Schools teach skills—reading, writing and arithmetic—as well as traits like punctuality and perseverance.

**Processes of Socialization** Socialization is based on three different processes. (1) Instrumental conditioning, the association of rewards and punishments with an act, is a basis for learning both behaviors and performance standards. Studies of the effectiveness of various child-rearing techniques indicate that rewards do not always make a desirable behavior more likely to occur and punishments do not always eliminate an undesirable act. Through instrumental learning, children develop the ability to judge their own behaviors and to engage in self-reinforcement. (2) We learn many behaviors and skills by observation of models. We may not perform these behaviors, however, until we are in the appropriate situation. (3) Socialization also involves internalization, the acquisition of behavioral standards and making them part of the self. This process enables the child to engage in self-control.

**Outcomes of Socialization**  (1) The child gradually learns a gender role—the expectations associated with being male or female. Whether the child is independent or dependent, aggressive or passive, depends on the expectations communicated by parents, relatives, and peers. (2) Language skill is another outcome of socialization; it involves learning both words and the rules for combining them into meaningful sentences. Related to the learning of language is the development of thought and the ability to group objects and persons into meaningful categories. (3) The learning of social norms involves parents, peers, and teachers as socializing agents. More generally, children learn that conformity to norms facilitates social interaction. Children also develop the ability to make moral judgments. (4) Children acquire motives—dispositions that produce sustained, goal-directed behavior. One such motive—the achievement motive—is the desire to reach high standards of excellence. Orientations toward work are influenced primarily by parents; middle-class families seem to emphasize self-direction, while working-class families emphasize conformity.

**Adult Socialization** Socialization continues throughout life. In adulthood, it involves preparing the person to successfully enact major roles such as intimate partner and parent. (1) Role acquisition is often accompanied by on-the-job training. In addition, time spent in the role of trainee may be prerequisite to entry into a new role. (2) Anticipatory socialization involves unintentional role preparation. Its effectiveness depends on the visibility and accuracy of the portrayal of the new role and consensus about the expectations for the role. (3) Role discontinuity makes role acquisition especially difficult. Discontinuity can be eased by anticipatory socialization and by rites of passage.

## Key Terms

| | |
|---|---|
| ACHIEVEMENT MOTIVE | 74 |
| ANTICIPATORY SOCIALIZATION | 77 |
| ATTACHMENT | 57 |
| EXTRINSICALLY MOTIVATED BEHAVIOR | 63 |
| GENDER ROLE | 67 |
| INSTRUMENTAL CONDITIONING | 61 |
| INTERNALIZATION | 66 |
| INTRINSICALLY MOTIVATED BEHAVIOR | 63 |
| MORAL DEVELOPMENT | 72 |
| NORM | 70 |
| OBSERVATIONAL LEARNING | 65 |
| PUNISHMENT | 63 |
| ROLE DISCONTINUITY | 79 |
| SELF-REINFORCEMENT | 64 |
| SHAPING | 61 |
| SOCIALIZATION | 53 |

chapter

# Self and Identity

# ■ Introduction

An amnesia victim who walked into a pizza parlor six weeks ago and said "Help me, I don't know who I am," now faces a new identity crisis.

"I've gotten lots of calls saying I'm two entirely different people," said the man, who dubbed himself John Jackson after he was taken to a homeless shelter.

Jackson's ordeal began May 3, when he woke up beside railroad tracks and Interstate 64 near where the borders of West Virginia, Ohio and Kentucky meet. "I woke up in a field and the first thing I thought was 'It's cold.' And then I wondered, 'Where am I?' And then, 'Who am I?' Then I realized I had no answers and it got very frightening."

"He was bewildered more than anything else," said mission cook Virginia Berry. "He was dressed OK, had on blue jeans. He has a beard but it's well-trimmed and shaped and been taken care of. His hair's that way, too."

The man asked mission staff to call him Jackson because he remembers roads and a baseball park in Jacksonville, Fla. Mission workers asked Jacksonville newspapers to publish the man's photograph and on Friday he was inundated by calls from people claiming to know him.

Jackson did not recognize any of the callers, although some of the details they provided matched the few pieces he has to his puzzle.

One group of people who saw Jackson's photograph told him they believe he is Roy Moses, 33, who frequently vanished for months at a time "on a whim." Moses' stepmother, stepsister and brother-in-law all talked to Jackson on the telephone "and they seem convinced that's who I am." But Jackson said the family could provide few details, "so I'm not so sure."

Another Jacksonville resident said he recognized Jackson as Michael Shawper, 40, a career Navy man. The caller knew several little known naval bases that Jackson remembers, "and the knowledge of history and geography, that matches. But he said the last time he had seen Shawper was 11 years ago in Guam."

Shawper and Moses both attended the same college in Jacksonville. One has been married two or three times, the other once. "So

I'm either an educated responsible man or an educated irresponsible man," Jackson said. "I can't be both, that's for sure. One of them has to be wrong. They both may be wrong." (*Wisconsin State Journal,* June 12, 1988, p. 4A)

"Who am I?" Few human beings in Western societies live out their lives without pondering this question. The search for self-knowledge and for a meaningful identity is pursued eagerly by some, desperately by others. College students, in particular, are often preoccupied with discovering who they are. Few, however, have experienced the existential uncertainty faced by John Jackson.

Each of us has unique answers to this question, answers that reflect our **self-concepts**— the various thoughts and feelings we have about ourselves. Self-concepts are sometimes assessed by having people answer the question "Who am I?" This "test" is the focus of Box 4-1. Before you read on, take a few moments and respond to this question yourself in the space provided in the box. For comparison, read the responses of a 9-year-old boy and a female college sophomore to the question "Who am I?" listed at the bottom of the box.

The following five major questions will be addressed in this chapter.

1. What is the self and how does it arise?

2. How do we acquire unique identities, the categories we use to specify who we are? How do we use them to locate ourselves in the world relative to others?

3. How do our identities guide our plans and behavior?

4. How does self-awareness influence the ways we think and feel?

5. Where do the feelings of evaluation that inevitably accompany thoughts about ourselves come from, and how do they affect our behavior? How do we protect our self-esteem against attack?

# ■ The Nature and Genesis of Self

## The Self as Source and Object of Action

We can behave in a wide variety of ways toward other persons. For example, if Bob is having coffee with Carol, he can perceive her, evaluate her, communicate with her, motivate her to action, attempt to control her, and so on. What is interesting to note, however, is that Bob can also act in the same fashion toward himself —that is, he can engage in self-perception, self-evaluation, self-communication, self-motivation, and self-control. Behavior of this type, in which the individual who acts and the individual toward whom the action is directed are one and the same, is termed *reflexive behavior.*

For example, if Bob, a student, has an important term paper due Friday, he engages in the reflexive process of self-control when he pushes himself ("Work on that history paper now"). He engages in self-motivation when he makes a promise to himself ("You can go out for pizza and beer Friday night"). Both of these processes are part of the self. To have a self is to have the capacity to engage in reflexive actions, to plan, observe, guide, and respond to our own behavior (Mead, 1934; Bandura, 1982).

By definition, the **self** is the individual viewed as both the source and the object of reflexive behavior. Clearly, the self is both active (the source that initiates reflexive behavior) and passive (the object toward which reflexive behavior is directed). The active aspect of the self is labeled the "I" and the object of self-action is labeled the "Me" (James, 1890; Mead, 1934).

It is useful to think of the self as a continuing process. Action involving the self begins with the "I," with an impulse to act. For example, Bob wants to see Carol. In the next moment that impulse becomes the object of self-reflection and, hence, part of the "Me" ("If I don't work on that paper tonight, I won't get it done on

Box 4-1

## MEASURING SELF-CONCEPTS

In order to study self-concepts, we need ways to measure them. Many methods have been used. For example, one approach asks people to check those adjectives on a list (intelligent, aggressive, trusting, and so on) that describe themselves (Sarbin & Rosenberg, 1955). In another approach (Osgood, Suci, & Tannenbaum, 1957), people rate themselves on pairs of adjectives (strong-weak, good-bad, active-passive): Are they more like one of the adjectives in the pair or more like its opposite? Another technique developed by Miyamoto and Dornbusch (1956) asks people whether they have more or less of a characteristic (self-confidence, likeableness) than members of a particular group (such as fraternities, sororities, and so on). In yet another technique, people sort cards containing descriptive phrases (interested in sports, concerned with achievement) into piles according to how accurately they think the phrases describe them (Stephenson, 1953).

Each of these popular methods provides respondents with a single, standard set of categories to use in describing themselves. Using the same categories for all respondents makes it easy to compare the self-concepts of different people. These methods have a weakness, however. They do not reveal the unique dimensions that individuals use in spontaneously thinking about themselves. For this purpose, techniques that ask people simply to describe themselves in their own words are especially effective (Kuhn & McPartland, 1954; McGuire & McGuire, 1982).

Instructions for the "Who Am I?" technique for measuring self-concepts (Gordon, 1968) are provided below. You can try this test yourself.

In the 15 numbered blanks write 15 different answers to the simple question "Who am I?" Answer as if you were giving the answers to yourself, not to somebody else. Write the answers in the order they occur to you. Don't worry about "logic" or "importance."

### I AM

1. _____     6. _____     11. _____
2. _____     7. _____     12. _____
3. _____     8. _____     13. _____
4. _____     9. _____     14. _____
5. _____     10. _____    15. _____

The following responses have been obtained from two persons, Josh and Arlene.

**Josh: A 9-year-old male**
a boy
do what my mother says, mostly
Louis's little brother
Josh
have big ears
can beat up Andy
play soccer
sometimes a good sport
a skater
make a lot of noise
like to eat
talk good
go to third grade
bad at drawing

**Arlene: A female college sophomore**
a person
member of the human race
daughter and sister
a student
people lover
people watcher
creator of written, drawn, and spoken (things) creations
music enthusiast
enjoyer of nature
partly the sum of my experiences
always changing
lonely
all the characters in the books I read
a small part of the universe, but I can change it
I'm not sure?! (Gordon, 1968)

time"). Next, Bob responds actively to this self-awareness, again an "I" phase ("But I want to see Carol, so I won't write the paper"). This, in turn, becomes the object to be judged, again a "Me" phase ("That would really hurt my grade"). So Bob exercises self-control and sits down at his desk to write. The "I" and "Me" phases continue to alternate as every new action (I) becomes in the next moment the object of self-scrutiny (Me). Through these alternating phases of self, we initiate, judge, and continuously guide our own behavior.

Mead portrays action as guided by an internal dialogue. People engage in conversations in their minds as they regulate their behavior. They use words and images to symbolize their ideas about themselves, other persons, their own actions, and others' probable responses to them. This description of the internal dialogue suggests that there are three capacities human beings must acquire in order to engage successfully in action. They must (1) develop an ability to differentiate themselves from other persons, (2) learn to see themselves and their own actions as if through others' eyes, and (3) learn to use a symbol system or language for inner thought. In this section, we will examine, in turn, how children come to differentiate themselves and how they learn to view themselves from others' perspectives. We will also note how language learning is intertwined with acquiring these two capacities.

## Self-Differentiation

In order to take the self as the object of action, we must—at a minimum—be able to recognize ourselves. That is, we must distinguish our own faces and bodies from those of others. This may seem elementary, but infants are not born with this ability. At first they do not even discriminate the boundaries between their own bodies and the environment. With cognitive growth, continuing tactile exploration of their bodies, and social experience with caregivers who treat them as distinct beings, infants gradually discover their physical uniqueness. Studies of

when children can recognize themselves in a mirror suggest that most children are able to discriminate their own image from others' by about 18 months (Bertenthal & Fischer, 1978).

Children must not only learn to discriminate their physical selves from others, but they must also learn to discriminate themselves as a social object. Mastery of language is critical in childrens' efforts to learn the latter (Denzin, 1977). Learning one's own name is one of the earliest and most important steps in acquiring a self. As Allport (1961) put it: "By hearing his name repeatedly the child gradually sees himself as a distinct and recurrent point of reference. The name acquires significance for him in the second year of life. With it comes awareness of independent status in the social group" (p. 115).

A mature sense of self entails recognizing that our thoughts and feelings are our private possessions. Young children often confuse processes that go on in their own minds with external events (Piaget, 1954). They locate their own dreams and nightmares, for example, in the world around them. The distinction between self and nonself sharpens as social experience and cognitive growth bring children to realize that their own private awareness of self is not directly accessible to others. By about age 4, children report that their thinking and knowing goes on inside their heads. Asked further, "Can I see you thinking in there?" they generally answer "No," demonstrating their awareness that self-processes are private (Flavell, Shipstead, & Croft, 1978).

Changes in the way children talk also reveal their dawning realization that the self has access to private information. During their first years of talking, children's speech patterns are the same whether they are talking aloud to themselves or directing their words to others. Gradually, however, they begin to distinguish speech for self from speech for others (Vygotsky, 1962). Speech for self becomes abbreviated until it is virtually incomprehensible to the outside listener, while speech for others becomes more elaborated over time. "Cold" suffices for Amy to tell herself she

To take the self as the object of our action, observing and modifying our own behavior, we must be able to recognize ourselves. Although infants are not born with this ability, they acquire it quickly.

wants to take off her wet socks. But no one else would understand this without access to her private knowledge. When addressing others, Amy would expand her speech to include whatever private information they would need to understand ("Gotta change my wet socks. They're making me cold."). This reflects her growing awareness that each self has its own unique store of knowledge.

Access to private information about the self leads to systematic differences in adults' self-descriptions compared to descriptions of others (McGuire & McGuire, 1986). Descriptions of the self are likely to focus on what one does, on physical action, and on affective reactions to others.

Descriptions of others focus on who the person is, on social interactions and on his or her cognitive reactions. Did your responses in Box 4-1 reflect these characteristics of self-descriptions?

## Role Taking

Recognizing that one is physically and mentally differentiated from others is only one step in the genesis of self. Once we can differentiate ourselves from others we can also recognize that each person sees the world from a different perspective. The second crucial step in the genesis of self is **role taking**—the process of imaginatively occupying the position of another person and viewing the situation from that person's perspective (Hewitt, 1988).

Role taking is crucial to the genesis of self, because through it the child learns to respond reflexively. Imagining others' responses to the self, children acquire the capacity to look at themselves as if from the outside. Recognizing that others see them as objects, children can become objects (Me) to themselves (Mead, 1934). They can then act toward themselves to praise ("That's a good girl"), to reprimand ("Stop that!"), and to control their own behavior ("Wait your own turn").

Long ago, C. H. Cooley (1908) noted the close tie between role taking and language skills. One of the earliest signs of role-taking skills is the correct use of the pronouns "you" and "I." To master the use of these pronouns requires taking the role of self and of the other simultaneously. Almost all children firmly grasp the use of "you" and "I" by 2½ (Clark, 1976). This suggests that children are well on their way to effective role taking at this age. Observations of preschoolers at play reveal complex uses of pronouns and perspectives (Denzin, 1977).

Contrast the thoughts of a child of 1 and a child of 9 as they seek their objectives. The 1-year-old is incapable of adopting the perspectives of others, whereas the 9-year-old has an expanding capacity for role taking. When 1-year-old Lisa sees an ice-cream cone, she is likely to tug at her mother's skirt and point. If ignored, she may cry or scream. Years later, at age 9, Lisa might generate a strategy based on the following internal dialogue: "Pointing won't help, since Mom's busy talking. I'll tug her sweater. No, she'll say supper's in an hour. I'll suggest buying ice cream for dessert. She'll like that." Lisa's greater effectiveness at 9 derives from her ability to adopt her mother's perspective as well as her own. Of course, role taking is often not perfect. Sometimes we misjudge other's probable responses to us because we are ignorant of or misinformed about their perspective.

## The Social Origins of Self

Our self-concepts are produced in our social relationships. Throughout life, as we meet new people and enter new groups, our self-concepts are modified by the feedback we receive from others. This feedback is not an objective reality that we can grasp directly. Rather, we must interpret others' responses in order to figure out how we appear to them. We then incorporate others' imagined views of us into our self-concepts.

To dramatize the idea that the origins of self are social, Cooley (1902) coined the term *look-ing-glass self.* The most important looking glasses for children are their parents and immediate family and, later, their playmates. They are the child's **significant others,** the people whose reflected views have greatest influence on the child's self-concepts. As we grow older, the widening circle of friends and relatives, school teachers, clergy, and fellow workers provides our significant others. The changing self-concepts we acquire through our lives depend on the social relationships we develop (see Table 4-1).

**Play and the Game**  Mead (1934) identified two sequential stages of social experience leading to the emergence of the self in children. He called these stages play and the game. Each stage is characterized by its own form of role taking.

In the play stage, young children imitate the activities of people around them. Through such play, children learn to organize different activities into meaningful roles (nurse, doctor, fire fighter). For example, using their imaginations, children carry sacks of mail, drop letters into mailboxes, greet homeowners, and learn to label these activities as fitting the role "mail carrier." At this stage, children take the roles of others one at a time. They do not recognize that each

■ **TABLE 4-1**

SIGNIFICANT OTHERS MENTIONED IN SELF-DESCRIPTIONS, BY AGE

| | Ratio of the Frequency of Mentioning | | |
| | 1 | 2 | 3 |
| Age | Parents versus Teachers | Brothers and Sisters versus Friends and Fellow Students | Nonfamily Members versus Extended Family |
|---|---|---|---|
| 7 years | 1.7 to 1 | 1.7 to 1 | 4 to 1 |
| 9 years | 1 to 1.4 | 1 to 1.4 | 8 to 1 |
| 13 years | 1 to 1 | 1 to 1 | 13 to 1 |
| 17 years | 1 to 2.3 | 1 to 2.3 | 49 to 1 |

Note: In this study, 560 boys and girls were asked, "Tell us about yourself." The children's responses suggest that their self-definitions in terms of other people tend to shift away from family members with age—from parents to teachers (column 1), from brothers and sisters to friends and fellow students (column 2), from extended family members (cousins, aunts, uncles) to nonfamily members (column 3). For example, 7-year-olds mentioned parents almost twice as often as teachers, while 17-year-olds mentioned teachers more than twice as often as parents (column 1).

Source: Adapted from McGuire and McGuire, 1982.

By playing complex games, children learn to organize their actions into meaningful roles and to imagine the viewpoints of the different game players at the same time. These experiences teach them how to coordinate effectively with others and to follow social rules.

role is intertwined with others. Playing mail carrier, for example, the child does not realize that mail carriers also have bosses to whom they must relate. Nor do children in this stage understand that the same person simultaneously holds several roles—that mail carriers are also parents, store customers, and golf partners.

The game stage comes later, when children enter into organized activities such as complex games of house, school, and team sports. These activities demand interpersonal coordination because the various roles are differentiated. Role taking at the game stage requires children to imagine the viewpoints of several others at the same time. For Ellen to play shortstop effectively,

for example, she must adopt the perspectives of the infielders and of the base runners as she fields the ball and decides where to throw. In the game, children also learn that different roles relate to each other in specified ways. Ellen must understand the specialized functions of each position, the ways the players in different positions coordinate their actions, and the rules that regulate baseball.

**The Generalized Other** Repeated involvement in organized activities lets children see that their own actions are part of a pattern of interdependent group activity. This experience teaches children that organized groups of people share common perspectives and attitudes. With this new knowledge, children construct a **generalized other**—a conception of the attitudes and expectations held in common by the members of the organized groups with whom they interact. When we imagine what "the group" expects of us, we are taking the role of the generalized other. We are also concerned with the generalized other when we wonder what "people" would say, or what "society's standards" demand. As children grow older, they control their own behavior more and more from the perspective of the generalized other. This helps them to resist influence from specific others who just happen to be present in the situation at the moment.

Over time, children internalize the attitudes and expectations of the generalized other, incorporating them into their self-concepts. But building up self-concepts involves more than accepting the reflected views of others. It is often fraught with confusion and turmoil. We may misperceive or misinterpret the responses that others direct to us, for example, due to our less-than-perfect role-taking skills. Others' responses may themselves be contradictory or inconsistent. We may also resist the reflected views we perceive because they conflict with our prior self-concepts or with our direct experience. A boy may reject his peers' view that he is a sissy, for example, because he previously thought of himself as brave and could still visualize his experience of beating up a bully.

**Self-Evaluation** The views of ourselves that we perceive from others usually imply positive or negative evaluations. These evaluations also become part of the self-concepts we construct. Actions that others judge favorably produce self-concepts about which we feel good. In contrast, when others disapprove or punish our actions, the self-concepts we derive are negative. Our self-evaluations are based on our interpretations of the feedback we receive from others.

# ■ Identities: The Self We Know

In Box 4-1 Arlene described herself as a person, member of the human race, daughter, and people lover. This is the self she knows, a self that includes specific identities. **Identities** are the categories people use to specify who they are, to locate themselves relative to other people. When we think of our identities, we are actually thinking of various plans of action that we expect to carry out. When Arlene identifies herself as a student, for example, she has in mind that she plans to attend classes, write papers, take exams, and so on. If Arlene does not engage in these behaviors, she will have to relinquish her student identity.

In this section, we will consider several questions about the self we know. (1) How do our experiences in society influence the identities we include in this self? (2) What aspects of self do people note in their actual self-descriptions? (3) How do the aspects of self that people note vary from one situation to another? (4) What evidence is there that the self we know is based on the reactions we perceive from others?

## Role Identities

Each of us occupies numerous positions in society—student, friend, son or daughter, customer. Each of us, therefore, enacts many different social roles. We construct identities by observing our own behavior and the responses of others to us as we enact these roles. For each role we enact, we develop a somewhat different view of who we are, an identity. Because these identities are concepts of self in specific roles, they are called **role identities.** The role identities we develop depend on the social positions available to us in society. As a result, the self we know is linked to society fundamentally through the roles we play. It reflects the structure of our society and our place in it (Gordon, 1976; McCall & Simmons, 1978; Stryker, 1980).

Do societal role expectations strictly dictate the contents of our role identities? Apparently not. Consider, for example, the role expectations for the college instructor. Some instructors deliver lectures, while others lead discussions; some encourage questions, while others discourage them; some assign papers, and others do not. As this example indicates, role expectations usually leave individuals room to improvise their own role performances. It is probably more accurate to think of people as "making" their roles—that is, shaping them—rather than as conforming rigidly to role expectations (Turner, 1978).

Several influences affect the way we make the roles we enact. A general framework is set by the conventional role expectations in society. In the role of student, for example, you must submit assigned papers. Within this general framework, you can fashion your actual role performances to reflect your personal characteristics and competencies (selecting topics that interest you and are likely to highlight your strengths and cover your weaknesses). You also mold your role performances to impress your audience (writing in the style the instructor prefers). Finally, you adjust your different performances to maintain some consistency among them (trying for a level of quality consistent with your other course work). Since each person makes roles in a unique, personal fashion, we each derive somewhat different role identities even if we occupy similar social positions. Consequently, our role identities as student, team player, and so on differ from the role identities of others who also occupy these same positions.

## Actual Self-Descriptions

If the self we know consists of role identities, people's actual self-descriptions should reflect their major social roles. Do they? To answer this question, we examine responses people give to the "Who am I?" questionnaire. These responses reveal three general types of self-description: role identities, personal qualities, and self-evaluation (Gordon, 1968).

**Role Identities**  In describing themselves, most people mention their major social roles (age group, student, or work role). This is demonstrated in the responses of high school students (see Table 4-2), as well as responses from other large samples. Most respondents spontaneously list their role identities as members of occupational, educational, or family groups among their first self-descriptions. They also frequently mention identities as members of various social categories such as religion, activity, group, race, and so on. There is usually substantial public agreement about the role identities a person claims (Kuhn & McPartland, 1954).

The other two types of self-description—personal qualities and self-evaluations—refer to conceptions of self that others may dispute or ignore. Our conceptions of how moody we are,

■ TABLE 4-2

WHO AM I? SELF-DESCRIPTIONS OF HIGH SCHOOL STUDENTS

| Type of Response (example) | Percentage of Students Who Mentioned |
|---|---|
| *Role Identities* | |
| Age group (teenager, 15-year-old) | 82 |
| Student role (freshman, student) | 80 |
| Sex (a boy, daughter) | 74 |
| Abstract category (a person, human) | 41 |
| Participant in activities (hiker, poet) | 31 |
| Membership in actual group (on the football team) | 17 |
| Kinship role (sister, husband) | 17 |
| Name (Claire, Mark) | 17 |
| Territoriality, citizenship (a Bostonian) | 16 |
| Religion (Catholic, Lutheran) | 11 |
| *Personal Qualities* | |
| Interpersonal style (introverted, cool, friendly) | 59 |
| Emotional, psychological style (moody, optimistic, in love) | 52 |
| Body image (good-looking, 112 lbs.) | 36 |
| Preferences: judgments, tastes, likes (loves Bach, hates abstract art) | 27 |
| Material possessions (owns a Porsche) | 5 |
| *Self-Evaluations* | |
| Sense of competence (creative, always making mistakes) | 36 |
| Sense of self-determination (ambitious, self-starter) | 23 |
| Sense of moral worth (trustworthy, evil) | 22 |
| Sense of unity (mixed up, a whole person) | 5 |

Note: Responses of 157 high school students to the "Who Am I?" questionnaire illustrate the types of self-concepts that constitute the self we know.

*Source: Adapted from* The self in social interaction, I: Classic and contemporary perspectives, C. Gordon and K.J. Gergen(eds.) © 1968 John Wiley & Sons.

how moral, or how competent are not matters of public consensus. These self-concepts express the personal, idiosyncratic variation in our self-descriptions.

**Personal Qualities** In describing their personal qualities, people mention most frequently the styles of interpersonal behavior (introverted, cool) that distinguish the way they fashion their unique role performances. People also mention the emotional or psychological styles (optimistic, moody) that characterize these performances. Individual preferences point to specific ways people express their role identities. For example, a person who sees herself as a musician expresses this role identity differently depending on whether she prefers Bach or rock. Body image, the aspect of the self we recognize earliest, remains important throughout life. Beyond this, our self extends to include our material possessions, such as our clothing, house, car, records, and so on (James, 1890).

**Self-Evaluations** The third type of self-description refers to the ways we evaluate ourselves. We form these self-evaluations when reflecting on the adequacy of our role performances, on the extent to which we live up to the role identities to which we aspire. Our evaluations most commonly focus on our competence, self-determination, moral worth, or unity. Self-evaluations also influence the ways we express our role identities. A musician, for example, will pursue opportunities to perform in public more persistently if she sees herself as competent than if she thinks she is never quite good enough. Self-evaluations are so important that the concluding section of this chapter will be devoted to them.

## The Situated Self

If we were to describe ourselves on several different occasions, the identities, personal qualities, and self-evaluations mentioned would not remain the same. This is not due to errors of reporting. Rather, it demonstrates that the aspects of self that enter our awareness and matter to us most depend on the situation. The **situated self** is the subset of self-concepts chosen from our identities, qualities, and self-evaluations that constitutes the self we know in a particular situation (Hewitt, 1988).

The self-concepts most likely to enter into the situated self are those distinctive in the setting and relevant to the ongoing activities. Consider a black woman for whom being black and being a woman are both important self-concepts. When she interacts with black men, she is more likely to think of herself as a woman; when she interacts with white women, she is more likely to be aware that she is black. Similarly, whether gender is part of your situated self depends in part on the gender composition of those present (Cota & Dion, 1986). Male and female college students placed in a group with two students of the opposite gender were more likely to list gender in their self-descriptions than members of all male or all female groups. Thus self-concepts that are distinctive or peculiar in the social setting tend to enter into the situated self (McGuire & McGuire, 1982).

The activities in which we are engaged also determine the self-concepts that constitute the situated self. A job interview, for example, draws attention to your competence; a party makes your body image more salient. The self we experience in our imaginings and in our interactions is always situated, because setting characteristics and activity requirements make particular self-concepts distinctive and relevant.

## Research on Self-Concept Formation

Two of the key theoretical ideas discussed so far are: (1) A person's self-concept is shaped by the reactions that he or she receives from significant others during social interaction and (2) A person's self-concept is more likely to include roles or personal traits that are distinctive rather than those that are not. Each of these has been the focus of empirical research.

**Reflected Appraisals** The idea that the person bases his or her self-concept on the reactions he or she perceives from others during social interaction is captured by the term *reflected appraisals*. Studies of this process (Miyamoto & Dornbusch, 1956; Marsh, Barnes, & Hocevar, 1985) typically compare people's self-ratings on various qualities (intelligence, self-confidence, physical attractiveness) with the views of themselves that they perceive from others. The studies also compare self-ratings with actual views of others. Results of these studies support the hypothesis that it is the perceived reactions of others, rather than their actual reactions, that are crucial for self-concept formation.

Recent research has focused on the differential effect of various significant others on one's appraisal of self in particular roles/domains. Felson (1985; Felson & Reed, 1986) has studied the relative influence of parents and peers on the self-perceptions of fourth through eighth graders with regard to academic ability, athletic ability, and physical attractiveness. The results indicate that parents affect self-appraisals in the areas of academic and athletic ability, whereas peers are an important influence on perceived attractiveness. One aspect of attractiveness is weight. Although there is an objective measure of weight (pounds, or pounds in relation to height), it is the social judgment ("too fat," "too thin," or "just right") that is incorporated into the self-concept. A study of adolescent health obtained self-appraisals of weight from 6,500 adolescents, as well as appraisals from their parents and a physician (Levinson, Powell, & Steelman, 1986). These young people were generally unhappy with their weight, with males judging themselves to be too thin and females judging themselves to be overweight. For both, parental appraisal was significantly related to the young person's judgment, while the physician's rating was not.

Typically, a person's self-ratings are related more closely to his or her perceived ratings by others than to the actual ratings by others. Why is this so? Three reasons are especially important. First, others rarely provide full, honest feedback about their reactions to us. Second, the feedback we do receive is often inconsistent and even contradictory. Third, the feedback is frequently ambiguous and difficult to interpret. It may be in the form of gestures (shrugs), facial expressions (smiles), or remarks that can be understood in many different ways. For these reasons, we may know little about others' actual reactions to us. Instead, we must rely on our perceptions of others' reactions to construct our self-concepts (Schrauger & Schoeneman, 1979).

Evidence that self-concepts are related to the perceived reactions of others does not, of itself, demonstrate that self-concepts are actually formed in response to these perceived reactions. However, one study (Mannheim, 1966) does suggest such an impact of the perceived reactions of others on self-concepts. The investigators in this study asked college dormitory residents to describe themselves and also to report how they thought others viewed them. Several months later, self-concepts were measured again. In the interim, students' self-concepts had moved closer to the views they had originally thought that others held. Change toward the perceived reactions of others had indeed occurred.

**Distinctiveness**   As noted above, those of our features that attract attention by others should become part of our self-concepts. Other persons are most likely to notice and respond to features that distinguish us from our social groups—our weight if we are obese or skinny, our ethnic heritage if we are foreign born. These distinctive features are the very ones we are most likely to note when we compare ourselves to others. Thus, when describing ourselves, we should be more likely to mention a feature that is distinctive.

Studies using the "Who am I?" technique in India and in the U.S. support this prediction. In both countries, members of minority religious groups (Christians, Muslims, Sikhs in India; Catholics and Jews in the U.S.) mentioned their religious identity three times as often as members of majority religious groups (Kuhn & McPartland, 1954; Driver, 1969).

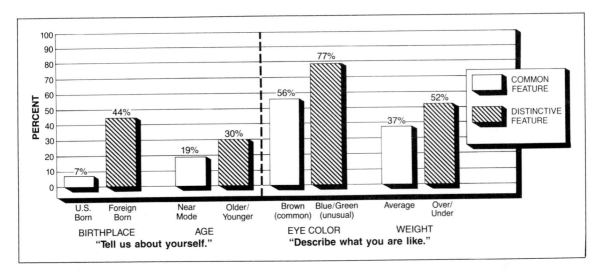

■ **FIGURE 4-1**

PERCENTAGE OF STUDENTS WHO MENTION A FEATURE SPONTANEOUSLY AS PART OF THEIR SELF-CONCEPT

A group of 252 sixth graders from ten classrooms were asked to describe themselves. Students mentioned a particular feature (for example, birthplace) more often if that feature distinguished them from their classmates. Because these are characteristics on which we stand out from our social groups, attracting more notice and social comment, we are more likely to build them into our self-concepts.

*Source: Adapted from McGuire and Padawer-Singer, 1976.*

Sixth graders in the U.S. also tend to pick features that distinguish them when asked to describe themselves (McGuire & Padawer-Singer, 1976). Students mention their age, birthplace, sex, hair color, eye color, or weight more frequently when that feature places them in the minority of their class or when they differ substantially on it from their class average (see Figure 4-1). These findings support the symbolic interactionist approach to self-concept, because they demonstrate that socially distinctive features are especially important in self-identification.

## ■ Identities: The Self We Enact

How does the self influence the planning and regulation of social behavior? The general answer to this question is that we are motivated to plan and to perform behaviors that will confirm and reinforce the identities we wish to claim for ourselves (Backman & Secord, 1968; Burke &

Reitzes, 1981). In elaborating on this answer, we will examine three more specific questions: (1) How are behaviors linked to particular identities? (2) Of the different identities available to us, what determines which ones we choose to enact in a situation? (3) How do our identities lend unity and consistency to our behavior?

### Identities and Behavior

Earlier we noted that self-conceptions include both role identities and personal qualities. Some people place greater emphasis on one than on the other. This is reflected, for instance, in responses to the question "Who am I?" Some people emphasize socially based characteristics (group membership, reputation), while others list primarily personal qualities (characteristic emotions, personal goals). Leary and colleagues (1986) predicted that the behavioral preferences of these two groups would differ. In the area of recreation, people who emphasize social aspects would prefer team sports (softball, volleyball, basketball), while those who emphasize personal qualities would prefer individual sports (swimming, running, aerobics). Furthermore, those whose self-concept is predominantly social should prefer occupations that offer social rewards, such as status and friendship, while those whose self-concept is predominantly personal should prefer jobs that offer rewards such as opportunities for self-expression and personal growth. Both predictions were verified.

The link between identities and behaviors is through their common meanings (Burke & Reitzes, 1981). If members of a group agree on the meanings of particular identities and behaviors, they can regulate their own behavior effectively. They can plan, initiate, and control behavior to generate the meanings that establish the identities they wish to claim. If the meanings of an identity or of particular behaviors are not agreed on, however, people have difficulty establishing their preferred identities. If Roberta sees no connection between competitiveness and femininity, for example, she will have trouble establishing a feminine identity in the eyes of friends who think being feminine means being noncompetitive.

When we travel, change schools or jobs, or move into new groups, we come into contact with individuals who do not share all the same meanings with us because their backgrounds or cultures differ from ours. We may then be unable to predict whether performing a particular behavior will confirm an identity or undermine it. Only by learning the meanings that prevail in our surroundings can we select behaviors that will confirm the identities we wish to claim.

## Choosing an Identity to Enact

Each of us has many different identities. Each identity suggests its own lines of action. These lines of action are not all compatible, however, nor can they all be pursued simultaneously in a single situation. If you are at a family reunion in your parents' home, for example, you might wish to claim an identity as a helpful son/daughter, an aspiring poet, or a witty conversationalist. These identities suggest different, even conflicting, ways of relating to the other guests. What influences the decision to enact one rather than another identity? Below are several factors affecting such choices.

**The Hierarchy of Identities**  The many different role identities we enact do not have equal importance for us. Rather, we organize them into a hierarchy according to their **salience**— that is, their relative importance to the self-concept. This hierarchy exerts a major influence on our decision to enact one or another identity (McCall & Simmons, 1978; Stryker, 1980). First, the more salient an identity is to us, the more frequently we choose to perform activities that express that identity (Stryker & Serpe, 1981). Second, the more salient an identity, the more likely we are to perceive that situations offer opportunities to enact that identity. Only a person aspiring to the identity of poet, for example, will perceive a family reunion as a chance to recite his or her poems. Third, we are more active in seeking out opportunities to behave in terms of salient identities (searching for guests willing to listen to poems). Fourth, we conform more with the role expectations attached to the identities that we consider the most important.

What determines whether a particular identity occupies a central or a peripheral position in the identity hierarchy? In general, several factors affect the importance we attach to a role identity: (1) the resources we have invested in constructing the identity (time, effort, and money expended, for example, in learning to be a sculptor); (2) the extrinsic rewards that enacting the identity has brought (purchases by collectors, acclaim by critics); (3) the intrinsic gratifications derived from performing it (the sense of competence and aesthetic pleasure obtained when sculpting a human figure); and (4) the amount of self-esteem staked on enacting the identity well (the extent to which a positive self-evaluation has become tied to being a good sculptor). As we engage in interaction and experience greater or lesser success in performing our different identities, their salience shifts.

**Social Networks** Each of us is part of a network of social relationships. These relationships may stand or fall on whether we continue to enact particular role identities. The more numerous and significant the relationships that depend on enacting an identity, the more committed we become to that identity (Callero, 1985). Consider, for example, your role as a student. Chances are that many of your relationships—with roommate(s), friends, instructors, and perhaps a lover—depend on your continued occupancy of the student role. If you left school, you could lose a major part of your life. Given this high level of commitment, it isn't surprising that, for many students, being forced to leave school is traumatic.

The more commitment we have to a role identity, the more important that identity will be in our hierarchy. For instance, adults for whom participating in religious activities was crucial for maintaining everyday social relationships ranked their religious identity as relatively important compared with their parent, spouse, and worker identities (Stryker & Serpe, 1981). Similarly, the importance rank that undergraduates gave to various identities (student, friend, son/daughter, athlete, religious person, and dating partner) depended on the importance to them of the social relationships maintained by enacting each identity (Hoelter, 1983).

**Need for Identity Support** We are likely to enact those of our identities that most need support at the moment because they have recently been challenged. For instance, suppose that someone has recently had difficulty getting a date. That person may now choose actions calculated to elicit responses indicating she is an attractive dating partner. We also tend to enact identities likely to bring intrinsic gratifications (such as a sense of accomplishment) and extrinsic rewards (such as praise) that we especially need

or miss at the moment. For example, if, after hours of solitary study, you feel a need for relaxed social contact, you might seek gratification by going to a student lounge or union to find someone to chat with.

**Situational Opportunities** Social situations are restrictive; they let us enact only some identities profitably, not others. Thus, in a particular situation, the identity we choose to enact depends partly on whether the situation offers opportunities for profitable enactment. Regardless of the salience of your identity as poet, if no one wants to listen to your poems, there will be no opportunity to enact that identity.

A study of racial values, attitudes, and behavior among white Canadian college students vividly demonstrates that, given the situational opportunity, we enact the identities that are most in need of support (Dutton & Lake, 1973). Researchers selected subjects on the basis of their earlier statements that they valued equality highly and that they viewed themselves as free of racial prejudice. They invited students individually to the research site and connected them to a rigged lie detector. The lie detector purportedly revealed to the students their true attitudes by measuring psychological changes in response to seeing slides of interracial scenes. For half the students, the lie detector "revealed" strong negative reaction, implying that they were prejudiced against blacks and threatening their identity as egalitarian. The other students were told their responses were typical of unprejudiced individuals.

Upon leaving the research site, each student was approached either by a white panhandler or a black panhandler. The black panhandler received substantially more money from students whose egalitarian identity had been threatened than from those whose egalitarianism had not been threatened. The white panhandler received the same moderate amount of money from both groups of students. These results, shown in Figure 4-2, suggest that the threatened students went out of their way to prove they were not

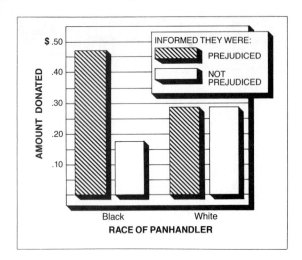

■ **FIGURE 4-2**

ACTING TO REAFFIRM THREATENED IDENTITIES

White students' self-identities as egalitarians were either threatened by false feedback from a lie detector, indicating that they were racially prejudiced, or confirmed by feedback that they were not prejudiced. Following the lie detector test, students were solicited either by a black or a white panhandler. Those who were approached by the black panhandler after their egalitarian identity had been threatened donated the most money. In this way they "proved" they were not racially prejudiced and reaffirmed their recently threatened egalitarian identity. Donations to the white panhandler were unaffected by the feedback about prejudice. These results demonstrate how the desire to affirm threatened identities influences behavior.

*Source: Adapted from Dutton and Lake, 1973.*

racially prejudiced when given the chance to help a black. They perceived the situation as an opportunity to reaffirm their recently challenged egalitarian identity.

## Identities as Sources of Consistency

Although the self includes multiple identities, people usually experience themselves as a unified entity. Our most salient identities provide consistent styles of behavior and priorities that lend continuity and unity to our behavior. In this way, the importance hierarchy helps construct a unified sense of self from our multiple identities.

The hierarchy of identities influences consistency in three ways. First, the hierarchy provides us with a basis for choosing which situations we should enter and which ones we should avoid. For instance, in one study of the everyday activities of college students, subjects recorded information about each situation they encountered over a 30-day period (Emmons, Diener, & Larson, 1986). There were clear patterns of choice and avoidance in each student's interactions; these patterns were consistent with the person's characteristics, such as sociability.

Second, the hierarchy influences the consistency of behavior across different situations. In another study of situation selection, each person was asked to report the extent to which each of ten affective states and ten behavioral responses occurred in various situations, again over a 30-day period (Emmons & Diener, 1986). The results indicated a significant degree of consistency in affective responses across situations the person chose to be in and a significant degree of consistency in behavioral responses across situations that were imposed on the person.

Third, the hierarchy influences consistency in behavior across time. Serpe (1987) studied a sample of 310 freshmen, collecting data at three points during their first semester in college. The survey measured the salience at each point of five identities: academic ability, athletic/recreational involvement, extracurricular involvement, personal involvement (friendships), and dating. There was a general pattern of stability in salience. Change in salience was more likely for those identities where there was greater opportunity for change (for example, dating).

To be consistent across time and situations, we must monitor our behavior to ensure that it confirms the chosen identity. We are especially likely to monitor and regulate behavior intended to express our important identities (Santee & Jackson, 1979). As a result, people expect to behave more consistently with their important identities, and they generally do so (Bem & Allen, 1974; Markus, 1977). Imagine that an identity as "expert driver" is important to Steve,

The more important an identity is to us, the more consistently we act to express it, regardless of others' reactions. Are any of your identities so important that you would express them the way these Hassidic Jews do?

for example, but unimportant to Craig. Steve, but not Craig, would insist on driving whatever the occasion.

Behavior is more likely to be inconsistent when it is based on unimportant identities. We make no persistent efforts to behave consistently with identities low in our hierarchy. When unimportant identities are at stake, we readily switch from one to another to maximize profit in the situation. Craig might volunteer to drive, for example, if driving lets him choose a destination he especially wants. He might read the map for someone else, if he believes a navigator identity will win admiration.

Major life changes sometimes upset or destabilize the importance hierarchy of our identi-

ties. When this happens, we lose a crucial basis for self-organization and for planning our behavior. This occurs especially during periods such as adolescence and retirement, when our social relationships, role opportunities, and abilities change rapidly. We are then likely to feel a weakened sense of unity and confusion about how to behave. This has been called an *identity crisis* (Erikson, 1968). In order to overcome such confusion, we must reorganize our identity hierarchy, giving greater importance to identities based on our newly available or remaining social positions. A retiree may successfully reorganize the hierarchy, for example, by upgrading identities based on new hobbies (gardener) and on continuing social ties (witty conversationalist).

# ■ The Self in Thought and Feeling

Many of us are preoccupied by our own thoughts and feelings, as well as by information that is especially relevant to us. At a noisy, crowded party you can often barely hear the conversation you are directly involved in. But should someone mention your name, even halfway across the room, you are likely to hear it and immediately shift your attention in that direction. Experimental research has confirmed this power of self-relevant information to grab our attention (Moray, 1959).

In this section we will discuss three ways in which the self affects our thoughts and feelings. These include (1) the impact of information's relevance to the self on memory and judgment, (2) ways that focusing attention on the self influences the relationship between our identities and our behavior, and (3) ways the self influences the emotions we experience.

## Self-Schema

The influence of self on thought occurs through the operation of the self-schema (Greenwald & Pratkanis, 1984). **Self-schema** refers to the organized structure of information about the self possessed by the individual (Markus, 1977). It influences what we perceive and how we interpret it, organizes information in memory, and shapes the inferences we make about ourselves. It includes important identities and personal qualities. It also includes generalizations about the self based on past experience. Repeated thinking about the self creates detailed and elaborate mental connections among those identities, qualities, and experiences. The self-schema is readily activated whenever we encounter relevant information or events.

The self-schema influences the processing of incoming information. Compare Sara, for whom an athlete identity is important, with Maggie, who does not think about herself as either athletic or nonathletic. Sara will judge more quickly and confidently whether traits like agile, clumsy, muscular, and puny apply to her than will Maggie. She will also reject more strongly information purporting to show that she is either more coordinated than she had previously thought or less. In short, people are quicker and more certain when judging and interpreting information related to their important identities or qualities.

The self-schema helps to maintain one's self-concept. When discussing reflected appraisals, we noted that the feedback one receives from others is always incomplete, frequently ambiguous, and sometimes inconsistent. The self-schema determines how we receive and process this feedback. We pay more attention to relevant information and selectively focus on information that confirms our self-concepts, especially highly salient identities. It also provides us with a basis for interpreting the responses of others to us. Because of the influence of self-schema, we typically perceive more confirmation about our self-concept than actually exists (Schwann, 1987).

The self-schema also influences memory. In one study, a group of dating couples met for a discussion. Afterward, each member of the group was asked to recall what was said during their meeting. Answers indicated that they remembered their own remarks best, the remarks of their partners (an extension of the self) next best, and the words of strangers least (Brenner, 1976). Thus, memory for events is better the more they relate to the self. Relating information to the self has also been shown to enhance learning and memory for factual material. We remember a list of personality traits better, for instance, if we think about how each trait applies to us while trying to memorize the list (Rogers, 1977).

Thus, the important identities and other interrelated self-concepts constituting our self-schema provide a finely tuned set of mental categories that we can use to process new information. By employing these self-related categories, we seek out, perceive, judge, and remember information more effectively.

## Effects of Self-Awareness

While eating with friends, reading a book, or participating in conversation, your attention is usually directed toward the objects, people, and events that surround you. But what happens if—upon looking up—you discover a photographer with his lens focused on you, snapping away? Or what if you suddenly notice your image reflected in a large mirror? In such circumstances most of us become self-conscious. We enter a state of **self-awareness**—that is, we take the self as the object of our attention and focus on our own appearance, actions, and thoughts. This corresponds to the "Me" phase of action (Mead, 1934).

Numerous circumstances cause people to become self-aware: seeing our reflection in a mirror, facing a camera, hearing our voice on tape, encountering unfamiliar tasks and surroundings, or blundering in public. Mirrors, cameras, and recordings of our own voice cause self-awareness because they directly present the self to us as an object. Unfamiliarity and blundering also cause self-awareness because they disrupt the smooth flow of action and interaction. When this happens, we must attend to our own behavior more closely, monitoring its appropriateness and bringing it into line with the demands of the situation. In general, anything that reminds us that we are objects of others' attention will increase our self-awareness.

How does self-awareness influence behavior? When people are highly self-aware, they are more likely to be honest and to more accurately report on their mood state, psychiatric problems, and hospitalizations (Gibbons et al., 1985). High self-awareness is also associated with behaving helpfully and industriously. In general, people who are self-aware act in ways more consistent with personal and social standards (Wicklund & Frey, 1980; Wicklund 1982). Their behavior is controlled more by the self. In the absence of self-awareness, behavior is more automatic, habitual, or driven. Society gains control over its members through the self-control individuals exercise when they are self-aware

(Shibutani, 1961). This is because the standards to which people conform are largely learned from significant groups in society. Self-awareness is thus often a civilizing influence.

A study of Halloween trick-or-treaters illustrates the impact of self-awareness on conformity to standards (Beaman et al., 1979). Children arriving at various homes on Halloween were sent into the living room alone to collect candy from a bowl. They were first asked their name and age and told, "You may take one of the candies." This instruction defined taking extra candies as "cheating." Self-awareness was raised for half the children. While reaching for the candy, they saw themselves in a large mirror. The presence of the mirror reduced cheating to one-tenth the level of violations exhibited in the absence of the mirror (see Figure 4-3). Evidence on cheating by college students also suggests that self-awareness increases conformity (Diener & Wallbom, 1976). These and other studies show that important aspects of our identities—such as being honest—have more impact on our behavior when we are self-aware.

The most widely endorsed theory to explain these effects of self-awareness assumes that attention to self leads to self-evaluation (Wicklund, 1975). We evaluate ourselves by comparing our current or planned behavior to our ideal standards or to the standards salient in the situation. Because our behavior rarely matches these standards, this comparison arouses self-dissatisfaction. In order to reduce this dissatisfaction, we first attempt to escape self-awareness (fleeing the scene, covering our face or ears, and so on). If this is impossible, we try to reduce dissatisfaction by controlling and modifying our behavior to make it more consistent with desirable standards.

When social standards and internalized standards correspond, an increase in self-awareness brings behavior closer to ideals. But what if social standards conflict with internalized standards, as when groups pressure individuals to change their attitudes or to violate their personal standards? When this happens, the impact of increased self-awareness depends on whether

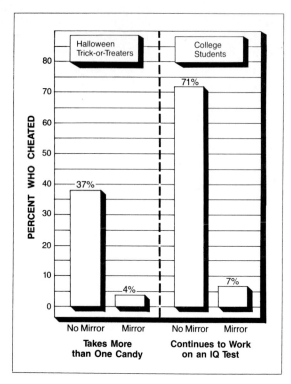

■ **FIGURE 4-3**

SELF-AWARENESS: A CIVILIZING INFLUENCE ON BEHAVIOR

When we are in a state of self-awareness, we are more likely to behave in a manner that corresponds with prevailing social standards or with the personal standards built into our identities. This is illustrated by the effects of the presence of a mirror—which increases self-awareness—on cheating. Halloween trick-or-treaters over age 8, instructed to take only one candy, cheated much less if exposed to a mirror while reaching into the candy bowl. The presence of a mirror also reduced cheating by college students who were given a chance to cheat by continuing work on an intelligence test beyond the time limit.

*Sources: Adapted from Beaman et al., 1979; Diener and Wallbom, 1976.*

our attention is drawn to the public, social aspects of self or to the private, covert aspects of self (Scheier & Carver, 1981).

If we attend to public aspects of self (the way our behavior appears to others, our mannerisms, our public image), we respond more to group influence and adhere less to our personal stan-

dards. On the other hand, if we attend to private aspects of self (our personal values, attitudes, and internalized standards), the effect is the opposite: we guide our behavior more by personal standards and yield less to group influence. In studies of these opposing effects, it was shown that exposure to cameras could induce public self-awareness and increase social responsiveness at the expense of self-direction. In contrast, exposure to mirrors was shown to induce private self-awareness and to increase self-direction at the expense of social conformity (Scheier & Carver, 1983).

These findings suggest that groups enhance their social control over individual behavior when they expose individuals to conditions like an attentive audience, unfamiliar circumstances, and socially awkward tasks that increase awareness of the public self. Interestingly, these are precisely the conditions used so effectively by cults. Self-awareness is thus a double-edged sword: if we focus on our internalized standards, it increases independence and self-directed behavior, whereas if we focus on our public image, it increases conformity.

## Influences of Self on Emotions

At first, knowing how we feel appears to be a straightforward matter. We know whether we are happy or sad because we feel these emotions directly. But the matter is more complex than this. When you feel tightness in the stomach and sweaty palms, how do you know immediately whether you are excited about your date for tonight or fearful of the exam you have to take tomorrow? Closer examination suggests that the emotions we recognize are often merely plausible explanations we generate for physiological reactions. Whether we interpret our clenched teeth as a sign of fear or of sexual excitement depends on whether we are watching a horror film or an erotic movie.

**Cognitive Labeling Theory** According to the cognitive labeling theory, emotions do not happen to the self. Rather, we ourselves actively construct our emotions (Schachter, 1964). The theory proposes that emotional experience is the result of the following three-step sequence.

1. An event in the environment produces a physiological reaction.
2. We notice the physiological reaction and search for an appropriate explanation.
3. By examining situational cues (What was happening when I reacted?), we find an emotion label (joy, disgust) for the reaction.

The theory further assumes that arousal is a generalized state, so that different emotions are not physiologically distinguishable. This implies that general arousal can signify virtually any emotion, depending on the situation.

In one study of this issue (Schachter & Singer, 1962), researchers gave students an injection of epinephrine, a drug that produces heightened physiological arousal. They informed one group (the Informed Group) of students that this injection would probably cause them to experience a pounding heart, flushed face, and trembling. They told a second group (the Ignorant Group) nothing about the drug's side effects. All students then waited with a confederate who, while appearing to be another student, was actually employed by the researchers. Depending on the experimental treatment, the confederate behaved either euphorically (shooting crumpled paper at a waste basket, flying paper airplanes, playing with a hula hoop) or angrily (reacting with hostility to items on a questionnaire and finally tearing it up).

According to the theory, students in the Informed Group would not need to seek an explanation for their arousal, because they knew their symptoms were drug induced. Students in the Ignorant Group, however, lacked an adequate explanation for their symptoms and, thus, would need to search the environment for cues to help them label their feelings. Results confirmed these predictions. Students in the Ignorant Group adopted the label for their arousal suggested by the environment. That is, those in the Ignorant Group who waited with the euphoric confederate described themselves as hap-

py, while those who waited with the angry confederate described themselves as angry. The self-descriptions of the Informed Group, on the other hand, were largely unaffected by the confederate's behavior.

Numerous later studies have expanded these findings to additional emotions (Kelley & Michela, 1980). They show that people who are unaware of the true cause of their physiological arousal can be induced to view themselves as anxious, guilty, amused, or sexually excited by placing them in environments that suggest these emotions (Dutton & Aron, 1974; Zillman, 1978). As the theory predicts, environmental conditions influence people to mislabel their physiological arousal only when they are not aware of its true origins.

Later research suggests that the emotional label sometimes precedes the awareness of arousal (Kemper, 1978; Leventhal, 1980; Pennebaker, 1980). We begin with a belief that we are experiencing a particular emotion and, only then, search our bodily sensations for signs that will verify our belief. If environmental cues give us reason to believe we are angry, we attend to our flushed face and racing heart and verify our anger. If the cues suggest we are happy, we attend to our feelings of alertness and trembling and confirm our happiness. At any given time, our physiological state may afford evidence to support several emotion labels.

This extension of cognitive labeling theory suggests two ways in which role identities may influence the emotions and sensations we experience. First, particular role identities direct attention selectively to particular sensations. Hypochondriacs verify their identity as sickly by attending to their stuffed noses, raspy throats, and uneven pulse; while people who view themselves as healthy tend to overlook these same physical signs. Second, role identities imply how sensations should be interpreted. The pounding heart that accompanies a drop to earth from an airplane signifies thrill to a veteran skydiver but terror to a novice paratrooper. Each interprets the same physical sign in line with his or her role identity.

**Sentiments**  Most of the emotions we have discussed (such as fear, joy, anger) are a response to immediate situations and tend to fade quickly when arousal subsides. Complex feelings like grief, love, jealousy, or indignation, however, arise out of our enduring social relationships. These socially significant feelings are called **sentiments** (Gordon, 1981); each is a pattern of sensations, emotions, actions, and cultural beliefs appropriate to a social relationship. Sentiments such as grief, loyalty, envy, and patriotism develop around our attachments to family, friends, fellow workers, and country.

Sentiments reflect the nature of our social relationships and the changes in them. Grief and nostalgia reflect social losses. Jealousy and envy reflect problems over control of possessions. Indignation and resentment reflect betrayal of commitments. We label our feelings with the culturally appropriate sentiment to make sense of our diverse emotional responses. For example, Mark's joy in Laurie's presence, his sorrow in her absence, his anger when she is criticized, and his fear when threatened with her loss make sense if he labels his feeling "love." Like simpler emotions, sentiments are produced by cognitive labeling. In choosing a sentiment label, however, we take into account all the information we have about our enduring relationship.

As we develop our role identities, we also learn which sentiments are appropriate to them (Denzin, 1983). By expressing appropriate sentiments, we affirm or modify our identities and the social relationships in which they are embedded. To affirm our identity as a romantic partner, for example, we must express love, jealousy, anxiety, tenderness, and ecstasy, all at the appropriate times and places.

**Emotion Work**  On occasion, the self takes an especially direct, active role in the control of emotions. At one time or another most of us have psyched ourselves up, forced ourselves to have a good time even when we were tired, tried to feel grateful in the face of an unwanted gift, and displayed a stiff upper lip despite severe disappointment. These are all instances of **emotion work**—attempts to change the intensity or quality of our feelings to bring them into line with the requirements of the occasion (Hochschild, 1983). Emotion work is needed when we find that we are in violation of **feeling rules**—rules that dictate what people with our role identities ought to feel in a given situation.

There are two basic kinds of emotion work: (1) evocation of feelings that are not present but should be and (2) suppression of feelings that are present but should not be. For example, an airline flight attendant is expected to feel calm and cheerful as she interacts with passengers. But suppose the flight attendant has been working for ten hours, serving hundreds of people on three different flights. Fatigue and irritation may be her main feelings. If so, she must then work directly on her own emotions to evoke feelings of cheerfulness and suppress feelings of irritation.

Among methods used to evoke suitable emotions and suppress unsuitable ones are adopting appropriate postures, shaping our facial expressions, breathing quickly or deeply, and imagining a situation that produces the required feeling. To recapture some cheerfulness, the flight attendant may take a deep breath and relax to reduce her irritation and imagine how good it will feel to be home tonight. Emotion work modifies our actual feelings. An alternative strategy, employed when we are not personally committed to the identity to which the feeling rules apply, is to control only the outward expression of our feelings (Gordon, 1981).

## ■ Self-Esteem

Do you have a positive attitude about yourself, or do you feel you do not have much to be proud of? On the whole, how capable, successful, significant, and worthy are you? Answers to these questions reflect our overall self-esteem, our general evaluation of ourselves. The term **self-esteem** refers to the evaluative component of the self-concept (Gergen, 1971; Gecas, 1972).

This section will address the following four questions. (1) How is self-esteem assessed? (2) What are the major sources of self-esteem? (3) How is self-esteem related to behavior? (4) What techniques do we employ to protect and enhance our self-esteem?

## Assessment of Self-Esteem

Individuals evaluate themselves, and researchers attempt to measure those evaluations. In this section, we will consider these two processes.

**Evaluating the Self** Our overall self-esteem depends on how we evaluate our numerous specific role identities. We evaluate each as relatively positive or negative. For instance, you may consider yourself a competent athlete and a worthy friend but an incompetent debater and an unreliable employee. According to theory, our overall level of self-esteem is the product of these individual evaluations, with each identity weighted according to its importance (Rosenberg, 1965; Sherwood, 1965).

Ordinarily, we are unaware of precisely how we combine and weight the evaluations of our specific identities. If we weight our positively evaluated identities as more important, we can maintain a high level of overall self-esteem while still admitting to certain weaknesses. If we weight our negatively evaluated identities heavily, we will have low overall self-esteem, even though we have many valuable qualities.

**Measurement Techniques** We can further clarify this concept by looking at three methods for measuring self-esteem. One approach, using a questionnaire (Rosenberg, 1965), asks respondents whether they agree or disagree with statements such as: (1) I certainly feel useless at times. (2) On the whole I am satisfied with myself. (3) I have a number of good qualities. (4) I wish I could have more respect for myself. Each answer reveals direct evaluations of competence, power, or control on the one hand or of moral and social worth or acceptance on the other.

A less direct approach to measuring self-esteem asks people to accept or reject factual self-descriptions. For example, consider the following items from the Coopersmith Self-Esteem Inventory (1967): (1) I'm pretty sure of myself. (2) I have a low opinion of myself. (3) I'm easy to like. (4) I give in easily. On the surface, these descriptions are factual, not evaluative. This approach assumes, however, that all respondents evaluate particular descriptions either positively (items 1 and 3) or negatively (items 2 and 4). The more a person accepts positive descriptions and rejects negative ones, the higher his or her self-esteem is considered to be.

A third approach derives from the idea that self-esteem depends on the correspondence between actual self-perceptions and ideals (Sherwood, 1965). In line with this idea, respondents furnish two separate but parallel self-descriptions: one of their actual self and one of the self they aspire to be. A self-esteem score is then derived from the discrepancy between these self-descriptions. Small actual-ideal discrepancies point to high self-esteem, while large discrepancies point to lower self-esteem.

These three self-report approaches are the most widely used because they measure self-esteem as a reflexive process of self-evaluation. A review of various measures of self-esteem concludes that these approaches are valid measures of experienced self-esteem (Demo, 1985).

## Sources of Self-Esteem

Why do some of us enjoy high self-esteem while others suffer low self-esteem? To help answer this question, consider three major sources of self-esteem—family experience, performance feedback, and social comparisons.

**Family Experience** As you might expect, parent-child relationships are important for the development of self-esteem. From an extensive study of the family experiences of fifth and sixth graders, Coopersmith (1967) concluded that four types of parental behavior promote higher self-esteem: (1) showing acceptance, affection, interest, and involvement in children's affairs; (2) firmly and consistently enforcing clear limits on children's behavior; (3) allowing children latitude within these limits and respecting initiative (children setting their own bedtime and participating in making family plans); (4) favoring noncoercive forms of discipline (denying privileges and discussing reasons rather than punishing physically). Findings from a representative sample of 5,024 New York high school students corroborate these conclusions (Rosenberg, 1965).

Family influences on self-esteem confirm the idea that the self-concepts we develop mirror the view of ourselves communicated by significant others. Children who see that their parents love, accept, care about, trust, and reason with them come to think of themselves as worthy of affection, care, trust, and respect. But why is discipline related to high self-esteem? Perhaps because parental limits paired with reasoning provide evidence of parental concern and interest. Firm management may also promote self-esteem because it creates a more orderly, predictable environment where children can interact more successfully and feel more capable and in control.

Research also suggests that self-esteem is produced by the reciprocal influence of parents and their children on each other. Children with higher self-esteem exhibit more self-confidence, competence, and self-control. Such children are probably easier to love, accept, reason with, and trust. Consequently, they are likely to elicit responses from their parents that further promote self-esteem.

As young people move into adolescence, their overall or global self-esteem becomes linked to the self-evaluations tied to specific role identities. A study of 416 sixth graders found that evaluations of self as athlete, son/daughter, and student were positively related to global self-esteem (Hoelter, 1986). Also, the number of significant others expands to include friends and teachers, .in addition to parents. The relative importance of these others appears to vary by gender. A study of 1,367 high school seniors

found that the perceived appraisals of friends had the biggest impact on females' self-esteem, whereas perceived appraisals of parents had the biggest impact on males' self-esteem (Hoelter, 1984). For both males and females, teachers' appraisals were second in importance.

**Performance Feedback** Everyday feedback about the quality of our performances—our successes and failures—influences our self-esteem. We derive self-esteem from experiencing ourselves as active causal agents who make things happen in the world, who attain goals and overcome obstacles (Franks & Marolla, 1976). In other words, self-esteem is based partly on our sense of efficacy—of competence and power to control events (Bandura, 1982). People who hold low power positions (such as clerks, unskilled workers) have fewer opportunities to develop efficacy-based self-esteem because such positions limit their freedom of action. Even so, people seek ways to convert almost any kind of activity into a task against which to test their efficacy and prove their competence (Gecas & Schwalbe, 1983). In this way, they obtain performance feedback useful for building self-esteem.

**Social Comparison** In order to interpret whether performances represent success or failure, we must often compare them with our own goals and self-expectations or with the performances of others. Getting a B on a math exam, for example, would raise your sense of math competence if you had hoped for a C at best, but it would shake you if you were counting on an A. The impact of the B on your self-esteem would also vary depending on whether most of your friends got As or Cs.

Social comparison is crucial to self-esteem, because the feelings of competence or worth we derive from a performance depend in large part on whom we are compared with, both by ourselves and others. Even our own personal goals are largely derived from our aspirations to succeed in comparison with people whom we admire. We are most likely to receive evaluative feedback from others in our immediate social

Not everyone can win an Olympic gold medal. But, for all of us, an inner sense of self-esteem depends on experiencing ourselves as causal agents who make things happen, overcome obstacles, and attain goals.

context—our family, peers, teachers, and work associates. We are also most likely to compare ourselves with these people and with others who are similar to us (Festinger, 1954; Rosenberg & Simmons, 1972).

A study of job applicants clearly demonstrates the effect of social comparison on self-esteem (Morse & Gergen, 1970). After each applicant had completed a set of forms, including a self-esteem scale, another applicant entered the waiting room. For half the participants, the second applicant wore a dark business suit, carried an attache case, and communicated an aura of competence. The remaining participants each waited with an applicant who wore a smelly sweatshirt and no socks and appeared dazed. Several minutes later, while still in the presence of the highly impressive or unimpressive compet-

itor, applicants completed additional forms, including a second self-esteem scale. Applicants exposed to the obviously inferior competitor revealed a substantial increase in self-esteem from the first to the second self-esteem measurement; among those faced with the impressive competitor, self-esteem dropped substantially.

Losing one's job is generally interpreted as a serious failure in our society. A national survey of American employees reveals that job loss undermined self-esteem, but the size of the drop in self-esteem depended on social comparison (Cohn, 1978). In neighborhoods with little unemployment, persons who lost their jobs suffered a large drop in self-esteem. In neighborhoods where many others were unemployed too, the drop was less. This difference points to the importance of the immediate social context for defining success and failure.

### Self-Esteem and Behavior

People with high self-esteem often behave quite differently from those with low self-esteem. Research findings indicate that high self-esteem is generally associated with active and comfortable social involvement, whereas low self-esteem is a depressing and debilitating state (Coopersmith, 1967; Rosenberg, 1979; Wylie, 1979).

Compared with those having low self-esteem, children, teenagers, and adults with higher self-esteem are socially at ease and popular with their peers. They are more confident of their own opinions and judgments and expect they will be well received and successful. They are more vigorous and assertive in their social relations, more ambitious, and more academically successful. During their school years, those with higher self-esteem participate more in extracurricular activities, are elected more frequently to leadership roles, show greater interest in public affairs, and have higher occupational aspirations. On psychiatric examinations and psychological tests, they appear healthier, better adjusted, and relatively free of symptoms. Adults with high self-esteem experience less stress following the death of a spouse and cope with the

Box 4-2

## MINORITY STATUS AND SELF-ESTEEM

Members of racial, religious, and ethnic minorities may have special problems in developing a reasonable level of self-esteem. Because of prejudice, minority group members are likely to see a negative image of themselves reflected in others' appraisals. When they make social comparisons of their own economic, occupational, and educational success with that of the majority, they are bound to compare unfavorably. Therefore, we might assume that members of minority groups will interpret their performances and achievements as evidence of their basic lack of worth and competence.

For example, note the following statement to the United States Supreme Court:

"As minority-group children learn the inferior status to which they are assigned and observe that they are usually segregated and isolated from the more privileged members of their society they react with deep feelings of inferiority and with a sense of personal humiliation" (Clark, 1963).

Is this hypothesis true? Hundreds of studies have sought to determine whether minority status undermines self-esteem in America (Wylie, 1979). Surprisingly, the vast majority of studies offer little support for the conclusion that minorities (racial, religious, or ethnic) have appreciably lower self-esteem. A number of the best-controlled studies suggest that minorities may even have somewhat higher self-esteem (Bachman, 1970; Rosenberg & Simmons, 1972; Trowbridge, 1972; Simmons et al., 1978).

Close examination of these studies reveals that reflected appraisals from significant others and social comparisons of success do indeed affect minority group members. For example, the self-esteem of black school children is strongly related to their perception of what their parents, teachers, and friends think of them. In everyday life, however, appraisals and social comparisons are usually not negative (Barnes, 1972; Yancey, Rigsby, & McCarthy, 1972; Rosenberg, 1973, 1981). Living in segregated neighborhoods, minority-group children usually see themselves through the unprejudiced eyes of their own group, not the eyes of the dominant majority. In contrast, blacks attending largely white schools and Catholics growing up in non-Catholic neighborhoods—circumstances likely to produce negative reflected appraisals—do tend to have lower self-esteem than majority groups (Rosenberg & Simmons, 1972). In effect, segregation insulates minority children from regularly comparing themselves with more successful majorities.

When social comparisons with majorities are made, their impact on self-esteem depends on who is blamed for failures. Minorities can protect their self-esteem by blaming the system of discrimination rather than themselves for their lesser accomplishments. Indeed, minority statuses such as race, religion, and ethnicity show virtually no association with overall self-esteem (Jacques & Chason, 1977). Social failure affects self-esteem only when people attribute it to poor individual achievement (Rosenberg & Pearlin, 1978).

resulting problems more effectively (Johnson, Lund, & Dimond, 1986).

The picture of people with low self-esteem forms an unhappy contrast. People low in self-esteem tend to be socially anxious and ineffective. They view interpersonal relationships as threatening, feel less positively toward others, and are easily hurt by criticism. Lacking confidence in their own judgments and opinions, they yield more readily in the face of opposition. They expect others to reject them and their ideas and have little faith in their ability to achieve. In school, they set lower goals for themselves, are less successful academically, less active in the classroom and in extracurricular activities, and less popular. People with lower self-esteem appear more depressed and express more feelings of unhappiness and discouragement. They more frequently manifest symptoms of anxiety, poor adjustment, and psychosomatic illness.

Most of these contrasts are drawn from comparisons between naturally occurring groups of people who report high or low self-esteem. It is, therefore, difficult to determine whether self-esteem causes these behavior differences or vice versa. For example, high self-esteem may enable people to assert their opinions more forcefully and, thus, to convince others. But the experience of influencing others, in turn, may increase self-esteem. Thus, reciprocal influence, rather than causality from self-esteem to behavior, is probably most common.

## Protecting and Enhancing Self-Esteem

Ordinarily, we seek to maintain a positive evaluation of our own worth and competence. To this end, we strive—while enacting our various role identities—to meet or surpass our own standards for achievement and to elicit supportive responses from others. Inevitably, however, we fail to live up to all our idealized self-expectations and to impress others sufficiently to obtain the full role support we desire. Here, four different techniques used to protect or enhance self-esteem will be examined (see McCall & Simmons, 1978; Rosenberg, 1979).

**Manipulating Appraisals** We choose to associate with people who support our role performances and communicate favorable appraisals, and we avoid people who do not see us the way we see ourselves. This enables us to conclude, "Others think highly of me." For example, a study of interaction in a college sorority revealed that women associated most frequently with those who they believed saw them as they saw themselves (Backman & Secord, 1962). Such selective interaction practically ensures that we will receive favorable appraisals.

Another way to enhance our self-esteem is by interpreting others' appraisals as more favorable than they actually are. For instance, college students took an analogies test and subsequently were given positive, negative, or no feedback about their performance (Jussim, Coleman, & Nassan, 1987). Each student then completed a questionnaire. Students who were high in self-esteem evaluated their performance more favorably and perceived the feedback as more positive than students low in self-esteem. Such distortion occurs even among people who interact frequently (Sherwood, 1965). Similarly, people often overestimate how highly others regard their own racial, religious, ethnic, and occupational groups (Rosenberg, 1979).

**Selective Information Processing** Another way we enhance our self-esteem is by attending more to those occurrences that reflect favorably on our merits and virtues. This is known as selective information processing. Psychiatric attendants focus more on their contributions to alleviating patients' suffering, for example, than on the cleanup and housekeeping activities that occupy most of their time (Simpson & Simpson, 1959). We also protect our self-esteem by taking credit for successes (such as good grades) while denying responsibility for failures (such as poor grades) and for the ill-effects of our actions (Bradley, 1978). In order to maintain high self-esteem, people will unrealistically attribute positive traits to the self and deny negative traits (Roth, Snyder, & Price, 1986). Memory also acts to protect self-esteem. We recall our good, responsible, and successful activities more often than our bad, irresponsible, and unsuccessful ones. We also enhance esteem by remembering our own contributions to joint efforts (preparing a project, preserving a relationship) as greater than others' contributions (Ross & Sicoly, 1979).

**Selective Social Comparison**  When we lack objective standards for evaluating ourselves, we engage in social comparison (Festinger, 1954). By carefully selecting others with whom to compare ourselves, we can further protect our self-esteem. We usually compare ourselves with persons who are similar in age, sex, occupation, economic status, abilities, and attitudes (Suls & Miller, 1977; Walsh & Taylor, 1982). We tend to avoid comparing ourselves with the class valedictorian, homecoming queen, or star athlete, thereby forestalling a negative self-evaluation.

Once people make a social comparison, they tend to overrate their relative standing (Felson, 1981). This is illustrated by self-ratings obtained from a large sample of American adults (Heiss & Owens, 1972). Only 2 percent rated themselves "below average" as parents, spouses, sons or daughters or in the qualities of trustworthiness, intelligence, and willingness to work.

**Selective Commitment to Identities**  Still another technique involves committing ourselves more to those self-concepts in which we excel and downgrading those in which we fall short. This protects overall self-esteem because self-evaluation is based most heavily on those identities and personal qualities we consider most important. In one study, adults estimated how good they were on 16 qualities and stated how much they cared about each (see Table 4-3). Results show that, in general, people were more likely to consider a quality important if they thought they excelled in it (Rosenberg, 1979).

People tend to enhance self-esteem by assigning more importance to those identities (religious, racial, occupational, family) they consider particularly admirable (Hoelter, 1983). They also increase or decrease identification with a social group when the group becomes a greater or lesser potential source of esteem (Tesser & Campbell, 1983). In one study, students were part of a group that either succeeded or failed at a task (Snyder, Lassegard, & Ford, 1986). On measures of identification with the group, students belonging to a successful group claimed closer association (that is, basked in the reflected glory), while those in an unsuccessful group distanced themselves from the group. Similarly, students are more apt to wear clothing that displays university affiliation following a football victory rather than a defeat. They also identify more with their school when describing victories

---

■ **TABLE 4-3**

IMPORTANCE TO SELF AND SELF-ESTIMATED GOODNESS OF "WORKING WITH HANDS"

| Self-Estimated Goodness | Importance to Self | | Number |
|---|---|---|---|
| | *Great Deal* | *Some or Little* | |
| Very good | 68% | 32% | 533 |
| Fairly good | 27% | 73% | 392 |
| Poor | 6% | 94% | 224 |

Note: A sample of 1,149 Chicago adults were asked to rate such qualities as honesty, kindness, likeability, and being realistic. The responses for 16 qualities showed a pattern similar to that illustrated here for "working with hands." People rated those qualities they believed they excelled in as more important to self. In this way they protected and enhanced their self-esteem.

*Source: Adapted from Rosenberg, 1979.*

Elated with their victory, these fans are proud to be students at the university. They boost their own self-esteem by identifying with their school team. But following a defeat, fans are likely to protect their self-esteem by identifying less with their team.

("We won") than defeats ("They lost"), thereby enhancing or protecting self-esteem (Cialdini et al., 1976).

All four techniques for protecting self-esteem described here portray human beings as active processors of social events. People do not accept social evaluations passively or allow self-esteem to be buffeted by the cruelties and kindnesses of the social environment. Nor do successes and failures directly affect self-esteem. The techniques described here testify to human ingenuity in selecting and modifying the meanings of events in the service of self-esteem.

## ■ Summary

The self is the individual viewed both as the source and the object of reflexive behavior.

**The Nature and Genesis of Self** (1) The self is the source of action when we plan, observe, and control our own behavior. The self is the object of action when we think about who we are. (2) Newborns lack a sense of self. Later, they come to recognize that they are physically separate from others. As they acquire language, they learn that their own thoughts and feelings are also separate. (3) Through role taking, children come to see themselves through others' eyes. They can then observe, judge, and regulate their own behavior. (4) Children construct their identities based on how they imagine they appear to others. They also develop self-evaluations based on the perceived judgments of others.

**Identities: The Self We Know** The self we know includes multiple identities. (1) Each identity is linked to one of the social roles we enact. At the same time, we shape it to fit our own unique qualities. (2) Individuals mention role identities very often when describing themselves. They also mention personal qualities and self-evaluations. (3) The self we know varies with the situation. We attend most to those aspects of our selves that are distinctive and relevant to the ongoing activity. (4) We form self-concepts primarily in response to the perceived reactions of others. Because we notice and react to features that distinguish people from their social groups, such features are especially likely to become important self-concepts.

**Identities: The Self We Enact** The self we enact expresses our identities. (1) We choose behaviors in order to evoke responses from others that will confirm particular identities. To confirm identities successfully, we must share with others our understanding of what these behaviors and identities mean. (2) We choose which identity to express based on that identity's salience, need for support, and situational opportunities for enacting it. The salience of an identity depends on the resources invested in constructing that identity and the benefits tied to enacting it. (3) We gain consistency in our behav-

ior over time by striving to enact important identities.

**The Self in Thought and Feeling** The self affects both thought and feeling. (1) We learn and remember information better if it relates to the self. We seek out, perceive, and judge information more effectively if it relates to our important identities. (2) When attention is drawn to the self, we become self-aware and take greater control over our behavior. We then conform more with our own personal standards and with salient social standards. (3) The emotions we experience also depend on the self. Our identities cause us to notice particular sensations and influence the interpretations and emotional labels we apply to them. When our emotions seem socially inappropriate, we try to modify our feelings.

**Self-Esteem** Self-esteem is the evaluative component of self. Most people try to maintain positive self-esteem. (1) Overall self-esteem depends on the evaluations of our specific role identities. Self-reports are a general method for measuring self-esteem. (2) Self-esteem derives from three sources: family experiences of acceptance and discipline, direct feedback on the effectiveness of actions, and comparisons of our own successes and failures with those of others. (3) People with higher self-esteem tend to be more popular, assertive, ambitious, academically successful, better adjusted, and happier. (4) We employ numerous techniques to protect and enhance self-esteem. Specifically, we seek favorable reflected appraisals, process information selectively, overestimate our relative worth compared to similar others, and attribute greater importance to qualities in which we excel.

## Key Terms

chapter

# Social Perception and Attribution

# ■ Introduction

It's 10 P.M. and the mental hospital admitting physician is interviewing a respectable-looking man asking for treatment. "You see," the patient says, "I keep on hearing voices." After taking a full history, the physician diagnoses the man as a schizophrenic and assigns him to the psychiatric ward.

In one unusual study (Rosenhan, 1973), eight pseudopatients who were actually research investigators gained entry into mental hospitals by claiming to hear voices. During the intake interviews, the pseudopatients gave true accounts of their backgrounds, life experiences, and present (quite ordinary) psychological condition. They falsified only their names and their complaint of hearing voices. Once in the psychiatric ward, the pseudopatients stopped simulating symptoms of abnormality. They reported that the voices had stopped, talked normally with other patients, and made observations in their notebooks. Although some of the other patients suspected that the investigators were not really ill, the staff did not. Even upon discharge, the pseudopatients were still diagnosed as schizophrenic, though now it was "schizophrenia in remission."

A person who presents himself to a mental hospital for admission may pose a difficult perceptual problem for the hospital staff. Is he really "mentally ill" and in need of hospitalization, or is he basically "healthy?" Is he no longer able to function in the outside world, or is he merely seeking a break from his work or his family? Or might he be faking?

The staff doctor must determine what kind of person the potential patient really is by gathering information, classifying it as indicating illness or health, combining this information into a general impression, and picking a diagnostic label (paranoid, schizophrenic, mildly depressed) that suggests what form of treatment should be administered. While performing these actions, the staff doctor is engaging in social perception. Broadly defined, **social perception** refers to constructing an understanding of the social world out of the data we obtain through our senses. More narrowly defined, social perception refers to the processes through which we use available information to form impressions of other people, to assess what traits they have and what they are like.

In making his diagnosis, the physician not only forms an impression regarding the traits and characteristics of the new patient but also engages in attribution. That is, the physician tries to understand the causes of the observed behavior. He tries, for instance, to figure out whether a given patient acts as he does because of some internal dispositions or because of external pressures from the environment. **Attribution** is the process through which observers infer the causes of others' behavior. That is, in attribution we observe others' behavior and then infer backward to its causes—to the intentions, abilities, traits, motives, and situational pressures that explain why people act as they do.

Social perception and attribution involve more than passively registering the stimuli that impinge upon our senses. Our current expectations and our prior experiences influence what we perceive and how we interpret it. The intake physician, for example, expects to meet people who are mentally ill, not research investigators, and so he gathers information and interprets it in ways likely to confirm that expectation.

In many cases, our everyday social perceptions and attributions work fairly well. The impressions we form of others and the judgments we make about the causes of behavior are usually accurate enough to permit smooth interaction. Yet, as our example shows, social perception and attribution can be unreliable. Even a skilled observer, such as the intake physician, can misperceive, misjudge, and reach the wrong conclusions. Once we form a wrong impression, we are likely to persist in our misperceptions and misjudgments.

In this chapter we will take a close look at the processes of social perception and attribution. The following questions will be addressed.

1. How do we use concepts to make sense of the flood of information that surrounds us? What are schemas and what functions do they serve?

2. What are group stereotypes? Why do we use them and what problems do they create?

3. How do we form impressions of others? That is, how do we combine the diverse (or even contradictory) information we receive about someone into a coherent, overall impression?

4. How do we judge the causes of other people's behavior and interpret the origins of actions we observe? More specifically, when we judge someone's behavior, how do we know when to attribute the behavior to different causes, such as the person's internal dispositions or the external situation that the person faces?

5. What sorts of errors do we commonly make in judging the behavior of others, and why do we make such errors?

## ■ Foundations of Social Perception

### Concepts

During most of our waking moments we are bombarded with an immense number of stimuli —sounds, colors, patterns of light, smells, flavors, pressures against our skin, and so on. Yet this bombardment is seldom the buzzing confusion it might be, because we impose some order on the stimuli in our environment. We do this by using concepts to encode stimuli and make sense of them (Bruner, 1958; Rosch, 1978). Thus, we perceive and interpret sounds as music or noise, smells as perfume or food cooking, pressures as handshakes or kisses.

When a perceiver can group together certain stimulus objects and exclude others, he is said to have a concept. A **concept** is an idea that specifies in what way various objects or stimuli are similar to each other or related to each other. We employ concepts to organize the complex flow of incoming information into useful forms. By using concepts to group stimuli, we are able to simplify the world and make it more manageable.

Concepts may refer to objects that are natural (caves) or manufactured (skyscrapers), physically real (hammers) or imaginary (tooth fairy). They may refer to actions (flirting) or to feelings (compassion), to categories of people (hairdressers) or to specific individuals (Charlie Chaplin). They may be as clearly defined as a "cheeseburger" or as loosely defined as "justice."

The particular concepts that we use greatly affect how we perceive our environments (and how we interpret the threats and opportunities they pose), as well as how we react to them. For example, a woman driving a car and seeing a male hitchhiker on the road ahead will respond quite differently depending on whether she encodes the hitchhiker as a "friendly college student who needs a ride" or as a "potential murderer or rapist." Similarly, a young child, upon bringing home a cute black-and-white "doggy" that he found in a nearby field, might be astonished by his parents' reaction to what they insist is a "skunk."

## Schemas

The human mind may be viewed as, among other things, a highly sophisticated system for processing information. Humans constantly scan their environments, select stimuli to attend to, and code or interpret those stimuli in terms of whatever concepts they choose. They then use the encoded information as a basis for action and decision making, or they store the information in memory so that it can be retrieved and used at a later time.

Concepts are not isolated from one another. Rather, humans organize concepts into related clusters or groups. For instance, we may think of an object ("Jonathan") not only as having various attributes ("tall," "wealthy") but also as bearing certain relations with other objects ("friend of Caroline," "stronger than Bill"). And the attributes themselves may also bear relations with other attributes. In this way, an entire cognitive organization (consisting of concepts, attributes, and relations among them) will be constructed. Such an organization greatly simplifies our task in perceiving the world by grouping together similar experiences. It also enables us to make inferences beyond the information immediately available in a situation. When we apply a concept to an object or event, we readily infer additional facts about it. For example, if we conceptualize a neighborhood as "safe" (rather than as "dangerous"), it affects how we will relate to people and objects in that neighborhood. In a "safe" neighborhood, we can smile at strangers and walk through the streets with confidence, and we need not keep a hand on our wallet or clutch our purse tightly.

Social psychologists use the term **schema** to designate an organized structure of concepts about some social stimulus, such as a person, group, role, or event. Schemas are cognitive structures that organize and guide the processing of social information. (In this book, we will use *schemas* as the plural of schema, although some prefer the plural form *schemata*.) A given individual may know and use many different schemas. Some schemas are simple, others very complex.

To illustrate, suppose that Mike is a member of a committee and that it is time for him and the other members to select a new chairperson. Mike may not know exactly which individual to select, but he does have a schema regarding the "chairperson" role. In Mike's schema, anyone serving as chairperson should exhibit strong leadership qualities, which means they should be both goal-oriented and affiliative. For Mike, a goal-oriented person is one who has certain attributes —he or she is ambitious, takes calculated risks, is dominant and issues orders, and has a clear image of the goals to be achieved by the committee. An affiliative person, in Mike's schema, is one who is sociable, friendly, joins groups, and talks with people. Moreover, Mike's schema includes some attributes that a chairperson should not have; a chairperson is not shy and retiring, not biased in his or her treatment of members, and not waffling and indecisive. Mike's schema for chairperson will help him make judgments regarding which committee members are the best candidates for that role.

There are several important classes of schemas. The example involving Mike details a schema for a social role (chairperson). Schemas of this type are sometimes called *prototypes* or role schemas. But schemas are used to organize other types of social stimuli as well (Taylor & Crocker, 1981). For instance, our conceptions of other people's personalities may be organized into schemas. Schemas of this type are called *implicit personality theories*. Then, too, we may use schemas regarding the characteristics of the members of a particular group or social category. Schemas of this type are usually referred to as *stereotypes* (Hamilton, 1981). Even our perceptions of our own qualities and characteristics are organized in terms of schemas (self-conceptions or *self-schemas*).

Schemas serve three important functions: (1) They guide what we perceive in our environment and how we interpret it. (2) They organize information in memory and, thus, affect what we remember and what we forget (Sherman, Judd, & Park, 1989). (3) They guide our inferences and judgments about people and things (Fiske & Taylor, 1984).

In the following sections, we will illustrate these functions by discussing in more detail two important types of schemas—schemas regarding group members (stereotypes) and schemas regarding persons (implicit personality theories).

## ■ Stereotypes

"Blacks are lazy."
"Southerners are bigots."
"Jocks are dumb."
"Polacks are dumb, too."
"Engineering students are bores."
"Sociologists are left-wing radicals."

We can hardly avoid making a snap judgment about the personalities of these individuals, but are we right? Stereotypes enable us to form impressions about people merely by knowing the group to which they belong.

Each of these statements is an example of what is generally termed a stereotype. Originally, the term referred to a rigid and simplistic "picture in the head" (Lippman, 1922). In current usage, a **stereotype** is a fixed set of characteristics that observers attribute to all members of some specified social group (McCauley, Stitt, & Segal, 1980; Taylor, 1981). As we noted above, stereotypes are one type of schema (Hamilton, 1981).

Stereotypes are useful because they help us simplify the overwhelmingly complex social world we face. Based on the simple expedient of categorizing people into groups, stereotypes enable us to make quick judgments about people when we have only minimal information. They allow us to form impressions of people and to predict their behavior merely by knowing the groups to which they belong.

Yet, there is something quite invidious about stereotypes. Most people are unaware of the impact of stereotypic beliefs on their judgments of others (Hepburn & Locksley, 1983). And, unfortunately, most stereotypes are negative. Although one can point to a few that are positive ("Orientals excel at math"), the overwhelming majority of them disparage the persons belonging to the group stereotyped. Then, too, because they involve overgeneralization, stereotypes are frequently very inaccurate. This unhappy possibility, however, does not seem to stop us from using them.

## Common Stereotypes

In American society, some of the most widely known—and widely used—stereotypes are those that pertain to ethnic, racial, and gender groups. Some ethnic stereotypes held by Americans might include, for example, the view that Germans are industrious and scientifically minded; Irish, quick-tempered; Italians, passionate; and Americans, materialistic (Karlins, Coffman, & Walters, 1969). Ethnic and racial stereotypes have been studied intensively for many years, and the research shows clearly that these stereotypes change over time. Few of us now believe, as many once did, that the typical Native American Indian is a drunk, that the typical black is superstitious and musical, or that the typical Chinese American is conservative and inscrutable. The fact that stereotypes have changed over time, however, does not mean that they have vanished or faded away.

Just as with those for ethnic and racial groups, stereotypes regarding gender groups are among the most commonly used in our society (see Chapter 16 for more details on gender stereotypes in society). Usually our first observation upon meeting or even just seeing people at a distance is to classify them as male or female. In spite of the women's movement, this classification is likely to elicit a rich—though questionable—stereotype. Many men and women consider males more independent, dominant, competent, rational, assertive, and stable in handling crises. They see females as more emotional, sensitive, expressive, gentle, helpful, and patient (Minnigerode & Lee, 1978; Ashmore, 1981; Martin, 1987).

## Origins of Stereotypes

In addition to the ethnic, racial, and gender stereotypes we have described, people regularly attribute characteristics to members of groups defined by occupation, age, mental illness, hobbies, school attended, and so on (Miller, 1982). How do various stereotypes originate? One possibility is that stereotypes arise out of direct experience with a member of the stereotyped group (Campbell, 1967). We may once have known Italians who were passionate, blacks who were musical, or Japanese who were polite. We then build a stereotype by overgeneralizing; we infer that all members of a group share the attribute that we know to be characteristic of particular members. Thus, stereotypes may be based on a "kernel of truth."

Some theorists (Eagly & Steffen, 1984) have argued that stereotypes derive in part from a biased distribution of group members into social roles. Roles have associated characteristics, and eventually those characteristics come to be linked to the persons occupying the roles. If members of some social group disproportionately occupy roles with associated negative characteristics, an unflattering stereotype of that group may eventually emerge. For instance, if members of a given racial group disproportionately occupy jobs that entail despised work, the negative characteristics of the job eventually may be ascribed to members of the racial group.

Another basis of stereotypes may be the need to boost our own self-esteem (Lemyre & Smith, 1985; Crocker et al., 1987). This would explain why most stereotypes are negative. By comparing ourselves with groups of others whom we stereotype as inferior, we can assert our own superiority. People are also motivated to increase the solidarity of their own group by developing negative stereotypes of groups against whom they compete, whether on the battlefield or in the economic arena.

Recent research suggests that stereotypes arise even in the absence of a kernel of truth or of self-interested motivation. Stereotyping appears to be a natural outcome of social perception. When people have to process and remember a great deal of information about others, they tend to store this information in terms of group categories rather than in terms of individuals (Taylor et al., 1978). In trying to remember what went on in a classroom discussion, you may recall that several women spoke and that a black expressed a strong opinion, even though you cannot remember which women spoke or who the black was. Because people remember behavior by group rather than by individual, they are likely to form stereotypes of these groups (Rothbart et al., 1978). Remembering that *women* spoke and that a *black* expressed a strong opinion, you might infer that women are talkative in general and that blacks are opinionated. You would not form these stereotypes if you remembered these characteristics as belonging to individuals.

## Errors Caused by Stereotypes

Although some stereotypes may contain a kernel of truth, they are always overgeneralizations. This means they inevitably lead to various errors in social perception and judgment. First, stereotypes lead us to assume that all members of a group possess certain traits. Yet—obviously—individual members of a group often differ greatly on any trait. One hard hat may shoulder you into the stairwell on a crowded bus; another may offer you his seat. Second, stereotypes lead us to assume that all the members of one group differ greatly from all the members of other groups. Football players and ballet dancers, for instance, may be thought to have nothing in common. In fact, among football players as well as among ballet dancers there are individuals who are patient, neurotic, hardworking, intelligent, and so on.

Stereotypes also promote inaccurate perceptions because people mistakenly assume that

the salient features they use to distinguish between groups are also the *causes* of all differences between the groups. For example, skin color is a salient feature that leads people to distinguish racial groups. Once people distinguish groups by skin color, they also tend to assume that all other differences between these groups are caused by race. Thus, people attribute the fact that whites obtain higher average scores than blacks on standard intelligence tests to race. By focusing on this one salient feature, they ignore other, more likely causes such as socioeconomic opportunities, education, and cultural bias in tests.

Although stereotypes can produce inaccurate perceptions and judgments in simple situations, they are especially likely to do so in complex situations. When the judgment to be made is multifaceted and the situation involves a lot of complex data, reliance on stereotypes can prove particularly misleading. This happens because perceivers use a stereotype as a central theme around which they organize presented information that is consistent with it, and they neglect or ignore information that is inconsistent with it (Bodenhausen & Lichtenstein, 1987). Thus, the use of stereotypes in complex judgmental situations can cause a perceiver to overlook a lot of information that may not fit the stereotype and, hence, to reach a poor judgment.

Because stereotypes can cause serious errors in perception, one might think that people would avoid using them altogether. But this is not the case. Stereotypes are resistant to change even in the face of concrete evidence that contradicts them. This occurs because people tend to welcome evidence that confirms their stereotypes and to ignore or explain away disconfirming evidence (Snyder, 1981; Weber & Crocker, 1983; Lord, Lepper, & Mackie, 1984). Suppose, for example, that Stan stereotypes homosexual males as effeminate, nonathletic, and artistic. If he stumbles on a homosexual bar, he is especially likely to notice those males who fit this description, thereby confirming his stereotype. What does he make of the rough-looking, athletic

# ARE ETHNIC STEREOTYPES CHANGING?

Contact between ethnic groups has greatly increased over the past 50 years. Has this increased contact been accompanied by a reduction in stereotypes? Studies of stereotypes among white Princeton undergraduates over four decades shed some light on this question. In 1932, students were asked to check off the traits on a list that they thought were typical of various groups (Katz & Braly, 1933). This same checklist was presented to Princeton students in 1950 (Gilbert, 1951) and in 1967 (Karlins, Coffman, & Walters, 1969). The last survey also asked students to rate each trait in terms of its favorableness. The students' responses for four different groups—Americans, Italians, Negroes, and Jews—are shown in Table B-5-1.

At first glance, it may seem that stereotypes have faded over the last 35 years. Compared with 1932 undergraduates, few 1967 undergraduates characterized Americans as industrious or intelligent, Italians as artistic or impulsive, Negroes as superstitious or lazy, and Jews as shrewd or mercenary. Overall, the percentage of undergraduates who agreed with any of the most common stereotypes in 1932 declined substantially.

A closer look at the data, however, reveals that the percentages for some of the traits viewed as "typical" of a group actually increased during this period. For instance, more undergraduates in 1967 said Americans are materialistic and ambitious than did undergraduates in 1932. Likewise, Italians were more frequently viewed as passionate in 1967 than in 1932, Negroes as more musical, and Jews as more intelligent and ambitious. If we pay attention to the type of traits assigned to the various groups, an interesting trend appears. Students have become more positive in their stereotypes of Negroes and Jews, but more negative in their self-stereotypes as Americans.

The trend toward expressing less negative stereotypes of minority groups suggests a possible problem with the checklist method for measuring stereotypes. There has been a growing sensitivity in America to prejudiced statements about minorities. The idea that negative stereotyping of minority groups is bigoted and socially undesirable has become especially strong on college campuses. Perhaps, then, negative stereotypes have not faded; they have merely gone underground.

To address this issue, researchers developed an ingenious method for measuring stereotypes (Sigall & Page, 1971). Half their respondents evaluated how characteristic various traits were of white Americans and of blacks using a standard questionnaire procedure. The remaining respondents were wired to a sophisticated-looking machine through bogus electrodes on their forearms. They were persuaded that this "EMG" machine could measure their true, undistorted feelings. Their task, when evaluating the traits typical of groups, was to see how accurately they could sense their own feelings by predicting the EMG readings. Students wired to the EMG were under pressure to respond honestly, because they believed any lying would be detected. Students using the standard questionnaire, however, had no reason to fear detection if they responded in a socially desirable manner.

The results of this study indicate that responses to standard questionnaires or interviews may be biased due to attempts to hide bigotry. In the EMG condition, blacks were characterized as having positive traits (honest, sensitive, intelligent) less often than whites and as having negative traits (lazy, unreliable, stupid, physically dirty) more often than whites. In the standard rating (checklist) conditions, blacks were characterized as more honest and sensitive than whites and whites as more lazy, unreliable, and stupid than blacks. Thus, a negative stereotype of blacks emerged only in the honesty-eliciting EMG condition. The stereotype was

Box 5-1

even partly reversed in the standard rating condition, as if students were bending over backward to present themselves as unprejudiced.

The picture is not entirely discouraging, however. The black stereotype revealed in the EMG condition was still less negative than the Negro stereotype expressed by students in 1932. The researchers conclude that the apparent changes in stereotypes of blacks (Negroes) over four decades probably reflect a "little fading and a little faking." What are the current stereotypes on your campus today? How sure can you be?

▪ **TABLE B-5-1**
STEREOTYPES OF FOUR ETHNIC GROUPS

| | Percent Who Rate the Trait as Typical | | |
|---|---|---|---|
| | 1932 | 1950 | 1967 |
| *Americans* | | | |
| Industrious | 48 | 30 | 23 |
| Intelligent | 47 | 32 | 20 |
| Materialistic | 33 | 37 | 67 |
| Ambitious | 33 | 21 | 42 |
| Progressive | 27 | 5 | 17 |
| Pleasure-loving | 26 | 27 | 28 |
| *Italians* | | | |
| Artistic | 53 | 28 | 30 |
| Impulsive | 44 | 19 | 28 |
| Passionate | 37 | 25 | 44 |
| Quick-tempered | 35 | 15 | 28 |
| Musical | 32 | 22 | 9 |
| Imaginative | 30 | 20 | 7 |
| *Negroes* | | | |
| Superstitious | 84 | 41 | 13 |
| Lazy | 75 | 31 | 26 |
| Happy-go-lucky | 38 | 17 | 27 |
| Ignorant | 38 | 24 | 11 |
| Musical | 26 | 33 | 47 |
| Ostentatious | 26 | 11 | 25 |
| *Jews* | | | |
| Shrewd | 79 | 47 | 30 |
| Mercenary | 49 | 28 | 15 |
| Industrious | 48 | 29 | 33 |
| Grasping | 34 | 17 | 17 |
| Intelligent | 29 | 37 | 37 |
| Ambitious | 21 | 28 | 48 |

*Source: Adapted from Karlins, Coffman, and Walters, 1969.*

males who are there? There are several ways he can prevent their presence from threatening his stereotype. He might scrutinize them closely for hidden signs of effeminacy, underestimate their number and say they are atypical, comment that they are the exceptions that prove the rule, or even assume that they, like he, are straight. Through responses like these, people explain away contradictory information and preserve their stereotypes.

## ■ Impression Formation

### The Warm-Cold Variable

In a classic study of first impressions, Solomon Asch (1946) used a very straightforward procedure. A group of undergraduate students were given a list of seven traits of a hypothetical person and asked to write a character sketch about that person. The seven traits describing that person were *intelligent, skillful, industrious, warm, determined, practical,* and *cautious.* A second group of students was given the same task and the same list of traits, but with one critical difference: The trait *warm* was replaced by *cold.*

Several findings emerged from this study. First, the students had no difficulty performing the task. They were quite able to weave the trait information into a coherent whole and to construct a composite sketch of the stimulus person. Second, substituting the trait *warm* for the trait *cold* produced a large difference in the overall sketch produced by the students. When the stimulus person was characterized as *warm,* the students typically described him as happy, successful, popular, humorous, and so on. But when he was characterized as *cold,* they described him as self-centered, unsociable, and unhappy. Third, the terms *warm* and *cold* seemed to have an especially large impact on the overall impression formed of the stimulus person. For instance, if the terms *polite* and *blunt* were inserted in the attribute list in place of *warm* and *cold,* there was considerably less difference in the impressions formed by the students.

A subsequent study by Kelley (1950) replicated this finding in a more realistic setting. Students in sections of a psychology course were given descriptions of a guest lecturer before he spoke. Two different descriptions were used. The experimenter gave one description to half the class and the other description to the other half. Both descriptions contained adjectives similar to those Asch used (*industrious, critical, practical, determined*), but one characterized the guest lecturer as *warm,* whereas the other characterized him as *cold.* This one difference profoundly influenced the impressions students formed following the discussion. Those who had read that the guest instructor was *cold* rated him as less considerate, sociable, popular, good-natured, humorous, and humane than those who had read he was *warm.* Because the students all simultaneously observed the same guest instructor engaging in the same classroom behavior, their different impressions could only have been based on the warm/cold trait in the profile they had read earlier.

How could a single variation embedded in otherwise identical profiles have such an impact on impressions of someone's actual behavior? There are various explanations, but one is that the students had a theory about which traits go with being warm and which go with being cold. The students in the Asch and Kelley studies may have believed, for example, that people who are cold are also inconsiderate and antisocial, whereas people who are warm are also popular and good-natured. The impressions the students reported regarding the stimulus person would then have reflected their assumptions about which personality traits typically go together.

### Implicit Personality Theories

Each person makes assumptions about how personality traits are related—which ones go together and which do not. This set of assumptions is called an **implicit personality theory** (Schneider, 1973; Cantor & Mischel, 1979; Sternberg, 1985). As we noted above, an implicit personality theory is one kind of schema (Grant & Holmes, 1981). Because people make different assumptions about how traits cluster together, there are many different implicit personality theories in use every day. People usually do not subject their theories about personality to explicit, systematic examination, nor are they typically aware of the contents of these theories—hence, the label "implicit."

As do all schemas, implicit personality theories enable us to make inferences well beyond the information given. Often without realizing it, we draw upon them to flesh out our impressions of a person based on just a few bits of information. Instead of withholding judgment until we know more about their relevant traits, we jump to conclusions about others' personalities using our implicit personality theories. If we learn that someone is warm (not cold), we conclude that she is also likely to be sociable, popular, good-natured, and so on. If we hear that someone else is pessimistic, we tend to assume that he is also humorless, irritable, and unpopular, even though we lack evidence that he in fact possesses these traits.

**Mental Maps**   An implicit personality theory is best pictured as a "mental map" of the way we believe traits are related to one another. Traits we believe to be similar to each other are located close together in our mental map, meaning that people who possess one probably possess the other. Traits we believe to be dissimilar are located far apart, meaning that they rarely occur together in one person. The mental map in Figure 5-1 shows relationships among various traits, or attributes, based on judgments made by college students (Rosenberg, Nelson, & Vivekananthan, 1968).

If your own mental map is similar to the one portrayed in Figure 5-1, you are likely to think that people who are wasteful are also unintelligent and irresponsible (see the lower left part of the map) and that people who are persistent are likely to be determined and skillful (the upper right part of the map). When you observe that a person has a particular trait, you infer that the person possesses traits that are close to it on your mental map.

Figure 5-1 neatly clarifies the warm/cold findings in the studies by Asch and Kelley. Note the locations of "warm" and "cold" on the map. Now note the other attributes close by. The warm person was judged as more sociable, popular, good-natured, and humorous than the cold person, because these traits are all located close to "warm" and far from "cold" on many people's mental maps.

Attempts to map people's implicit personality theories reveal that traits are organized along two distinct positive-negative dimensions—a social dimension and an intellectual dimension. These dimensions are represented by the lines shown in Figure 5-1. "Warm" and "cold" differ mainly on the social dimension, for example, whereas "lazy" and "industrious" differ on the intellectual dimension (Rosenberg & Sedlak, 1972). Some traits (such as "important") are good on both the social and the intellectual dimensions, while other traits (such as "unreliable") are bad on both. In fact, there is a general tendency for traits to be either good or bad on both dimensions.

The fact that many traits are arranged along good-bad dimensions in implicit personality theories explains a common bias in impression formation. We tend to judge persons who have several good traits as generally good, and those who have several bad traits as generally bad. Once we believe a person has a specific trait, we assume (in the absence of explicit information to the contrary) that the other traits nearby in our mental map also apply to that person. This tendency to perceive personalities as clusters of either good or bad traits is called the **halo effect** (Thorndike, 1920).

**Origins of Implicit Personality Theories**   From where do our implicit personality theories come? First, they come from the distilled wisdom of our socializers, who use expressions like, "power corrupts," "ignorance is bliss," and "tall, dark, and (fill in the blank)." These expressions tell us which traits go together. Second, and probably more important, the very meanings of words in our language suggest certain rela-

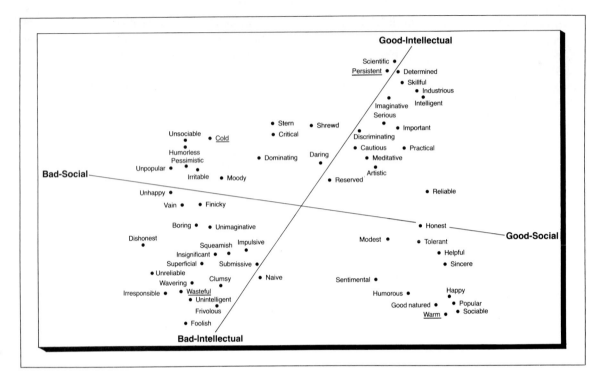

■ **FIGURE 5-1**

RELATIONSHIPS AMONG ATTRIBUTES: A MENTAL MAP

Each of us has an implicit theory of personality—a theory about which personality attributes tend to go together and which do not. We can represent our theories of personality in the form of a mental map. The closer attributes are located to each other on our mental map, the more we assume that these attributes will appear in the same person. The mental map shown above is based on the mental maps of many American college students.

*Source: Adapted from Rosenberg, Nelson, and Vivekananthan, 1968.*

tionships. The word "generous" implies both "helpful" and "sympathetic"; "fickle" implies "unreliable" and "unpredictable." As a result, when one word applies to a person, we assume that other words that share similar meanings also apply (Shweder, 1977). Third, people we know actually do possess sets of personality characteristics that occur together; we observe their actions, infer the traits underlying these actions, and then formulate our implicit personality theories accordingly (Stricker, Jacobs, & Kogan, 1974; Borkenau & Ostendorf, 1987).

**Individual Differences** Within any single culture and language group, individuals share many socialization experiences and word meanings.

Thus, the implicit personality theories they hold tend to be similar and share a common core. Yet each of us also has unique experiences that contribute a unique component to our implicit personality theories (Rosenberg, 1977).

Our unique experiences direct our attention to particular trait categories when we form impressions. For example, when meeting someone, some of us may pay particular attention to physical attractiveness; others, to intelligence, friendliness, and so on. Suppose that two people meet the same intelligent, friendly individual. If one attends more to intelligence, she is likely to form an impression that the individual is industrious, imaginative, and skillful—all traits associated with intelligence in most people's mental maps. If the other attends more to friendliness, he is likely to form an impression that the individual is popular, good-natured, and warm —traits associated with friendliness. Both impressions may be valid, and they are not necessarily contradictory, but they are very different. Thus, people's impressions reflect as much about their own modes of perception as they do about the characteristics of the person being perceived (Higgins, King, & Mavin, 1982).

## Integrating Information about Others

In everyday life, we often receive a lot of information at once. New experiences with other persons typically add information to impressions we already hold. Our task as perceivers is to process all this information, to put it together into an overall impression. If your uncle—whom you view as loving and tolerant—criticizes your cousin for his sloppy clothes and stringy hair, how do you make sense of this new information? You might conclude that your uncle's personality is not well integrated. More likely, however, you would try to understand how this contradictory information fits with your earlier impression of your uncle. In general, when forming an impression of a person, perceivers try to integrate information about many, seemingly contradictory attributes in order to create a unified impression of that individual (Asch & Zukier, 1984). For example, you might conclude that your uncle's apparent intolerance was motivated by his love for your cousin and desire to protect him from others' criticism.

**Models of Information Integration** Most studies of how perceivers combine information are based on a key assumption—namely, that the most important aspect of a perceiver's impression of a person is the overall positive or negative *evaluation* of that person. Although this assumption is restrictive, it seems justified by two facts. First, studies of implicit personality theories show that evaluation is the most important dimension on our mental maps of personality traits. Second, this evaluative dimension is crucial when we make practical judgments. For

instance, we turn down a job because overall we think it is poor, and we invite a new acquaintance to a party because we like her.

Assume that a theorist wishes to predict the overall evaluation that a perceiver would make of a stranger who is "sincere, friendly, cautious, and dishonest." To do this, the theorist needs to know several things. First, he must know how positively or negatively perceivers evaluate each one of these traits. Most college students assign highly positive values to such traits as "sincere" (say +3), less positive values to "friendly" (+2) and "cautious" (+1), and negative values to "dishonest" (−3) (Anderson, 1968).

Next, the theorist needs to know how perceivers combine the trait values to form an overall evaluation of the stranger. Social psychologists have proposed several different models to predict the overall favorableness of impressions formed from diverse traits. One of these, termed the **additive model,** is based on the idea that perceivers form an overall evaluation by summing the values of all the single traits. An example of the additive model is shown in the top panel of Table 5-1 with four different combinations of traits. A key feature of the additive model is that when we add traits with a positive value, we increase the favorableness of our overall impression (column I versus column II), whereas when we add traits with a negative value, we decrease favorableness (column III versus column IV).

The additive model is not the only one that a theorist might use. An alternative theory—termed the **averaging model**—is based on the idea that perceivers form an overall evaluation by averaging the values of all the single traits (see the bottom panel of Table 5-1). In this averaging model, the effect of adding new information depends on whether it is more or less favorable than the overall impression we already have. Thus, incorporating a new, mildly positive trait into a strongly positive impression makes it less positive (column I versus column II), whereas incorporating a mildly negative trait to a strongly negative impression makes it less negative (column III versus column IV).

■ **TABLE 5-1**

A COMPARISON OF ADDITIVE AND AVERAGING MODELS FOR FORMING IMPRESSIONS

| Model | Trait Combinations | | | |
|---|---|---|---|---|
| | I | II | III | IV |
| *Additive* | sincere +3 | sincere +3 | serious +1 | serious +1 |
| | friendly +2 | friendly +2 | irresponsible −3 | irresponsible −3 |
| | tolerant +1 | tolerant +1 | dishonest −3 | dishonest −3 |
| | | cautious +1 | | unimaginative −1 |
| Overall impression: | +6 | +7 | −5 | −6 |
| *Averaging* | sincere +3 | sincere +3 | serious +1 | serious +1 |
| | friendly +2 | friendly +2 | irresponsible −3 | irresponsible −3 |
| | tolerant +1 | tolerant +1 | dishonest −3 | dishonest −3 |
| | | cautious +1 | | unimaginative −1 |
| Overall impression: | +2.00 | +1.75 | −1.67 | −1.50 |

Although both the additive and averaging models have some appeal, the bulk of evidence from empirical studies of impression formation supports a refined version of the averaging model called the **weighted averaging model** (Anderson, 1981). According to this model, perceivers average the values of the single traits, but they also give more weight to certain types of information and less to others. For instance, if a stranger is described as "handsome, talkative, blue-eyed, armed, and murderous," the last two traits on this list will certainly overshadow the others and receive much more weight in the perceiver's overall evaluation of the stranger.

Several factors influence the weights that perceivers assign to trait information. First, they give greater weight to information obtained from credible sources than to information from less reliable sources. Second, they tend to weight negative attributes more heavily than positive attributes (Hamilton & Zanna, 1972; Ronis & Lipinski, 1985), perhaps because negative information is more striking in a world where people present favorable fronts. Third, they attend more to attributes that are particularly relevant to the purpose or judgment at hand. Fourth, they discount information that is very inconsistent with previous impressions and stereotypes

or information that is totally redundant with what they already know. Finally, they weight early impressions more heavily than subsequent impressions. We consider the importance of first impressions in more detail below.

**First Impressions** You have surely noticed the special effort that individuals make to create a good impression when starting a new job, entering a new group, or meeting an attractive potential date. This effort reflects the widely held belief that first impressions are especially important. In fact, this belief is supported by a body of systematic research. Information presented early in a sequence is weighted more heavily in impression formation than information presented later; this phenomenon is called the **primacy effect** (Luchins, 1957).

What accounts for the power of first impressions? One explanation is that after forming an initial impression of a person, we interpret information received subsequently in a way that makes it consistent with our initial impression. Having concluded that your new roommate is neat and considerate, you interpret the dirty socks on the floor the next day as a sign of temporary forgetfulness rather than a sign of sloppiness and lack of concern. This explanation holds that the

This man makes a first impression as strong, stern, self-controlled, and daring. First impressions are hard to change because people pay less attention to later information. Told the man is fearful, for example, observers are likely to ignore this information or to interpret it as meaning he shows healthy fear in extremely dangerous situations.

existing schema into which new information is assimilated influences the interpretation of that new information (Zanna & Hamilton, 1977).

A second explanation for the primacy effect is that we attend most carefully to the first bits of information we obtain about a person but that our attention wanes once we have enough information to make a judgment. It is not that we interpret later information differently; we simply use it less. This explanation assumes that whatever information we attend to most will have the greatest effect on our impressions (Dreben, Fiske, & Hastie, 1979). There is evidence that both reinterpretation of later information and waning of attention contribute to the power of first impressions.

Primacy effects do not occur all the time. Sometimes, the direct opposite occurs. That is, under certain conditions, the most recent information we acquire exerts the strongest influence on our impressions—an occurrence known as the **recency effect** (Luchins, 1957; Jones & Goethals, 1971). A recency effect is likely to occur when so much time has passed that we have largely forgotten our first impression or when we are judging characteristics that we expect to change over time, like moods or certain attitudes. In laboratory settings, a recency effect can be induced by asking perceivers to make a separate evaluation after each new piece of information is received (Stewart, 1965).

In most everyday interactions, primacy effects are more prevalent than recency effects. Primacy effects can have a substantial impact on the impressions people form of one another. In a study demonstrating this impact (Jones et al., 1968), subjects observed the performance of a college student on an SAT-type aptitude test. In one condition, the student started off successfully on the first few items; after this his performance deteriorated steadily. In a second condition, the student started poorly then gradually improved. In both conditions, the students answered 15 out of 30 test items correctly. After observing one or the other performance, subjects were asked to rate the student's intelligence and to predict how well he would do on the next 30 items. Remember that overall performance was the same in both conditions. Nonetheless, subjects rated the student who started well and then tailed off as more intelligent than the student who started poorly and improved. They also predicted that the former would do better on the next series. Apparently, subjects gave more weight to the student's performance on the first few items—a clear primacy effect.

Consider the consequences such a primacy effect might have in school. Students who start the semester poorly but then improve will probably be judged more negatively than they deserve to be by teachers and peers. Students with relatively weak backgrounds, who are bound to perform poorly at first compared with their better prepared classmates, are especially disadvantaged by the primacy effect. And what if a teacher forms an impression about a student before he enters the classroom because he is a member of a stereotyped minority group? The primacy effect will make it difficult for that student to shake the negative first impression others hold.

### Impressions as Self-fulfilling Prophecies

Whether correct or not, the impressions we form of people influence our behavior toward them. Recall, for instance, the study in which students read that their guest instructor was "warm" or "cold" before meeting him (Kelley, 1950). Not only did the students form different impressions of the instructor, they also behaved differently toward him. Those who believed the instructor was "warm" participated more in the class discussion than those who believed he was "cold."

When our behavior toward people reflects our impressions of them, we may cause them to react in ways that confirm our original impressions. For example, if we ignore someone because we think he is dull, he will probably withdraw and add nothing interesting to the conversation. Because our own actions evoke appropriate reactions from others, our initial impressions—correct or incorrect—are often confirmed by the reactions of others. Thus, our impressions may become *self-fulfilling prophecies* (Merton, 1948; Darley & Fazio, 1980).

A study of "getting acquainted" conversations between male and female college students demonstrates vividly how impressions may become self-fulfilling (Snyder, Tanke, & Berscheid, 1977). The study provided males with a snapshot of either an attractive coed or an unattractive coed and asked them to rate her personality. Consistent with a physical-attractiveness stereotype, the snapshot of the attractive woman generated more favorable personality impressions. Each man then engaged in a "get acquainted" phone conversation, presumably with the woman in his picture. Men who believed their partner was attractive, spoke with more animation, sociability, and warmth than those who thought their partner was unattractive—evidence that their impressions influenced their own behavior. Although the women knew nothing about their

partners' impressions, they responded in a more poised, confident, animated, sociable, sexually warm, and outgoing manner when they were speaking with men who thought they were attractive rather than unattractive. The responses of the women demonstrate the self-fulfilling impact of the men's impressions on their behavior. They also demonstrate the continuous, reciprocal impact that people have on each others' behavior.

Table 5-2 shows the sequence of steps in self-fulfilling prophecies in two important settings—schools and the military. In many studies, students whose teachers believed that they had high potential responded by growing intellectually (Rosenthal & Rubin, 1978). Teachers expressed their positive impressions of students by showing more warmth, closer attentiveness, more frequent contact, and greater persistence in seeking responses to questions (Snodgrass & Rosenthal, 1982). Another study involving thousands of American soldiers shows that impressions that become self-fulfilling prophecies can also do harm (Hart, 1978). In this case, groups of enlisted men, especially blacks, behaved according to the "lawless" stereotype that their company leaders conveyed to them by imposing unjustified punishments. Thus punishment led to crime.

## ■ Attribution Theory

When we interact with other people, they seldom announce their traits or provide us with a list of their qualities. Rather, we observe only words and actions, as well as their effects. The task of constructing an interpretation is left to us. When a woman performs a favor, does it mean she is generous or manipulative? When she teases a man, does it mean she likes him or that she thinks he's ridiculous? In order to build impressions, we must first figure out why people act as they do. Once we have done this, we may be able to predict their future behavior and choose how to act effectively toward them.

As defined earlier, *attribution* refers to the

■ **TABLE 5-2**

IMPRESSIONS AS SELF-FULFILLING PROPHECIES IN SCHOOL AND IN THE MILITARY

| | Sequence of Steps | | |
|---|---|---|---|
| | *Initial Impression* ⟶ | *Action* ⟶ | *Target's Confirming Reaction* |
| *School* | Teachers believe particular students have unusually high potential for intellectual growth, even though these students do not differ from others. | Teachers show greater warmth to students and allow them more time to think when they are unsure of answers. | These particular students raise their scores on intelligence tests over time. |
| *U.S. Army* | Company leaders believe groups of enlisted men—especially blacks—are lawless, although they have no objective evidence of lawbreaking. | Company leaders deal harshly with the enlisted men and punish them. | The enlisted men react to punishment with a sense of injustice that spurs them to defiant lawbreaking. |

process through which observers infer the causes of others' behavior. That is, the observer attempts to arrive at an answer to the question "Why did the stimulus person act as he (she) did?" In attribution, we observe behavior and infer backward to its causes—to the intentions, abilities, traits, motives, and situational pressures that explain why people act as they do. Theories of attribution focus on the methods we use to interpret other people's behavior and its sources. They describe how the average person comes to attribute behavior to some possible causes rather than others (Ross & Fletcher, 1985).

## Dispositional versus Situational Attributions

Most people use commonsense reasoning to understand the causes of behavior. Whether or not their beliefs about behavior are scientifically valid, people act on the basis of these beliefs. Fritz Heider (1944, 1958), whose work stimulated the study of attribution, argued that we must study people's ordinary, commonsense explanations regarding the causes of events if we are to explain their own behavior.

Central to Heider's theoretical view is the notion that people, in their efforts to understand the causes of a behavior, subject the events they observe to a kind of psychological analysis. The most crucial decision that observers make is whether to attribute a behavior to the person who performed it (a *dispositional attribution*) or to the surrounding environment (a *situational attribution*).

As an example, consider the attributions you might make upon observing that your neighbor is unemployed. You might judge that he is out of work because he is lazy, irresponsible, or lacking in ability. These are dispositional attributions, because they attribute the causes of someone's behavior to the internal qualities of that person. Alternatively, you might attribute his unemployment to the poor state of the economy, to employment discrimination, or to the evils of the capitalist system. These are situational attributions, because they attribute the person's behavior to external causes.

What determines whether an observer attributes an act to a person's disposition or to the situation? Heider suggests that people consider two factors in making this decision: (1) the strength of situational pressures impacting on the person and (2) the person's own intentions. Consider a circumstance in which a man and a woman have sexual intercourse, and the woman subsequently states that she was raped. If the woman engaged in the act at knifepoint, observers would call this rape, attributing her behavior to overwhelming situational pressure. But suppose the act occurs in her apartment, following a candlelight dinner with a man she is dating? How do observers decide whether she was raped or consented to have sexual intercourse? What attribution will be made?

Even when situational pressures are weak (as in the absence of a weapon), an attribution to personal dispositions is uncertain unless observers believe the act is intentional. How do ordinary observers decide whether or not a behavior is intentional? According to Heider, the logic followed by the naive observer is that a person (in this case, the rape victim) cannot be perceived as having intent unless she could foresee the effects produced by her behavior and had the ability to produce those effects. To assess intentionality, observers might wish to know such things as: Did the woman initiate physical contact with the man, or, in contrast, did she explicitly discourage him? Did she make a special effort to be alone with the man, or, on the other hand, did she work hard to avoid being alone with him? Ordinary observers infer intention when the act is freely chosen and goal-directed and when a great deal of effort is exerted to perform the act. When observers conclude that an act is intentional, they usually attribute it to personal dispositions rather than to the situation.

**Social Consequences** Under certain conditions, the choice between dispositional and situational attribution may have important social consequences. Dispositional attributions define suffering as due to personal problems and prescribe solutions that involve treating individuals (therapy, counseling, education, handouts). Situational attributions, however, define suffering as a social problem and prescribe solutions that involve changing the social structure (revolu-

tions, new laws, job retraining programs, subsidized employment).

Consider the fact that most women in Western societies have lower status and lower paying jobs than most men. If we attribute this fact to women's personal dispositions (fear of success, lack of assertiveness, poorer skills), we would be implying that women should receive assertiveness training, go into psychotherapy, or strive harder to improve their skills. If we attribute this fact to sexual prejudice and discrimination in society, we would be implying that societal institutions and attitudes are at fault and that reducing discrimination against women in the workplace, providing adequate day-care facilities, changing societal definitions of sex-appropriate roles, and so on are the appropriate solutions.

## Correspondent Inferences

Although Heider's analysis is useful in identifying some of the conditions under which observers make dispositional attributions, it does not explain which specific dispositions they will make. Suppose, for instance, that you are on a city street during the Christmas season, and you see a young, well-dressed man walking with a woman. Suddenly, the man stops and tosses several coins into a Salvation Army pot. From this observed act, what can you infer about the man's dispositions? Is he generous, altruistic? Or was he trying to impress the woman? Or was he perhaps just trying to clear out some small change from his coat pocket? Or what?

When we try to infer a person's specific dispositions, our perspective is much like that of a detective. All we can observe is an act (a man gives coins to the Salvation Army) and the effects of that act (the Salvation Army receives more resources, the woman smiles at the man, the man's pocket is no longer cluttered with coins). From this observed act and its effects, we must infer back to the man's dispositions.

According to one theory (Jones, 1979; Jones & Davis, 1965), there are two major steps in the process of inferring personal dispositions.

First, we try to deduce a person's specific *intentions* from his actions. In other words, we try to figure out what the person originally intended to accomplish by performing the act. Second, from these intentions we then try to infer what prior personal *disposition* would cause a person to have such intentions. If we conclude that the man's act was intended to help, for example, we infer the disposition "helpful" or "generous." On the other hand, if the man is seen to have some other intention(s), we do not infer that he has the disposition "helpful." In other words, we attribute a disposition that reflects the presumed intention.

Note that in some cases the inferred disposition has a direct, one-to-one correspondence with the observed act, whereas in other cases it does not. When the inferred disposition corresponds directly to the nature of the act, we call it a **correspondent inference.** Correspondence is a matter of degree, ranging from high correspondence to low or no correspondence.

One problem in inferring dispositions from acts, of course, is that any given act may have multiple effects. A different disposition may correspond to each intended effect. In order to make confident attributions, perceivers must decide which effect(s) the person is really pursuing and which effects are merely incidental. This is not always easy to do. When the man donated money to the Salvation Army, for example, was his intention to perform a charitable act or to impress the woman accompanying him? His act accomplished both effects. Before making the correspondent inference that the man is generous and helpful by disposition, an observer must know which effect(s) the man intended the act to produce.

Several factors influence observers' decisions regarding which effect(s) the person is really pursuing and, hence, whether a highly correspondent inference is appropriate. These factors are (1) the commonality of effects, (2) the social desirability of effects, and (3) the normativeness of effects (Jones & Davis, 1965).

**Commonality** In most cases, any act that a person performs will produce a number of effects. It is the multiplicity of effects that makes it difficult for observers to infer dispositions from acts. If any given act produced one, and only one, effect, then inferences of dispositions from acts would always be correspondent. Due to the multiplicity of effects, however, observers interested in attributing specific intentions and dispositions to a person often find it informative to observe that person in situations where he is making a choice.

Suppose, for example, that a person can choose to engage either in action 1 or in action 2. Action 1, if chosen, will produce effects a, b, and c. Action 2 will produce effects b, c, d, and e. As we can see, two of these effects (b and c) are *common* to actions 1 and 2. The remaining effects (a, d, and e) are unique to a particular alternative; these are *noncommon* effects.

Now suppose the person chooses action 2. What can we infer about his intentions and dispositions? The main inference is that while he may or may not have intended to produce effects b and c, he certainly intended to produce either effect d or effect e (or both) or to avoid effect a. It is the unique (noncommon) effects of acts that enable observers to make correspondent inferences regarding intentions and dispositions (Jones & Davis, 1965). The common effects of acts provide little or no basis for inferences.

Thus, observers who wish to discern the specific dispositions of a person will try to identify effects that are unique to the action chosen. Research shows that the fewer noncommon effects associated with the chosen alternative, the more confident observers are about their attributions (Ajzen & Holmes, 1976). Moreover, when a person makes choices that eliminate some effects, observers make increasingly more correspondent inferences each time an effect is eliminated; the smaller the number of effects that remain to be considered, the greater the correspondence of inferred dispositions (Newtson, 1974).

Is this man really as tough as he looks? Or is he striking a pose to fit his position as a guard? Actions that conform to role expectations reveal little about personality dispositions.

**Social Desirability**   In many situations, people engage in particular behaviors because they are considered socially desirable. Yet, a person who performs a normative, socially desirable act shows us only that he is "normal" and reveals nothing about his distinctive dispositions. Suppose, for instance, that you observe a guest at a party politely thank the hostess upon leaving. What does this tell you about the guest? Did he really enjoy the party? Or was he merely behaving in a socially desirable fashion (that is, polite behavior)? You cannot be sure—either inference could be correct. Now, suppose instead that upon leaving the guest complained loudly to the hostess that he had a miserable time at her dull party. This act would tell you quite a bit more about him, wouldn't it? Acts low in social desirability produce highly correspondent inferences

and are interpreted by observers as clear indicators of underlying dispositions (Miller, 1976).

**Normative Expectations**   When inferring dispositions from acts, observers consider the normativeness of behavior. *Normativeness* is the extent to which we expect the average person to perform a behavior in a particular setting. This includes conformity to role expectations in groups (Jones & McGillis, 1976). Actions that conform to role expectations are uninformative about personal dispositions, whereas actions that violate role expectations are informative and lead to correspondent inferences by observers.

This effect of out-of-role behavior on inferences is illustrated in a study by Jones, Davis, and Gergen (1961). Subjects in this study listened to a tape-recorded interview of an individual seeking employment either as a submariner or as an astronaut. The first part of the tape provided a description (by the interviewer) of the characteristics of the ideal job applicant. For the submariner (who had to work long hours in cramped quarters with other people), the ideal characteristics were friendliness, cooperativeness, obedience, and gregariousness (an "other-directed" person). In contrast, for the astronaut (who at that time travelled alone in space), the ideal characteristics were resourcefulness, thoughtfulness, independence, and a capacity to perform without help or the company of others (an "inner-directed" person).

Next, subjects heard a tape of the applicant presenting himself for the job. Depending on experimental treatment, he sought employment either as a submariner or as an astronaut, and he presented himself either as an outer-directed person or as an inner-directed person. The important point is that two of these combinations are role appropriate (the outer-directed person applying for the submariner job, and the inner-directed person applying for the astronaut job), whereas the other two combinations are role inappropriate (the outer-directed person applying for the astronaut job, and the inner-directed person applying for the submariner

■ **TABLE 5-3**

MEAN RATINGS BY SUBJECTS OF INTERVIEWEES

| | Role | | | | |
| --- | --- | --- | --- | --- | --- |
| | Astronaut | | | Submariner | |
| Trait | Inner-Directed | Other-Directed | | Inner-Directed | Other-Directed |
| Conformity | 13.09 | 15.91 | | 9.41 | 12.58 |
| Affiliation | 11.12 | 15.27 | | 8.64 | 12.00 |
| Candor | 9.68 | 12.42 | | 12.08 | 10.09 |

*Source: Adapted from Jones, Davis, and Gergen, 1961.*

job). After listening to the taped interview, the subjects rated the applicant on various trait measures and also indicated how much confidence they had in their ratings.

Results are displayed in Table 5-3. The two applicants whose behavior was role appropriate (the outer-directed submariner and the inner-directed astronaut) were rated by subjects as conforming and affiliative. The subjects had relatively low confidence in their own ratings of these applicants. Since these applicants knew the job requirements and presented themselves accordingly during the interview, subjects could not infer much about what they were really like as persons. (Were the applicants merely posing in order to get the job, or were they truly what they claimed to be?)

On the other hand, when the subjects rated the two applicants whose behavior was role inappropriate (the outer-directed astronaut and the inner-directed submariner), the results were different. Subjects rated the inner-directed submariner as quite independent (nonconforming) and nonaffiliative, and they reported a high level of confidence in their ratings. His behavior was contrary to role requirements, and the subjects made strong attributions. Subjects also were confident of their ratings for the outer-directed astronaut; they rated him as very conforming and affiliative. This candidate's behavior was also contrary to role requirements, and subjects once

again made strong attributions. Subjects knew that these candidates were behaving contrary to expectations (and that the candidates themselves were aware of this), and they concluded the reason was that the candidates were presenting themselves as they actually were. In general, behavior that violates role expectations leads to highly correspondent inferences by observers.

## Covariation Model of Attribution

Up to this point we have examined how people make attributions based on single observations of behavior in specific situations. Often, however, we have multiple observations that provide more information. Specifically, we may obtain information about the way a given person behaves in a variety of situations. Moreover, if we have information about other persons in the same situation(s), we can compare their behavior against that of the given person. Thus, multiple observations make possible many different comparisons, and these, in turn, make possible many attributions. How do perceivers use multiple observations to arrive at a conclusion about the cause of a behavior? In an extension of Heider's theoretical view, Kelley (1967, 1973) addresses this question.

Kelley (1967) argues that when we make causal attributions and when we have information about the behavior of more than one person in more than one situation, we analyze information essentially the same way that a scientist would. In other words, we assess whether the behavior occurs in the presence or absence of various potential causes (actors, objects, contexts). In doing so, we use the **principle of covariation:** We attribute the behavior to the potential cause that is both present when the behavior occurs and absent when the behavior fails to occur—the cause that "covaries" with the behavior.

Suppose you hear a presidential candidate deliver a rousing speech opposing proliferation of nuclear weapons? To what would you attribute his behavior? There are at least three potential causes: (1) the *actor*, (2) the *object* of the behavior, and (3) the *context* or setting in which the behavior occurs. For example, you might attribute the speech to the candidate's powerful personality (a characteristic of the actor), to the intrinsic threat posed by nuclear proliferation (a characteristic of the object), or to the audience's known opposition to nuclear weapons (a characteristic of the context).

Kelley (1967) suggests that, in using the principle of covariation to determine whether a behavior is caused by the actor, object, or context, we rely on three types of information: consensus, consistency, and distinctiveness information.

**Consensus** refers to whether all or only a few people perform the same behavior. For example, do all the other political candidates speak against proliferation of nuclear weapons (high consensus), or is this the only candidate who does so (low consensus)?

**Consistency** refers to whether the actor behaves the same way at different times and in different settings. If the candidate speaks against nuclear proliferation throughout the campaign before many different audiences, his behavior is high in consistency; in contrast, if the candidate opposes proliferation in some appearances and favors it in others, his behavior is low in consistency.

**Distinctiveness** refers to whether the actor behaves differently toward a particular object than toward other objects. If the candidate

rarely takes a strong public stand on controversial issues, his outspoken opposition to nuclear proliferation is a behavior high in distinctiveness. If he speaks out in the same way on many topics, his behavior is low in distinctiveness.

While people sometimes examine these three types of information deliberately and consciously, they often do so without self-awareness or reflection. The particular combinations of consensus, consistency, and distinctiveness information that people associate with a behavior determine their attributions. To illustrate, Table 5-4 displays three combinations of information that are particularly important, because they produce clear attributions to (1) the actor, (2) the object, and (3) the context.

As Table 5-4 indicates, people usually attribute a behavior to the actor (the candidate) when the behavior is low in consensus (only this candidate speaks in opposition to nuclear proliferation), low in distinctiveness (the candidate also speaks in opposition on many other issues), and high in consistency (the candidate gives speeches opposing nuclear proliferation whenever and wherever he appears). The candidate apparently believes that nuclear proliferation is highly undesirable, which is why he gave such a rousing speech. In short, the combination of low consensus, low distinctiveness, and high consistency is understood to imply that something about the actor (his beliefs) caused the behavior.

We usually attribute a behavior to the object when the behavior is high in consensus, high in distinctiveness, and high in consistency. In this case we might infer that the nuclear proliferation issue is popular; that is why various candidates all give speeches opposing it. Thus, the high-high-high combination is understood to imply that something about the object (the evils of nuclear proliferation) rather than something about the actor caused the behavior.

Finally, we tend to attribute a behavior to context when there is a combination of low consensus, high distinctiveness, and low consistency. Perhaps this particular audience or the latest turn in the campaign make a speech opposing proliferation desirable. Thus, a low-high-low

■ **TABLE 5-4**

ATTRIBUTING CAUSALITY FROM COMBINATIONS OF INFORMATION

*Why does a candidate deliver a rousing speech against nuclear proliferation?*

| Combination | Type of Information | | | Attribution |
|---|---|---|---|---|
| | *Consensus* | *Distinctiveness* | *Consistency* | |
| (1) LLH | *Low:* Only this candidate speaks against nuclear proliferation. | *Low:* This candidate speaks out on many different objectives. | *High:* Whenever this candidate appears, he speaks against nuclear proliferation. | Actor (candidate's beliefs) |
| (2) HHH | *High:* All the candidates speak against nuclear proliferation. | *High:* This candidate speaks out only against nuclear proliferation but not other objectives. | *High:* Whenever this candidate appears, he speaks against nuclear proliferation. | Object (evils of nuclear proliferation) |
| (3) LHL | *Low:* Only this candidate speaks against nuclear proliferation. | *High:* This candidate speaks out only against nuclear proliferation but not other objectives. | *Low:* This candidate speaks against nuclear proliferation before this audience but not before other audiences. | Context (audience's view or situational demands) |

combination implies that something about the context (the audience's view or situational demands) rather than something about the actor or the object caused the behavior.

Numerous studies confirm that people use consensus, consistency, and distinctiveness information much the way Kelley theorized (McArthur, 1972; Pruitt & Insko, 1980; Hewstone & Jaspars, 1987). Of course, combinations of information available sometimes differ from those shown in Table 5-4. In these cases, attributions tend to be more complicated, more ambiguous, and less confident.

## Attributions for Success and Failure

For students, football coaches, elected officials, and anyone else whose fate is determined by evaluations of their achievements, attributions for success and failure are vital. Attributions of this type are problematic, however, because whenever someone succeeds at a task, a variety of alternative reasons can be advanced as explanations for the outcome. For example, a student who passes a test could credit her own intrinsic ability ("I have a lot of intelligence"), her effort ("I really studied for that exam"), the easiness of the task ("The exam could have been much more difficult"), or even luck ("They just happened to test us on the few articles I read").

These four factors—ability, effort, task difficulty, and luck—are quite general and apply in many settings. For this reason, an important question is how do observers decide which factor is the "real" cause of an outcome (for example, success or failure)? When observers look at an event and try to figure out the cause for success or failure, they must consider two things. First, they must decide whether the outcome (success or failure) is due to sources within the person (an internal or dispositional attribution) or due to forces in the environment (an external or situational attribution). Second, they must decide whether the outcome is a stable or unstable occurrence. That is, they must determine whether the cause is a relatively permanent feature (of the person or the environment) or whether it is labile and changing. (Was Bob's game-winning home run the sort of thing that he has done frequently before and could do again, or was it the sort of thing he has done only once every ten years?) Only when observers have made judgments regarding internality-externality and stability-instability can they reach conclusions regarding the cause of the success or failure.

As various theorists (Heider, 1958; Weiner et al., 1971) have pointed out, the four factors mentioned above—ability, effort, task difficulty, and luck—can be grouped according to internality-externality and stability-instability. Ability, for instance, is usually considered to be internal and stable; that is, ability is usually taken to be a property of the person (not the environment), and it is stable in the sense that it does not change from moment to moment. In contrast, effort is internal and unstable; it is a property of the person but can change depending on how hard he or she tries. Task difficulty is usually considered to be external and stable. Luck is external and unstable (fickle). These relations are displayed in Table 5-5.

**Determinants of Attributed Causes**  Whether we attribute a performance to internal or external causes depends on how the actor's performance compares with those of others. Extreme or unusual performances will typically be attributed to internal causes. For example, we would judge a tennis player who wins a major tournament as extraordinarily able or possibly as highly motivated; and, similarly, we would view a player who turns in an unusually poor performance (for

■ TABLE 5-5
PERCEIVED CAUSES OF SUCCESS AND FAILURE

| Degree of Stability | Locus of Control | |
|---|---|---|
| | Internal | External |
| Stable | Ability | Task difficulty |
| Unstable | Effort | Luck |

Source: Adapted from Weiner et al., 1972.

example, loses 6–0 6–0 in the first round to an unseeded competitor) as weak and/or unmotivated. In contrast, average or common performances will usually be attributed by observers to external causes. If defeat comes to a player halfway through the tournament, we are likely to attribute it to tough competition or perhaps bad luck.

Whether we attribute a performance to stable or unstable causes depends on how consistent the actor's performance is over time. When performances are very consistent, observers are likely to attribute the outcome to stable causes. Thus, if a tennis player wins consistently, we are likely to attribute this success to her great talent (ability) or perhaps to the uniformly low level of her opponents (task difficulty). On the other hand, when performances are very inconsistent, observers are likely to attribute outcomes to unstable causes, rather than stable ones. Suppose, for example, that our tennis player is unbeatable one day and a pushover the next. In this case, we might attribute the outcomes to fluctuations in motivation (effort) or to random external factors such as wind speed, condition of the court, and so on (luck).

These effects were illustrated clearly in an experimental study by Frieze and Weiner (1971). Subjects were given information about an individual's performance (success or failure) at a given task. They were also given information about that person's past success rate on the same and similar tasks as well as information about other persons' success rates on that task. The purpose of these data was to influence whether

Failing is never pleasant, but its impact may vary. If we attribute our poor performance to lack of effort, we are likely to feel guilt or shame and to renew our efforts. If we attribute it to lack of ability, we may despair and quit trying.

the subjects viewed the actor's performance as consistent (or inconsistent) with his past performance, and as similar to (or different from) the performances of other persons on the same task.

Subjects were then asked to report their judgments about the impact of internal factors (ability, effort) and external factors (task difficulty, luck) in causing the actor's performance outcome (success or failure on the immediate task). The results showed that, first, success was more likely to be attributed to internal factors (ability, effort) than was failure. Second, performance similar to that of others was attributed to external factors (task difficulty), whereas performance different from that of others (such as success while others fail, or failure while others succeed) was attributed to internal factors (ability, effort). Third, performance consistent with one's own past record (such as success when one has succeeded in the past, or failure when one has failed in the past) was attributed to stable factors (ability, task difficulty), whereas perfor-

mance inconsistent with one's past was attributed to unstable factors (luck, effort).

**Consequences of Attributions** Attributions for performance are important because they influence our emotional reactions to success and failure as well as our future expectations and strivings. For instance, if we attribute a poor exam performance to lack of ability, we may despair of future success and give up studying; this is especially likely if we view ability as given and not controllable by us. Alternatively, if we attribute the poor exam performance to lack of effort, we may feel shame or guilt, but we are likely to study harder and expect improvement. If we attribute the poor exam performance to bad luck, we may experience feelings of surprise or bewilderment, but we are not likely to change our study habits, because the situation will not seem controllable; despite this lack of change, we may, nevertheless, hold expectations of improved grades in the future. Finally, if we attribute our poor exam performance to the fact that the test was too difficult, we tend to become angry, but we do not strive for improvement (Valle & Frieze, 1976; McFarland & Ross, 1982; Weiner, 1985, 1986).

## ■ Bias and Distortion in Attribution

According to the picture we have drawn so far, observers scrutinize their environment, gather information, form impressions, and interpret behavior in rational, if sometimes unconscious, ways. In actuality, however, observers often deviate from the logical methods described by attribution theories. They fall prey to biases that distort their judgment. Biases may lead observers to misinterpret events and, hence, to behave in ways that are personally damaging and socially harmful. In this section we will consider several major types of bias and distortion in attribution.

### Overattributing to Dispositions

Suppose that someone asks you to read an essay that supports Fidel Castro—or one that opposes him. The person explains that the essay was written by a student who was assigned by the instructor to take either a pro-Castro or anti-Castro stand. Your task is to infer the writer's true underlying attitude. If you are like most people given this task, you will attribute a strong pro-Castro attitude to the writer of the supporting essay and a strong anti-Castro attitude to the writer of the opposing essay. But that attribution is biased rather than rational, for it ignores the situational pressures on the writer. Specifically, it ignores the fact that the essay writer had no choice: he was assigned the position he took in the essay. His true opinion may bear little relation to the content of the essay (Jones & Harris, 1967).

The tendency to overestimate the importance of personal (dispositional) factors and to underestimate situational influences as causes of behavior is so common that it is called the **fundamental attribution error** (Ross, 1977;

Small & Peterson, 1981; Higgins & Bryant, 1982). The tendency was first identified by Heider (1958), who noted that most observers ignore or discount the impact of role pressures and situational constraints on others and see behavior as caused by people's intentions, motives, or attitudes. This bias is especially dangerous when it causes us to overlook the advantages or disadvantages of power built into social roles. We may incorrectly attribute the failures of persons without power to their personal weaknesses and the successes of the powerful to their superior personal capabilities.

This bias was clearly demonstrated in a study (Ross, Amabile, & Steinmetz, 1977) in which university students participated in a quiz and were randomly assigned to serve in the role of either questioner or contestant. Questioners were instructed to make up ten challenging questions. They were encouraged to show off their knowledge by composing difficult questions from their own areas of expertise. Following the quiz, participants as well as observers were asked to rate the general knowledge of the questioners and the contestants. Their responses are shown in Table 5-6.

The different performances of contestants and questioners were obviously due to the differences in power built into the roles to which they were randomly assigned. Whereas questioners controlled the outcome, contestants were virtually doomed to fail by the powerlessness of their roles. Still, contestants and observers rated questioners as much more knowledgeable than contestants. They made the fundamental attribution error of overlooking the impact of situational influences (roles) and attributing the performance to the actor's characteristics (knowledgeability). Only questioners understood their advantage in choosing the questions and were aware of their own ignorance on other topics.

### Focus of Attention Bias

Another common tendency, or bias, is to overestimate the causal impact of whomever or whatever our attention is focused on. This tendency is known as the **focus of attention bias.**

■ **TABLE 5-6**
THE FUNDAMENTAL ATTRIBUTION ERROR: GENERAL KNOWLEDGE RATINGS FOR PARTICIPANTS IN A QUIZ

|  | General Knowledge Rating | | Questioner–Contestant Difference |
|  | Questioners | Contestants |  |
| --- | --- | --- | --- |
| Observers | 82.1 | 48.9 | 33.2 |
| Contestants | 66.8 | 41.3 | 25.5 |
| Questioners | 53.5 | 50.6 | 2.9 |

Note: Ratings are based on a 100-point scale of general knowledge.
Source: Adapted from Ross, Amabile, and Steinmetz, 1977.

A striking demonstration of this bias appears in a study by Taylor and Fiske (1978). The study involved six subjects who observed a conversation between two persons (Speaker A and Speaker B). Although all six subjects heard the same dialogue, they differed in the focus of their visual attention. Two observers sat behind Speaker A, facing Speaker B; two sat behind Speaker B, facing Speaker A; and two sat on the sides, equally focused on both speakers (see Figure 5-2). Measures taken after completion of the conversation showed that observers thought the speaker whom they faced not only had more influence on the tone and content of the conversation but also had a greater causal impact on the other speaker's behavior. Observers who sat on the sides and were able to focus equally on both speakers attributed equal influence to them.

We perceive the stimuli that are most salient in the environment—those that attract our attention—as most causally influential. Thus, we tend to attribute most causal influence to people who are noisy, colorful, vivid, or in motion. We tend to credit the person who talks the most with leadership abilities, the same way we tend to blame the person who runs past us when we hear a rock shatter a car window. Although salient stimuli may be causally important in some cases, we tend to overestimate their importance (McArthur & Post, 1977).

The focus of attention bias suggests one theoretical explanation for the fundamental attribution error. The person who behaves is the active element in the environment; therefore, that person is likely to capture our attention. Because our attention is directed more at people who act than at the surrounding situation, we tend to attribute more causal importance to people than to situations.

**Actor-Observer Difference**

Actors and observers make somewhat different attributions for behavior. Observers tend to attribute actors' behavior to internal personal characteristics, whereas actors tend to see their own behavior as due more to characteristics of the external situation (Jones & Nisbett, 1972; Watson, 1982). This tendency is known as the **actor-observer difference.** Thus, while other customers in a market may attribute the selection of items in your grocery cart (beer, vegetables, candy bars) to your personal characteristics (alcoholic, vegetarian, chocolate freak), you will probably attribute the selection to the requirements of your situation (preparing for a party) or the qualities of the items (nutritional value or special treat).

In one systematic demonstration of the actor-observer difference (Nisbett et al., 1973), male students wrote descriptions explaining why they liked their girlfriends and why they chose their majors. Then, as observers, they explained why their best friend liked his girlfriend and chose his major. When explaining their own actions, students emphasized external characteristics, like attractive qualities of their girlfriends and the interesting aspects of their majors. When

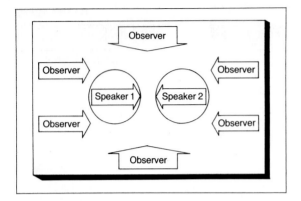

■ **FIGURE 5-2**

THE FOCUS OF ATTENTION BIAS

This diagram depicts the seating arrangement for speakers and observers in a study investigating the effect of the focus of visual attention on observers' attributions. Arrows indicate visual focus of attention. Following a conversation between both speakers, observers attributed more influence to the speaker whom they faced than to the other speaker. Observers on the side, however, attributed equal influence to both speakers. This illustrates our tendency to attribute more causal impact to the object of our attention.

*Source: Adapted from Taylor and Fiske, 1978.*

explaining the behavior they observed in their friends, they emphasized external characteristics less but mentioned internal dispositions, like their friends' own preferences and personalities, more.

Two explanations for the actor-observer difference in attribution are (1) that actors and observers have different visual perspectives and (2) that they have different information.

**Visual Perspectives** The actor's natural visual perspective is to look at the situation, whereas the observer's natural perspective is to look at the actor. Thus, the actor-observer difference reflects a difference in the focus of attention. Both the actor and observer attribute more causal influence to the source they focus on.

Storms (1973) reasoned that, if the actor-observer difference in attributions was due simply to a difference in perspective, that it might be possible to reverse the actor-observer difference by making the actor see behavior from the observer's viewpoint and the observer see the same behavior from the actor's viewpoint. To give each the other's point of view, Storms videotaped a conversation between two people, using two separate cameras. One camera recorded the interaction from the visual perspective of the actor, the other from the perspective of the observer. Storms then showed actors the videotape made from the observer's perspective, and he showed observers the videotape made from the actor's perspective. As predicted, reversing visual perspectives reversed the actor-observer difference in attribution.

**Information** A second explanation for the actor-observer difference is that actors have information about their own past behavior that observers lack. Consequently, observers may make dispositional attributions for actions they mistakenly assume to be typical of the actor's behavior. For example, observers who see a clerk return an overpayment to a customer may assume that the clerk always behaves this way. They may then use this behavior as the basis for a dispositional attribution of honesty. However, if the clerk knows he has often cheated customers in the past, he would probably not interpret his current behavior as evidence of his honest nature.

Even when observers have some information about an actor's past behavior, they often do not know how changes in context influence the actor's behavior. This is because observers are usually exposed to an actor in limited contexts. Students may observe a professor deliver witty, entertaining lectures in class week after week, for example. Yet the professor may know that in most other social situations she is shy and withdrawn. As a result, observers (students) may infer dispositions from apparently consistent behavior that the actor (the professor) knows to be inconsistent across a wider range of contexts.

**Motivational Biases**

Up to this point, we have considered attribution biases that are based on cognitive factors. That is, we have traced biases to the types of

What are these people looking at? Has an accident occurred or has someone been shot? Is someone demonstrating or giving a speech? The world around us does not make sense automatically. We must construct an understanding of social events and figure out the causes of behavior.

information that observers have available, acquire, and process. Motivational factors—one's needs, interests, and goals—are a second source of bias in attributions. When events affect one's self-interests, biased attribution is likely. Specific motives that influence attribution include the desire to (1) maintain and defend deep-seated beliefs, (2) protect and enhance one's self-esteem, (3) increase one's sense of control over the environment, and (4) strengthen the favorable impression of oneself that others have.

The defense of stereotypes illustrates how people use biased attribution to maintain cherished beliefs. People tend to perceive actions that correspond with their stereotypes as caused by the actor's personal dispositions. They may attribute a woman executive's outburst of tears in a crisis to her emotional instability, for example, because that corresponds to a female stereotype. At the same time, people tend to attribute actions that contradict stereotypes to situational causes. If the woman executive manages the crisis smoothly, people may credit this to the effectiveness of her male assistants. When they selectively attribute behaviors that contradict stereotypes to situational influences, these behaviors reveal nothing new about the persons who perform them. As a result, stereotypes persist even in the face of contradictory evidence (Hamilton, 1979).

Attributions for success and failure are also influenced by motivational biases. People tend to take personal credit for acts that yield positive outcomes and to deflect blame for bad outcomes, attributing them to external causes (Bradley, 1978; Ross & Fletcher, 1985). This phenomenon, which is referred to as the **self-serving bias,** is illustrated clearly in a study where college students were asked to explain the grades they received on three examinations (Bernstein, Stephan, & Davis, 1979). Students who received As and Bs attributed their grades much more to their own effort and ability than to good luck or easy tests. However, students who received Cs, Ds, and Fs attributed their grades largely to bad luck and the difficulty of the tests. In a similar way, athletes show the self-serving bias in reporting the results of competitions (Lau & Russell, 1980; Ross & Lumsden, 1982). Whereas members of winning teams take credit for winning ("We won"), members of losing teams are more likely to attribute the outcome to an external cause—their opponent—("*They* won," not "We lost").

Various motives may contribute to this self-serving bias in attributions of performance. For instance, attributing success to personal qualities and failure to external factors enables people to enhance or protect their self-esteem. Regardless of the outcome, they can continue to see themselves as competent and worthy. Then, too, by avoiding the attribution of failure to personal qualities they maximize their sense of control. This, in turn, supports the belief that they can master challenges successfully if they choose to apply themselves, because they possess the necessary ability. Finally, biased attributions enable people to present a favorable public image and to make a good impression on others.

# ■ Summary

Social perception is the process of using information we acquire to construct understandings of the social world and form impressions of people.

**Foundations of Social Perception** (1) A concept is an idea that specifies in what way various objects or stimuli are similar or related to each other. We employ concepts to organize the complex flow of incoming environmental data into useful forms. (2) Schemas are organized structures of concepts about some social stimulus, such as a person, group, role, or event. Schemas serve several functions. They regulate what we perceive in our environment, they organize information in memory, and they guide our inferences and judgments about people and things.

**Stereotypes** A stereotype is a fixed set of characteristics we tend to attribute to all members of a given group. Stereotypes permit us to make quick judgments of people. (1) In American society, the most widespread stereotypes are those that pertain to ethnic, racial, gender, and occupational groups. (2) They arise out of direct experience, the need to boost self-esteem, or the tendency to remember information in terms of group categories. (3) Because they are overgeneralizations, stereotypes cause errors in perception; this is especially true in complex situations. Stereotypes are often negative and resistant to change.

**Impression Formation** (1) Research on the "warm-cold" variable illustrates how variations in a single trait can produce a large difference in the impression formed by observers of a stimulus person. (2) Perceivers use implicit personality theories to flesh out impressions of others based on limited information. Most perceivers organize traits along two distinct positive-negative dimensions—social and intellectual. Impressions may reflect as much about the perceiver's preferences for trait categories as about the person perceived. (3) Perceivers try to integrate the bits of information they receive to create a unified impression of others. They apparently average these bits of information together after weighting certain types of information more heavily than other types. Information received early on usually has a more significant impact on impressions than information received later. (4) Impressions become self-fulfilling prophecies when we behave toward others according to our impressions and evoke appropriate reactions from them.

**Attribution Theory** Through attribution, people infer an action's causes from its effects. (1) Observers tend to attribute intentional behavior to actors' dispositions rather than to the situation. They infer intentionality if a behavior appears to be voluntary, goal-directed, and effortful. (2) To attribute specific dispositions to an actor, observers first observe an act and its effects and then try to infer the actor's intention with respect to that act. Observers then attribute the disposition that corresponds best with the actor's inferred intention. (3) Observers who have information about many of an actor's behaviors make attributions to the actor, object of action, or context, depending on which of these causes covaries with the behavior in question. They assess covariation by considering consensus, consistency, and distinctiveness information. (4) Observers attribute success or failure to four basic causes—ability, effort, task difficulty, and luck. Consistent performances are attributed to stable rather than unstable causes; average performances, to external rather than internal causes. Attributions for performance influence both emotional reactions to success and failure as well as future expectations and strivings.

**Bias and Distortion in Attribution** (1) A fundamental attribution error is to overestimate the importance of dispositional causes of behavior and to underestimate the impact of situational pressures. (2) People also tend to overestimate the causal impact of whatever their attention is focused on. (3) Actors and observers have different attribution tendencies. Actors attribute their own behavior more to characteristics of the external situation, whereas observers attribute the same behavior more to the actor's personal characteristics. (4) Motivations—needs, interests, and goals—lead people to make self-serving, biased attributions. People defend deep-seated beliefs by attributing behavior that contradicts their beliefs to situational influences. People defend and enhance their self-esteem, sense of control, and public image by attributing their failures to external causes and taking personal credit for their successes.

## Key Terms

6

chapter

# Attitudes

# ■ Introduction

"Woody Allen movies are great."
"My human sexuality class is really boring."
"I like my job."
"I think government spending causes inflation."
"The law requiring 18-year-olds to register is a lousy law."
"Guns don't kill people; people kill people."

What do all of these statements have in common? Each represents an **attitude**—a predisposition to respond to a particular object in a generally favorable or unfavorable way (Ajzen, 1982). A person's attitudes influence the way in which he or she perceives and responds to the world (Allport, 1935; Thomas & Znaniecki, 1918). Attitudes influence attention: the person who likes Woody Allen movies is more likely to notice news stories about Allen's activities. Attitudes influence behavior: the young man who opposes the draft is more likely to participate in a demonstration against the draft law.

Because attitudes are an important influence on people, they occupy a central place in social psychology. But what exactly is an attitude? What do we mean by a "predisposition to respond"? Further, how do we measure a person's attitudes? We cannot study predispositions directly; instead, we must rely on various measures that reflect a person's attitudes. Moreover, a particular attitude does not exist in isolation. The person who believes that government spending causes inflation has a whole set of beliefs about the role of government in the economy, and this attitude about spending is related to those other beliefs. If attitudes influence behavior, perhaps we can change behavior by changing attitudes. This leads us to ask, "How do attitudes change?" Politicians, lobbyists, auto manufacturers, and brewers spend billions of dollars every year trying to create favorable attitudes. Even if they succeed, do these attitudes affect behavior?

In this chapter we will consider the four following questions.

1. What is an attitude? Where do attitudes come from, and how are they formed?
2. How do we find out someone's attitudes? How are attitudes measured?
3. How are attitudes linked to other attitudes? How does this organization influence attitude change?
4. What is the relationship between attitudes and behavior?

## ■ The Nature of Attitudes

An attitude exists in a person's mind; it is a mental state. Every attitude is about something, an object. In this section we will consider the components of an attitude as well as the sources of attitudes and their functions.

### The Components of an Attitude

Consider the following statement. "I don't want to go to my human sexuality class. It's boring." This attitude has three components: (1) beliefs or cognitions; (2) a favorable or unfavorable evaluation; and (3) a behavioral disposition.

**Cognition**  Our cognitions, or beliefs, are the ways in which we perceive objects. The person who doesn't like his or her human sexuality class perceives it as involving certain content, taught by a particular person. We cannot always prove whether particular beliefs are true or false. For example, economists and government officials disagree on whether government spending causes inflation, with both sides equally convinced they are right.

**Evaluation**  An attitude also has an evaluative (or affective) component. "It's boring" indicates that the course arouses a mildly unpleasant emotion in the speaker. Stronger negative emotions include dislike, hatred, or even loathing: "I

can't stand punk rock." Of course, the evaluation may be positive: "I like Woody Allen movies," or "This food is terrific." Generally, the evaluative component can be thought of as having both a direction (either positive or negative) and an intensity (ranging from very weak to very strong feelings).

**Behavior**  An attitude involves a predisposition to respond or a behavioral tendency toward the object. "I don't want to go" represents a tendency to avoid the class. "I'm going to vote for Steve Smith" indicates an intention to engage in a behavior. Persons having a specific attitude are inclined to behave in some ways toward an object and not in other ways.

**Relationships between the Components**  Cognitive, evaluative, and behavioral components all have the same object, so we would expect them to be organized into a single, relatively consistent whole. At the same time, these three components are distinct; if they were identical we would not need to distinguish among them. This implies that we should be able to measure each component and that we should be able to find a relationship between them.

This relationship is demonstrated in a study where researchers took a survey of women's attitudes toward contraceptives (Kothandapani, 1971). They classified statements about birth control as representing feelings, beliefs and opinions, or actions. For example, "Birth control causes birth defects" is a belief; "The very thought of birth control disgusts me" is a feeling; and "I would volunteer to speak about the merits of birth control" is an action. Researchers interviewed a group of married black women, none of whom were pregnant at the time. From the women's responses, they constructed measures of each of the components. Results indicated that there was a positive correlation among items that were designed to measure the same component. Measures of the three components were also positively associated, but the relationships were not as close.

The degree of consistency between components is related to other characteristics of the attitude. Greater consistency between the cognitive and affective components is associated with greater attitude stability and resistance to persuasion (Chaiken & Yates, 1985). Greater consistency is also associated with a stronger relationship between attitude and behavior, as we will discuss later in this chapter.

## Attitude Formation

"Woody Allen movies are great."
"I like my job."
"Blacks are lazy."
"Southerners are bigots."
"The law requiring 18-year-olds to register is a lousy law."
"Guns don't kill people; people kill people."

Where do attitudes like these come from? How are they formed? The answer lies in the processes of social learning, or socialization (discussed in Chapter 3). Attitudes may be formed through reinforcement (instrumental conditioning), through associations of stimuli and responses (classical conditioning), or by observing others (observational learning).

We acquire an attitude toward Woody Allen's films or our jobs through *instrumental conditioning*—that is, learning based on direct experience with the object. If you experience rewards related to some object, your attitude will be favorable. Thus, if your work provides you with good pay, a sense of accomplishment, and compliments from your co-workers, your attitude toward it will be quite positive. Conversely, if you associate negative emotions or unpleasant outcomes with some object, you will dislike it. For example, repeated exposure to bland, overcooked food leads many students to have a very negative attitude toward dormitory food.

Only a small portion of our attitudes are based on direct contact with some object, however. We have attitudes about many political figures whom we have never met. We may have attitudes

toward members of certain ethnic or religious groups even though we have never been face to face with a member of those groups. Attitudes of this type are acquired through our interactions with third parties. We learn some attitudes from our parents as part of the socialization process. Research shows that children's attitudes toward male–female relations (gender roles), divorce, and politics frequently are similar to those held by their parents (Thornton, 1984; Glass, Bengston, & Dunham, 1986). This influence also involves instrumental learning; parents typically reward their children for adopting the same or similar attitudes.

Friends are another important source of our attitudes. The attitude that the law requiring selective service registration is bad, for example, may be learned through interaction with peers. A classic study of Bennington College women by Newcomb (1943) demonstrated the impact of peers on the political attitudes of college students. Although most of these women were raised in wealthy, politically conservative families, the faculty of Bennington had very liberal political attitudes. The study demonstrated that freshmen who maintained close ties with their families and did not become involved in campus activities remained conservative, whereas women who became active in the college community and who interacted more frequently with other students gradually became more liberal. Presumably, the students at Bennington rewarded the liberal attitudes of their peers.

We acquire attitudes and prejudice toward a particular group through *classical conditioning*, in which a neutral stimulus gradually acquires the ability to elicit a response through repeated association with other stimuli that elicit that response. Children learn at an early age that "lazy," "dirty," "stupid," and many other characteristics are undesirable. Children themselves are often punished for being dirty or hear adults say "Don't be stupid!" If they hear their parents (or others) refer to members of a particular group as lazy or stupid, children will increasingly associate the group name with the negative reactions initially elicited by these terms. A number of experiments have shown that classical conditioning can produce negative attitudes toward groups (Staats & Staats, 1958; Lohr & Staats, 1973). A recent study demonstrates that even a brief contact between a disliked or disgusting object and a neutral one creates a negative attitude toward the latter (Rozin et al., 1986). For example, a drink touched by a sterilized, dead cockroach was rated as undesirable by people even though no diseases could be transmitted by the contact.

Some attitudes are held primarily or exclusively by members of particular groups or subcultures. Attitudes like "Guns don't kill people; people kill people" and "When guns are outlawed, only outlaws will have guns" are widespread among members of the National Rifle Association. Holding such an attitude may be both a prerequisite to acceptance by other group members and a symbol of one's loyalty to the group.

Another source of attitudes is the media, especially television and films. Here, the mechanism may be *observational learning.* The attitude that "southerners are bigots" may result primarily from exposure to the media. Because racism and bigotry are unacceptable in our current social and political climate, people who hold such attitudes often attract the attention of the media. Many persons living in the U.S. have had little contact with southerners. Consequently, residents of other regions who see or read of incidents in which southerners express racist or prejudiced attitudes may (erroneously) conclude that all southerners are bigots.

## The Functions of Attitudes

We acquire attitudes through learning. But why do we retain them? That is, why do attitudes stay with us for extended periods of time? One answer is that they serve important functions for us. Each attitude fulfills at least one of four functions or purposes for the individual (Katz, 1960). One way of measuring the function(s) of an attitude is to have people write essays about an attitude object and analyze the essays for themes or patterns (Herek, 1987).

Some attitudes serve an instrumental function. Obviously, we develop favorable attitudes toward objects that aid or reward us and unfavorable attitudes toward objects that thwart or punish us. People in business generally have favorable attitudes toward Republican candidates because Republican politicians frequently propose legislation that benefits business. Similarly, we like those co-workers and fellow students who help us with our own work, and we dislike persons who compete against us for scarce resources, whether grades, jobs, or lovers.

Attitudes often serve a knowledge function—that is, they provide us with a meaningful and structured environment. Because the world is too complex for us to understand, we group objects and events into categories and develop simplified (stereotyped) attitudes that allow us to treat individuals as members of a category. Our attitudes about that category (object) provide us with meaning, with a basis for making inferences about the person (Bodenhauser & Wyer, 1986). The belief that blacks are untrustworthy leads some whites to be guarded in their interaction with blacks. Reacting to every member of the group in the same way is more efficient, if less satisfying, than trying to learn about each as an individual.

Stereotypes of groups are often associated with intense emotions. A strong like or dislike for members of a specific group is referred to as **prejudice.** Prejudice and stereotyping go together, with people using their stereotypic beliefs to justify prejudice toward members of the group. (Stereotypes are discussed in Chapter 5.) The emotional component of prejudice can lend to intergroup conflict (Chapter 15) or violence toward members of another group (Chapter 18).

Some attitudes express the individual's basic values and reinforce self-image. Many conservatives in our society have negative attitudes toward abortion, racial integration, and equal rights for women. Thus, a person whose self-concept includes conservatism may adopt these attitudes because they express that self-image. The Catholic church positively values having children and opposes the use of mechanical or chemical birth control. A Catholic man may favor large families because that attitude is consistent with these religious values.

Finally, some attitudes protect the person from recognizing certain thoughts or feelings that threaten his self-image or adjustment. For instance, an individual may have feelings that he cannot fully acknowledge or accept, such as hostility toward his father. If he recognized this hostility, he would feel very guilty because such sentiments are contrary to his upbringing. So, instead of acknowledging that he hates his father, he may direct anger and hatred toward members of a minority group or authority figures such as policemen or teachers. Research indicates that experiences that threaten one's self-esteem, such as failing a test, lead to a more negative evaluation of other groups (Crocker et al., 1987), particularly among persons whose self-esteem was initially high.

In sum, attitudes are psychological entities comprised of cognitive, evaluative, and behavioral components. These components are interrelated but not identical. Attitudes are learned through direct experience or through interaction with other persons. An individual holds attitudes because they serve one or more functions for him or her.

## ■ The Measurement of Attitudes

When you meet someone, you need information about the person in order to interact smoothly. You need to find out her attitudes about objects that are relevant to your interaction—the class or workplace where you meet, the persons whom you both know, or the current local or national events. Because attitudes are mental states, they cannot be directly observed. Sometimes we can infer someone's attitudes from some form of display associated with the person. A button that

It can be difficult to find out another person's attitudes. But some people help us by displaying what they believe.

says "No Nukes" indicates that the person wearing it is opposed to nuclear power plants and/or weapons. Slogans printed on T-shirts are another source of information about attitudes, as are bumper stickers (Wrightsman, 1969). But we cannot rely on these sources, because most people with whom we interact do not put their attitudes on display. To find out someone's attitude, we usually ask them.

Social psychologists, advertisers, and politicians also are interested in finding out people's attitudes. Often they want to systematically assess the attitudes held by the members of some population. Social psychologists have developed a variety of methods for measuring attitudes, some direct and others indirect.

### Direct Methods

The most direct way of finding out someone's attitude is to ask a direct question and record the person's answer. This is the way most of us "study" the attitudes of persons with whom we interact. It is also the technique used by newspaper and television reporters. In order to make the process more systematic, social psychologists employ several methods, including the single-item measure, Likert scales, and semantic differential techniques (see Box 6-1). These measures are obtained by surveys.

**Single Items**   The use of just one question to assess attitudes is very common in surveys. The *single-item scale* usually consists of a direct positive or negative statement about the object, and the respondent indicates whether he or she agrees, disagrees, or is unsure. Such a measure is economical; it takes a minimum of time or space to present. It is also easy to score. The major drawback of the single item is that it is not very precise. Of necessity, it must be general and detects only gross differences in attitude. Using the single item measure in Box 6-1, we could only separate people into two groups: those who favor premarital abstinence and everybody else.

**Likert Scales**   Often we want to know not only how each person feels about an object but also how each respondent's attitude compares with the attitudes of others. The **Likert scale,** a technique based on summated ratings, provides such information (Likert, 1932).

Box 6-1 includes a two-item Likert scale. Each possible response is given a numerical score, indicated in parentheses. We would assess the respondent's attitude by adding his scores for all the items. For example, suppose you strongly agree with item 1 ($+2$) and strongly disagree with item 2 ($+2$), your score would be $+4$, indicating strong opposition to premarital intercourse. Your roommate might strongly disagree with the statement that people should wait until they are married ($-2$) and might also disagree that premarital sex strengthens a marriage ($+1$); the resulting score of $-1$ indicates a slightly negative view. Finally, someone who strongly disagrees with item 1 ($-2$) and agrees with item 2 ($-1$) would get a score of $-3$ and

could be differentiated from a person who received a $-4$.

Typically, a Likert scale includes at least four items. The items should be counterbalanced—that is, some should be written as positive statements, and others should be written as negative ones. Our two-item scale in Box 6-1 has this property; one item is positive, and the other is negative. The Likert scale allows us to order respondents fairly precisely; items of this type are commonly used in public opinion polls. Such a scale takes more time to administer, however, and involves a scoring stage as well.

**Semantic Differential Scales**   Like most attitude scales, the single-item and Likert scales measure the *denotative* or dictionary meanings of the object to the respondent. However, objects also have a *connotative* meaning—a set of psychological meanings that vary from one respondent to another. For instance, one person may have had very positive experiences with sexual intercourse, whereas another person's experiences may have been very frustrating.

The **semantic differential scale** (Osgood, Suci, & Tannenbaum, 1957) is a technique for measuring connotative meaning. In using it, an investigator presents the respondents with a series of bipolar adjective scales. Each of these is a scale whose ends are two adjectives having opposite meanings. The respondent rates the attitude object on each scale. After the data are collected, the researcher can analyze them by various statistical techniques. Analyses of such ratings frequently identify three aspects of connotative meaning: evaluation, potency, and activity. Evaluation is measured by adjective pairs such as good-bad and positive-negative; potency, by weak-strong and light-heavy; and activity, by fast-slow and exciting-boring.

The example in Box 6-1 includes two bipolar scales measuring each of the three dimensions. Scores are assigned to each scale from $+3$ to $-3$; they are then summed across scales of each type to arrive at evaluation, potency, and

## THE MEASUREMENT OF ATTITUDES

Suppose you want to assess attitudes toward premarital sexual behavior. Here are three techniques you could employ.

### Single Item

The single item is probably the most common measure of attitudes. An example of this type is:

**I think people should wait until they are married to have sex.**

—— Yes
—— No
—— Not sure

### Likert Scale

The Likert scale consists of a series of statements about the object of interest. The statements may be positive or negative. The respondent indicates how much he or she agrees with each statement. For example:

**1. I think people should wait until they are married to have sex.**

| | |
|---|---|
| —— Strongly agree | (+2) |
| —— Agree | (+1) |
| —— Undecided | (0) |
| —— Disagree | (−1) |
| —— Strongly disagree | (−2) |

activity scores. In the example shown, scores on each dimension could range from −6 (bad, weak, and slow) to +6 (good, strong, and fast).

One advantage of the semantic differential technique is that researchers can compare an individual's attitudes on three dimensions, allowing more complex differentiation among persons. Another advantage is that because the meaning it measures is connotative, it can be used with any object, from a specific person to an entire nation. Its disadvantages include the fact that it often requires more time to administer and to score.

### Indirect Methods

All of the methods discussed so far involve asking direct questions. They assume that people will honestly report their attitudes toward the object of interest. But is this assumption valid? Some persons might be uneasy if asked questions

Box 6-1

**2. I think having sex before marriage strengthens the marriage.**

| | |
|---|---|
| ___ Strongly agree | (−2) |
| ___ Agree | (−1) |
| ___ Undecided | (0) |
| ___ Disagree | (+1) |
| ___ Strongly disagree | (+2) |

## Semantic Differential Scale

The semantic differential scale consists of a number of dimensions on which the respondent rates the attitude object. For example:

**Rate how you feel about premarital sexual intercourse on each of the following dimensions.**

| | | | | | | | | |
|---|---|---|---|---|---|---|---|---|
| good | ___ (+3) | ___ (+2) | ___ (+1) | ___ (0) | ___ (−1) | ___ (−2) | ___ (−3) | bad |
| weak | ___ (−3) | ___ (−2) | ___ (−1) | ___ (0) | ___ (+1) | ___ (+2) | ___ (+3) | strong |
| fast | ___ (+3) | ___ (+2) | ___ (+1) | ___ (0) | ___ (−1) | ___ (−2) | ___ (−3) | slow |
| negative | ___ (−3) | ___ (−2) | ___ (−1) | ___ (0) | ___ (+1) | ___ (+2) | ___ (+3) | positive |
| light | ___ (−3) | ___ (−2) | ___ (−1) | ___ (0) | ___ (+1) | ___ (+2) | ___ (+3) | heavy |
| exciting | ___ (+3) | ___ (+2) | ___ (+1) | ___ (0) | ___ (−1) | ___ (−2) | ___ (−3) | boring |

about their sexual behavior. Many people with strong prejudices toward other racial or ethnic groups might be unwilling to express those attitudes to a stranger, the interviewer. Furthermore, these methods assume that every member of the sample has an attitude; some may not, and yet they may answer direct questions as if they did. If either of these assumptions is false, direct methods will yield erroneous results.

In order to avoid such error, we can measure attitudes indirectly by observing overt behavior. Several researchers have obtained behavioral measures from representative samples.

Suppose your phone rings about 9:00 P.M. and the caller says, "Hello, Ralph's Garage? This is George Williams. Listen, I'm stuck out here. I'm wondering if you'd be able to come out here and take a look at my car?"

You would probably say, "Sorry, this isn't Ralph's Garage."

The caller would reply: "This isn't Ralph's Garage! Listen, I'm terribly sorry to have disturbed you. But listen, I'm stuck out here on the highway, and that was the last change I had. Now I'm really stuck out here. What am I going to do? Listen, do you think you could do me the favor of calling the garage and letting them know where I am? I'll give you the number. They know me over there."

Would you call? Would it make any difference if the caller sounded white or sounded like a southern black?

Although this phone call appears to have come from a person in need, in actuality it was an instance of the *wrong-number technique,* a procedure for measuring attitudes. Since white persons may be reluctant to report prejudice toward blacks if they are asked direct questions, the wrong-number technique was developed to provide an indirect method of studying attitudes toward blacks. In one study conducted in Brooklyn, New York (Gaertner & Bickman, 1971), researchers selected subjects from the telephone directory who were living in mostly black or white residential areas. They called over 500 whites and 500 blacks and recorded how many of these subjects contacted the garage (actually a confederate of the experimenter, waiting to receive calls). The results showed that whites were more likely to help other whites (65 percent) than blacks (53 percent), whereas blacks were equally likely to help both blacks and whites (63 percent).

A second indirect technique involves littering. As you walk across campus, you are often given a leaflet about a candidate for campus or political office. If it is for a candidate you support, you will probably read and perhaps keep the flier. But if you dislike the candidate, you will probably get rid of the flier quickly, perhaps by simply dropping it on the ground. In one study, handbills were placed on windshields of parked cars. As the drivers returned, observers noted whether they kept the flier or dropped it. As each car left the lot, the interviewer stopped the car and then measured the driver's attitude by asking direct questions. Results showed there was a strong relationship between dislike of the candidate and littering (Cialdini & Bauman, 1981).

Researchers have studied not only what people throw away but also what they pick up, using the *lost-letter technique* (Schwartz & Ames, 1977). The researcher prepares letters and places them in stamped envelopes addressed to organizations with known positions on major issues. The researcher then drops them in areas of considerable pedestrian traffic so they appear to have been lost before they were mailed. When addressing the letters, the researcher can vary the organization named on the envelope; for example, he can address one half of the letters to the National Abortion Rights League and the other half to Parents Opposed to Abortion. Or, he can address all letters to the same organization and vary the location where they are dropped. The behavioral measure is the percentage of letters returned (the letters are addressed to a post office box controlled by the researcher). Mailing a lost letter to an organization presumably indicates a favorable attitude toward that organization. Variation in the return rate is assumed to reflect variation in the attitudes of a particular neighborhood, campus, or other area where the letters were "lost."

Few studies have compared these indirect methods with direct procedures. There is evidence, however, that indirect procedures are less reliable and less valid than the direct ones (Lemon, 1973). In addition, direct techniques are more sensitive to the differences in attitudes between individuals (Petty & Caccioppo, 1981). Finally, the available evidence suggests that most people do respond honestly to direct questions, even when the questions involve sensitive topics (DeLamater & McKinney, 1982). Unobtrusive methods, such as determining whether "lost" letters are mailed, are a useful approach when it is impossible to ask direct questions or when there is reason to believe that some persons do not have an attitude on the issue under study.

# ■ Attitude Organization and Change

## Attitude Structure

Have you ever tried to change another person's attitude toward an object (such as a political candidate or a racial group) or a behavior (such as premarital sex)? If you have, you probably discovered that he had a counter argument for most every argument you made. He seemed to have quite a few reasons why he felt his attitude was correct. An individual's attitude toward some object usually is not an isolated psychological unit. It is embedded in a cognitive structure, linked with a variety of other attitudes. We can often find out what other cognitive elements are related to a particular attitude by asking the person why he holds that attitude. Consider the following interview.

INTERVIEWER: "Why do you think premarital sexual intercourse is bad?"

BILL: "Because sex outside of marriage is wrong; it is against the teachings of God, in the Bible."

INTERVIEWER: "Are there any other reasons?"

BILL: "Well, I think people who have sex before marriage are usually promiscuous, and they could spread AIDS."

INTERVIEWER: "Any other reasons?"

BILL: "Um . . . yeah. They may get pregnant, and teenage pregnancy is really bad."

This exchange indicates Bill's reasons for his attitude. More than that, it illustrates the two basic dimensions of attitude organization, vertical and horizontal structure (Bem, 1970).

**Vertical Structure** Bill is opposed to premarital sex because it violates his religious beliefs. Specifically, it violates the biblical injunction against intercourse before marriage. Bill accepts this injunction because he views the Bible as a statement of God's teachings. Bill's attitude toward premarital intercourse ultimately rests on his belief in God. The unquestioning acceptance of the credibility of some authority, such as God, is termed a *primitive belief* (Bem, 1970).

Attitudes are organized hierarchically. Some attitudes (primitive beliefs) are more basic or fundamental than others. The linkages between fundamental beliefs and minor beliefs in cognitive structure are termed *vertical*. Vertical linkages signify that a minor belief is derived from or dependent on a fundamental belief. Such a structure is portrayed in the center of Figure 6-1.

A fundamental or primitive belief, such as a belief in God, is often the basis for a large number of specific or minor attitudes (Bem, 1970). For example, Bill probably is opposed to murder, adultery, and other sins mentioned in the Bible. Changing a primitive belief may result in widespread changes in the person's attitudes. If Bill comes in contact with, say, members of the Unification church ("Moonies"), they may attempt to persuade him that the Reverend Moon is the only legitimate religious authority. If he is converted, the resulting change in primitive beliefs will lead to changed attitudes toward many objects, including family and friends.

**Horizontal Structure** When the interviewer asked Bill why he was opposed to sex before marriage, Bill gave two other reasons. One was his belief that people who engage in premarital sexual intercourse are promiscuous and that promiscuity spreads AIDS. The other reason was that premarital sex leads to teenage pregnancy and such undesirable consequences as birth defects. These belief structures are represented in the right-hand and left-hand columns of Figure 6-1. When an attitude is linked to more than one set of underlying beliefs—that is, when there are two or more different justifications for it—the linkages are termed *horizontal*.

An attitude with two or more horizontal linkages, or justifications, is more difficult to change than one based on a single set of more fundamental attitudes. Even if you show Bill statistical evidence that AIDS is not associated

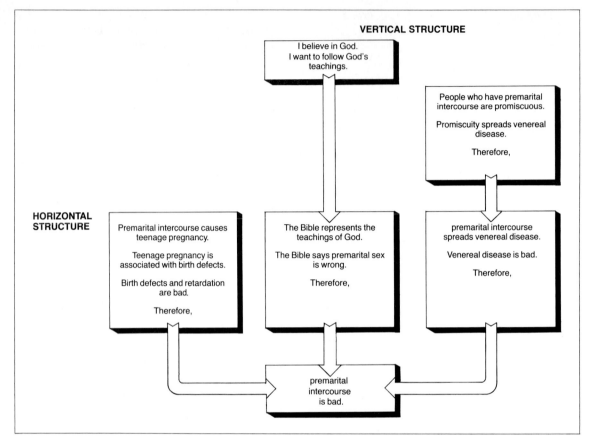

**VERTICAL STRUCTURE**

I believe in God.
I want to follow God's teachings.

People who have premarital intercourse are promiscuous.

Promiscuity spreads venereal disease.

Therefore,

**HORIZONTAL STRUCTURE**

Premarital intercourse causes teenage pregnancy.

Teenage pregnancy is associated with birth defects.

Birth defects and retardation are bad.

Therefore,

The Bible represents the teachings of God.

The Bible says premarital sex is wrong.

Therefore,

premarital intercourse spreads venereal disease.

Venereal disease is bad.

Therefore,

premarital intercourse is bad.

■ **FIGURE 6-1**
THE STRUCTURE OF ATTITUDES

with premarital intercourse, his religious beliefs and his concern about teenage pregnancy make it unlikely that this attitude will change.

Thus, attitudes are often embedded within a larger cognitive structure. A particular attitude may rest on one or more vertical linkages to primitive beliefs and may be linked to several other attitudes to form a horizontal structure. In general, attitudes embedded in such structures will be more resistant to change.

## Balance Theory

An important theory about the nature of linkages between attitudes is balance theory. This theory assumes that we prefer consisten-

cy and that cognitive systems generally are consistent.

**The Drive toward Consistency** The elements of cognitive structure are called **cognitions.** A cognition is an individual's perception of his own attitudes, beliefs, and behaviors. Bill perceives himself as someone who believes in God and follows God's teachings. These two cognitions seem to go together; we are not surprised that Bill perceives both as applying to him. Many of his attitudes are consistent with what he perceives as God's teachings; for example, he has very negative attitudes toward adultery and murder. Given his attitude toward premarital sex, we would expect Bill to abstain from intercourse until he marries; and, indeed, he has never engaged in that behavior. Thus, Bill's behavior is consistent with his attitudes.

Consistency among a person's cognitions—that is, beliefs and attitudes—seems to be widespread. If you have liberal political values, you probably favor medical assistance programs for the poor. If you value equal rights for all persons, you probably support school integration, and you may try hard to behave in nonsexist ways when you interact with members of the opposite sex. The observation that most people's cognitions are consistent with one another implies that individuals are motivated to maintain that consistency. Several theories of attitude organization are based on this principle. In general, these *consistency theories* hypothesize that, if inconsistency develops between cognitive elements, people are motivated to restore harmony between elements.

**Balanced Cognitive Systems**  One important theory based on the consistency principle is **balance theory,** which was formulated by Heider (1958) and elaborated by Rosenberg and Abelson (1960). To see how balance theory works, consider the following statement: "I'm going to vote for Steve Smith; he's in favor of reducing taxes." Balance theory is concerned with cognitive systems with multiple elements like this one. This system contains three elements—the speaker, P; another person (candidate Steve Smith), O; and an impersonal object (taxes), X. From the standpoint of balance theory, two types of relationships may exist between elements. **Sentiment relations** refer to sentiments or evaluations directed toward objects and people; a sentiment may be either positive (liking, endorsing) or negative (disliking, opposing), symbolized as + or −. **Unit relations** refer to the extent of perceived association between elements. For example, a positive unit relation may result from ownership, a relationship (such as friendship or marriage), or causality. A negative relation indicates dissociation, like that between

ex-spouses or members of groups with opposing interests. A null relation exists when there is no association between elements.

With these terms in mind, we will analyze our example. One can depict this system as a triangle (see Figure 6-2). Balance theory is concerned with the elements and their interrelations from P's viewpoint. In our first example (Figure 6-2A), the speaker favors reduced taxes, perceives Steve Smith as favoring reduced taxes, and intends to vote for Steve. This system is balanced. By definition, a **balanced state** is one in which all three sentiment relations are positive or in which one is positive and the other two are negative. Consider another example (Figure 6-2B). Suppose you favor legalizing possession of marijuana, and candidate Mary Jones wants mandatory prison sentences for its possession. Your cognitions would be balanced if you disliked Mary Jones.

**Imbalance and Change**  From the standpoint of balance theory, an **imbalanced state** is one in which two of the relationships between elements are positive and one is negative or in which all three are negative. Consider Judy and Mike, who are seniors in college. They have been going together for three years and are in love with each other. Mike is thinking about going to law school. Judy doesn't want him to stay in school after he gets his Bachelor's degree. She wants him to get a job so that they can get married. Figure 6-2C illustrates the situation from Mike's viewpoint. It is imbalanced; there are two positive relations and one negative relation.

In general, an imbalanced situation like this is unpleasant. When subjects are presented hypothetical triads like those shown in Figure 6-2 and asked to rate each triad, imbalanced triads are rated less pleasant than balanced ones (Price, Harburg, & Newcomb, 1966). Balance theory assumes that people will try to restore balance—that is, they will eliminate perceived imbalance among cognitions by changing one or another of their attitudes.

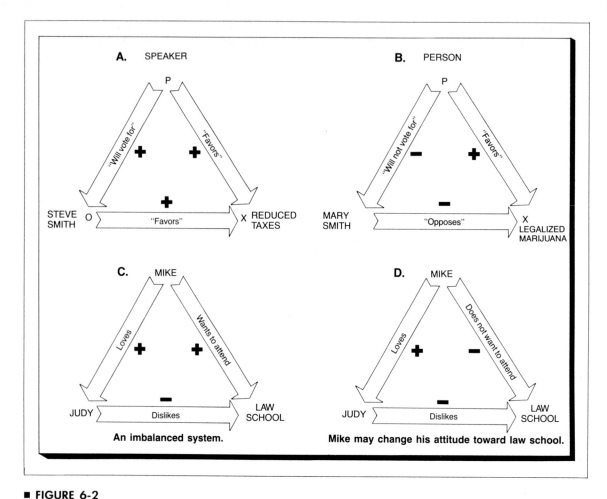

**A. SPEAKER**

P

"Will vote for" +

"Favors" +

+

STEVE SMITH O "Favors" X REDUCED TAXES

**B. PERSON**

P

"Will not vote for" −

"Favors" +

−

MARY SMITH "Opposes" X LEGALIZED MARIJUANA

**C. MIKE**

Loves +

Wants to attend +

−

JUDY Dislikes LAW SCHOOL

**An imbalanced system.**

**D. MIKE**

Loves +

Does not want to attend −

−

JUDY Dislikes LAW SCHOOL

**Mike may change his attitude toward law school.**

■ **FIGURE 6-2**

BALANCED COGNITIVE SYSTEMS AND RESOLUTION OF IMBALANCED SYSTEMS

When the relationships among all three cognitive elements are positive (A), or when one relationship is positive and the other two are negative (B), the cognitive system is balanced. When two relationships are positive and one is negative, the cognitive system is imbalanced. Judy's negative attitude toward law school (C) creates an unpleasant psychological state for Mike. He can resolve the imbalance by deciding he does not want to go to law school (D), by deciding he does not love Judy, or by persuading Judy to like law school.

There are three basic ways to restore balance. First, Mike may change his attitudes so that the sign of one of the relations is reversed (Tyler & Sems, 1977). For instance, Mike may decide he does not want to attend law school (Figure 6-2D). Alternatively, Mike may decide he does not love Judy, or he may persuade Judy that it is a good idea for him to go to law school. Each of these involves changing one relationship so that the system of beliefs contains either zero or two negative relationships.

Second, Mike can restore balance by changing a positive or negative relation to a null relation (Steiner & Rogers, 1963). Mike may decide that Judy doesn't know anything about law school and that her attitude toward it is irrelevant. Third, Mike can restore balance by differentiating the attributes of the other person or object (Stroebe et al., 1970). For instance, Mike may distinguish between major law schools, which require all the time and energy of their students, and less prestigious ones, which require less work. Although Judy is correct in her belief that they would have to postpone marriage if he goes to Yale Law School, Mike believes that he can go to a local school part-time and also support a wife.

Which technique will a person use to remove imbalance? Balance is usually restored in whichever way is easiest (Rosenberg & Abelson, 1960). If one relationship is weaker than the other two, the easiest mode of restoring balance is to change the weaker relationship (Feather, 1967). Because Mike and Judy have been seeing each other for three years, it would be very difficult for Mike to change his sentiments toward Judy. It would be easier for him to change his attitude toward law school than to get a new fiancee. However, Mike would prefer to maintain their relationship and go to law school. In this case, he may attempt to change Judy's attitude, perhaps by differentiating the object (law schools). If this influence attempt fails, Mike will probably change his own attitude toward law school.

## Theory of Cognitive Dissonance

Another major consistency theory is the **theory of cognitive dissonance.** Whereas balance theory deals with the relationship between three cognitions, dissonance theory deals with consistency between two or more elements (behaviors and attitudes). There are three situations in which dissonance commonly occurs: (1) after a decision, (2) when one acts in a way that is inconsistent with his beliefs, or (3) when an important belief is disconfirmed.

**Postdecisional Dissonance** Although Susan will begin her junior year in college next week, she will need to work part-time in order to pay for school. After two weeks of searching for

work, she receives two offers. One is a part-time job typing for a faculty member whom she likes that pays $4 per hour with flexible working hours. The other is a job in a restaurant as a cashier that pays $6 per hour but has working hours from 5:00 P.M. to 9:00 P.M., Thursdays, Fridays, and Saturdays. Susan has a hard time choosing between jobs. Both are located near campus, and she thinks she would like either one. Whereas the typing job offers flexible hours and easier work, the cashier's job pays more and offers her the opportunity to meet interesting people. In the end, Susan chooses the cashier's job, but she is experiencing dissonance.

Dissonance theory (Festinger, 1957) assumes that there are three possible relationships between any two cognitions. Cognitions are consistent, or *consonant*, if one naturally or logically follows from the other; they are *dissonant* when one implies the opposite of the other. The logic involved is *psycho logic* (Rosenberg & Abelson, 1960)—logic as it appears to the individual, not logic in a formal sense. Two cognitive elements may also be irrelevant; one may have nothing to do with the other. In Susan's case the decision to take the cashier's position is consonant with its convenient location, the higher pay, and the opportunities to meet people, but it is dissonant with the fact that she will be responsible for hundreds of dollars and has to work weekend nights (see Figure 6-3).

Having made the choice, Susan is experiencing **cognitive dissonance**—a state of psychological tension induced by dissonant relationships between cognitive elements. Although dissonance is a cognitive concept, it has physiological correlates. A recent study used the galvanic skin response (GSR)—a widely utilized measure of physiological arousal—as a measure of the tension induced by dissonance (Elkin & Lieppe, 1986). As expected, subjects who were placed in a dissonance-arousing situation showed an elevated GSR.

Some decisions produce a large amount of cognitive dissonance, others very little. The magnitude of dissonance experienced is based on the

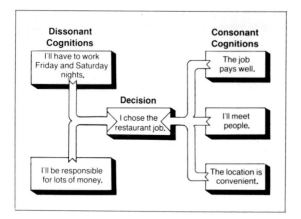

Dissonant Cognitions

I'll have to work Friday and Saturday nights.

Consonant Cognitions

The job pays well.

Decision

I chose the restaurant job.

I'll meet people.

I'll be responsible for lots of money.

The location is convenient.

■ **FIGURE 6-3**

POSTDECISIONAL DISSONANCE

Whenever we make a decision, there are some cognitions—attitudes, beliefs, knowledge—that are consonant with that decision and other cognitions that are dissonant with it. Dissonant cognitions create an unpleasant psychological state that we are motivated to reduce or eliminate. In this example, Susan has chosen a job and is experiencing dissonance. Although three cognitions are consistent with her decision, two other dissonant cognitions are creating psychological tension.

proportion of elements that are dissonant with one's decision. In Susan's case, there are three consonant and only two dissonant cognitions, so she will experience moderate dissonance. The magnitude is also influenced by the importance of the elements; she will experience less dissonance if it is not important that she has to work every Friday and Saturday, more dissonance if an active social life on weekends is important to her.

Dissonance is an uncomfortable state. To reduce dissonance, the theory predicts that Susan will change her attitudes. She can either change the cognitive elements themselves or change the importance associated with the elements. It is hard to change cognitions. She chose the restaurant job, and she is, therefore, committed to working weekend nights and to being responsible for large sums of money. Alternatively, Susan can change the relative importance of her cognitions. She can emphasize the importance of one (or more) of the consonant cogni-

tions and perceive one (or more) of the dissonant cognitions as less important. Even though she has to work to earn money, she can emphasize the fact that the cashier's job pays well. Although she would prefer to be able to go out on weekends, this is less important than otherwise because the cashier's job will allow her to meet people.

In a laboratory study of postdecision dissonance, undergraduate women were given a choice between two products, such as a toaster and a coffee maker. Subjects rated the attractiveness of each item before and after their choice. Researchers predicted that to reduce dissonance, the women would minimize the importance of cognitions dissonant with the decision. That is, after the choice is made, the attractiveness of the item chosen would increase and the attractiveness of the item not chosen would decrease. Results verified these hypotheses (Brehm, 1956).

**Counterattitudinal Behavior** A second circumstance that produces dissonance is when a person behaves in a way that is inconsistent with his or her attitudes. Such situations may involve *forced compliance*—that is, pressures on a person to comply with a request to engage in counterattitudinal behavior.

Imagine that you have volunteered to serve in a psychology experiment. You arrive at the lab and are told you are participating in a study of performance. You are given a pegboard and told to carefully turn each peg exactly one-quarter turn. After you have turned the last peg, you are told to start over, to turn each peg another one-quarter turn. Later you are told to carefully remove each peg from the pegboard and then to put each peg back. After an hour of such activity, the experimenter indicates that you are finished. The experimenter says, "We are comparing the performance of subjects who are briefed in advance with that of others who are not briefed. You did not receive a briefing. The next subject is supposed to be briefed, but my assistant who usually does this couldn't come to work today."

The plain, unfiltered fact is that people who smoke cigarettes get lung cancer a lot more frequently than nonsmokers.
And lung cancer can finish you. Before your time.

We'd rather have you stay alive and in good health. Because even if you do gain a few pounds, you'll have the time to take them off.
**American Cancer Society**

People use various strategies for handling the dissonance aroused by messages that are inconsistent with their behavior. Faced with these two ads, nonsmokers resolve the inconsistencies by emphasizing the importance of health and denying that smoking leads to fun; smokers resolve the inconsistencies by emphasizing fun and denying their risk of cancer.

He then asks you to help out by telling a waiting subject that the tasks you have just completed were fun and exciting. For your help, he offers you either $1 or $20.

In effect, you are being asked to lie, to say that the boring and monotonous tasks you performed are enjoyable. If you actually tell the next subject the tasks are fun, you may experience cognitive dissonance afterward. Your behavior is inconsistent with your cognitions that the tasks are boring. In addition, lying to the next subject is dissonant with your beliefs about yourself— that you are moral and honest. To reduce dissonance, you can change one of the cognitions. Which will you change? You can't really change your awareness that you told the next subject the

task is fun. The only cognition open to change is your attitude toward the tasks, which can change in the direction of greater liking for the tasks.

The theory of cognitive dissonance predicts, first, that you will change your attitudes toward the tasks (like them better) and, second, that the amount of change will depend on the incentive you were paid to tell the lie. Specifically, the theory predicts that greater attitude change will occur when the incentive to tell the lie is low ($1) rather than high ($20), because you will experience greater dissonance under low incentive than you would under high incentive.

These predictions were tested in a classic experiment by Festinger and Carlsmith (1959). In this study, most of the subjects agreed to brief the next subject. They told him that the tasks were interesting and that they had fun doing them. A secretary then asked each subject to rate the experiment and the tasks. These ratings provide the measures of the dependent variable. As expected, control subjects who did not brief anyone and were not offered money rated the tasks as very unenjoyable and did not want to participate in the experiment again.

What about the experimental subjects who were paid money to tell a lie? For those receiving $20, the situation was not very dissonant. The money provided ample justification for engaging in counterattitudinal behavior (lying). Therefore, these subjects rated the task as unenjoyable and were unwilling to participate again. In the $1 condition, however, the subjects experienced more dissonance because they did not have the justification for lying provided by a large amount of money. These subjects could not deny that they lied, so they reduced dissonance by changing their attitude—that is, by increasing their liking for the task and the experiment. Results of this study confirmed the predictions from dissonance theory. Subjects in the high-incentive ($20) condition experienced little dissonance and rated the task negatively, whereas those in the low-incentive ($1) condition experienced more dissonance and rated the task and experiment positively.

These results reflect what is often termed the **dissonance effect:** the greater the reward or incentive for engaging in counterattitudinal behavior, the less the resulting attitude change. The opposite of this is the **incentive effect:** the greater the incentive for engaging in counterattitudinal behavior, the greater the resulting attitude change.

Under what conditions does each of these effects occur? Research suggests that the dissonance effect is more likely when subjects choose (or have the illusion of choosing) whether or not to engage in the behavior. In one study (Sherman, 1970), subjects were asked to write essays taking a position on current issues that contradicted their own attitudes. They were paid either .50 or $2.50 for the essay. In one condition, subjects were allowed to choose whether or not to write the essay; in the second condition, students were offered no choice. For those subjects given a choice, the results showed a dissonance effect: subjects given little incentive (.50) wrote longer, more persuasive essays and showed more attitude change than those given larger incentive ($2.50). On the other hand, for those subjects given no choice, the results showed an incentive effect: subjects given more incentive ($2.50) wrote longer, more persuasive essays and showed more attitude change than those given a small incentive (.50).

**Disconfirmation of a Belief**   To this point, we have discussed how cognitive dissonance can occur in two types of situations—after a person makes a decision and when a person acts in a way that is inconsistent with his or her beliefs. Dissonance can also arise in a third situation—when an important belief is disconfirmed. Consider again the case of Susan, who decided to work as a cashier. After she accepted the job, Susan realized she would need some nice clothes to wear to work. She bought three outfits, which cost her over $200. She told her friends that she would be working on weekend evenings. Three weeks after she started work, the manager called her in and told her that business had fallen off sharply and the restaurant had to cut expenses. Accordingly, he cut back her hours from twelve to six, from 6:30 to 9:30 P.M. on Friday and Saturday. Susan was very upset. Her initial decision was based on the belief that she would work three nights per week and would earn more working in the restaurant than at the typing job. She had also bought three outfits, believing she would be working three nights a week.

This case illustrates how the disconfirmation of a belief can produce dissonance. Four conditions must exist in order for disconfirmation to produce dissonance (Petty & Cacioppo, 1981). First, the belief must be firmly held. Susan was certain she would be working Thursday, Friday, and Saturday. Second, the person must be committed to the belief; he or she must take action based on it. In this case, Susan chose one job over another and spent about $200 on clothes. Third, the belief must be specific enough so that events can disconfirm it. Finally, the disconfirmation must occur, and the person must perceive it. The perception that one took action based on the belief is dissonant with the cognition that the belief was disconfirmed.

Three modes of resolving dissonance are common in situations of this type. First, Susan could quit her job as a cashier. This action would remove the discrepancy between her acts and her present cognitions that the job involves six hours per week. This mode of resolution is most likely to be used if the costs of maintaining the original decision are high (Frey, 1982). Second, she could change her assessment of how much money she needs to earn; she might decide, for instance, that working six hours per week will provide her with enough income after all. This changes one of the cognitions that is dissonant with her act. Third, she can become a more committed believer. She may believe that the reduction in hours is just temporary, that business will pick up when the weather gets cold, and that the manager will soon ask her to work Thursday nights as well. This mode of dissonance reduction is likely only when there is support from others who share this belief (Festinger, Riecken, & Schachter, 1956); the fact that others made the same choice or share the same belief provides information consonant with one's own choice/belief (Stroebe & Diehl, 1981). For example, if Susan spends time with other employees who have been adversely affected by the decline in business, they may support each other in the belief that the cutback is temporary.

Thus, dissonance occurs only in some situations (Wicklund & Brehm, 1976). To experience dissonance, a person must be committed to a belief or course of action (Brehm & Cohen, 1962). In addition, the person must believe that he or she chose to act voluntarily and is thus responsible for the outcome of the decision (Linder, Cooper, & Jones, 1967). This is shown in the case of Susan, who chose the cashier's job. If the owner of the restaurant were Susan's father and he demanded she work for him, she would have had little or no postdecision dissonance.

## Is Consistency Inevitable?

If our beliefs and behavior were always consistent, all of our cognitions would be in harmony. Obviously, that is not the case. Practically every adult in the U.S. knows that cigarette smoking is related to lung cancer, yet millions continue to smoke. Most of us overindulge in a favorite food (pizza, chocolate) or beverage (soda, beer) at least occasionally, even though we know it is not healthy to do so. When we do, our behavior is inconsistent with the belief that overindulgence is unhealthy. How is it that people can hold mutually inconsistent cognitions?

For one thing, many of our cognitions never come into contact with each other; we may never become aware that our contradictory cognitions are in fact related. Thus, many people who have a very positive attitude toward nature flock to Yosemite National Park and are unaware that their behavior is overtaxing the park. One reason this happens is that some of our behavior is mindless. Because we do not think about our actions, we are unaware that they are inconsistent with our beliefs (Triandis, 1980). The cigarette smoker often lights up without consciously thinking about the act; sometimes he is surprised to find a lit cigarette in the ashtray. Thus, the relationship between the act and one's knowledge is often not salient to the person.

Another reason why inconsistency occurs is that each belief, attitude, or self-perception is embedded in a larger structure of consistent, related attitudes, beliefs, and self-perceptions. For example, Bill's attitude toward premarital sexual intercourse discussed earlier was embedded in a structure of other attitudes and values. Although two attitudes may be inconsistent, each may be related to several other consonant attitudes. To change one or the other would create new inconsistencies. In effect, people tolerate some inconsistencies in order to avoid others.

In sum, we prefer consistency among our cognitions. Should inconsistency arise, we are motivated to change our attitudes or behavior in order to restore harmony. Not all of our cognitions are consistent, however. Inconsistency persists for a variety of reasons, because we may not be aware of imbalance or dissonance, or because restoring consistency between some cognitions would create other inconsistencies.

# ■ The Relationship between Attitudes and Behavior

## Do Attitudes Predict Behavior?

We have seen how behavior can affect our attitudes and how people sometimes change their attitudes when their behavior appears to contradict them. However, most people think of attitudes as the source of behavior. For example, we often assume that when we know a person's attitude toward an object (another person, volleyball, or Woody Allen movies), we can predict how that person will behave toward the object. If you know someone enjoys volleyball, you would expect her to accept your invitation to play volleyball with friends. When we are able to predict another person's responses, we can decide how to behave toward that person in order to achieve our own goals. But can we truly predict someone's behavior if we know their attitudes?

In 1930, the social scientist Richard LaPiere traveled around the U.S. by car with a Chinese couple. At that time, there was considerable prejudice against the Chinese, particularly in the western part of the country. The three travelers stopped at more than 60 hotels, auto camps, and tourist homes and more than 180 restaurants. They kept careful notes about how they were treated. In only one place were they denied service. Later, LaPiere sent a questionnaire to each place asking whether they would accept Chinese guests. He received responses

from 128 establishments; 92 percent of them indicated that they would not serve Chinese guests (LaPiere, 1934). Evidently there can be a great discrepancy between what people do and what they say.

Many studies on the topic have found only a modest correlation between attitude and behavior (Wicker, 1969). The correlation (*r*) is a measure of the relationship between two variables and may range from −1.00 to +1.00. If there is no relationship between variables, the correlation is zero. If two variables increase together (or decrease together), the correlation is positive. A survey of 33 studies of attitudes and behavior found that the average correlation between these two variables is +.30 or less, a relatively weak correlation. Several reasons why the relationship is not stronger have been suggested. In this section, we will consider four variables that influence the relationship between attitudes and behavior: (1) the correspondence between attitude and behavior, (2) the characteristics of the attitude, (3) the activation of the attitude, and (4) situational constraints on behavior. Each of these variables is considered below.

## Correspondence

Attitudes are more likely to predict behavior when the two are at the same level of specificity (Schuman & Johnson, 1976). For example, suppose you have invited a casual acquaintance to dinner and you want to plan the menu. You know that she is Italian, so she probably likes Italian food. But can you predict with confidence that she will eat green noodles with red clam sauce? Probably not. A favorable attitude toward a type of cuisine does not necessarily mean that the person will eat every dish of that type. Many studies have attempted to predict from general attitudes to specific behaviors. For instance, some studies of racial prejudice have tried to predict from people's general attitudes toward blacks to specific behaviors, such as willingness to have one's photograph taken with particular blacks in particular settings (Green, 1972). Not surprisingly, the relationship between attitude and behavior was weak.

A general attitude is a summary of many feelings about an object under a variety of conditions or about a whole class of objects. Logically, it should not necessarily predict behavior in any particular single situation. On the other hand, it might predict a composite measure of a number of relevant behaviors. For example, even though a white person with pro-black attitudes might object to being photographed with a black person in a way that implies that they were on a date, she might engage in other pro-black behaviors.

This point can also be illustrated by the results of a study of environmental behavior. Can we predict whether people will engage in various environmentally oriented behaviors if we know their general attitudes toward environmental quality? To answer this question, researchers distributed a questionnaire, including a 16-item Likert scale, to measure attitudes toward conservation and pollution (Weigel & Newman, 1976). Between three and eight months later, they contacted subjects three times and asked them to participate in various environmental projects. The projects included signing and circulating copies of three petitions, participating in a litter pickup program and recruiting a friend to do so, and participating for up to eight weeks in a recycling program. The results are shown in Table 6-1. The first column of numbers displays the relationship between the general attitude measure and individual behaviors. These correlations range from .12 to .57; seven of the fourteen correlations are about .30, the magnitude noted by Wicker (1969). When the researchers constructed aggregate measures by combining the fourteen individual behavioral measures into three categories, the attitude-

■ **TABLE 6-1**

CORRELATION BETWEEN SUBJECTS' ENVIRONMENTAL ATTITUDES AND BEHAVIORAL CRITERIA

| Single Behaviors | r | Categories of Behavior | r | Behavioral Index | r |
|---|---|---|---|---|---|
| Offshore oil | .41 | | | | |
| Nuclear power | .36 | Petitioning behavior | | | |
| Auto exhaust | .39 | scale (0–4) | .50 | | |
| Circulate petitions | .27 | | | | |
| Individual participation | .34 | Litter pickup | | Comprehensive | .62 |
| Recruit friend | .22 | scale (0–2) | .36 | behavioral index | |
| Week 1 | .34 | | | | |
| Week 2 | .57 | | | | |
| Week 3 | .34 | | | | |
| Week 4 | .33 | Recycling behavior | | | |
| Week 5 | .12 | scale (0–8) | .39 | | |
| Week 6 | .20 | | | | |
| Week 7 | .20 | | | | |
| Week 8 | .34 | | | | |

Note: N = 44; r = correlation.

*Source: Adapted from Weigel and Newman, 1976.*

behavior correlations improved somewhat (middle column). When the researchers created a composite "behavioral index" and correlated this with the general attitude measure, the relationship increased even more. Thus, general attitudes can predict a general measure of relevant behaviors (Weigel & Newman, 1976).

What about predicting a specific behavior, such as whether your Italian guest will eat green noodles and red clam sauce? Just as general attitudes best predict a composite index of behavior, we need a specific measure of attitude to predict a specific behavior. We can think of an attitude and a behavior as having four elements: an action (eating), object or target (green noodles and red clam sauce), context (in your home), and time (tomorrow night). The greater the degree of **correspondence**—that is, the number of elements that are the same in the two measures—the better we can predict behavior from attitudes (Ajzen & Fishbein, 1977).

A study of birth control use by 244 women (Davidson & Jaccard, 1979) demonstrated that attitudinal measures that exhibit correspondence with the behavioral measure are better predictors of behavior. In this study, the behavior of interest was whether women used birth control pills during a particular two-year period. Attitude was measured in four ways. The measure of the women's general attitude toward birth control had only one element in common with the behavior (object); the correlation between this attitude measure and behavior was a modest .323, as shown in Figure 6-4. When the attitude measure had two elements in common with the behavior (object and action), the correlation rose to .525. Finally, an attitude measure that included three elements (object, action, and time)—"Do you plan to use birth control pills in the next two years?"—was most highly correlated with the behavioral measure. Thus, attitudinal measures having high correspondence with the behavioral measure were better predictors of behavior than attitudinal measures having low correspondence.

Earlier in this chapter we mentioned that in LaPiere's study most establishments that served

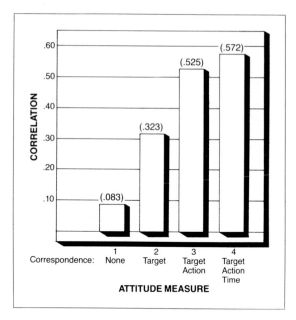

■ **FIGURE 6-4**

CORRELATIONS OF ATTITUDE MEASURES THAT VARY IN CORRESPONDENCE WITH BEHAVIOR

Every behavior involves a target, action, context, and time. In order to predict behavior from attitude, the measures of attitude and behavior should correspond—that is, involve the same elements. The larger the number of elements in common, the greater the correlation between attitude and behavior. Researchers obtained four measures of attitudes toward birth control from 244 women: (1) general attitude toward birth control, (2) attitude toward birth control pills, (3) attitude toward using pills, and (4) attitude toward using pills in the next two years. The behavioral measure was actual use of pills during the two-year period. Note that as correspondence increased from zero to three elements, the correlation between attitude and behavior increased.

*Source: Adapted from Davidson and Jaccard, 1979.*

the Chinese couple later said they would not. The lack of a relationship between attitude and behavior in LaPiere's study may be due to lack of correspondence. The behavioral measure was whether a particular Oriental couple (object) was served (action) in a particular restaurant or hotel (context) on a particular day (time). However, LaPiere's questionnaire measuring attitudes simply asked whether Chinese guests would be served. Thus, there was correspondence between the measures on only one element—the action—which may account for the discrepancy LaPiere found between attitude and behavior.

## Characteristics of the Attitude

The relationship between attitude and behavior is also affected by the nature of the attitude itself. Four characteristics of attitudes that may influence the relationship are (1) the degree of consistency between the affective (evaluative) and the cognitive components, (2) whether the attitude is grounded in personal experience, (3) the certainty with which the person holds it, and (4) whether it is stable over time.

**Affective-Cognitive Consistency** At the beginning of the chapter, we identified three components of an attitude: cognition, evaluation (affect), and behavior. When we consider the relation between attitude and behavior, we are looking at the relationship between the first two components and the third. Not surprisingly, it has been shown that the degree of consistency between the affective and cognitive components affects the attitude-behavior relationship. That is, the greater the consistency between cognition and evaluation, the greater the strength of the attitude-behavior relation.

Recall that the cognitive component is a belief about the attitude object (for example, "Capital punishment is necessary to protect society"). The affective component is the emotion associated with the object ("I am strongly in favor of capital punishment"). In this case, there is a high degree of affective-cognitive consistency. Now suppose another person endorses the belief but at the same time is opposed to capital punishment. Whose behavior could you confidently predict? The first person is much more likely to write letters to legislators supporting the death penalty and to vote for candidates who advocate its use. In fact, greater consistency has been shown to be associated with a closer relationship between attitude and behavior (Norman, 1975).

In a recent experiment, subjects' beliefs and evaluations regarding capital punishment were assessed by questionnaires (Chaiken & Yates, 1985). Next, subjects who were either high or low in consistency were asked to engage in a behavior—to write two essays, one on the death penalty and one on an unrelated topic. The relevant essays written by high-consistency subjects were much more internally consistent; that is, their attitudes were part of an internally consistent structure. Furthermore, high-consistency subjects dealt with discrepant information by discrediting it or minimizing its importance, making their attitudes more resistant to change.

**Direct Experience** If you have a positive attitude toward an activity based on having done it once, and your roommate has a positive attitude based on hearing you rave about it, which of you is more likely to accept an invitation to engage in it again?

One study (Regan & Fazio, 1977) provides an answer to this question. The behavior of interest was the proportion of time spent playing with several kinds of puzzles. Subjects in the direct experience condition played with sample puzzles, while those in the indirect experience condition were given only descriptions of the puzzles. Researchers then asked subjects to respond to some attitude measures and subsequently gave them an opportunity to play with the puzzles. They discovered that the average correlation between attitude and behavior was much higher for subjects who had direct experience than for those who did not.

There are several reasons why attitudes based on direct experience are more predictive of subsequent behavior (Fazio & Zanna, 1981). The best predictor of behavior is past behavior; the more frequently you have played tennis in the past, the more likely you are to play it in the future (Fredricks & Dossett, 1983). An attitude is a summary of a person's past experience; thus, one grounded in direct experience predicts future behavior more accurately. In addition, direct experience makes more information available about the object itself (Kelman, 1974). In a test of the hypothesis that the attitude-behavior

Although most Americans have attitudes about abortion, only a minority act on their beliefs like these demonstrators. People who are more certain of their attitudes, whether pro or con, are more likely to engage in such behavior.

relation will increase as the amount of information increases, researchers studied three different behaviors, including voting for specific candidates in an election (Davidson et al., 1985). The results indicated that both the amount of information and direct experience increase the relationships.

**Certainty**   Suppose you ask two friends which candidate they like in the upcoming presidential election. One replies with certainty, "I'm voting for X"; the other hedges a bit and says, "Well, maybe I'll vote for Y." Which person's behavior do you think you could predict? In general, the more certain the person is of his attitude the more likely it is to influence behavior. Studies of voting behavior find that many of the errors in

predictions occur among those who report indifference to the election—that is, people who have weak or uncertain attitudes (Schuman & Johnson, 1976). In one study, researchers measured people's attitudes before an upcoming election. Subjects completed a 15-item Likert scale and also indicated how certain they were of each response (Sample & Warland, 1973). For each subject, two measures were constructed: one of attitude and one of certainty. The 243 subjects were divided into high- and low-certainty groups. Researchers noted how subjects voted in the election fifteen days after they had completed the questionnaire. Among subjects who were more certain of their attitudinal responses, the attitude-behavior correlation was .47, whereas among those who were less certain the correlation was .06. The average for all subjects was in the usual range (.29).

The certainty with which an attitude is held depends partly on whether it is based on direct experience. Attitudes based on direct experience with the object, as opposed to information obtained from others, may be held with greater certainty. The person who has direct experience has more information about the object and, thus, a more clearly defined attitude. The attitude may also be more resistant to change (Fazio & Zanna, 1981).

**Temporal Stability**   Most studies attempting to predict behavior from attitudes measure people's attitude first and their behavior weeks or months later. A modest or small correlation may mean there is a slight attitude-behavior relationship. Or it could mean that people's attitudes have changed in the interim period. If the attitude changes after it is measured, the person's behavior may be consistent with his present attitude, even though it appears inconsistent with our measure of his attitude. Thus, in order to predict behavior from attitudes, the attitudes must be stable over time.

In general, we would expect that the longer the time period between the measurement of attitude and of behavior, the more likely the attitude will change and the smaller the attitude-behavior relationship will be. In a study designed to test this possibility (Schwartz, 1978), an appeal was mailed to almost 300 students to volunteer as tutors for blind children. Earlier, students had filled out a questionnaire measuring general attitudes toward helping as well as questions about tutoring blind children. Some students had filled out the questionnaire six months earlier; some, three months earlier; some, both three and six months earlier; and still others had not seen the questionnaire. The correlation between attitude toward tutoring and actually volunteering was greater over the three-month period than over the six-month period. Thus, to avoid problems of temporal instability, the amount of time between the measurement of attitudes and of behavior should be brief.

On the other hand, some attitudes evidence a remarkable degree of stability. Thornton (1984) studied the attitudes of 458 women toward divorce; their attitudes were measured in 1962, 1977, and 1980. He found substantial stability over the eighteen-year period, particularly among women who attended church regularly. Marwell, Aiken, and Demerath (1987) studied the political attitudes of 220 white young people who spent the summer of 1964 organizing blacks in the South to vote. They measured the same attitudes of two thirds of these activists in 1984, two decades later. The extreme, radical attitudes these people held in 1965 softened in the following twenty years; but, in general, these people remained liberal and committed to the needs of disadvantaged groups (see Box 6-2).

Thus, characteristics of attitudes influence the degree to which we can predict behavior. Our ability to predict will be greater if the attitude is characterized by affective-cognitive consistency, is based on direct experience with the object, is held with certainty, and is stable over time.

## Activation of the Attitude

Each of us has thousands of attitudes. Most of the time a particular attitude is not within our conscious awareness. In order for an attitude to influence behavior, it must be **activated**—that is, brought from memory into conscious awareness (Zanna & Fazio, 1982).

An attitude is usually activated by exposure of the person to the object, particularly if the attitude was originally formed through direct experience with the object (Fazio, Powell, & Herr, 1983). Earlier sections of this chapter may have activated your attitudes toward many objects, such as Woody Allen's films, premarital sexual activity, birth control, Italian food, and cigarette smoking. If you are at home and watching television, your attitudes toward various programs may be activated. Thus, one way to activate attitudes is to arrange situations in which persons are exposed to relevant objects. Soft lighting, a cozy fire, and glasses of wine are all associated with seduction; we often set up these

## WHERE ARE THEY NOW? PERSISTENCE AND CHANGE IN POLITICAL ATTITUDES

The decade from 1961 to 1971 was a period of political activism, when tens and perhaps hundreds of thousands of Americans engaged in activities such as demonstrations, sit-ins, voter registration drives, and community organizing in an attempt to bring about social and political change. College campuses were frequently the setting for these activities, and some of the most active and visible participants were college students. Both the news media and social scientists referred to these people as "radicals" because they were calling for large-scale changes: real freedom and civil rights for blacks, the withdrawal of American troops from Vietnam, and the restructuring of colleges and universities to give students more power. In the early 1970s, the student movement subsided, and the faces of its leaders faded from front pages and television screens.

Where are they now? What happened to those student activists? What happened to their sometimes radical attitudes as they grew older and conditions in American society changed? Some suggest that their radical attitudes reflected youthful idealism and that as they became involved in family and work their attitudes mellowed (DeMartini, 1983). In fact we might find that yesterday's activists are today's establishment, that their education gave them access to positions as lawyers, bankers, and stockbrokers in the 1980s. Others argue that the activists of the 1960s were the product of unique historical experiences—the presidency of John F. Kennedy, the war in Vietnam—that created radical attitudes that will last throughout their lives. According to this view, we may find the former activists working in social service and community action settings.

An unusual longitudinal study provides us with some answers. In early 1965, 300 young people—95 percent of them students—volunteered to spend ten weeks in six southern states, working to get blacks registered to vote. During their training, 231 of the activists completed a questionnaire that measured various characteristics and attitudes. The volunteers were young, from metropolitan and urban areas, and from liberal families (Demerath, Marwell, & Aiken, 1971). Their attitudes were very liberal, if not radical: 90 percent favored federal intervention to secure civil rights for blacks in the South; 71 percent felt that a large portion of the federal

cues in the hope of activating our partner's positive attitudes toward romantic and sexual activity.

Attitudes differ in the ease with which they are activated—that is, they differ in *accessibility*. Some attitudes are highly accessible and are activated automatically by the presentation of the object. One indication (measure) of the degree to which an attitude is accessible is the speed of activation. Attitudes activated instantaneously are, by definition, highly accessible. Other attitudes are activated more slowly—that is, they are less accessible (Fazio & colleagues, 1986).

There is some evidence that the more accessible an attitude, the more it is likely to guide future behavior. This was shown, for example, in a study by Fazio and Williams (1986) that looked at the impact of accessibility on voting in the 1984 presidential election. In June and July of 1984, a sample of 245 people were questioned about their attitudes toward Ronald Reagan and Walter Mondale. The latency of the answer—how quickly the person replied to the question

## Box 6-2

budget should be devoted to social programs; half of the volunteers thought there was a 50–50 chance of a nuclear war in the next ten years; and about one third expressed pessimism regarding American institutions, saying that they had grave doubts about democracy as a political system and that the system had proven itself incapable of coping with racial discrimination.

In 1984 and 1985, the researchers attempted to locate those who participated in the original study. They were able to find and obtain data from 145 of the 231 (63 percent). They compared those in the follow-up sample with the original group on thirteen background characteristics. The only significant difference was on gender; there were proportionately more males in the 1984–85 group, because the researchers had greater difficulty in locating the women who had been members of the original group.

The follow-up study yields an interesting picture of student activists nineteen years later (Marwell, Aiken, & Demerath, 1987). In 1984–85, they were all in their mid-40s and often embedded in work and family roles. What hap-

pened to their attitudes? First, the researchers compared selected political attitudes of the former activists with those of a national sample surveyed in 1982. The activists differed on all nine items, generally picking the most liberal response possible. Next, the researchers compared the respondent's attitudes in 1984–85 with his or her attitudes in 1963. Changes were noted in four areas. First, the activists commitment to nonviolence as a strategy was significantly lower in 1984–85. Second, their attitudes toward the South had become less hostile in the intervening years; this may in part reflect the fact that conditions in the South have changed. Third, the former activists were more pessimistic about nuclear war. Finally, they had less trust in federal political institutions than they had in 1965. These results suggest a softening of the extreme attitudes held by these people in the 1960s. At the same time, the former activists were still very liberal. There is more persistence than change in their political attitudes.

---

about each candidate—was used as a measure of accessibility. After the election, each person was asked for whom he or she voted. The more accessible the attitude—that is, the more quickly the person replied to the question about the candidate—the more likely the person was to vote for that candidate in November.

In general, once an attitude has been activated, the person needs to decide what that attitude implies as a guide for behavior. There appear to be systematic differences in the kind of information that people use to guide their be-

havior. Some persons rely heavily on situational cues, whereas others rely heavily on information about their internal states. The term *self-monitoring* refers to the extent to which people use information about the environment as a basis for modifying their own behavior to meet the expectations of others (Snyder, 1979). Persons who rely primarily on environmental cues as a guide to behavior are considered to be high in self-monitoring, whereas others who rely primarily on information about their inner states, including attitudes, are low in self-monitoring. The interesting point is that those persons who rely on internal information—that is, persons low in self-monitoring—should exhibit greater attitude-behavior consistency than those who utilize situational cues. Experimental data support this prediction (Snyder & Tanke, 1976; Ajzen, Timko, & White, 1982).

Thus, whether an attitude influences behavior depends in part on (1) whether the attitude is activated, and (2) whether the person uses attitudes as guidelines for behavior.

## Situational Constraints

If you believe in L. Ron Hubbard—author and founder of the Church of Scientology—as your spiritual leader and you attend a service of the church, your behavior reflects your attitude. If you believe that Scientology, the Unification church, and The Way are all dangerous cults and you attend a meeting of the Citizens Freedom Foundation—a national group opposed to such cults—your behavior is consistent with your attitude. Suppose, however, that you are opposed to cults but find yourself in a conversation with three followers of Mr. Hubbard. Would you voice your opposition—that is, behave in a manner consistent with your attitudes—or tactfully end the interaction? Your reaction would probably depend partly on the strength and certainty of your attitude. If you are strongly opposed to cults, you may speak your mind. But if situational constraints prevent you from expressing your attitude, you may behave in a way that is inconsistent with your beliefs. In LaPiere's study, for instance, hotel and restaurant employees confronted by a white man and a Chinese couple may have felt compelled to serve them rather than run the risk of creating a scene by refusing to do so.

**Situational constraint** refers to an influence on behavior due to the likelihood that other persons will learn about a behavior and respond positively or negatively to it. Situational constraints often determine whether our behavior is consistent with our attitudes. In fact, how we behave is frequently a result of the interaction between our attitudes and constraints present in the situation (Warner & DeFleur, 1969). This relationship is summarized in Figure 6-5, using attitudes toward cults as an example. A conversation between someone weakly opposed to cults and followers of Mr. Hubbard (weak pressures) would be a situation of conflict for the individual, whereas someone strongly opposed to cults is more likely to voice his opposition.

Sometimes we feel constrained by the possibility that others may learn of our behavior. At other times those around us exert direct social influence; they communicate specific expectations about how we should behave. The greater the agreement among others˙ about how we should behave, the greater the situational constraint on persons whose attitudes are inconsistent with the situational norms (Schutte, Kendrick, & Sadalla, 1985). Under these conditions, there is a weaker relationship between attitudes and behavior. Consequently, the less visible our behavior is to others, the more likely it is that our behavior and attitudes will be consistent (Acock & Scott, 1980).

But what if persons whose opinions we value are not actually present? Several studies have assessed the impact of reference groups on the attitude-behavior relationship. Such research involves measuring the subject's attitudes toward some object and then asking him to indicate the position of various social groups with regard to that object. One survey of a sample of adults assessed the respondent's attitudes toward drinking alcoholic beverages and the degree to which his or her friends approved of drinking (Rabow, Newman, & Hernandez, 1987). When attitudes and social support were congruent—that is, when the respondent's and friends' views were the same—there was a much stronger relation between attitudes and behavior than when attitudes and social support were not congruent. Several other studies report similar findings (Schuman & Johnson, 1976).

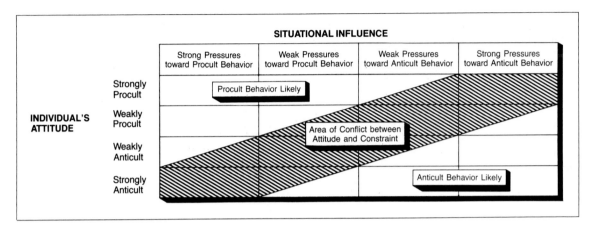

SITUATIONAL INFLUENCE

■ **FIGURE 6-5**

THE INFLUENCE OF ATTITUDE AND SITUATIONAL CONSTRAINTS ON BEHAVIOR

Our behavior is influenced not only by our attitudes but also by situational constraints, the behavior of others, or the likelihood that others will find out what we do. When the individual has a strongly held attitude and situational influences encourage behavior consistent with the attitude, there will be a strong relationship between attitudes and behavior. But when situational influences produce pressure to behave in ways inconsistent with one's attitude or when the attitude is weak, behavior and attitude are less likely to be consistent.

*Source: Adapted from Warner and DeFleur, 1969, Figure 3.*

## ■ The Reasoned Action Model

In the preceding section, we identified several influences on the relationship between attitudes and behavior. Obviously, the relationship is not a simple one. One important attempt to specify this relationship is the **theory of reasoned action,** developed by Fishbein and Ajzen (1975; Ajzen & Fishbein, 1980). This model is based on the assumption that behavior is rational, and it incorporates several factors that have been shown to affect the consistency between attitudes and behavior (see Figure 6-6).

According to the reasoned action model, behavior is determined by behavioral intention. Behavioral intention is primarily influenced by two factors: attitude ($A_B$) and subjective norm ($S_N$). Attitude refers to positive or negative feelings about engaging in a behavior. **Subjective norm** is the individual's perception of others' beliefs about whether or not a behavior is appropriate. (In other words, subjective norm is one

form of situational constraint.) The reasoned action model also specifies the determinants of attitude and of subjective norm. Attitude is influenced by one's beliefs about the likely consequences of the behavior and one's evaluation —positive or negative—of each of those outcomes. Subjective norm is influenced by the person's beliefs about the reactions of other persons or groups to the behavior and his or her motivation to comply with their expectations.

### Formal Model

The model can be summarized using the following expression.

$$\text{Behavior} = \text{Behavioral intention}$$
$$= \text{Attitude} + \text{Subjective norm}$$

Suppose Bill has two good friends who are Moonies. They have been giving him literature and encouraging him to join the Unification church. In order to predict what Bill will do, we need to know his attitude ($A_B$) and his perception

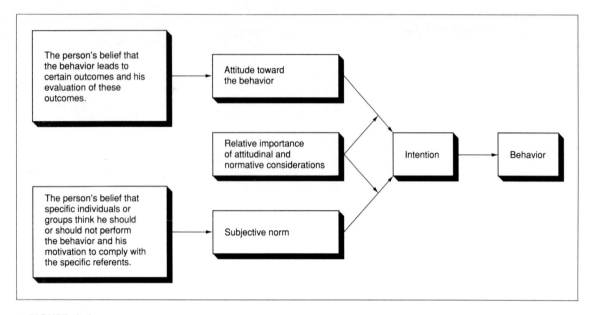

■ **FIGURE 6-6**

THE REASONED ACTION MODEL

Note: Arrows indicate the direction of influence.

Source: Ajzen and Fishbein, 1980, Figure 1-1.

of how others will react to his joining this cult. Attitude is the sum ($\Sigma$) of beliefs ($b$) about the likelihood of various consequences of the act and the evaluation ($e$)—positive or negative—of each consequence.

$$A_B = \Sigma\, b\, e$$

Bill has several beliefs about joining the Moonies. If he joins, he will gain a greater sense of purpose in his life, and his physical needs will be met. At the same time, he will have to end his relationship with Cindy, his girlfriend. Also, Bill has read that excult members claim they had to relinquish their personal freedom when they joined. These beliefs and their evaluations are shown in Table 6-2. Bill is certain consequences (1) and (2) would occur; hence, their value ($b$) is +3. His evaluation of consequence (1) is very positive (+3), whereas his evaluation of having his physical needs cared for is less positive (+1).

He believes it is likely that he will have to give up Cindy (+2), which would be unpleasant (−2). He is skeptical about the claim that he will lose his freedom (+1), although he would be very upset if that occurred (−3). Bill's attitude is +5, the value of $b \times e$.

Subjective norm ($S_N$) is the product of normative beliefs ($NB$)—expectations about how significant others will react—and motivation to comply ($MC$) with each:

$$S_N = \Sigma\, (NB \times MC)$$

For Bill, the significant others are his parents, his peers, and his girlfriend, Cindy. His parents are strongly opposed to his joining a cult (−3), and he is moderately motivated to comply with their views (+2). He is equally motivated to comply with his friends' views (+2), who strongly favor his joining (+3). And he is highly motivated to comply with Cindy, who opposes his becoming a

This Hari Krishna member is passing out literature and seeking new recruits. A person who considers joining the cult will be influenced not only by his or her attitudes, but also by subjective norms—the anticipated reaction of family and friends.

Moonie. This is summarized in Table 6-3. Thus, the value of $S_N$ is the product of $NB \times MC$, or $-6$. Behavioral intention is simply attitude multiplied by subjective norm, or $(+5)(-6) = -30$. The model predicts that Bill will not join the Unification Church. While his attitude toward the behavior is positive, the social pressures are negative.

## Assessment of the Model

The reasoned action model combines several elements discussed earlier in this chapter. It has been used to predict behaviors such as signing up for a treatment program for alcoholics (McArdle, 1972), using birth control pills (Davidson & Jaccard, 1979), and smoking (Fishbein, 1980). When combined with quantitative

■ **TABLE 6-2**
DETERMINING ATTITUDE $(A_B)$ FROM BELIEFS $(b)$ AND EVALUATIONS $(e)$

| Consequences of Joining the Unification Church | Belief (b) | Evaluation (e) | Product (b × e) |
|---|---|---|---|
| (1) Gain a sense of purpose | +3 | +3 | +9 |
| (2) Have one's physical needs provided for | +3 | +1 | +3 |
| (3) Loss of relationship with Cindy | +2 | −2 | −4 |
| (4) Loss of some personal freedom | +1 | −3 | −3 |
| | | **Attitude** $(A_B) = \Sigma\, b \times e = +5$ | |

■ **TABLE 6-3**

DETERMINING SUBJECTIVE NORM ($S_N$) FROM NORMATIVE BELIEFS (*NB*) AND MOTIVATION TO COMPLY (*MC*)

| Significant Others | Normative Beliefs (*NB*) | Motivation to Comply (*MC*) | Product (*NB* × *MC*) |
|---|---|---|---|
| (1) Parents | −3 | +2 | −6 |
| (2) Friends | +3 | +2 | +6 |
| (3) Cindy | −2 | +3 | −6 |
| | | Subjective Norm ($S_N$) = Σ (*NB*)(*MC*) = −6 | |

measures of the components of attitudes, this model can predict a specific behavior under specific circumstances. For instance, one recent study attempted to predict weight loss among college women (Schifter & Ajzen, 1985). The participants' subjective intention, attitude, and subjective norm with respect to losing weight were measured. Several other variables were also assessed, including whether or not the woman had a detailed plan regarding weight loss. Six weeks later measurements were taken regarding the amount of weight actually lost. The amount of weight lost was associated with intention and with having a detailed plan; intention to lose weight was determined by attitude and subjective norm.

This model has also been the target of some criticism (Liska, 1984) because it assumes that our behavior is determined largely by our intentions. This assumption is not always correct; in some situations, our own past behavior may be even more influential than our intentions. For example, whether one has donated blood in the past is a much better predictor of whether a person will donate blood in the next four months than his statement about whether he intends to do so (Bagozzi, 1981).

In effect, much of our behavior is habitual and may not match our conscious intentions. Then, too, our behavior may be affected not only by intentions but also by whether we have a vested interest in the outcome of that behavior (Sivacek & Crano, 1982). In a study of 79 adult women committed to a six-week weight loss program, the amount of weight actually lost was influenced both by intentions and by the importance each woman placed on physical appearance and on good health (Saltzer, 1981). Finally, the reasoned action model applies primarily to behavior under the individual's volitional control. Thus, the reasoned action model does not apply to all behaviors, but it is useful for explaining and predicting behaviors under the person's conscious, volitional control.

## ■ Summary

**The Nature of Attitudes**    Attitudes have three characteristics. (1) Every attitude has three components: cognition, an evaluation, and a behavioral predisposition toward some object. (2) We learn attitudes through reinforcement, through repeated associations of stimuli and responses, and by observing others. (3) Attitudes are useful; they may serve instrumental and knowledge functions, express a person's values, or protect a person's self-image.

**The Measurement of Attitudes**    There are two major types of attitude measures. (1) Direct methods involve asking a direct question and recording the answer. They include the use of single items, Likert scales, and semantic differential techniques. (2) Indirect methods involve observing overt behavior. Such methods are useful when a direct question might elicit a false response. Examples include the wrong-number and lost-letter techniques.

**Attitude Organization and Change** (1) An attitude is usually embedded in a larger cognitive structure and is based on one or more fundamental or primitive beliefs. Consistency theories assume that when cognitive elements are inconsistent, individuals will be motivated to change their attitudes or behavior in order to restore harmony. (2) Balance theory assesses the relationship between three cognitive elements and suggests ways to resolve imbalance. (3) Dissonance theory cites three situations in which inconsistency often occurs: after a choice between alternatives, when people engage in behavior that is inconsistent with their attitudes, and when an important belief is disconfirmed. The theory also cites three ways to reduce dissonance: by changing one of the elements, by adding consonant cognitions, or by changing the importance of the cognitions involved.

**The Relationship between Attitudes and Behavior** The attitude-behavior relationship is influenced by four variables: correspondence, characteristics of the attitude, activation of the attitude, and situational constraints. (1) The relationship is stronger when the measures of attitude and behavior correspond in action, object, context, and time. (2) The relationship is also stronger if affective-cognitive consistency is high and if the attitude is based on direct experience, is held with certainty, and is stable over time. (3) Several factors determine whether an attitude will influence behavior, including whether it is activated and whether the person uses it as a guide for behavior. (4) Situational constraints may facilitate or prevent the expression of attitudes in behavior.

**The Reasoned Action Model** This model suggests that behavior is determined by behavioral intention. In turn, intention is determined by one's attitude and perception of social norms. This model allows precise predictions of behavior, and a number of studies report results consistent with such predictions. At the same time, it may not apply to some types of behavior.

## Key Terms

| | |
|---|---|
| ACTIVATION (OF AN ATTITUDE) | 171 |
| ATTITUDE | 148 |
| BALANCE THEORY | 159 |
| BALANCED STATE | 159 |
| COGNITION | 149 |
| COGNITIVE DISSONANCE | 161 |
| CORRESPONDENCE | 168 |
| DISSONANCE EFFECT | 164 |
| IMBALANCED STATE | 159 |
| INCENTIVE EFFECT | 164 |
| LIKERT SCALE | 153 |
| PREJUDICE | 151 |
| SEMANTIC DIFFERENTIAL SCALE | 153 |
| SENTIMENT RELATIONS | 159 |
| SITUATIONAL CONSTRAINT | 174 |
| SUBJECTIVE NORM | 175 |
| THEORY OF COGNITIVE DISSONANCE | 161 |
| THEORY OF REASONED ACTION | 175 |
| UNIT RELATIONS | 159 |

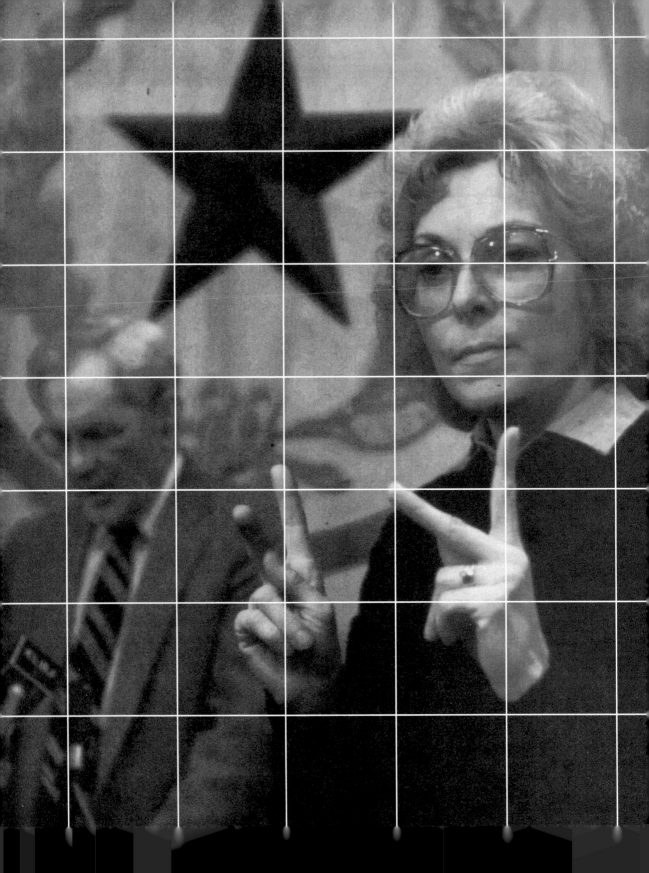

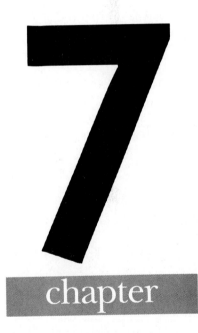

chapter

# Symbolic Communication and Language

# ■ Introduction

Communication is a basic ingredient of every social situation. Imagine playing a game of basketball or buying a pair of shoes without some form of verbal or nonverbal communication. Without it, interaction breaks down, and the goals of any social encounter are foiled. Indeed, it would be impossible to arrange commercial transactions, trials, birthday parties, or any other social occasion without communication. In its absence, people have no sense of what a social situation is all about. What then, is this crucial social behavior?

**Communication** is the process whereby people transmit information about their ideas and feelings to one another. We communicate through spoken and written words, through voice qualities and physical closeness, through gestures and posture. Often communication is deliberate: we smile, clasp our beloved in our arms and whisper, "I love you." Other behavior communicates unintentionally. If we forget a birthday or become totally absorbed in a book, for example, a sensitive partner may interpret our behavior as communicating lack of affection.

Because people do not share each other's experiences directly, they must convey their ideas and feelings in ways that others will notice and understand. We often do this by means of symbols. **Symbols** are arbitrary forms that are used to refer to ideas, feelings, intentions, or any other object.

Symbols can represent our experiences in a way that others can perceive with their sensory organs—through sounds, gestures, pictures, even fragrances. But for us to interpret symbols as they are intended, their meanings must be socially shared. To communicate successfully, we must master the ways for expressing ideas and feelings that are accepted in our community.

Symbols are arbitrary stand-ins for what they represent. Green could stand as reasonably for "stop" as for "go"; the sound *luv*, as reasonably for negative as for positive feelings. The arbitrariness of symbols may become painfully obvious when we travel in foreign countries. We are then likely to discover that the words and even the gestures we take for granted fail to communicate accurately. A North American who makes a circle with thumb and index finger to express satisfaction to a waiter may be in for a rude surprise if he is eating at a restaurant in Ghana, where the waiter may interpret his gesture as a sexual invitation. In Venezuela, it may be interpreted as a sexual insult! The traveler may then have serious difficulties straightening out these misunderstandings, because he and the waiter lack a shared language of verbal symbols to discuss them.

Communication is rarely a process of consciously translating ideas and feelings into symbols and then transmitting these symbols in hopes that others will interpret them correctly. Most communication occurs without any self-conscious planning. As we communicate with others, we usually produce our ideas and thoughts as we go along. In fact, we often learn about our own thoughts and feelings only when we express them to others. Sometimes we are surprised to discover what we ourselves think and feel as we communicate.

Language and nonverbal forms of communication are amazingly complicated. They must be understood and used with flexibility and creativity. Most of us fail, on occasion, to communicate our ideas and feelings with accuracy or to understand others' communications as well as we might wish. Yet, considering the problems a communicator must solve, most people do surprisingly well. This chapter begins with an examination of language, moves on to nonverbal communication, then analyzes the mutual impacts of communication and social relationships on each other. Finally, this chapter considers the delicate coordination involved in our most common social activity—conversation. In doing so, it addresses the following questions:

1. What is the nature of language, and how is it used to grasp meanings and intentions?

2. What are the major types of nonverbal communication, and how do they combine with language to convey emotions and ideas?

3. How do social relationships shape communication, and how does it in turn express or modify those relationships?

4. What rules and skills do people employ to maintain a smooth flow of conversation and to avoid disruptive blunders?

## ■ Language and Verbal Communication

Although people have created numerous symbols (mathematics, music, painting), language is the main vehicle of human communication. All people possess a spoken language. There are thousands of different languages in the world. This section addresses several crucial questions regarding the role of language in communication: What is the nature of language? How do people attain mutual understanding through language use? How are language and thought related to one another?

### Language as a Symbol System

Little is known about the origins of language (Lieberman, 1975), but humans have possessed complex spoken languages since earliest times (Kiparsky, 1976). **Spoken language** is a socially acquired system of sound patterns with meanings agreed on by the members of a group. We will examine the basic components of spoken language as well as some of the advantages of language use.

**Basic Components** Consider the following statement of one roommate to another: "Wherewereyoulastnight?" What the listener hears is a string of sounds much like this, rather than the sentence, "Where were you last night?" Spoken languages include sounds, words, meanings, and grammatical rules. To understand a string of sounds and to produce an appropriate response, people must recognize the following components: (1) the distinct sounds of which the language is composed (the *phonetic* component), (2) the combination of sounds into words (the *morphologic* component), (3) the common meaning of the words (the *semantic* component), and (4) the conventions built into the language for putting words together (the *syntactic* component, or *grammar*). We are rarely conscious of manipulating all these components during conversation, though we do so regularly and with impressive speed.

Unspoken languages, such as Morse code, computer languages, and the sign languages of the deaf, lack a phonetic component; although they do possess the remaining components of spoken language. People who use sign languages, for example, use upper-body movements to signal words (morphology) with shared meanings (semantics), and they combine these words into sentences according to rules of order (syntax). For a communication system to be considered a language, morphology, semantics, and syntax are all essential. Linguists study these components, seeking to uncover the rules that give structure to language. Social psychologists are more interested in how language fits into social interaction and influences it and in how language expresses and modifies social relationships (Giles, Hewstone, & St. Clair, 1981).

**Advantages of Language Use** Words—the symbols around which languages are constructed—provide abundant resources with which to represent ideas and feelings. The average adult native speaker of English knows the meanings of some 35,000 words and actively uses close to 5,000. Because it is a symbol system, language enhances our capacity for social action in several ways.

First, language frees us from the constraints of the here and now. Using words to symbolize objects, events, or relationships, we can communicate about things that happened last week or last year, and we can discuss things that may happen in the future. The ability to do the latter

allows us to coordinate our behavior with the activities of others.

Second, language allows us to communicate with others about experiences we do not share directly. You cannot know directly the joy and hope your friend feels at bearing a child nor her grief and despair at her mother's death. Yet she can convey a good sense of her emotions and concerns to you through words, even in writing, because these shared symbols elicit the same meanings for you both.

Third, language enables us to transmit, preserve, and create culture. Through the spoken and written word, vast quantities of information pass from person to person and from generation to generation. Language also enhances our ability to go beyond what is already known and to add to the store of cultural ideas and objects. Working with linguistic symbols, people generate theories, design and build new products, and invent social institutions.

Finally, words used to name an object let us know how to behave toward that object; one might call a person "friend," "stranger," or "enemy," for example. The behaviors we recognize as appropriate are drastically different depending on which of these verbal labels is applied. So are the behaviors implied by calling a remark an "insult" or a "joke" or by naming a yellow liquid "poison" or "Scotch." The label applied to an object tells us how to relate to it.

## Linguistic Meaning

"That's a great deal to make one word mean," Alice said in a thoughtful tone.

"When I make a word do a lot of work like that," said Humpty Dumpty, "I always pay it extra."

"Oh!" said Alice. She was too much puzzled to make any other remark. (Lewis Carroll, 1872)

Meaning is one of the most puzzling and controversial terms in the study of language, yet also one of the most central. We talk in order to express the meanings of our thoughts, and we listen in order to discover the meanings of what

These apes seem to be carrying on a lively conversation. How similar is their communication to human language? Although researchers hotly debate this question, they do agree that animal communication lacks the flexibility and creativity of human language.

others say. A commonsense theory is that the meaning of a word is simply what the word refers to. But this theory of meaning is inadequate. First, if meaning equals reference, then all words that refer to the same object should have the same meaning. This is not necessarily so. For example, "the large building," "Merton Hall," and "the library," could all refer to the same place, yet these words obviously have different meanings. Second, a single word may refer to different objects at different times yet retain the same meaning. One such word is "I." "I" refers to Jonathan when *he* uses it but to Judy when *she* uses it. Finally, some words ("the," "although," "ought") are quite meaningful, yet they do not refer to any object.

Most students of *semantics*—the study of meanings—agree that the sense of a word is somehow related to the attributes associated with

that word. Based on two types of associations, a useful distinction can be made between two types of meanings: denotative and connotative.

**Denotative Meaning** The literal, explicit properties associated with a word as defined in a dictionary are called **denotative meanings.** These are "objective" in the sense that they are shared by many people. The way dictionaries are written provides insight into the nature of denotative meaning. Dictionary writers collect samples of language from literature, the media, and speech and then figure out what meanings are being employed from the context of a given word. In short, they derive the meanings of words in part from the ways people commonly use them. At the same time, once they establish the definition of a word, they tend to resist changing it.

## ANIMAL AND HUMAN COMMUNICATION: A COMPARISON

Speech is a uniquely human activity. Still, numerous animals have ways of communicating. Bees communicate about the locations of nectar sources by performing elaborate dances. Ants communicate chemically by laying down trails for other ants to follow.

Against the backdrop of animal communication, the qualities of human language stand out more clearly. Unlike language, these varieties of animal communication are not symbolic. Each one is linked genetically to specific, physically present events. These nonsymbolic modes of communication cannot be used either to think about present or future, or to communicate new ideas.

Some animals use sounds as one means of communication. Recordings of the sounds animals make under natural conditions suggest that birds and mammals have vocabularies ranging between five and thirty distinct calls. The small number of natural animal calls contrasts sharply with the thousands of words in human speech. Most scientists believe that animal calls are largely instinctive, automatic responses to pleasant or fearful situations—expressions of pleasure, alarm, warning, and so on (Miller, 1981). With rare exceptions, animal calls, unlike human vocabularies, are not learned. Nor can they be applied to novel situations.

The absence of developed speech among animals does not necessarily signify a lack of complex communication. Communicating mainly through sight, smell, and touch, for example, the intelligent apes maintain an elaborate social organization. The possibility that they could be taught to communicate with symbols captured the imagination of scientists and stimulated numerous frustrating attempts to teach the apes language. Eventually, researchers recognized that these attempts were doomed to failure because chimps lack the vocal apparatus necessary to produce human speech.

Noting that chimps in the wild are responsive to gestures, scientists next tried to teach them American Sign Language (ASL), a gestural language of the deaf. These efforts were much more successful. Several chimps have mastered and used hundreds of ASL signs (Gardner & Gardner, 1980). More important, chimps have generalized these signs to objects other than the ones used in training and even to absent objects. For example, the chimp Washoe learned the sign for "open" in connection with a particular door, later generalized it to other doors, then to closed containers, and eventually to a water faucet. This certainly looked like symbol use.

The next step was to ask whether chimps could combine signs into meaningful sequences,

We can convey part of the denotative meaning of words that refer to concrete, material objects ("elephants," "books") by pointing out examples of these objects. But a word means more than any single example can express. "Book," for example, is a symbol that stands for a whole category of related objects that vary in size, shape, color, contents, and so on. Thus, even the words for concrete objects indicate categories that mean much more than one example can convey.

For more abstract words ("justice," "authority"), it is difficult or impossible to find clear examples to which we can point. Such words consist of complex relationships among people, events, activities, and so on. "Authority," for example, means the right and power of one party in a relationship to judge and exact obedience from another party.

**Connotative Meaning** A word has **connotative meaning** when it includes all the personal

Box 7-1

another crucial feature of human language. Washoe's first combination of signs occurred at the age of 20 months. By the time she was 3, the Gardners had recorded well over 300 combinations. This rate of development is much slower than for children. Still, it suggests that chimps may have some ability to produce new meanings by combining signs.

Could chimps master the even more developed language skill of ordering signs according to rules of syntax? This problem was tackled using a different method (Premack & Premack, 1984). A chimp named Sarah was taught to form sequences with various plastic tokens she had learned to associate with different objects (foods, colors, actions, relations, and so on). During these experiments, Sarah also followed instructions presented in complex sequences of tokens. For example, in response to "Red on green if-then Sarah take apple; green on red if-then Sarah take banana," Sarah took the apple rather than the banana that was available when a red card was on top of a green card.

What can we conclude from such studies? Enthusiasts conclude that apes are really capable of understanding human language and engaging in symbolic communication. Apes will never speak, but they are capable of manipulating abstract symbols, if only at an elementary level. Human symbolic communication is a great advance, but it separates us from other organisms only in degree (Meddin, 1979).

Critics, on the other hand, are not convinced (Sebeok & Umiker-Sebeok, 1980; Terrace, 1984). They argue that ingenious stimulus-response training and constant signalling by their teachers enables apes to imitate some of the outward forms of language. But, they say, the key features of language are missing. Critics see no evidence that apes use signs as symbols—recognizing their arbitrary relationship to objects. They view the apes' responses to sequences of signs as based on rote learning not on applying syntactic rules. Finally, critics claim that apes do not combine signs spontaneously to produce new meanings; they only imitate or respond to prompting by their trainers.

Whichever argument is correct, all agree that animal communication lacks the infinite flexibility and creativity of human language. Moreover, field studies indicate that animals make no natural use of symbols to communicate, whereas the use of symbols is central to natural human communication.

associations and emotional responses that an individual gives to that word. Warm, happy feelings that some people associate with the word "home," for example, are connotative meanings. For other people, "home" may conjure up coldness or suffocation. Connotative meanings are personal; they are not part of the dictionary definitions of words. Words with similar denotative meanings may have very different connotative meanings. "Separate" and "segregate," for example, both denote keeping objects apart. Yet "segregate" has negative connotations associated with racism, whereas "separate" is emotionally neutral.

When the connotative meanings people associate with words are at variance, communication problems may result (Kuhlman, Miller, & Gungor, 1973). For instance, the different connotative meanings of the phrase "gun control" produce tension and misunderstanding between members of the National Rifle Association and gun-control activists. For members of the NRA, gun control means an infringement on their freedom and attempts to disarm law-abiding citizens. To supporters of laws limiting access to handguns, gun control means making it harder for criminals to buy cheap handguns.

**Meaning and Context** Words have multiple denotative meanings. The word "pen," for example, is both an instrument for writing with ink and a fenced enclosure for pigs. One of the striking characteristics of language is that the more frequently words are used, the more meanings they seem to have. One dictionary, for example, lists 80 meanings for "take," 39 for "charge," but only 3 for "rudder," and 1 for "surrey." Although some meanings of a word may be related, other meanings can be very distinct. The senses of "charge" in "Please charge it!" and in "The charge is $20" are related; but its meanings in "Charge the battery," "Who's in charge?" and "Fire as you charge" are quite different.

Given the large number of meanings packed into a single word, the listener must decipher the correct meaning in a particular sentence. The key to selecting the correct meaning of a word is context. Three types of context are especially important: (1) the context of other words in the sentence, (2) the context of other sentences in the conversation, and (3) the social context of people, situations, and events in which the word is used.

Suppose you heard the sentence "I can't take the bike." What is the meaning of "take" here? If the sentence context included the words "but I'll get there by car," most of us would infer that "take" meant "ride on." If, instead, the words that followed were "because it's too heavy," we might infer that "take" meant "lift up and carry." The context of sentences could also clarify the meaning of "take." Consider the following example: "He's pulling away. I can't take the bike. I'm too slow." In this context, "take" seems to mean "catch up with." Finally, the social context may clarify word meaning. If one thief mutters to another, "I can't take the bike," as he vainly struggles to cut the chain lock, "take" would seem to mean "get into one's possession."

Thus, successful communication is much more than transmitting and receiving words with fixed, shared meanings. Conversationalists must select and discover the meanings of words through their context. In ordinary social interaction, the meanings of whole sentences and conversations may be ambiguous. Speakers and listeners must jointly work out these meanings as they go along. This next section will consider how people manage to attain mutual understanding through the use of language.

## Language Use as a Social Accomplishment

Virtually all adults have linguistic competence in their native language. They know how sounds and meanings go together, and they know the implicit rules that enable them to generate and to understand grammatically acceptable sentences. But linguistic competence does not guarantee that a person will use language well. In normal speech we hesitate, repeat ourselves, mispronounce words, and make grammatical errors. If you listen carefully to a natural conversation, you will notice many errors of performance. Despite these errors, people usually succeed in understanding each other. It is failure to use language in socially appropriate ways that most frequently disrupts communications during interaction. We will therefore examine several social requirements for effective language use.

**Sociolinguistic Competence** To attain mutual understanding, language performance must be appropriate to the social and cultural context. Otherwise, even grammatically acceptable sentences will not make sense. "My mother eats raw termites" is grammatically correct and meaningful; it reflects linguistic competence. But as a serious assertion by a North American, this utterance would probably draw amazed looks. It expresses an idea that is totally incongruous with

American culture, and listeners would have difficulty interpreting it. In a termite-eating culture, however, the same utterance would be quite sensible. This demonstrates that successful communication requires **sociolinguistic competence**—that is, knowledge of the implicit rules for generating socially appropriate sentences. Such sentences make sense to listeners because they fit with the listeners' social knowledge (Hymes, 1974).

Speech that clashes with what is known about the social relationship to which it refers suggests that a speaker is not socially competent (Grimshaw, 1981). Speakers are expected to use language that is appropriate to the status of the individuals they are discussing and to their relationship of intimacy. For example, socially competent speakers would not state seriously "The janitor ordered the president to turn off the lights in the Oval Office." They know that low-status persons do not "order" those of much higher status; at most they "hint" or "suggest." Referring to a relationship of true intimacy, socially competent speakers would not say, "The lover bullied his beloved." Rather, they would select such socially appropriate verbs as "coaxed" or "persuaded." In short, socially competent speakers recognize that social and cultural constraints make some statements interpretable in a specific situation and others uninterpretable.

**Speech Acts** We speak in order to accomplish a purpose. To express the idea that speech is a form of social action, we use the term **speech act,** which refers to the smallest unit of verbal social behavior (Austin, 1962; Searle, 1979). Speakers intend whatever they say to be interpreted as some type of social act: to warn listeners, inform, question, order, accuse, thank, complain, invite, and so on. We attain mutual understanding only when listeners recognize the purposes of that speaker's speech acts. When a question is used, for example, the speaker wants listeners to recognize that the purpose is to request information.

The purpose of a speech act may not be obvious; taken out of context, most statements are ambiguous (Grimshaw, 1987). For example, what is intended when a speaker says "John drinks"? Is this statement meant to inform us about John, to issue a warning, to pronounce an evaluative judgment, or to suggest a possible drinking partner? No matter how well listeners grasp the literal meaning of a speaker's words, communication has not succeeded unless the listeners also recognize the speaker's intended purpose.

To make sure purpose is understood, we can explicitly state our intent ("I must warn you that John drinks"). It is sometimes difficult to make our intentions clear without pointing to them explicitly. That is why we begin statements with phrases, such as "I authorize" (appoint, nominate, challenge, bet, and so on), that designate our intention. For the most common speech acts, however, each language has standard forms that designate the intended act implicitly. These standard forms reduce the time and effort needed for communication. In English, for example, there are standard forms for telling ("John is waiting."), asking ("Is John waiting?"), and ordering ("Wait, John!"). Standard forms employ syntax (word order and verb forms) and intonation to convey their purpose.

So far we have discussed speech acts in which people indicate their intentions directly, either by stating them explicitly or by using standard forms. Most of the time, however, speech is more subtle. We indicate our intentions only indirectly. Suppose, for example, you want someone to open the door. The standard form that directly signals an order is "Open the door!" But this same intention is often expressed in various indirect ways, including the following: "Would you mind opening the door?" "The door should be open." "Haven't you forgotten something?" "I'm more comfortable with the door open." Under certain circumstances, each of these indirect speech acts might accomplish the purpose of getting someone to recognize that you want them to open the door.

One situation in which we commonly use indirect speech acts is to express a thought or feeling that threatens the "face"—self-presentation—of another person (Holtgraves, 1986). Imagine meeting a good friend who is wearing an obviously new sweater, and your immediate reaction is that it is ugly. A direct speech act ("I think it looks terrible on you") would challenge your friend's claim to being nicely dressed. At the same time, assume that you don't want to lie. In this situation, a sociolinguistically competent speaker might select an indirect speech act ("I like that color"). If your friend is also competent, you will both understand that your intent is to communicate your dislike in a tactful way.

**Understanding and Cooperation** Mutual understanding is a cooperative enterprise. Because language does not convey thoughts and feelings in an unambiguous manner, people must work together to attain a shared understanding of each others' utterances (Goffman, 1983). A speaker must cooperate with a listener by formulating the content of speech acts in a manner that reflects the listener's way of thinking about objects, events, and relationships. The speaker must also take into consideration the listener's current knowledge. For example, the indirect request "Could you pick up my laundry?" shouted while racing to class will be effective only if the listener knows where the laundry was left. Is it at the laundromat or strewn all over the room?

Listeners must cooperate by actively trying to understand. They must go beyond the denotative meanings of what they hear to determine what the speaker is really trying to say. Only by making a creative effort to understand can listeners cope successfully with the fact that we often formulate our speech acts indirectly, leave out words ("Paper come?"), abbreviate familiar terms ("See ya in calc."), and make vague references ("He told him he would come later.").

According to a theory proposed by Grice (1975), listeners assume that much talk is based on a **cooperative principle.** In other words,

conversationalists ordinarily assume that the speaker is behaving cooperatively by trying to be (1) informative—giving as much information as is necessary and no more, (2) truthful, (3) relevant to the aims of the ongoing conversation, and (4) clear—avoiding both ambiguity and wordiness.

The principle of cooperation is more than a code of conversational etiquette. Often we can reach a correct understanding of otherwise ambiguous talk only by assuming that speakers are indeed trying to satisfy this principle. Consider, for example, how the relevance assumption (item 3 above) enables the conversationalists to understand each other in the following exchange:

TONY:   I'm exhausted.

CAROLYN:   Fred will be back next Monday.

On the surface, Carolyn's statement seems unrelated to Tony's declaration. In fact, in some contexts, we would infer that she has changed the subject, indirectly sending the message that she does not care about Tony's physical state. In fact, however, Carolyn is stating that she and Tony won't have to work as hard after their colleague Fred returns to the office. But why does she expect that Tony will understand this? Because she expects him to assume that she is adhering to the relevance maxim, that her comment relates to what he said.

The cooperative principle is also crucial for speech forms like sarcasm or understatement to succeed. In sarcasm or understatement, speakers want listeners to recognize that their words mean something quite different from what they seem to convey. One way we signal listeners that we intend our words to imply something different is by obviously violating one or two maxims of the cooperative principle while holding to the rest. Consider Carrie's sarcastic reply when asked what she thought of the lecturer: "He was so exciting that he came close to keeping most of us awake the first half hour." By flouting the maxim of clarity (responding in an unclear, wordy way) while still being informative, truthful, and relevant, Carrie implies that the lecturer was, in fact, a bore. Speakers add force and interest to their conversation by such violations of the cooperative principle. Listeners understand the speakers' indirect meanings because they can recognize intended violations against the backdrop of general adherence to the principle.

Thus, the successful use of language during interaction is an impressive social accomplishment. To attain mutual understanding, conversationalists must demonstrate sociolinguistic competence—that is, use language appropriate to the cultural and social setting. They must select speech acts that will best express their intentions under the circumstances, directly or indirectly. And they must cooperate in their roles as speakers and listeners by talking in ways that take account of others' current knowledge and perspectives and listening creatively for what others wish to imply through their frequently ambiguous words.

## Language and Thought

Language has developed to serve a purpose: to communicate thoughts and feelings. The features of language are molded by its uses. To be useful, a language must enable adults to speak and understand it easily and efficiently. Children must be able to learn it. It must express the ideas people normally want to convey, including ideas that are special to a particular social and cultural system. In short, the features of language are shaped by human capacities to process information and by the ideas that arise through exposure to the physical and social environments. This leads to the first question addressed in this section: What features are shared by all languages and why?

There is another side to the relationship between language and thought. Once people have learned a language, it wields a power of its own. It aids people in thinking about some ideas and hinders them in thinking about others. In fact, language may mold many aspects of behavior. This leads to the second question: What impacts do the distinctive features of different languages have on the ways speakers think?

**Linguistic Universals**   If languages are molded in part by the capacities, ideas, and experiences all people share, languages should also have certain features in common. **Linguistic universals** are features common to all languages. Every language has nouns and verbs, for example, because people must refer to objects and to actions. Certain sets of concepts are so basic that every language has words to express them. Some universal lexical concepts are shown in Table 7-1. The 100 basic concepts listed are likely to be used in everyday speech. They are learned by children at an early age and are seldom borrowed by one language from another.

Every language appears to have terms that express height, length, and distance and directions like up/down, front/back, and left/right. The terms for spatial dimensions are universal probably because all human beings use the same basic perceptual capacities to orient themselves in the physical world (Clark & Clark, 1977). The universality of experience with time, number, and negativity produces terms to express these abstract concepts in virtually every language. All languages have ways of distinguishing between present, past, and future; between single and multiple objects; between ideas and their negation—"go" versus "not go"—(Greenberg, 1966). The origins of these universals lie in basic human capacities for thought about abstract events and relations.

Of most interest to social psychologists are universals rooted in the social and cultural conditions of life. For example, the universal characteristics of families and of human conversations give rise to two sets of linguistic universals: kin terms and pronouns.

Languages invariably enable speakers to distinguish at least three characteristics of relatives: generation, blood relationship, and sex. But the precision with which a language designates particular relatives may vary. In English, "mother-in-law" indicates all these characteristics, while "nephew" designates generation and sex but not blood relationship. We can specify whether or not a nephew is related by blood by using

■ **TABLE 7-1**

SOME UNIVERSAL LEXICAL CONCEPTS

| | | | | |
|---|---|---|---|---|
| 1. I | 21. dog | 41. nose | 61. die | 81. smoke |
| 2. thou | 22. house | 42. mouth | 62. kill | 82. fire |
| 3. we | 23. tree | 43. tooth | 63. swim | 83. ash |
| 4. this | 24. seed | 44. tongue | 64. fly | 84. burn |
| 5. that | 25. leaf | 45. claw | 65. walk | 85. path |
| 6. who | 26. root | 46. foot | 66. come | 86. mountain |
| 7. what | 27. bark | 47. knee | 67. lie | 87. red |
| 8. not | 28. skin | 48. hand | 68. sit | 88. green |
| 9. all | 29. flesh | 49. belly | 69. stand | 89. yellow |
| 10. many | 30. blood | 50. neck | 70. give | 90. white |
| 11. one | 31. bone | 51. breasts | 71. say | 91. black |
| 12. two | 32. grease | 52. heart | 72. sun | 92. night |
| 13. big | 33. egg | 53. liver | 73. moon | 93. hot |
| 14. long | 34. horn | 54. drink | 74. star | 94. cold |
| 15. small | 35. tail | 55. eat | 75. water | 95. full |
| 16. woman | 36. leather | 56. bite | 76. rain | 96. new |
| 17. man | 37. hair | 57. see | 77. stone | 97. good |
| 18. person | 38. head | 58. hear | 78. sand | 98. round |
| 19. fish | 39. ear | 59. know | 79. earth | 99. dry |
| 20. bird | 40. eye | 60. sleep | 80. cloud | 100. name |

Source: Adapted from Swadesh, 1971.

more complicated constructions ("my nephew through my wife"). In all languages, precise, simple terms are used for the relatives who spend most time together because of caregiving and biological ties—parents, children, siblings, and spouses. This suggests that languages develop kin terms in response to the universal requirements of family interactions.

Similarly, the demands of conversation give rise to a universal set of pronouns. All languages have pronouns that designate the roles of speaker (I), of those addressed (you), and of other participants (he, she, they). Pronoun systems also invariably distinguish singular from plural (I/we, he/they). These features are critical for efficient conversation. Speakers must constantly refer to themselves and to one or more addressees during conversations. Pronouns eliminate the need to repeat names, roles, or other designations of participants each time they are mentioned. "You," for example, is much more efficient than "John, Mary, Steve, and Ellen." Thus, pro-

nouns, like kin terms, take their universal features from the need for effective social interaction.

Just as the common capacities and shared experiences of humankind give rise to linguistic universals, so the different social and environmental experiences of groups lead to differences between languages. Most obvious are variations in vocabulary. The concepts for which a language provides words depend on what speakers like and need to talk about. There are precise words for frequently needed concepts. The Hanunoo, for whom rice is a staple, for instance, have 92 names for rice (Brown, 1965). Each name conveys the shape, color, texture, state, and so on of a different type of rice. This makes communication accurate and easy. To convey the same information in English would not be impossible, but it would be inefficient. Instead of one basic word, English speakers would have to say something like "the long-grained, brown-freckled, firm, cooked rice."

## Box 7-2

## OPERATING ON THE ENGLISH LANGUAGE

Though Latin names for common body parts may seem bad enough; the special form of English employed within the medical community can be almost as perplexing.

Doctors, for example, don't call each other doctors. They say "physician," to distinguish themselves from that lesser species of doctor, the Ph.D.

For-profit hospitals don't call themselves for profit. They say "proprietary" or "investor owned," two terms with soothing neutral timbres.

Neither hospitals nor physicians call their charges a "price." Instead they speak genteelly of "reimbursement."

In the new world of medical marketing, hospitals refer to departments, like orthopedics or radiology, as "product lines." Package concepts clearly tied to one hospital are "branded products"; services arranged through the hospital but delivered elsewhere are "product-line extensions." "High-touch products" are those requiring physical contact with patients.

The process of getting more business is "patient accrual." People who pay with private insurance are "retail customers." Patients in general are now referred to as "consumers."

Anything a doctor does that requires cutting, jabbing or injecting is a "procedure." Anything a doctor does that requires thinking, talking or counseling of patients is "cognitive services." Procedures pay much better than cognitive services.

Colleges have begun conferring doctorates of pharmacology. As a result there are now Ph.D. pharmacists roaming hospital corridors sporting little name tags prefaced with the magic abbreviation Dr. This is driving M.D.'s, "medical doctors," crazy.

When spoken by an official of the Health Care Financing Administration, "realistic fees" means low fees. When spoken by a doctor, "realistic fees" means high fees.

The American Medical Association does not use the word "malpractice." It speaks of "physician liability."

"General medicine" is now considered a specialty.

*Source: Newsweek,* January 26, 1987, p. 44.

---

All languages multiply terms for concepts that are central to daily activities. This principle also applies to subgroups within larger language groups. Groups like surgeons, farmers, sociologists, and cooks each have special vocabularies that are relatively unknown to others. We sometimes refer to these vocabularies disparagingly as *jargons.* But jargons enable speakers to perform their tasks more effectively. Some contemporary medical jargon is presented in Box 7-2.

**Linguistic Reality** So far we have discussed the influence of thought and experience on language. What of the reverse? Does the language we speak influence the way we think about and

experience the world? The most famous theory on this question—the Sapir-Whorf **linguistic relativity hypothesis**—holds that language "is not merely a reproducing instrument for voicing ideas, but is itself a shaper of ideas, the program and guide for the individual's mental activity" (Whorf, 1956). Both strong and weak forms of this hypothesis have been proposed.

According to the strong form of the linguistic relativity hypothesis, language determines our perceptions of reality, so we cannot perceive or comprehend distinctions that do not exist in our own language. Orwell's description of "Newspeak," the language developed by the totalitarian rulers in his novel *1984,* gives frightening expression to the impact of language on thought.

> Don't you see that the whole aim of Newspeak is to narrow the range of thought? In the end we shall make thought crime literally impossible because there will be no words in which to express it . . . Every year fewer and fewer words, and the range of consciousness always a little smaller . . . The revolution will be complete when the language is perfect. (Orwell, 1949, pp. 46–47)

Orwell's description suggests that language determines thought through the words it makes available to people. We cannot talk about objects or ideas for which we lack words. The ways we think about the world are determined by the way our language slices up reality.

This strong form of the linguistic relativity hypothesis has not fared well in research. Consider some of the evidence. Some languages have only two basic words ("dark" and "white") to cover the whole spectrum of colors. Yet people from these and all other known language groups can discriminate and communicate about whatever large numbers of colors they are shown (Heider & Olivier, 1972). Most likely any concept can be expressed in any language, though not with the same degree of ease and efficiency. Before either the object or the word "television" existed, for example, someone undoubtedly referred to the concept of "a device that can transmit pictures and sounds over a distance." When new concepts are encountered, people invent words ("laser") or borrow them from other languages ("sabotage" from French, "goulash" from Hungarian).

Thus, the strict hypothesis that language determines thought has found little support. But there is considerable evidence for a weaker form of this hypothesis: Each language facilitates particular forms of thinking because it makes some events and objects more easily codable or symbolized. The availability of linguistic symbols for objects or events has been shown to have two clear effects: (1) it improves the efficiency of communication about these objects and events;

and (2) it enhances success in remembering them.

Regarding communication efficiency, recall that the availability of 92 names for rice enabled the Hanunoo to communicate quickly and precisely about their staple food. To demonstrate the value of language labels for memory, consider a study that measured the effect of language on memory for information about persons (Hoffman, Lau, & Johnson, 1986). English- and Chinese-language descriptions were created of two persons whose traits could be easily labeled in English but not in Chinese and of two persons whose traits could be easily labeled in Chinese but not in English. Three groups of subjects read the descriptions: English monolinguals, Chinese-English bilinguals who read in Chinese, and Chinese-English bilinguals who read in English. Subjects' memory of the descriptions was assessed; memory was much better when the information about the target conformed to labels in the subject's language of processing.

Based on the accumulated research we can conclude that language is more a reflection of human capacities and culture than a determinant of thought. But language also influences memory and perhaps the efficiency of thought.

## ■ Nonverbal Communication

Have you ever been in a situation in which you tried to communicate without using words? Perhaps you were interacting with someone who was deaf or someone who was too far away for your words to be heard. Imagine a situation where you are looking out of a window of your third-floor dorm room or apartment. You notice a man on the sidewalk below, dressed immaculately in a three-piece suit, pacing back and forth. He looks up and sees you and immediately begins to gesture. He points to you, then to some other window, and then to his watch. His movements are quick and sharp. His face is tense. What is he trying to communicate to you?

Even without the use of words, most of us can make some correct inferences about the

Nonverbal cues suffuse words with life, emphasizing them and clarifying their meaning. Nonverbal cues such as the posture and direct gaze of these two individuals also carry their own message.

man's message and emotional state. We do so by interpreting his nonverbal communication. But some ambiguity remains. This section examines three questions concerning nonverbal communication: (1) What are the major types of nonverbal communication? (2) How is emotion communicated through facial expressions? (3) What is gained and what problems arise by combining nonverbal and verbal communication in ordinary interaction?

## Types of Nonverbal Communication

By one estimate, the human face can make some 250,000 different expressions (Birdwhistell, 1970). Combining these with other nonverbal cues, the number of nonverbal communication possibilities is infinite. Four major types of nonverbal cues are described below and summarized in Table 7-2.

**Paralanguage** Speaking involves a great deal more than the production of words. Vocal behavior includes loudness, pitch, speed, emphasis, inflection, breathiness, stretching or clipping of words, pauses, and so on. All the vocal aspects of speech other than words are called **paralanguage.** This includes such highly communicative vocalizations as moaning, sighing, laughing, and even crying. Shrillness of voice and rapid delivery communicate tension and excitement in most situations (Scherer, 1979). Combined with other nonverbal communications, paralinguistic cues reinforce the expression of anger and superior status. Various uses and

■ **TABLE 7-2**

TYPES OF NONVERBAL COMMUNICATION

| Type of Cue | Definition | Examples | Channel |
| --- | --- | --- | --- |
| Paralanguage | Vocal (but nonverbal) behavior involved in speaking | Loudness, speed, pauses in speech | Auditory |
| Body language (kinesics) | Silent motions of the body | Gestures, facial expressions, eye gaze | Visual |
| Interpersonal spacing (proxemics) | Positioning of body at varying distances and angles from others | Intimate closeness, facing head-on, looking away, turning one's back | Primarily visual, also touch, smell, and auditory |
| Choice of personal effects | Selecting and displaying objects that others will associate with you | Clothing, makeup, room decorations | Primarily visual, also auditory and smell |

interpretations of paralinguistic and other non-verbal cues will be examined later in this chapter. For now, see how many distinct meanings you can give to the sentence "George is on the phone again" by varying the paralinguistic cues you use.

**Body Language** The silent motion of body parts—scowls, smiles, nods, gazes, gestures, leg movements, postural shifts, caresses, slaps, and so on—all constitute **body language.** Because body language entails movement, it is known as *kinesics* (from the Greek *kinein,* meaning "to move"). While paralinguistic cues are auditory, we perceive kinesic cues visually. The body movements of the man in our example were probably particularly useful to you in interpreting his feelings and intentions.

**Interpersonal Spacing** We also communicate nonverbally by using **interpersonal spacing** cues—that is, positioning ourselves at varying distances and angles from others (standing close or far away, facing head-on or to one side, adopting various postures, and creating barriers with books or other objects). Because proximity

is a major means of communication between people, this type of cue is called *proxemics.* When there is very close positioning, proxemics can convey information through smell and touch as well.

**Choice of Personal Effects** Though we usually think of communication as expressed through our bodies, people also communicate nonverbally through the personal effects they select: their choice of cars, home decoration, clothing, contact lenses, and—if one believes the commercials—beer. A uniform, for example, may communicate social status, political opinion, life style, and occupation, revealing a great deal about how its wearer is likely to behave (Joseph & Alex, 1972). You may have made assumptions about the status and life style of the man in our sketch based on the fact that he wore a three-piece suit. The deliberate use of personal effects to communicate impressions is discussed in Chapter 9.

For the most part, nonverbal cues—like language—are learned rather than innate. As a result, the meanings of particular nonverbal cues may vary from culture to culture. Other features of nonverbal communication may have universal meanings, however. These universals are based in our biological nature. The nonverbal communication of emotion reveals an interesting combination of learned and innate features.

## Facial Communication of Emotion

More than 2,000 years ago, the Roman scholar Pliny the Elder said, "The face of man is the index to joy and mirth, to severity and sadness." When we want to hide our feelings, we look away or cover our face. If we must show our face, we try carefully to compose our expressions (Goffman, 1959). Our sense that we must shield our natural facial expressions to conceal our true emotions from others suggests that we share two beliefs: (1) people express their emotions in distinctive ways on their faces; and (2) observers can accurately recognize the emotions others are experiencing. Are these two beliefs correct? This section will address this question and discuss the interplay of innate and cultural influences on the facial communication of emotion.

**Emotions Expressed by the Face**  Research indicates that people communicate six different emotions by distinctive facial expressions: happiness, sadness, surprise, fear, anger, and disgust (Ekman & Friesen, 1975; Harper, Wiens, & Matarazzo, 1978). Certain facial features are crucial for the expression of each emotion. Using only the lower face, for instance, nearly everyone in one study was able to identify happiness but was unsuccessful in identifying fear (Ekman, Friesen, & Tomkins, 1971). Both fear and sadness are judged best from the area around the eyes (Boucher & Ekman, 1975).

**Learned or Innate?**  Social psychologists are especially interested in the communication of emotion because it reflects a meeting between biological, social, and cultural influences. A century ago, Darwin first proposed that facial expressions of emotion are an innately determined part of our biological heritage. If this is so, all peoples of the world should exhibit very similar expressions when experiencing the same emotion. Individuals should also be able to recognize emotions expressed by members of other cultures.

To test the universality of expression, photos of individuals from many different cultural groups were shown to members of other groups (Ekman & Friesen, 1975). Members of each group successfully recognized the facial emotions expressed by members of other cultural groups. This suggests that all were responding to a common set of facial expressions that represents the primary emotions across cultures. Observations of children who were born blind provide convincing, added support for the universality of emotional expression. These blind children, who cannot learn how to express emotions from seeing others, still smile, laugh, and frown much like sighted children (Eibl-Eibesfeldt, 1979).

A recent study provides an impressive demonstration of the universality of recognition of the six basic emotions (Ekman et al., 1987). Three photographs were selected that depicted each of the six emotions. The photos were black-and-white pictures of the head and shoulders of a caucasian man or woman. The photographs were shown to college student observers in ten countries, including Western and non-Western nations. The results are displayed in Table 7-3. In every country, from 65 to 98 percent of the observers correctly identified the emotion portrayed in the pictures.

**Cultural Influences**  What is universal in facial expressions is the particular combination of facial muscles that move when we experience a given emotion. There are strong cultural influences, however, on the actual expression of emotion in everyday interaction. These influences take two main forms. First, through learning, culture helps determine which stimuli evoke particular emotions. Thus, cultural groups differ in the emotions that various odors, sounds, events, and so on evoke in their members. For example, cultural learning influences whether

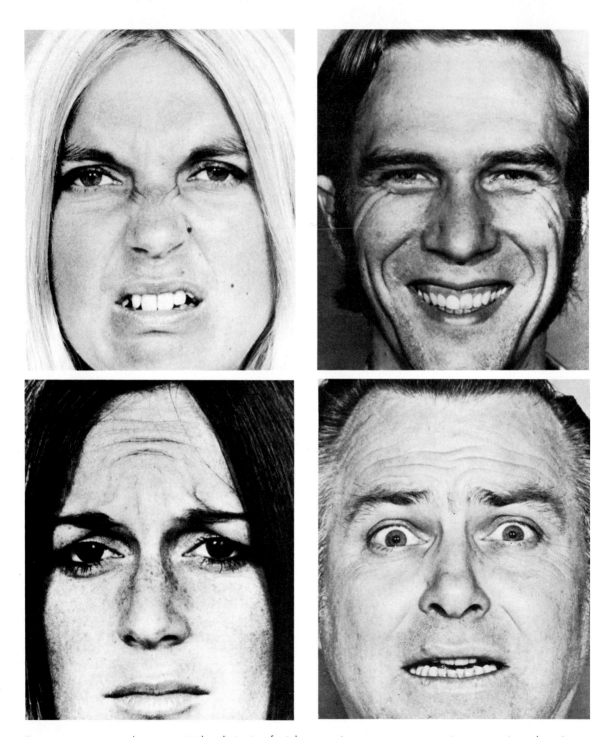

People from many cultures use similar distinctive facial expressions to express six primary emotions: happiness, sadness, surprise, fear, anger, and disgust. Can you identify the emotion expressed in each of the pictures above? Cross-cultural agreement in the expression and recognition of emotions suggests that these basic facial expressions are innate.

■ **TABLE 7-3**

SINGLE-EMOTION JUDGMENT TASK: PERCENTAGE OF SUBJECTS WITHIN EACH CULTURE WHO CHOSE THE PREDICTED EMOTION

| Nation | Happiness | Surprise | Sadness | Fear | Disgust | Anger |
|---|---|---|---|---|---|---|
| Estonia | 90 | 94 | 86 | 91 | 71 | 67 |
| Germany | 93 | 87 | 83 | 86 | 61 | 71 |
| Greece | 93 | 91 | 80 | 74 | 77 | 77 |
| Hong Kong | 92 | 91 | 91 | 84 | 65 | 73 |
| Italy | 97 | 92 | 81 | 82 | 89 | 72 |
| Japan | 90 | 94 | 87 | 65 | 60 | 67 |
| Scotland | 98 | 88 | 86 | 86 | 79 | 84 |
| Sumatra | 69 | 78 | 91 | 70 | 70 | 70 |
| Turkey | 87 | 90 | 76 | 76 | 74 | 79 |
| United States | 95 | 92 | 92 | 84 | 86 | 81 |

*Source: Adapted from Ekman et al., 1987.*

the event of death will elicit sadness or happiness for the deceased.

Second, culture influences the expression of emotion through display rules. **Display rules** are culture-specific norms for modifying facial expressions of emotion to make them fit with the social situation (Ekman, 1972). Display rules are typically learned in childhood. They become habits that automatically control facial muscles. Display rules may require modifying facial expressions of emotion in one of several ways. They may require (1) greater intensity in the expression of an emotion, (2) less intensity, (3) complete neutralization of the emotional expression, or (4) masking one emotion with a different one. If cultures vary in the intensity of emotional displays considered appropriate, we would expect observers to have a hard time assessing the intensity of an emotional display by someone from another culture. In the study by Ekman and his colleagues (1987) using observers from ten cultures, each observer was asked to rate how intensely the person in the photograph was experiencing happiness, anger, and so on. Whereas judgments of the emotion showed high levels of agreement across cultures, the judgments of intensity showed lower levels of agreement from one culture to another.

In response to our earlier question, we can conclude that people do indeed express primary emotions in distinct ways on their faces, and that others can accurately recognize these emotions. These universal features of the facial expression of emotion are innate. In order to communicate emotions effectively in everyday interaction, however, people must learn and employ the display rules of their own culture.

## Combining Nonverbal and Verbal Communication

When we speak on the telephone or shout to a friend in another room we are limited to communicating through verbal and paralinguistic channels. When we wave to arriving or departing passengers at the airport we use only the visual channel. Ordinarily, however, communication is multichanneled. Information is conveyed simultaneously through verbal, paralinguistic, kinesic, and proxemic cues. What is gained and what problems are caused by combining different communication channels? If they appear to convey consistent information, they reinforce each other, and communication becomes more accurate. But sometimes the information conveyed through different channels appears inconsistent. This produces confusion and may even arouse suspicion of deception. This section examines outcomes of apparent consistency and inconsistency among channels.

**Reinforcement and Increased Accuracy** The multiple cues we receive often seem redundant, each carrying the same message. A smile accompanies a compliment delivered in a warm tone of voice; a scowl accompanies a vehemently shouted threat. But multiple cues are seldom entirely redundant. They are better viewed as complementary (Poyatos, 1983). The smile and warm tone convey that the compliment is sincere; the scowl and vehement shout imply that the threat will be carried out. Thus, multiple cues convey added information, reduce ambiguity, and increase the accuracy of communication.

Taken alone, each channel lacks the capacity to carry the entire weight of the messages exchanged in the course of a conversation. Paralinguistic and kinesic cues, in particular, suffuse words with life by supporting and emphasizing them. The word "maybe," for example, is a weak way to express doubt unless it is delivered in a slow, deliberate manner and accompanied by raised eyebrows or a shrug. The words "come here" take on meaning either as a coy invitation, frantic demand, or matter-of-fact request depending on the tone of voice and beckoning gesture that accompany them.

By themselves, the verbal aspects of language are insufficient for accurate communication. The importance of paralinguistic cues is illustrated in a study of students attending a Nigerian secondary school and teachers college (Grayshon, 1980). Although these students took courses in English and knew the verbal language well, they did not know the paralinguistic cues of British native speakers. The students listened to two British recordings with identical verbal content. In one recording, paralinguistic cues indicated that the speaker was giving the listener the brush-off. In the other recording, paralinguistic cues indicated that the speaker was apologizing. Of 251 students, 97 percent failed to perceive any difference in the meanings the speaker was conveying. Failure to distinguish a brush-off from an apology could be disastrous in everyday communication. Accurate understanding requires paralinguistic as well as verbal knowledge.

Our accuracy in interpreting events is greatly enhanced if we have multiple communication cues, rather than verbal information alone. The value of a full set of cues was demonstrated in a study of students' interpretations of various scenes (Archer & Akert, 1977). Students observed scenes of social interaction that were either displayed in a video broadcast or described verbally in a transcript of the video broadcast. Thus, students received either full, multichannel communication or verbal cues alone. Afterward, students were asked to answer questions about what was going on in each scene, questions that required going beyond the obvious facts. Observers who received the full set of verbal and nonverbal cues were substantially more accurate in interpreting social interactions. For instance, of those provided multichannel cues, 56 percent correctly identified which of three women engaged in a conversation had no children, compared with only 17 percent of those limited to verbal cues. These and other findings convincingly demonstrate the gain in accuracy from multichannel communication.

**Resolving Inconsistency** At times the messages conveyed by different communication channels may appear inconsistent. This makes interaction problematic. What would you do, for example, if your instructor welcomed you during office hours with warm words, a frowning face, and an annoyed tone of voice? You might well react with uncertainty and caution, puzzled by the apparent inconsistency among the verbal and nonverbal cues you were receiving. You would certainly try to figure out the instructor's "true" feelings and desires, and you might also try to guess why the instructor was sending such confusing cues.

The strategies people use to resolve apparently inconsistent cues depend upon their inferences about the reasons for the apparent inconsistency (Zuckerman, DePaulo, & Rosenthal,

1981). It could be due to the communicator's ambivalent feelings, to poor communication skills, or to an intention to deceive. A large body of research has compared the relative weight we give to messages in different channels when we do not suspect deception.

In one set of studies, people judged the emotion expressed by actors who posed contradictory verbal, paralinguistic, and facial signals (Mehrabian, 1972). These studies showed that facial cues were most important in determining which feelings were interpreted as true. Paralinguistic cues were second, and verbal cues were much less important. Later research, exposing receivers to more complete combinations of visual and auditory cues, replicated the finding that people rely more on facial than on paralinguistic cues when the two conflict. This preference for facial cues increases with age from childhood to adulthood, indicating that it is a learned strategy (DePaulo et al., 1978).

People also use social context to help them to judge which of the channels is more credible (Bugenthal, 1974). They consider whether the facial expression, tone of voice, or verbal content are appropriate to the particular social situation. If people recognize a situation as highly stressful, for example, they rely more on the cues that seem consistent with a stressful context (a strained tone of voice) and less on cues that seem to contradict it (a happy face or verbal assertion of calmness). In short, people tend to resolve apparent inconsistencies between channels by favoring the channels whose message seems most appropriate to the social context.

# ■ Social Structure and Communication

So far, this chapter has examined the nature of verbal and nonverbal communication and some consequences of the fact that everyday communication usually combines the two. But how do social relationships shape communication? And how does communication express, maintain, or modify social relationships? These questions pinpoint social psychology's concern with the reciprocal impacts of social structure and communication on each other. This section examines three aspects of these impacts. First, it discusses the links between styles of speech and position in the stratification system. Second, it analyzes ways that communication expresses and modifies the two central dimensions of relationships: status and intimacy. Third, it examines social norms that regulate interaction distances and some of the outcomes when these norms are violated.

## Social Stratification and Speech Style

The way we speak reflects and regulates our social relationships. A major proponent of this view, Basil Bernstein, studied the connections between speech styles and modes of interaction in various social classes, families, and schools in England (1974, 1975). He identified two major speech styles or sociolinguistic codes: a restricted code and an elaborated code. An example of speech in each code will help us to define these two speech styles.

Suppose a child demands to stay up late to watch television. Using the restricted code, a parent might say, "No! I say you can't." Using the elaborated code, a parent might reply, "That's probably not a good idea. If you don't sleep now you'll be exhausted tomorrow and you probably won't enjoy the trip we planned."

The **restricted code,** as these examples demonstrate, is a concrete and egocentric speech style. It is direct, rooted in the here and now, lacking in qualifications, and emotionally expressive. In contrast, the **elaborated code** is a relatively abstract speech style attuned to the characteristics of the particular listener. It allows for subtle differences in meaning by employing qualifications and extended perspectives on time and events. It encourages listeners to reason for themselves. The restricted language code is used in settings that emphasize rigid role definitions and respect for formal social positions. In these settings, people exercise social control through issuing direct commands, demanding obedience, and applying the same rules to everyone. The restricted code does an efficient job of communicating when the aim is to express straightforward expectations authoritatively or with strong feeling. The elaborated code is used in settings that emphasize unique individual characteristics and needs. In these settings, social control is person oriented. People tailor their demands to one another's capabilities and provide reasons and explanations. Such social relations are found in families or classrooms with flexible or loose role expectations and definitions.

According to this analysis, the mode of social control that prevails in relationships in a cultural group determines the type of language code the group members are likely to acquire. In working-class families, schools, and work settings in Great Britain, for instance, social relationships tend toward rigid, positional control; and the working class socialize toward restricted codes. Middle-class relationships tend toward more flexible, personal control and socialize toward elaborated codes (Bernstein, 1974, 1975). Class differences in speech style have been found in studies of the way mothers talk to their children and answer their children's questions and in studies of children's speech (Robinson & Rackstraw, 1972; Turner, 1974).

In interviews about a devastating tornado, American adults also revealed class-connected preferences for restricted versus elaborated codes. Lower-class persons tended to describe the disaster that hit their community entirely through their own eyes. They rarely qualified their statements or provided an abstract overview. Upper-middle-class persons tended to describe the disaster from others' viewpoints as well as their own. They set their observations in context for the listener and explained and qualified their reactions (Schatzman & Strauss, 1955).

These ideas and findings have given rise to sharp, sometimes heated controversy (Edwards, 1976). Advocates of so-called *deficit theories* claim that people who use restricted codes are less capable of logical reasoning and abstract thought. They also claim that restricted speech styles are typical of lower-class, black, and other culturally disadvantaged groups in America as well as in Great Britain. Combining these two claims, deficit theorists argue that the children from disadvantaged groups perform poorly in school because their restricted language makes them cognitively inferior.

The strongest criticism of deficit theories has come from Labov (1972b). Based on interviews in natural environments, he demonstrates that "black English," which has been described as restricted and impoverished, is every bit as rich and subtle as standard English. It differs mainly in surface details like pronunciation ("ax" = ask) and grammatical forms ("He be busy" = He's always busy). Nonstandard speech may appear impoverished because nonstandard speakers feel less relaxed in the social contexts where they are typically observed (schools, interviews). Social researchers or other "outsiders" who observe them may also inhibit their language (Grimshaw, 1973). When interviewed by a member of their own race, for instance, black job applicants used longer sentences and richer vocabularies and employed words more creatively (Ledvinka, 1971).

Thus, speech differences between groups have not been shown to reflect differences in cognitive ability (Thorlundsson, 1987). Nonetheless, members of socially less valued groups do experience communication problems in some social situations. Recognizing this, they often "correct" their speech to conform with higher prestige usages in these situations (Labov, 1972a). The struggle over what language use is "correct" or "preferred" is part of the larger intergroup struggle for power in society. In Quebec, for example, the previously disadvantaged French-speaking community passed laws making French the only official language for use in commerce and education. Through legislating language use they have asserted their power over the formerly dominant English speakers.

## Communicating Status and Intimacy

The two central dimensions of social relationships are status and intimacy. Status is concerned with the exercise of power and control. Intimacy is concerned with the expression of affiliation and affection that creates social solidarity (Kemper, 1973). Verbal and nonverbal communication express and maintain particular levels of intimacy and relative status in relationships. Moreover, through communication we may challenge existing levels of intimacy and relative status and negotiate new ones (Scotton, 1983).

Communication can signal our view of a relationship only if we recognize which communication behaviors are appropriate for an expected level of intimacy or status and which are inappropriate. The following examples suggest that we easily recognize when communication behaviors are inappropriate. What if you repeatedly addressed your mother as Mrs. _____?; used vulgar slang during a job interview?; draped your arm on your professor's shoulder as she explained how to improve your test answers?; or looked away each time your beloved gazed into your eyes? Each of these communication behaviors would probably make you uncomfortable. They would doubtlessly also cause others to think you were inept, disturbed, or hostile. Each behavior expresses levels of intimacy or relative status easily recognized as inappropriate to the relationship. Let us survey systematically how specific communication behaviors express, maintain, and change status and intimacy in relationships.

**Status**  Forms of address clearly communicate relative status in relationships. Inferiors use formal address (title and last name) for their superiors ("When is the exam, Professor Levine?"), whereas superiors address inferiors with familiar forms (first name or nickname: "On Friday, Daphne"). Status equals use the same form of address with one another. Both use either formal (Ms./Mr.) or familiar forms (Carol/Bill), de-

pending on the degree of intimacy between them (Brown, 1965). When status differences are ambiguous, individuals may even avoid addressing each other directly. They shy away from choosing an address form because it might grant too much or too little status.

A shift in forms of address signals a change in social relationships, or at least an attempted change. During the French Revolution, in order to promote equality and fraternity, the revolutionaries demanded that everyone use only the familiar (*tu*) and not the formal (*vous*) form of the second-person pronoun, regardless of past status differences. Presidential candidates try to reduce their differences with voters by inviting the use of familiar names ("George," "Mike"). In cases where there is a clear status difference between people, the right to initiate the use of the more familiar or equal forms of address belongs to the superior ("Why don't you drop that 'Dr.' stuff?"). This principle also applies to other communication behaviors. It is the higher-status person who usually initiates changes toward more familiar behaviors, such as greater eye contact, physical proximity, touch, or self-disclosure.

We each have a repertoire of different pronunciations and dialects and a varied vocabulary from which to choose when speaking. Our choices of language to use with other people express a view of our relative status and may influence our relationships. People usually make language choices smoothly, easily expressing status differences appropriate to the situation (Gumperz, 1976; Stiles et al., 1984). Teachers in a Norwegian town, for instance, were observed to lecture to their students in the standard language (Blom & Gumperz, 1972). When they wished to encourage student discussion, however, they switched to the local dialect, thereby reducing status differences. Note how your teachers also switch to more informal language when trying to promote student participation.

Language choice by bilinguals also expresses status in relationships. Paraguayans speak

Spanish with persons of higher status but switch to the tribal language, Guarani, among equals (Rubin, 1962). Bilingual Puerto Ricans in New York use English in the formal, status-differentiated settings of school and work. They typically switch to Spanish with their friends who are status equals (Greenfield, 1972).

Paralinguistic cues also communicate and reinforce status in relationships. People of higher status interrupt their partners more during conversations and talk more themselves. Inferiors grant status to others by not interrupting and by responding with "M'hm" more frequently at appropriate intervals to indicate they are following the conversation. Among status equals, these paralinguistic behaviors are distributed more equally (LaFrance & Mayo, 1978; Leffler, Gillespie, & Conaty, 1982). An experimental study of influence in small, three-person groups systematically varied the paralanguage of one member (Ridgeway, 1987). This member, a confederate, was most influential when she spoke rapidly, in a confident tone and gave quick responses. She was less influential when she behaved dominantly (spoke loudly, gave orders) or submissively (spoke softly, in a pleading tone). Thus, engaging in the paralinguistic behaviors appropriate to the statuses of group members, in this case equals, enhances one's influence.

Body language also serves to express status. When status is unequal, people of higher status tend to adopt relatively relaxed postures with their arms and legs in asymmetrical positions. Those of lower status stand or sit in more tense and symmetrical positions. The amount of time we spend looking at our partners and the timing also indicate status. Higher-status persons look more when speaking than when listening, whereas lower-status persons look more when listening than when speaking. Overall, inferiors look more at their partners, but they are also first to break the gaze between partners. Finally, superiors are much more likely to intrude physically on inferiors by touching or pointing at them (LaFrance & Mayo, 1978; Dovidio & Ellyson, 1982; Leffler, Gillespie, & Conaty, 1982).

**Intimacy** This second, central dimension of relationships is also expressed through communication. One way we signal intimacy or solidarity is by addressing each other by our first names. The exchange of titles and last names is common for strangers. In other languages, speakers express intimacy by their choice of familiar versus formal second-person pronouns. As noted above, the French can choose between the familiar *tu* or formal *vous;* the Spanish have the familiar *tu* or formal *usted.*

Our choice of language is another way to express intimacy. For example, Paraguayan men usually court women in Spanish; after marriage, they converse with their wives in the tribal language as a sign of increased intimacy (Rubin, 1962). Similarly, residents of a Norwegian town were found to use the formal version of their language with strangers and the local dialect with friends. They spoke the formal version when transacting official business in government offices and then switched to dialect for a personal chat with the clerk after completing their business (Blom & Gumperz, 1972).

The intimacy of a relationship is clearly reflected in and reinforced by the content of conversation. As a relationship becomes more intimate, we disclose more personal information about ourselves. Intimacy is also conveyed by conversational style. In one study (Hornstein, 1985), telephone conversations were recorded and later analyzed; the conversations were between strangers, acquaintances, and friends. Compared to strangers, friends used more implicit openings ("Hi." or "Hi. It's me."), raised more topics, and were more responsive to the other conversationalist (for example, asked more questions). Friends also used more complex closings (for example, making concrete arrangements for the next contact). Conversations of acquaintances were more like those of strangers.

The **theory of speech accommodation** (Giles, 1980; Beebe & Giles, 1984) illustrates an important way that people use verbal and paralinguistic behavior to express intimacy or liking. According to this theory, people express or reject intimacy by adjusting their speech behavior during interaction to converge with or diverge from their partner's. To express liking or evoke approval, they make their own speech behavior more similar to their partner's. To reject intimacy or communicate disapproval, they accentuate the differences between their own speech and their partner's.

Adjustments of paralinguistic behavior demonstrate speech accommodation during conversations (Taylor & Royer, 1980; Thakerar, Giles, & Cheshire, 1982). Individuals who wish to express liking tend to shift their own pronunciation, speech rate, vocal intensity, pause lengths, and utterance lengths during conversation to match those of their partner. Individuals who wish to communicate disapproval modify these vocal behaviors in ways that make them diverge more from their partner's. Among bilinguals, speech accommodation may also determine the choice of language (Bourhis et al., 1979). To increase intimacy, bilinguals choose the language they believe their partner would prefer to speak. To reject intimacy, they choose their partner's less preferred language.

The use of slang gives strong expression to in-group intimacy and solidarity. Through slang, group members assert their own shared social identity and express their alienation from and rejection of the out-group of slang illiterates. At the same time, society tends to adopt slang in the standard language. This blunts its usefulness for affirming in-group solidarity and makes necessary the constant manufacture of new slang. Middle-class white youths, for example, adopt black slang as a sign of their own rebellion and out of a desire to identify with the rebellion of alienated blacks. As a result, black slang must continuously undergo renewal to express black solidarity (Drake, 1980).

The ways we express intimacy through body language and interpersonal spacing are well recognized. For instance, research supports the folklore that lovers gaze more into each other's eyes (Rubin, 1970). In fact, we tend to interpret a high level of eye contact from others as a sign of

intimacy. We communicate liking by assuming moderately relaxed postures, moving closer and leaning toward others, orienting ourselves face-to-face, and touching others (Mehrabian, 1972). There is an important qualification to these generalizations, however. Mutual gaze, close distance, and touch reflect intimacy and promote it only when the interaction has a positive cast. If the interaction is generally negative—if the setting is competitive, the verbal content unpleasant, or the past relationship antagonistic—these same nonverbal behaviors intensify negative feelings (Schiffenbauer & Schiavo, 1976).

### Normative Distances for Interaction

American and Northern European tourists in Cairo are often surprised to see men touching and staring intently into each other's eyes as they converse in public. Surprise may turn to discomfort if the tourist engages an Arab male in conversation. Bathed in the warmth of his breath, the tourist may feel sexually threatened. In our own communities, in contrast, we are rarely made uncomfortable by the overly close approach of another. People apparently know the norms for interaction distances in their own cultures, and they conform to them. What are these norms, and what happens when they are violated?

**Normative Distances**   Edward Hall (1966) has described four spatial zones that are normatively prescribed for interaction among middle-class Americans. Each zone is considered appropriate for particular types of activities and relationships. *Public distance* (12–25 feet) is prescribed for interaction in formal encounters, lectures, trials, and other public events. At this distance, communication is often one-way, sensory stimulation is very weak, people speak loudly, and they choose language carefully. *Social distance* (4–12 feet) is prescribed for most casual social and business transactions. Here, people speak at a normal volume, they do not touch, frequent eye contact is needed to maintain smooth communication, and sensory stimulation is low. *Personal distance* (1½–4 feet) is prescribed for interaction among friends and relatives. Here, people speak softly, touch, and receive substantial sensory stimulation by sight, sound, and smell. *Intimate distance* (0–18 inches) is prescribed for giving comfort, making love, and aggressing physically. This distance provides intense stimulation from touch, smell, breath, and body heat. It signals unmistakable involvement.

Many studies support the idea that people know and conform to the normatively prescribed distances for particular kinds of encounters (LaFrance & Mayo, 1978). When we compare different cultural and social groups, both similarities and differences in distance norms emerge. All cultures prescribe closer distances for friends than for strangers, for example. The specific distances for preferred interactions vary widely, however. Latin Americans, Arabs, Greeks, and French use shorter interaction distances than Americans, British, Swiss, and Swedes (Sommer, 1969). Females tend to interact with one another at closer distances than males do in various cultures (Sussman & Rosenfeld, 1982). Social class may also influence interpersonal spacing. In Canadian school yards, the lower-class primary school children were observed to interact at closer distances than the middle-class children, regardless of race (Scherer, 1974).

Differences in distance norms may cause discomfort in cross-cultural interaction. People from different countries or social classes may have difficulty in interpreting the amount of intimacy implied by each other's interpersonal spacing and in finding mutually comfortable interaction distances. Cross-cultural training in nonverbal communication can reduce such discomfort. For instance, Englishmen were liked more by Arabs with whom they interacted when the Englishmen had been trained to behave nonverbally like Arabs—to stand closer, smile more, look more, and touch more (Collett, 1971).

Two aspects of interpersonal spacing that clearly influence and reflect status are physical distance and the amount of space each person occupies. Equal-status individuals jointly determine comfortable interaction distances and tend to occupy approximately equal amounts of space with their bodies and with the possessions that surround them. When status is unequal, superiors tend to control interaction distances, keeping greater physical distance than equals would choose. Superiors also claim more direct space with their bodies and possessions than inferiors (Hayduk, 1978; Gifford, 1982; Leffler, Gillespie, & Conaty, 1982).

**Violations of Personal Space**  What happens when people violate distance norms by coming too close? In particular, what do we do when strangers intrude on our personal space?

The earliest systematic examination of this question included two parallel studies (Felipe & Sommer, 1966). In one, strangers approached lone male patients in mental hospitals to a point only 6 inches away. In the other, strangers sat down 12 inches away from lone female students in a university library. The mental patients and the students who were approached left the scene much more quickly than the patients and students who were not approached. After only two minutes, 30 percent of the patients who were intruded on had fled, compared with none of the others. Among the students, 70 percent of those whose space was violated had fled by the end of 30 minutes, compared with only 13 percent of the others. Results of this study are shown in Figure 7-1.

Fleeing is not the only response to space violation. In addition, people protect their privacy by turning their backs on intruders, leaning away, and placing barriers such as books, purses, or elbows in the intervening space. Only very rarely do people react verbally to space violations (Patterson, Mullens, & Romano, 1971). The reaction to violations depends in part on the setting in which they occur. While violations of space norms at library tables lead to flight, violations in library aisles lead to the person spending more time in the aisle (Ruback, 1987). It is possible that when you are looking for a book, a violation of distance norms is distracting, so it takes you longer to find the book.

Despite the crowded circumstances, the eight people on this bench have maintained some privacy. Strangers feel uncomfortable when they must intrude on each others' personal space. To overcome this discomfort, they studiously ignore each other, avoiding touch, eye contact, and verbal exchanges.

Violating personal space is uncomfortable for intruders, too. Individuals required to pass through the personal space of others who are engaged in conversation feel more awkward and display more unpleasant facial expressions than individuals who merely pass nearby (Efran & Cheyne, 1974). Due to their discomfort, people will avoid intruding. They will drink from a water fountain much less frequently, for instance, if someone is standing within a foot of it (Baum, Riess, & O'Hara, 1974).

Staring is a powerful way to violate another's privacy without direct physical intrusion. Staring by strangers elicits avoidance responses, indicating that it is experienced as an intense negative stimulus. When stared at by strangers, for instance, pedestrians cross the street faster, and drivers speed away from intersections more quickly (Ellsworth, Carlsmith, & Henson, 1972; Greenbaum & Rosenfeld, 1978).

## ■ Conversational Analysis

Conversation is a regular daily activity. Yet we all have trouble communicating at times. The list of what can go wrong is long and painful: inability to get started, irritating interruptions, awkward silences, failure to give others a chance to talk, failure to notice that listeners are bored or have lost interest, changing topics inappropriately, assuming incorrectly that others understand, and so on. This section examines the ways people avoid these embarrassing and annoying blunders. To maintain smooth-flowing conversation requires knowledge of certain rules and communication skills that are often taken for granted.

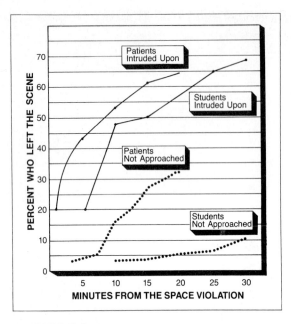

■ **FIGURE 7-1**

REACTIONS TO VIOLATIONS OF PERSONAL SPACE

How do people react when strangers violate norms of interpersonal distance and intrude on their personal space? A common reaction is illustrated here. Strangers sat down 12 inches away from lone female students in a library or approached lone male patients in a mental hospital to within 6 inches. Those who were approached left the scene much more quickly than control subjects who were not approached. Violations of personal space often produce flight.

*Source: Adapted from Felipe and Sommer, 1966.*

We will discuss some of the rules and skills that are crucial for initiating conversations, regulating turn taking, and coordinating conversation through verbal and nonverbal feedback.

## Initiating Conversations

Conversations must be initiated with an attention-getting device, a summons to interaction. Greetings, questions, or the ringing of a telephone can serve as the summons. But conversations do not get underway until potential partners signal that they are attending and willing to converse. Eye contact is the crucial nonverbal signal of availability for face-to-face inter-

action. Goffman (1963) suggests that eye contact places a person under an obligation to interact: when a waitress permits eye contact, she places herself under the power of the eye-catcher.

The most common verbal lead into a conversation is the **summons-answer sequence** (Schegloff, 1968). Response to a summons ("Jack, you home?" . . . "Yeah.") indicates availability. More importantly, this response initiates the mutual obligation to speak and to listen that produces conversational turn taking. The summoner is expected to provide the first topic—a conversational rule little children exasperatingly overlook. Our reactions when people violate the summons-answer sequence demonstrate its widespread acceptance as an obligatory rule. When people ignore a summons, we conclude either that they are intentionally insulting us, socially incompetent, or psychologically absent (sleeping, drunk, crazy).

Telephone conversations exhibit a common sequential organization. Consider the following conversation between a caller (C) and a recipient (R).

0.      (ring)

1.   R:   Hello?

2.   C:   This is John.

3.   R:   Hi.

4.   C:   How are you?

5.   R:   Fine. How are you?

6.   C:   Good. Listen, I'm calling about . . .

The conversation begins with a summons-answer sequence (lines 0 and 1). This is followed by an identification-recognition sequence (lines 2 and 3); in this example, the recipient knows that John recognizes his voice, so he does not state his name. Next, there is a trading of "how are you" sequences (lines 4–6). Finally, at line 6, John states the reason for the call. This organization is found in many types of telephone calls. However, in emergency phone calls, when seconds count, the organization is quite different (Whalen & Zimmerman, 1987). Consider the following example.

0.  (ring)
1.  R:  Mid-City Emergency.
2.  C:  Um yeah. Somebody jus' vandalized my car.
3.  R:  What's your address?

Notice that the opening sequence is shortened; both the greeting and the "how are you" sequences are omitted. In emergency calls, the reason for the call is stated sooner. Note also that the recognition element of the identification-recognition sequence is moved forward, to line 1. Both of these changes facilitate communication in an anonymous, urgent situation. However, if the dispatcher answers a call and the caller says, "This is John," that signals an ordinary call. Thus, the organization of conversation clearly reflects situational contingencies.

Examination of how new topics are initiated reveals yet another rule. Strangers are expected to introduce new topics into conversation as questions that invite a response ("You going to the game tonight?"). People who know each other may simply announce a topic ("I'm going to the game tonight!"), claiming the conversational floor and the right of a response (Maynard, 1978). Using a socially inappropriate method is likely to fail. A stranger who announces "I'm going to Chicago for the weekend" is likely to evoke embarrassed silence from others. The same remark by an old friend, in contrast, would produce a flood of suggestions of things to do, places to go, and tales of past visits.

## Regulating Turn Taking

A pervasive rule of conversation is to avoid bumping into someone verbally. To regulate turn taking, people use many verbal and nonverbal cues, singly and together, with varying degrees of success (Kendon, Harris, & Key, 1975; Duncan & Fiske, 1977; Sacks, Schegloff, & Jefferson, 1978).

**Signaling Turns**  Speakers indicate their willingness to yield the floor by looking directly at a listener with a sustained gaze toward the end of an utterance. People also signal readiness to give over the speaking role by pausing, stretching the final syllable of their speech in a drawl, terminating hand gestures, dropping voice volume, and tacking relatively meaningless expressions ("you know") onto the end of their utterances. Listeners indicate their desire to talk by inhaling audibly as if preparing to speak. They also tense and move their hands, shift their head away from the speaker, and emit especially loud vocal signs of interest ("Yeah," "M'hm").

Speakers retain their turn by avoiding eye contact with listeners, tensing their hands and gesticulating, and increasing voice volume to overpower others when simultaneous speech occurs. People who persist in these behaviors are soon viewed by others as egocentric and domineering. They have violated an implicit social rule: "It's all right to hold a conversation, but you should let go of it now and then."

Verbal content and grammatical form of speech also provide important cues for turn taking. People usually exchange turns at the end of meaningful speech units after an idea has been completed. First priority for the next turn goes to any person explicitly addressed by the current speaker with a question, complaint, or other invitation to talk. People expect turn changes to occur after almost every question but not necessarily after other pauses in conversation (Hanni, 1980). It is difficult to exchange turns without using questions. When speakers in one study were permitted to use all methods except questions for signaling their desire to gain or relinquish the floor, the length of each speaking turn virtually doubled (Kent, Davis, & Shapiro, 1978).

**Turn Allocation**  Much of our conversation takes place in settings where turn taking is more organized than in spontaneous conversations. In class discussions, meetings, interviews, and therapy sessions, for example, responsibility for allocating turns tends to be controlled by one person, and turns are often allocated in advance. Prior allocation of turns reduces strains that arise from people either competing for speaking time or avoiding their responsibilities to speak. Allocation of turns also increases the efficiency of talk. It can arrange a distribution of turns that best fits the task or situation—a precisely equal distribution (as in a debate) or the assignment of just one speaker (as in a football huddle).

Prior allocation of turns influences both how long people talk and the methods they use to extend their turns (Sacks, Schegloff, & Jefferson, 1978). With prior allocation, people tend to speak longer during each turn, which they lengthen by stringing many sentences together. In spontaneous conversation, turn lengths are ordinarily shorter. When people wish to extend their turns, they increase the complexity of what they say within single sentences. These observations on turn allocating illustrate some of the subtle, taken-for-granted rules that govern everyday conversation.

## Feedback and Coordination

We engage in conversation to attain interpersonal goals—to inform, persuade, impress, control, and so on. To do this effectively, we must assess how what we say is affecting our partner's interest and understanding as we go along. Both verbal and nonverbal feedback help conversationalists make this assessment. Through feedback, conversationalists coordinate what they are saying to each other from moment to moment. Responses called **back channel feedback** are especially important for regulating speech as it is happening. These are the small vocal and visual comments a listener makes while a speaker is talking, without taking over the speaking turn. They include such responses as "Yeah" and "M'hm," short clarifying questions ("What?" "Huh?"), brief repetitions of the speaker's words or completions of his or her utterances, head nods, and brief smiles. When conversations are proceeding smoothly, the fine rhythmic body movements of listeners (swaying, rocking, blinking) are precisely synchronized with the speech sounds of speakers who address them (Condon & Ogston, 1967). These automatic listener move-

ments are another source of feedback that indicates to speakers whether they are being properly "tracked" and understood (Kendon, 1970).

Both the presence or absence and the timing of back channel feedback influence speakers. In smooth conversation, listeners time their signs of interest, agreement, or understanding to occur at the end of long utterances or when the speaker turns his head toward them. The absence or mistiming of such listener feedback undermines coordination. It makes speakers uneasy or upset and causes them to hesitate or stop talking (Rosenfeld, 1978). Alerted to the possible loss of listener attention and involvement by the absence of feedback, speakers employ attention-getting devices to evoke feedback. One such attention-getting device is the phrase "You know." Speakers frequently insert "You know" into long speaking turns immediately prior to or following pauses if their partner seems to be ignoring their invitation to provide feedback or to accept a speaking turn (Fishman, 1980).

Another device a speaker can use to regain the attention of another participant is to ask her a question. Such displays of uncertainty—for example, "What was the name of that guy on the Carson show?"—restructure the interaction by getting listeners more involved (Goodwin, 1987). If the speaker shifts his gaze to a specific person as he asks the question, it will draw that person into the conversation.

When speakers are denied feedback, the quality of their speech deteriorates. They become less coherent and communicate less accurately. Their speech becomes more wordy, less efficient, and more poorly fitted to the specific information needs of their partner (Kraut, Lewis, & Swezey, 1982). Lack of feedback causes such deterioration because it prevents speakers from learning several things about their partners. They cannot discern whether their partners (1) have relevant prior knowledge they need not repeat, (2) understand already so they can wrap up the point or abbreviate, (3) have misinformation they should correct, (4) feel confused so they should backtrack and clarify or (5) feel bored so they should stop talking or change topics.

The fact that feedback influences the quality of speech has another interesting consequence. Listeners who frequently provide their conversational partners with feedback also understand their partner's communication more fully and accurately. Through their feedback, active listeners help shape the conversation to fit their own information needs. This finding reinforces a central theme of this chapter: Communication is a shared social accomplishment.

Feedback is important not only in conversations but also in formal lectures. Lecturers usually monitor members of the audience for feedback. If listeners are looking at the speaker attentively and nodding their heads in agreement, the lecturer infers that her message is understood. On the other hand, quizzical or "out of focus" expressions suggest failure to understand. Similarly, members of the audience use feedback from the lecturer to regulate their own behavior; a penetrating look from the speaker may be sufficient to end a whispered conversation between listeners.

An important form of feedback in many lectures is applause. Speakers may want applause for a variety of reasons, not just ego gratification. Sometimes lecturers subtly signal the audience when to applaud; audiences watch for such signals in order to maintain their involvement. As an illustration, analysis of 42 hours of recorded political speeches suggests that there is a narrow range of message content that stimulates applause (Heritage & Greatbatch, 1986). Attacks on political opponents, foreign persons and collectivities, statements of support for one's own positions, record, or party, and commendations of individuals or groups generate applause. When these messages are framed within particular rhetorical devices, applause is from two to eight times more likely. For example:

SPEAKER: Governments will argue (pause) that resources are not available (short pause) to help disabled people. (long pause) The fact is that too much is spent on the munitions of war, (long pause) and too little is spent (applause begins) on the munitions of peace.

In this example, the speaker uses the rhetorical device of contrast or antithesis. Using this device, the speaker's point is made twice. Audiences can anticipate the completion point of the statement by mentally matching the second half with the first. This rhetorical device is an "invitation to applaud," and in the example the audience begins to applaud even before the speaker completes the second half.

## ■ Summary

Communication is the process whereby people transmit information about their ideas and feelings to one another.

**Language and Verbal Communication** Language is the main vehicle of human communication. (1) All spoken languages consist of sounds that are combined into words with arbitrary meanings and put together according to grammatical rules. Language enhances our capacity for social action. It frees us from the here and now and allows communication about nonshared experiences. (2) The meanings of words consist of both the properties most people associate with them and the personal responses they evoke in each individual. Most words have multiple meanings. (3) The successful use of language during interaction is a social accomplishment. To attain mutual understanding, conversationalists must use socially appropriate language, express their intentions in ways listeners can recognize, take account of others' current knowledge, and actively work to decipher meanings. (4) All languages share universal features based on human capacities for thought and on common social experience. Each language also has distinctive features that make it easier or more difficult to think about and remember particular ideas.

**Nonverbal Communication** A great deal of information is communicated nonverbally during interaction. (1) Four major types of nonverbal communication are paralanguage, body language, interpersonal spacing, and choice of personal effects. (2) People from all cultures express and recognize the same six primary emotions in distinctive facial expressions. During interaction, however, people modify facial expressions of emotion according to cultural display rules. (3) Information is usually conveyed simultaneously through nonverbal and verbal channels. Multiple cues may add information to each other, reduce ambiguity, and increase accuracy. But if cues appear inconsistent, people must determine which cues reveal the speaker's true intentions.

**Social Structure and Communication** The ways we communicate with others reflect and influence our relationships with them. (1) The more rigid control patterns observed among the lower class foster concrete, direct, egocentric, and emotionally expressive styles of speech. The more personalized control observed among the middle class fosters speech styles that are more abstract, qualified, and attuned to the listeners' unique qualities. (2) We express, maintain, or challenge the levels of relative status and intimacy in our relationships through our verbal and nonverbal behavior. Status and intimacy influence and are influenced by forms of address, choice of dialect or language, interruptions, matching of speech styles, gestures, eye contact, posture, and interaction distances. (3) The appropriate interaction distances for particular types of activities and relationships are normatively prescribed. Cultural groups differ in their distance norms; this makes it difficult in cross-cultural interaction to find mutually comfortable interaction distances. When strangers violate distance norms, people flee the scene or use other devices to protect their privacy.

**Conversational Analysis** Smooth conversation depends on conversational rules and communication skills that are often taken for granted. (1) Conversations are initiated by a summons to interaction. They get underway only if potential partners signal availability, usually through eye contact or verbal response. (2) Conversationalists avoid verbal collisions by taking turns. They signal either a willingness to yield the floor or a desire to talk through verbal and nonverbal cues. In some situations turns are allocated in advance. (3) Effective conversationalists assess their partner's understanding and interest as they go along through vocal and visual feedback. If feedback is absent or poorly timed, the quality of communication deteriorates. An effective speech also involves coordination between speaker and audience; the timing of applause is a joint accomplishment.

## Key Terms

8

chapter

# Social Influence and Persuasion

## ■ Introduction

Carol faces a problem. One year ago, shortly after graduating from college, she began work as a technical assistant in a small company. She is reasonably happy with her job—it challenges her abilities and she likes the people—but she is troubled by her low salary. The job did not pay very well at the start, and she still has received no raise. Carol realizes that under company policy she may not get a pay raise for another 12 to 18 months. Carol's boss, Martha, has the authority to grant her a higher salary immediately. Carol broached the issue in a tentative way about a week ago, but Martha did not respond favorably. Carol wonders what to do now.

Thinking the problem over, Carol realizes there are many things she could do. She could go out of her way to be nice to Martha, hoping that Martha might like her better and eventually offer her a raise. She could try to persuade Martha that she is doing more for the company than Martha realized and is therefore underpaid at her present salary. She could threaten to leave the company and find work elsewhere. Or she could promise to assume additional job responsibilities in the hope that this would justify a salary increase.

Which of these techniques, if any, should Carol use? She wants a raise, and these influence techniques may help her accomplish that goal. Which will be effective and which will not?

### Techniques of Social Influence

To obtain a raise, Carol will have to exert social influence. By definition, **social influence** occurs when one person's behavior causes another person to change an opinion or to perform an action. That is, in social influence, one person (the **source**) engages in some behavior, such as persuading, threatening, or promising, that causes another person (the **target**) to behave differently from what he or she would otherwise do. In our example, the source (Carol) is contemplating various techniques she might use to influence the target (Martha) to give her a pay raise. Influence, then, is a causal relationship between the source's behavior and the target's response (Tedeschi, Bonoma, & Schlenker, 1972).

There are many forms of social influence. Influence attempts may be either open or manipulative (Tedeschi, Schlenker, & Lindskold, 1972). Open influence attempts are based on techniques that are apparent to the target—that is, when the target understands that someone is attempting to change his or her attitudes or behavior. Manipulative influence attempts are hidden from the target. Ingratiation and tactical self-presentation are examples of manipulative influence. In this chapter, we will focus on open influence; we will take up manipulative influence in Chapter 9.

There are three basic techniques of open influence—persuasion, threats, and promises. In persuasion, the source uses information and argument to change the target's beliefs and attitudes about a situation. For example, a financial executive might try to persuade an oil-drilling contractor by saying, "Why don't you change plans and try to finish drilling the Ewing-3 well as soon as possible? Otherwise, you may have to wait a long time before getting approval to connect it to a pipeline, and that will cause you further headaches." The executive is trying to change the contractor's outlook by presenting special facts that show the situation in a new light.

When using promises and threats, the source does not rely primarily on information and argument. Instead, he manipulates the rewards and punishments received by the target. For example, the financial executive might promise the oil-drilling contractor: "If you complete the Ewing-3 well 30 days ahead of schedule, I'll give you a 5-percent bonus." Alternatively, the executive might threaten: "If you don't complete the well on schedule, I'll deduct 1 percent of your payment for every day that you are late." In making these statements, the source (executive) introduces rewards (5-percent bonus) or punishments (1-percent daily penalty) into the situa-

tion. He does not try to convince the target to view the situation differently. Instead, he changes the situation itself by adding rewards or imposing punishments contingent on the target's performance.

Influence attempts vary in their degree of success—some are successful while others are not. In fact, our everyday attempts at persuasion probably fail more often than they succeed. Even threats and promises do not always achieve their intended effect. Because we all engage in, and are exposed to, influence attempts, we need to understand the conditions under which they prove effective. We will devote the remainder of this chapter to a discussion of open influence and the effectiveness of various influence attempts. The following issues are considered:

1. What factors determine whether a persuasive communication will be effective in changing a target's beliefs and attitudes? In what ways, for instance, do characteristics of the source and of the target, as well as properties of the message itself, determine whether the persuasion attempt will be effective?

2. Under what conditions do threats and promises prove successful in gaining compliance from the target?

3. When two persons have the capacity to exchange rewards, what processes are involved in bargaining and negotiating between them? Likewise, when two persons have the capacity to threaten and punish one another, what processes are involved in the negotiations? What tactics can be used in bargaining to secure superior outcomes?

## ■ Communication and Persuasion

### Effectiveness of Persuasion

Day in and day out we are bombarded with persuasive messages. As an example, consider what happens to Steve Maxwell on a typical day.

First thing in the morning, Steve's clock radio comes on. Before he can even get out of bed, the cheerful voice of the announcer is trying to sell him a new mouthwash. On the way into work, one of the members of his car pool attempts to persuade him to vote for a particular candidate in the upcoming city election. At lunch, a friend describes plans to attend a concert the following weekend and urges him to come along. Later in the day, he listens to an argument from a co-worker who wants to change some of the paper-work procedures in the office. At 5:30 that afternoon, he stops in a store to look for a new pair of shoes. The clerk recommends one type over another and encourages him to try them on. When he finally arrives home, Steve opens his mail. One letter is a carefully worded appeal from a charitable organization asking him to volunteer his time. Other letters are junk mail flyers asking for money. Later that evening, when Steve is watching television, advertisers bombard him with ads for their products—laundry soap, beer, shampoo, and new cars.

All of these communications have something in common. In each and every case, the message seeks to persuade. **Persuasion** may be defined as an effort (by a source) to change the beliefs or attitudes of a target person through the use of information or argument. Persuasion is widespread in social interaction and assumes many different forms (McGuire, 1985).

While some attempts at persuasion succeed, many do not. When a target is exposed to a persuasive message, he or she can respond in a variety of ways, some of which involve no change in attitudes or beliefs. For instance, one possible response by the target is simply to ignore the message. Rather than heed the automobile advertisement or the beer commercial, the television viewer can switch channels. Another possible response is to listen to the message but to suspend judgment on the issue; in this case, the target might decide to seek additional information before concluding that one viewpoint or another is correct. A third possible response to a

persuasive communication is to derogate the communicator. Instead of changing attitudes, the target may dismiss the communicator as poorly informed, illogical, or even stupid. A fourth response is to engage in counterpersuasion—that is, the target may react by attempting to change the source's attitude rather than his or her own.

Because a target can respond to a persuasive communication in so many ways, the fundamental question becomes, Under what conditions will a persuasive message succeed in changing the target's beliefs or attitudes? Or, stated another way, What factors lead to success in persuasion, and what factors lead to failure?

**Central and Peripheral Routes**   According to a model proposed by Petty and Cacioppo (1981, 1986a, 1986b), there are two basic routes through which a message may alter a target's existing attitudes—a *central route* and a *peripheral route*. The central route emphasizes the way the target processes the specific information and arguments in a message, whereas the peripheral route emphasizes the impact on persuasion of peripheral cues that are independent of the message's content.

Persuasion via the central route occurs when a target scrutinizes the arguments contained in a persuasive message, interprets and evaluates them, and then integrates them into a reasoned and coherent position. Information held by the target prior to the message, as well as new information in the message itself, comes into play during this integration. Petty and Cacioppo use the term *elaboration* to describe the process whereby the target thinks through the implications of the arguments contained in a message. In elaboration, what matters is whether the arguments are strong, consistent, factually correct, and so on.

In contrast, the peripheral route to persuasion emphasizes the importance for attitude change of peripheral cues that are independent of the content of the message. Such cues might

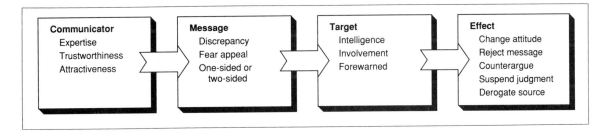

■ **FIGURE 8-1**
THE COMMUNICATION PARADIGM

include characteristics of the communicator (expertise, likability, trustworthiness) or of the target (mood, self-concept, and so on). Peripheral cues are linked to the message but are not involved directly in the elaboration of the message. Yet, they may have a significant impact on a message's effectiveness.

**Communication Paradigm** Much of modern persuasion research is organized in terms of the question "Who says what to whom with what effect?" Both central and peripheral routes to persuasion are reflected in this question. The effectiveness of a message depends both on central factors—what is said (the message)—and on peripheral factors—who sends the message (the communicative source) and who receives it (the target audience).

Figure 8-1 presents a diagram showing how these factors interrelate. First, properties of the communicative source can affect how the target person will construe the message. For instance, the expertise of the source can make a difference in persuasion. Other characteristics of the source (such as trustworthiness and likability) may also affect whether the target is persuaded. Second, properties of the message itself can have a significant impact on its persuasiveness. For instance, messages that entail a fear-arousing appeal are sometimes more effective than those that do not. Other properties of the message (such as whether it is highly discrepant from what the target believes prior to receiving the

message or whether it presents only one-sided arguments rather than multisided ones) can also affect whether a persuasive attempt is successful. Third, the characteristics of the target person also enter into the picture. For instance, what a target person already believes about an issue as well as the extent of that person's involvement and commitment on an issue can significantly affect whether a message leads to attitude change or merely to rejection and derogation of the communicator.

In the discussion that follows, we will consider in turn each of these factors—properties of the communicator, the message, and the target person—and look more closely at their impact on a persuasive message's effectiveness.

### The Communicator

Suppose we ask 25 persons to read a persuasive communication (such as a newspaper editorial) advocating a position on a nutrition-related topic. We tell this group that the message is written by a Nobel Prize-winning biologist. Suppose we ask another 25 persons to read the same message but we tell them it was written by a cook at a local fast-food establishment. Then we ask both groups to indicate their attitude toward the position advocated in the message. The question is, Which group of persons will be more persuaded by the communication?

Assuming that we assembled the groups at random, the only difference between them is that they ascribe the message to a different source.

Thus, any difference in their reaction to the message must be due to the identity of the source. Most likely, the persons who received the message from the prize-winning biologist will be more persuaded than those who received it from the cook.

Why will this occur? In most cases, the identity of the source provides the target with information above and beyond the content of the message itself. The target uses this information to assess the communicator's credibility. This, in turn, affects the target's willingness to accept the message. By definition, **communicator credibility** is the extent to which the communicator is perceived by the target audience as a believable source of information. Note that the communicator's credibility is "in the eye of the beholder." A source is credible if the target believes that he or she is credible. Thus, a communicator may be viewed as credible for some audiences but not for others.

Many factors influence a communicator's level of credibility, but several are of special importance. These include the communicator's expertise, trustworthiness, and attractiveness.

**Expertise**   In general, a message from a source having a high level of expertise on the issue at hand will bring about greater attitude change than a similar message from a source having a lower level of expertise (Maddux & Rogers, 1980; Haas, 1981; Petty, Cacioppo, & Goldman, 1981). One possible explanation for this is that recipients may be more accepting and uncritical of messages from high-credibility sources, whereas they may consider the issue more carefully when receiving a message from a low-credibility source.

This is illustrated in one study (Sternthal, Dholakia, & Leavitt, 1978) in which subjects were exposed to a message advocating the passage of a consumer protection bill by the U.S. Senate. For some subjects, the message was ascribed to a lawyer educated at Harvard with extensive experience in consumer issues (a high-

credibility source). For other subjects, the message was ascribed to a citizen interested in consumer affairs (a low-credibility source). Afterward, subjects were asked to express their reaction to the message. In general, subjects who initially opposed the position advocated in the message expressed more unquestioning agreement and less counterargument when the message came from the high-credibility source than when it came from the low-credibility source.

One important factor in communicator expertise pertains to the target's prior involvement with and knowledge about the issue at hand. When the target has little involvement or prior knowledge on some issue, messages from highly expert sources produce more attitude change than those from less expert sources. But the more involving the issue or the more knowledge the target has about the issue, the less likely that communicator expertise will make any difference in persuasion (Rhine & Severance, 1970). This occurs because when involvement and knowledge are high, the content of the message itself becomes the overriding determinant of attitude change (Petty & Cacioppo, 1979).

**Trustworthiness**   Although expertise is an important factor in credibility, it is not the only one. Under some conditions, a communicator can be highly expert but still not very credible. As an example, suppose that your car has been running poorly and you take it into a garage for a checkup. A mechanic you have never met inspects your car. He identifies several problems, one of which involves major repair work on the engine. The mechanic offers to complete this work for $380. The mechanic may have a high level of expertise, but can you accept his word that the expensive repair is necessary? How much does he stand to gain if you believe his message?

As this example shows, the target person pays attention not only to a communicator's expertise but also to his or her motives. If the message appears highly self-serving and benefi-

As the automobile owner listens to the message from the garage mechanic, she assesses not only the quality of the argument but also the credibility of the communicator. He may have expertise, but can he be trusted?

cial to the communicator, the recipient may distrust the communicator and discount the message.

The opposite effect also holds true. Communicators who argue against their own vested interests seem especially trustworthy. For example, if an employee of a local business quietly told you that you should not purchase a product made by her company but one made by a Japanese competitor instead, the message would be surprising but also persuasive. Her comment would have more impact than if she had argued for purchasing her own American-made model. A communicator who violates our initial expectations by arguing against her own vested interest will appear particularly trustworthy and, therefore, will be especially persuasive (Walster, Aronson, & Abrahams, 1966; Eagly, Wood, & Chaiken, 1978).

Another factor that affects trustworthiness (and hence credibility) is the social identity of the communicator. A communicator's identity provides many clues regarding his underlying goals and values. A communicator who is perceived as having goals similar to the audience will be more persuasive than a communicator perceived as having dissimilar goals (Berscheid, 1966; Cantor, Alfonso, & Zillmann, 1976). For example, a given policy proposal will probably be received differently by conservative Republicans depending on whether it was made by George Bush or Jesse Jackson. The political identity of the source reveals much about the source's goals and intentions, and these, in turn, affect his perceived trustworthiness.

**Attractiveness** The physical attractiveness of a communicator can affect a message's persuasiveness. Although attractiveness is not as important as expertise or trustworthiness, it nevertheless has some impact on the target. Political parties, for example, have a tendency to select candidates who are physically attractive. Whether or not their abilities and experience suit them for public office, the candidates' good looks may increase their persuasiveness and get them elected. Likewise, television and magazine advertisements for skin cream, panty hose, and so on, employ attractive models as communicators. Advertisers realize that such models are able to hold an audience's attention. This is especially important with messages that are not highly involving and that might otherwise be ignored by the audience.

Physically attractive communicators are influential not only on television but in face-to-face interactions as well. Attractive communicators are better liked than unattractive ones. Consequently, they will be more effective in changing the target's beliefs and attitudes (Horai, Naccari, & Fatoullah, 1974; Chaiken, 1979). Attractiveness is especially important when the message is not what the target audience would like to hear. If the message is unpopular, the difference in persuasiveness between an attractive and unattractive communicator can be substantial (Eagly & Chaiken, 1975).

**Effect of Multiple Sources**  Factors other than the expertise or the attractiveness of a communicator can affect whether a message is persuasive. One of these is the sheer number of communicative sources that send the persuasive arguments to the target. Social impact theory (Latane, 1981) predicts that a message will be more persuasive if it is received from multiple sources rather than from a single source. Consistent with this view, several recent experiments have shown a clear tendency for a message presented by three different communicators to be more persuasive than the same information presented by a single communicator (Harkins & Petty, 1981b, 1983, 1987).

This effect is particularly marked when the arguments presented in the message are strong as opposed to weak. Strong messages coming from multiple sources receive greater scrutiny and foster more issue-relevant thinking by the target. This increase in information processing by the target tends to produce greater agreement with the message and more attitude change. In contrast, weak messages from multiple sources may receive added scrutiny but produce no extra agreement and persuasion (Harkins & Petty, 1981a).

Certain qualifications regarding this effect should be noted. First, for multiple sources to have more effect than a single source, the target person must perceive the multiple sources to be unrelated and independent of one another. If the target believes that the sources colluded in sending their messages, the added impact of multiple sources will vanish, and the communication will have no more effect than if it came from a single source (Harkins & Petty, 1983).

Second, there is certainly some upper limit to increases in persuasion from the multiple-source effect (Tanford & Penrod, 1984). Adding more and more sources will increase persuasion, but only up to a point. For example, a message with strong arguments coming from three independent sources will likely be more persuasive than the same message coming from a single source, but a message coming from, say, thirteen sources may not be appreciably more persuasive than the same message coming from eleven sources. After some threshold, no additional impact will be achieved by adding more sources.

## The Message

Persuasive communications differ dramatically in their content, and most of these differences are relevant to the central route to persuasion rather than the peripheral route. Some messages contain arguments that are highly factual and rational, whereas others contain emotional appeals that motivate action by arousing fear or greed. Moreover, messages may differ in terms of their detail and complexity (simple versus complex arguments), their strength of presentation (strong versus weak arguments), and their balance of presentation (one-sided versus two-sided arguments). In this section we will discuss the impact of these properties on persuasion, beginning with the question of message discrepancy.

**Message Discrepancy**  Suppose a woman told you that Elizabeth II, the queen of England, is five feet four inches tall. Would you believe her? What if she said five feet ten inches tall—would you believe that? How about six feet three inches? Or seven feet four inches? You may not know how tall the queen actually is, but you probably have a rough idea. Although you might believe five feet ten inches, you would probably doubt six feet three inches and certainly doubt seven feet four inches. The message asserting that the queen is seven feet four inches tall is highly discrepant from your beliefs.

By definition, a **discrepant message** is one advocating a position that is different from what the target believes. To bring about a change in beliefs, a message must be at least somewhat discrepant from the target's current position; otherwise, it would merely reaffirm what the target already believes. Up to a certain point, greater levels of message discrepancy will lead to greater change in attitudes and beliefs (Jaccard, 1981). A message that is moderately discrepant will be more effective in changing a target's beliefs than a message that is only slightly discrepant. Of course, it is possible for a message to be so discrepant that the target will simply dismiss it. To say that the queen of England is seven feet four inches tall is just not believable.

There is an important interaction between message discrepancy and source credibility. If a message is only slightly discrepant, it will be persuasive only when it comes from a high-credibility source. Highly discrepant messages from a low-credibility source are likely to be ineffective, because the target will derogate the source. Figure 8-2 summarizes these relationships between message discrepancy, communicator credibility, and attitude change.

Numerous empirical studies report findings consistent with the relationships shown in Figure 8-2 (Aronson, Turner, & Carlsmith, 1963; Rhine & Severance, 1970; Fink, Kaplowitz, & Bauer, 1983). In one experiment, for instance, subjects were given a written message on the number of hours of sleep that people need each night to function effectively (Bochner & Insko, 1966). In some cases, the message was attributed to a Nobel Prize-winning physiologist (high credibility), and in other cases it was attributed to a YMCA director (medium credibility). The arguments contained in the message were identical for all subjects, with one important exception. In some cases the message proposed that people need eight hours of sleep per night; in others, seven hours; in others, six hours; and so on down to zero hours of sleep per night. Most subjects began the experiment with the idea that approximately eight hours of sleep were needed each night. Therefore, these messages had different levels of discrepancy.

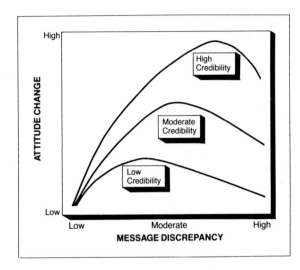

■ **FIGURE 8-2**

ATTITUDE CHANGE AS A FUNCTION OF COMMUNICATOR CREDIBILITY AND MESSAGE DISCREPANCY

These three curves summarize the relationship between message discrepancy and attitude change conditional on the credibility of a source. Note that messages from low-credibility communicators produce maximum attitude change at moderate levels of discrepancy, whereas messages from high-credibility communicators produce maximum attitude change at high levels of discrepancy.

Results of this study show that the more discrepant the position advocated by the highly expert source (the Nobel Prize winner), the greater the amount of attitude change. Only when the source argued for the most extreme position (zero hours of sleep) did the subjects refuse to believe the message. This same pattern was noted for the medium-expert source (the YMCA director), except that his effectiveness peaked out at moderate levels of discrepancy (three hours of sleep per night). For very extreme positions (two hours of sleep or less), the medium-expert source was less effective. Thus, this study demonstrates that communicators with higher credibility produce maximum attitude change at higher levels of discrepancy.

**Fear Arousal**   Most persuasive communications are based on either rational appeals or emotional appeals. Rational appeals are factual in nature; they present specific, verifiable evidence to support claims. Rational messages are frequently drive reducing—that is, they address a need already felt by the audience and provide the missing solution. An emotional appeal, in contrast, tries to arouse basic drives and creates a need where none was present.

Perhaps the most common emotional appeals are those involving fear. Fear-arousing messages are especially useful when the source is trying to motivate the target to take some specific action. A political candidate, for example, might warn that if his opponent is elected to office, the nation will find itself embroiled in international conflict. Likewise, an industrial leader may warn that unless steps are taken to impose trade restrictions, American jobs will be lost to foreign competition. In each of these cases, the source is using a fear-arousing communication. Messages of this type direct the target's attention to some negative or undesired outcome that is likely to occur unless the target takes certain actions advocated by the source (Higbee, 1969).

In most cases, research shows that communications arousing high levels of fear produce more change in attitude than communications arousing low levels of fear (Leventhal, 1970; Dembroski, Lasater, & Ramires, 1978). For example, fear-arousing communications have been effective in persuading people to reduce their cigarette smoking, to drive more safely, to improve their dental hygiene practices, and to change their attitudes toward Communist China (Insko, Arkoff, & Insko, 1965; Leventhal & Singer, 1966; Leventhal, 1970).

The impact of fear-arousing communications is shown clearly in a study in which college students received messages advocating inoculations against tetanus (Dabbs & Leventhal, 1966). These messages described tetanus as being easy to catch and as producing serious, even fatal consequences. The message also indicated that inoculation against tetanus, which could be obtained easily, provided effective protection against the disease. Depending on experimental treatment, subjects received either high-fear, low-fear, or control communications. In the high-fear condition, the messages described tetanus in extremely vivid terms in order to create a high level of apprehension and fear. In the low-fear condition, the messages carried a less detailed description so that no more than moderate fear would be produced. In the control condition, the message provided little detail about the disease, and correspondingly little fear was aroused.

To determine the message's effectiveness, the students were asked how important they thought it was to get a tetanus inoculation and whether they actually intended to get one. Responses showed that students exposed to the high-fear message had stronger intentions to get shots than those exposed to the other messages. Moreover, records kept at the university health service indicated that students receiving the high-fear message were more likely to obtain inoculations during the following month than were students receiving the other messages.

This study demonstrates that fear-arousing messages are effective in changing attitudes. In general, however, fear-arousing messages are effective only when certain conditions are met. First, the message must assert that if no changes are made in behavior, the target will suffer serious negative consequences. Second, the message must show convincingly that these negative consequences are highly probable. Third, the message must recommend a specific course of action that, if adopted, will enable the target to avoid the negative consequences. Messages that predict negative consequences but fail to assure the target that the consequences can be avoided through action will produce little attitude change or subsequent action. Instead, they will leave the target feeling that the negative consequences are inevitable regardless of what he or she may do (Rogers & Mewborn, 1976; Maddux & Rogers, 1983; Patterson & Neufeld, 1987).

**One-sided versus Two-sided Messages** When a source uses rational rather than emotional appeals, other message characteristics come into play. A common technique in persuasion is the one-sided message. Such a message emphasizes only those points that explicitly support the position advocated by the source. A two-sided message, in contrast, acknowledges that there are opposing viewpoints and then attempts to refute or downplay one of them. For example, if a man used a one-sided message to persuade his wife to spend their vacation at the seashore, he would mention only the reasons for going to the shore. If he used a two-sided message, he would mention both the reasons for going and the reasons for not going, but stress the reasons for going.

Which is more effective, a one-sided message or a two-sided message? The answer depends heavily on the audience. One-sided messages have the advantage of being uncomplicated and easy to grasp. They are more effective when the audience already agrees with the speaker or is not well informed about the issue. Two-sided messages have the advantage of making the speaker appear less biased and more trustworthy. They are more effective when the audience is initially opposed to the speaker's viewpoint or is well informed about alternative positions (Karlins & Abelson, 1970; Sawyer, 1973).

## The Audience

Some persons are easier to persuade than others. That is, a given message from a specific communicator will have more impact on some targets than on others. Although the attributes of the source and the content of the message are important in persuasion, so are the characteristics of the target. Among these characteristics are the level of the target's intelligence, the degree of the target's involvement with the issue, and the extent to which the target

has been prepared to counterargue against the message.

**Intelligence** Who would be easier to persuade, a genius or an imbecile? Or—to rephrase the question in less extreme terms—who would be easier to persuade, someone having high intelligence or someone having low intelligence? The person with low intelligence might yield more readily to an argument, and, in this sense, he might be easier to persuade than the target with high intelligence. But what if the low-intelligence person is not able to comprehend the argument in the first place? Without adequate comprehension, no attitude change can occur. The person with high intelligence may not yield readily to an argument, but at least she will comprehend it.

As this example shows, intelligence and persuasion are related in two distinct ways. Although high intelligence may increase comprehension of an argument and induce higher levels of attitude change, it may also heighten resistance and therefore inhibit or block attitude change (McGuire, 1972).

Thus, any consideration of the relation between intelligence and persuasion must take into account message properties. This is illustrated in a study (Eagly & Warren, 1976) involving high school students who took a standard intelligence test (verbal ability). The students were exposed to a persuasive communication. For some, this message included several complex (and apparently valid) arguments. For others, the message was simple and included no arguments. The results showed that the high-intelligence students displayed greater comprehension for the complex messages and were slightly more persuaded by these arguments than the low-intelligence students. At the same time, the high-intelligence students were more prone to reject the simple messages than the low-intelligence students.

Thus, a person's intelligence does not of itself determine whether he will be more or less easy to persuade. Much depends on the message itself. Although persons of higher intelligence are better able to resist persuasive communications, they are also better able to understand complex messages and, therefore, may be more influenced by such messages than are persons of lower intelligence.

**Involvement with the Issue**  Another important aspect of targets of persuasion is the extent of their involvement with a particular issue. Suppose, for example, that someone advocates a fundamental change at your college, such as increasing the number of comprehensive exams required for graduation. The proposed change would take effect in September of next year. Most undergraduates at your college would probably be very involved with this issue, because the change would affect their chances of getting a college degree. Now, suppose the source advocated that the change take place ten years in the future rather than next September. Students would probably have little interest in this proposal, simply because they would be finishing college long before any changes take effect.

Thus, involvement with the issue affects the way a target person processes a message. When highly involved, a target will be motivated to scrutinize the message closely and to think carefully about its content. Strong arguments will likely produce substantial attitude change, whereas weak arguments will produce little or no attitude change. In contrast, the target who is less involved will be less motivated to scrutinize the message or to think about it carefully. If any change in attitude occurs, it will probably depend more on peripheral factors (such as communicator attractiveness or credibility) rather than on the arguments themselves (Chaiken, 1980; Petty, Cacioppo, & Heesacker, 1981; Leippe & Elkin, 1987).

In one study, a message similar to that described above was in fact presented to a group of college students (Petty, Cacioppo, & Goldman, 1981). The message proposed that college seniors be required to take a comprehensive exam prior to graduation. Three variables were

manipulated in this study. The first variable was personal involvement with the issue. Half the subjects were told that the new policy would take effect next year at their college (high involvement), while the other half were told that the policy would take effect ten years in the future (low involvement). The second variable was the strength of the message's argument. Half the subjects received eight strong and cogent arguments in favor of the proposal; the other subjects received eight weak and specious arguments. The third variable was the expertise of the source. Half of the subjects were told that the source of the message was a professor of education at Princeton University (high-expert source); the other half were told that the source was a student at a local high school (low-expert source).

The results of this study are illustrated in Figure 8-3. In the high-involvement condition, the target's attitude toward comprehensive exams was determined primarily by the strength of the arguments. Strong arguments produced significantly more attitude change than weak ones. The expertise of the communicative source had no significant impact on attitude change. In the low-involvement condition, attitudes were determined primarily by source expertise; the high-expert source produced more attitude change than the low-expert source. The strength of the arguments had relatively little effect on this group. Thus, the primary determinant of attitude change depended on the target's involvement with the issue. For subjects with high involvement, the strength of the argument was more important because subjects cared about the issue. For those with low involvement, peripheral factors (such as communicator expertise) were more important because subjects had little motivation to scrutinize the arguments critically.

**Immunization against Persuasion**  Most people are motivated to defend their attitudes and beliefs from attack, especially with respect to important issues. If attitudes are based on a large

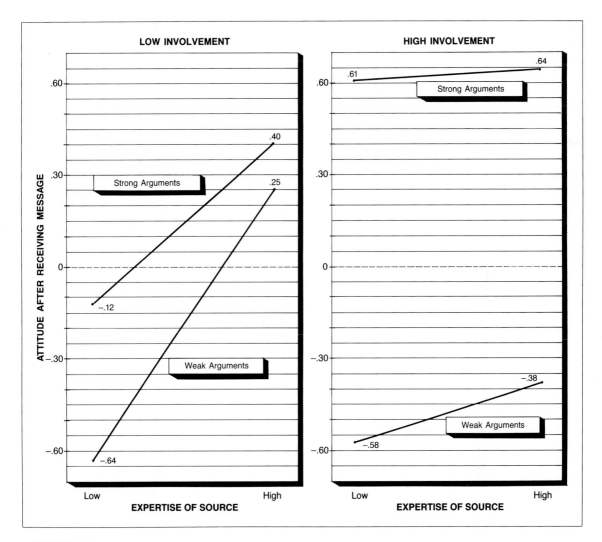

■ FIGURE 8-3

THE EFFECTS OF PERSONAL INVOLVEMENT ON PERSUASION

In this study, students received a message advocating that college seniors be required to take a comprehensive exam prior to graduation. Half the students were told that the new policy would take effect next year (high involvement), while the others were told the policy would take effect ten years later (low involvement). Results show that students in the high-involvement condition were affected primarily by the strength of the arguments rather than the expertise of the source, whereas students in the low-involvement condition were more affected by the expertise of the source rather than the strength of the arguments.

*Source: Adapted from Petty, Cacioppo, and Goldman, 1981.*

SHE'S VERY CHARLIE.

Charlie
REVLON

This magazine advertisement is part of a media campaign to sell perfume. The ad uses the technique of reversing traditional gender roles to define the intended market for the product.

amount of information, they usually are relatively easy to defend. However, if attitudes are based on little information, they may be vulnerable to attack.

One class of beliefs that target persons find hard to defend is the *cultural truism.* Truisms are beliefs widely accepted by the members of a given culture. Examples of American truisms are: "Everyone should brush his teeth after every meal if at all possible"; "The effects of penicillin have been, almost without exception, of great benefit to mankind"; and "Mental illness is not contagious."

Because truisms are rarely questioned, people have little practice in defending them against

attack. However, McGuire (1964) proposed that a target person can be "immunized" against persuasion on an issue of this type. The immunization treatment consists of giving the target (1) information that is discrepant with the truism and (2) arguments that refute this discrepant information and that support the truism. By exposing a target to weak attacks and allowing the target to refute them, the immunization treatment prepares the target for stronger attacks against the truism at a future time. This treatment, called a *refutational defense,* is analogous to medical immunization in which a patient is given a small dose of a pathogen so that he can develop antibodies. By exposing the target to small attacks, the immunization builds up the target's resistance to subsequent, stronger attacks.

A study by McGuire and Papageorgis (1961) demonstrated the effectiveness of a refutational defense against persuasion attempts. College students were exposed to messages attacking three different cultural truisms. Two days prior to the attack, the students had received an immunization treatment designed to create resistance to persuasion. For one truism, they received a refutational defense. For a second truism, they received a *supportive defense*—information containing elaborate arguments in favor of the truism. For the third truism, they received *no defense.* Following exposure to attacks on their attitudes, students rated the extent of their agreement with each of the truisms. They also rated their agreement with a fourth truism, one that had not been attacked and for which no defense had been previously provided. The fourth truism served as a *control.*

Students' final ratings regarding the truisms are summarized in Table 8-1. Results show that the refutational defense provided a high level of resistance to persuasion, whereas the supportive defense provided somewhat less resistance. When no defense was present, there was still less resistance to persuasion. In general, the findings demonstrate the effectiveness of

■ **TABLE 8-1**

AGREEMENT WITH TRUISMS AFTER ATTACK AS A
FUNCTION OF THE TYPE OF DEFENSE PROVIDED
BEFORE THE ATTACK

| Type of Defense Provided | Agreement with Truisms |
|---|---|
| Control (no exposure to attack) | 12.62 |
| Refutational defense prior to attack | 10.33 |
| Supportive defense prior to attack | 7.39 |
| No defense prior to attack | 6.64 |

Note: Higher scores indicate more agreement with the truisms
after attack and hence less attitude change.

Source: Adapted from McGuire and Papageorgis, 1961.

the refutational defense in creating resistance
to persuasion.

# ■ Threats and Promises

On many occasions, influence attempts are intended to persuade the target person—that is,
to change the target's beliefs and attitudes. On
other occasions, however, a source may not care
about changing the target's beliefs and attitudes
but only about influencing how the target behaves. In other words, the source may care only
about securing **compliance**—that is, conformity by the target to the source's requests or
demands.

Sometimes compliance is obtained via persuasion. By changing what the target believes,
the source might succeed in changing how the
target behaves. Frequently, however, compliance
is obtained not through persuasion but through
the use of threats and promises.

As an example, consider how Richard
Sorenson exercises influence by means of prom-

ises. Sorenson, a home owner, lives in the upper
peninsula of Michigan, where it snows heavily
each winter. One cold January day, a blizzard
dumps eight inches of snow on his driveway and
sidewalk. Not wanting to shovel snow, he notices
that his neighbor's 14-year-old son is outside
clearing his own driveway with a snowblower.
Sorenson approaches the boy and says, "If you
use your snowblower to clear my driveway and
sidewalk, I will pay you $5." This is an influence
attempt in the form of a promise. Sorenson
promises to pay the boy $5 in return for specific
behaviors by the boy.

Influence based on promises and threats
differs from persuasion in a fundamental way. In
using persuasion, the source tries to change the
way a target person views the situation.
Sorenson, for example, might have attempted to
persuade the boy that shoveling snow is fun or
that clearing out the driveway would make him a
good neighbor. Neither of these appeals restructures the situation; if successful, they would
change only how the boy looks at the situation.
In contrast, when using threats and/or promises,
the source directly restructures the situation
itself. By promising to pay money for a clear
driveway, Sorenson has added a new contingency
to the situation—money in return for snow
removal. He hopes this approach will achieve
compliance from the boy.

## Effectiveness of Threats and Promises

Before we consider the effectiveness of
threats and promises, it may be helpful to define
these terms more precisely. A **threat** is a communication from one person (the source) to
another (the target) that takes the general form,
"If you don't do X (which I want), then I will do
Y (which you don't want)" (Tedeschi, Schlenker,
& Lindskold, 1972; Boulding, 1981). For example, a boss might say to his employee, "If you
don't obtain a new advertising contract from that
customer, I'll fire you." If the employee needs
his salary to keep food on the table and has no
other job prospects, he will certainly take the

## PERSUASION VIA MASS MEDIA

Although many of the persuasive messages we receive daily come to us through face-to-face interaction with other persons, a significant portion of them arrive another way—via the mass media. The term **mass media** refers to those channels of communication that enable a source to reach a large audience. Whereas face-to-face communications can reach only a small audience, mass media can potentially influence a large number of people. The most influential mass medium in the U.S. is television, followed by newspapers, radio, and magazines (Atkin, 1981).

Not everyone has equal exposure to the mass media. For example, in the U.S. women view more television than do men. Children and retirees view more television than do adolescents and working adults. Viewing is negatively related to the level of education, income, and occupational status (Comstock et al., 1978; Newspaper Advertising Bureau, 1980).

Consideration of the mass media immediately raises a central question: To what extent are communications transmitted by the mass media effective in changing the beliefs and attitudes of large numbers of people? We will look at this issue from the standpoint of media campaigns.

### Media Campaigns

A **media campaign** is a systematic attempt by a source to use the mass media to change attitudes and beliefs of a target audience. Media campaigns are common in the industrialized world. They are used by advertisers to sell new products or services, by political parties to sway voters' sentiment, and by public officials to change citizens' behavior—"Only you can prevent forest fires," "Take a bite out of crime," "Friends don't let friends drive drunk," and so on (Farhar-Pilgrim & Shoemaker, 1981; Nimmo & Sanders, 1981; Solomon, 1982).

A huge amount of money is spent on media campaigns. Each year tens of billions of dollars are spent on persuasive communications delivered through the mass media (McCombs & Eyal, 1980). Nevertheless, most media campaigns do not produce large amounts of attitude change. In general, persuasive messages sent via mass media have only a small impact on their target audience's attitudes (Bauer, 1964). Consider, for example, what occurs during presidential campaigns. Soon after the political conventions nominate their candidates in midsummer, most Americans know how they intend to vote in the upcoming November election. Although the parties spend millions of dollars on political advertising during the fall campaign, they will not change many voters' attitudes. In most presidential elections, only about 7 to 10 percent of the voters change their candidate preferences during the campaign. It is difficult to change political attitudes via the mass media.

Of course, from another perspective, a shift of 7 to 10 percent may be quite significant. Many professional advertisers and politicians

threat seriously and do his best to comply with his boss' demand.

Threats can be issued with respect to virtually any behavior. Then, too, the sanction that is threatened can be almost anything—a physical beating, the loss of a job, a monetary fine, the loss of love. The important point is that for a threat to be effective, the target must want to avoid the sanction. If the employee threatened by his boss happens to have a new job lined up elsewhere, he may not care whether he is fired. In this case, the threat will have little impact because the target has no real need to avoid the sanction.

## Box 8-1

are quite satisfied if their media messages can shift public opinion a few percentage points in the intended direction. In a close political race, a net gain of 1 or 2 percent might be sufficient to win the election. Thus, even though a media campaign might be considered a disappointment in terms of producing widespread attitude change, it may also be considered a clear success from the standpoint of getting a candidate elected (Mendelsohn, 1973). Even a small amount of attitude change may be sufficient to justify the cost of the media campaign.

Why are media campaigns usually able to produce only small amounts of attitude change? There are several reasons. First, there is the phenomenon of *selective exposure:* Many messages do not even reach the audience they are intended to influence. Instead of reaching persons who disagree with the message (and whose opinions might therefore be changeable), many media communications are heard by persons who already agree with the message (and whose opinions will therefore be reinforced, not changed). In most routine media exposure, persons encounter more messages supporting than not supporting their preexisting attitudes (Sears & Freedman, 1967).

Second, even if the intended targets receive messages from the media, they may resist them in the same way they reject messages in face-to-face contacts. A target person might disbelieve the media message or derogate the source. Recipients of media communications are not passive. The impact of a message depends heavily on the uses and gratifications that the audience can obtain from the information (Swanson, 1979; Dervin, 1981). For example, in selling consumer products, persuasion is more effective when the target person's involvement with the decision is low and when he perceives relatively small differences among alternative products. In contrast, the impact of the media will be slight when involvement with the decision is high and the differences between products appear clear cut (Ray, 1973; Chaffee, 1981).

Third, even when a target person finds a media message compelling, he or she may be subject to counterpressures that inhibit attitude change (Atkin, 1981). Some of these pressures come from social groups (such as family, friends, and co-workers); these groups may exert influence that nullifies the media's impact. In addition, target persons are exposed to conflicting persuasive communications and cross-pressures transmitted over the media. For example, beer advertisements would be very successful if only one manufacturer advertised his product. But because dozens of brands advertise, media messages largely offset one another.

### Other Effects of Media Campaigns

Although media campaigns do not usually bring about a massive change in attitudes, they do exert other significant influences on audiences. First, they are effective in strengthening

---

A **promise** is similar to a threat, except that it involves contingent rewards, not punishments. A person using a promise says, "If you do X (which I want), then I will do Y (which you want)." Notice that a promise involves a reward controlled by the source. Richard Sorenson promises a payment of $5, provided that his 14-year-old neighbor clears the driveway and sidewalk. Promises are frequently used in exchanges, both monetary and nonmonetary.

Threats and promises are often—but not always—effective in gaining compliance from the target. In issuing a promise, the source creates a set of options for the target. Suppose,

Box 8-1

(continued)

preexisting attitudes. In other words, they reinforce and buttress preferences already held by the target audience. Televised debates between presidential candidates, for example, usually strengthen existing attitudes rather than change them. This was demonstrated in the famous Kennedy-Nixon debates of 1960. Viewers who were pro-Kennedy prior to the first debate concluded that Kennedy had "won" the debate, whereas those who were pro-Nixon concluded that Nixon had "won" (Kraus, 1962; Sears & Whitney, 1973).

In addition to strengthening preexisting attitudes, mass media are also successful in creating new attitudes toward previously unknown objects. Many media campaigns have cultivated new attitudes toward objects that previously were unknown or unimportant to the audience. For example, the "Smokey the Bear" campaign and related media efforts have been effective in raising peoples' consciousness regarding the prevention of forest fires (McNamara, Kurth, & Hansen, 1981). Likewise, businesses use media campaigns to create positive attitudes toward new products, such as special-ingredient breakfast cereals or new-design toys. The average child, for example, sees over 20,000 commercials per year (Adler et al., 1977). This is worrisome because young children are more trusting toward and less skeptical of television advertisements than are older children and adults (Rossiter, 1981; Christenson, 1982). A media blitz in the weeks immediately preceding Christmas may be quite sufficient to create positive attitudes toward new toys even among children having comparatively strong defenses.

Media campaigns to create positive attitudes toward new objects are also common in politics. One example is the campaign created for Jimmy Carter shortly before his nomination in 1976 (Patterson, 1980). Although Carter had been governor of Georgia, he was almost entirely unknown outside the South. To win the Democratic nomination for president, Carter needed more name recognition among voters. The solution to this problem was a media campaign based on the theme "Jimmy Who?" By asking this question, the campaign sought first to pique voters' curiosity and later to create a positive view of this previously unknown candidate. To some degree, the same problem arose for the democratic candidate Michael Dukakis in the 1988 campaign against George Bush. The Republican campaign managers, noting that Dukakis was not well known to many voters, moved quickly to brand Dukakis as a soft-on-crime, tax-raising "liberal" before Dukakis could create advertisements defining himself in other terms. Dukakis was never able to shed the label, a fact that many believe hurt him in the election.

for example, that the source makes the promise, "If you clear the snow from my driveway with your snowblower, I will pay you $5." In response, the target can (1) comply with the source's request and clear the driveway, (2) refuse to comply and let the matter drop, or (3) propose a counteroffer ("How about $7? It's a long driveway, and the snow is very deep").

In similar fashion, a threat also creates a choice for the target. Once a threat is issued, the target can (1) comply with the threat, (2) refuse to comply, or (3) issue a counterthreat (Boulding, 1981).

The range of possible responses to threats and promises raises a fundamental question: Under what conditions will threats and promises be successful in gaining compliance, and under what conditions will they fail? Certain characteristics of threats and promises, such as their magnitude and credibility, affect the probability that the target will comply.

**Magnitude of Threats and Promises**   Late at night, on an isolated street corner, a bandit brandishes a pistol and issues the threat "Hand over your money or I'll blow you away!" The victim is forced to choose between two undesirable alternatives: losing his money or losing his life. Facing a negative outcome of enormous magnitude, the victim will almost certainly hand over his wallet. However, if the threat were of smaller magnitude, his reaction might be different. A bandit would not get the victim's wallet if, instead of using a gun, he said, "Hand over your money or I'll zap you with this rubber band!" Obviously, the effectiveness of a threat varies directly with the magnitude of the punishment involved.

A similar principle holds true for promises. The greater the magnitude of the reward promised, the greater the probability of compliance (Lindskold & Tedeschi, 1971). A factory supervisor, for example, might obtain compliance from a worker by saying, "If you are willing to work the late shift next month, I'll approve your request for four extra days of vacation in September." The worker's reaction might be less accommodating, however, if his supervisor offered only a trivial reward: "If you work the late shift next month, I'll let you take your coffee break five minutes earlier today." The greater the magnitude of the reward promised, the higher the probability that the worker will comply with the supervisor's request.

**Credibility of Threats and Promises**   Suppose someone says to you, "If you don't do X (where X is something I want), then I will do Y (where Y is something you don't want)." Does the person making the threat really mean it, or is he merely bluffing? You might comply if the threat is credible, but you certainly do not want to comply if the threat is merely a bluff. Unfortunately, there is no easy or risk-free way to find out whether a threat is credible. The only true way to test credibility is to call the source's bluff—that is, to refuse to comply. Then, if the threat was merely a bluff, that fact will quickly become evident. Of course, if the threat was real, you will suffer the consequences.

Bluffing or not, any threatener wants the target to believe the threat is credible and to comply with his demand. He doesn't want the target to call his bluff. After all, a successful threat is one that achieves compliance without actually having to be carried out. If the target refuses to comply, the threatener must either admit that he is bluffing or incur some costs in carrying out the threat.

To some degree, the credibility of a threat depends on the social identity of the source. A threat involving physical violence, for example, will be more credible if it comes from a karate expert wearing a black belt than if it comes from the proverbial 97-pound weakling. In some instances, the source may attempt to manipulate how he is perceived by the target in order to increase credibility. Thus, bank robbers try to increase compliance with their threats by adopting a sinister appearance, perhaps by wearing trench coats and face masks.

**The SEV Model**   Reinforcement theory provides some insight regarding the conditions under which a threat will be effective in gaining compliance from a target person. According to this view, compliance with a threat will increase directly with the threat's **subjective expected value (SEV)** for the target. SEV, in turn, depends on three factors. First, SEV increases as the threat's credibility increases. Second, SEV increases as the magnitude of punishment to be inflicted increases. Third, SEV increases as the product of credibility and punishment magnitude (that is, credibility × punishment magnitude) increases (Tedeschi, Bonoma, & Schlenker,

1972; Stafford et al., 1986). This theory holds that in deciding whether to comply with the threat, a target will (consciously or unconsciously) calculate the SEV. Thus, the higher the SEV, the greater the likelihood the target will comply with the threat.

Note that when both credibility and punishment magnitude are high, SEV is also high, and consequently the target will feel a lot of pressure to comply. When both credibility and punishment magnitude are low, SEV is low, and consequently the target will feel little or no pressure to comply. When one is high and the other is low (as, for example, when magnitude is high but credibility is low), the threat will have a low SEV and, therefore, is not likely to produce much compliance by the target. (It is to handle situations like this that the theory includes the product term, *credibility* × *punishment magnitude*.)

Various empirical studies support the SEV model of threat effectiveness (Faley & Tedeschi, 1971; Grasmick & Bryjak, 1980; Stafford et al., 1986). For example, in one laboratory study (Horai & Tedeschi, 1969), several pairs of persons were asked to play a game repeatedly over a number of trials. One member of each pair was a naive subject, whereas the other member, while appearing to be a subject, was actually a confederate following a programmed strategy. The object of the game was to gain points. Both the subject and the confederate were required to make certain choices during each trial.

Threats could be issued by the confederate but not by the subject. To gain extra points, the confederate occasionally delivered threats to influence the subject's choice. ("If you do not pick Choice 1 on the next trial, I will take [some number of] points away from your counter.") The magnitude of the punishment threatened by the confederate was either high (20 points), medium (10 points), or low (5 points), depending on experimental treatment. The credibility of the confederate was manipulated by varying the frequency with which he carried out his threats if the subject did not comply. In the high-credibil-

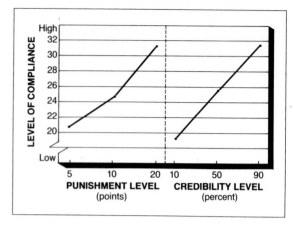

■ FIGURE 8-4

COMPLIANCE WITH THREATS

Compliance is a function of both the magnitude of the punishment threatened and the credibility of the threat. The greater the punishment threatened, the greater the compliance with the threat. Likewise, the greater the credibility of the threat, the greater the compliance.

*Source: Adapted from Horai and Tedeschi, 1969.*

ity condition, the confederate carried out his threats 90 percent of the time; in the medium-credibility condition, 50 percent of the time; and in the low-credibility condition, 10 percent of the time.

The results of this study are shown in Figure 8-4. They indicate that both the magnitude of punishment threatened and the credibility of the source affect the subject's level of compliance. Compliance increased as a direct function of both variables. The greatest amount of compliance occurred under the highest magnitude of punishment and the highest level of credibility. Results of this study support the SEV model.

Although we have described the SEV model as pertaining to threats, it may also be applied to promises. Of course, the relevant variable with respect to promises is the magnitude of reward promised. Both reward magnitude and promise credibility affect compliance to promises. Results of one study showed that target persons were more influenced when the reward prom-

A terrorist issues a threat at gunpoint during an airplane hijacking. Passengers are more likely to comply when threats are both large and credible.

ised was large rather than small and when the source had some credibility in following through on his promises (Lindskold et al., 1970). Consistent with the SEV view, these conclusions hold true only when the reward promised for compliance is greater than the rewards that might be gained from refusing to comply.

## Problems in Using Threats and Promises

Threats and promises pose certain problems for their user. First, the person using threats or promises must maintain surveillance over the target's behavior. That is, the source must watch closely to determine whether compliance has occurred or not. Typically, this entails some cost for the source in the form of time and trouble to monitor the target's behavior. The problems involved in maintaining surveillance can be especially troublesome if threats rather than promises are being used. When threats are used, the targets may not only fail to comply, but may also attempt to conceal their noncompliance to avoid punishment (Ring & Kelley, 1963).

The cost of surveillance is one reason why the source might prefer to use persuasion rather than threats or promises. If successful, persuasion requires no surveillance. For example, if an employer uses persuasion and succeeds in changing an employee's mind, she can expect that the employee will carry out the task on his own, without surveillance. This is certainly more convenient and less costly for the employer.

A second problem facing the source is that threats may backfire and cause resentment or

Box 8-2

## "YOUR MONEY OR YOUR LIFE"

Perhaps the strongest threat that can be made is one against your life. Threats of this type frequently occur during robbery or holdup attempts involving a lethal weapon, such as a knife or firearm. How do robbery victims react to threats made against their lives?

To answer this question, Luckenbill (1982) studied 201 cases of robbery and attempted robbery that occurred in a Texas city between February 1976 and March 1977. Police records indicated that victims take several factors into account before deciding whether to comply with a threat. First, they consider the source's capacity to inflict serious injury or death. For example, compliance was high (75 percent) when the source had a lethal weapon (a knife or a gun) and was in a position to use it. As two victims noted:

> "I wasn't going to try anything because he had a piece (firearm). When he's got a piece, you give him the money. That's all there is to it. If you try anything, he might shoot you."

> "I would've killed him if I could've gotten to my pistol. But . . . I couldn't do a thing, you know. He was standing right there watching me. And his pistol was loaded. I could see the steel bullets in the cylinder. And he was shaking so bad he might've shot me if I tried anything. If I had a clear chance, you know, I would've

nailed him. But hell, I'm not going to risk my life for a few measly bucks." (Luckenbill, 1982, p. 814)

On the other hand, when the source had no lethal weapon, compliance by the targets was low (5 percent). According to police records, the target opposed the demand for money in almost all cases in which the robber did not have a weapon.

A second factor considered by the victims was the source's intent regarding the use of force—that is, whether the source's threat was contingent or noncontingent. If the threat was contingent ("If you don't give me your money, then I'll shoot you"), victims assumed they would not be hurt if they complied. However, if the threat was noncontingent ("I'm going to shoot you and take the money"), most victims concluded that the robber intended to inflict harm indiscriminately. Luckenbill's data show that in cases where the robber issued a noncontingent threat, victims actively opposed the robber, primarily because they feared imminent death.

In general, these findings support the view that compliance to a threat is likely when the subjective expected value of the threat is large, provided that compliance with the threat provides an assured means of avoiding punishment or harm.

---

hostility by the target toward the influencer (Zipf, 1960). Promises usually do not entail this difficulty, because they are based on rewards and not punishments. In general, a target will have more positive feelings toward someone who uses promises than toward someone who uses threats (Rubin & Lewecki, 1973). For this reason, in situations where the source has an ongoing relationship with the target and where maintain-

ing a good relationship is important, the source may prefer to use promises, not threats.

## ■ Bargaining and Negotiation

Up to this point we have considered various techniques of influence—persuasion, threats, and promises. Throughout our discussion we have assumed that one person (the source) attempts to influence another person (the target) in order to change the behavior of that person. In many real-life situations, however, it happens that not one but both persons can exercise influence. When person A (the source) attempts to influence person B (the target), person B (now the source) may respond by attempting to influence person A (now the target).

Influence in these two-way situations may be based on persuasion as, for example, when a persuasion attempt is met not by attitude change but by counterpersuasion. Alternatively, it may be based on promises of rewards or on threats of punishment. For example, if Hugo threatens Norman, Norman may respond by threatening Hugo:

HUGO: "Get out of my way or I'll rearrange your face."

NORMAN: "If you touch me, I'll sue you for every penny you've got."

Situations in which both persons exercise influence are often complex. For instance, when each person controls rewards valued by the other, the interaction between them may entail bargaining over prices or exchange rates. When each person controls punishments feared by the other, the potential for bargaining likewise exists, but there is also the further possibility of escalation of conflict. In this section we will discuss both of these processes.

## Bargaining in Exchange Relations

An **exchange** is a transaction in which person A gives person B something that B values in return for B giving A something that A values (Kelley & Thibaut, 1978; Emerson, 1981). The items exchanged can be virtually anything—goods, services, money, information, or whatever. We see exchanges occurring every day. If a consumer makes a purchase in a grocery store, she exchanges money for food. Workers exchange their labor for a paycheck. Politicians often exchange their votes on one issue for the votes of colleagues on another issue. Prostitutes conduct a sex-for-money exchange with their clients. In holding a conversation, two persons exchange information; if the news is fresh, both may gain.

Some exchange relations involve explicit bargaining, whereas others do not. In some exchanges, the terms and the rate of exchange (that is, the price) are institutionally established or fixed, and little or no bargaining occurs. The price of food in American grocery stores, for instance, is largely fixed; food store customers do not often haggle with the checkout clerk regarding the price of a loaf of bread. In other exchanges, however, terms and prices are negotiable and hence open to bargaining. When you go to a used car lot to purchase an auto, the price is almost always subject to discussion. Likewise, when you are offered a new job, the price of your services (that is, your salary) may be a topic for bargaining.

**Bargaining Situations**  The term **bargaining** refers to an interaction in which two (or more) persons have divergent (conflicting) preferences regarding some outcome, are able to communicate with one another, and can make concessions in an attempt to reach an agreement that is mutually acceptable. This agreement, if achieved, specifies what each person will give and what each will receive (Rubin & Brown, 1975; Druckman, 1977; Bacharach & Lawler, 1981).

People can bargain over anything. Although the rate of exchange (price) is a very common issue, in some cases the terms of an agreement may be more important than the price itself. A labor union, for example, in bargaining against management, may be more concerned with work rules than with wage increases. Similarly, a new employee may bargain not over salary, but over the terms and details of the employment—the rights and responsibilities of the new job, the duration of employment, and so on.

Bargaining behavior usually consists of tactics whereby each party tries to obtain concessions from the other. A bargainer who obtains concessions from another while making few concessions himself will consider the outcome relatively favorable. Thus, the bargaining process usually involves a series of concessions leading to a final resolution that both parties find acceptable. Since the resolution is almost always a compromise, neither party may find it 100 percent satisfying.

**The Bargaining Range**  From a tactical standpoint, the most fundamental question in bargaining is, What actions will lead to a good outcome? In other words, What moves should a person make during bargaining to obtain a favorable agreement? Although there is no simple formula that will always produce a good outcome, a bargainer's actions can certainly influence the result for better or worse.

Suppose, for example, that Juan has recently taken a job and finds that he needs a car to commute to work. Late one afternoon Juan goes down to Friendly Al's Used Car Lot and takes a look around. Scrutinizing all the cars on the lot, Juan finds one he likes. After checking out the car mechanically, Juan decides that he wants to buy it. The only stumbling block is Friendly Al's asking price, $2,995.

Juan wants the car, but not at Friendly Al's price. He checks his finances closely and figures that the most he could pay for a car without borrowing a lot of money is $2,500. He decides that $2,500 is the maximum price he will pay for the car under any conditions. This amount is termed Juan's *limit*. Of course, Juan would like to get the car for even less. He has scrutinized the local newspapers to see what prices other dealers are asking for similar automobiles and has noticed that some sell for as little as $1,500, although these are a year or two older and in poorer condition. Juan hopes that he can bargain Friendly Al down to $1,500. This lower amount is referred to as Juan's *level of aspiration* —that is, the lowest price for which he thinks he might realistically get the car.

The seller, Friendly Al, also has a limit and a level of aspiration, although these differ from Juan's. Friendly Al paid $1,500 for the car when it originally came onto his lot, and he also paid for some minor repairs. Thus, Friendly Al's limit—the lowest price he'll accept for the car— is $1,800. Friendly Al's level of aspiration is the $2,995 asking price; this is the highest price that he thinks he might realistically get.

Thus, Juan has a limit of $2,500 and an aspiration level of $1,500. Friendly Al has a limit of $1,800 and an aspiration level of $2,995. Since Friendly Al will not accept less than $1,800 and Juan will not pay more than $2,500, any deal made will have to fall within this range. This $700 range—the difference between the two limits—is termed the **bargaining range.** Of course, at the start of discussions Juan and Al may not know the exact size of the bargaining range, because Juan doesn't know Friendly Al's limit and Al does not know Juan's. Nevertheless, if Juan and Al are to strike a deal, the price of the car has to fall somewhere between the two limits, $1,800 and $2,500.

**Initial Offer**  In the opening discussions with Friendly Al, what initial offer should Juan make? Juan could either make an extremely low offer ($1,500) or he could make his highest possible offer ($2,500). Which offer will lead to the most favorable outcome for Juan?

If Juan opens with a high offer ($2,500), an agreement will be easy to reach. On the other hand, if Juan makes a low initial offer ($1,500), an agreement may be more difficult to reach. Experimental evidence for situations of this type shows that a low initial offer is more likely to lead to favorable outcomes for the buyer (Juan) than a high initial offer, provided that the initial offer is not so meager that it causes the seller (Friendly Al) to become disgusted and break off the bargaining (Benton, Kelley, & Liebling, 1972; Harnett & Vincelette, 1978). A low initial offer from one bargainer tends to produce a favorable outcome because it reduces the aspiration level

In negotiating the purchase of a used car, the customer must estimate not only the salesman's level of aspiration but also his limit. Only then does the customer have any real chance of striking a deal.

of the other bargainer (Liebert et al., 1968; Yukl, 1974a, 1974b). After receiving an extremely low initial offer, the other bargainer will reconsider the situation and conclude that his original level of aspiration was simply out of reach. Typically, he will react by lowering his demand or making larger concessions than he otherwise might. If this happens, the person making the low initial offer will obtain a relatively favorable outcome.

**Concessions**   If Juan makes a low initial offer of $1,500 for the automobile, with Friendly Al asking $2,995, someone will obviously have to

make some concessions if an agreement is to be reached. Quite possibly both Juan and Al will have to make concessions. Bargaining is generally a give-and-take process during which each person reduces his demands and moves in the direction of a compromise.

A bargainer such as Juan could take any of several approaches with respect to concessions. At one extreme, Juan could hang tough and refuse to make any concessions whatever. At the other extreme, Juan could consistently initiate unilateral concessions irrespective of Friendly Al's response (or lack of response). Between these extremes, Juan could adopt a *matching* strategy and make concessions that depend expressly on Al's response. For instance, Juan might make it clear to Al that he will match any concessions in a tit-for-tat manner. If Al lowers his asking price, Juan would respond by raising his offer; if Al refuses to lower his price, Juan would respond by refusing to raise his offer.

Several studies have shown that a matching strategy can be effective in inducing an opposing bargainer to make concessions. Many bargainers will make concessions in direct response to a concession made by the other. Bargainers are especially likely to make a concession immediately after one has been offered. (Bartos, 1974; Hopmann & Smith, 1977). Moreover, when one bargainer initiates concessions frequently during negotiations, the other bargainer is likely to match with frequent concessions (Chertkoff & Conley, 1967; Yukl, 1974b).

There is, however, an important distinction between the frequency of concessions and the size of concessions. Analysts use the term **concession magnitude** to denote the average size of a bargainer's concessions. For example, if Juan offered a concession by raising his bid from $1,500 to $1,700 and subsequently offered a second concession by raising from $1,700 to $2,000, Juan's concession magnitude would be $250—that is, a total concession of $500 over two moves. Concession magnitude is an important concept because, although bargainers may match in the number of concessions made, they frequently mismatch in terms of magnitude. To illustrate, suppose that Juan offered his first concession, raising $1,500 to $1,700, and Friendly Al responded by lowering his asking price from $2,995 to $2,850. Then Juan offered a second concession, raising $1,700 to $2,000, and Friendly Al responded by lowering $2,850 to $2,700. It is apparent that although Friendly Al is matching the frequency of Juan's concessions, he is trying to gain an advantage by refusing to match their magnitude.

Bargainers will attempt to match the concession magnitude of their opponent only under certain circumstances. Concession magnitude is more likely to be matched when a bargainer perceives that the other party is engaging in cooperative behavior because he wants to rather than because he has to. In other words, if one bargainer enjoys a position of strength or is operating in terms of some principle, he will be perceived as impervious to pressure or competitive tactics; thus, any concessions from him will be construed as a conciliatory gesture. These are likely to be met with concessions of similar magnitude (Michener et al., 1975; Wall, 1977).

**Bargaining as Problem Solving**  Because bargaining involves opposing interests, many people view it as a battle to be won rather than as a problem to be solved. Yet, it is often more effective to adopt a problem-solving orientation when bargaining because this increases the chances that all participants will come out ahead. Given this viewpoint, the basic objective during bargaining is to invent **integrative proposals** —alternatives that reconcile bargainers' divergent interests by providing high benefits to both of them (Pruitt, 1981; Pruitt & Rubin, 1985).

As an illustration, consider the case of Dave and Cathy, a husband and wife who disagreed on where to spend their summer vacation (Pruitt, 1981). Although they both preferred some kind of vacation to none at all, Dave wanted to go to the mountains, while Cathy wanted to go to the seashore. Rather than adopting rigid positions, Dave and Cathy agreed to discuss the factors underlying their different preferences. This revealed that Dave preferred the mountains because he liked fishing, hiking, and the breathtaking mountaintop views. Of these, fishing and hiking had the highest value for him. Cathy preferred the seashore because she liked swimming, sunning on the beach, seafood dinners, and the salt air, in that order.

Initially, Dave and Cathy considered only three options: the mountain, the seashore, and no vacation at all. After the conflict became apparent, they also began to consider a fourth option: spend one week in the mountains and one week at the seashore. This fourth option was a compromise, but not an integrative proposal. Because it was positioned halfway between Dave's and Cathy's initial preferences, it required each of them to sacrifice value. Although attractive in some respects, this compromise was not entirely satisfactory. Rather than stop here, the couple continued to discuss the problem. By analyzing the factors underlying their conflict, they tried to create new integrative proposals that were superior to any of the options on the table. The breakthrough came when Cathy suggested a fifth option: spend their vacation at an inland lake. This location would give Dave a chance to fish and hike, and Cathy an opportunity to swim, sun, and eat fresh fish. By meeting the underlying needs of both persons, this integrative proposal resolved the conflict successfully.

By adopting a problem-solving orientation and striving to develop integrative proposals, bargainers may be able to avoid open conflict and achieve a constructive resolution. Such an outcome is most likely to emerge when bargainers take the following steps: (1) Separate the people from the problem. To avoid an entanglement of egos, the bargainers should view themselves as working side by side, attacking the problem, not one another. (2) Focus on interests, not positions. By exchanging information on underlying values and interests, bargainers can determine what conditions must be met by an integrative proposal. (3) Invent new options that are mutually beneficial. Integrative proposals tailored to the underlying values and interests of the bargainers are frequently superior to superficial compromises between initial positions. This is shown clearly in the preceding example where Dave and Cathy preferred the new option (inland lake) to the halfway compromise between initial positions (one week in the mountains and one week at the seashore). (4) In those extreme cases where interests are directly opposed and integrative proposals cannot be developed, attempt logrolling. Bargainers should make concessions on low-priority issues in exchange for concessions on high-priority issues. Alternatively, bargainers should insist on objective criteria as a basis for concessions. Rather than rewarding intransigent and stubborn behavior with concessions, bargainers should insist that certain criteria (for example, market value, expert opinion, custom, or law) be met as a condition for concessions. By placing emphasis on objective criteria rather than on what the bargainers are willing or unwilling to do, neither party need give in to the other. Both can defer to an objectively determined resolution without losing face (Pruitt & Carnevale, 1980; Fisher & Ury, 1981).

## Bilateral Threat and Escalation

Thus far, we have characterized bargaining as a process of reciprocal influence between two persons. This process may involve a sequence of rigid demands and counterdemands between individuals, or it may involve joint attempts at integrative problem solving. Either way, the influence techniques used are persuasion and promises.

Beyond this, however, the process of bargaining may also involve threats. If one (and only one) bargainer has the capacity to issue threats, this may give him or her some extra leverage in the negotiations. But if both persons can issue threats and inflict punishments on one another, the situation is quite different because it can degenerate into open hostility and aggression. A case of this type—where both persons can issue threats and inflict punishment—is referred to as a situation of **bilateral threat.**

For example, suppose that two persons, Alex and Paul, have reached a deadlock in their negotiations. To coerce a favorable settlement, Alex threatens to inflict punishment if Paul does not make some further concessions. How might Paul respond to this threat? He could decide to comply and make the concessions. On the other hand, suppose Paul decides to issue a counterthreat. Paul's counterthreat will probably be larger in magnitude than Alex's because he wishes to deter Alex from carrying out the original threat. The result will be an escalation of the conflict. At this point Alex might back down, but if he feels committed to his position, he will have no alternative but to respond with yet another, bigger threat. These actions constitute a **threat-counterthreat spiral,** a situation in which bargainers stand firm in their positions and issue increasingly larger threats to inflict more and more damage on one another. (Deutsch, 1973; Smith & Anderson, 1975; Lawler, 1986).

**Trucking Game Studies** A classic laboratory study by Deutsch and Krauss (1960, 1962) demonstrates what can happen when both bargainers have threat capability. This study was a two-person bargaining simulation in which participants took the roles of the chief officers of two trucking companies called Acme and Bolt. Acme's objective was to move its cargo to a destination over a road displayed on a board in front of the subjects. Bolt's objective was the same, except that it moved its cargo in the opposite direction (see Figure 8-5). Each participant's profits depended on the speed with which he or she moved the truck from start to destination. The experiment involved a total of 20 trips. For each trip, a player earned 60 cents minus 1 cent for each second the cargo was in transit between points.

As Figure 8-5 indicates, Acme and Bolt each had two routes: a long one and a short one. The short route permitted the faster trip. This situation created a sharp opposition of interests between Acme and Bolt—and, hence, a need for bargaining—because the short route was only one lane wide and could accommodate only one truck at a time. If both participants were to select the one-lane road as their route, they would not be able to move past one another and would waste valuable time. The logical solution to this problem was for the participants to take turns using the one-lane road, thereby alternating the loss involved in traveling the longer route.

Depending on experimental conditions, participants in this study had different threat capabilities. In one condition (bilateral threat), Acme and Bolt each controlled gates that, when lowered, blocked the other's movement along the one-lane road. In another condition (unilateral threat), only one participant, Acme, controlled a gate that could be used to prevent Bolt from taking the short route. In a third condition (no threat), neither Acme nor Bolt controlled a gate.

In the basic version of this study, the subjects communicated their threats by means of electronic controls (lights); no verbal communication was permitted among participants. Bargaining effectiveness in this study was indexed by the sum of the payoffs to the two players, Acme and Bolt. By definition, effective bargainers were those who required less time to reach a joint agreement and, therefore, achieved higher summed payoffs. The results of this study, displayed in Table 8-2, show that bargaining effectiveness decreased when participants had threat capability. Specifically, the best summed payoffs (that is, the quickest resolution) occurred in the no-threat condition. The next-best summed payoff occurred in the unilateral-threat condition (one gate), while the worst summed payoff (that is, the slowest resolution) occurred in the bilateral-threat condition. Thus, the presence of

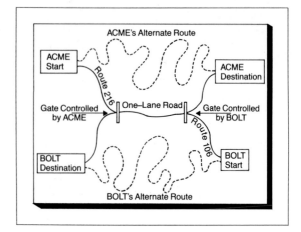

**■ FIGURE 8-5**
ROAD MAP OF THE TRUCKING GAME

In this game, the objective of each player (Acme and Bolt) is to move a truck from start to destination as quickly as possible. The one-lane road (center) provides the fastest route, although both players cannot use it simultaneously. When both players control gates, each player can threaten to block the other's access to the one-lane road. Under these conditions (bilateral threat), conflict between players may escalate.

*Source: Adapted from Deutsch and Krauss, 1960.*

threat capability made the conflict more difficult to resolve and coordination difficult to establish. In fact, when both participants could issue threats, a threat-counterthreat spiral frequently occurred. This created a deadlock that wasted valuable time and resulted in a large negative joint payoff to participants.

Subsequent trucking game experiments investigated threat under conditions where the subjects were able to communicate verbally with each other. In one study (Deutsch & Krauss, 1962), subjects were either required to talk to one another on every play (compulsory communication) or they were given the option to talk if they wished to do so (permissive communication). The findings were similar to those in the original experiment. Once again, even with communication, the bilateral-threat condition led to

■ **TABLE 8-2**

MEAN PAYOFFS TO BARGAINERS IN THE TRUCKING GAME EXPERIMENT

| Payoffs | Threat Condition | | |
| --- | --- | --- | --- |
| | No Threat | Unilateral Threat | Bilateral Threat |
| Acme's Payoff | 122.44 | −118.56 | −406.56 |
| Bolt's Payoff | 80.88 | −287.31 | −468.56 |
| Summed Payoffs (Acme + Bolt) | 203.31 | −405.88 | −875.12 |

Note: In the unilateral threat condition, Acme had the threat capability.

Source: Adapted from Deutsch and Krauss, 1962.

worse outcomes than the unilateral-threat or no-threat conditions. However, a further study (Stech & McClintock, 1981) showed that within the bilateral-threat condition itself, subjects usually obtained more favorable outcomes when they were able to communicate verbally than when they could not.

**Escalation and Deterrence** In bilateral situations where both parties are armed, there is a tension between the temptation to use punishment against the other and the fear of having punishment used against oneself. When the former prevails, a threat-counterthreat spiral ensues, and the conflict escalates; when the latter prevails, a situation of deterrence exists, and the conflict stabilizes.

Therefore, an interesting question is, When does escalation occur, and when does deterrence occur? Several studies have investigated the effects of power differences on escalation in a bilateral situation. Specifically, these studies investigated the case where both persons in a bilateral situation have threat and punishment capability, but one person has more than the other. The outcomes of theoretical interest are threat use, concessions, and deterrence.

To begin, there is evidence that situations of unequal power are less stable than situations of equal power. In other words, situations of unequal power tend to produce greater use of damage tactics (threat, punishments) than do situations of equal power (Lawler, Ford, & Blegen, 1988).

In unequal-power situations, bargainers with more power than their opponets generally tend to behave exploitatively, whereas those with less power tend to behave submissively (Rubin & Brown, 1975). For example, persons who have more power are reluctant to make concessions. Not only may they refuse to initiate concessions, but they may even refuse to match concessions from the low-power bargainer (Smith & Leginski, 1970; Michener et al., 1975). In addition, bargainers in a position of high power tend to have high aspirations and to make few concessions, whereas bargainers with low power tend to match the concessions made by the other.

The power of the strong bargainer also has a deterrent effect on his weaker opponent. Because he fears retaliation, the weaker opponent hesitates to use threats or behave aggressively. The deterrence effect is unmistakable—as the punishment magnitude of the high-power person increases, the aggressive behavior of the low-power person decreases (Michener & Cohen, 1973).

Despite this deterrence effect, situations in which both bargainers have threat capability are potentially explosive. Even a single threat or punitive act may set off a threat-counterthreat spiral. Once a threat has been issued by one party, a matching response (that is, a counter-threat by the other) becomes increasingly likely. Bargainers, in fact, often overmatch or exceed the opponent's threats at low levels of conflict, a tendency that produces conflict spirals (Youngs, 1986). One threat provokes another of larger size, making an agreement increasingly difficult to achieve.

## ■ Summary

Social influence occurs when action by one person (the source) causes another person (the target) to change an opinion or to comply with a directive.

**Communication and Persuasion**  A common form of social influence is persuasion. (1) While some attempts at persuasion succeed, many do not. Persuasive messages can change attitudes and beliefs through either the central route or the peripheral route. A large number of factors —properties of the source, the message, and the target—affect whether a persuasion attempt will succeed in changing beliefs and attitudes. (2) Certain characteristics of the source determine a message's effectiveness. Communicators who are credible (that is, highly expert, trustworthy, and/ or attractive) will generally be more persuasive than communicators who are not. A message with strong arguments coming from multiple, independent sources will have more persuasive impact than the same message from a single source. (3) Message characteristics also determine a message's effectiveness. Highly discrepant messages are more persuasive when they come from a source having high credibility.

Fear-arousing communications (warnings) are most effective when they specify a course of action that can avert impending negative consequences. One-sided messages have more impact than two-sided messages when the audience already agrees with the speaker's viewpoint or is not well informed. (4) Characteristics of the target also determine a message's effectiveness. Intelligent targets are most responsive to complex messages containing strong arguments. Targets who are highly involved with an issue scrutinize messages closely and are more influenced by the strength of the arguments than by peripheral factors, such as communicator attractiveness or credibility. Targets exposed to a refutational defense show a high level of resistance to persuasion, whereas those exposed to a supportive defense show somewhat less resistance. When no defense is present, there was still less resistance to persuasion.

**Threats and Promises**  Threats and promises are influence techniques used primarily to achieve compliance (not persuasion) from the target person. In using threats and promises, the source alters the environment of the target by directly manipulating reward contingencies. (1) The effectiveness of a threat depends on both the magnitude of the punishment involved and the probability that it will be carried out. Greater compliance results from high magnitude and high probability. Similar effects hold true for promises, although of course these involve rewards rather than punishments. (2) Threats and promises pose several problems for the source. The source must maintain surveillance over the target. In addition, threats often arouse resentment or hostility toward the source, and hence can disrupt established relationships.

**Bargaining and Negotiation**  In many situations, influence is bilateral rather than unilateral. (1) When two persons are able to reward one another, bargaining and negotiation frequently result. Certain tactics may lead to favorable outcomes in bargaining. Among these are the use of a low initial offer and the use of mismatching responses to concessions by the opponent. These tactics are risky, however, for they may create a stalemate or a failure to reach agreement. A superior approach to bargaining, where possible, is to develop integrative proposals that benefit both parties. (2) When each party controls punishments, bargaining may escalate into bilateral threat and conflict. If the two persons have very unequal power, the weaker person will probably hesitate to issue threats and will accommodate the demands of the stronger person. But bilateral situations are volatile, in part because threats by one party are often matched with increasingly larger threats by the other.

## Key Terms

| | |
|---|---|
| BARGAINING | 237 |
| BARGAINING RANGE | 238 |
| BILATERAL THREAT | 241 |
| COMMUNICATOR CREDIBILITY | 220 |
| COMPLIANCE | 229 |
| CONCESSION MAGNITUDE | 239 |
| DISCREPANT MESSAGE | 222 |
| EXCHANGE | 237 |
| INTEGRATIVE PROPOSAL | 240 |
| MASS MEDIA | 230 |
| MEDIA CAMPAIGN | 230 |
| PERSUASION | 218 |
| PROMISE | 231 |
| SOCIAL INFLUENCE | 216 |
| SOURCE | 216 |
| SUBJECTIVE EXPECTED VALUE (SEV) | 233 |
| TARGET | 216 |
| THREAT | 229 |
| THREAT-COUNTERTHREAT SPIRAL | 241 |

9

chapter

# Self-Presentation and Impression Management

# ■ Introduction

Remember the shabbily dressed street violinist you saw downtown? Your friend remarked on his artistic demeanor and sensitive eyes, and you agreed it was a shame a guy with such talent couldn't afford to study at Juilliard. When people dropped coins into his battered violin case you both felt pleased.

If you lived in New York City, you may have seen 29-year-old Richard Wexler (*Newsweek*, 1978). By day, Richard lives the good life in his fashionable Manhattan apartment. But come nightfall, Wexler kicks off his expensive clothes and Italian-made loafers, slips into torn sneakers, tattered jeans, and a frayed shirt, and is transformed into Richie, the ragged street artist. He hails a cab for Broadway, where he sets up a sign on the sidewalk ("Violinist Needs Money to Further Studies") and serenades theatergoers with sentimental show tunes. On a typical night, Richie pulls in more than $300 for an hour's fiddling—enough to buy himself $400 watches and Bermuda vacations.

Richie is a true artist—an artist at managing impressions. He masterfully creates the perception of himself he desires. Richie capitalizes on an important principle in social psychology: The way we perceive people's behavior and the impressions we form based on their behavior largely determine how we act toward them. Because people respond to one another on the basis of social identities, they find it advantageous to control the self-images they present (Schlenker, 1980). Thus, by presenting himself as a "struggling violinist," Richie influences the behavior of passersby and achieves his goal of making lots of money easily.

Although few of us make our living by intentionally creating a false impression, we all engage in presenting particular images of who we are. When we shout or whisper, dress up or dress down, smile or frown, we actively influence the impressions others form of us. In fact, presenting some image of ourselves to others is an inescapable aspect of all social interaction.

The term **self-presentation** refers to all our attempts, both conscious and unconscious, to control the images we project in social interaction. Whenever we care about the impressions others have of us, self-presentation is intertwined with other aspects of behavior (Jones & Pittman, 1982). Self-presentation may involve carefully calculated tactics designed to make a particular impression. Such intentional use of tactics to manipulate the impressions others form of us is called **impression management**.

This chapter considers the ways in which people actively determine how others perceive them. It examines the following questions:

1. What is the content of self-presentation, what are its goals, and what are the obstacles that prevent achievement of these goals?

2. What special impression-management tactics can we use when we want to claim a particular identity such as "overworked employee," "attractive date," or "competent student"? Under what conditions do we choose one impression-management tactic over another?

3. How can people detect when impression-management tactics are being used against them? That is, how do they unmask the deceptive impression manager?

4. What are the consequences when people fail to project the social identities they desire?

# ■ Self-Presentation in Everyday Life

The main goal of self-presentation is to project a particular social identity in a given situation. Thus, Richie was concerned with convincing passersby of his identity as a struggling violinist who deserved contributions. But social identities cannot be isolated from the social context. Richie would have failed, for example, had passersby interpreted his performance as part of an advertising campaign for a newly opened Broadway play.

To project social identities successfully, people must share understandings about the situation in which they are participating. Successful self-presentation involves efforts to control the definition of the situation, to disclose information about the self consistent with the claimed identity and the actor's goals, and to avoid certain obstacles. Each of these will be discussed in this section.

## The Definition of the Situation

In order for social interaction to proceed smoothly, people must share a common perception of what is happening between them. In other words, people who are interacting must achieve a shared **definition of the situation**— an agreement about who they are, what actions are appropriate, and what their behaviors mean. The view that interaction depends on shared understandings of social reality is central to the theory known as *symbolic interactionism* (Blumer, 1962; Stryker, 1980). According to this theory, people create shared definitions during interaction by negotiating the meaning of events (McCall & Simmons, 1978; Stryker & Gottlieb, 1981). In these negotiations people must answer three questions: (1) What type of social occasion is at hand? (2) What identities will be granted? and (3) How much leeway will be given to enact roles in unique ways?

**Frames** The first requirement in defining the situation is for people to agree on the type of social occasion at hand. Is it a wedding? fishing trip? family reunion? job interview? The type of social occasion people recognize themselves to be in is called the **frame** of interaction (Goffman, 1974). (A frame can be viewed as one type of cognitive schema, as discussed in Chapter 5.) Each frame is comprised of a set of stable, widely known rules or conventions that indicate what roles are operative and what kinds of behavior are appropriate. When people recognize a social occasion as a wedding, for example, they immediately know that there will be a bride, a groom, and a clergyperson, that the others present are mostly friends and relatives of the couple, and that it is acceptable (indeed, appropriate) to kiss the bride.

The frame of interaction is usually known in advance or else quickly discovered. Sometimes, however, the frame of interaction must be negotiated. When a family sends their wayward teenager to a physician for a talk, for example, the discussion may begin with subtle negotiations about whether this is a "psychiatric interview" or merely a "friendly chat." Once negotiated, the frame limits the potential meanings that any particular behavior can have (Gonos, 1977). If the situation is defined as a "psychiatric interview," for example, the jokes the teenager tells may now be interpreted as symptoms of illness, not as friendly banter.

**Identities**  The second requirement in defining a situation is for people to agree on the identities they will grant one another. That is, people must agree on the type of person they will treat each other as being. To a large extent, each person's identity is determined by the frame. For example, a teenager in a psychiatric interview can no longer claim an identity as a "normal, well-adjusted kid."

Each person participating in an interaction has a **situated social identity**—a conception of who he or she is in relation to the other people involved in the situation (Alexander & Wiley, 1981; Alexander & Rudd, 1984). Identities are "situated" in the sense that they are specific to a particular situation. The identity projected while discussing a film (insightful critic) differs from the identity projected when asking for a small loan (reliable friend). These identities are "social" in the sense that they are adopted for purposes of smooth social interaction, although they are not necessarily accepted privately. In order to avoid unpleasant arguments, for example, you and your friend might relate to one another as if you were more insightful or more reliable than either of you truly believed yourselves to be.

Much of the time, we want perceivers to grant to us the same identities that we ascribe to ourselves (Swann, 1987). Yet, many of our identities are not self-evident to others. This is because their perceptions of us are based on stereotypes, implicit personality theories, expectancies, and motivations brought to any situation. Their biases influence the identities they perceive and grant to us and to others. Thus, even if the situated identity that a person wishes to claim is "true," it may be necessary to dramatize it with self-presentation (Goffman, 1959). For instance, if some adolescents display their usual nonchalant, defiant image when stopped by police, they might be arrested even if innocent. They are much more likely to avoid arrest if they dramatize their innocence by presenting a polite, regretful demeanor (Piliavin & Briar, 1964). Thus, the "true" identity of innocent adolescent must be presented appropriately.

**Enacted Roles**  The third requirement in defining the situation is for people to agree on the roles to be enacted. Of course, the frames and identities on which people agree when defining the situation place some limits on the roles they may appropriately enact. Still, it is necessary to negotiate the ways in which people will actually enact these roles and to what extent idiosyncratic behavior will be accepted (Turner, 1962; Cicourel, 1972).

People tend to enact roles in ways that express their unique personalities. Consider, for example, how different people enact the role of student in a classroom setting. One person may participate actively in the discussion, another may listen attentively, a third may take notes, a fourth may read a newspaper, and a fifth may fall asleep. These different role enactments reflect each individual's interests, styles of learning, and reasons for attending the class. Although some of these idiosyncratic role enactments are accepted, others may be contested. For example, some instructors prohibit the reading of newspapers or sleeping in their classes. If the class is to

Even if they have done nothing wrong, these teenagers had best dramatize their innocence by presenting themselves to this policeman with polite deference. True identitities may not be self-evident because perceivers are biased by the stereotypes and expectations they bring to a situation.

proceed smoothly, the participants must reach an agreement about the boundaries of acceptable role behavior.

## Self-Disclosure

A primary means we use to make identity claims is to reveal information about ourselves. When we first meet someone, we may discuss only superficial things; but eventually, as we get to know the other better, we reveal more personal and intimate details about ourselves. This may include information about our needs, attitudes, values, background, worries, and aspirations (Archer, 1980). This process of revealing personal aspects of one's feelings and behavior to others is termed **self-disclosure** (Jourard, 1971; Derlega & Chaiken, 1975).

Although self-disclosure can be unilateral, it usually is not. There is a widely accepted social norm that one person's disclosures should be met with disclosures at a similar level of intimacy by another (Ehrlich & Graven, 1971; Cohn & Strassberg, 1983). Because people are responsive to this norm of *reciprocity in disclosure,* self-disclosure is usually two-sided and gradual. One qualification, however, is that strict reciprocity in disclosure is more prevalent in newly established relationships or developing friendships than in established ones (Davis, 1976; Won-Doornink, 1979).

We do not disclose ourselves to just anyone; that is, we are more prone to engage in self-disclosure with some persons than with others. For instance, we are more likely to disclose personal information to individuals we find attractive and desirable as friends (Kleinke & Kahn, 1980). Then, too, males and females differ in self-disclosure. Males find it more difficult than females to disclose information about themselves and their relationships (Hacker, 1981; Davidson & Duberman, 1982), in part because men see excessive disclosure as a sign of weakness (Cunningham, 1981). Men and women disclose different things about themselves. When men engage in self-disclosure, they tend to discuss such topics as aggressive behavior or risk taking, whereas women tend to disclose with respect to such topics as immature behavior or flaws in their appearance (Derlega, Durham, Gockel, & Sholis, 1981).

Self-disclosure often increases liking and gains social approval. People who reveal private information about themselves often increase their partner's liking for them, especially if the content of the self-disclosure complements what their partner has revealed (Daher & Banikiotes, 1976; Davis & Perkowitz, 1979).

Under some conditions, however, self-disclosure may inspire disliking rather than liking. If it is too intimate for the depth of the relationship (a new acquaintance discussing her deepest anxieties and emotions), or if it uncovers profound dissimilarities with the partner (revealing strong religious commitment to a nonbeliever),

These chess players seem to be building up trust and liking through reciprocal self-disclosure. But self-disclosure may also be used in tactical impression management to obtain information and to create a relationship that will later be exploited for personal advantage.

self-disclosure may produce disliking (Cozby, 1973; Derlega & Grzelak, 1979).

Because we have substantial control over what we do (and do not) reveal about ourselves, self-disclosure is an important process through which we not only promote friendship and liking but also make identity claims. For instance, if a young woman is on a first date with a man and wishes to make a good first impression, she will be fairly selective regarding what she does and does not disclose because she wishes to establish a certain identity in his eyes.

## Goals of Self-Presentation

Self-presentation permeates most social interaction, at least to some degree. Usually, either of two major goals will underlie self-presentation (Baumeister, 1982; Baumeister & Hutton, 1987). First, an individual may wish to make a specific impression (usually a positive impres-

sion) on others. He will try to do this because he wants the others to approve of him or to behave in certain ways, such as give him rewards at some future time (Arkin, 1980). For example, during a job interview the objective is to appear responsible and competent in order to get the job and gain the rewards that go with it.

Second, an individual may wish to construct and maintain a public image that is congruent with his or her own self-concept (ideal self). For instance, a woman may have an ideal image of the kind of person she would like to be, and she may strive to come as close to the ideal in her actual behavior as she possibly can. In each of these cases, the individual either may or may not be conscious of the intention behind self-presentation.

If an individual is trying to make a positive impression on an audience, he or she typically aims at creating the most highly approved situated identity possible. Selecting the appropriate behaviors to create a favorable identity often requires careful planning. First, the individual must discover those qualities that participants in the interaction are most likely to approve. A partygoer, for example, must recognize that friendliness is appreciated by fellow bon vivants, whereas a saleswoman must understand that honesty is considered important by customers. Second, the individual must anticipate which of various behaviors will be seen as indicating that he or she possesses the qualities desired in that situation. For example, the partygoer must anticipate whether or not telling an off-color joke or complimenting the hostess will make him appear friendly. Only after correctly assessing the situation can the individual confidently choose behaviors to gain the desired situated identity.

Often there are norms that dictate how we should present ourselves in order to gain approval. When such norms are well defined and obvious, there will be widespread agreement about which behaviors will lead to positive or negative evaluations. Under these circumstances, individuals find it easy to decide how to behave and to predict others' reactions. When the norms are unclear or ambiguous, however, individuals do not know how to present themselves, and their behavior is therefore more erratic and unpredictable (Alexander & Lauderdale, 1977).

## Obstacles to Successful Self-Presentation

People do not always succeed in establishing the situated social identity that they desire. Successful self-presentation occurs when participants in the interaction respond in a way that meshes with the actions used to claim a particular identity. Success occurs, for example, when they accept offers of help (thereby conferring the identity of "altruist" on the person offering), when they reciprocate romantic embraces (conferring the identity of "lover"), and when they accept commands and leadership directives (conferring the identity of "leader"). Thus, one's success or failure in establishing a situated social

identity depends on the complementary responses of others.

This process was demonstrated in a study in which students read a paragraph describing a conversation among three persons (Turner & Shosid, 1976). One person in the conversation was giving directions to two others about planning a party. When the others accepted these directions—thereby providing a complementary response to his attempted leadership—almost all of the students characterized the person who gave the directions as a "leader" or "organizer." When identical statements by the would-be leader were not met with a complementary response by the other two persons, almost half of the subjects failed to recognize this same person as a leader. The results of this study are shown in Table 9-1.

Why do people sometimes reject the identity claims of others? Two major reasons are: (1) the identity claims may conflict with the interests of interaction partners, and (2) the claims may be inappropriate to the frame of interaction.

**Conflict of Interest** One obstacle to successful self-presentation is conflict between the goals of the participants in the interaction. In a court of law, for example, a witness for the defense may have difficulty maintaining a social identity as a trustworthy person when faced with cross-examination by a hostile prosecuting attorney. Less extreme but still troublesome conflicts may occur in many everyday interactions. Consider the student who discusses his poor examination results with his instructor. By questioning the fairness of the exam or the quality of the grading, the student can present himself as knowledgeable and hard working, despite his grades. But this self-presentation may conflict with the instructor's identity as a competent preparer of exams and an impartial, careful grader. Hence, the student and the instructor may challenge each other's self-presentations.

**Inappropriate Claims** A second obstacle to successful self-presentation arises when an individual claims an identity that conflicts with the

■ **TABLE 9-1**

PERCENTAGE OF OBSERVERS WHO GRANTED DIFFERENT IDENTITIES TO A PERSON WHO DIRECTED OTHERS

| Responses of Others | Identity Granted to Person Who Gave Directions | | |
| --- | --- | --- | --- |
| | "Leader" or "Organizer" | No Clear Identity | Negative Identity ("Troublemaker") |
| Complementary (accept direction) | 92% | 6% | 2% |
| Not complementary (reject direction) | 56% | 36% | 8% |

*Source: Adapted from Turner and Shosid, 1976.*

frame of interaction. Flirting with a member of the opposite sex during a funeral, for example, is more likely to create a situated social identity of "uncultured boor" than of "attractive potential date." Claims inappropriate to the interaction frame usually occur when a person is ignorant of the rules specifying appropriate behavior on such occasions (this might happen, for example, if one has never attended a funeral before), or when an individual's need to reaffirm a recently challenged social identity blinds him or her to the rules (for example, the recently jilted suitor).

## ■ Tactical Impression Management

As noted above, self-presentation is an everyday occurrence. It is inherent in most, if not all, social situations. Probably most people use self-presentation to create impressions of themselves that are "true"—that is, consistent with their own self-concept. Nevertheless, it must be recognized that individuals can intentionally present themselves in such a way as to create false, exaggerated, or misleading images in the eyes of others. The process of creating false images is referred to as tactical *impression management.*

There are many specific reasons why we might engage in tactical impression management

(Jones & Pittman, 1982; Tetlock & Manstead, 1985). One major reason is to make others like us more than would otherwise be the case (ingratiation). Other reasons for impression management are to make others fear us (intimidation), respect our abilities (self-promotion), respect our morals (exemplification), or feel sorry for us (supplication). Fundamentally, however, when we engage in impression management, the main motive usually is to increase our power and control over valued outcomes mediated by other persons (Tedeschi, 1981).

Social interaction may be viewed as a kind of drama in which each person performs a "line" —a set of carefully chosen verbal and nonverbal acts (Goffman, 1963, 1967). These acts are selected to communicate the person's definition of the situation and to establish the desired social identity. In order to manipulate impressions, people attempt to control the information others obtain about them. This section examines in detail a variety of tactics used in impression management.

### Managing Appearances

Consider the planning and control of appearances. Appearances refer to everything about a person that others may observe— clothes, grooming, habits such as smoking or chewing gum, choice and arrangement of personal possessions, verbal communications (accents, vocabulary), and nonverbal communications. Through the appearances we present, we indicate to others the line of action we intend to pursue and the kind of persons we are (Stone, 1962).

**Physical Appearances and Props** In many situations, we consciously arrange our clothing and physical appearance to achieve some effect. This would be true, for example, if we were attending a dance or going on a date. It is also true when we go for a job interview, as illustrated in a study of female job applicants (von Baeyer, Sherk, & Zanna, 1981). In this study, some applicants were led to believe that their male interviewer felt the ideal female employee should conform closely to the traditional female stereotype (passive, gentle, and so on). Other applicants were led to believe that he felt the ideal female employee should be nontraditional (independent, assertive, and so on). Results indicated that applicants managed their physical appearance to match their interviewer's stereotyped expectations. Those expecting to meet the traditionalist wore more makeup and used a greater number of accessories, such as earrings, than those planning to meet the nontraditional interviewer.

A job interview is a special situation, but even in routine daily interaction our appearances are hardly left to chance. Managing appearances to present a desired identity affects everyday choices of which clothes to wear, how to arrange our hair, whether to shave, whether to use perfume, and so on.

The impression an individual makes on others depends not only on clothes, makeup, and grooming but also on physical props in the environment. For instance, the impression a woman makes on her friends and acquaintances will depend in part on the props she uses—the big pile of books she places on her desk, the music she selects to play on her CD, the wine she serves. It can even include the art hanging on her wall. In many circles, modern art in an individual's home conveys to visitors that the owner is a person of taste and discrimination and should be treated in those terms. Thus, managing impressions includes not only clothes and grooming, but also the manipulation of environmental props.

**Regions** Goffman (1959) draws a parallel with the notion of a theater's frontstage and backstage to illuminate other ways we manage appearances. **Front regions** are settings in which people carry out interaction performances and exert efforts to maintain appropriate appearances. In a restaurant dining room, for example, waiters smile and courteously offer to help customers. **Back regions** are settings inaccessible to outsiders in which people routinely and knowingly violate the lines they present in front regions. Behind the kitchen doors, those same waiters shout, slop food on plates, and even mimic their customers.

Back regions are often used to prepare, rehearse, and rehash performances. For example, anticipating dinner with the boss, a couple will take advantage of the privacy of their home to rehearse the topics to discuss during the meal. Then, at home after their dinner performance, they will let down their hair in a raucous postmortem. Front and back regions are separated by barriers to perception, like the restaurant's kitchen door and the walls of a home. When performers cross these barriers, their behavior changes—smiles sprout or fade, and postures straighten or relax.

The barriers between front and back regions are crucial to successful impression management because they block access to the inevitable violations of impressions that occur during preparation and relaxation. The breakdown of such barriers, for instance, has undermined the ability of national figures to project a dignified, leaderlike image in recent years (Meyrowitz, 1985). Because the media now expose almost every detail of a politician's life, for example,

An undertaker employs a range of impression management techniques to create an atmosphere of quiet comfort. The opulent front region of a casket showroom is carefully separated from the back region where a team of workers prepares the body for burial.

presidents and other officials are often caught off guard. They are seen expressing views and performing actions they would prefer to keep hidden from the public. The slips and inconsistencies that are inevitably revealed undercut the leaders' stature. American presidents may find it difficult to project a heroic identity when the media show one bumping his head on a door, another nearly collapsing while jogging, and a third nodding off during an audience with the Pope. It was much easier to be a hero in the days of Jefferson or Lincoln. Then, reporters were barred from the White House, and the electronic media that penetrate the barriers to the president had yet to be invented.

## Ingratiation

On most occasions people want to be liked. Not only do we find it inherently pleasant, but liking may gain us a promotion, a better grade, or a date, and it may save us from being fired, failed, or ignored. How do we persuade others to like us? Often we are sincere in our relations with others; at other times we engage in the impression-management technique called ingratiation (Jones, 1964; Jones & Wortman, 1973). **Ingratiation** refers to the deliberate use of deception to increase a target person's liking for us in hopes of gaining tangible benefits that he or she controls. Techniques such as flattery or exaggerating one's admirable qualities are often used to ingratiate oneself with others.

Ingratiation is based on the idea that target persons are more likely to grant benefits to someone they like. People ingratiate themselves with a target only when they think that he or she has the power to decide whom to benefit. That is, people tend to ingratiate themselves with targets who are not constrained by preestablished regulations but who set their own standards for allocating benefits (Jones et al., 1965). For example, students are less likely to try to ingratiate themselves with instructors who base grades on objective multiple-choice exams than with instructors who assign papers and whose grading may be influenced by their liking for a student.

There are three major ingratiation tactics, each of which involves the use of deception. They include opinion conformity (pretending to share the target person's views on important issues), other enhancement (outright flattery or complimenting the target person), and selective self-presentation (exaggerating one's own admirable qualities).

**Opinion Conformity** Faced with a target person who has discretionary power, people may ingratiate themselves by expressing insincere agreement on important issues. This tactic, termed *opinion conformity,* is based on the underlying fact that individuals tend to like others more if they hold opinions similar to their own (Byrne, 1971). But obvious or excessive opinion conformity is likely to arouse suspicion. In order to increase their credibility and minimize detection, ingratiators may attempt to mix their opinion conformity on important issues with disagreement on unimportant issues.

Opinion conformity sometimes requires us to tailor the content of the opinions we express to match a target person's general values rather than any specific opinions he or she may hold. Such clever modification of expressed opinions was demonstrated by students who were failing at an experimental task of judging advertising slo-gans (Jones et al., 1965). Some of the students were led to believe that their supervisor valued friendliness and social compatibility, whereas others heard that the supervisor valued efficiency and independence. The students were told that a positive evaluation from their supervisor would help them earn $10. As a result, students shifted their publicly expressed opinions to conform with their supervisor's values. By agreeing with the supervisor, students in the first group implied that they too valued friendliness and compatibility. In contrast, students in the second group expressed less agreement with the supervisor in order to show their independence and to increase their instructor's liking for them.

**Other Enhancement** People may attempt to ingratiate themselves with a target person by using *other enhancement*—that is, flattery. To be effective, flattery cannot be indiscriminate. More than two centuries ago, Lord Chesterfield (1774) asserted that people are best flattered in those areas in which they wish to excel, particularly if they are somewhat unsure of themselves. This assertion was tested in a study similar to that cited earlier in which students were told that their supervisor valued either efficiency or sociability (Michener, Plazewski, & Vaske, 1979). The supervisor was an appropriate target for ingratiation, because the students' possible earnings depended on the evaluations they would receive from her. Before these evaluations were made, the students had a chance to flatter their supervisor. The experimenter asked them to rate her efficiency and sociability on scales that their supervisor would be shown. The results showed that the supervisor's values channeled the form of flattery the students used. Students who believed the supervisor valued efficiency rated her higher on efficiency than on sociability, whereas students who believed she valued sociability rated her higher on sociability than on efficiency. Thus, the students were discriminating in their use of praise. In addition, they avoided extreme ratings that might suggest insincerity.

Box 9-1

## PLAYING DUMB

"Playing dumb" is one widely employed ingratiation tactic. By playing dumb, people give the target person a sense of superiority by presenting themselves as inferior. Thus, playing dumb is a form of other enhancement. Although popular belief and early research suggest that women are more prone to play dumb (Wallin, 1950), a national survey of American adults indicates otherwise (Gove, Hughes, & Geerken, 1980).

Significantly more males than females agreed that they had pretended, at least once, to be less intelligent or knowledgeable than they really were. As shown in Table B-9-1, men play dumb more often than women in many situations.

What leads people to play dumb? The data indicate that people who use this method are often young, highly educated, of high occupational status, and male. These are certainly not the characteristics usually associated with people who are really inferior. But such people are the ones who are most likely to find themselves in settings where playing dumb may be necessary or appropriate (Gove et al., 1980). Many of these people are located near the bottom of an occupational ladder they aspire to climb, in a setting where intelligence and knowledge are prized. Under these circumstances, a person's relatively low status may require deferring to one's superiors despite one's own abilities.

In contemporary American society, young, well-educated males are more likely to be channeled into lower-status positions in competitive occupations where knowledge is valued and where they are expected to defer to their elders. This is often the fate of junior executives, law clerks, and graduate students, for example. In these situations, people of lower status stand to gain by hiding any intellectual superiority they feel—that is, by playing dumb.

The survey also shows that women play dumb significantly more often than men in relating to their spouses. Perhaps this reflects the continuing cultural expectations that women should refrain from displaying superior knowledge that might challenge their husband's assumed superiority. Our reasoning about the causes of playing dumb suggests that college-educated women probably play dumb more often than women with less education. This is because highly educated women are especially likely to find themselves in school and career settings where knowledge is valued but deference to superiors is expected.

■ TABLE B-9-1

PERCENTAGE OF PEOPLE WHO REPORTED "PLAYING DUMB"

| | Overall | Target Person | | | | | |
| | | Date | Spouse | Boss | Co-workers | Friends | Strangers |
|---|---|---|---|---|---|---|---|
| Males (1,065) | 31.1 | 8.2 | 6.0 | 13.1 | 14.6 | 12.6 | 15.8 |
| Females (1,182) | 22.9 | 9.0 | 10.0 | 5.2 | 6.3 | 8.7 | 7.7 |

*Source: Adapted from Gove et al., 1980.*

**Selective Self-Presentation** A third major tactic of ingratiation is *selective self-presentation,* which involves the explicit presentation or description of one's own attributes to increase the likelihood of being judged attractive by the target. There are two quite distinct forms of self-presentation. On the one hand, a person may present herself in such a way as to advertise her strengths, virtues, and admirable qualities. When successful, this tactic is a quick and easy way to generate a desirable public identity and gain liking by others. On the other hand, she may present herself in very humble or modest terms. A modest presentation of self can be an effective way to increase approval and liking, especially when it aligns the ingratiator with such important cultural values as honesty and objectivity in self-appraisal.

Of these two variations, the overly positive (self-aggrandizing) presentation of self is the more common. For instance, if subjects in experimental settings are asked to describe themselves in such a way as to make a good impression, the general tendency is for them to emphasize their positive attributes and play down their weaknesses (Jones, 1964). People may mention their honesty, competence, and friendliness directly, or they may imply their admirable qualities indirectly, as by publicly attributing their behavior to appropriate motives (Tetlock, 1981).

Although frequently effective, the tactic of selectively emphasizing our admirable qualities can be quite risky, especially if the target knows enough about us to suspect we are boasting. There is also the danger that future events may prove our claims invalid. Ingratiators, therefore, prefer to present self-enhancing descriptions only when these risks are minimal—that is, when the target person does not know them well and has no way to check their future performances (Schlenker, 1975; Frey, 1978). For example, people are more likely to boast about themselves to strangers who have no way of checking their stories than to close acquaintances who may learn the truth. People who believe their performances will support boastful claims tend to risk selective self-presentation even when others can check up on their future performances. Bowlers who believe they have a good chance of excelling in a tournament, for example, will often talk about their prowess during prematch discussions. Caution prevails, however, when people believe they will not perform successfully in the future and when they know others already have negative information about them (Schlenker, 1975; Ungar, 1980).

Due to the risks inherent in self-enhancement, the opposite approach—a self-deprecating or modest self-presentation—is often a safer ingratiation tactic. But does modesty increase others' liking for us? Consider the options available to a person who has succeeded or failed at a task. One who succeeds has the choice between claiming full credit for the success (a self-enhancing response) or disclaiming credit (a modest response). One who fails may deny blame for the failure (self-enhancing) or accept responsibility for it (modesty). Which strategy is more effective?

In a study addressing this question, subjects read about a teacher who had either succeeded or failed in teaching his pupil and who gave either a self-enhancing or a modest explanation for the outcome (Tetlock, 1980). Subjects liked the teachers more when teachers responded to their success or failure with modest claims. They viewed these teachers as more concerned for others than teachers whose responses were self-enhancing. This finding demonstrates that modesty can be a successful ingratiating technique.

In another relevant study, members of a group were asked to evaluate other members following the group's success or failure at a task (Forsyth, Berger, & Mitchell, 1981). Group members reported greater liking for those who took blame for the group's failure or credited others for the group's success (modesty) than for those who blamed others for failure and claimed credit themselves for the group's success (self-enhancement). Taken together, these studies suggest that when observers have clear-cut, objective evidence about someone's performance

—whether favorable or unfavorable—modesty is the more effective ingratiation tactic. Under these conditions, self-enhancement tends to reduce others' liking because it is perceived as deceitful impression management (Carlston & Shovar, 1983).

## Aligning Actions

In the course of interaction, occasional failures of impression management are inevitable. In pursuit of our goals we may sometimes be caught performing actions that violate group norms (missing an appointment) or contravene laws (running a red light). Such actions potentially undermine the social identities we have been claiming, challenge the definition of the situation we have negotiated, and disrupt smooth interaction. When this occurs, people engage in a variety of **aligning actions**—attempts to define their apparently questionable conduct as actually in line with cultural norms. Aligning actions are tactics intended to repair cherished social identities, restore meaning to the situation, and reestablish smooth interaction (Hewitt & Stokes, 1975; Spencer, 1987). Three major types of aligning actions are considered below.

**Vocabularies of Motive** Our initial reaction to failures of impression management is to request an explanation for someone's unsuitable behavior. "Why are you late handing in this paper?" an instructor may ask. Explanations of the motives that supposedly underlie behavior are termed *motive talk*. "I was taking care of my parents who are recovering from a car crash," the student may reply. The explanation of our motives is intended to deny that the behavior has negative implications for the identity in question. Not every motive is equally acceptable. Rather, we must learn **vocabularies of motive**—the sets of explanations regarded as appropriate by particular groups in specific situations (Mills, 1970; Karp & Yoels, 1986). For example, "God's will" is a good explanation for a priest's celibacy, but it would raise eyebrows if it were used by a labor leader to explain his compromises on a union contract.

**Disclaimers** When people anticipate that the actions they will take may be disruptive, they often employ **disclaimers,** which are verbal assertions intended to ward off any negative implications of these actions in advance (Hewitt & Stokes, 1975). Different social conditions require different types of disclaimers.

When individuals are certain that an intended act is discrediting, they use disclaimers to acknowledge that although the act ordinarily implies a negative identity, theirs is an extraordinary case. For example, before making a bigoted remark, a person may point to his extraordinary credentials ("My best friend is Hispanic, but . . ."). Similar disclaimers are used prior to acts that would normally undermine one's identity as moral ("I know I'm breaking the rules, but . . .") or as mentally competent ("This may seem crazy to you, but . . ."). These disclaimers emphasize that the person is aware that the act could threaten his identity, but that he is appealing to a higher morality or to a superior competence. Still other disclaimers plead for a suspension of judgment until the whole event is clear: "Please hear me out before you jump to conclusions."

When individuals are not certain how others will react and care little about the potentially threatened identity, they are more likely to preface their actions with hedging remarks ("I'm no expert, but . . ." or "I could be wrong, but . . ."). Such remarks proclaim in advance that possible mistakes or failures should not reflect on one's crucial identities.

**Accounts** When disclaimers are not given or are rejected, accounts are necessary to repair the damage. **Accounts** are the explanations people offer after they have performed acts that threaten their social identities (Scott & Lyman, 1968; Schlenker, 1980). There are two types of accounts: those that excuse the unsuitable behavior and those that justify it. Excuses minimize one's responsibility by citing uncontrollable events ("My car broke down"), coercive external pressures ("She made me do it"), or compelling internal pressures ("I suddenly felt dizzy"). Presenting an excuse reduces the observer's tendency to hold the individual responsible or to make negative inferences about his character (Riordan, Marlin, & Kellogg, 1983; Weiner et al., 1987). Justifications admit responsibility, but, at the same time, they define the behavior as appropriate under the circumstances ("Sure I hit him, but he hit me first") or as prompted by praiseworthy motives ("It was for his own good"). Justifications reduce the perceived wrongness of the behavior.

Accounts are more likely to be accepted when their content appears truthful and conforms with the explanations commonly used for such behavior (Riordan et al., 1983). Even so, accounts are sometimes rejected. Accounts are honored more readily when the person who gives them is trustworthy, penitent, and of superior status and when the identity violation is not serious (Blumstein, 1974). Thus, we are more likely to accept a psychiatrist's quiet explanation that he struck an elderly mental patient because she kept shouting and would not talk with him than to accept a delinquent's defiant use of the same excuse to explain why he struck an elderly woman.

## Altercasting

The tactics discussed so far demonstrate how people claim identities and protect them from being discredited. But the actions of one person in an encounter also place limits on who the others can claim to be. Therefore, it makes sense to impose identities on others that complement those we claim and to pressure others to enact roles that mesh with the roles we wish to construct for ourselves. **Altercasting** is the use of tactics to impose roles and identities on others. Through altercasting, we cast others into situated roles and identities that are to our advantage (Weinstein & Deutschberger, 1963).

In general, altercasting involves treating others as if they already have the roles and identities we wish to impose on them. "After all I've done for you . . ." is a typical altercasting remark. It treats the other as indebted and thereby sets up the obligation to reciprocate a favor. Teachers engage in altercasting when they tell a student, "I know you can do better than that." This remark pressures the student to live up to an imposed identity of competence. Altercasting can also entail carefully planned duplicity. An employer may invite subordinates to dinner, for example, casting them as personal friends in hopes of eliciting employee secrets.

Putting others on the defensive is an especially common form of altercasting: "Explain to the voters why you can't control the runaway national debt," says the challenger, altercasting the incumbent official as incompetent in running the economy. Should the incumbent rise to his own defense, he admits that the charge merits discussion and that the negative identity may be correct. Should he remain silent, he implies acceptance of the altercast identity because he may be unable to refute it. Putting others on the defensive is a powerful technique used against rivals, because the negative identity it imposes is so difficult to escape.

In order to interact smoothly with one another, people must reach a working consensus regarding their situated identities (Goffman, 1959). Bargaining with respect to identities is therefore an essential feature of social interaction. In the bargaining process, we concede to others the identities they desire in return for their acceptance of our own identity claims. We also try to deny the identity claims of others that would prevent us from achieving or maintaining our own prized social identities.

The use of altercasting in the give-and-take of identity bargaining is nicely demonstrated in a study of interaction among dates (Blumstein, 1975). Women were instructed to claim an identity of "healthily assertive" by altercasting their dates into a more submissive identity. The women did this by making remarks to their dates such as, "Must you insist on making all the decisions?" Some of the men conceded the assertive identity claimed by their dates by presenting a self consistent with the altercast ("Sorry I've been so pushy. Whatever you say goes."). Other men rejected their date's assertive identity by altercasting the woman in return, pressuring her to return to a submissive identity ("You always liked me to make the decisions before. What's up?").

This research reveals one of the factors determining whether people resist altercasting or whether they give in to it. Men who had indicated earlier that dominance was an important part of their self-concept tended to resist their date's altercasting, whereas men who had rated dominance as unimportant tended to accept the submissive identity imposed by the women. In general, we tend to concede identities that are unimportant to our overall self-concept, while we reject altercasting aimed at our more central identities.

## ■ Unmasking the Impression Manager

Tactical impression management is inherently deceitful because it tries to create a false image. Sometimes the people being deceived will accept the false image because they have little to gain by questioning the sincerity of others. For instance, morticians strive to convey an air of sympathy and concern even though they did not know the deceased person and are bored by their client's grief. But mourning relatives ask very few questions because they would only be more upset by discovering the mortician's true feelings. In other cases, however, unmasking the impression manager is vital for protecting our own interests. In attempting to win a contract, for example, house builders may claim to be reliable businessmen and skilled artisans even when they are total frauds. To the homeowner, it may be worth thousands of dollars to determine whether the builder's hearty handshake belongs to a fly-by-night operator before making a down payment.

How do people go about trying to unmask the impression manager? In general, they attend to two major types of information: the ulterior motives the other person has for an action and the nonverbal cues that accompany the action.

### Ulterior Motives

Recognition that another person has a strong ulterior motive for his behavior usually colors an interaction. For example, when a used-car salesman tells us that a battered vehicle with sagging springs was driven only on Sundays by his retired aunt, his ulterior motive is transparent, and we are likely to suspect deceit. In such a case, we may come to distrust the salesman, discount what he says about *any* car on his lot, and decide not to do business with him.

When ulterior motives become apparent to a target, they undermine the success of ingratiation tactics. For instance, in an early study (Dickoff, 1961), an experimenter expressed praise for the performances of her subjects under two different conditions. In one condition, where the experimenter had no apparent ulterior motive, this flattery worked. Subjects liked her better the more she praised them. In the second condition, however, when the experimenter had an obvious ulterior motive (she wanted them to volunteer for another experiment), subjects discounted her remarks, and the flattery failed to increase their liking for her.

Target persons have some understanding of the conditions under which such tactics as flattery or opinion conformity are likely to be used against them. One important clue is the nature of the power relation between individuals—who controls what and who wants what? If there are great differences in power between people, the low-power person may be tempted to use ingratiation to achieve his or her ends. High power persons may not need to ingratiate to achieve their ends because they can deploy resources. But low-power persons may find that ingratiation is one of the few ways they can influence the powerful targets on whom they depend. Other important modes of influence are usually unavailable to the powerless, because they lack material goods (to exchange for benefits) or clout (to back up threats).

Thus, situations of large power differences pose a dilemma for would-be ingratiators. When large power differences exist, the low-power person is very tempted to resort to ingratiation tactics to achieve his or her ends. However, it is precisely under these conditions (that is, situations involving large power differences where the target has a lot of discretion) that the target is most vigilant and suspicious of deception. In other words, the greater the need for ingratiation, the less likely it is to succeed. This irony, which is termed the *ingratiator's dilemma,* means that any ingratiation tactics used under conditions of large power disparity must be very subtle to conceal ulterior motives and avoid detection. Indeed, research shows that ingratiators tend to avoid using tactics such as opinion conformity under these conditions; they are more prone to use ingratiation tactics only under conditions when they will be less salient to the target (Kauffman & Steiner, 1968).

## Nonverbal Cues

It is widely believed that nonverbal cues, such as the look in people's eyes, are telltale signs of deception. Trial lawyer Louis Nizer (1973), for example, asserts that there are identifiable cues that reveal when witnesses are attempting to deceive jurors. Sigmund Freud asserted years ago that "He who has eyes to see and ears to hear may convince himself that no mortal can keep a secret. If his lips are silent, he chatters with his fingertips; betrayal oozes out of him at every pore" (1905, p. 78). Although recent research shows that these views are somewhat overblown and exaggerated, nonverbal cues do in fact provide a modest basis for detecting deception by impression managers.

### Controllability-Leakage Hierarchy When trying to deceive someone in face-to-face interaction, an impression manager generally will send messages through both verbal and nonverbal channels. The messages transmitted through some of these channels are more controllable than others. An impression manager will generally have a high level of intentional control over her own verbal expression (choice of words), and a fair amount of control over her facial expressions (smiles, frowns, and so on). She may have somewhat less control, however, over all her body movements (arms, hands, legs, feet) and still less control over her voice quality and vocal inflections (the pitch and waver of her voice). Thus, these channels can be ranked from more controllable to less controllable in this order: verbal expression, facial expression, body movement, and voice quality. It is the less controllable channels that pose the greatest risk for the impression manager of leaking and conveying unintended information. In fact, various studies show that the order of leakage among channels is the exact reverse of controllability (Krauss, Geller, & Olson, 1976; DePaulo & Rosenthal, 1979; Blanck & Rosenthal, 1982).

Investigators use the term **controllability-leakage hierarchy** to refer to the fact that there is a ranking among communication channels, with the order in terms of controllability being the direct opposite of the order in terms of leakage (Ekman & Friesen, 1969; Brown, 1986). The verbal channel is generally considered to carry the greatest amount of information. However, when deception is occurring, it is the leakier and less controllable nonverbal channels that are likely to be most revealing.

### Nonverbal Cues Indicating Deception Deception by an impression manager is usually signalled in one of two ways. First, there are some behavioral cues that frequently accompany an effort to deceive. For the most part, these cues come through the less controllable, leakier channels of communication. They include such behaviors as avoidance of eye contact, shifts in posture, excessive gestures, little smiling, slow speech or slips in speech, long delays in response, and high vocal pitch (Zuckerman,

DePaulo, & Rosenthal, 1981; DePaulo, Stone, & Lassiter, 1985).

A second way in which deception is signalled is when a clear discrepancy or inconsistency occurs between the messages carried in different channels. One common discrepancy is that between the verbal message and vocal quality. A job applicant lying about his work background may suddenly be revealed as false if his voice cracks or reaches a strange pitch. Another common discrepancy is that between verbal message and body movement. Nervous movements or the tapping of fingers may seem incongruous with a salesman's confident spiel, alerting a customer that something is amiss.

**Accuracy in Detection**    How good are observers at detecting acts of deception? Although some people believe they can always detect deception when it is used by an impression manager, recent research suggests the contrary. Results of most experiments reveal that observers are not especially adept at correctly identifying when deception is being used. Rates of detection are generally somewhat better than chance but not especially good in absolute terms (Zuckerman, DePaulo, & Rosenthal, 1981).

Difficulty in detection is illustrated by a study in which airline travelers at an airport in New York were asked to participate in a mock inspection procedure (Kraut & Poe, 1980). Some of these travelers were given "contraband" to smuggle past inspection, while others carried only their own legitimate luggage. All subjects were instructed to present themselves as honest persons. (As motivation, the researchers offered travelers prizes up to $100 for appearing honest.) Later on, professional customs inspectors and laymen judges watched videotaped playbacks of each of the travelers and were asked to decide which ought to be searched. Results showed that both the customs inspectors and the inexperienced judges failed to spot a substantial proportion of the travelers who were smuggling the contraband. The rate of detection, even by the customs inspectors, was no better than

chance. Interestingly, however, the inspectors and judges agreed on which travelers should be searched and which not. This occurred because the inspectors and judges used the same behavioral cues by travelers as indicative of deception. Travelers were more likely to be selected for search if they were young and lower class, appeared nervous, hesitated before answering questions, gave short answers, avoided eye contact, and shifted their posture frequently. Unfortunately for the inspectors, these cues were imperfect indicators of deception.

Why aren't observers better at detecting deception? First of all, nonverbal behaviors that are taken as telltale signs of deception, such as high vocal pitch and avoidance of eye contact, are imperfect indicators. They may reflect deception, but they can also result from conditions unrelated to deception, such as excitement or anxiety. In such circumstances, the innocent will appear guilty, and mistakes in detection will be made by observers. Second, certain skilled impression managers are able to give near-flawless performances when deceiving. One study (Riggio & Friedman, 1983) finds evidence that there is a class of people who are able to give off what seem to be honest emotional cues (facial animation, some exhibitionism, few nervous behaviors) even when they are deceiving. In such cases, the guilty will appear innocent, again causing mistakes in detection by observers. Third, it must be noted that face-to-face interaction is a two-way street: impression managers not only behave, but also observe the reactions of their audiences. The feedback that impression managers receive from their audiences in face-to-face situations is fairly rich and often provides them with a clear indication whether their attempts at deception are succeeding. If it appears that they are not succeeding, they may be able to adjust or fine-tune their deceptive communications to be more convincing.

The picture is not entirely bleak, however. There are some indications that observers' success in detecting deception can be increased by special discrimination training (Zuckerman,

Box 9-2

## BEHAVIOR THAT REVEALS DECEPTION

Although some observers might wish otherwise, there are no telltale signs of deception that apply to all situations. It is true that paralinguistic cues such as hesitation in speech and tone of voice are frequently interpretable correctly as revealing deception (DePaulo, Lanier, & Davis, 1983; Zuckerman & Driver, 1984). However, even these nonverbal behaviors must be interpreted within their social context (Krauss, 1981). For instance, the following study of speech hesitation (Kraut, 1978) illustrates how interpretations of deception cues vary according to context.

Half of the subjects in this study heard an applicant for the position of dorm counselor admit to an interviewer that she used marijuana regularly, whereas the other half heard the applicant deny that she ever used marijuana. Before making this statement, the applicant either hesitated, or did not. Subjects judged the applicant as more truthful when she hesitated before admitting marijuana use than when she did not hesitate. In contrast, hesitation before denial of marijuana use led listeners to judge her as more dishonest. Thus, hesitations increased or decreased the perception of deception, depending on the context.

In general, observers are better able to interpret various nonverbal cues (pitch, speed of talking, smiling, and so on) when they attend to context. High vocal pitch, for example, often indicates deception, but it may also reflect nervousness if the person is honestly discussing a taboo topic or a self-incriminating event. Thus, like hesitation, high vocal pitch is an unreliable sign of deception unless taken in context. Research has only now begun to identify the conditions under which different behavioral cues signify deception or honest self-presentation.

Koestner, & Alton, 1984). Moreover, success in detecting deception can be affected by instructions given to observers. For instance, in one study (DePaulo, Lassiter, & Stone, 1982), when observers in face-to-face interaction were given instructions to pay particular attention to auditory cues and to downplay visual cues, they were more successful in discriminating truth from deception than when they paid attention to both visual and auditory cues. Evidently, the visual cues were distracting to the observers. By emphasizing auditory cues, observers avoided the mistake of overattending the moderately controllable visual channels and, hence, ignoring the leaky cues that are least under an impression manager's intentional control, such as voice quality. In general, lack of attention to verbal content and paralinguistic cues seriously impairs the ability to detect deception (Geller, 1977; Littlepage & Pineault, 1978).

## ■ Ineffective Self-Presentation and Spoiled Identities

Social interaction is a perilous undertaking, one that is likely to be disrupted at any moment by challenges to one's identity. Some of us manage to recover from ineffective self-presentation; others are permanently saddled with spoiled identities. This section will discuss what happens when impression management fails. First, it considers embarrassment, a spontaneous reaction to sudden or transitory challenges to our identities. Second, it analyzes two deliberate actions aimed at destroying or debasing the identities of persons who fail repeatedly: cooling-out and identity degradation. Finally, this section examines the fate of those afflicted with stigmas—physical, moral, or social handicaps that may spoil their identities permanently.

## Embarrassment and Saving Face

**Embarrassment** is the feeling we experience when the identity we claim in an encounter is discredited. Many people describe it as an uncomfortable feeling of mortification, awkwardness, exposure, and chagrin. We feel embarrassed not only when our own identity is sabotaged but also when the identities of people with whom we are interacting are discredited (Miller, 1987). Our embarrassment at others' spoiled identities arises from the knowledge that we have been duped about the identities on which we built our interaction (Goffman, 1967; Edelmann, 1985). For example, someone who claims to be an outstanding ballplayer will feel embarrassed when he drops the first three pop flies to center field. But the team members who accepted his claims and let him play the outfield in a crucial game will also be embarrassed because they were foolish enough to believe him.

**Sources of Embarrassment**  A study of several hundred cases of embarrassment revealed three conditions that produce this feeling (Gross & Stone, 1970). First, people feel embarrassed when they fail to maintain an appropriate identity. This is the plight of the math professor when he discovers that he cannot solve the demonstration problem he has written on the chalkboard. His carefully nurtured identities as a competent mathematician and a well-prepared teacher collapse into embarrassed confusion, and those students who admired his competence share in his discomfort.

Second, embarrassment ensues when people display a lack of poise. Poise is lost if they stumble, spill coffee, barge unaware into places they don't belong, or can't find the wedding ring during the ceremony. Poise may also vanish if they lose control of their equipment (a dentist dropping his drill), of their clothing (a speaker splitting his pants), or even of their own bodies (trembling, burping, and worse). In general, poise is lost whenever we lose control over those aspects of our self-presentation that we ordinarily manage routinely.

We can read the embarrassment on the face of Police Chief D. R. Sinclair as he announces that one of his own trusted officers has been arrested on a drug charge. People experience embarrassment when an important social identity they claim for themselves or accept in others is discredited.

Third, people become embarrassed when something happens to destroy their confidence in their own expectations regarding interaction with others. This may happen when an interaction partner deliberately redefines the situation, destroying the identities she and the other have built together. Imagine, for example, two friends (male and female) on a romantic dinner date, each responding to the other's identity as a potential lover. Suddenly one launches into a description of her job-hunting plans and asks the other to recommend her for a good-paying job with his company. In switching identities abruptly, she redefines the situation as a career-planning session, not a romantic interlude. This deliberate contradiction of carefully built assumptions embarrasses her partner and destroys his confidence in the predictability of their future interactions.

These three sources of embarrassment are all occasions in which a central assumption in a transaction is unexpectedly and unqualifiedly discredited for at least one participant. Thus, embarrassment arises during social interaction, challenges the identities of the participants and the definition of the situation, and makes continued interaction difficult (Gross & Stone, 1970; Miller, 1986).

**Responses to Embarrassment** A continuous state of embarrassment is uncomfortable for everyone involved. For this reason, it is usually in everyone's interest to cooperate in eliminating embarrassment quickly. Unless their goal is to harm one another, interaction partners typically try to help the embarrassed person restore face. When a party guest trips and falls while demonstrating his dancing prowess, for instance, his partner might help him save face by remarking that the floor tiles seem to have worked their way loose. Mutual commitment to supporting each other's social identities is a fundamental rule of social interaction (Goffman, 1967). That is, we are committed to protecting and restoring the identities of others, just as we are committed to protecting our own identities.

The major responsibility for restoring order lies with the person whose actions produced the embarrassment. As noted earlier, the first line of defense is to provide accounts that realign one's actions with the normative order. People offer excuses that minimize their responsibility or justifications that define their behavior as acceptable under the circumstances. If the audience accepts these accounts, an appropriate identity is restored, and interaction can proceed smoothly. In the interests of eliminating embarrassment, interaction partners may accept accounts that would seem like a lame excuse to an uninvolved observer.

When accounts are unavailable or insufficient, people often apologize for their discrediting behavior. By apologizing, people admit that they view their own behavior as wrong. In this way they reaffirm threatened norms and reassure others that they will not violate the norms again. Most important, sincere apologies imply that the discrediting behavior does not fairly represent what the individual is really like as a person (Schlenker, 1980). When an apology is accepted, both the individual and the audience dismiss the discrediting behavior as irrelevant to the individual's "true" social identity. Embarrassment then recedes, and the person's identity is reestablished.

When our behavior discredits a particular, narrow identity, we can often save face through an exaggerated reassertion of that identity. A man whose masculine identity is threatened by behavior suggesting he is infantile, for example, might make substantial efforts to reassert his courage and strength. In a test of this hypothesis (Holmes, 1971), male subjects were asked to suck on a rubber nipple, a pacifier, and a breast shield—all embarrassing experiences. Others were asked to touch surfaces such as sandpaper and cloth. The men were next asked how intense an electric shock they would be willing to endure later in the experiment. Men who anticipated the embarrassing experiences indicated willingness to endure more intense shocks than men who anticipated no threat to their masculinity. By taking the intense shocks, the embarrassed men could present themselves as tough and courageous, thereby reasserting their masculinity.

Sometimes people embarrass others intentionally and make no effort to help them to save face. In such circumstances, embarrassed persons are likely to respond aggressively. They may vigorously attack the judgment of those who embarrassed them. Alternatively, they may assert that the task on which they failed is worthless or absurd (Modigliani, 1971). Retaliation against those who embarrass us is frequently a component of efforts to recover a positive identity. Retaliation asserts an image of strength and dignity. It also gains revenge and may forestall future embarrassment by demonstrating the resolve to punish those who would discredit us.

## Cooling-Out and Identity Degradation

When people repeatedly or glaringly fail to meet performance standards or to present appropriate identities, others cease to help them save face. Instead, they may act deliberately to modify the offenders' identities or to remove them from their positions in interaction. Failing students are dropped from school, unreliable employees are let go, tiresome suitors are rebuffed, schizophrenics are institutionalized. Attempts to modify an offender's identity assume the form either of cooling-out (Goffman, 1952) or of degradation (Garfinkel, 1956), depending on the social conditions surrounding the failure.

The term **cooling-out** refers to gently persuading an offender to accept a less desirable, though still reasonable, alternative identity. A counselor at a community college may cool-out a weak student by advising him to switch from premed to an easier major, for example, or by recommending that he seek employment after completing community college rather than transfer to the university. Persons engaged in cooling-out seek to persuade offenders, not to force them. Cooling-out actions usually protect the privacy of offenders, console them, and try to minimize their distress. Thus, the counselor meets privately with the student, emphasizes the attractiveness of the alternative, listens sympathetically to the student's concerns, and leaves the final choice up to him.

**Identity degradation** is the process of destroying the offender's current identity and transforming him or her into a "lower" social type. Degradation establishes the offender as a nonperson, an individual who cannot be trusted to perform as a normal member of the social group because of reprehensible motives. This is the fate of a political dissident who is fired from his job, declared a threat to society, and relegated to isolation in a prison or work camp.

Because the offender's loss is severe, it is usually imposed forcibly. Identity degradation often involves a dramatic ceremony, such as a criminal trial or sanity hearing, in which a denouncer acts in the name of the larger society or the law (Scheff, 1966). In such ceremonies, persons who had previously been treated as free, competent citizens are brought before a group or individual legally empowered to determine their "true" identity. They are then denounced for serious offenses against the moral order. If the degradation succeeds, offenders are forced to give up their former identities and to take on new ones as "criminal" or "insane."

Two social conditions strongly influence the choice between cooling-out and degradation: the offender's prior relationships with others and the availability of alternative identities (Ball, 1976). Cooling-out is preferred when the offender has had prior relations of empathy and solidarity with others and when alternative identity options are available. Lovers, for example, who have been close in the past can cool their partners out by offering to remain friends. Prior relationships entailing little intimacy and the absence of respectable identity alternatives foster degradation. Thus, strangers found guilty of molesting children are degraded and transformed into immoral, subhuman creatures.

Professional football and baseball present an interesting contrast in their treatment of players who fail (Ball, 1976). Professional football players who do not perform well are usually cooled-out. When cut by their team, they often have acceptable alternatives available to them. They can join another major team or parlay their college education into a reasonable nonfootball job. Degradation is more common for baseball players because they lack respectable alternatives. A drop into the minor leagues produces a sharp loss of status, and baseball players rarely have sufficient education to find a prestigious job outside sports. Cut by their team, baseball players experience social isolation and loss of identity, and they sometimes refer to themselves as "dead men."

Relationships among peers, families, and within bureaucratic organizations are usually more conducive to cooling-out than to identity degradation. Feelings of solidarity tend to prevail in everyday contacts, and alternative identities are usually available for those who no longer enact their cherished identities successfully. In large organizations, the cooling-out of unsuccessful employees is crucial for smooth functioning. As a result, most of the leading corporations in the U.S. retain professional consulting firms that specialize in cooling-out employees whom the company wishes to fire. Using "directional counseling," consultants help unsuccessful executives explore alternative careers and find new jobs.

## Stigma

A **stigma** is a characteristic widely viewed as an insurmountable handicap that prevents competent or morally trustworthy behavior (Goffman, 1963; Jones et al., 1984). There are three types of stigma. First, there are physical challenges and deformities—missing or paralyzed limbs, ugly scars, blindness, deafness. Second, there are character defects—dishonesty, unnatural passions, psychological derangements, or treacherous beliefs. These may be inferred from a known record of crime, imprisonment, sexual abuse of children, mental illness, or radical political activity, for example. Third, there are characteristics such as race, sex, and religion that, in particular societies, are believed to contaminate or morally debilitate all members of a group.

Once recognized during interaction, all types of stigma spoil the identities of the persons tainted by them. No matter what their other attributes, stigmatized individuals are likely to find that others will not view them as fully competent or moral. As a result, social interaction between "normal" and stigmatized persons is shaky and frequently experienced as uncomfortable.

**Sources of Discomfort** Discomfort arises during interaction between "normals" and stigmatized individuals because both are uncertain what behavior is appropriate. "Normals" may fear, for example, that if they show direct sympathy or interest in the stigmatized person's condition, they will be intrusive ("Is it difficult to write with your artificial hand?"); yet if they ignore the defect, they may make impossible demands ("Would you help me move the refrigerator?"). To avoid being hurt, stigmatized individuals may vacillate between shamefaced withdrawal (avoiding social contact) and aggressive bravado ("I can do anything anyone else can!"). Another source of discomfort for "normals" is fear that being seen with a stigmatized person may discredit them ("If I befriend a convicted criminal, people may wonder about my trustworthiness"). Coming face-to-face with a stigmatized individual may also arouse anxiety because it may cause "normals" to realize that they themselves are vulnerable ("I too may lose my sanity one day").

**Effects on Behavior and Perceptions** "Normals" tend to react toward stigmatized persons with an attitude of *ambivalence* (Katz, 1981; Katz, Wackenhut, & Glass, 1986). Toward a quadriplegic, for instance, "normals" have feelings of aversion and revulsion on the one hand but feelings of sympathy and compassion on the other. This ambivalence creates a tendency toward behavior instability, in which extremely positive or extremely negative responses may occur toward the stigmatized person, depending on how the specific situation is structured.

When relating to stigmatized individuals, "normals" tend to alter their usual behavior. They gesture less than usual, express opinions that reflect their actual beliefs less accurately, and terminate their encounters sooner when interacting with stigmatized others (Kleck, 1968). There is also evidence that "normals" speak faster in such interactions, ask fewer questions, agree less, make more directive remarks, and allow the stigmatized fewer opportunities to speak (Bord, 1976). These behaviors reflect the underlying ambivalence of "normals," who diminish discomfort by limiting the responses of the stigmatized and reducing uncertainty.

For their part, stigmatized persons have difficulty interacting with "normals." Remarkably, the mere belief that we have a stigma—even when we do not—leads us to perceive others as relating to us in a negative manner. In a dramatic demonstration of this principle (Kleck

& Strenta, 1980), some female students were led to believe that a woman with whom they would interact had learned that they had a mild allergy (a nonstigmatizing attribute). Other students believed that the woman would view them as disfigured by an authentic-looking scar that had been applied to their faces with stage makeup (a stigmatizing attribute). In fact, the interaction partner had no knowledge of either attribute. In the allergy condition, the partner had received no medical information whatever. In the scar condition, there was actually no scar to be seen, though the subject did not realize this. (The experimenter had removed the scar surreptitiously just prior to the discussion while presumably applying a moisturizer to it.)

After holding a six-minute discussion with the interaction partner, the students described their partners' behavior and attitudes. Those students who believed that they had a facial scar remarked more frequently that their partners had stared at them. They also perceived their partners as more tense, more patronizing, and less attracted to them than the nonstigmatized students did. Judges who viewed videotapes of the interaction perceived none of these differences. This is not surprising, since the partner knew nothing about either disability. However, these results indicate that people who believe they are stigmatized tend to perceive others as relating negatively toward them; this occurs even if the others are not in fact doing anything negative or irregular. The findings are illustrated in Figure 9-1.

When people believe they are stigmatized, they tend not only to perceive the social world differently but also to behave differently. In one study, for instance, one group of mental patients believed the person with whom they were interacting knew their psychiatric history, whereas another group thought their stigma was safely hidden (Farina et al., 1971). Patients in the first group performed more poorly on a cooperative test and found the task more difficult. Moreover, objective observers of the interaction perceived these patients to be more anxious, more tense, and less well adjusted.

**Coping Strategies** Stigmatized persons adopt various strategies to avoid awkwardness in their interactions with "normals" and to establish the most favorable social identities possible. Persons who are handicapped or physically challenged often must choose between engaging in interaction or concealing their stigma. A stutterer, for instance, may refrain from introducing himself to strangers; were he to introduce himself, he could do so only at the risk of stumbling over his own name and drawing attention to his stigma (Petrunik & Shearing, 1983). People whose speech reveals their stigmatized foreign origin or lack of education face a similar dilemma when meeting strangers.

Stigmatized persons usually try to induce "normals" to behave tactfully toward them and to build relationships around the untainted aspects of their selves. Their strategies depend on whether their stigma can be defined as temporary, such as a broken leg on the mend or a passing bout of depression, or whether it must be accepted as permanent, such as blindness or stigmatized racial identity (Levitin, 1975).

Persons who are temporarily stigmatized focus attention on their handicap, recounting how it befell them, detailing their favorable prognosis, and encouraging others to talk about their own past injuries. In dealings with doctors, nurses, or wardens, they demand treatment as if they were already cured, rejecting their stigmatized roles. They expect to share experiences and intimacies with others as if they were peers. To this end, they employ tactical self-disclosure to elicit reciprocation, and they altercast others as equals by asking personal questions.

In contrast, people who are permanently stigmatized try to focus attention on attributes unrelated to their stigma (Davis, 1961). They often use props to highlight aspects of the self that are unblemished, such as proclaiming their intellectual interests (carrying a heavy book), their political involvements (campaign buttons), or their hobbies (a knitting bag). Often the permanently stigmatized try to strike a deal with "normals." They will behave in a nondemanding and nondisruptive manner in exchange for being

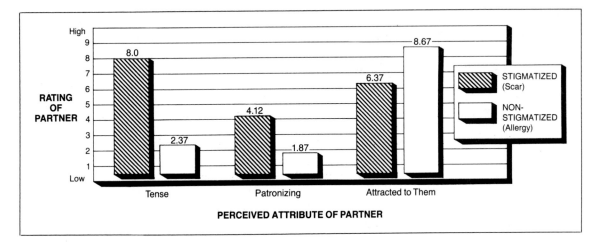

High
9
8
7
6
RATING
OF 5
PARTNER 4
3
2
1
Low

8.0
2.37
4.12
1.87
6.37
8.67

STIGMATIZED
(Scar)

NON-
STIGMATIZED
(Allergy)

Tense        Patronizing        Attracted to Them

**PERCEIVED ATTRIBUTE OF PARTNER**

■ **FIGURE 9-1**

PERCEPTIONS OF INTERACTION PARTNERS BY STIGMATIZED AND NONSTIGMATIZED INDIVIDUALS

In this study, some female students were led to believe that a large facial scar stigmatized them in the eyes of their female interaction partner. Others were led to believe their partner knew they had a mild allergy—a nonstigmatized characteristic. In fact, interaction partners were unaware of either of these characteristics. Nonetheless, students who believed they were stigmatized perceived their partners as substantially more tense and patronizing and as less attracted to them. This suggests that the mere belief that we are stigmatized leads us to perceive others as behaving negatively toward us.

*Source: Adapted from Kleck and Strenta, 1980.*

treated as trustworthy human beings despite their handicaps. Interaction is most comfortable when they acknowledge their stigma as a minor attribute with which they are coping successfully and about which they are not overly sensitive (Hastorf, Wildfogel, & Cassman, 1979).

Studies of interaction between "normal" and stigmatized individuals yield a consistent picture. "Normals" tend to pressure the stigmatized—whether intentionally or unintentionally—to accept and conform to their inferior identities. For example, sighted persons tend to discourage the blind from undertaking such pursuits as sports, politics, and entertaining. Stigmatized individuals are expected to cultivate a cheerful manner, whatever their limitations, and to avoid bitterness and self-pity. Many blind persons conform to these expectations, refraining from overt complaint.

Everyone gains some benefit from coping with stigmas in these ways. Stigmatized persons avoid the constant embarrassment of indelicate questions, inconsiderateness, and awkward offers of help. They gain substantial acceptance and manage to enjoy relatively satisfying interaction in most encounters. "Normals" gain because this resolution assuages the ambivalence they feel toward the stigmatized. They can be spared facing the true pain and unfairness the stigmatized suffer. In addition, they need not admit to themselves how limited their own tactfulness and tolerance really are. Thus, even identities spoiled by physical, moral, and social handicaps are managed in a way that preserves social order.

## ■ Summary

Self-presentation refers to our attempts, both conscious and unconscious, to control the images we project of ourselves in social interaction.

**Self-Presentation in Everyday Life** Successful self-presentation requires efforts to control how others define the interaction situation and accord identities to participants. (1) In defining the situation, people negotiate the type of social occasion they will agree is at hand, the identities they will grant each other, and the amount of leeway they will allow for enacting roles in unique, individual ways. (2) Self-disclosure is a process through which we not only promote friendship and liking but also make identity claims. Self-disclosure is usually two-sided and gradual, and it follows a norm of reciprocity. (3) There are two primary goals of self-presentation: to create a positive impression in the eyes of other people and to construct and maintain a public image that is congruent with one's own self-concept (ideal self). Self-presentation typically aims at creating a highly approved social identity because favorable outcomes are usually associated with social approval. (4) Success in self-presentation depends on inducing others to behave in ways that complement our chosen identity. Obstacles to success include conflicts of interest over interaction goals and claiming an identity inappropriate to the setting.

**Tactical Impression Management** People employ various tactics to manipulate the impressions others form of them. (1) They manage appearances (clothes, habits, possessions, and so on) in order to indicate the kind of person they claim to be. (2) They ingratiate themselves with others through opinion conformity, other enhancement, and selective presentation of their admirable qualities. (3) When caught performing socially unacceptable actions, people try to repair their identities through aligning actions—attempts to align their questionable conduct with cultural norms. They explain their motives, disclaim the implications of their conduct, or offer accounts that excuse or justify their actions. (4) They altercast others, imposing roles and identities that mesh with the identities they claim for themselves.

**Unmasking the Impression Manager** Observers attend to two major types of information in detecting deceitful impression management. (1) They assess others' possible ulterior motives. If a large difference in power is present, an impression manager's ulterior motives may become transparent to observers, making techniques such as ingratiation difficult. (2) They scrutinize others' nonverbal behavior. Although detection of deceit is difficult, observers are more accurate when they concentrate on leaky cues, such as tone of voice, and discrepancies between messages transmitted through different channels.

**Ineffective Self-Presentation and Spoiled Identities** Ineffective self-presentation has several consequences. (1) People experience embarrassment when their identity is discredited. Interaction partners usually help the embarrassed person to restore an acceptable identity. Otherwise, embarrassed persons tend to reassert their identity in an exaggerated manner or to attack those who discredited them. (2) Repeated or glaring failures lead others to modify the offender's identity through deliberate actions. Others may attempt to cool-out offenders by persuading them to accept less desirable alternative identities, or they may degrade offenders' current identities and transform them into "lower" social types. (3) Numerous physical, moral, and social handicaps stigmatize individuals and permanently spoil their identities. Interaction between stigmatized and "normal" persons is marked by ambivalence and is frequently awkward and uncomfortable. In general, "normals" pressure stigmatized individuals to accept inferior identities. At the same time, stigmatized individuals seek to build relationships around the untainted aspects of their selves.

## Key Terms

**10**

chapter

# Altruism and Aggression

## ■ Introduction

*Item:* "A hundred points earns you full membership in the Studded Stompers," explained three male youths who had been arrested for viciously kicking an elderly man into unconsciousness. "Each unprovoked attack on a stranger is worth 10 points." Contrast these aggressive youths with another group of urban teenagers who organize voluntary senior citizen escort programs; they take the elderly shopping and sit in the park or visit museums or theaters with them. "Until these youngsters came along, I almost never left my apartment," one smiling lady reported. "I didn't dare take the subway alone."

*Item:* On January 17, 1989, Patrick Purdy walked up to a Stockton, California, elementary school. It was recess time, and hundreds of children were playing in the school yard. Purdy was carrying a rapid-fire assault rifle; he opened fire, spraying 100 or more bullets into the crowd of children. His shots killed five students, and wounded twenty-nine children and one teacher. Several adults risked their lives to protect the children near them.

*Item:* PHYSICIANS EMPLOYED TO MAKE TORTURE MORE EFFECTIVE. Every year, newspaper headlines like these bring new reports of doctors who dispense drugs that allow torturers to administer more intense shock treatment to their victims. By preventing heart attacks, these doctors block death as an escape. But other physicians leave personal comfort and lucrative practices behind in efforts to relieve human suffering in underdeveloped nations. These doctors subject themselves to primitive living conditions in deserts, jungles, and rain forests in order to treat people afflicted with leprosy, malnutrition, parasitic diseases, and the depredations of war.

The above examples of helping and hurting, altruism and aggression, illustrate an important fact of life: Human beings are capable of vastly different, even opposing, social behaviors. In any situation, some people may help, whereas others may cause pain or show indifference. And the same people who behave aggressively in one situation may show compassion in another. The same Nazi doctors who performed cruel experiments on concentration camp inmates labored selflessly to save hospital victims of bombings.

The challenge for social psychologists is to explain this variation in human social behavior. When will people help, when will they do harm, and why? Drawing on research and theory, this chapter addresses the following questions:

1. What motivates people to help others?
2. How do characteristics of the situation influence helping?
3. How do emotions affect helping?
4. What motivates people to aggress against others?
5. How do characteristics of the situation influence aggression?
6. How do emotions affect aggression?

In order to address these questions, we need clear definitions of helping and aggression. At first glance, two simple definitions seem to fit the behavior in the examples mentioned above: (1) Helping is any behavior that benefits another; and (2) Aggression is any behavior that hurts another. But these definitions acknowledge only the observable consequences of behavior, ignoring the actor's intentions. Consequently, they often lead to absurd conclusions. For example, according to these definitions, a surgeon might be considered an aggressor if a heart transplant patient died on the operating table, despite heroic efforts to preserve the patient's life. At the same time, a would-be assassin might be considered a helper if the bullet he intended for the president killed the president's chief political rival instead.

Clearly, intentions are important in defining an act as helping or aggression (Krebs, 1982).

Intentions are particularly important for social psychologists, who are interested in how personal and situational factors together influence decisions to behave. Given this viewpoint, we will use the following definitions of helping and aggression.

**Helping** is any behavior intended to benefit another person. According to this definition, attempts to benefit others that fail (an unsuccessful heart transplant) are help; unintended benefits (an accidental assassination) are not. Helping is also sometimes labeled *prosocial behavior,* because helping has positive social consequences and is approved and encouraged by prevailing social standards. Of course, behavior intended to benefit others is often guided by additional motives. The surgeon, for example, is also motivated to earn a living, to demonstrate strong professional competence, and perhaps to achieve social recognition.

Does helping ever occur in the absence of these additional motives? This question is controversial. Reinforcement-based theories maintain that human nature is basically selfish and that individuals are motivated to maximize their own net gains (Phelps, 1975; Gelfand & Hartmann, 1982); others argue that helping does occur in the absence of external rewards. This type of helping—voluntary behavior intended to benefit another with no expectation of external reward—is called **altruism** (Macaulay & Berkowitz, 1970).

**Aggression** is any behavior intended to harm another person. According to this definition, a bungled assassination is an act of aggression; it involves intended harm that the target certainly would wish to avoid. Heart surgery—approved by the patient and intended to improve her health—is clearly not aggression, even if the patient dies.

Note that helping and aggression, as defined here, share four features: (1) Both refer to actual behavior, not merely to good or evil wishes or hopes. (2) Both may entail verbal as well as physical benefits or harm. (3) The benefits or harm may affect the target either directly or indirectly (as an example, by enhancing or damaging the target's possessions or dear ones). (4) Most important, both helping and aggression are intentional—the goal of one is to benefit; of the other, to harm.

## ■ Altruism and the Motivation to Help

There are three main types of motivation for helping. First, people may help because social norms define helping as the right thing to do. Second, people help in response to their own feelings of compassion or discomfort, feelings aroused by seeing others in distress. Third, people help because they can obtain rewards and avoid costs. These three types of motivation may operate singly or in combination. Each type causes some people to help more than others, and each is more important in some situations than in others.

A young businessman gives money to a needy street person. Altruistic behavior of this type may depend in part on personal attitudes, but it can also result from situational events that make salient the social responsibility norm.

### Normative Motivation

People expect each other to help in a wide variety of situations. Thus, helping is normative in the sense that it is an approved behavior, supported by social sanctions. From early childhood we are urged repeatedly to act in ways that benefit others. Our helping is praised, and our selfishness is condemned. Prevailing social norms enable us to anticipate how others will respond to our behavior and to calculate in advance what the likely social costs and rewards for helping might be. We may also internalize social norms, adopting them as standards for our own conduct. We then feel good or bad about ourselves, depending on whether we live up to our internalized standards. People often mention their sense of what they "ought to do"—their norms—when asked why they offer help (Berkowitz, 1972).

**Social Responsibility**  A widely accepted, very general norm that motivates helping is the **social responsibility norm.** It states that individuals should help people who are dependent on them. Several studies show that simply informing individuals that another person, even a stranger, is dependent on them is enough to elicit helping. For example, students worked harder for a supervisor when told that the chances of his winning a prize depended on how hard they worked for him (Berkowitz, Klanderman, & Harris, 1964). In this study, all other motives for hard work by students were eliminated by the researchers. Students were offered no material rewards for their work. They also were offered no social rewards, because no one would learn whether they had helped for many weeks. Their supervisor also showed no distress that might arouse the students' feelings of compassion or discomfort. Thus, the internalized social responsibility norm was the only apparent motivation for students to help.

Although mere awareness of a stranger's dependency sometimes elicits help, it often fails to do so. Speeding passersby frequently disregard stranded motorists they notice on the roadside; bystanders watch fascinated, but immobile, during rapes and other assaults; and thousands of people reject charity appeals every day. Thus, people regularly ignore the social responsibility norm.

Some theorists have suggested that the social responsibility norm effectively motivates helping only when people are reminded of it. In a test of this hypothesis (Darley & Batson, 1973), theological students were asked to prepare a talk on the parable of the Good Samaritan. On the way to record their talk, the students passed a man slumped in a doorway. Although these students were presumably thinking about the virtues of altruism, they helped the stranger only slightly more than a similar group of students who had prepared a talk on an unrelated topic (careers). A second variable—being in a hurry—had a much stronger impact on the amount of help offered. Students who were in a hurry offered much less help than those not in a hurry. These findings suggest that the social responsibility norm is a weak source of motivation to help, one easily negated by the costs of helping.

**Reciprocity** The **reciprocity norm** states that people should (1) help those who help them and (2) not hurt those who help them (Gouldner, 1960; Trivers, 1983). This norm applies to a person who has already received some benefit from another. Small kindnesses that create the conditions for reciprocity are a common feature of family, friendship, and work relationships. People typically report that the reciprocity norm influences their behavior (Muir & Weinstein, 1962). But does it really?

Many studies involving both children and adults demonstrate that people are inclined to help those who helped them earlier, in accord with the reciprocity norm (Bar-Tal, 1976). Moreover, people try to match the amount of help they give to the quantity they received earlier (Wilke & Lanzetta, 1970), and people are less likely to ask for help when they believe they will not be able to repay the aid in some form (Fisher,

Nadler, & Whitcher-Alagna, 1982). By matching benefits, people maintain balance in their relationships and avoid becoming overly indebted to others.

People do not reciprocate every benefit they receive, however. Whether we feel obligated to reciprocate depends on the intentions we attribute to the person who helped us. We feel more obligated to reciprocate if we perceive that the original help was given voluntarily rather than coerced and that it was chosen consciously rather than accidentally (Greenberg & Frisch, 1972; Gergen et al., 1975). Help that is voluntary increases reciprocity because it implies good intentions.

Reciprocity is also more likely when the original benefit appears to have been tailored specifically to the recipient's preferences. We are more inclined to reciprocate if a friend specially prepared a meal of our favorite dishes, for example, than if he invited us to take potluck. Help that is general, inappropriate, or merely what is required by the helpers' role gives no evidence of strong positive intentions. Recipients also consider how much of a sacrifice the helper is making on their behalf. They consider not only the size of the benefit but, even more important, the portion of the donor's own resources devoted to the gift. In an experimental demonstration of this idea, students received a gift of eighty cents from their partner in the first round of a game. When they were given the opportunity to reciprocate, students offered much more when they believed their partner had only one dollar before making the donation than when they believed their partner had four dollars before the donation (Pruitt, 1968).

If recipients infer that their benefactors' motives are selfish, they are unlikely to reciprocate. We are especially suspicious of people who are overly generous or extend help beyond our ability to reciprocate. Exaggerated generosity obligates us to submit to our benefactors' wishes (Greenberg, 1980). Gifts we cannot reciprocate threaten our freedom of action. They arouse an unpleasant emotional state called *psychological reactance,* which motivates people to restore their freedom and regain control (Brehm, 1972). Not surprisingly, people dislike benefactors they cannot repay. They attribute their benefactors' generosity to selfish, manipulative motives and thereby avoid indebtedness and regain a sense of freedom.

**Critique**  Several critics contend that broad social norms like social responsibility and reciprocity are inadequate to explain helping (Latane & Darley, 1970; Schwartz, 1977). Given the variety of situations people encounter, they claim that such norms are too general to dictate our behavior. Second, if such norms are accepted by almost everyone in society, why do individuals differ in the extent to which they help? Third, the social norms that apply to any given situation often conflict. Social responsibility may obligate us to help an abused wife, for example, but the widely accepted norm against meddling tells us not to intervene. Finally, while it is always possible to think of some norm that endorses a helping act, how do we know if that norm actually motivated the helper?

**Personal Norms Theory**  In response to these criticisms, a different normative theory has been developed (Schwartz, 1977; Schwartz & Howard, 1981). This theory explains not only the conditions under which norms are likely to motivate helping but also individual differences in helping in particular situations. Instead of dealing with broad social norms, this theory focuses on **personal norms**—feelings of moral obligation to perform specific actions that stem from individuals' internalized systems of values.

To investigate whether feelings of moral obligation motivate helping, researchers measure an individual's personal norms by means of a survey questionnaire. For example, a survey on medical transplants might ask, "If a stranger needed a bone-marrow transplant and you were a suitable donor, would you feel a moral obligation to donate bone marrow?" This survey would usually be followed by an apparently unrelated encounter with a representative of an organization who would ask these individuals for help. In various studies, individuals' personal norms have predicted differences in their willingness to donate bone marrow, to tutor blind children, to take class notes for students called up for army reserve duty, to work for increased welfare payments for the elderly, and to donate blood.

According to the personal norms theory, three phases of decision making precede overt helping: (1) *Awareness*—We notice another's need, think of relevant helping acts, and recognize our own ability to perform some of these acts. (2) *Motivation*—We arrive at a sense of how much we want to perform relevant acts by weighing how well the acts express our internalized moral values and the social and material costs and benefits of these acts. (3) *Defense*—If our motivations for and against helping are nearly equal, we try to reduce this conflict and avoid helping by neutralizing our personal norms. We do this by denying the reality or seriousness of the other's need, by denying the feasibility of helping, or by denying our own personal responsibility.

Helping is most likely to occur when conditions favor the activation of personal norms and oppose defenses that might neutralize personal norms. For example, personal norms predicted quite accurately how often college students who accepted responsibility for their actions would volunteer to tutor blind children. Among students who tended to deny such responsibility, however, there was no relationship between personal norms and behavior (Schwartz & Howard, 1980). Results of this study are illustrated in Figure 10-1. Several other studies (Schwartz & Howard, 1984) have revealed that the relationship between personal norms and helping depends on conditions that support norm activation (for example, the tendency to notice others' suffering) and oppose defensive denial of norms (for example, the focus of responsibility on the potential helper).

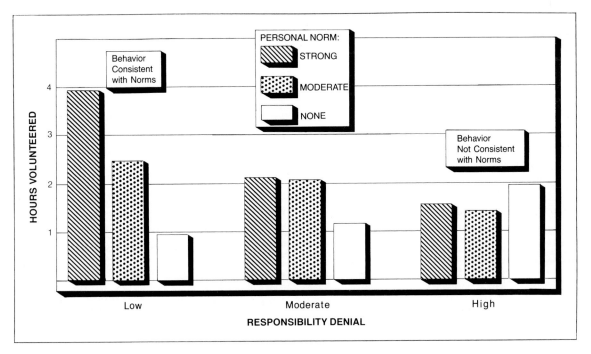

■ **FIGURE 10-1**

VOLUNTEERING AS A FUNCTION OF PERSONAL NORMS AND RESPONSIBILITY DENIAL

In a survey on social issues, university students indicated how much of a moral obligation they would feel (personal norm) to read texts to blind children. Three months later, the director of the Institute for the Blind wrote to the students, requesting that they volunteer time for just this purpose. Students who rarely denied responsibility for the consequences of their acts (low responsibility denial) behaved consistently with their personal norms: the stronger their moral obligation, the more they volunteered. Students moderate in responsibility denial showed weak consistency between personal norms and behavior. Students high in responsibility denial showed no consistency between personal norms and behavior. These findings indicate that the impact of our personal norms on helping behavior depends on whether we accept or deny our own responsibility.

*Source: Adapted from Schwartz and Howard, 1980.*

**Norms and Altruism**   Can helping that is normatively motivated be considered truly altruistic behavior? Normative explanations of helping point to people's feelings of what they ought to do. These feelings may derive from broad social norms or from a person's own internalized values. Helping that is motivated by the desire to gain the benefits or avoid the costs tied to social norms is not altruism. However, helping that is motivated by a person's internalized norms may be altruism, depending on the values that underlie these norms (Schwartz & Howard, 1984). If these are values that entail concern for the welfare of others—values like justice and compassion—then the help is considered altruistic.

**Empathy**

People often respond to the distress of others on an emotional level. **Empathy** occurs when we imagine ourselves in the same situation as someone else and experience vicariously some of the same emotions that the other person is feeling. Through empathy, another person's pleasure gives us pleasure, or another person's pain causes us pain. When we perceive that other people need help, the empathic response is to feel distress ourselves. This can motivate us to relieve others' suffering as well as our own.

The tendency to respond empathically to the perceived emotional states of others is probably biologically based (Hoffman, 1977). People of all ages, including 1-year-old infants, respond empathically when viewing others in distress. Their empathy is revealed in facial expressions, self-reports of emotion, and measures of physiological arousal. Empathic responses occur very rapidly, and they are largely automatic. Although empathy does not necessarily lead to helping, it does provide a basis for helping. Numerous studies have demonstrated that empathic arousal often increases helping (Piliavin, Dovidio, Gaertner, & Clark, 1981).

**Conditions that Promote Empathy**  Believing that a victim is similar to oneself—whether in race, sex, personality, or attitudes—promotes empathy. For example, students in one study observed another student receiving painful shocks when he lost at roulette. Those students who had been told that the victim had a personality similar to their own became more physically aroused than those who thought the victim had a personality different from their own. Those who were more aroused also donated more of their own money to help the victim (Krebs, 1975). When observers take the role of another—imagining how that person feels rather than simply observing how that person reacts—both empathy and helping behavior increase (Harvey et al., 1980; Toi & Batson, 1982).

**Empathy and Altruism**  When empathy motivates us to help, what is our real goal? Is our goal selfish (to relieve our own pain aroused by witnessing another's suffering), or is our goal altruistic (to benefit another person directly)? Among young children, empathically motivated helping is probably aimed at benefiting the self. As people grow older, however, feelings of sincere concern for victims may also emerge, mak-

ing help truly altruistic (Hoffman, 1981a, 1981b). Adults report two distinct states of emotional arousal while witnessing another's suffering: (1) personal distress, consisting of emotions such as shock, alarm, worry, and upset, and (2) empathic concern, consisting of emotions such as compassion, concern, warmth, and tenderness (Batson & Coke, 1981). Helping in response to emotional arousal is considered altruistic if motivated by empathic concern but not by personal distress.

Several experimental and natural field studies have investigated whether empathy can lead to true altruistic behavior. In general, these studies have examined situations where potential helpers can easily alleviate their own personal distress by leaving the situation and where there are no external rewards for helping. Under these circumstances, any helping behavior that occurs in response to empathic arousal would be truly altruistic, because a desire merely to alleviate one's own personal distress would motivate one to leave the situation rather than to help.

Studies of this type have investigated such helping behaviors as contributing to a muscular dystrophy telethon, aiding a master's candidate with her thesis, taking electric shocks in place of a fearful experimental subject, and helping a fellow student catch up on course work (Toi & Batson, 1982; Batson et al., 1983; Davis, 1983). In each of these studies, people who experienced high levels of empathic concern—feelings of compassion, concern, warmth, and tenderness—were the ones most likely to help. These results support the view that empathic concern motivates altruism.

## Cost-Reward Motivation

One reason why helping fascinates those who study human motivation is that it seems to defy a widely held assumption about human nature: People are motivated to act in ways that maximize their rewards and minimize their costs. Altruistic helping does defy this assumption. But most helping is not purely altruistic. And some theorists argue that individuals help only when they perceive that the rewards to them for helping outweigh the costs (Lynch & Cohen, 1978; Piliavin et al., 1981). The problem lies in revealing the rewards and costs that a potential helper takes into account. In general, the potential helper considers three types: rewards for helping the victim, costs for helping, and costs for not helping.

**Rewards for Helping**   The rewards that motivate potential helpers are many and varied. They include admiration and approval from others, thanks from the victim, financial awards and prizes, and recognition from others for competence. The rewards people seek through helping reflect their own needs. This is illustrated by a study in which undergraduates were invited to volunteer their assistance in projects such as studying unusual states of consciousness (ESP and hypnosis, for example) or counseling troubled high school students (Gergen, Gergen, & Meter, 1972). Most participants chose to help by means of activities that satisfied their personal preferences and needs. Those who enjoyed novelty volunteered more frequently to help with the project on unusual states of consciousness. Those who liked close social relationships volunteered more frequently to help troubled high school students. Thus, the rewards people seek through helping match their own personality needs.

Rescuers carry an injured climber down a mountain to safety. Many people satisfy their own needs, such as the need for approval or excitement, through helping.

**Costs for Helping**   Every helping act imposes costs on the helper. These may include the threat of danger, embarrassment, financial costs, time loss, effort expended, or exposure to repulsive people and objects. A study in the New York City subway (Allen, 1972) demonstrates the inhibiting impact of such costs on potential helpers. A confederate asked the subject (a passenger) whether the train was going uptown or downtown. The man in the neighboring seat (also a confederate) responded quickly but gave an obviously wrong answer. The subject could help by correcting this misinformation only at the risk of challenging the misinformer. Whether or not the

subjects helped depended on how threatening the misinformer appeared to be. When the misinformer had loudly threatened physical harm to a person who had just stumbled over his outstretched feet, only 16 percent helped. When the misinformer had insulted and embarrassed the stumbler, 28 percent helped. When the misinformer had made no reaction to the stumbler, 52 percent helped. Thus, the greater the anticipated cost of antagonizing the misinformer, the less likely people were to help.

**Costs for Not Helping**   Failure to help also imposes certain costs on potential helpers.

Among these are public disapproval or condemnation by the victim. Students may donate blood, for example, to avoid condemnation for failing to help their fraternity meet its quota. Employees may authorize payroll deductions for the United Fund to avoid censure by their employers. In some instances, costs are normatively imposed. In certain European countries, for example, bystanders who fail to help victims not only incur disapproval but may even face criminal prosecution. Much helping is simply compliance with social norms to avoid socially imposed costs.

## ■ Situational Impacts on Helping

"What time is it?" someone asks; and practically anyone whose watch is working answers helpfully. A child darts from your side into the path of an oncoming car, and you—as virtually anyone would—reach out to grab her. But would you give a dollar to a drunken stranger for beer? Few people would. And very few of us would intervene in a heated argument between a man and a woman we believe are married. These examples demonstrate that aspects of the situation have a strong impact on whether we do or do not help. Situations influence our reactions by activating norms, arousing empathy, and making salient various costs and rewards. In this section, we consider two aspects of situations that influence helping: characteristics of the needy person and interactions among bystanders.

### Characteristics of the Needy

Some people have a much better chance of receiving help than do others. Our willingness to help needy persons depends on various factors, such as whether we know them, whether they are similar to us or different from us, and whether we consider them truly deserving of help.

**Do We Know Them?** We are especially inclined to help people to whom we feel close. Studies of reactions following natural disasters, for example, indicate that people tend to give aid first to needy family members, then to friends and neighbors, and lastly to strangers in the stricken area (Form & Nosow, 1958). Relatedness increases helping because ongoing relationships entail stronger normative obligations, more intense empathy, and greater costs if we fail to help. While strong relations increase helping, so also can weak ones. For instance, even the briefest prior encounter with a stranger —merely glimpsing at another—has been found to increase helping. This probably occurs because the earlier encounter generates a visual image of the victim who is suffering, thereby increasing our empathy (Liebhart, 1972).

**Are They Similar?** Because we tend to feel closer to people who are similar to us, we are more likely to help those who resemble us in race, nationality, attitudes, political opinions, ideologies, and even in mode of dress. In one study, for example, a confederate dressed either in jeans and a work shirt or in conventional sports clothes and asked college students for a coin to make a phone call. Students complied with the requests more often if the confederate's mode of dress was similar to their own (Emswiller, Deaux, & Willis, 1971).

A series of field studies demonstrated that similarity of opinions and political ideologies increases helping (Hornstein, 1978). In these studies, New York pedestrians came across "lost" wallets or "lost" letters planted by researchers in conspicuous places. These objects contained information indicating the original owner's views on the Arab-Israeli conflict, on worthy or unpopular organizations, or on trivial opinion items. The owner's views on these topics either resembled or differed from the views known to characterize the neighborhoods in which the objects were dropped. Persons finding the wallets or letters took steps to return them to their owner much more frequently when the owner's views were similar to their own.

**Are They Deserving?** Suppose you received a call asking you to help elderly people who had just suffered a reduction in their income after losing their jobs. Would it matter whether they lost their jobs because they were caught stealing and lying or because their work program was being phased out? A study of Wisconsin homemakers who received such a call showed that respondents were more likely to help if the elderly people had become dependent because their program was cut than because they had been caught stealing (Schwartz & Fleishman, 1978). In general, potential helpers respond more when the needy persons' dependency is caused by circumstances beyond their control. Such people are true "innocent victims" who deserve help.

Need caused by a person's own misdeeds or failings elicits little help. Contributors to the *New York Times' 100 Neediest Cases* are much less generous, for example, in cases where need can be blamed on the individual's own failures— moral transgressions and psychological illness— than in cases where need clearly is not due to the individual's own failures—abused children and physical illness (Bryan & Davenport, 1968). Need viewed as illegitimate undermines helping in several ways. It inhibits empathic concern, blocks our sense of normative obligation, and increases the possibility of condemnation rather than social approval for helping (Brickman et al., 1982).

Even in emergencies, potential helpers are influenced by whether they consider a victim deserving. Consider responses to an emergency staged by experimenters in the New York subway (Piliavin, Rodin, & Piliavin, 1969). Shortly after the subway train left the station, a young man (confederate) collapsed to the floor of the car and lay staring at the ceiling during the 7½-minute trip to the next station. In one experimental condition, the man carried a cane and appeared crippled. In another condition, he carried a liquor bottle and smelled of whiskey. Bystanders helped the seemingly crippled victim quickly, usually leaping to his aid within seconds, but they often left the apparent drunk lying on the floor for several minutes. Much of this

difference probably reflects the fact that many people—rightly or wrongly—blame drunks for their own plight.

## Interactions among Bystanders

Prior to engaging in a helping act, bystanders go through a sequence of decisions. According to the decision-making theory proposed by Latané and Darley (1970), there are five steps in this sequence. A potential helper must (1) notice that something is happening, (2) decide what's going on (Is someone in trouble? Is it an emergency?), (3) decide whether he or she has the responsibility to act, (4) decide what action or behavior is appropriate, and (5) decide to implement the chosen behavior. At each step in this decision process, shown in Figure 10-2, potential helpers are influenced by their relations with other bystanders whom they believe to be present.

Consider the tragic murder of Kitty Genovese, a heavily publicized event that inspired the early research on bystander intervention. Shortly before 3:20 A.M. on March 13, 1964, Kitty Genovese parked her red Fiat in the lot of the Long Island Railroad station. Although she lived in Kew Gardens, a quiet, middle-class residential area, Kitty must have sensed something wrong. Instead of taking the short route home, she started running along well-lighted Lefferts Boulevard. She didn't get far.

Milton Hatch awoke at the first scream. Staring from his window, he saw a woman kneeling on the sidewalk directly across the street and a small man standing over her. "Help me! Help me! Oh God, he's stabbed me!" she cried. Leaning out his window, Hatch shouted, "Let that girl alone!" As other windows opened and lights went on, the assailant fled in his car. No one called the police.

With many eyes now following her, Kitty dragged herself along the street, but not fast enough. More than 10 minutes passed before the neighbors saw her assailant reappear, hunting for her. When he stabbed her a second time, she screamed, "I'm dying! I'm dying!" Still no

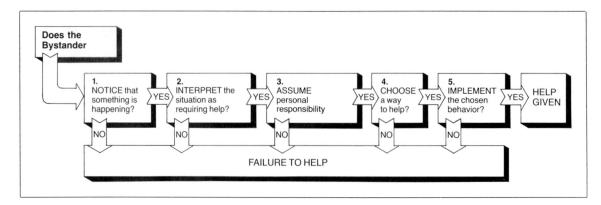

■ **FIGURE 10-2**

DECISIONS LEADING TO HELPING

When bystanders confront an ambiguous problem situation, they go through a sequence of decisions. At each step, a "yes" decision leads to the next decision on the way toward helping. A "no" decision at any step aborts the sequence and results in failure to help.

*Source: Adapted from Latané and Darley, 1970.*

one called the police. Emil Power would have, had his wife not insisted someone else must have already done so.

The third, fatal attack occurred in the vestibule of a building a few doors from Kitty's own entrance. Onlookers saw the assailant push open the building door, though few could hear the weak cry that greeted him. Finally, at 3:55 A.M., 35 minutes after Kitty's first scream, Harold Klein, who lived at the top of the stairs where Kitty was murdered, called the police. The first patrol car arrived within 2 minutes, but it was too late (Seedman & Hellman, 1975).

Why did no one react sooner? Newspaper editors and psychiatrists called it "bystander apathy" and wrote about urban alienation. Social psychologists, however, theorized that bystanders' failure to intervene in tragic emergencies may be due more to the social influence that bystanders have on each other than to individual callousness (Latané & Darley, 1970). To test this idea and to discover the nature of possible bystander influences, researchers conducted a series of laboratory studies.

Each of these studies involved a simulated emergency of one kind or another. For instance, in one experiment (Latané & Rodin, 1969), subjects heard a loud crash from the room next door, followed by a woman's screaming, "Oh my God, my foot I . . . I . . . can't move it. Oh my ankle. I . . . can't get this . . . thing off me." In another experiment (Darley & Latané, 1968), subjects participating in a discussion over an intercom suddenly heard someone in their group begin to choke, gasp, and call for help, apparently gripped by an epileptic seizure.

In each experiment, the number of people present when the emergency occurred was varied. Subjects believed either that they were alone with the victim or that one or more bystanders were present. Time and again the same finding emerged: As the number of bystanders increased, the likelihood that any one of them would help decreased. That is, bystanders helped most often and most quickly when they were alone. This finding, called the **bystander effect,** is illustrated in Figure 10-3, using data from the nervous-seizure experiment.

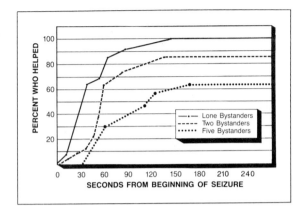

**■ FIGURE 10-3**

THE BYSTANDER EFFECT

Students who were engaged in a discussion via intercom of their adjustment to college life heard one participant begin to choke, then gasp and call for help, as if he were undergoing a serious nervous seizure. Students intervened to help the victim most quickly and most often when they believed they were the lone bystander to witness the emergency. More than 90 percent of lone bystanders helped within the first 90 seconds after the seizure. Among those who believed other bystanders were present, however, fewer than 90 percent intervened, even after 4 minutes. The bystander effect refers to the fact that the greater the number of bystanders in an emergency, the less likely any one bystander will help.

*Source: Adapted from Darley and Latané, 1968.*

Theorists have suggested that three distinct processes contribute to the bystander effect: social influence on the interpretation of the situation, evaluation apprehension, and diffusion of responsibility (Latané et al., 1981; Piliavin et al., 1981). Each of these processes affects specific steps in the decision-making sequence (see Figure 10-2).

**Interpreting the Situation** Emergencies and other situations requiring help are often ambiguous, at least initially. For instance, is a choking, gasping student in real trouble? When faced with ambiguity, people look to the reaction of others for cues about what is going on. Others' reactions influence three of the steps in the decision-making sequence leading to helping. If others appear calm, the bystander may decide that nothing special is happening (step 1) or that whatever is happening requires no help (step 2). Likewise, the failure of others to act in the situation may influence the bystander to decide that there is no appropriate way to help (step 4).

Bystanders often try to appear calm, avoiding overt signs of worry until they see whether others are alarmed. Through such cautiousness, they encourage each other to define the situation as nonproblematic. In that way they inhibit each other from helping. The larger the number of apparently unruffled bystanders, the stronger their inhibiting influence is on each other. However, consistent with this explanation, increasing the number of bystanders does not inhibit individual helping under two conditions: (1) when observation reveals that others are indeed alarmed (Darley, Teger, & Lewis, 1973) and (2) when the need for help is so unambiguous that others' reactions are unnecessary to define the situation (Clark & Word, 1972).

**Evaluation Apprehension** Bystanders are not only interested in others' reactions. They also realize that other bystanders are an audience for their own reactions. As a result, bystanders may feel **evaluation apprehension**—concern about what others expect of them and about how others will evaluate their behavior. Evaluation apprehension can either inhibit or promote helping. On one hand, evaluation apprehension inhibits helping when bystanders fear that others will view their intervention as foolish or wrong. When they see that other witnesses to an emergency are not reacting (as in the Genovese case), bystanders tend to infer that the others oppose intervention. Evaluation apprehension has its main impact at step 4 of the decision-making sequence (choosing a way to react) and at step 5 (deciding whether to implement the chosen behavior).

On the other hand, evaluation apprehension promotes helping if there are no cues to suggest that other witnesses oppose intervention. Bystanders then tend to assume that others approve intervention. In three laboratory studies demonstrating this effect, bystanders witnessed a convulsive nervous seizure or a violent assault (Schwartz & Gottlieb, 1976, 1980). Knowledge that an audience of other bystanders was present led to increased helping when the audiovisual system prevented each bystander from learning how others were reacting.

**Diffusion of Responsibility** The presence of multiple witnesses also affects the bystander's decision whether to assume personal responsibility for helping (step 3 in the decision sequence). When there is only a lone bystander, responsibility to intervene is focused wholly on that individual. But when there are multiple bystanders, responsibility to intervene is shared, as is the blame if the victim is not helped. Hence, a witness is less likely to intervene when other bystanders are present ("Why should I help? Let someone else do it."). This process of accepting less personal responsibility because responsibility is shared is called **diffusion of responsibility.** Bystanders sometimes diffuse their own responsibility by wishfully assuming that others have already taken action. Of the 38 witnesses to Kitty Genovese's murder, many claimed they thought someone else must surely have called the police.

Diffusion of responsibility occurs only when a bystander believes that the other witnesses are capable of helping. We do not diffuse responsibility to witnesses who are too far away to take effective action or too young to cope with the emergency (Bickman, 1971; Ross, 1971). The tendency to diffuse responsibility is particularly strong if a bystander feels less competent than others who are present. Bystanders helped less, for example, when one of the other witnesses to a seizure was a premed student with experience working in an emergency ward (Schwartz & Clausen, 1970).

# ■ Emotional States of the Helper

## Good Mood

Imagine a day like this: You find out that you did better than expected on your last exam. Your summer job comes through, paying enough to cover the vacation trip you were planning. The friend you've been trying to persuade to travel with you says "yes." The sun is shining. You feel great. According to many studies, you are more likely to help a needy person on this day than when your mood is just fair (Rosenhan et al., 1981). Good moods promote both spontaneous helping and compliance to requests. But moods can pass quickly; and as a person's good mood fades, helping also drops quickly back to normal (Isen, Clark, & Schwartz, 1976).

Almost every experience that puts us in a good mood increases the chance that we will help others. Consider a few examples. Suburban school teachers who learned they scored well on a battery of tests donated more to a school library fund than teachers who received no feedback on their performance (Isen, 1970). Pay phone users who found some change in the coin-return slots mailed lost unstamped letters they found more frequently than phone users who found no change (Levin & Isen, 1975). People were more likely to help a stranded caller who had dialed the wrong number from a pay phone if they had just received a small gift than if they had received no gift (Isen & Levin, 1972). Other mood-enhancing experiences that have been shown to increase helping include recalling happy experiences, reading statements describing pleasant feelings, hearing good news on the radio, listening to soothing music, and enjoying good weather. Some powerful effects of good mood on helping are shown in Table 10-1.

**How Good Moods Promote Helping** There are several theoretical explanations why good mood increases the propensity to help others.

■ **TABLE 10-1**

THE EFFECTS OF GOOD MOOD ON HELPING

| | Percent Who Helped | | |
|---|---|---|---|
| | Picking up Dropped Papers | Mailing a Lost Letter | Making a Phone Call for a Stranger |
| Ordinary mood | 4% | 10% | 12% |
| Good mood | 88% | 88% | 83% |

Sources: Adapted from Isen and Levin, 1972; Levin and Isen, 1975; Isen, Clark, and Schwartz, 1976.

First, people in a good mood are less preoccupied with themselves and less concerned with their own problems. This allows them to feel empathy for the needs and problems of others.

Second, people in a good mood feel relatively fortunate compared with persons who are deprived. They recognize that their good fortune is out of balance with others' needs. They restore a just balance by using their resources to help others (Rosenhan, Salovey, & Hargis, 1981).

Third, people in a good mood tend to see the world in a positive light and try to maintain a warm glow of happiness. If they can relieve others' suffering by helping, they can protect and even increase their own positive feelings. At the same time, people avoid helping that is unpleasant or embarrassing because it might ruin their good mood (Cunningham, Steinberg, & Grev, 1980).

## Bad Mood

Now imagine a very different day: You get your exam results back and your grade is lower than you expected. Your summer job falls through, forcing you to cancel your planned vacation. Your friend reacts in anger when you call. And a cold rain soaks you to the bone while you wait for the bus. You feel rotten. In this bad mood, are you more likely or less likely to help a needy person than when your mood is neutral?

Results of many studies make it clear that bad mood promotes helping under some conditions but inhibits it under others (Rosenhan et al., 1981).

What conditions determine whether bad mood promotes helping or inhibits helping? Research on this question has identified three important conditions. First, the salience of others' needs affects an individual's propensity to help. When others' needs do not stand out in a situation, help is less likely to be given (Rogers et al., 1982). This is significant, because people in a bad mood frequently are especially concerned about their own worries and problems. Focusing attention on their own needs, they are less likely to notice the needs of others. For this reason, people in a bad mood are less likely to offer help than those in a good mood or a neutral mood (Thompson, Cowan, & Rosenhan, 1980).

Second, the costs of helping affect an individual's propensity to actually give help. When the costs of helping are high, people are less inclined to help others. This fact is relevant to mood, because people in a bad mood often see themselves as less fortunate than others. Feeling relatively impoverished and underbenefitted, they may resist using their own resources to help others lest they become even more disadvantaged (Rosenhan et al., 1981). The net effect is that the feelings of relative impoverishment often associated with bad moods inhibit helping when the costs for helping are great.

Third, the benefits of helping affect an individual's propensity to give help. This is relevant, because people in a bad mood are often motivated to act in ways that make themselves feel better. One way to escape a bad mood is to help others in order to gain praise and approval in response. If praise of this type boosts their own morale and self-esteem, helping will serve as a route to relieve their negative mood. Thus, under these conditions, a bad mood can actually produce more rather than less helping (Cialdini et al., 1987).

## Guilt

Guilt is a negative emotional state aroused when we transgress, when we do something we consider wrong. Research has shown consistently that guilt increases helping. For example, people induced to feel guilty in various experiments helped more by making telephone calls for an ecology group, picking up scattered papers, volunteering to participate in further experiments, and donating blood than others not induced to feel guilty (Rosenhan et al., 1981). Guilt-producing transgressions—such as killing a laboratory animal, giving painful electric shocks, lying, damaging expensive machinery—have all led to increased helping. Transgressions of this type have led to helping regardless of whether they have been intentional or unintentional, public or private, in the laboratory or in the field. Such transgressions have also produced help both spontaneously and in response to requests, as well as help that benefits either the victim of the transgression or a third party.

We can explain these findings by assuming that helping allows transgressors to relieve their feelings of guilt in some way. As a socially approved behavior, helping can compensate or make up for the transgression; this in turn may relieve the guilt and boost the transgressors' damaged self-esteem. If this explanation is correct, one interesting implication is that transgressors who find an alternative way to relieve

their guilt might be expected to help less. Confessing is presumably such an alternative. A study of churchgoers' contributions to charity revealed precisely this effect; persons who had just gone to confession contributed less to charity than those who had not yet confessed (Harris, Benson, & Hall, 1975).

Similar effects of transgression, guilt, and confession on helping have been observed in experimental settings. In one laboratory study (Carlsmith, Ellsworth, & Whiteside, 1968), subjects were made to feel guilty for ruining an experiment. When these subjects had no opportunity to confess, they helped more on a later task than control subjects who had not transgressed. When the guilty subjects were given an opportunity to confess, however, the increase in helping was eliminated. These and other studies (Cunningham et al., 1980) support the view that transgression leads to guilt, which is then relieved through helping.

# ■ Aggression and the Motivation to Harm

Capable as we are of helping, human beings also show a remarkable capacity to harm others. The remainder of this chapter examines aggression. As we noted above, aggression is any behavior intended to harm another, behavior that the target would want to avoid. Our first question concerns the motivation for human aggression: Why do human beings turn against others? Consider three possible answers to this question: (1) People are instinctively aggressive. (2) People become aggressive in response to events that are painful or provoking. (3) People learn to use aggression as an effective means of obtaining what they want.

## Aggression as an Instinct

The best known proponent of an instinct theory of aggression was Sigmund Freud (1930, 1950). In Freud's view, we carry within us from the moment of conception both an urge to create and an urge to destroy. The innate urge to destroy, or **death instinct,** is as natural as our need to breathe. This instinct constantly generates hostile impulses that demand release. We release these hostile impulses either by hurting others (aggression), by turning violently against ourselves (suicide), or by suffering internal distress (physical or mental illness). Because social constraints prevent our releasing hostility in frequent, minor destructive acts, violent aggressive outbursts are virtually inevitable.

Many animal behavior studies conclude that aggression is rooted in instinct. According to Lorenz (1966, 1974), the aggressive instinct has developed in the course of evolution because it contributes to an animal's survival: animals motivated to fight succeed better in protecting their territory, obtaining desirable mates, and defending their young. Through evolution, animals have also developed an instinct to inhibit their aggression once their opponents signal submission, whereas humans have not. For this reason, humans are more dangerous and destructive than animals.

Because instinct theories postulate that the urge to harm others is genetically determined, they are pessimistic about the possibility of controlling human aggression. At best, aggression can be partly channeled into approved competitive activities such as athletics, academics, or business. Social rules that govern the expression of aggression are designed to prevent competition from degenerating into destructiveness. Quite often, however, socially approved competition stimulates aggression: soccer fans riot violently, medical students ruin others' laboratory experiments, and business people destroy competitors through ruthless practices. If aggression is instinctive, we should not be surprised that it is always with us.

Despite the popularity of instinct theories of aggression, most social psychologists find them neither persuasive nor particularly useful. Applying observations of animal behavior to humans is hazardous. Moreover, cross-cultural studies suggest that human aggression lacks two characteristics typical of instinctive behavior: universality and periodicity. The needs to eat and breathe, for example, are universal to all members of a species. They are also periodic, for they rise after deprivation and fall when satisfied. Aggression, in contrast, is not universal. It pervades some individuals and societies but is virtually absent in others. Moreover, aggression is not periodic. The occurrence of aggression seems largely governed by specific social circumstances; it shows no periodic rise when people have not aggressed for a long time or drop after they have recently aggressed. Thus, our biological makeup provides only the capacity for aggression, not an inevitable urge to aggress. We must look elsewhere to explain why particular people harm others in particular circumstances.

## Aggression as an Elicited Drive

In an early experiment (Barker, Dembo, & Lewin, 1941), researchers showed children a room full of attractive toys. They allowed some of the children to play with the toys immediately. Others were kept waiting about 20 minutes, looking at the toys, before they were allowed into the room. The children who were made to wait behaved much more destructively when given a chance to play, smashing the toys on the floor and against the walls. This study illustrates aggression as a direct response to **frustration**— that is, to the blocking of goal-directed activity. By blocking the children's access to the tempting toys, the researchers frustrated them. This frustration elicited an aggressive drive that the children expressed by destroying the researcher's toys.

The most famous view of aggression as an elicited drive is the **frustration-aggression hypothesis** (Dollard et al., 1939). This hypothesis makes two bold assertions: (1) every frustration leads to some form of aggression and (2) every aggressive act is due to some prior frustration. In contrast to instinct theories, this hypothesis states that aggression is instigated by environmental events external to the person. Several decades of research have led to a modification of

## HELP: A MIXED BLESSING

How does it feel to receive help? Help not only relieves need but also demonstrates to the recipient that someone cares. We might, therefore, assume that recipients will feel gratitude and joy. But help often brings resentment, hostility, and anxiety. Recipients of aid—both individuals and nations—may come to despise benefactors unless they can discharge their debt by reciprocating. Why does help have such negative consequences? Because it undermines recipients' self-esteem or national pride, increases their dependency, weakens their future self-reliance, and makes them vulnerable to manipulation by the donor. Help is a mixed blessing.

Ideally, help bolsters recipients' own abilities, increases their self-help, and leads to independence. The avowed purpose of foreign aid, for example, is to enable nations to overcome crises and to develop their own resources to cope with future problems independently. The avowed purpose of welfare is to aid impoverished individuals and to help families escape hunger while they establish themselves as self-supporting. Yet nations and individuals are often reluctant to accept aid because it is given in ways that do not promote these ideas. Instead,

help often communicates the message that recipients are inferior in status and competence (DePaulo & Fisher, 1980; Rosen, 1984). By accepting help, recipients fail to display the self-reliance and achievement admired in Western societies. The problem is to identify the conditions under which help will have desirable or undesirable impacts on recipients. Several conditions are considered below.

**Centrality**   Help that implies inferiority in intelligence, competence, morality, or other qualities that are central to recipients' self-conceptions is considered threatening. Help is not threatening when it does not imply any important personal or national inadequacy (Tessler & Schwartz, 1972; Nadler & Fisher, 1984a). For example, help can be supportive and nonthreatening if need is attributed to uncontrollable or chance factors like a drought, epidemic, or unprovoked attack or if the aid is defined as enabling one to overcome a trivial inadequacy in experience or effort.

**Similarity**   Help that implies an important inadequacy is more threatening when received

---

the original hypothesis (Berkowitz, 1978). First, frustration does not always produce aggressive responses. Frustration sometimes leads to other responses such as despair, depression, or withdrawal. Second, aggression often occurs without prior frustration. An insulting remark, for example, may elicit an aggressive response even though it blocks no goal-directed activity.

In light of research findings, frustration today is viewed as only one of a variety of aversive events that can elicit an aggressive drive. Pain, insults, and physical attacks are other aversive events that may elicit aggression. Current analyses also hold that aversive events do not lead to

aggression directly. Rather, they create a readiness for aggression by arousing anger. Whether this readiness translates into destructive behavior depends on the situation. Aggressive behavior is more likely, for example, when it is culturally approved, when an appropriate target is available, and when retaliation is unlikely.

### Aggression as a Rewarding Behavior

Often people behave aggressively because they anticipate that the aggressive act will be rewarding to them. Gang members mug an elderly woman, stealing her purse to obtain money. A child knocks down another to obtain the

## Box 10-1

from those who are similar to us rather than dissimilar in attitudes, knowledge, or background (Nadler & Fisher, 1984a). Similarity aggravates recipients' negative self-evaluations, because similar helpers are relevant targets for self-comparison ("If we are both alike, why do I need help while you can give it?"). People who receive aid from similar as compared to dissimilar helpers report lower self-esteem, less self-confidence, and more personal threat (DePaulo, Nadler, & Fisher, 1983).

**Self-Esteem**  People whose self-esteem is high are more threatened by help than those whose self-esteem is low (Nadler & Mayseless, 1983). The inferiority implied by accepting help contradicts the positive self-conceptions of people with high self-esteem ("How can a capable person like me need help?"). Inferiority is consistent with the unfavorable self-conceptions of people with low self-esteem.

**Threat as a Motivator**  Is the fact that aid threatens recipients' pride and self-esteem undesirable? Not if we consider the long-term consequences (Nadler & Fisher, 1984b). If recipients experience aid as nonthreatening, they feel favorably towards themselves and the helper, but they have little motivation to change. Consequently, they invest little in developing self-reliance and continue to seek help in the future. Nations become dependent satellites, and individuals become helpless parasites. In contrast, aid that threatens the self generates negative feelings toward the helper as well as the self, but it motivates recipients to change. This motivation can promote self-reliance. Both nations and individuals are more likely to invest in developing their own resources and regaining their independence if they feel threatened.

**Perceived Control**  Threat motivates positive change only when recipients perceive themselves as having some control over their environment. In an uncontrollable environment, threat produces long-term helplessness (Abramson, Seligman, & Teasdale, 1978). In short, recipients shake off their demeaning dependence on charity or foreign aid only when they believe they have a chance to succeed on their own.

---

toy he desires. Professional killers work for substantial fees. Students destroy library materials to improve their own chances on exams. These and other aggressive acts provide rewards to their perpetrators. According to social learning theory, the expectation of reward is the most important—perhaps the only—motivation for aggression (Bandura, 1973). The social learning view holds that aggressive responses are acquired and maintained like any other social behavior through experiences of reward.

If aggression is motivated by the expectation of any reward, which aggressive responses, if any, will people perform in a particular situation? This depends on two factors: the range of aggressive responses the person has acquired and the cost-reward consequences the person anticipates for performing these responses. A person may be skilled, for example, in using a switchblade, a Molotov cocktail, biting sarcasm, or subtle gossip to harm others. People also consider the likely consequences of enacting particular aggressive responses in the current situation. "Will this response gain me the rewards I seek? At what costs?" Answers to these questions largely determine which aggressive act(s), if any, people will perform.

If aggressive acts are performed primarily

because they are rewarding, then there is hope that aggression in society can be controlled. Because aggression is learned, for example, child-rearing and socialization techniques that withhold rewards for aggressive acts should reduce aggression. Because the performance of aggression depends on available rewards, social groups can structure situations to make nonaggressive responses more rewarding than aggressive ones.

# ■ Situational Impacts on Aggression

So far our analysis indicates that situational conditions largely determine whether the motivation to harm will arise and find expression in aggressive acts. In this section, we will examine four aspects of situations that influence aggression: reinforcements available, occurrence of attacks, modeling of aggression by others, and conditions that cause deindividuation.

## Reinforcements

Rewards and punishments are two very different types of reinforcement. Whereas rewards intended to encourage aggression are usually successful, punishments intended to suppress aggression often fail.

**Rewards**  Three rewards that promote aggression are direct material benefits, social approval, and attention. The material benefits that large-scale criminals, and even younger bullies, obtain through using violence support their aggression. If material benefits are reduced by vigorous law enforcement or by training the bullied children in karate, this aggressive violence will drop.

Despite the general condemnation of aggression, social approval is a second common reward for specific aggressive acts. Virtually every society has norms approving aggression against particular targets in particular circumstances. Soldiers are honored for shooting the enemy in war, children are praised for defending

their siblings in a fight, and almost all of us urge friends to respond aggressively to insults or exploitation on occasion.

Attention is a third source of positive reinforcement for aggressive acts. The teenager who aggressively breaks school rules basks in the spotlight of attention even as he is reproached. If we ignored aggressive behavior and rewarded cooperation with attention and praise, would this reduce aggressive acts? This is in fact what researchers found in a study of aggression among 27 male preschool children (Brown & Elliott, 1965). In this study, teachers selectively ignored aggressive acts and rewarded cooperation with attention and praise. After two weeks, the frequency of verbal and physical aggression declined substantially. Three weeks after these reinforcement schedules were discontinued, however, physical aggression rebounded, while verbal aggression continued to decline. The researchers then asked the teachers again to ignore aggression and reward cooperation. After this second treatment, acts of both physical and verbal aggression declined dramatically. (Even two of the most violent boys in the class became friendly and cooperative.) These findings demonstrate that the careful use of rewarding attention and approval can have powerful effects on aggression. Results of this study are shown in Table 10-2.

**Punishment**  Given the widespread use of punishment as a means of controlling aggression, we might assume that the threat of punishment and/or actual punishment are effective deterrents. In fact, however, threats are effective in eliminating aggression only under certain narrowly defined conditions (Baron, 1977). For threats to inhibit aggression, the anticipated punishment must be great and the probability that it will be delivered very high. Even so, threatened punishment is largely ineffective when potential aggressors are extremely angry and when they have relatively much to gain by being aggressive.

For actual punishment (in contrast to mere

■ **TABLE 10-2**

REDUCING AGGRESSION IN CHILDREN THROUGH SELECTIVE REINFORCEMENT

| Timing of Observations of Children's Behavior | Average Number of Aggressive Acts Observed | |
|---|---|---|
| | Verbal | Physical |
| Before selective reinforcement treatment | 22.8 | 41.2 |
| After first two-week treatment | 17.4 | 26.0 |
| After three weeks of no treatment | 13.8 | 37.8 |
| After second two-week treatment | 4.6 | 21.0 |

Source: Adapted from Brown and Elliott, 1965.

threats) to control aggression, certain other conditions must be met (Baron, 1977): (1) the punishment must follow the aggressive act promptly; (2) it must be seen as the logical outcome of that act; and (3) it must not violate legitimate social norms. Unless these conditions are met, people perceive punishment as unjustified, and they respond with anger.

The criminal justice system often fails to meet many of these conditions. In most cases, the probability that any single criminal act will be punished is low simply because most criminals are not caught, and punishment rarely follows the crime promptly. Moreover, few criminals see the punishment as a logical or legitimate outcome of their act, and they often have much to gain through their aggression. As a result, the justice system is relatively ineffective in deterring criminal aggression.

## Attacks

Direct verbal or physical attacks provoke aggressive behavior in a variety of situations. Individuals often react sharply to insulting remarks or to the impatient beeping of another driver's horn. Verbal and physical attacks arouse us physiologically and provoke anger. The typical sequence is: provoking attack ⟶ anger ⟶ aggressive response. Two factors that can modify this sequence are attributions the victim makes for the attack and the victim's belief in reciprocating harm. We consider these factors next.

**Attribution for the Attack**  Direct attacks, both verbal and physical, typically produce an aggressive reaction (Geen, 1968; White & Gruber, 1982). Nevertheless, we tend to withhold retaliation when we perceive that an attack was not intended to harm us. We are unlikely to respond aggressively, for example, if we see that a man is trying to save a child falling from his grocery cart when he smashes into us. Aggression following an attack is both stronger and more likely when we attribute the attack to the actor's intentions rather than to accidental or legitimate external pressures (Dyck & Rule, 1978).

Attributing an attack to mitigating circumstances could reduce aggressive reactions in one of two ways. It might forestall anger so that the victim does not become motivated to aggress. Alternatively, even if the victim becomes angry, it might block aggression because the victim realizes that social norms preclude retaliating under such circumstances.

In a study comparing these two possibilities, an experimenter provoked subjects by humiliating and insulting them while they performed various tasks (Zillmann & Cantor, 1976). A second experimenter attributed this attack to external pressures, mentioning that the first experimenter was "really uptight about a midterm he has tomorrow." This external justification was given either prior to the provocation, after provocation, or not at all. Subjects informed of the mitigating circumstances prior to provocation were much less angry and retaliated less than other subjects. These findings suggest that attribution to mitigating circumstances reduces retaliation by forestalling the arousal of anger. Victims who know that a provocation is

A young boy reacts aggressively to a physical attack by another. An attack typically produces anger which in turn leads to an aggressive response.

due to external pressures are less likely to become angry in the first place. External justifications brought to the victims' attention only after provocation have little power to prevent retaliation once a victim has resolved to strike back (Kremer & Stephens, 1983).

**Negative Reciprocity and Escalation** Corresponding to the positive norm of reciprocity, which supports helping, is a negative norm of reciprocity that encourages us to return harm for harm. This norm—"An eye for an eye, a tooth for a tooth"—justifies retaliation for attacks. In a national survey, more than 60 percent of U.S. males considered it proper to respond to an attack on one's family, property, or self by

killing the attacker (Blumenthal et al., 1972). Milder attacks called for milder retaliation.

The negative reciprocity norm requires that the retaliation be proportionate to the provocation. Numerous experiments indicate that people tend to match the level of their retaliation to the level of the attack (Taylor, 1967). In the heat of anger, however, we are likely to overestimate the strength of another's provocation and to underestimate the intensity of our own response. When angry, we are also more likely to misinterpret responses that have no aggressive intent as intentional provocation. Thus, even when people strive to match retaliation to provocation, aggression may escalate.

A study of 444 assaults against police officers revealed that escalation of retaliation due to mutual misunderstanding was the most common factor leading to violence (Toch, 1969). Typically, the police officer began with a routine request for information. The person confronted interpreted the officer's request as threatening, arbitrary, and unfair and refused to comply. The officer interpreted this noncompliance as an attack on his own authority and reacted by declaring the suspect under arrest. Angered further by the officer's seemingly illegitimate assertion of power, the suspect retaliated with verbal insults and obscenities. From there on the incident escalated quickly. The officer angrily grabbed the suspect, who retaliated by attacking physically. This sequence illustrates how a confrontation can spiral into violent aggression even when the angry participants feel they are merely matching their opponents' level of attack.

## Modeling

A third situational factor that promotes aggression is the presence of behavior models. For example, parents who punish their children physically serve (often inadvertently) as models of aggressive behavior. Children may not imitate this aggression at home for fear of retaliation. When this fear is removed, however, the child is likely to aggress against others. Children punished severely for aggression at home are indeed more aggressive outside the home than children punished less severely (Sears et al., 1953).

The following account of events at the Woods End protest demonstration illustrates the ways an aggressive model can influence others.

Many prominent officials have come to dedicate the Woods End nuclear plant. As the motorcade approaches, hundreds of respectable citizens—business people, professionals, students—surround the cars in protest. At a prearranged signal, the demonstrators abruptly sit down, forcing the motorcade to halt. Tension mounts as demonstrators and police await each others' responses. Suddenly a demonstrator wearing a dark suit leaps to his feet and begins to pound

the hood of the first car. A rumor sweeps the crowd: "It's the priest." The police hold back, shouting over their bullhorns for the demonstrators to disperse. Within seconds hundreds more fists smash against cars.

Aggressive models like the demonstrator in the dark suit provide three types of information that influence observers. First, through their actions, models demonstrate specific aggressive acts that are possible in the situation. The idea of pounding on the cars, for example, never crossed the minds of most demonstrators until they saw the model do it. The model's acts identified available opportunities for aggressive action.

Second, models provide information about the appropriateness of aggression, about whether it is normatively acceptable in the setting. Models who have high social status or who represent society's moral order are especially likely to be imitated. Their aggression gives legitimacy to the aggression of their imitators. The fact that many demonstrators thought that the model for aggression in our example was a priest encouraged them to imitate him.

Third, models provide information about the consequences of acting aggressively. Observers see whether the model succeeds in attaining goals, whether the behavior is punished or rewarded. Observers are more likely to imitate aggressive behaviors that yield reward and avoid punishment. Hesitation by the police in our example gave demonstrators the impression that smashing cars might go unpunished. Had the police immediately surrounded the aggressive model, observers would probably not have joined in, although they might have erupted in another way.

An aggressive model conveys the message that aggression is acceptable in a particular situation. This message matters little when observers are not motivated to do harm. But people who feel provoked and are suppressing their inclination to aggress, like the initially peaceful demonstrators in our example, tend to lose their inhibitions after observing an aggressive model. They are the most likely to imitate aggression.

Just as aggressive models may increase aggression, nonaggressive models may reduce it. The pacifist tactics of Mahatma Gandhi, who freed India of British colonialism, have been imitated by protestors around the world. Laboratory research has also demonstrated the restraining influence of nonaggressive models. In one study (Baron & Kepner, 1970), participants observed an aggressive model deliver many more shocks to a confederate than were required by the task. Other participants observed a nonaggressive model who gave the minimum number of shocks required. A control group observed no model. Results showed that subjects who observed the nonaggressive model displayed less subsequent aggression than subjects in the control group, whereas subjects who observed the aggressive model displayed more subsequent aggression than those in the control group. Other research has shown that nonaggressive models not only reduce aggression, they can also offset the influence of aggressive models who are present (Baron, 1971).

### Deindividuating Conditions

Did the respectable citizens who attended the Woods End nuclear demonstration anticipate that they would end up smashing cars, shouting obscene phrases, or defying the police? Surely few expected to behave in such a wild, uninhibited manner. There are times, however, when surrounding conditions reduce a person's self-awareness and lead to impulsive, counternormative behavior. People may then behave in a violently aggressive manner, and, when conditions return to normal, they may feel shocked or embarrassed by what they have done.

The temporary loss of self-awareness brought on by situational conditions is called *deindividuation*—the reduction of one's sense of individual identity and feelings of personal responsibility. Several conditions produce deindividuation. Among the most important are ano-

nymity, an undifferentiated crowd, darkness, and consciousness-altering drugs. When no one knows who we are, when we are swallowed up in a faceless crowd, when we cannot be seen by others, and when our critical faculties are numbed, our sense of uniqueness and individuality is weakened. In this state we are relatively unconcerned with how we are evaluated by ourselves and others. We are, therefore, less prone to feel the guilt, shame, or fear that ordinarily regulate our own behavior and make it conform with social and personal standards. Our reduced capacity for self-regulation, in turn, makes us more likely to perform behaviors we would ordinarily inhibit (Diener, 1980).

There is a substantial amount of evidence that deindividuating conditions promote aggression (Dipboye, 1977; Prentice-Dunn & Rogers, 1980). In an early laboratory study, for example, students who were anonymous expressed much more hostility toward their parents than students whose personal identity was stressed (Festinger, Pepitone, & Newcomb, 1952). Among cultures that disguise their warriors before battle (by painting the face and body, or by wearing masks or elaborate battle clothes), the warriors are much more likely to kill, torture, or mutilate their enemies in war (Watson, 1973). Warriors whose identities are disguised are more deindividuated; hence, they are less inhibited in the expression of violent aggression.

## ■ Emotional States of the Aggressor

The previous section demonstrates that situational conditions have a strong impact on aggression. But so far, little attention has been given to the emotional states aroused by these conditions. Aggression can be cold and calculating, but very often it is accompanied by emotional arousal. The next section will consider three aspects of the role of emotion in aggression, including (1) the factors that channel anger into aggressive acts and intensify aggression, (2) the process by

Swept away by the excitement of the crowd, these men are experiencing a reduced sense of self-awareness that enables them to express aggressive violence they would ordinarily inhibit. When we experience deindividuation, we are capable of uncontrolled behaviors we would never have thought possible.

which general arousal amplifies aggressive responses, and (3) the release of aggressive emotions through displacement and catharsis.

## Angry Arousal

Attack and frustration arouse anger. This angry emotional state in turn prepares the person to behave aggressively. Three factors in a given situation that intensify angry arousal are the strength of frustration, the arbitrariness of frustration, and the aggressive cues available. Each of these factors, discussed below, also increases aggressiveness.

**Strength of Frustration**  The more we desire a goal and the closer we are to achieving it, the more frustrated and aroused we become when we are blocked. If someone cuts ahead of us just as we reach the front of a long waiting line, for example, frustration is strong, and we are likely to become especially angry. According to theory, this intense anger should lead to aggressiveness.

A field experiment based on this idea demonstrated that stronger frustration elicits more aggression (Harris, 1974). Researchers recorded the reactions of people when a confederate cut ahead of them in waiting lines at theaters, restaurants, and grocery checkout counters. The confederate cut in front of either the second or the twelfth person in line. As predicted, people at the front of the line responded far more aggressively. They made more than twice as many abusive remarks to the intruder than people at the back of the line.

**Arbitrariness of Frustration**  People's perceptions of the reasons for frustration markedly influence the degree of hostility they feel. People are apt to feel much more hostile when they believe the frustration is arbitrary, unprovoked, or illegitimate than when they attribute it to a reasonable, accidental, or legitimate cause. As a result, arbitrary or illegitimate frustration elicits more aggression.

In a study demonstrating this principle, students were asked to make appeals for a charity over the telephone (Kulik & Brown, 1979). They were frustrated by refusals from all potential donors (in reality, confederates). In the legitimate-frustration condition, potential donors offered good reasons for refusing (such as "I just lost my job"). In the illegitimate-frustration condition, they offered weak, arbitrary reasons (such as "charities are a rip-off"). As shown in Figure 10-4, individuals exposed to illegitimate frustration were more aroused than those exposed to legitimate frustration. They also directed more verbal aggression against the potential donors.

**Aggressive Cues** Whether angry arousal actually leads to aggressive acts depends in part on the presence of suitable aggressive cues in the environment (Berkowitz, 1978). These cues may be necessary to bring the idea of aggression to mind. In their absence, for example, a frustrated, angry homemaker may simply despair and become resigned. Situational cues (sounds from the neighboring apartment of dishes being smashed) may suggest the idea of aggression as well as specific acts that serve as an appropriate means for expressing it (smashing one's spouse's pile of dirty dishes).

Aggressive cues have a second function. They intensify the arousal that an angry person experiences. Seeing a gun or viewing a violent film, for example, may further arouse a frustrated person and make him more aggressive. Substantial cross-cultural evidence shows that people who have been frustrated respond more aggressively when in the presence of a gun than in the presence of neutral objects (Berkowitz & LePage, 1967; Frodi, 1975). This so-called *weapons effect* occurs when people are already aroused. The sight of weapons intensifies their arousal. The presence of handguns in many homes can help to create the conditions for a weapons effect, if family members become angry.

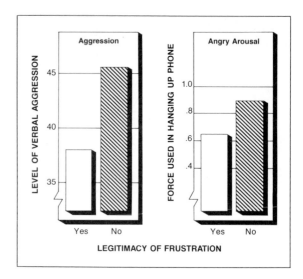

■ **FIGURE 10-4**

EFFECTS OF LEGITIMACY OF FRUSTRATION ON AGGRESSIVE RESPONSES

Students were frustrated when potential donors whom they telephoned all refused their appeal for a charity. Half the students heard legitimate reasons for refusal, whereas the others heard illegitimate and arbitrary reasons. Those exposed to illegitimate frustration showed a higher level of angry arousal and aggression than those exposed to legitimate frustration. They slammed the receiver down harder when hanging up and expressed more verbal aggression toward potential donors during their conversations. These findings demonstrate that the perceived reasons for frustration influence angry arousal and aggression.

*Source: Adapted from Kulick and Brown, 1979.*

## Nonspecific Arousal

It is no surprise that arousal due to provocation leads to aggression. But what of general physiological arousal that has no specific connection with anger—arousal caused by an ear-splitting rock band, for example, or by strenuous physical exercise? Although we might not expect such arousal to increase aggression, numerous studies reveal that virtually any source of strong physiological arousal can intensify harm-inflicting responses (Rule, Ferguson, & Nesdale,

1980). This includes arousal from vigorous exercise, loud noise, competition, stimulating drugs, hard-core pornography, and stimulating music. Because these sources of arousal are so common, their intensifying effect on aggression is important.

**Excitation Transfer Theory** General arousal does not cause aggression by itself. Rather, it amplifies aggression when aggression is already the dominant tendency in the situation—that is, when aggression is the natural response because the person has been provoked. If there is no strong tendency toward aggression in a situation, nonspecific arousal does not promote the infliction of harm. For example, in one study of the amplifying effect of nonspecific arousal (Zillmann, Katcher, & Milavsky, 1972), aggression was established as the dominant tendency for half the participants by having a confederate insult them. The others were treated politely. Participants engaged either in an arousing activity (pedaling a bicycle vigorously) or in a quiet activity (stringing discs on a wire) and were then given a chance to administer shocks to the confederate. Those who had been provoked by insults retaliated with more intense shocks if they had been exercising vigorously than if they had been stringing discs. For those participants who had not been provoked by insults, exercising had no effect on their responses. In short, arousal amplified the infliction of harm when aggression was the dominant response.

Why do effects of this type occur? Zillmann (1982) has theorized that it comes about through a process called *excitation transfer*. Physiological arousal, once it has occurred, does not dissipate immediately; rather, it declines slowly over time. Thus, some portion of the arousal may carry over from one situation to another or from one activity to another. This residual arousal, when transferred to a new situation, energizes whatever responses are made in that situation. Thus, while pedaling a bicycle has virtually nothing to do with being insulted by a confederate, the excitation from pedaling is transferred over and affects the intensity of the reaction to the insult.

**Arousal Misattributed to Provocation** Although effects of this type occur quite often, there is one important qualification of the theory. For nonspecific arousal to amplify aggression, the persons who are aroused must misinterpret their arousal as provocation. That is, they must misattribute the arousal caused by exercise, noise, and so on, to provocation or frustration (Zillmann, 1979). If people correctly attribute their exercise-induced arousal to riding a bicycle, the arousal will not intensify aggression in response to an insult. In short, provided that it is erroneously attributed to provocation, nonspecific arousal intensifies feelings of anger and amplifies aggression.

## Displacement and Catharsis

Infuriated by a day of catering to the whims of her boss, Ruth turned on her teenage son as he drove her home. "Why must you drive like a maniac?" she snapped. Ralph was stunned. He was driving 25 miles per hour and had done nothing to provoke his mother's aggression. Why was he being attacked?

**Displacement** According to instinct and drive theories of aggression, the everyday irritations and provocation we encounter cause a buildup of anger that demands release. We prefer to release this pent-up anger by directly attacking whoever caused it. If we are unable to do this, however, we seek another outlet for our aggressive energy. **Displacement** is the release of pent-up anger in the form of aggression against a target other than the original cause of anger. Displacement occurs when we cannot identify a tormentor or when the tormentor is unavailable or is so powerful that fear of retaliation inhibits us (Fitz, 1976). Ruth's boss was unavailable, and she would not have attacked him anyway for fear of losing her job.

## MEDIA VIOLENCE AND AGGRESSION IN SOCIETY

While carrying a two-gallon can of gasoline back to her stalled car, Evelyn Wagler was cornered by six young men who forced her to douse herself with the fuel. Then one of the men tossed a lighted match. She burned to death. Two nights earlier, a similar murder had been depicted on national television. Four days after the television screening of *Born Innocent,* a 9-year-old girl was raped by four teenagers who reenacted, in detail, a scene from the movie. Violence pervades television: stabbings, shootings, poisonings, and beatings—even cartoon characters torment each other in astonishingly creative ways. In an average week, a viewer can watch 50 to 100 violent deaths during prime time.

Do these presentations of violence encourage viewers to behave aggressively? Is our society more violent because of media violence? These are vexing questions, especially when the average U.S. child spends more time watching television than performing any other waking activity (Huston & Wright, 1982). By age 16, the average American child is likely to have seen about 13,000 homicides on television (Waters & Malamud, 1975)—and aggression on television is performed by heroes as well as villains. The two incidents cited above suggest that, in some cases, media violence induces aggression. Yet establishing a causal link between watching television and subsequent aggression has been difficult and highly controversial (Milavsky et al., 1983).

There are four theoretical processes that would explain why exposure to media violence might increase aggressive behavior: (1) *Imitation.* Viewers learn specific techniques of aggression from media models. Social learning through imitation evidently played a part in both of the violent attacks cited above. (2) *Legitimation.* Exposure to violence that successfully attains goals and has positive outcomes (as in police stories) legitimizes aggression and makes it more acceptable. (3) *Desensitization.* After observing violence repeatedly, viewers become less sensitive to aggression. This makes them less reluctant to hurt others and less inclined to ease others' suffering. (4) *Arousal.* Viewing violence on television produces excitement and physiological arousal, which may amplify aggressive responses in situations that would otherwise elicit milder anger. Research has shown that all four of these processes operate in linking media violence to aggression (Murray & Kippax, 1979).

In hundreds of laboratory and field experiments, children, adolescents, and adults have been exposed to scenarios of aggression, both live and on film. Some of these scenarios have been specifically prepared, whereas others have been taken directly from popular television shows. With rare exceptions, these experiments show that observing violence increases subsequent aggression by viewers (Geen, 1978; Comstock, 1984).

Three field experiments conducted in the U.S. and Belgium (Parke et al., 1977) illustrate the strengths and weaknesses of this kind of research. In each experiment, adolescent male delinquents who lived together in small groups were exposed for a one-week period either to several violent films (such as *Bonnie and Clyde* and *Ride Beyond Vengeance*) or to several nonviolent films (such as *Ride the Wild Surf* and *The Absent-Minded Professor*). Observers rated the amount of aggression exhibited by each adolescent during the film-viewing week and at least one week before and after this period. They recorded physical threats and attacks, verbal taunts and cursing, and attacks against property. The pattern of results from all three experiments shows that adolescent boys who viewed

Box 10-2

the violent films became more aggressive, especially during the film-viewing week. Physical attacks resembling those seen in the films were the type of aggression that increased most sharply but also faded most quickly. There were indications that the boys who were initially highly aggressive were most affected by viewing violence on television.

These experiments convincingly demonstrate a causal link between viewing violence and aggression. However, our ability to generalize from these experiments to the effects of media violence on society is clearly limited. For one thing, few members of society are exposed to a concentrated, week-long diet of exclusively violent films. For another, few of us are delinquent adolescents, predisposed to violence. Note also that the observed effect tended to fade away within a week or two. For these reasons, researchers have turned to *correlational studies* that complement the experimental approach in order to assess the impacts of media violence in society.

In one correlational study (Singer & Singer, 1981), researchers measured the television viewing experiences and spontaneous aggression of preschoolers several times over a period of one year. Consistently, aggressive behavior during this period was linked to heavy viewing of aggressive, action-adventure programs or cartoons. The level of aggressiveness these children exhibited four to five years later also correlated positively with the amount of violent television they had viewed as preschoolers. A second study (Milavsky et al., 1983) of elementary school children and teenagers covered a three-year period. This study revealed a weak positive correlation between watching television violence and subsequent aggression. A third study (Eron, 1980, 1982) examined children's preferences for violent television and their lev-

els of aggression in third grade and again ten and twenty-one years later. This study revealed that viewing violent television in third grade correlated positively with later aggression.

The correlations in these studies may reflect a causal impact of viewing violence on subsequent aggression, but correlations do not prove causality. The correlation might also suggest that aggressive children prefer violent programs. The growing body of evidence suggests that the relationship between aggression and television viewing is circular. Because aggressive children are relatively unpopular with their peers, they spend more time watching television. This exposes them to more violence, reassures them that their behavior is appropriate, and teaches them new aggressive techniques. When they then try to use these techniques in interaction with others, they become even more unpopular and are driven back to television—and the vicious cycle continues (Huesmann, 1982; Singer & Singer, 1983).

After twenty years of research, most observers agree that viewing media violence does cause later aggression. Although no single study proves this connection beyond a doubt, the combined evidence from hundreds of studies is overwhelming (Rubinstein, 1983; Comstock, 1984). The inevitable question is, Should media violence be censored? Some argue censorship of the media is a dangerous, antidemocratic step. Others say it depends on how strong the effects of media violence really are. For the general population, the increase in aggression due to media violence is apparently quite small. For those already inclined to violence, it may be greater. But note that even a very small increase spread across millions of viewers would add substantially to the violence in society. What then, should be done about media violence?

What determines the target of displaced aggression? First, we tend to displace aggression onto persons who resemble the source of our anger. Similarity to the original tormentor in appearance, sex, religion, age, and various other characteristics all increase the chances of someone becoming the target of displaced aggression. Second, we displace aggression more onto potential targets whom we perceive as weak and unable to retaliate. Target weakness and similarity influenced Ruth's displacement of aggression onto Ralph. She chose a target who was weaker than herself and similar in sex to her tormentor. Third, we displace aggression more onto socially sanctioned targets—people who are negatively stereotyped—against whom aggression is viewed as "justified." Because minority groups are often both negatively stereotyped and weak, they are tempting targets. People who are frustrated by economic or social stress may therefore displace their aggressive energy by inflicting violence on blacks, Hispanics, Jews, and other minorities (Harding et al., 1969).

**Catharsis** Did Ruth feel better after venting her anger on Ralph? She probably did. It is generally thought that letting off steam is better than bottling up hostility. A very old psychological concept, catharsis, captures this idea. **Catharsis** is the reduction of aggressive arousal brought about by performing aggressive acts. The catharsis hypothesis states that we can purge ourselves of hostile emotions by intensely experiencing these emotions while performing aggression. A broader view of catharsis suggests that by observing aggression as an involved spectator to drama, television, or sports we also release aggressive emotions.

Numerous studies support the basic catharsis hypothesis: Aggressive acts directed against the source of anger reduce physiological arousal fairly consistently (Geen & Quanty, 1977). We usually feel relieved after letting off steam against a tormentor. However, if the person who provoked our anger is relatively powerful, ag-

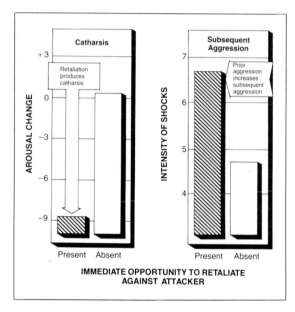

■ **FIGURE 10-5**

AGGRESSION, CATHARSIS, AND SUBSEQUENT AGGRESSION

Participants in a learning experiment who were antagonized by a confederate became physically aroused. Some participants had an opportunity to retaliate immediately, whereas others did not. Those who retaliated experienced catharsis—a sharp drop in arousal indicated by a drop in blood pressure. Those who did not retaliate remained aroused (left). Later, both groups of participants had a chance to attack the confederate (right). Those who had retaliated earlier delivered more intense shocks to the confederate than those who had not retaliated earlier. Thus, the group that experienced catharsis earlier subsequently behaved more aggressively. These findings contradict the idea that releasing pent-up anger through retaliation reduces future aggression against a target. Instead, as this study showed, catharsis increases subsequent aggression.

*Source: Adapted from Geen, Stonner, and Shope, 1975.*

gression may not bring catharsis because the fear of retaliation keeps us aroused. Aggression against a tormentor is also unlikely to reduce arousal under two other conditions: (1) if we feel our aggression is inappropriate to the situation and will make us look foolish or (2) if our internalized values oppose the aggression and make us feel guilty about our aggression.

A second hypothesis that extends the original catharsis idea asserts that catharsis reduces subsequent aggression (Dollard et al., 1939; Freud, 1950). That is, once people act aggressively and release their anger, they are less inclined to future aggression. This hypothesis underlies advice such as "Put on the gloves and get the fight over with once and for all." Yet, with few exceptions, research has shown that performing aggressive acts will increase future aggression, not reduce it. This is true whether the initial aggression is a verbal attack, a physical attack, or even aggressive play. One particularly telling study, illustrated in Figure 10-5, showed a clear catharsis effect after retaliating against a tormentor, followed by increased aggression against that same tormentor (Geen, Stonner, & Shope, 1975).

Initial aggression promotes further aggression in several ways. First, initial aggressive acts produce **disinhibition**—the loosing of ordinary internal controls against socially disapproved behavior. Disinhibition is reflected in the reports of murderers as well as soldiers who comment that killing was difficult the first time but became easier thereafter. Second, initial aggressive acts serve to arouse our anger even further. Third, they give us experience in harming others. Finally, the catharsis that follows aggression and the pleasure of thwarting our tormentor reward aggression; and rewarded behaviors are repeated more frequently.

# ■ Summary

Helping is behavior intended to benefit others, whereas aggression is behavior intended to harm others.

**Altruism and the Motivation to Help** There are three main types of motivation for helping. (1) Norms define helping as the correct thing to do. The social responsibility norm motivates us to help whoever is dependent on us. The reciprocity norm motivates us to reciprocate intentional benefits. Personal norms, based on internalized values, motivate help when we notice another's need and feel responsible to relieve it. (2) Empathic arousal in response to others' distress provides a biological basis for helping. Our similarity to victims increases our empathy. Empathic concern motivates altruism—helping with no expectation of reward. (3) Several cost-reward concerns motivate helping. For example, people help more when it leads to rewards, such as social approval, material gain, and satisfaction of personal needs. People often refuse to help when it entails costs, such as danger and effort. They also avoid costs imposed for not helping, such as public disapproval and self-condemnation.

**Situational Impacts on Helping** The characteristics of the needy and interactions among bystanders are two situational influences on helping. (1) Needy individuals have a better chance of receiving help if they are similar to potential helpers, liked by them, and seen as deserving. (2) Bystanders influence each others' behavior in three ways. First, the reactions of other bystanders affect the interpretation of a situation as requiring help or not. Second, the expectations of other bystanders create evaluation apprehension, which may inhibit or enhance helping. Third, the presence of other bystanders fosters diffusion of responsibility, which reduces helping.

**Emotional States of the Helper** (1) Good moods promote helping because they reduce self-preoccupation, increase feelings of relative good fortune, and are maintained through helping. (2) Bad moods inhibit helping because they increase self-preoccupation and reduce feelings of relative fortune. However, a bad mood may increase helping when the need is salient and the costs for helping are low. Under these conditions, people will attempt to relieve their bad feelings through helping. (3) Guilt promotes helping because helping boosts transgressors' self-esteem.

### Aggression and the Motivation to Harm

There are three main theories regarding motivation for aggression. (1) Aggression is based on a biological instinct that generates hostile impulses demanding release. Most social psychologists reject this theory. (2) Aggression is a drive elicited by frustration and other provocations that arouse anger. (3) Aggression is a learned behavior motivated by rewards.

### Situational Impacts on Aggression

Situational conditions largely determine whether motivation for aggression arises and finds expression. (1) Rewards that encourage aggression include material benefits, social approval, and attention. Punishment and threat effectively inhibit aggression only under narrowly defined conditions. (2) Verbal and physical attacks provoke aggression if they are perceived as intentional. (3) Aggressive models provide information about the available options, normative appropriateness, and consequences of aggressive acts. (4) Deindividuating conditions reduce self-awareness and self-regulation, thereby releasing aggression that people ordinarily inhibit.

### Emotional States of the Aggressor

(1) Factors that intensify angry arousal and increase aggression include the strength and arbitrariness of frustration and the presence of aggressive cues in a situation. (2) General physiological arousal amplifies aggression when two conditions occur jointly: when aggression is the dominant response in the situation and when people erroneously attribute their arousal to provocation. (3) People who cannot release anger against the source of provocation displace it as aggression against targets who are similar to the source but weaker; they also displace it against socially sanctioned targets, such as minorities. Retaliation usually provides catharsis—the relief of angry arousal—but it also promotes subsequent aggression against that same target.

## Key Terms

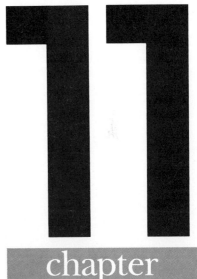

11

chapter

# Interpersonal Attraction and Relationships

# ■ Introduction

Dan was looking forward to the new semester. Now that he was a junior, he would be taking more interesting classes. He walked into the lecture hall and found a seat halfway down the aisle. As he looked toward the front, he noticed a very pretty young woman removing her coat; as he watched, she sat down in the front row.

Dan noticed her at every class; she always sat in the same seat. One morning he passed up his usual spot and sat down next to her.

"Hi," he said. "You must like this class. You never miss it."

"I do, but it sure is a lot of work."

As they talked, they discovered they were from the same city and both were economics majors. When the professor announced the first exam, Dan asked Sally if she would like to study for it with him. They worked together for several hours the night before the exam, along with Sally's roommate; Dan and Sally did very well on the test.

The next week he took her to a film at a campus theater. The week after she asked him to a party at her dormitory. That night, as they were walking back to her room, Sally told Dan that her roommate's parents had just separated and that her roommate was severely depressed. Dan replied that he knew how she felt because his older brother had just left his wife. Because it was late, they agreed to meet the next morning for breakfast. They spent all day Sunday talking, about love, marriage, parents, and their hopes for the future. By the end of the semester, Sally and Dan were seeing each other two or three times a week.

At its outset, the relationship between Dan and Sally was based on **interpersonal attraction**—a positive attitude held by one person toward another person. Over time, however, the development of their relationship involved increasing interdependence and increasing intimacy, or pair relatedness, as shown in Figure 11-1.

The development and outcome of personal relationships involves several stages. This chapter will discuss each of these stages. Specifically it will consider the following questions:

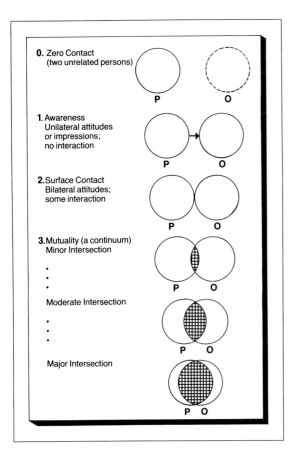

■ **FIGURE 11-1**

LEVELS OF PAIR RELATEDNESS

A relationship between two persons develops through a series of stages, or levels. Before it begins, the two are unrelated. The next level occurs when one person becomes aware of the other. The third level involves interaction. Relationships that develop beyond the surface contact stage involve progressively greater mutuality—disclosure of personal information, trust, and interdependence.

*Source: Adapted from Levinger and Snook, 1978.*

1. Who is available? What determines whom we come into contact with?

2. Who is desirable? Of those available, what determines whom we attempt to establish relationships with?

3. What are the determinants of attraction or liking?

4. How do friendship and love develop between two people?

5. What is love?

6. What determines whether love thrives or dies?

## ■ Who Is Available?

There may be dozens or thousands of persons who go to school or live or work where you do. Most of them remain strangers, persons with whom you have no contact. Those persons with whom we come into contact, no matter how fleeting, constitute the field of **availables**—the pool of potential friends and lovers (Kerckhoff, 1974). What determines who is available? Is it mere chance that George rather than Bill is your roommate or that Dan met Sally rather than Heather? The answer, of course, is no.

Two basic influences determine who is available. First, institutional structures influence our personal encounters. The admissions office of the school, faculty committees that decide on degree requirements, and the scheduling office all influence whether Dan and Sally are enrolled in the same class. Second, individuals' personal characteristics influence their choice of activities. Dan chose to take the economics class where he met Sally because of an interest in that field and the desire to go to graduate school in business. Thus, institutional and personal characteristics together determine who is available.

Given a set of persons who are available, how do we make contact with one or two of these persons? Three influences progressively narrow our choices: routine activities, proximity, and familiarity.

### Routine Activities

Much of our life consists of a routine of activities that we repeat daily or weekly. We attend the same classes and sit in the same seats, eat in the same places at the same tables, shop in the same stores, ride the same bus, and work with the same people. These activities provide opportunities to interact with some availables

When we think about where people meet, we often picture the singles bar. But one study of heterosexual relationships found that relatively few people met their partners at a bar. Much more common were meetings in classes, dorms, or at work.

but not with others. More importantly, the activity provides a focus for our initial interactions. We rarely establish a relationship by saying "Let's be friends" at our first meeting. To do so is risky, because the other person may decide to exploit us. Or that person may reject such an opening, which may damage our self-esteem. Instead, we begin by talking about something shared—a class, an ethnic background, a school, or the weather.

Most relationships are initiated in the context of routine activities. This was shown clearly in a study of a random sample of college sophomores who described their most recently established heterosexual relationships. Some of the relationships were brand new; others had existed for many months. Of these relationships, 36 percent began in classes, dorms, clubs, or at work; 38 percent began with an introduction by a

third person; 18 percent of the couples met at parties; and only 14 percent met at bars or cocktail lounges. Thus, routine activities, introductions, and socially sanctioned meeting places such as parties are legitimate places for people to initiate contact with each other (Marwell et al., 1982). Similar studies of the friendship patterns of city dwellers have found that friends are selected from relatives, co-workers, and neighbors (Fischer, 1984). Thus, routine activities set limits on the pool of availables.

## Proximity

Although routines bring us into the same classroom, dining hall, or work place, we are not equally likely to meet every person who is present. Rather, we are more likely to develop a relationship with someone who is in close physical proximity to us.

In classroom settings, seating patterns are an important influence on the development of friendships. One study (Byrne, 1961a) varied the seating arrangements for three classes of about 25 students each. At the beginning of the semester, students were assigned seats alphabetically. In one class, they remained in the same seats for the entire semester (14 weeks). In the second class, they were assigned new seats halfway through the semester. In the third class, they were reassigned new seats every 3½ weeks. The relationships between students were assessed at the beginning and end of the semester. Few relationships developed among the students in the class where seats were changed every 3½ weeks. In the other two classes, students in neighboring seats became acquainted in greater numbers than students in non-neighboring seats. Moreover, by the end of the semester, the relationships between students were closer in the class where seat assignments were not changed during the semester.

Similar positive associations between physical proximity and friendship have been found in a variety of natural settings, including dormitories (Priest & Sawyer, 1967), married student housing projects (Festinger, Schachter, & Back

1950), and business offices (Schutte & Light, 1978).

We are more likely to develop friendships with persons in close proximity because such relationships provide interpersonal rewards at the least cost. First, interaction is easier with those who are close by. It is less costly in terms of time and energy to interact with the person sitting next to you than with someone on the other side of the room. A second factor is the influence of social norms. In situations where people are physically close or interact frequently —such as in dormitories, classes, and offices— we are expected to behave in a polite, friendly way. Polite, friendly behavior provides increased rewards in these interactions. Failure to adhere to these norms might result in disapproval from others, which increases costs.

## Familiarity

As time passes, people who take the same classes, live in the same apartment building, or do their laundry in the same place become familiar with each other. Having seen a person several times, sooner or later we will smile or nod. Repeated exposure to the same stimulus is sufficient to produce a positive attitude toward it; this is called the **mere exposure effect** (Zajonc, 1968). In other words, familiarity breeds liking, not contempt. This effect is highly general and has been demonstrated with respect to a wide variety of stimuli—such as music, visual art, and comic strips—under many different conditions (Harrison, 1977).

Does mere exposure produce attraction? The answer appears to be yes. In one experiment, female undergraduates were asked to participate in an experiment on their sense of taste. They entered a series of booths in pairs and rated the taste of various liquids. The schedule was set up so that two subjects shared the same booth either once, twice, five times, ten times, or not at all. At the end of the experiment, each woman rated how much she liked each of the other subjects. As predicted, the more frequently a woman had been in the same booth with

another subject, the higher the rating (Saegert, Swap, & Zajonc, 1973).

## ■ Who Is Desirable?

We come into contact with many potential partners, but contact by itself does not ensure the development of a relationship. Whether a relationship of some type actually develops between two persons depends on whether each is attracted to the other. Initial attraction is influenced by social norms, physical attractiveness, and processes of interpersonal exchange.

### Social Norms

For each type of relationship between persons, norms specify what kinds of people are allowed to have such a relationship. These norms tell us which persons are appropriate as friends, lovers, and mentors. In U.S. society, there is a **norm of homogamy**—a norm requiring that friends, lovers, and spouses be characterized by similarity in age, race, religion, and socioeconomic status (Kerckhoff, 1974). Research shows that homogamy is characteristic of adolescent friendship pairs (Kandel, 1978) and married couples (Murstein, 1980). Differences on one or more of these dimensions make a person less appropriate as a friend and more appropriate for some other kind of relationship. Thus, a person who is substantially older but the same social class and ethnicity may be appropriate as a mentor, someone who can provide advice about how to manage your career. Potential dates are single persons of the opposite sex who are of similar age, class, ethnicity, and religion.

Norms that define appropriateness influence the development of relationships in several ways. First, each of us uses norms to monitor our own behavior. We hesitate to establish a relationship with someone who is normatively defined as an inappropriate partner. Thus, a low-status person is unlikely to approach a high-status person as a potential friend; the law clerk who just joined a firm would not discuss her hobbies

# Box 11-1

## LET'S MAKE A DEAL

Reasonably good-looking professional man, 34, wishes to meet attractive, professional, sensitive woman, 24–30, to share interest in music, hockey, people, and conversation. Reply to Box 3010.

Attractive Gemini, 36, blond, blue eyes, 5′7″, 125 lbs., tanned, loves sports, playing piano, guitar. Interested in meeting man 26–46, physically fit, financially secure, witty. Reply to Box 2407.

Ads like these can be found in most daily newspapers in the U.S. Other papers carry dozens or even hundreds of these ads in each issue. If we look at these ads in terms of what they offer and what they seek, we can see that people are looking for a complementary exchange.

One study analyzed 800 such ads in an attempt to identify patterns in the "revelations" and "stipulations" contained in them (Harrison & Saeed, 1977). The ads were taken from a national weekly over a six-month period. An ad was excluded if the person was under 20 or over 60 or if age was not given. Results of the study showed that most women offered physical attractiveness and were seeking financial security and an older companion. Conversely, most men offered financial security and were seeking an attractive, younger woman. Thus, each group offered complementary resources.

Results also indicated that people sought partners whose level of social desirability was approximately equal to their own. This supports the *matching hypothesis* (Berscheid et al., 1971). The researchers calculated an "overall desirability score" for both the offerer and the person being sought. Attractiveness, financial security, and sincerity were each worth one point on the offerer's index, whereas seeking each of these was worth one point on the other index. Points were added to obtain desirability scores, which

ranged from zero to three. There was a significant correlation between the level of desirability offered and the level of desirability sought. In addition, most persons who stated they were physically attractive were seeking good-looking partners—evidence of the importance of physical attractiveness.

A recent study of 400 ads from two weekly newspapers replicated the finding that women offered physical attractiveness and were seeking status (security), whereas men offered status and were seeking attractiveness (Koestner & Wheeler, 1988). In addition, women offered low body weight and were seeking tall men, whereas men offered height and were seeking light women. Finally, women offered instrumental traits—independent, competent, bright—valued by men, while men offered expressive traits—warm, gentle, emotional—valued by women. Advertisers seem to understand attraction, offering precisely the traits that are sought by the opposite sex.

What kinds of people advertise for partners? You might suspect that these are people who are unable to find partners appropriate to their needs or characteristics in their routine activities. Often they are people who have, or are seeking persons with, atypical characteristics and are thus less likely to meet eligibles in their immediate environment. Advertisers are often older (40+), members of racial or ethnic minorities (such as Asian-American), religiously devout or Pentecostal, vegetarian, divorced with young children, or physically handicapped. They are seeking persons with specific characteristics—that is, "old-fashioned," nondrinkers, or born-again Christians. By placing ads, people hope to transcend the limits imposed on who's available by proximity and routine activities.

with the senior partner (unless the senior partner asked). Second, if one person attempts to initiate a relationship with someone who is normatively defined as inappropriate, the other person will probably refuse to reciprocate. If the clerk did launch into an extended description of the joys of restoring antique model trains, the senior partner would probably politely end the interaction. Third, even if the other person is willing to interact, third parties often enforce the norms that prohibit the relationship (Kerckhoff, 1974). Another member of the firm might later chide the clerk for presuming that the senior partner cared about her personal interests.

## Physical Attractiveness

In addition to social norms that define who is appropriate, individuals also have personal preferences regarding desirability. Someone may be normatively appropriate but still not appeal to you. Personal attributes—physical attractiveness, competence—can have a significant impact on desirability. We will consider these two attributes in this section.

### The Impact of Physical Attractiveness

A great deal of evidence shows that given more than one potential partner, individuals will prefer the one who is more physically attractive. A study of 752 college freshmen, for example, demonstrates that most individuals prefer more attractive persons as dates (Walster [Hatfield] et al., 1966). As part of the study, freshmen were invited to attend a dance. Prior to the dance, each student had to purchase a ticket; as students waited in line, four ticket sellers secretly rated the person's physical attractiveness as low, average, or high. Before the dance, each student filled out several questionnaires and was told that dance partners would be selected by the computer. In fact, males and females were paired randomly. The couples arrived at the dance at about 8:00 P.M. and danced until intermission at 10:30 P.M. During the intermission, students filled out a questionnaire in which they gave their impression of their date.

This study tested the **matching hypothesis** —the idea that each of us looks for someone who is of approximately the same level of desirability. The hypothesis predicts that those students whose dates matched their own attractiveness would like their date most. Those whose dates were very different in attractiveness were expected to rate their date as less desirable and less considerate. Contrary to the hypothesis, the research showed that in this situation, students preferred a more attractive date, regardless of their own attractiveness. The more physically attractive their partners, the more the students were eager to continue their relationship.

The importance of physical attractiveness in the computer dance study could not have been predicted from reports by men and women about what they look for in a potential date. When asked directly, young people consistently rank character, emotional stability, and various other traits as more important than attractiveness (Berscheid & Walster [Hatfield], 1974b). How then can we explain the significance of attractiveness?

One factor is simply esthetic; generally, we prefer what is beautiful. Although beauty is to some extent "in the eye of the beholder," cultural standards influence our esthetic judgments. A recent study of female facial beauty found substantial agreement among male college students about which features are attractive (Cunningham, 1986). These men rated such features as large eyes, small nose, and small chin as more attractive than small eyes, large nose, and large chin. Research has also found a high level of agreement among men that certain female body shapes are more appealing than others (Wiggins, Wiggins, & Conger, 1968). Men prefer a figure with average bust, waist, and hip measurements. There is also agreement among women about which male body shapes are attractive (Beck, Ward-Hull, & McLear, 1976).

A second factor is that we anticipate more rewards when we associate with attractive persons. A man with an extremely attractive woman receives more attention and prestige from other

According to the matching hypothesis, people seek partners whose level of social desirability is about equal to their own. We frequently encounter couples who are matched—that is, who are similar in age, race, ethnicity, social class, and physical attractiveness.

persons than if he is seen with an unattractive female, and vice versa (Sigall & Landy, 1973).

A third factor is the **attractiveness stereotype**—the belief that "what is beautiful is good" (Dion, Berscheid, & Walster [Hatfield], 1972). We assume that an attractive person possesses other desirable qualities. This was demonstrated in one study in which male and female college students were shown three photographs and were asked to rate the personality of the person photographed on various dimensions. The people in the photos varied in physical attractiveness; half of the male and female subjects were shown pictures of men, whereas the other half were shown pictures of women. Physically attractive people were rated as more sensitive, kind, interesting, strong, modest, and sexually responsive than less attractive persons (Sigall & Landy, 1973). They were also considered to be more talented than unattractive people (Landy & Sigall, 1974).

When we believe another person possesses certain qualities, those beliefs influence our

behavior toward that person so that he behaves in ways that are consistent with our beliefs. In one experiment, males were shown photographs, randomly chosen, of either an attractive woman or an unattractive woman. They were then asked to interact with the woman in the photograph via intercom for ten minutes. (The woman over the intercom was actually one of several student volunteers.) Each conversation was tape-recorded and rated by judges. Women who were perceived as attractive by the men were rated as behaving in a more friendly, likable, and sociable way, compared to women who were perceived as unattractive. This happened in part because the men gave the target person opportunities to act in ways that would confirm their expectations based on the attractiveness stereotype (Snyder, Tanke, & Berscheid, 1977).

Each of us knows that physically attractive people may receive preferential treatment. As a result, we spend tremendous amounts of time and money in an effort to increase our attractiveness to others. Men and women purchase clothing, perfumes, colognes, and hair dyes in an effort to enhance their physical attractiveness. Our choice of products reflects current standards of what looks good. Increasingly, people are using cosmetic surgery to enhance their appearance. Plastic surgeons can "lift your face, bob your nose, unbag your eyes, strengthen your chin, pin your ears, tuck your tummy, enlarge your breasts, shape your fanny and smooth your thighs" ("Plastic Surgery Now Commonplace," 1979). In 1984, at least 1.5 million such operations were performed in the U.S., costing an estimated 4 billion dollars (Ubell, 1985).

Not everyone is favorably predisposed toward attractive persons. Many of us were taught, and some of us believe, that "Beauty is only skin deep" and "You can't judge a book by its cover." What kinds of people are influenced by another's attractiveness, and what kinds of people "read the book" before making a judgment?

People who hold traditional attitudes toward men and women are those whose judgments are much more likely to be influenced by beauty. In one study (Touhey, 1979), students were given the Macho scale, a series of questions used to assess to what degree a person endorses sexist/traditional attitudes. Men and women who obtained high and low scores (a high score reflecting more traditional attitudes) were given photographs and biographical descriptions of several members of the opposite sex. Each subject indicated probable liking for and willingness to date the person in the photograph and rated the person on various aspects of character. Overall, men and women with high Macho scores liked attractive persons more than unattractive ones, whereas low scorers liked attractive and unattractive persons equally well. High scorers rated the attractive person as more dutiful and respecting of authority, a better dancer, and a better potential parent than the unattractive person. The low scorers' overall ratings for attractive versus unattractive persons were not significantly different.

Another characteristic that influences the importance of attractiveness in judgments of others is *self-monitoring*. People who are high in self-monitoring look for external cues to tell them how to behave, whereas those who are low in self-monitoring use internal cues, such as their own attitudes and values, to guide their behavior. Two studies were conducted in which men who were high or low in self-monitoring were allowed to select one of two or more women to date (Snyder, Berscheid, & Glick, 1985). Men who were high in self-monitoring were more likely to select on the basis of physical attractiveness, whereas men who were low on this characteristic gave greater weight to the woman's personal attributes.

**Attractiveness Isn't Everything** Physical attractiveness may have a major influence on our judgments of others because it is readily observable. If other relevant information is available, it may reduce or eliminate the impact of attractiveness on our judgments. For example, in a classroom or work setting, we often prefer to establish contact with someone who is competent and can give us help. In such cases, information about competence should influence our liking for another person. To determine the influence of such information, researchers asked men and women to rate a stimulus person (female) whose physical attractiveness and competence varied (Solomon & Saxe, 1977). Subjects were shown a videotape of an unattractive or an attractive woman and were either told the woman was very intelligent (grade point average of 3.75, high test scores) or not very intelligent (grade point average 2.25, low test scores). Afterward, subjects were asked to rate the personality of the woman shown on the tape. Subjects gave a more positive rating to women who were described as more intelligent as well as more attractive. In fact, intelligence had greater influence on the personality ratings than did beauty. Moreover, when people are given information about a person's ability before they receive information about physical attractiveness, attractiveness does not influence their judgments of that person's performance (Benassi, 1982).

Thus, physical attractiveness greatly influences our initial impressions of other persons. If it is the only information we have about someone, the attractiveness stereotype leads us to assume that that person possesses other desirable traits. But if we also have information about competence, this may reduce and even eliminate the impact of physical attractiveness.

## Exchange Processes

How do we move from the stage of awareness of another person to the stage of contact? Recall that in our introduction Dan noticed Sally at every lecture. Because she was young and not wearing a wedding ring Dan hoped that she was available. She was certainly desirable—she was very pretty and seemed like a friendly person. What factors did Dan take into account when deciding whether to initiate contact? One important factor in this decision is the availability and desirability of alternative relationships (Backman, 1981). Thus, before Dan chose to initiate contact with Sally, he probably considered whether there was anyone else who might be a better choice.

**Choosing Your Friends**   We can view each actual or potential relationship—whether involving a friend, co-worker, roommate, or date—as promising rewards but entailing costs. Rewards are the pleasures or gratifications we derive from a relationship. These might include a gain in knowledge, enhanced self-esteem, satisfaction of emotional needs, or sexual gratification. Costs are the negative aspects of a relationship, such as physical or mental effort, embarrassment, and anxiety.

Exchange theory proposes that this is in fact the way people view their interactions (Homans, 1961; Blau, 1964). People evaluate interactions and relationships in terms of the rewards and costs that each is likely to entail. They calculate likely **outcomes** by subtracting the anticipated costs from the anticipated rewards. If the expected outcome is positive, people are inclined to initiate or maintain the relationship. If the ex-

pected outcome is negative, they are unlikely to initiate a new relationship or to stay in a relationship that is ongoing. Dan anticipated that dating Sally would be rewarding; she would be fun to do things with, and others would be impressed that he was dating such an attractive woman. At the same time, he anticipated that Sally would expect him to be committed to her and that he would have to spend money to take her to movies, plays, and restaurants.

It is not easy to evaluate the outcomes of a relationship. What standards can we use to make such an evaluation? Two standards have been proposed (Thibaut & Kelley, 1959; Kelley & Thibaut, 1978). One is the **comparison level** (CL), which is based on the average of a person's experience in past relevant relationships. Based on this experience, the person expects a certain minimum level of outcomes. Each relationship is evaluated in terms of whether it is above or below that person's CL—that is, better or worse than the average of past, relevant relationships. Relationships that fall above a person's CL are satisfying, whereas those that fall below it are unsatisfying.

If this were the only standard, we would always initiate relationships that appeared to promise outcomes better than those we already experienced and avoid relationships that appeared to promise poorer outcomes. Sometimes, however, we may use a second standard, called the **comparison level for alternatives** ($CL_{alt}$). This is the lowest level of outcomes a person will accept in light of available alternatives. A person's $CL_{alt}$ varies, depending on the outcomes that he or she believes can be obtained from the best of the available alternative relationships. The fact that behavior is based on $CL_{alt}$ helps explain why we may sometimes turn down interaction opportunities that appear promising or why we may remain in a relationship even though we feel that the other person is getting all the benefits.

Whether or not a person initiates a new relationship will depend on both the CL and the $CL_{alt}$. An individual usually avoids relationships whose anticipated outcomes fall below the comparison level. If a potential relationship appears likely to yield outcomes above a person's CL, then initiation will depend on whether the outcomes are expected to exceed the $CL_{alt}$. Dan believed that a relationship with Sally would be very satisfying. He was casually dating another woman, and that relationship was not gratifying. Thus, the potential relationship with Sally was above both CL and $CL_{alt}$, leading Dan to initiate contact.

Whereas CL is an absolute, relatively unchanging standard, several factors influence a person's $CL_{alt}$, including the extent to which one's routine activities provide opportunities to meet people, the size of the pool of eligibles, and one's skills in initiating relationships. Situational factors are another influence. At a dance, party, or bar, where people hope to initiate relationships, an important situational factor is time. When the party or dance ends or the bar closes, it will no longer be possible to establish contact. This suggests that one's $CL_{alt}$ declines as the evening passes. In fact, some research shows that as the alternatives become fewer, originally less attractive potential partners are seen as more attractive (Pennebaker et al., 1979).

**Making Contact** Once we decide to initiate interaction, the next step is to make contact. Sometimes we use technology, such as the telephone or computer mail. Often we arrange to get physically close to the person. At parties and bars, people often circulate, which brings them into physical proximity with many of the other guests.

Once in proximity, strangers frequently use a direct gaze as a signal that they are interested in conversation. In one study, pairs of undergraduates were brought together in a waiting room and videotaped for five minutes. In the room two chairs were placed side by side with a magazine table in between. Almost every subject gave an initial look at the other subject. If the gaze was mutual, continuous conversation during the five-minute period was likely; when the gaze was not

mutual, conversation was unlikely (Carey, 1978). This pattern is probably repeated many times every day in airplanes, on trains and buses, and in classes and waiting rooms.

In most cases, initiation requires an opening line. Usually it is about some feature of the situation. At the beginning of this chapter Dan initiated conversation by commenting that Sally never missed the class. Two people waiting to participate in a psychology experiment may begin talking by speculating over the purpose of the experiment. The weather is a widely used topic for openings. The opening line often includes an *identification display*—a signal that we believe the other person is a potential partner in a specific kind of relationship (Schiffrin, 1977). Approaching someone in a department store and asking for help clearly communicates your belief that the other person is an employee. When Dan commented about Sally's attitude toward the class, it conveyed an interest in friendship; a different message would have been sent if he had asked whether she knew the woman sitting next to her. The other person, in turn, decides whether she is interested in that type of relationship; if she is, she engages in an **access display**—a signal that further interaction is permissible. Thus, Sally responded warmly to Dan's opening line, encouraging continued conversation.

# ■ The Determinants of Liking

Once two people make contact and begin to interact, several factors will determine the extent to which each person will like the other. Four of these factors are considered in this section: similarity, complementarity, shared activities, and reciprocal liking.

## Similarity

How important is similarity? Do "birds of a feather flock together"? Or is it more the case that "opposites attract"? These two aphorisms about the determinants of liking are inconsistent and provide opposing predictions. A good deal of research has been devoted to finding out which one is more accurate. On the whole, evidence indicates that birds of a feather do flock together; that is, we are attracted to people who are similar to ourselves. Probably the most important kind of similarity is **attitudinal similarity**—the sharing of beliefs, opinions, likes, and dislikes.

**Attitudinal Similarity**   A widely employed technique for studying attitudinal similarity is the attraction-to-a-stranger paradigm, initially developed by Byrne (1961b). Potential subjects fill out an attitude questionnaire that measures their beliefs about various topics, such as life on a college campus. Later, subjects receive information about a stranger, as part of a seemingly unrelated study. The information they receive describes the stranger's personality or social background and may include a photograph. They also are given a copy of the same questionnaire they completed earlier, ostensibly filled out by the stranger. In fact, the stranger's questionnaire is completed by the experimenter, who systematically varies the degree to which the stranger's supposed responses match the subject's responses. After seeing the stranger's questionnaire, subjects are asked how much they like or dislike the stranger and how much they would enjoy working with that person.

In most cases, the subject's attraction to the stranger is positively associated with the percentage of attitude statements by the stranger that agree with the subject's own attitudes (Byrne & Nelson, 1965; Gonzales et al., 1983). We rarely agree with our friends about everything; what matters is that we agree on a high proportion of issues. This relationship between similarity of attitudes and liking is very general; it has been replicated in studies using both men and women as subjects and strangers under a variety of conditions (Berscheid & Walster [Hatfield], 1978).

In the attraction-to-a-stranger paradigm, the subject forms an impression of a stranger without any interaction. This allows researchers to determine the precise relationship between similarity and liking. But what do you think the relationship would be if two people were allowed to interact? Would similarity have as strong an effect?

A study attempting to answer this question arranged dates for 44 couples (Byrne, Ervin, & Lamberth, 1970). Researchers distributed a 50-item questionnaire measuring attitudes and personality to a large sample of undergraduates. From these questionnaires they selected 24 male-female couples whose answers were very similar (66–74 percent identical) and 20 couples whose answers were not similar (24–40 percent identical). Each couple was introduced, told they had been matched by a computer, and asked to spend the next 30 minutes together at the student union—they were even offered free sodas. The experimenter rated each subject's attractiveness before they left on their date. When they returned, the couple rated each other's sexual attractiveness, desirability as a date and marriage partner, and indicated how much they liked each other. The experimenter also recorded the physical distance between the two as they stood in front of his desk.

Results of this experiment showed that both attitudinal similarity and physical attractiveness influenced liking. Partners who were attractive and who held highly similar attitudes were rated as more likable. In addition, similar partners were rated as more intelligent and more desirable as a date and marriage partner (see Table 11-1). The couples high in similarity stood closer together after their date than couples low in similarity, another indication that similarity creates liking.

At the end of the semester, 74 of the 88 subjects in this study were contacted and asked whether they (1) could remember their date's name, (2) had talked to their date since their first meeting, (3) had dated their partner, or (4) wanted to date their partner. Researchers compared reports of subjects in the high attractiveness/high similarity condition with those in the low attractiveness/low similarity condition. Those in the former condition were more likely

■ **TABLE 11-1**

THE INFLUENCE OF SIMILARITY AND PHYSICAL ATTRACTIVENESS ON LIKING

| Physical Attractiveness of Date | Liking/ Desirability Rating | |
| --- | --- | --- |
| | Proportion of Similar Responses | |
| | Low | High |
| Male Subjects | | |
| Attractive date | 10.55 | 12.00 |
| Unattractive date | 9.89 | 10.43 |
| Female Subjects | | |
| Attractive date | 11.25 | 12.71 |
| Unattractive date | 9.50 | 11.00 |

A computer paired 24 couples with a high proportion of identical answers to a 50 item questionnaire and 20 couples with a low proportion of identical respones. Following a 30-minute date, each person rated the likability and desirability of their date on a scale from 1 to 7. The sum of the average of these two ratings is shown by gender and condition. Both men and women liked attractive dates more than unattractive ones, and both liked dates who were more similar.

*Source: Adapted from Byrne, Ervin, and Lamberth, 1970, Table 4.*

to remember their partner's name, to report having talked to their partner, and to report wanting to date their partner.

The story of Dan and Sally at the beginning of this chapter illustrates the importance of similarity in the early stages of a relationship. After their initial meeting, they discovered they had a number of things in common. They were from the same city. They had chosen the same major and held similar beliefs about their field and about how useful a bachelor's degree would be in that field. Each also found the other attractive; like the subjects in the high attraction/high similarity condition, Dan and Sally continued to talk after their first meeting.

**Why Is Similarity Important?** Why does attitudinal similarity produce liking? One reason is the desire for consistency between our attitudes and

perceptions. The other reason focuses on our preference for rewarding experiences.

Most people desire *cognitive consistency*—consistency between attitudes and perceptions of whom and what we like and dislike. If you have positive attitudes toward certain objects and discover that another person has favorable attitudes toward those same objects, your cognitions will be consistent if you like that person (Newcomb, 1971). When Dan discovered that Sally had a positive attitude toward his major, his desire for consistency produced a positive attitude toward Sally. Our desire for consistency attracts us to persons who hold the same attitudes toward important objects.

Similarity in attitudes is an important influence on the development of friendships. This was demonstrated in a study of the friendships that developed among 17 college men who shared a rooming house for a semester (Newcomb, 1961). Each man filled out several questionnaires prior to moving into the house. Roommates were assigned by the research staff. During the semester, the men expressed their liking for each other several times. The liking ratings stabilized by the middle of the semester, and they were best predicted by the degree of similarity in attitudes and values. Another study reported similar findings among women (Hill & Stull, 1981).

Another reason why we like persons with attitudes similar to our own is because interaction with them provides three kinds of reinforcement. First, interacting with persons who share similar attitudes usually leads to positive outcomes (Lott & Lott, 1974). At the beginning of this chapter, Dan anticipated that he and Sally would get along well because they shared similar likes and dislikes.

Second, similarity validates our own view of the world. We all want to evaluate and verify our attitudes and beliefs against some standard. Sometimes physical reality provides objective criteria for our beliefs. But often there is no physical standard, and so we must compare our attitudes with those of others (Festinger, 1954). Persons who hold similar attitudes provide us with support for our own opinions, which allows us to deal with the world more confidently (Byrne, 1971). Such support is particularly important in areas like political attitudes, where we realize others hold attitudes dissimilar to our own (Rosenbaum, 1986).

Third, we are attracted to others who share similar attitudes because we expect that they will approve of us. In one experiment, college students were recruited to join groups of strangers. Before the first meeting, students were privately informed that, based on personality tests, certain other members would probably like them. At the end of the meeting, each person indicated whom he preferred to work with subsequently. Most students chose the person whom they had been told would like them (Backman & Secord, 1959). In the absence of specific information about others' feelings, we prefer to develop friendships with those we think will evaluate us favorably (Santee & Jackson, 1978).

### Complementarity

We have seen that similarity produces liking. Does that mean that *dissimilarity* promotes *disliking*? What about the aphorism "Opposites attract"? This commonsense adage is based on the idea that people are attracted to those whose needs complement their own. Thus, a person with strong needs to dominate, to be in control, should be attracted to a submissive person who will take orders without complaint. A person needing to nurture others by providing love and support should be attracted to a person with a strong need to be nurtured.

This theory of complementary needs has been applied to mate selection (Winch, 1958), leading to the hypothesis that most people will choose as a mate someone whose needs complement their own. In several studies, results of personality tests given to a small sample of married couples support this hypothesis, although most of the research on mate selection does not (Berscheid & Walster [Hatfield], 1978; Backman, 1981).

More recently, it has been suggested that it is not *complementarity of personal needs* that is

important, but *role compatibility* (Murstein, 1976). As a couple develops their relationship, they must adopt roles that complement each other and fulfill each other's expectations. Thus, a couple with traditional values—in which the man is expected to support the family and take care of the house and the car and the woman is expected to do the housework and raise the children—will find it relatively easy to develop complementary roles. A woman who is committed to a career and hates housework will have a difficult time developing a role if her mate holds traditional expectations.

### Shared Activities

As people interact, they share activities. Recall that after Sally and Dan met, they began to sit together in class and to discuss course work. When the first exam was announced, Dan invited Sally to study for it with him. Sally's roommate was also in the class; the three of them reviewed the material together the night before the exam. Sally and Dan both got As on the exam, and each felt that studying together helped. The next week they went to a movie together, and several days after that Sally invited Dan to a party.

Shared activities provide opportunities for each person to experience reinforcement—that is, to have positive experiences with the other individual. Some of these reinforcements come from the other person; Sally finds Dan's interest in her very reinforcing. Often the other person is associated with a positive experience, which leads us to like the person (Byrne & Clore, 1970). Getting an A on the examination was a very positive experience for Dan and for Sally; the association of the other person with that experience led to increased liking for the other.

Thus, as a relationship develops, the sharing of activities contributes to increased liking. This was shown in a study in which pairs of friends of the same sex both filled out attitude questionnaires and listed their preferences for various activities (Werner & Parmalee, 1979). The duration of the friendships varied from 3 months to 20 years, with an average of 5 years. Both 13 male and 11 female pairs were included in the study. Results of the study showed similarity between friends in both activity preferences and attitudes. In addition, people were able to predict their friend's activity preferences more accurately than their friend's attitudes. This suggests that participation in mutually satisfying activities may be a strong influence on the development and maintenance of friendship. As Dan and Sally got to know each other, their shared experiences—studying, seeing movies, going to parties—became the basis for their relationship, supplementing the effect of similar attitudes.

### Reciprocal Liking

One of the most consistent research findings is the strong positive relationship between our liking someone and the perception that the other person will like us in return (Backman, 1981). In most relationships, we expect reciprocity of attraction; the greater the liking of one person for the other, the greater the other person's liking will be in return. But will the degree of reciprocity increase over time, as partners have greater opportunities to interact? To answer this question, one study obtained liking ratings from 48 persons (32 males and 16 females), who had been acquainted for 1, 2, 4, 6, or 8 weeks (Kenny & La Voie, 1982). Results showed a positive correlation between each person's liking rating for the other, and reciprocity of attraction increased somewhat from week 1 to week 8. However, because some of the subjects in this study were roommates rather than friends, they would be expected to like each other due to the proximity effect. When roommate pairs were eliminated from the results, the correlation between liking ratings increased substantially, as expected.

## ■ The Growth of Relationships

We have traced the development of relationships from the stage of zero contact through awareness (who is available), surface contact (who is desirable), to mutuality (liking). At the beginning of this chapter Dan and Sally met, discovered that they had similar attitudes and interests, and shared pleasant experiences—such as doing well on an examination, going to a movie, and later, to a party.

Many of our relationships remain at the "minor" level of mutuality (see Figure 11-1)—that is, we have numerous acquaintances, neighbors, and co-workers whom we like and interact with regularly but to whom we do not feel especially close. A few of our relationships grow closer; they proceed through "moderate" to "major" mutuality. Three aspects of the growth of relationships are examined in this section: self-disclosure, trust, and interdependence. As the degree of mutuality increases between friends, roommates, and co-workers, so self-disclosure, trust, and interdependence will also increase.

### Self-Disclosure

Recall that when Dan and Sally returned from the party, Sally told Dan that her roommate's parents had just separated and that her roommate was very depressed. Sally said that she didn't know how to help her roommate, that she felt unable to deal with the situation. At this point, Sally was engaging in *self-disclosure*—the act of revealing personal information about oneself to another person. Self-disclosure usually increases along with mutuality in a relationship. Initially, people reveal things about themselves that are not especially intimate and that they believe the other will readily accept. Over time, they disclose increasingly intimate details about their beliefs or behavior, as well as information that they are less certain the other will accept (Backman, 1981).

Self-disclosure increases as a relationship grows. In one study, same-sex pairs of previously unacquainted college students were brought into a laboratory setting and asked to get acquainted (Davis, 1976). They were given a list of 72 topics; each topic had been rated earlier by other students on a scale of intimacy from 1 to 11. Subjects were asked to select topics from this list and to take turns talking about each topic for at least one minute's duration while their partner remained silent. The interaction continued until each partner had spoken on 12 of the 72 topics. Results showed that the intimacy of the topic selected increased steadily from the first to the twelfth topic chosen. The average intimacy of topics discussed by each couple increased from 3.9 to 5.4 over the 12 disclosures.

When Sally told Dan about the situation with her roommate, Dan replied that he knew how she felt because his older brother had just separated from his wife. This exchange reflects reciprocity in self-disclosure; as one person reveals an intimate detail, the other person usually discloses information at about the same level of intimacy (Altman & Taylor, 1973). In the Davis study discussed above (1976), each subject selected a topic at the same level of intimacy as the preceding one or at the next level of intimacy. However, reciprocity decreases as a relationship develops. In one study, the researcher recruited students to be subjects and asked each one to bring an acquaintance, a friend, or a best friend to the laboratory (Won-Doornink, 1985). Each dyad was given a list of topics that varied in degree of intimacy and was instructed to take at least four turns choosing and discussing a topic. Each conversation was tape-recorded and later analyzed for evidence of reciprocity. The association between the stage of relationship and the reciprocity of intimate disclosures was curvilinear; there was greater reciprocity of intimate disclosures among friends than among acquaintances and less reciprocity among best friends than among friends (see Figure 11-2).

Dan and Sally had known each other several weeks before they told each other about the separations involving someone close to them. What would have happened if one of them had divulged this information the day they met? The fact that disclosures become more intimate over time suggests that an intimate disclosure very early in a relationship may be inappropriate and reduce the discloser's attractiveness. In a laboratory experiment (Wortman et al., 1976), a confederate revealed to some male subjects just two

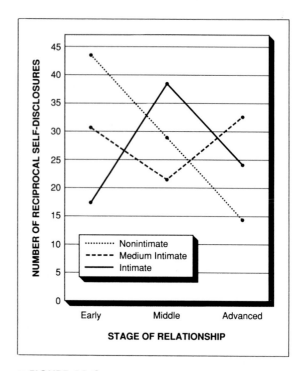

■ **FIGURE 11-2**

THE RELATIONSHIP BETWEEN RECIPROCITY AND INTIMACY

Reciprocity—picking a topic of conversation that is as intimate as the last topic introduced by your partner—is the process by which relationships become more intimate. The extent of reciprocity depends on the intimacy of the topic and the stage of the relationship. Students talked with an acquaintance (early stage), friend (middle stage), or best friend (advanced stage). With topics that were not intimate (such as the weather), reciprocity declined steadily as the stage increased. With intimate topics, on the other hand, reciprocity was greatest at the middle stage, less at the advanced stage, and least at the early stage of a relationship.

*Source: Adapted from Won-Doornink, 1985, Figure 4.*

minutes after they had met that his girlfriend was pregnant. Most subjects rated the confederate as phony and immature. When the same disclosure was made after eight minutes of interaction, the confederate was rated as sincere and likable. Thus, the timing of disclosure may affect the other person's liking for an individual.

Not all people divulge increasingly personal

information as you get to know them. You have probably known people who were very open—who readily disclosed information about themselves—and others who said little about themselves. In this regard, we often think of men as less likely to discuss their feelings than women. However, research has shown that self-disclosure depends not only on gender but also on the nature of the relationship. In casual relationships (with men or women), men are less likely to disclose personal information than women (Reis, Senchak, & Solomon, 1985). In intimate heterosexual relationships, men and women do not differ in the degree of self-disclosure (Hatfield, 1982). Among dating couples, the amount of disclosure is related more to sex-role orientation rather than gender. Men and women with traditional orientations tend to disclose less to their partners than those with egalitarian orientations (Rubin et al., 1980). Traditional sex roles are more segregated, with each person responsible for certain tasks, whereas egalitarian orientations emphasize sharing. An emphasis on joint activity leads to greater communication and self-disclosure.

The degree of self-disclosure also depends on the personal characteristics of the people involved. Some people are *openers*—individuals who elicit intimate disclosures from others. In one study, women were given a questionnaire measuring the degree to which they were openers. After completing the questionnaire, each woman was asked to interact with a female stranger. Women who were characterized as openers were more likely to elicit personal information from the stranger. The same study found differences in the degree to which women disclose personal information (Miller, Berg, & Archer, 1983).

Thus, as we get to know someone, we are likely to reveal information about ourselves that increases in intimacy over time. Initially there is reciprocity of self-disclosure, as the other person responds to our disclosures with revelations about himself or herself; but this declines over time. The amount of information revealed depends on the type of relationship and on our beliefs about the appropriateness of disclosure in such relationships.

## Trust

Why did Dan confide in Sally that his brother had just left his wife? Perhaps he was offering reciprocity in self-disclosure. Because Sally had confided in Dan, she expected him to reciprocate. But had he been suspicious of Sally's motives, he might not have. This suggests the importance of trust in the development of a relationship.

When we **trust** someone, we believe that person is both honest and benevolent (Larzelere & Huston, 1980). We believe that the person tells us the truth—or at least does not lie to us—and that his intentions toward us are positive. One measure of interpersonal trust is the interpersonal trust scale reproduced in Table 11-2. The questions focus on whether the other person is selfish, honest, sincere, fair, or considerate. It has been suggested that we are more likely to disclose personal information to someone we trust. How much do you trust *your* partner? Answer the questions on the scale and determine your score. Higher scores indicate greater trust.

To study the relationship between trust and self-disclosure, researchers recruited men and women from university classes, from a list of people who had recently obtained marriage licenses, and by calling persons randomly selected from the telephone directory. Each person was asked to complete a questionnaire concerning his or her spouse or current or most recent date. The survey included the interpersonal trust scale reproduced in Table 11-2. Researchers averaged the trust scores for seven types of relationships as shown in Figure 11-3. Note that as the relationship becomes more exclusive, trust scores increase significantly. Is there a relationship between trust and self-disclosure? Each person was also asked how much he or she had disclosed to the partner in each of six areas—religion, family, emotions, relationships with others, school or work, and marriage. Trust scores were positively

■ **TABLE 11-2**
INTERPERSONAL TRUST SCALE

| | Strongly Agree | Agree | Slightly Agree | ? | Slightly Disagree | Disagree | Strongly Disagree |
|---|---|---|---|---|---|---|---|
| 1. My partner is primarily interested in his/her own welfare. | ___ | ___ | ___ | – | ___ | ___ | ___ |
| 2. There are times when my partner cannot be trusted. | ___ | ___ | ___ | | ___ | ___ | ___ |
| 3. My partner is perfectly honest and truthful with me. | ___ | ___ | ___ | | ___ | ___ | ___ |
| 4. I feel I can trust my partner completely. | ___ | ___ | ___ | | ___ | ___ | ___ |
| 5. My partner is truly sincere in his/her promises. | ___ | ___ | ___ | | ___ | ___ | ___ |
| 6. I feel my partner does not show me enough consideration. | ___ | ___ | ___ | | ___ | ___ | ___ |
| 7. My partner treats me fairly and justly. | ___ | ___ | ___ | – | ___ | ___ | ___ |
| 8. I feel my partner can be counted on to help me. | ___ | ___ | ___ | – | ___ | ___ | ___ |

Note: For items 1, 2, and 6, Strongly Agree = 1, Agree = 2, Slightly Agree = 3, and so on. For items 3, 4, 5, 7, and 8 the scale is reversed.

Source: Adapted from Larzelere and Huston, 1980.

correlated with disclosure—that is, the more the person trusted the partner, the greater the degree of disclosure was.

Other research on interpersonal trust suggests that, in addition to honesty and benevolence, reliability is an important aspect of trust. We are more likely to trust someone who we feel is reliable—on whom we can count (Johnson-George & Swap, 1982)—and predictable (Rempel, Holmes, & Zanna, 1985).

## Interdependence

Earlier in this chapter we noted that people evaluate potential and actual relationships in terms of the outcomes (rewards minus costs) they expect to receive. Dan initiated contact with Sally because he anticipated that he would experience positive outcomes. Sally encouraged the development of a relationship because she, also, expected the rewards to exceed the costs. As their relationship developed, each discovered that the relationship was rewarding. Consequently, they increased the time and energy devoted to their relationship and decreased their involvement in alternative ones. As their relationship became increasingly mutual, Sally and Dan became increasingly dependent on each other for various rewards (Backman, 1981). The result is strong, frequent, and diverse interdependence (Kelley et al., 1983).

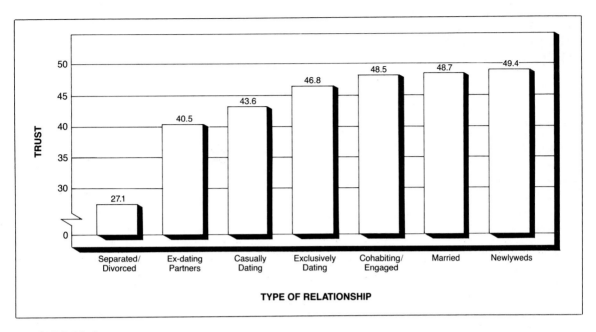

■ **FIGURE 11-3**

AVERAGE INTERPERSONAL TRUST SCORES FOR SEVEN TYPES OF HETEROSEXUAL RELATIONSHIPS

Trust involves two components: the belief that a person is honest and that his or her intentions are benevolent. More than 300 persons completed the interpersonal trust scale (see Table 11-2) for their current or most recent heterosexual partner. Results showed that there was a strong relationship between the degree of intimacy in a relationship and the degree of trust.

*Source: Adapted from Larzelere and Huston, 1980, Table 3.*

Increasing reliance on one person for gratifications and decreasing reliance on others is called **dyadic withdrawal** (Slater, 1963). One study of 750 men and women illustrates the extent to which such withdrawal occurs. Students identified the intensity of their current heterosexual relationships, then listed the names of persons whose opinions they considered to be important. They also indicated how important each person's opinions were and how much they had disclosed to that person (Johnson & Leslie, 1982). As predicted, the more intimate their current heterosexual relationship, the smaller the number of friends listed by the subjects; there was no difference in the number of relatives listed. In addition, as the degree of involve-

ment increased, the proportion of mutual friends of the couple also increased (Milardo, 1982). Other studies have found that as heterosexual relationships become more intimate, each partner spends less time interacting with friends and relatives (Milardo, Johnson, & Huston, 1983).

A study of the development of interdependence asked newlyweds to reconstruct their relationship from the time they met until the time they married (Surra, 1985). Each person reported the extent to which he or she shared affectional, instrumental, and leisure activities with the future spouse as their relationship grew. The length of time between meeting and marrying (from a few months to 6½ years) was closely related to how quickly the two people reduced their dependence on others for these activities and became interdependent.

Interdependence evolves out of the process of negotiation (Backman, 1981). Each person offers various potential rewards to the partner; the partner accepts some and rejects others. As the relationship develops, the exchanges stabilize. If a couple shares traditional sex-role orientations, the male will probably pay for their joint activities and drive the car when they are together; a couple that shares egalitarian orientations may contribute equally to the cost of joint activities and alternate driving (Rubin et al., 1980).

A potential reward in many relationships is sexual gratification. As relationships develop and become more mutual, physical intimacy increases as well. The couple negotiates the extent of sexual intimacy, with the woman's preferences having a greater effect on the outcome (Peplau, Rubin, & Hill, 1977). How important is sexual gratification in dating relationships? A study of 149 couples assessed the importance of various rewards in relationships of increasing intimacy (preferred date, going steady, engaged, living together, and married). Among intimate couples, sexual gratification was much more likely to be cited as a major basis for the relationship (Centers, 1975). Other surveys indicate that the longer couples have been dating, the more likely they are to engage in sexual intimacy (DeLamater & MacCorquodale, 1979).

# ■ Love and Loving

It is fair to say that what we feel for our friends, roommates, co-workers, and some of the people we date is attraction. But is that all we feel? Occasionally, at least, we experience something more intense than a positive attitude toward others. Sometimes we feel and even say "I love you."

How does loving differ from liking? Much of the research in social psychology on attraction or liking is summarized earlier in this chapter. By contrast, there has been much less research on love. Three views of love are considered in this section: the distinction between liking and loving, passionate love, and romantic love.

## Liking versus Loving

One of the first empirical studies of love distinguished between liking and loving (Rubin, 1970). *Love* is something more than intense liking: it is the attachment to and caring about another person (Rubin, 1974). Attachment involves a powerful desire to be with and be cared about by another person. Caring involves making the satisfaction of another person's needs as significant as the satisfaction of your own.

Based on this distinction, Rubin developed scales to measure both liking and love. These are reproduced in Box 11-2. The liking scale (scale A) evaluates one's dating partner, lover, or spouse on various dimensions—such as adjustment, maturity, responsibility, and likability. The love scale (scale B) deals with attachment (items 3, 4, and 9), caring (items 2, 5, and 7), and intimacy, discussed earlier as self-disclosure (items 1 and 8). These scales were completed by each member of 182 dating couples, both for their partner and their best friend of the same sex (Rubin, 1970). Results showed a high degree of internal consistency within each scale and a low correlation between scales. Thus, the two scales measure different things.

If the distinction between liking and loving is valid, how do you think you would rate a dating partner and your best friend on these scales? Rubin predicted there would be high scores on both liking and love for the dating partner, lover, or spouse and a high liking score and lower love score for the (platonic) friend. The average scores of the 182 couples confirmed these predictions. Recent work by Davis (1985) also distinguishes between friendship and love. Friendship involves several qualities, including trust, understanding, and mutual assistance. Love involves all of these plus caring (giving the utmost to and being an advocate for the other) and passion (obsessive thought, sexual desire).

# Box 11-2

## LIKING AND LOVING IN PERSONAL RELATIONSHIPS

Liking and loving are different emotions. Think about your closest friend of the same sex (SSF) and your closest dating partner or lover or spouse (OSF). Then use this scale to answer each of the statements below.

| 1 | 2 | 3 | 4 | 5 | 6 | 7 | 8 | 9 |

Not at all true (Disagree completely)  Moderately true (Agree to some extent)  Definitely true (Agree completely)

In the column headed SSF, enter the number that best reflects your answer for your closest friend of the same sex. Then, in the column headed OSF, enter the appropriate number for your closest friend of the opposite sex.

### Scale A                                                     SSF    OSF

1. I think that _____ is usually well-adjusted.  ___ ___
2. I would highly recommend _____ for a responsible job.  ___ ___
3. In my opinion _____ is an exceptionally mature person.  ___ ___
4. I have great confidence in _____'s good judgment.  ___ ___
5. Most people would react favorably to _____ after a brief acquaintance.  ___ ___
6. I think _____ is one of those people who quickly win respect.  ___ ___
7. _____ is one of the most likable people I know.  ___ ___
8. _____ is the sort of person who I myself would like to be.  ___ ___
9. It seems to me that it is very easy for _____ to gain admiration.  ___ ___
TOTAL  ___ ___

### Scale B                                                     SSF    OSF

1. I feel that I can confide in _____ about virtually everything.  ___ ___
2. I would do almost anything for _____.  ___ ___
3. If I could never be with _____, I would feel miserable.  ___ ___
4. If I were lonely, my first thought would be to seek _____ out.  ___ ___
5. One of my primary concerns is _____'s welfare.  ___ ___
6. I would forgive _____ for practically anything.  ___ ___
7. I feel responsible for _____'s well being.  ___ ___
8. I would greatly enjoy being confided in by _____.  ___ ___
9. It would be hard for me to get along without _____.  ___ ___
TOTAL  ___ ___

Now add up the numbers on each scale.

Scale A is a measure of liking. Notice that it asks the extent to which you respect, admire, and have confidence in another person. Often people have about the same (high) degree of liking for their best friends and lovers. Your totals for Scale A may be very similar. Scale B is a measure of love and inquires about your dependence on, trust in, and feelings of responsibility for another. Usually people report greater love for their opposite-sex friends than for their same-sex friends.

*Source: Adapted from Z. A. Rubin, 1970.*

The distinction between liking and loving is important, because social norms allow many behaviors between lovers that are not allowed between mere friends. One study (Mack, cited in Levinger, 1974) presented 84 college students with a list of activities and asked how likely a couple was to engage in each one. Four types of relationships were specified: casual acquaintances, good friends, romantically attracted, and in love. The results are shown in Table 11-3. The students felt that most of these activities became more appropriate as the emotional involvement of the couple increased. Thus, as the couple became more emotionally intimate, physical intimacy—holding hands, back rubs, and sexual intercourse—became more appropriate. These differences in norms undoubtedly produce differences in behavior.

## Passionate Love

Love certainly involves attachment and caring. But is that all? What about the agony of jealousy and the ecstasy of being loved by another person? An alternative view of love emphasizes emotions such as these. It focuses on **passionate love**—a state of intense absorption and of intense physiological arousal (Hatfield & Walster, 1983).

The concept of passionate love is based on the two-factor theory of emotion (Schachter, 1964). According to this theory, the experience of emotion depends on: (1) perceptible physiological arousal and (2) environmental cues, which we use to determine what emotion is being experienced. Arousal itself may be caused by a variety of factors, including fear, social rejection, frustration, excitement, or sexual desire. Because arousal is generalized, and its causes complex or ambiguous, an individual must rely on situational cues to determine what emotion is being felt. Thus, we experience passionate love when two conditions occur simultaneously: when we experience a state of intense physiological arousal and when situational cues indicate that love is the appropriate label for that arousal (Berscheid & Walster [Hatfield], 1974a).

There has been no direct empirical test of the two-factor theory of passionate love. However, several studies provide supportive evidence by investigating behavior or feelings related to love or sexual attraction (Murstein, 1980). In these studies, physiological arousal in subjects is elicited through fear, frustration, or excitement. At the same time, situational cues are presented. When the situational cue suggests that passionate love might be the source of their arousal, subjects' feelings and behavior often reflect romantic or sexual attraction.

Imagine you are hiking in a state park on a warm, sunny day. You come out of a wooded area, and in front of you is a bridge. It is about 450 feet long, and it crosses a canyon more than 200 feet deep. The bridge doesn't look very stable; it is 5 feet wide and is constructed of boards attached to cables with low, wire handrails. As you cross it, the bridge tends to wobble and sway. Chances are that by the time you reach solid ground on the other side you would be experiencing perceptible arousal—at least, that is what two social psychologists hoped when they picked such a site as the location for their experiment (Dutton & Aron, 1974).

The control site in this study was another bridge in the same park made of solid wood. It was 10 feet wide with only a 10-foot drop below. The subjects were men between ages 18 and 35 who had just crossed either bridge and were not accompanied by a woman. Researchers varied the gender of the person approaching the subject to manipulate the environmental cue available to the subject. Thus, there were aroused (high-bridge) and unaroused (low-bridge) men approached by an attractive female or a male. Following this encounter, each subject completed a short questionnaire, which included writing a story about a drawing of a woman; these stories were scored for sexual imagery. In addition, the experimenter gave each subject his or her phone number, and noted the percentage of subjects who later phoned the experimenter.

As predicted, subjects who crossed the high bridge (presumed to be aroused) and were ap-

■ TABLE 11-3
THE RELATIONSHIP BETWEEN ROMANTIC INVOLVEMENT AND BEHAVIOR

| Activity | Type of Couple | | | |
|---|---|---|---|---|
| | Casually Acquainted | Good Friends, Not in Love | Romantically Attracted | Much in Love, Fully Committed |
| Communication | | | | |
| Smile at each other | 89* | 98 | 95 | 98 |
| Stay up late and talk | 52 | 78 | 79 | 79 |
| Confide in each other | 26 | 69 | 76 | 94 |
| Physical Contact | | | | |
| Stand close to one another | 51 | 79 | 90 | 93 |
| Hold hands | 18 | 44 | 82 | 93 |
| Give back rubs | 12 | 36 | 76 | 86 |
| Joint Actions | | | | |
| Study together | 55 | 65 | 74 | 82 |
| Watch TV together | 57 | 82 | 79 | 89 |
| Go to parties together | 29 | 61 | 84 | 93 |
| Prepare meals together | 21 | 47 | 70 | 85 |
| Go camping together | 13 | 43 | 54 | 74 |
| Live together | 10 | 33 | 61 | 88 |
| Have sexual intercourse | 3 | 28 | 61 | 79 |
| Shared Ownership | | | | |
| Collect items together | 36 | 51 | 61 | 75 |
| Exchange clothing | 20 | 43 | 53 | 75 |
| Exclusive Commitment | | | | |
| Try to be alone with each other | 13 | 25 | 76 | 85 |
| Refuse to date other persons | 3 | 7 | 29 | 90 |

*Mean percentage of couples at each degree of involvement estimated "to engage in each of the listed behaviors." The raters were 84 college students.

Source: Adapted from Levinger, 1974.

proached by the female wrote stories with sexual content and later phoned the experimenter more often than subjects in the other conditions. Subjects who crossed the high bridge and were approached by the male wrote stories with little sexual imagery. In contrast, the stories of men who crossed the low bridge were much less affected by the sex of the experimenter, suggesting that arousal and the cue of an attractive female were both necessary to produce sexual attraction.

## The Romantic Love Ideal

The studies and theories of love discussed so far assume that love consists of a particular set of feelings and behaviors. Furthermore, most of us assume that we will experience this emotion toward a member of the opposite sex at least once in our lives. But these are very culture-bound assumptions. There are societies in which the state or experience we call love is unheard of. In fact, U.S. society is almost unique in accepting love as a major basis for marriage.

The United States is one of a small number of societies in which love is widely accepted as a basis for getting married. In many other societies, marriages reflect political and economic influences, not romance.

In U.S. society, we are socialized to accept a set of beliefs about love—beliefs that guide much of our behavior. The following five beliefs are known collectively as the **romantic love ideal:**

1. True love can strike without prior interaction ("love at first sight").
2. For each of us, there is only one other person who will inspire true love.
3. True love can overcome any obstacle ("love conquers all").
4. Our beloved is (nearly) perfect.
5. We should follow our feelings—that is, we should base our choice of partners on love rather than on other (more rational) considerations. (Lantz, Keyes, & Schultz, 1975)

Not everyone fully accepts this romantic love ideal, of course. But it is often expressed in films and novels. Many people admire King Edward VIII of England, who in 1936 gave up the throne, the royal life style, and many of his lifelong friends to marry the American divorcée he loved. Like many others, his mate-seeking behavior was influenced by the romantic love ideal.

This ideal has not always been popular in the U.S. A group of researchers conducted an analysis of best-selling magazines published during four historical periods (Lantz, Keyes, & Schultz, 1975; Lantz, Schultz, & O'Hara, 1977). They counted the number of times the magazines mentioned one or more of the five beliefs that make up the romantic ideal. The number of times the ideal was discussed increased steadily over time, as shown in Figure 11-4. Their findings suggest that American acceptance of the romantic ideal occurred gradually during the period from 1741 to 1865. The romantic ideal first really came into its own about the time of the Civil War.

# ■ Breaking Up

Few things last forever. Roommates who once did everything together lose touch after they finish school. Two women who were best friends gradually stop talking. Couples fall out of love, break up, and divorce. What causes the dissolution of relationships? Research suggests two answers: unequal outcomes and unequal commitment.

## Unequal Outcomes and Instability

Earlier this chapter discussed the importance of outcomes in establishing and maintaining relationships. Our decision to initiate a rela-

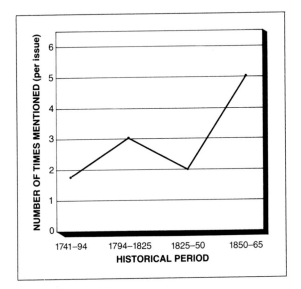

■ **FIGURE 11-4**

OCCURRENCE OF THE ROMANTIC LOVE IDEAL IN MAJOR AMERICAN MAGAZINES, 1741–1865

One way to measure the influence of the romantic love ideal on American society is to determine the frequency with which it is mentioned in popular magazines. A team of researchers selected a sample of the best-selling magazines from four different historical periods and counted the number of times each of the five romantic ideals was mentioned—including (1) idealization of the beloved, (2) love at first sight, (3) love conquers all, (4) there is one and only one for each of us, and (5) we should follow our hearts. They discovered that the number of times the ideals were mentioned increased more than 300 percent from 1741–94 to 1850–65.

*Sources: Adapted from Lantz, Keyes, and Schultz, 1975; Lantz, Schultz, and O'Hara, 1977.*

tionship is based on what we expect to get out of it. In ongoing relationships, we can assess our actual outcomes; we can evaluate the rewards we are obtaining relative to the costs of maintaining the relationship. A survey of college students examined the impact of several factors on satisfaction with a relationship; one factor was the value of overall outcomes as compared to a person's comparison level (CL) (Michaels, Edwards, & Acock, 1984). Analyzing the reports of both men and women involved in exclusive

relationships, the outcomes being experienced were most closely related to satisfaction with the relationship.

The comparison level for alternatives ($CL_{alt}$) is also an important standard used in evaluating outcomes—Are the outcomes from this relationship better than those obtainable from the best available alternative? One dimension on which people may evaluate relationships is physical appearance. A relationship with a physically attractive person may be rewarding. Two people who are equally attractive physically will experience similar outcomes on this dimension. What about two people who differ in attractiveness? The less attractive person will benefit from associating with the more attractive one, whereas the more attractive person will experience less positive outcomes. Because attractiveness is a valued and highly visible asset, the more physically attractive person is likely to find alternative relationships available and to expect some of them to yield more positive outcomes.

This reasoning was tested in a study of 123 dating couples. Photographs of each person in the study were rated by five men and five women for physical attractiveness, and a relative attractiveness score was calculated for each member of each couple. Both men and women who were more attractive than their partners reported having more friends of the opposite sex (that is, alternatives) than men and women who were not more attractive than their partners. Follow-up data collected nine months later indicated that dating couples who were rated as similar in attractiveness were more likely to be still dating each other (White, 1980). These results are consistent with the hypothesis that persons experiencing outcomes below $CL_{alt}$ are more likely to terminate the relationship.

But not everyone compares their current outcomes with those available in alternative relationships. Individuals in the study who were committed—that is, cohabiting, engaged, or married—did not vary in the number of alternatives they reported, and their relative attractiveness was not related to whether they were still in the relationship nine months later. Persons who are committed to each other may be more concerned with equity than alternatives.

Equity theory (Walster [Hatfield], Berscheid, & Walster, 1973) postulates that each of us compares the rewards we receive from a relationship to our costs or contributions. In general, we expect to get more out of the relationship if we put more into it. Thus, we compare our outcomes (rewards minus costs) to the outcomes our partner is receiving. The theory predicts that **equitable relationships**—those where the outcomes are equivalent—will be stable, whereas inequitable ones will be unstable.

This prediction was tested in a study involving 511 college students who were dating someone at the time (Walster [Hatfield], Walster, & Berscheid, 1977). Each student received a list of things that someone might contribute to a relationship, including good looks, intelligence, being loving, being understanding, and helping the other make decisions. Each student then rated the degree to which they and their partners actually contributed each of these things in their relationship. Each student also received a list of potential benefits from relationships and indicated the degree to which they and their partner received each benefit. Researchers calculated each person's benefits and contributions and obtained a measure of outcomes by dividing the benefit score by the contribution score. They divided the total benefits a person perceived the partner was receiving by the total contributions a person perceived the partner was making in order to obtain a perceived outcome-of-partner measure. By comparing a person's outcomes with a perceived partner's outcomes, the researcher determined whether the relationship was equitable.

Students were interviewed 3½ months later in order to assess the stability of their relationships. Stability was determined by whether or not they were still dating their partner, how long they had been going together (or how long they

had gone together), and how certain they were that they would be going together the following year. The results clearly demonstrated that inequitable relationships were unstable. The less equitable the relationship was at the start, the less likely the couple was to be still dating 14 weeks later. Furthermore, students who perceived that their outcomes did not equal their partner's outcomes reported that their relationships were of shorter duration and were much less certain that their relationship would last another year.

Thus, the outcomes experienced in a relationship are one important influence on its stability. If these outcomes are better than a person anticipates getting from alternative relationships and if they are believed equitable relative to those the partner is obtaining, then the person is more likely to remain in the relationship.

## Differential Commitment and Dissolution

Are outcomes (rewards minus costs) the only thing we consider when deciding whether or not to continue a relationship? What about emotional attachment or involvement? We often continue a relationship because we love someone, because through interaction over a long period we have developed an emotional commitment to the person and feel a sense of responsibility for that person's welfare. The importance of commitment is illustrated by the results of a survey of 234 college students (Simpson, 1987). Each student was involved in a dating relationship and answered questions about ten aspects of the relationship. Three months later, each person was recontacted to determine whether he or she was still dating the partner. The characteristics that were most closely related to stability included length and exclusivity of the relationship and having engaged in sexual intimacy; all three are aspects of commitment.

When both persons are equally committed, the relationship may be quite stable. But if one person is less involved than the other, the relationship may break up. The importance of equal

■ **TABLE 11-4**

DIFFERENTIAL INVOLVEMENT AND DISSOLUTION OF A RELATIONSHIP

| Characteristics of Relationships in 1972 | Status Two Years Later | | | |
|---|---|---|---|---|
| | Women's Reports | | Men's Reports | |
| | Together | Breakups | Together | Breakups |
| Mean Ratings | | | | |
| Self-report of closeness (9-pt. scale) | 7.9 | 7.3** | 8.0 | 7.2** |
| Estimate of marriage probability (as percentage) | 65.4 | 46.4** | 63.1 | 42.7** |
| Love scale (max. = 100) | 81.2 | 70.2** | 77.8 | 71.5** |
| Liking scale (max. = 100) | 78.5 | 74.0* | 73.2 | 69.6 |
| Number of months dated | 13.1 | 9.9* | 12.7 | 9.9* |
| Percentages | | | | |
| Couple is "in love" | 80.0 | 55.3** | 81.2 | 58.0** |
| Dating exclusively | 92.3 | 68.0** | 92.2 | 77.5** |
| Seeing partner daily | 67.5 | 52.0 | 60.7 | 53.4 |
| Had sexual intercourse | 79.6 | 78.6 | 80.6 | 78.6 |
| Living together | 24.8 | 20.4 | 23.1 | 20.4 |

Note: $N = 117$ together, 103 breakups for both men and women. Significance by $t$ tests or chi-square for together-breakup differences. $p$ = probability of obtaining such a result by chance.

*$p < .05$.

**$p < .01$.

*Source: Adapted from Hill, Rubin, and Peplau, 1976.*

degrees of involvement is illustrated in another study in which couples were recruited from four colleges and universities in the Boston area (Hill, Rubin, & Peplau, 1976). Each member of 231 couples filled out an initial questionnaire and completed three follow-up questionnaires, six months, one year, and two years later. At the time the initial data were collected, couples had been dating an average of eight months; most were dating exclusively, and 10 percent were engaged. Two years later, researchers were able to determine the status of 221 of the couples. Some were still together, whereas others had broken up.

What distinguished couples who were together two years later from those who had broken up? Some of the major differences are summarized in Table 11-4. Couples who were more involved initially—those who were dating

exclusively, who rated themselves as very close, who said they were "in love," and who estimated a high probability that they would get married— were more likely to be together two years later. Of those couples who reported equal involvement initially, only 23 percent broke up in the following two years. But of the couples who reported unequal involvement initially, 54 percent were no longer seeing each other two years later.

Earlier in this chapter we discussed the importance of similarity in establishing relationships. How important is similarity in determining whether a relationship persists over time? Among the 221 couples, both couples who stayed together and those who broke up were initially similar in their sex-role attitudes, approval of premarital sexuality, importance of religion, and in the number of children they wanted. Thus, although attitudinal similarity seems to determine the formation of relationships, it does not distinguish couples whose relationships persist from those whose relationships dissolve.

Not surprisingly, the breakup of a couple was usually initiated by the person who was less involved. Of those whose relationships ended, 85 percent reported that one person wanted to break up more than the other. There was also a distinct pattern in the timing of breakups; they were much more likely to occur in May–June, September, and December–January. This suggests that factors outside the relationship, such as graduation, moving, and arriving at school, led one person to initiate the breakup.

The dissolution of a relationship is often painful. But breaking up is not necessarily undesirable. It can be thought of as a part of a filter process through which people who are not suited to each other terminate their relationships.

## Responses to Dissatisfaction

Not all relationships that involve unequal outcomes or differential commitment break up. What makes the difference? The answer is, in part, the person's reaction to these situations. The level of outcomes a person experiences and his or her commitment to the relationship are the main influences on *satisfaction* with that relationship (Rusbult, Johnson, & Morrow, 1986). As long as the person is satisfied, whatever the level of rewards or commitment, he or she will want to continue the relationship.

An individual in an unsatisfactory relationship has four basic alternatives (Rusbult, Zembrodt, & Gunn, 1982): exit (termination), voice (discuss it with your partner), loyalty (grin and bear it), and neglect (stay in the relationship but don't contribute much). Which of these alternatives the person selects depends on the anticipated costs of breaking up, the availability of alternative relationships, and the level of rewards previously obtained from the relationship.

To assess the costs of breaking up, the individual weighs the costs of an unsatisfactory

relationship against the costs of ending that relationship. There are three types of barriers or costs to leaving a relationship: material, symbolic, and affectual (Levinger, 1976). Material costs are especially significant for persons who have pooled their financial resources. Breaking up will require agreeing on who gets what, and it may produce a lower standard of living for each person. Symbolic costs include the reactions of others. Will close friends and family members support or criticize the termination of the relationship? These costs probably increase as time and investment in the relationship increase (Backman, 1981). Affectual costs involve changes in one's relationships with others. Breaking up may cause the loss of friends and reduce or eliminate contact with relatives—that is, it may result in *loneliness.* One may conclude that the costs of breaking up are too great and stay in the relationship.

A second factor in this assessment is the availability of alternatives. The absence of an attractive alternative may lead the individual to maintain an unrewarding relationship, whereas the appearance of an attractive alternative may trigger the dissatisfied person to dissolve the relationship.

A third factor is the level of rewards experienced before the relationship became dissatisfying. If the relationship was particularly rewarding in the past, the individual is less likely to decide to terminate it.

How important are each of these three factors—that is, which factors are most important in determining whether a dissatisfied person responds by discussing the situation with his or her partner, by waiting for things to improve, by neglecting the partner, or by terminating the relationship? In one study, subjects were given short stories describing relationships in which these three factors varied. They were asked what they would do in each situation (Rusbult, Zembrodt, & Gunn, 1982). Results showed that the lower the prior satisfaction—that is, the less satisfied and the less positive their feelings and caring for their partner—the more likely they

Box 11-3

## ARE YOU LONELY TONIGHT?

Did you feel lonely when you first entered school here? If you did, you weren't alone. People entering a college or university are likely to feel lonely for the first several weeks or months (Cutrona, 1982). In fact, most people have experienced loneliness at some time during their lives.

**Loneliness** is an unpleasant, subjective experience that results from the lack of satisfying social relationships in either quantity or quality (Perlman, 1988). Loneliness is different from being alone, or social isolation. Social isolation is an objective situation, whereas loneliness is a subjective, internal experience. You can feel lonely in the midst of a family reunion, and you can be alone in your room and yet feel connected to others.

Loneliness is different from shyness; the latter is a personality trait that reflects characteristics of the person rather than the state of one's social ties. Shyness is defined as "discomfort and inhibition in the presence of others" (Jones, Briggs, & Smith, 1986). An extensive study of several measures of shyness found that the common element in these measures is distress in and avoidance of interpersonal situations. A researcher asked 40 subjects with high scores and 30 subjects with low scores on a shyness scale to interact with another person (Asendorpf, 1987). The interactions were videotaped, and each subject later watched the tape of his or her interaction and commented on specific events. Analyses of the comments identified several differences between the shy and nonshy subjects. Shy people recalled more fear that they were being evaluated by the other person and were more likely to think that they had made a negative impression on the other person.

There are two types of loneliness (Weiss, 1973), which differ in their cause. One is social loneliness, which results from a lack of social relationships or ties to others. Several studies have found that people with few or no friends and few or no ties to family are more likely to feel lonely (Stokes, 1985). The other type is emotional loneliness, which results from the lack of emotionally intimate relationships. One study of adolescents found a strong association between self-disclosure and loneliness; greater self-disclosure to others was associated with reduced loneliness (Davis & Franzoi, 1986). There is some evidence that loneliness in men is the result of having few or no relationships with others, while in women it is the result of having no intimate relationships (Stokes & Levin, 1986). Clearly, loneliness is tied to the state of one's interpersonal relationships.

Since loneliness is related to the number and quality of interpersonal relationships, we can predict that people in some circumstances are more likely to experience it. In general, people undergoing a major social transition are at greater risk of loneliness. The transition from school to work may be accompanied by feelings of loneliness, especially when this transition involves a geographic move. Second, living arrangements are related to feeling lonely. A study of 554 adult men and women found that living alone was the most important determinant of these feelings (de Jong-Gierveld, 1987). Third, one's marital status is important. We described earlier in this chapter the increasing self-disclosure and interdependence that accompanies the development of romantic relationships; people who are engaged or married should be less likely to experience emotional loneliness. Conversely, people who have recently gone through the termination of an intimate relationship—through breaking up, divorce, or death of a partner—may be especially vulnerable to loneliness.

were to neglect or terminate the relationship. The less the investment—that is, degree of disclosure and how much a person stands to lose—the more likely subjects were to engage in neglect or termination. Finally, the presence of attractive alternatives increased the probability of terminating the relationship. A subsequent study of ongoing relationships yielded the same results (Rusbult, 1983).

## ■ Summary

Interpersonal attraction is a positive attitude held by one person toward another person. It is the basis for the development, maintenance, and dissolution of close personal relationships.

**Who Is Available?** Institutional structures and personal characteristics influence who is available to us as potential friends, roommates, co-workers, and lovers. Three factors influence whom we select from this pool. (1) Our daily routines make some persons more accessible. (2) Proximity makes it more rewarding and less costly to interact with some people rather than others. (3) Familiarity produces a positive attitude toward those with whom we repeatedly come in contact.

**Who Is Desirable?** Among the availables, we choose on the basis of several criteria. (1) Social norms tell us what kinds of people are appropriate as friends, lovers, and mentors. (2) We prefer a more physically attractive person, both for esthetic reasons and because we expect to obtain rewards from associating with that person. Attractiveness is more influential when we have no other information about a person. (3) We choose based on our expectations about the rewards and costs of potential relationships. We will choose to develop those relationships whose outcomes we expect will exceed both comparison level (CL) and comparison level for alternatives ($CL_{alt}$). We implement our choices by making contact—using an opening line that often

indicates the kind of relationship in which we are interested.

**The Determinants of Liking** Many relationships, between friends, roommates, co-workers, or lovers, involve liking. The extent to which we like someone is determined by four factors. (1) The major influence is the degree to which two people have similar attitudes. The greater the proportion of similar attitudes, the more they like each other. Similarity produces liking because we prefer cognitive consistency and because we expect interaction with similar others to be reinforcing. (2) Complementarity—the possession of personal or social characteristics that complement our own—may influence liking in some instances. (3) Shared activities become an important influence on our liking for another person as we spend time with them. (4) We like those who like us; as we experience positive feedback from another, it increases our liking for them.

**The Growth of Relationships** As relationships grow, they change on three dimensions. (1) There may be a gradual increase in the disclosure of intimate information about the self. Self-disclosure is usually reciprocal, with each person revealing something about themselves in response to revelations by the other. (2) Trust in the other person—a belief in his or her honesty, benevolence, and reliability—also increases as relationships develop. (3) Interdependence for various gratifications also increases, often accompanied by a decline in reliance on and number of relationships with others.

**Love and Loving** (1) Whereas liking refers to a positive attitude toward an object, love involves attachment to and caring for another person. Love may also involve passionate feelings—a state of intense absorption in the other and of intense physiological arousal. (2) According to the two-factor theory of emotion, the experience of passionate love depends on perceptible physiological arousal and environmental cues that indicate that arousal is due to love. (3) The concept of love does not exist in all societies; the romantic love ideal emerged gradually in the U.S. and came into its own about the time of the Civil War.

**Breaking Up**   There are three major influences on whether a relationship dissolves. (1) Breaking up may result if one person feels that outcomes (rewards minus costs) are inadequate. A person may evaluate present outcomes against what could be obtained from an alternative relationship. Alternatively, a person may look at the outcomes the partner is experiencing and assess whether the relationship is equitable. (2) The degree of commitment to a relationship is an important influence on whether it continues. Someone who feels a low level of emotional attachment to, and concern for, the partner is more likely to break up with that person. (3) Responses to dissatisfaction with a relationship include exit, voice, loyalty, or neglect. Which response occurs depends on the anticipated economic and emotional costs, whether there are attractive alternatives, and the level of prior satisfaction in the relationship.

## Key Terms

| | |
|---|---|
| ACCESS DISPLAY | 319 |
| ATTITUDINAL SIMILARITY | 320 |
| ATTRACTIVENESS STEREOTYPE | 316 |
| AVAILABLES | 311 |
| COMPARISON LEVEL | 318 |
| COMPARISON LEVEL FOR ALTERNATIVES | 318 |
| DYADIC WITHDRAWAL | 327 |
| EQUITABLE RELATIONSHIP | 334 |
| INTERPERSONAL ATTRACTION | 310 |
| LONELINESS | 337 |
| MATCHING HYPOTHESIS | 315 |
| MERE EXPOSURE EFFECT | 313 |
| NORM OF HOMOGAMY | 313 |
| OUTCOMES | 318 |
| PASSIONATE LOVE | 330 |
| ROMANTIC LOVE IDEAL | 332 |
| TRUST | 325 |

12

chapter

# Group Cohesiveness and Conformity

# ■ Introduction

Groups are everywhere. Their existence is one of the most common facts of social life. We all participate in them, and many of us spend a large part of each day engaging in group activities. Families, work groups, sports teams, street gangs, classes and seminars, therapy and rehabilitation groups, classical quartets and rock groups, small military units, neighborhood social clubs, church groups—these are only some of the many groups we encounter.

Groups are important to our life. They provide social support, a cultural framework to guide activity, and rewards of all kinds. Without them an individual would be isolated, unloved, less productive, out of touch with the world, and perhaps even disoriented to the point of suicide.

## What Is a Group?

A **group** is a set of persons who relate to one another as parts of a system. Group members communicate with and influence one another, share at least one common goal, and perceive themselves as forming a unit. Groups are not mere collections of individuals, each pursuing his or her own objectives. Rather, groups are organized systems in which relations among persons are structured and patterned.

Although groups can differ with respect to their organization, they share certain properties or characteristics. These include the following:

1. *Shared goals.* Group members share specific goals and rely on one another's performances for collective success. Frequently, the goals are ones that individual members cannot attain alone.

2. *Communication.* Communication among members of a group may be either unlimited or restricted. But the capacity to communicate— either face-to-face or by other means—must be present for a set of persons to constitute a group.

3. *A set of normative expectations.* Members share certain expectations, called norms, that regu-

late interaction within the group. Norms coordinate behavior and allocate tasks to individuals within the group.

4. *Conscious identification of members with the group.* Individuals who belong to a group develop a concept of themselves as group members. Sometimes these concepts become central to their personal identities.

Given this definition, it is clear that groups are not mere collections of individuals. Not all behaviors in the presence of other people are group interactions; many times individuals interact without common goals or without a set of specific norms to regulate their behavior. For instance, a conversation between yourself and an insurance agent who is trying to sell you a policy you don't want would not qualify as a group interaction. There is no shared goal or conscious group identification. Likewise, a theater crowd escaping in panic from a fire is not a functioning group. Although there may be some communication among the individuals in the crowd, there are no shared, normative expectations or conscious identification with the others present.

Again, persons who interact in a group context will relate to one another as group members, not as unaffiliated individuals. Group interaction is based on shared goals, occurs via symbolic communication, is governed by norms, and involves a conscious identification with the group.

## Types of Groups

The term **small group** refers to any group sufficiently limited in size that its members are able to interact directly (that is, face-to-face). A group is not considered small if its size precludes interaction among all its members. Thus, small groups usually include from 2 to 20 members (Crosbie, 1975), although a discussion group with 30 or even 40 members would qualify as a small group if members were positioned so they could interact directly with each other. Most of the groups discussed in this chapter are small groups.

Small groups impact our lives in numerous ways. We spend many hours each day interacting with other persons in such groups. Most of us belong to family and work groups (or, if in school, to educational groups such as classes and seminars). Groups of this type provide the contacts with friends, relatives, and co-workers that are crucial to our personal welfare.

The small groups to which we belong exert a great amount of control over our behavior. Direct pressures to conform to group norms lead us to engage in some activities and to shun others. Moreover, many of the concepts and assumptions we use in perceiving our world are taken from the culture of groups to which we belong; thus, these groups often have a significant effect on how we perceive the world. Members of political or religious groups, for example, may accept the beliefs and stereotypes held by these groups as an objective view of reality.

There are various types of small groups. One useful distinction is that between natural groups and composed groups (McGrath, 1984). *Natural groups* are those that exist in natural (nonlaboratory) settings; natural groups usually form for their own purposes and may endure for extended periods of time. In contrast, *composed groups* are those put together for a particular purpose, as, for example, a group put together in a laboratory setting by a researcher interested in studying group productivity or problem solving. Composed groups typically endure only a short time (for example, a single session) and may work on restricted tasks not of the members' own choosing. In this chapter (as well as in chapters 13 and 14), we will discuss both natural groups and composed groups.

Another useful distinction among small (natural) groups is that between informal groups and formal groups. *Informal groups* are just that—informal. Interaction among members is not very structured or regulated, collective tasks are often ill defined or even nonexistent, and formation may be spontaneous and short lived. *Formal groups,* in contrast, are more structured. They usually have a well-defined set of roles and norms, with clear lines of authority and control.

Usually they have well-specified tasks and a division of labor. They often exist within the confines of a larger organization or bureaucracy, and they frequently endure for extended periods of time (sometimes years). Both formal and informal groups are important—most of us belong to both types—and we shall encounter them in this and subsequent chapters.

### Questions about Cohesiveness and Conformity

This chapter will examine the forces that give unity to a group. It addresses the following questions:

1. What factors hold a group together, and what factors split it apart? That is, what produces cohesiveness—or the lack of it—in groups?

2. What are group goals? How are they established, and how do they relate to group functioning? What are group norms, and how do they operate to regulate behavior?

3. A group cannot function without some minimum level of conformity by its members. How do groups obtain conformity? What factors cause group members to conform in greater or lesser degree?

4. Deviant behavior can threaten group cohesiveness and goal attainment. How do group members react when another member deviates from established norms? When does deviant behavior lead to change in group structure?

## ■ Group Cohesiveness

The Jaguars are a recreational baseball club with a long record of league championships. The ball players are proud of their performance and very committed to their team. At practice and in games they are a model of enthusiasm and coordination. On the rare occasion when they have a losing streak, the whole team voluntarily holds extra practice sessions to sharpen their skills and teamwork. The players like each other, and they enjoy playing together and celebrating their victories together. Though they do not always agree on strategy, the Jaguars seem to resolve their differences quickly. Several of the players consider their teammates best friends, and they even spend much of their time off the field together. The Jaguars rarely lose any of their players—even their second-stringers—and everyone turns out for practices and other meetings enthusiastically.

Another team in the league is called the Penguins. The Penguins are a different story. The players always seem to be busy with other activities and hate to spend much time practicing. Something less than a model of competence, the Penguins have finished in last place in the league standings for three seasons running. Occasionally, the team forfeits a game because it cannot even field nine players. When the players do agree to hold a practice, many forget or don't bother to show up; perhaps this is because they seldom run into one another outside of team activities. Last spring, the Penguins' planning session dissolved into chaos when the players could not agree on how to pay for some new equipment. Ever since then, there has been friction among certain members, and the team has not even decided whether to participate in the league next year.

Perhaps you've experienced groups like these two teams. Being a member of the Jaguars is more demanding but also much more rewarding. What is the basic difference between the teams? In a word, the Jaguars have a higher level of cohesiveness than the Penguins.

The term **group cohesiveness** refers to the degree to which the members of a group are attracted to and desire to remain in the group (Cartwright, 1968; Evans & Jarvis, 1980). If members place a high value on belonging to the group, the group is said to be cohesive; otherwise, it is not. Although the reasons for this attraction may vary, a highly cohesive group will maintain a firm hold over its members' time, energy, and commitment.

Interaction among happy sorority sisters on pledge day reflects a high level of group cohesiveness. Many groups safeguard cohesiveness by selecting new members carefully.

## Sources of Group Cohesiveness

A group will be highly cohesive only if its members find it attractive. Not only must it be attractive in absolute terms, it must also be relatively more attractive than other groups its members might join. Were this not so, its members would eventually leave for other groups.

What characteristics make a group attractive to its members? Among the most important are (1) the extent to which members find other group members personally likable and attractive, (2) the extent to which members feel that others in the group are similar to themselves, (3) the extent to which members feel that the group's tasks and goals are consonant with their own personal values and interests, and (4) the extent to which the group provides other rewards (such as prestige or money) for its members. Each of these characteristics helps to make a group attractive to its members and contributes to its cohesiveness (Lott & Lott, 1965; Anderson, 1974; Stokes, 1983).

In addition to these characteristics, several situational factors affect the level of group cohesiveness. One of these is whether the group succeeds or fails in achieving its major goals. Groups that succeed at achieving goals are generally more cohesive than groups that fail repeatedly, like the Jaguars. Failure to achieve important goals may not only cause members to reevaluate the group's worth but may also produce disagreement among members regarding which strategies will be effective and appropriate. This, in turn, can split the group into coalitions or factions, reducing the overall cohesiveness.

Another situational factor that affects cohesiveness is whether the group is involved in competition against other groups. Members' commitment to their own group and its goals is frequently heightened when it is engaged in competition or conflict with another group. Heightened commitment means greater cohesiveness. Thus, as competition or conflict between two groups increases, the level of cohesiveness within each group will increase (Dion, 1979). At the same time, the level of hostility of group members toward the outside group will also increase (Mulder & Stermerding, 1963; Sherif & Sherif, 1982).

## Consequences of Group Cohesiveness

What difference does it make whether a group is highly cohesive or not? That is, what are the consequences of group cohesiveness?

The patterns of interaction among group members demonstrate the impact of cohesiveness. Given the opportunity, members of highly cohesive groups communicate more than members of less cohesive groups. This has been observed in a wide variety of groups, ranging from student organizations to industrial training groups (Moran, 1966). Cohesiveness affects not only the amount of interaction but also its form. Interaction in highly cohesive groups is friendlier, more cooperative, and shows more attempts to reach agreements and to strengthen coordination among members (Shaw & Shaw, 1962).

Since a cohesive group is one that members find attractive and wish to belong to, members should be willing to do whatever is necessary to remain in the group. For this reason, we might expect a highly cohesive group to exert more influence over its members than a less cohesive group. Indeed, research findings indicate that members of highly cohesive groups try to influence other members more and are themselves more influenced (Lott & Lott, 1965). Members of highly cohesive groups are also more conforming to norms and expectations than members of less cohesive groups (Wyer, 1966; Sakurai, 1975).

The relationship between group cohesiveness and group productivity may be less obvious. On the one hand, there is some evidence that highly cohesive groups work harder and are more productive than less cohesive groups. This has been substantiated in military, classroom, and industrial settings (Van Zelst, 1952; Shaw & Shaw, 1962). On the other hand, there is also some evidence that highly cohesive groups are *less* productive (Cartwright & Robertson, 1961; Warwick, 1964). Although these findings appear contradictory, the explanation is logical. Highly cohesive groups obtain more conformity from their members, including conformity to standards of production. However, highly cohesive groups do not always establish high production standards. If "productivity" is not a goal, members of a highly cohesive group will spend all their time socializing with one another rather than producing.

## ■ Group Goals and Norms

The term **group goal** refers to a desirable outcome that group members strive collectively to accomplish or bring about. Most groups have at least one major goal, although some groups pursue several simultaneously. Group goals are an important basis for guiding and coordinating the activity of members. They also provide a framework for the development of norms and standards within the group.

We can find examples of groups goals in everyday life. A high school basketball team might hold the group goal of winning the state championship in its division. This is clearly a goal that could be pursued only by a group. That is, the championship itself can be won only by a team, not by an individual. Another example is four workmen pouring a concrete sidewalk. This task might be accomplished by a single individual, but it would take longer to complete the job, and the quality of work might not be as good.

If a group is to pursue its goals successfully, there must be goal consensus among its members. **Goal consensus** exists when a substantial proportion of members agrees on the group's goal and accepts the desirability of trying to achieve it. Goal consensus does not occur automatically in groups. Goal consensus often develops only after time-consuming, detailed discussions among members. Still, a goal for which there is little consensus has scant chance of being attained.

### Group Goals and Personal Goals

It is important to distinguish group goals from personal goals (Mills, 1967; Cartwright & Zander, 1968). *Personal goals* are outcomes desired by individuals for themselves. For example, one individual may want to earn a lot of money; a second may want an opportunity to acquire work

## Box 12-1

## PRIMARY GROUPS

A **primary group** is a small group in which there exist strong emotional ties among members. Members of primary groups usually know one another very well and like each other. Primary groups frequently have a distinctive subculture with informal rules that control the actions of members (Dunphy, 1972; Kimberly, 1984).

Primary groups exist in a variety of forms. The most common in our society are (1) nuclear families; (2) peer groups, such as children's play groups, teen delinquent gangs, and Wednesday night poker clubs; (3) informal work groups existing in organizational settings, such as factory work groups, classroom groups, and small military units; and (4) groups whose goal is to resocialize their members, such as rehabilitation groups, therapy groups, and self-analytic groups.

Members of a primary group usually possess similar values and attitudes, and they care about their mutual friendships. Interaction in primary groups is spontaneous, informal, and personal. This is not to say that it always runs a smooth course. The members of a family can certainly have a heated argument, and children often fight with others in their peer group. Still, the crucial characteristic of primary groups is that members care about one another as individuals.

Primary groups are often distinguished from **secondary groups,** which are groups whose members have few emotional ties with one another and relate in terms of limited roles. Interaction in secondary groups is formal, impersonal, and nonspontaneous. One example would be a formally structured work group in a bureaucratic setting with limited opportunities for personal contact.

Although the distinction between secondary and primary groups resembles that between formal and informal groups, these distinctions are in fact different. The primary-secondary distinction is based on differences in emotional bonding among members, whereas the formal-informal distinction is based on differences in the degree to which norms are explicit and codified formally within the group. Although most secondary groups are also formal groups, not all primary groups are informal groups.

Primary groups serve an important function in our urban, bureaucratic society. They provide an arena for friendship and deep personal commitments. They also serve to integrate the individual with larger organizations in the society.

---

experience and job skills. Very often, individuals join a group precisely because it helps them achieve their personal goals. Membership in a work group may pay a good salary and provide useful experience. Nevertheless, personal goals and group goals are different. A *group goal* is not the simple sum of personal goals, nor can it always be directly inferred from them. It is an outcome regarded by members as desirable for the group as a whole, not merely for the individual members.

Most groups function best when there is substantial similarity, or isomorphism, between group goals and the personal goals of its members. In some cases, of course, discrepancies or conflicts may occur between group goals and the individual goals held by members. This might occur, for example, when the individual's view of reality differs from the group's view of reality or when the individual's identity and group's role structure do not mesh smoothly. Some groups tolerate a certain amount of discrepancy between group goals and individual goals. If this discrepancy grows very large, however, group functioning will suffer, and the group may have to take steps to establish greater isomorphism between individual and group goals. Greater isomorphism usually can be created by exerting more social influence to establish goal consensus among members, by establishing strict norms to regulate behavior, and by appropriately excluding members from the group (Mackie & Goethals, 1987).

## Group Aspiration Level

Groups differ markedly in the difficulty of the goals they establish. Some groups aspire to high levels of accomplishment, whereas others pursue lesser goals. In other words, groups differ in their **level of aspiration**—the level of difficulty of the goals that members agree to pursue. When a group pursues a goal that is hard to achieve, it is said to have a high level of aspiration.

Choosing an appropriate level of aspiration for a group can be a difficult problem. Members must decide what level is suitable, based on their needs and capabilities. A team of mountain climbers, for example, might consider climbing several peaks in a mountain range. Group members view some peaks as more important and worthwhile than others, but the most desirable peaks are also the most difficult and dangerous to climb. (This is generally true in many other arenas as well: The most desired outcomes are usually the most difficult to attain. Were this not so, there would be little difficulty in choosing a level of aspiration.)

To illustrate, suppose that the climbers are considering one of two mountains, A or B, as a possible goal. Mountain A is very high and snow capped (hence, desirable to climb), but it is also very dangerous due to ice falls. Several members of an earlier expedition died in an avalanche while attempting to climb to the summit. Mountain B is less imposing, and certainly less dangerous, but still considered worth climbing by the members. None of the mountaineers have ever

climbed mountain A or B. Although they value success at climbing mountain A somewhat more than success at climbing mountain B, they believe that given the level of skill in their group, mountain B offers them a much higher probability of reaching the top and returning safely than does mountain A.

Zander (1971, 1977) proposed a theory describing how groups evaluate alternatives when establishing a group level of aspiration. According to Zander, we can express the expected value of an alternative as follows:

Expected value of an alternative

$$= \left( \begin{matrix} \text{perceived probability} \\ \text{of success} \end{matrix} \times \begin{matrix} \text{incentive value} \\ \text{of success} \end{matrix} \right)$$

$$- \left( \begin{matrix} \text{perceived probability} \\ \text{of failure} \end{matrix} \times \begin{matrix} \text{loss incurred} \\ \text{from failure} \end{matrix} \right)$$

If the climbers used this approach to assess the two alternatives (mountain A and mountain B), they would find that mountain A has a low "expected value." Although the mountaineers agree that climbing A is valuable—that is, has high "incentive value of success"—they also perceive that the probability of success is low. Moreover, there is a substantial "probability of failure"—a real chance that some climbers might fall and get hurt—and the "loss incurred from failure" could be severe. In contrast, mountain B has a higher expected value. Although climbing B has slightly lower incentive value than climbing A, there is a much higher probability of reaching the top of B. Moreover, any loss incurred from failure in climbing B would be slight; there is little chance that members would fall and get seriously hurt.

The theory holds that group members will, in effect, calculate the expected values of both alternatives (mountains A and B) and then choose that alternative having the highest expected value. Thus, in this instance, the group will choose mountain B as the appropriate goal for its expedition.

A team of mountaineers has climbed halfway toward their goal. Careful, realistic choices regarding level of aspiration are crucial in groups such as this one.

## Pursuit of Group Goals

Once a group establishes an appropriate level of aspiration, it can achieve this goal only if its members exercise continued effort and striving. This raises two important questions: (1) What factors cause a group to strive persistently toward its goals? and (2) How do groups react if they fail to attain their goals?

**Goal Striving and Feedback**  Several factors affect the extent to which a group will persist in its efforts to achieve a goal, even in the face of setbacks. One of these obviously is the value and importance of the goal itself. Provided that members believe the goal is realistic, groups will

work harder to attain a goal that is highly valued than a goal that is less valued.

A second factor that affects goal striving is the clarity of the path toward the goal. Although group members may know what they want, they may not always know the best way of obtaining it. Group motivation is impaired when the path to the goal is unclear. Facing this uncertainty, group members become less enthusiastic, less attracted to their task, and less efficient in their performance (Raven & Rietsema, 1957).

A third factor that affects group goal striving is *feedback*—information received by a group indicating whether it is progressing satisfactorily toward its goal. If feedback is negative and the group seems headed for failure, group members are likely to change their behavior. Either they will try harder to achieve their goal, or they will abandon it for a lesser one. The "try harder" reaction is most probable when the feedback indicates slight or moderate failure rather than extreme failure (Streufert, Streufert, & Castore, 1969). If the feedback indicates that the group is failing utterly, the members may simply give up.

The energizing effect of moderately negative feedback is illustrated in a study of a U.S. expedition climbing Mount Everest (Emerson, 1966). The expedition took 92 days to climb Everest, which, at 29,028 feet above sea level, is the world's tallest peak. At this altitude, the air is very thin, and even the simplest actions require enormous effort and will. A climber can honestly believe he wants to climb Everest very much but still fail to get out of his sleeping bag soon enough to get important work underway.

As a member of this expedition, the researcher was able to directly observe the group's efforts to reach its goal. His findings suggest that feedback from group leaders to other members was crucial in sustaining the group's motivation. Leaders were selective (sometimes even misleading) in passing along information. To dampen overoptimism, they occasionally gave negative feedback to members on the group's progress. This kept the team challenged, so that members exerted their best efforts under the harsh and uncertain conditions. Negative feedback at moderate levels served to intensify group goal striving.

**Reactions to Group Failure**    If a group fails to achieve its goals on the first attempt, it may elect to try again; members will reason that perhaps success will come on the second try. However, if a group fails again and again to achieve its goals, the risk increases that group cohesiveness will deteriorate. Progress in achieving group goals is essential in maintaining positive relations among members (Streufert & Streufert, 1969). Recurring failure often increases frustration and causes members to criticize one another. Under failure, members reduce their evaluation of the group, and this, in turn, may cause them to withdraw from the group or to pursue personal goals that do not contribute to the group's performance.

Because recurring failure erodes group cohesiveness, a group that fails repeatedly to achieve its central goals will usually react and make some fundamental changes. One typical change is to lower its level of aspiration and to pursue more attainable objectives (Zander, 1977). Alternatively, the group may retain its original, difficult goals but adopt a new, hopefully more effective strategy. In some cases, the group may even reject its old leaders and find new leadership in hopes of attaining success.

Other, less adaptive reactions to failure are also possible. For instance, members may attempt to avoid embarrassment and to protect themselves by blaming someone else for the group's failure (Zander, 1971; Deschamps, 1972). This method of deflecting responsibility for failure is known as *scapegoating*. It preserves good relations within the group at the expense of others.

The ideal scapegoat is someone who is powerless, different, and easily distinguishable. For this reason, persons chosen as scapegoats are usually outsiders in weak positions. On occasion, however, the blame may turn inward, and the majority may hold certain highly visible members responsible for the failure of the entire group.

Every autumn, for example, about a month before the end of the baseball season, the owner of some losing ball club holds a press conference and announces that he is firing the manager. The reason the team is 37 games out of first place, reporters are given to understand, is that the manager is incompetent. The fans, of course, are delighted to see him go. And, because everyone believes that a new manager cannot possibly lose any more games than his predecessor, tension abates within the group: "Just wait 'til next year."

## Structure of Norms

In many groups, the members perform different tasks from one another, and their activities must be highly coordinated if the group is to achieve its goals. To achieve this coordination, members must carry out the appropriate actions at the right time. But coordination does not happen without design; it can occur only when members conform to group norms.

A **norm** is a rule or standard that specifies how group members are expected to behave under given circumstances. Most groups have many norms, explicit or implicit. These norms regulate not only group members' actions and decisions but also their attitudes and beliefs. Norms are anchored in a group's values and goals, and they typically prescribe behaviors that will lead to attainment of these goals and prohibit behaviors that will hinder their attainment. In this sense, norms channel behavior so that group goals will be achieved.

Norms can apply to almost any behavior imaginable. A factory work group, for example, may have norms specifying what time workers are expected to show up on the assembly line and how much they are expected to produce; norms of this type will obviously have a large impact on the productivity of the entire work group. A group of college admissions officers may have norms that govern judgments by the officers; in effect, these norms will determine which applicants will be admitted. A family may have norms regulating who washes the dishes as well as norms specifying who can and cannot have sexu-

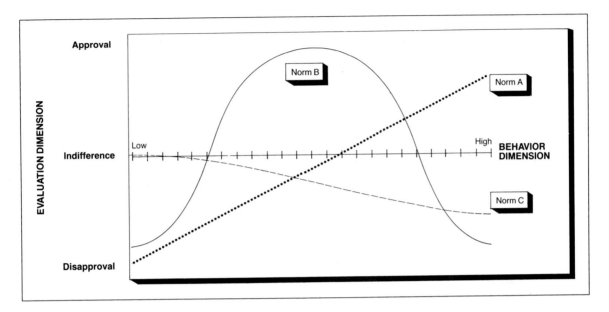

▪ **FIGURE 12-1**

RETURN POTENTIAL CURVES

The curves in this figure represent three group norms. In each case, varying amounts of behavior are met with some specified amount of approval or disapproval from group members. Under norm A, for example, low amounts of behavior are disapproved, whereas high amounts are approved. Under norm B, intermediate amounts of behavior are approved, whereas very low and very high amounts are disapproved. Under norm C, high amounts of behavior are disapproved, whereas low amounts are neither approved nor disapproved.

*Source: Adapted from Jackson, 1965.*

al relations with whom (for example, the incest taboo). A teenage street gang may have a wide variety of norms regulating the actions of its members. For instance, a study of several gangs of adolescent boys in the Southwest found that they had norms governing what kinds of information members could reveal to parents and police, how members should behave during street fights with rival gangs, and which members were permitted to steal someone else's girlfriend (Sherif & Sherif, 1964).

**The Return Potential Model** We can think of group norms as having structure and shape. One useful approach to the structure of norms is the **return potential model** (Jackson, 1965). This model treats norms as having two dimensions— the behavior dimension and the evaluation dimension. The *behavior dimension* specifies the frequency or amount of behavior regulated by

the group norm, whereas the *evaluation dimension* refers to the response to that behavior by other group members. The evaluation of behavior can be positive, indifferent, or negative.

Figure 12-1 displays three norms. These are drawn as curves in a space defined by the behavior dimension and the evaluation dimension. The curves represent different norms and, thus, have different shapes; yet in each case they show that evaluation is a function of behavior.

Let's suppose that norm A (in Figure 12-1) pertains to productive activity in a group of editors working on a magazine. Under norm A, group members encourage and reward higher levels of a given behavior, such as editing more and more pages of manuscript per day. Under this norm, the more pages a member edits, the more she is rewarded. Any member who produces at a very low level on this behavior dimension will receive negative evaluation from other group members. She may be criticized or castigated or perhaps face a monetary fine. Only by producing at a higher level does she receive a positive evaluation.

Norm B is different, although we can contrast it with the pattern of expectations shown in norm A. Whereas norm A basically said "the more the better," norm B says that there is such a thing as producing too much. Under norm B, a member will be most heavily rewarded for producing in the middle range on the behavior dimension. Productivity at very low levels or at very high levels will be negatively evaluated and discouraged. When group members notice that one among them is working much harder than the others, they may disapprove of this "overproduction."

A norm of this type was illustrated in a classic early study of 14 men, known as the Bank Wiring Group, who assembled banks of telephone switching equipment for the Western Electric Company (Roethlisberger & Dickson, 1939). The management of the company had recently instituted a wage incentive plan rewarding higher levels of productivity by individual workers. Managers believed that this plan would bring about an increase in the productivity of the Bank Wiring Group. They soon observed, however, that the workers continued to produce at about the same rate as before, despite the opportunity to increase their earnings. Subsequent investigation showed that this resulted because the group had a firmly established norm (like norm B in Figure 12-1) concerning what was "a fair day's work." Group members feared that if they began to produce at higher levels, the company might switch production standards and require higher levels of productivity for lower pay. Therefore, the group enforced its own production norm on individual members. If a man produced too much he was ridiculed by other members as a "rate buster." If he produced too little, he was disparaged as a "chiseler." If an offender did not respond to these verbal reprimands, he was soon subjected to another form of sanction termed "binging," whereby other members would hit him in the arm. Although this behavior might sound unusual, groups often develop their own special sanctions. What matters to the group is the sanction's effectiveness in regulating the behavior of its members.

The norms we have discussed up until this point—norms A and B—involve both positive and negative evaluation. Other norms, however, may pertain to behaviors that group members consider wholly undesirable—represented by norm C in Figure 12-1. Although group members may tolerate low levels of the undesirable behavior, their reaction becomes more negative as the behavior becomes more frequent or intense. Although a norm like C is unlikely to pertain to productive activities, it might well apply to such behaviors as talking a lot, importuning others, making a nuisance of oneself, and so on. Much of our everyday activity is regulated by norms like this one.

Every norm may be said to specify a **range of tolerable behavior**—that portion of the behavior dimension that members of a group approve and evaluate positively. For example, in Figure 12-1, norm A delimits a range of tolerable behavior at the higher end of the behavior dimension, whereas norm B delineates a range of tolerable behavior in the middle of the behavior dimension. Behaviors outside this range are not approved or tolerated; behaviors inside this range (that is, conforming behaviors) are tolerated, indeed approved.

Another characteristic of a norm is its intensity. The *intensity of a norm* is indicated by the height of the return potential curve from its highest to its lowest point, both above and below the point of indifference. The intensity of a norm reflects the strength of group feelings (whether approval or disapproval) regarding that behavior. In Figure 12-1, group members feel more strongly about the behavior regulated by norm B than about that regulated by norm C. Most norms that affect attainment of important group goals have high intensity, whereas norms about personal discretion or taste (such as style of dress or manner of speaking) typically have moderate or low intensity.

Group norms can extend to any aspect of behavior, including dress. The Savage Skulls have a different dress code from corporate executives, but conformity is high within each group.

## ■ Conformity in Groups

The term **conformity** means adherence by an individual to group norms. A group member is conforming to a given norm when his or her behavior falls within the range of tolerable behavior for that norm. Since most norms are related either directly or indirectly to important group goals, conformity to norms by members is essential if groups are to reliably achieve their goals.

Nevertheless, conformity by individuals cannot be taken for granted. Group members sometimes conform, sometimes not. In this section, we discuss two forms of influence that operate within groups to produce conforming behavior. Then we discuss some classic studies of conformity based on the Asch paradigm. These studies address the basic question of under what conditions will conformity occur? And finally, we will discuss some of the factors that affect the amount of conformity occurring in groups.

## Majority Influence in Groups

In many groups, influence flows in all directions. Individual members influence other individual members. Individual members influence the group's established majority. Most importantly, however, the group's majority influences the behavior of individual members. It is this impact of the majority on the individual that gives a group its integrity and continuity over time. The amount of influence exerted by the majority on individual members varies from group to group. Often it is slight to moderate. In some extreme cases, however, it can be nearly total (such as in religious cults, elite military units, and so on).

The influence exerted by a majority on individual members can assume various forms. One useful distinction is that between normative influence and informational influence (Deutsch & Gerard, 1955; Kaplan, 1987; Kaplan & Miller, 1987). **Normative influence** occurs when a member conforms to group norms in order to receive the rewards and/or avoid the punishments that are contingent on adherence to these expectations held by others. Usually the rewards and punishments are mediated by the group's majority. To administer these, the majority must be able to exercise some degree of surveillance. The effect of normative influence is heightened, for instance, when members respond publicly rather than anonymously (Insko et al., 1983, 1985). Normative influence is perhaps the fundamental source of uniformity within groups.

Another important form of influence exerted in groups is informational influence. **Informational influence** occurs when a group member accepts information from others as valid evidence about reality. Influence of this type is particularly likely to occur in situations involving uncertainty or in situations lacking external or objective standards of reference. Under these conditions, members will often rely on the group's majority to gain a sense of what is valid or true. The majority exerts influence on individual members by providing information that defines reality and that may serve as a basis for formulating a judgment or making a decision.

The operation of informational influence is illustrated in a famous early study by Sherif (1936). This study was based on a physical phenomenon known as the *autokinetic effect* (meaning "moves by itself"). The autokinetic effect occurs when a person stares at a stationary pinpoint of light located at a distance in a completely dark room. For most people, the light will appear to move in an erratic fashion. Sherif used the autokinetic effect as a basis for studying informational influence. First, he placed subjects in a laboratory setting by themselves and asked them to estimate how far the light moved. In making these judgments, subjects were quite literally in the dark—they had no external frame of reference. From their individual estimates, the researcher was able to determine a stable range for each subject. (Remember, however, that the pinpoint of light was not actually moving.)

Next, Sherif put two or three of these subjects together in the autokinetic situation at the same time. Although the estimates they had made earlier when alone were discrepant, the estimates they made in groups converged on a common judgment. Thus, lacking an external frame of reference and being uncertain about their own judgment, group members began to use one another's estimates as a basis for defining reality. Each group established its own (faulty) concept of reality, which members used as a frame of reference. A week or two later, when these subjects were again placed alone in the autokinetic situation, they used the group's norm as their new reference for individual judgments.

Subsequent research (Hood & Sherif, 1962) showed that subjects involved in an autokinetic experiment were quite unaware that their judgments were being influenced by other members. Informational influence is often indirect and subtle, but it can be an important source of uniformity in groups.

## The Asch Conformity Paradigm

The impact of a group's majority on individual members was well illustrated in a series of classic experiments by Asch (1951, 1955, 1957). Specifically, these studies investigated the process by which a majority pressures an individual to adopt a position on some issue, even when that position is self-evidently wrong. Using a laboratory setting, Asch created a situation in which an individual was confronted by a majority that agreed unanimously on a factual matter (spatial judgments) but was obviously in error. These studies showed that, within limits, groups can pressure their members to change their judgments and conform with the majority's erroneous position.

In a typical Asch experiment, a group of eight persons participated in an investigation of "visual discrimination." In fact, all but one of these persons were confederates working for the experimenter. The remaining individual was a naive subject. In front of the experimental room, large cards displayed a standard line and three comparison lines, as shown in Figure 12-2.

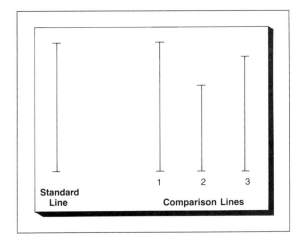

■ **FIGURE 12-2**

JUDGMENTAL TASK EMPLOYED IN ASCH
CONFORMITY STUDIES

In the Asch paradigm, naive subjects are shown one
standard line and three comparison lines. The task is to
judge which of the three comparison lines is closest in
length to the standard line. By itself, this task appears
easy. However, subjects are surrounded by other persons (supposedly also naive subjects, but actually experimental confederates) who publicly announce erroneous judgments regarding the match between lines.
Such a situation imposes pressure on the subject to
conform to their erroneous judgments.

*Source: Adapted from Asch, 1952.*

The subject's objective was to decide which of
the comparison lines (1, 2, or 3) was the same
length as the standard line.

As you can see, the task was very simple and
straightforward. One of the comparison lines
was the same length as the standard, while the
other two were very different. The group repeated this task 18 times, using a different set of lines
each time. On each trial, the standard line
matched one of the three comparison lines.
Although this task appears easy, it turned out to
be quite difficult for the naive subject because of
the confederates' behavior. During each trial of
the experiment, the confederates announced
their judgments publicly, one after another. The
subject also announced his opinion publicly. In
12 of the 18 trials, confederates gave an incor-

rect response. In the remaining 6 trials, which
were neutral, the confederates all responded
accurately.

The purpose of the study was to observe
how the naive subject would behave during the
12 critical trials. The group was seated so that all
of the confederates were asked to respond prior
to the real subject. This put the subject in a
difficult position. On the one hand, he knew (or
thought that he knew) the correct response
based on his own perception of the lines. On the
other hand, he heard all the other persons
(whom he believed to be sincere) announcing a
different and unanimous judgment.

Results of this and related experiments indicate that the incorrect opinion expressed by the
majority strongly influenced the judgments of
the naive subjects. In the 12 critical trials, nearly
one third of the responses by subjects were
incorrect (Asch, 1957). This compares with an
error rate of less than 1 percent in a control
condition in which no confederates were present
and subjects recorded their judgments privately
on paper. Most of the subjects were quite aware
of the discrepancy between the group's judgments and their own. They felt puzzled or under
pressure, and they tried to figure out what might
be happening. Some wondered whether they had
misunderstood the experimental instructions;
others began to question their eyesight. Even
those subjects who did not conform to the
majority felt some apprehension, but they eventually decided that the problem rested more with
the group than with themselves.

In a situation like this, most people find it
difficult to remain completely independent of
group pressure. Only about 25 percent of Asch's
subjects never yielded on any of the critical trials.
The rest did conform, either a little or a great
deal. From interviews conducted after the experiment, it became clear that this conformity was of
a particular type. For the most part, it involved
public compliance without private acceptance:
although many subjects conformed publicly,
they did not believe or accept the majority's
judgment privately. Instead, they viewed public

compliance as the best choice in a difficult situation.

## Factors Affecting Conformity

We have seen that majority pressures exert substantial influence on the behavior of individual members. But an individual's tendency to conform will be greater under some conditions than under others. Below are some of the factors that affect the amount of conformity in groups.

**Unanimity of the Majority** Consider the Asch experiment described earlier in which a single subject is confronted by a majority. If the majority is unanimous—that is, if all the members of the majority are united in their position—then the size of the majority will have an impact on the behavior of the subject. As the size of the majority increases, the pressure for conformity by this subject increases (Asch, 1955; Rosenberg, 1961). For example, a subject confronted by one other person in an Asch-type situation will conform very little; that is, he or she will answer independently and correctly on nearly all trials. However, when confronted by two persons, the subject will experience more pressure and will agree with the majority's erroneous answer more of the time. Confronted by three persons, the subject will conform at a still higher rate. In his early studies, Asch (1951) found that conformity to unanimous false judgments increased with majority size up to three members and then remained essentially constant beyond that point. Although some research (Gerard, Wilhelmy, & Conolley, 1968) has questioned the exact point at which the effect of majority size begins to level off, increases in the size of the majority beyond three persons generally have very little added impact.

What happens when the group's majority is not unanimous? Basically, lack of unanimity has a liberating effect on behavior. A subject will be less likely to conform if another member breaks away from the majority (Gorfein, 1964; Morris & Miller, 1975). One explanation for this is that the member who abandons the majority provides validation and social support for the subject. In an Asch experiment, for example, if one or several members abandon the majority and announce correct judgments, their behavior will reaffirm the subject's own perception of reality and reduce his or her tendency to conform to the majority.

It appears, moreover, that *any* breach in the majority—whether it provides social support or not—will reduce the pressure to conform (Allen & Levine, 1971; Levine, Saxe, & Ranelli, 1975). In one study (Allen & Levine, 1969), subjects participated in groups of five persons, four of whom were confederates. The subjects made judgments on a variety of items. These included visual tasks similar to those used by Asch, informational items (for example, "In thousands of miles, how far is it from San Francisco to New York?"), and opinion items for which there were no correct answers (Agree or disagree: "Most young people get too much education.") Depending on experimental conditions, subjects were confronted with either a unanimous majority of four persons (the control condition), a majority of three persons and a fourth person who broke from the majority and gave the correct answer (the social support condition), or a majority of three persons and a fourth person who broke from the majority but gave an answer that was more erroneous than that of the majority (the extreme erroneous dissent condition).

The results of this study are shown in Figure 12-3. Note that the control condition, which involved a unanimous majority, produced a very high level of conformity. The social support condition, in which the dissenter joined the subject, significantly reduced conformity. Even in the extreme erroneous dissent condition, when the dissenter gave an answer that was more extreme and incorrect than the majority's, conformity of the subjects declined. Thus, any breach in the majority reduced conformity. However, subjects in the social support and extreme erroneous dissent conditions had very different impressions of dissenters. Under the social support condition, subjects held a positive impression of that person, whereas under the extreme erroneous dissent condition, subjects

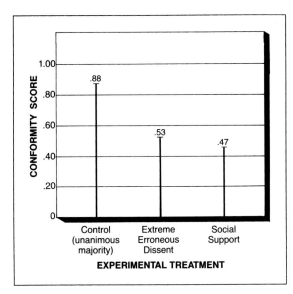

■ **FIGURE 12-3**

CONFORMITY SCORES AND DISSENT FROM
THE MAJORITY

In an Asch-type experiment, subjects conformed less
when one member broke from the majority and gave the
same answer as the subject (the social support condition) than when faced with a unanimous majority (the
control condition). Conformity was also lower when one
member broke from the majority but gave an answer
even more erroneous than the majority's (the extreme
erroneous dissent condition). Thus, a breach in the
majority of any kind had a liberating effect on the
subjects and reduced conformity.

*Source: Adapted from Allen and Levine, 1969.*

thought the dissenter was basically a jerk—
unlikable, stupid, insincere, and badly adjusted.
Either way, the breach in the majority had the
same effect: it called into question the correctness of the majority's position and reduced the
subject's tendency to conform.

**Attraction to a Group**   Members who are highly attracted to a group will conform more to
group norms than members who are less attracted to it (Kiesler & Kiesler, 1969; Mehrabian &
Ksionzky, 1970). One explanation for this is that
when individuals are attracted to a group, they
also wish to be accepted personally by its members. Because acceptance and friendship are

strengthened when members hold similar attitudes and standards, individuals highly attracted
to a group conform more to the views held by the
others (McLeod, Price, & Harburg, 1966; Feather & Armstrong, 1967). However, attraction to a
group will increase conformity only if that conformity is rewarded by group acceptance (Walker
& Heyns, 1962).

**Commitment to Future Interaction**   Members
are more likely to conform to group norms when
they anticipate that their relationship with the
group will be relatively permanent, as opposed
to short term. This was demonstrated in a study
that involved groups of five persons who were
asked to discuss such problems as urban affairs,
international relations, pollution, and population (Lewis, Langan, & Hollander, 1972). Some
of the participants were led to believe that they
would review these problems with group members again in the future, whereas others were not
given this expectation. Those who anticipated
future interaction conformed more to the majority's opinion than those who did not anticipate
future interaction.

Commitment to future interaction affects
conformity whether members are attracted to a
group or not. For instance, one investigation
showed that even if members do not like the
others in a group, they will conform at a high
level, provided they are already committed to
continuing in the group and cannot readily leave
or opt out. Such persons are likely to experience
distress or dissonance at having to interact with
others they do not like, and they may resolve this
problem by bringing their own attitudes and
beliefs into line with group standards (Kiesler &
Corbin, 1965; Kiesler, Zanna, & deSalvo, 1966).

**Competence**   Conformity to group norms is
also affected by an individual member's level of
expertise relative to that of other members. If a
member skilled at the task at hand differs from
the majority's view, he will resist pressure to the
degree that he believes himself to be more
competent than the other group members
(Ettinger et al., 1971). Interestingly, the extent

# Box 12-2

## BARRIERS TO INDEPENDENT BEHAVIOR

Conformity to norms is essential to group functioning. Without it, there will be a serious lack of coordination and risk of failure to achieve group goals. Of course, there are many situations in which an individual might desire to act independently rather than to conform to group norms. A person might disagree with the majority's position and prefer an innovative departure. Nevertheless, groups discourage independent behavior by a variety of means. One review (Hollander, 1975) suggests that there are six major barriers that inhibit independent behavior by group members:

1. *Risk of disapproval from other group members.* Most groups establish norms governing central activities. These norms specify a range of tolerable behavior. To the extent that a member's independent behavior falls outside that range, the member risks disapproval from others in the group. By deviating too far, a member may face outright rejection.

2. *Lack of perceived alternatives.* Individuals frequently are unaware of alternatives other than those specified by group norms. Unless someone else speaks out and proposes alternatives, a member may not realize that there is any choice other than conforming to the majority.

3. *Fear of disrupting the group's operations.* Because group norms usually reflect underlying group goals, an individual knows that departing from established norms may "rock the boat." Thus, he or she avoids acting independently due to fear that it may block the attainment of group goals. This effect is especially strong in cases where a member is

afraid of being held personally responsible for group failure, should it occur.

4. *Absence of communication among group members.* Even though a number of members may privately dissent from group standards, each may hesitate to express his or her reservations publicly. But without such communication, they do not learn that others are thinking along the same lines. Collective ignorance prevails. Lacking information that others would join in the nonconforming action, each avoids going out on a limb alone.

5. *No feeling of responsibility for group outcomes.* In some cases, where a group's established norms and procedures are deficient, members may actually cause a group to fail by conforming to the norms. Although some may realize that conformist behavior is producing poor outcomes, they may hesitate to take the initiative and turn the situation around. This is especially common when individuals feel that they are not personally responsible for the group's success or failure.

6. *A sense of powerlessness.* If a member feels unable to change a situation, he or she is unlikely to try anything new. The apathy becomes self-fulfilling. No one tries anything different, and, consequently, nothing improves.

These barriers to independent behavior are substantial contributors to conformity. If several are operating simultaneously, the probability of conformity is further increased.

to which a person *believes* that he is competent may be more important than his actual level of competence (Stang, 1972). Persons who in fact are not competent will still resist conformity pressure if they believe they are more skilled than other members. This is because group members who believe themselves to be competent rely less on the judgments of others. When confronted by a majority, they will usually try to persuade others to change their positions.

**Gender** The gender composition of a group also affects the amount of conformity of its members. The weight of available evidence indicates that, within same-sex groups, females yield somewhat more to conformity pressures than males. That is, females yield somewhat more to pressures from other females than males yield to pressures from other males. Although controversial, this finding is based on many different studies conducted during the last four decades. Some studies have shown no difference between the sexes in conformity, but the preponderance of evidence indicates that females conform slightly more than males (Cooper, 1979; Eagly & Carli, 1981; Becker, 1986). It is important, however, not to exaggerate this difference; the effect, while real, is small in size.

Various theoretical explanations have been advanced for this difference. One explanation holds that this difference is based on sex-role expectations (Eagly, 1987). The traditional female role in the U.S. rewards submissiveness, nurturance, passivity, and person orientation—all qualities related to conformity. The traditional male role rewards aggressiveness, assertiveness, dominance, and task orientation (Wiley, 1973). Consequently, standard sex-role expectations may lead to greater conformity among females than males.

Some direct support for the hypothesis that sex-role orientation affects conformity comes from research by Goldberg (1974, 1975). In these studies, women who accepted traditional sex roles conformed more than women who rejected traditional roles or who actively sup-

ported the women's movement. This pattern was particularly marked when the test items were traditionally feminine in nature.

A second explanation for gender differences in conformity lies in the nature of the tasks employed in conformity experiments themselves. At least some of the earlier studies employed tasks and opinion items involving male-related activities. Thus, differences in conformity that appeared to result from differences in sex may actually have resulted from differences in familiarity or competence with the tasks used in some of the experiments.

Several studies have shown this effect (Sistrunk & McDavid, 1971; Karabenick, 1983). Sistrunk and McDavid reasoned that if women are more conformist on activities regarded as traditionally male, then men might be more conformist on activities regarded as traditionally female. For example, whereas women may be more conformist about such things as cars and politics, men might be more conformist about child care and clothing fashions. Researchers compiled a list of statements, some of which were opinions and others of which were everyday matters of fact. Some of these test items had previously been judged to be of greater interest and familiarity to females, while other items had been judged to be of greater interest and familiarity to males. They presented these items (along with some neutral filler items) to several groups of males and females. Adjacent to each item there appeared some data indicating how the majority of the subjects' peers had responded to each item. This number was manipulated by researchers.

The results of this study showed that, on an overall basis, there was no significant difference between males and females in the amount of conformity. However, males conformed more than females on the feminine items, whereas females conformed more than males on the masculine items. Both sexes conformed about the same amount on the neutral items. These findings suggest that the amount of conformity depends on the nature of the task as well as the gender of the subject. Both females and males yield to pressure more when the task is one with which they are unfamiliar.

Thus, the overall conclusion regarding gender and conformity is that both sex-role orientation and task competence can serve as mediating factors in conformist behavior. Many studies show that women conform somewhat more than men in same-sex groups, but this holds true primarily when the women accept traditional sex roles or when they lack competence and familiarity with the task at hand.

## ■ Reactions to Deviance within Groups

Up to this point, we have been looking at causes affecting the amount of conformity within groups. We have seen that conformity is affected by various factors, such as the size of the majority and attraction to the group. We will now consider the opposite of conformity, which is known as *deviance*.

Consider the following scene. It is 11:00 P.M.—dark outside—and seven college students are holding a meeting. Everyone is drinking coffee by the gallon and trying to stay awake. The students have organized the group as part of one of their courses; the group is responsible for completing a large project. The school term is nearly over, and the project is due in three days. Unfortunately, the group is far behind schedule, and a major effort will be needed to meet the project deadline. No one wanted a meeting at this late hour, but there was no other time when everyone could fit it into their busy schedules.

The meeting is not progressing well. Two group members have offered a proposal regarding the project. Some other members like the idea, and a consensus is starting to develop. The proposal is better than anything else the group has considered. Everybody except a single member, Ben, thinks it has merit. Although Ben's own suggestions have not been especially useful, he has nothing but criticism for the idea. Support for the proposal is strong, but Ben's objections

grow more vigorous. Other members begin to argue with him, because unanimity is necessary before a decision can be reached. The group is running out of time. The discussion goes on and on, but Ben won't give an inch. Soon the clock in the bell tower strikes midnight. The majority of the group supports the proposal, but Ben still deviates. Something has to give.

As this example shows, deviant behavior can constitute a serious problem for a group's majority. Not only can deviance challenge a group's conception of reality and disrupt normal operations, but if allowed to continue unchecked, it may eventually cause a group to perform poorly or even to collapse and disband. How can a group's majority cope with deviant behavior? Fundamentally, there are three options available: (1) The majority can try to influence the deviant (using persuasion, threats, and so on) and bring him or her back within the range of tolerable behavior; (2) The majority can reject the deviant, either by means of outright expulsion from the group or by means of psychological exclusion; or (3) The majority can respond to the deviant by changing its own position and moving into line with the deviant's position.

Although these responses are diverse, they all constitute attempts by the majority to protect the group's integrity and effectiveness (Levine, 1980). If most of the members are committed to the group, they will want the group to continue to exist, to maintain its definition of social reality, and to succeed at its goals. Whatever the nature of their reaction to the deviant, members will try to offset the imbalance created by the deviant's behavior and to establish a new equilibrium within the group. This section will consider each of the possible responses of the majority.

### Pressure to Restore Conformity

When a group's majority first recognizes deviance in a member, it typically will increase communication with that person in hopes of persuading him or her to conform. Members of the majority will speak to the deviant, remind the deviant of the group's expectations, explain and justify these expectations, and urge him or her to comply. If the deviant responds by conforming, the problem will be solved and equilibrium will be restored.

On the other hand, if the deviant refuses to conform, the majority will likely react by applying more pressure. This may involve angry outbursts or threats of direct punishment. The willingness of a majority to punish deviance, once it has occurred, depends on a number of factors. Serious deviant acts are punished more harshly than minor deviant acts (Gouran & Andrews, 1984). Particularly relevant is the extent to which the deviant's behavior interferes with the attainment of important group goals. The majority will be very reluctant to let a deviant member disrupt the group's progress toward important objectives. Behavior that unequivocally prevents the group from attaining its goals—such as violating a central norm—will incite punishment that is both swift and severe (Michener & Burt, 1975).

This generalization must be qualified, however, by taking into account the status of the deviant. The term *status* refers to a member's relative standing within a group. Status is usually based, at least in part, on the extent to which a member has contributed to the attainment of important group goals. Members who contribute more are typically accorded higher status.

The reaction of a group's majority to a deviant act by a member depends both on the severity of the deviance and on the status of the deviant. According to one theoretical viewpoint (Sherif & Sherif, 1967), high-status persons are relatively free to violate minor norms, provided they do not interfere with the attainment of the group's goals. They will not be punished as severely as low-status members for minor violations. However, high-status persons will be punished more severely than low-status members if they breach important norms and impede progress toward the group's goals.

This phenomenon was demonstrated in a laboratory experiment involving 30 groups of four persons (Wiggins, Dill, & Schwartz, 1965). Each group attempted to solve a series of intellectual problems and competed against the other groups in the study for a $50 prize. One member within each group was made to appear to have violated work rules, which cost the group a penalty in points. Depending on experimental treatment, the point penalty was either high, medium, or low in magnitude. When the penalty was high, the violation virtually wiped out the group's chance of winning the $50 prize. When the penalty was medium or low, the violation lessened the group's chance of winning, but it did not put success entirely out of reach.

The study also varied the status of the supposed deviant. Depending on the condition, the deviant was either high status or medium status. The high-status member was a person who had performed very competently earlier in the session and, therefore, had contributed substantially toward the attainment of the group's goal. The medium-status member was one whose earlier performance had been of average competency.

After completing all the problems, group members were asked to indicate how they would distribute the $50 among themselves if they won the prize. Because each group in this study included four members, the average member might expect to win 25 percent of the prize. However, the actual distribution, which is shown in Figure 12-4, indicates that group members sanctioned the deviant by distributing rewards unequally. The severity of the sanction depended both on the deviant's status and the extent to which the deviant act interfered with the group goal. When the deviant act did not completely block the group's chance of winning, the high-status deviant was rewarded more (sanctioned less) than the medium-status deviant. However, when the violation seriously interfered with the attainment of the group's goal, the high-status deviant was rewarded less (sanctioned more) than the medium-status deviant.

These results are consistent with the hypothesis that group members are generally reluctant to label a high-status member as deviant. When the consequences of the deviation are minor, they prefer to overlook the matter. However, if the deviant act blocks attainment of the

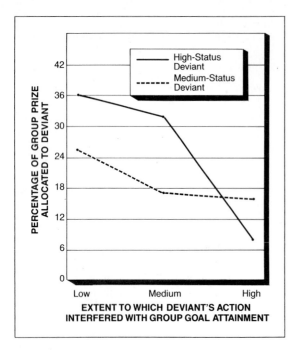

**■ FIGURE 12-4**

REACTION TO A DEVIANT AS A FUNCTION OF THE
DEVIANT'S STATUS AND THE INTERFERENCE
INTENSITY OF THE ACT

A deviant group member was punished more severely—
that is, given a smaller share of the group's prize—
when the deviant act interfered substantially with the
attainment of the group's goal than when it interfered
only slightly. The status of the deviant within the group
was also taken into account by members in determining
the punishment. Compared to the medium-status deviant, a high-status deviant was punished less severely for
acts of low or medium interference but more severely for
acts that greatly interfered with the attainment of group
goals.

*Source: Adapted from Wiggins, Dill, and Schwartz, 1965.*

group's goal, the members simply cannot overlook the matter; they react by labeling the high-
status member a deviant and applying severe
sanctions. This phenomenon is called *status liability*—that is, high-status members are more
liable than low-status members to be punished
for major offenses. Although some studies have
not found differences in sanctions for serious
offenses to be due to the level of status (Wahrman, 1977), other studies have replicated the
status-liability phenomenon (Alvarez, 1968).

## Rejection of the Deviant

If the majority members in a group lack the
capability or the inclination to apply pressure to
a deviant, they have another option available:
they can reject the deviant. Rejection can assume
various forms. One of these is expulsion from
membership. In some groups, for instance, the
majority may simply not invite the deviant back
to their next meeting. In other groups, expulsion involves a formal ceremony of status degradation. In still other groups, expulsion can occur
quite unceremoniously, as when members of the
majority pick up the deviant and throw him
physically out the door. Whatever the form, the
effect is the same: to remove the deviant from
membership and thereby "purify" the group.

Another form of rejection is psychological
isolation. In this case, the deviant member is
physically present, but the majority will ignore
and refuse to interact with him or her. Although
communication with the deviant may drop off
suddenly, it is more typical for communication to
decline gradually, as one member after another
comes to view the deviant as a lost cause (Sampson & Brandon, 1964). Even though the amount
of interaction is reduced, the level of covert
hostility toward the deviant will be high (Orcutt,
1973).

Rejection of the deviant is triggered by
many of the same factors that produce sanctioning. In fact, rejection and sanctioning often
occur at the same time. In the study of status
liability discussed above, the majority not only
punished the deviant monetarily but also did not
want him to participate in their future meetings.
Rejection was greatest when the deviant was a
high-status member who severely blocked the
group's progress. Other studies have shown that
rejection, like sanctioning, becomes more intense as a deviant takes an increasingly more
extreme position from that of the majority
(Hensley & Duval, 1976; Levine, Saxe, & Harris,
1976).

Rejection of the deviant is a means of reestablishing equilibrium within the group, because
it increases the homogeneity of the members.

That is, after a deviant is ostracized, only conforming members remain within the system. This enables the group to increase its integrity as a social unit.

## Minority Influence and Innovation

In many cases, behavior that violates the established norms constitutes a challenge to the existing social order. This might happen, for instance, when a group member engages in independent behavior, not from a desire to achieve selfish objectives but from a wish to change the rules or the procedures used in pursuing group goals. An important change of this type is referred to as an *innovation*. Although advocacy of change is sometimes received favorably by a group's majority, it just as readily may be viewed as deviant activity that threatens the group's existing way of life.

This raises a basic question: When is independent, counternormative behavior likely to produce innovations in group structure rather than meet with suppression by the majority? To answer, we must distinguish between group norms and the goals that underlie those norms. If a member's behavior departs from established norms but points to new ways to more fully realize the group's goals, the changes may be accepted by group members. On the other hand, if a member attempts to force a change in the nature of the group's goals, he or she will likely meet with greater resistance. Only if a group is drifting aimlessly or has failed repeatedly to achieve its goals will an effort to change its goals be favorably received.

Innovative suggestions are more likely to be adopted by a group if they are proposed by a well-regarded or high-status member. High-status members encounter less opposition and disapproval than lower-status members when suggesting innovations because high-status members are usually considered more skillful and more committed to group goals and norms than other members (Hollander & Julian, 1978).

Perhaps the most interesting case arises when not just one individual but several group members behave contrary to established norms and expectations. The term **active minority** is used to designate a subgroup of members that adopts a distinct viewpoint on some important issue(s) and that tries to persuade the majority to change its own position. While agitating for change, an active minority resists pressure to conform to the majority's position. Very often this push toward innovation creates tension and opposition between the majority and the active minority (Moscovici & Mugny, 1983; Moscovici, 1985; Levine & Russo, 1987).

Obviously, an active minority is not always successful in its efforts to modify group policies or goals. Success is more likely under some conditions than under others. One factor that affects the probability of success is the extent to which an active minority is consistent in its viewpoint. Several studies have shown that an active minority is more likely to win over group members to its distinctive position if it is consistent over time (Moscovici & Lage, 1976; Maass & Clark, 1984). When an active minority is consistent in its position, the majority is less likely to dismiss it as stupid or wayward. To be effective, an active minority must truly persuade other members regarding its viewpoint. A consistent position is persuasive because it implies that the minority is clearheaded, confident, and purposive.

A second factor that affects the probability of success of an active minority is the extent to which it is flexible (as opposed to rigid) in its negotiating style. An active minority that is flexible in its approach is more likely to be successful than one that adopts a rigid, hard-line approach (Mugny, 1982, 1984; Papastamou & Mugny, 1985). A rigid negotiation style, with its attendant refusal to make any concessions, is not likely to be successful, because members of the majority will perceive the minority's consistency as dogmatism (and, hence, as an idiosyncrasy of the minority).

A third factor that affects the probability of success of an active minority is the extent to which the group's majority perceives the active minority members as being similar to themselves rather than dissimilar (Wolf, 1985). Similarity increases the odds that the majority will view the members of the minority as being from the same social category as themselves, and it prevents the majority from attributing the minority's position to idiosyncratic personal characteristics. Thus, similarity makes it more difficult for the majority to engage in outright or wholesale rejection of the minority.

One of the interesting differences between a majority and an active minority concerns the types of influence they typically exert. A majority is in a position to compel compliance from group members. A minority, in contrast, is generally not able to force compliance, and so it relies on persuasion to exert influence. Group members confronted by the views of an active minority tend to think through the problem more fully than they otherwise would. Although they may not be persuaded to adopt exactly the position advocated by the minority, they do think in more divergent terms, attend to more aspects of the situation, and seek novel solutions (Moscovici, 1980; Nemeth, 1986; Maass, West, & Cialdini, 1987). The effectiveness of an active minority is best measured in terms of its capacity to change opinions and viewpoints, not its capacity to force behavioral compliance.

# ■ Summary

A group is a set of persons who relate to one another as parts of a system. Groups have certain hallmarks, including goals shared by group members, communication among members, norms that guide members' behavior, and identification of members with the group. The term small group refers to any group sufficiently limited in size that its members are able to interact directly. Various types of small groups include natural groups, composed groups, informal groups, and formal groups.

**Group Cohesiveness** A highly cohesive group is one that is able to attract and hold its mem-

bers. (1) Several factors affect a group's cohesiveness. A group will be more cohesive if its members like one another, are similar to one another, believe that the group's tasks and goals are consonant with their own personal goals, and gain rewards (prestige, money) by belonging to the group. (2) The level of a group's cohesiveness affects the interaction among members. Members in highly cohesive groups communicate more than those in less cohesive groups; they also exert more influence over one another, and their interaction is friendlier and more cooperative. Highly cohesive groups can be more productive than less cohesive groups, but only if they establish norms requiring a high level of productivity from members.

**Group Goals and Norms** A group goal is a desirable outcome that members strive collectively to bring about. (1) Group goals differ from personal goals, which are outcomes desired by individuals for themselves. (2) Groups differ with respect to the difficulty of the goals they choose. The level of aspiration chosen by a group is based on several considerations, including the value of the goal (if attained), the probability of actually reaching the goal, and the loss incurred if the goal is pursued but not achieved. (3) Several factors determine the extent to which a group will persist in the pursuit of its goals. Goal striving will be greater if the members see a direct path toward their goal rather than an ambiguous one. Moderate negative feedback strengthens goal striving, provided that members continue to believe their goal is attainable. Extreme negative feedback usually causes members to change or abandon their goal. (4) A norm is a rule or standard that specifies how group members are expected to behave under given circumstances. The return potential model of norms describes the relationship between behavior and subsequent reward or punishment.

**Conformity in Groups** Conformity means adherence by an individual to group norms and expectations. (1) Group majorities exert pressures on individual members to conform. These pressures take the form of both normative influence and informational influence. (2) The Asch conformity studies use a simple visual-discrimination paradigm to investigate conditions that produce conformity by individuals to the majority. (3) Many factors affect the amount of conformity in groups. First, more conformity occurs when the majority is unanimous than when it is not. Conformity is also greater when members are highly attracted to a group and especially when conformity leads to liking and acceptance by other members. Commitment to future interaction affects conformity; conformity is greater when members believe that their relationship with the group will be relatively permanent. Task competence affects conformity; members who oppose the majority's view will resist conformity pressures to the extent that they believe themselves to be more competent than other members. Finally, gender affects conformity. Females conform more than males, although the magnitude of this difference is small. Recent studies suggest that these differences may be based not merely on gender but on a person's sex-role orientation and familiarity with the task at hand.

**Reactions to Deviance within Groups**   Deviant behavior violates group norms and usually provokes a reaction from the group's majority. (1) The majority may apply pressure on the deviant to conform. This usually involves persuasion or sanctions. The amount of pressure applied will depend on several factors, including the severity of the violation and the status of the deviant. (2) The majority may also reject the deviant through psychological isolation or even expulsion. (3) Finally, an active minority may succeed in influencing a majority and, thus, innovate changes in the group's structure or procedures. This is more likely to occur when an active minority is consistent yet flexible in its approach.

## Key Terms

13

chapter

# Status and Interaction in Groups

## ■ Introduction

Consider a group of salespersons working for a small California company that manufactures laboratory equipment. This group consists of eleven people—one vice-president, two managers, and eight salespersons. Their basic goal is to market equipment to industrial laboratories in the northern and southern regions of the state.

The members of this group have different backgrounds and personalities. More important, they contribute in different ways to the group's functioning. The vice-president is the dominant member. He has a lot of experience in the industry, having established an outstanding record as a salesman. He is responsible for maintaining the group's sales performance at a high level. In addition, he communicates with executives in other divisions of the company regarding ideas for new products that might be developed by the company in the future. All important decisions made within the sales group must be approved by the vice-president.

Next in importance are the two managers. Both are experienced and have excellent skills in sales. One manager is responsible for the group's sales performance in the northern region of the state; the other, for the southern region. The managers are well paid, although they receive lower salaries than the vice president receives.

The least important members of the group are the eight salespersons. They are responsible for making direct contact with customers in their area and for providing products that meet their customers' needs. The salespersons are younger and less experienced than the managers and vice-president. They are paid partly on salary and partly on commission, but they still earn less than the managers.

### Status in Groups

As the above description indicates, the eleven persons in the group occupy different positions, and these positions differ substantially from one another. Each position carries a distinct set of role expectations. Moreover, each position can be evaluated by members in terms of its prestige, importance, or value to the group. This evaluation is referred to as the **social status** of the position. (Since the evaluation of a position often transfers to the occupant of that position, it is natural to talk in elliptical fashion of the social status of a group member— rather than of the position per se—and we shall sometimes use this form.) Within the sales group, the vice-president occupies the position of highest status, the two managers occupy positions of intermediate status, and the eight salespersons occupy positions of low status. The status structure of the sales group is hierarchical —that is, there are fewer members occupying positions of high status than members occupying positions of low status. The structure of this group is diagramed in Figure 13-1.

Note that the concept of status is multidimensional (Kimberly, 1970). In other words, positions within a group (and the occupants of these positions) can be ranked in terms of many different criteria. Three of special importance here include (1) the contribution that a member occupying a given position makes toward the group's overall performance, (2) the level of rewards received by that member, and (3) the level of control exercised by that member over group decisions and activities. Although there are exceptions, a member's standing on one of these dimensions usually correlates with his or her standing on the others. Thus, the vice-president contributes the most important skills to the group, receives the greatest rewards, and exercises the highest level of control over important group decisions.

This chapter examines the causes and consequences of status differences among group members. The questions to be addressed include the following:

1. How do differences in status arise among members in a newly forming group? That is, what processes are involved in status emergence?

2. In what ways do personal characteristics (such as sex, occupation, education, and race) affect a member's status?

3. In what ways do differences among members in skill and performance affect the rewards received by members? How will group members react if there is a mismatch between contributions and rewards, so that rewards are allocated inequitably?

4. What factors affect the willingness of members in formal groups to support and endorse the group's leader? What circumstances lead to the formation of coalitions that seek to overturn leadership?

5. To what degree are members responsive to legitimate power exercised by group leaders? Under what conditions will members obey orders from legitimate authorities, and under what conditions will they refuse to obey?

# ■ Status Emergence in Newly Formed Groups

The group of salespersons working for the small California company, as discussed above, obviously is a formal, structured group that has been operating continuously for an extended period of time. Groups of this kind are interesting (and will be considered in more detail later in this chapter). We begin our discussion here, however, with a more elementary case—that of groups that are newly forming.

## Communication in Task Groups

Consider a group of five undergraduates meeting for the first time. They are members of a task force discussing the case of Johnny Rocco, a delinquent juvenile who has committed a serious crime but who comes from an underprivileged background. The group members have been instructed to read a short summary of Johnny Rocco's history, discuss it, and reach a group decision regarding the handling of his case. At one extreme, they might decide that he should

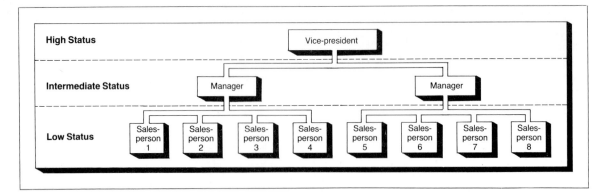

**■ FIGURE 13-1**

STATUS STRUCTURE IN A WORK GROUP

This work group has a three-level status hierarchy. The high-status member (vice-president) exercises the most control and receives the greatest rewards. Reporting to the vice-president are two intermediate-status members (managers). The eight low-status members (salespersons) exercise the least control and receive the least rewards.

be punished severely for the crime, while at the other extreme they might opt to treat him leniently. The case is complex and ambiguous, and it has no obviously right answer.

The five members in this group have been carefully chosen so that they are all "equal"— that is, the same age, same race, and same sex and of similar social background. The group has been assigned no formal leader, and the five members all enter the group on equal footing.

When these persons begin their discussion, something interesting happens. The initial equality disappears, and distinctions quickly arise among the members. Some participate more than others and exercise more influence regarding the group's decision. Members develop different expectations regarding their own and others' roles in the group, and one or more persons start to provide the group with leadership.

Suppose we wanted to investigate the process through which these differences emerge among group members. While there are many ways to do this, one is to monitor the flow of communication among group members. By analyzing the patterns that occur (rate, direction, and types of messages) we could obtain a picture

of a group's structure. One problem with this approach is that groups naturally have different concerns. A group of civil engineers, for example, might want to discuss the best way to repair an abandoned bridge; a group of high school teachers might be concerned with techniques for teaching algebra to students; a group of basketball players might ponder a way to increase rebounds and reduce turnovers. The content of communication will differ from group to group, depending on the problems or topics of concern. What we need, then, is a coding system that is general enough to apply to all groups but specific enough to break down the flow of communication into discrete and interpretable acts.

One approach to this is the coding system called **interaction process analysis (IPA)**, which was developed in various forms by Bales (1950, 1970). The IPA system uses a limited number of categories that apply to all problem-solving groups. Although the content may differ from group to group, any communicative act can be classified into one of the IPA categories.

The IPA system consists of 12 categories shown in Figure 13-2. The 12 IPA categories may be grouped into clusters. The broadest and most important distinction is that between social-emo-

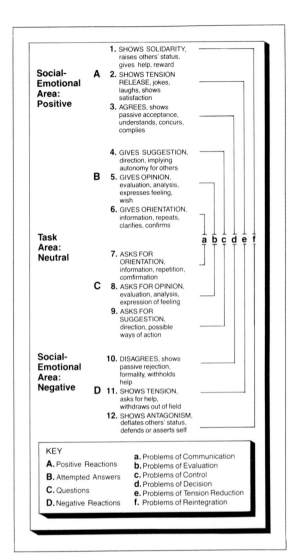

■ **FIGURE 13-2**

CODING CATEGORIES IN INTERACTION
PROCESS ANALYSIS

Interaction process analysis is a method of analyzing
patterns of communication among group members. Researchers observe a group in action and code members' acts into 12 distinct categories. Because these
categories are highly general, they apply regardless of
the specific topic or issue under discussion by members.
Half of the categories pertain to task, or instrumental,
acts, whereas the other half pertain to social-emotional,
or affective, acts.

*Source: Adapted from Bales, 1950.*

tional acts and task acts. **Social-emotional acts**
(categories 1–3 and 10–12) are emotional reactions, both positive and negative, directed toward other members in a group. **Task acts**
(categories 4–9) are instrumental behaviors that
push a group toward the realization of its goals.
Another breakdown of IPA categories is designated by the letters a–f in Figure 13-2. These
letters represent problems that confront any
newly forming group. Within the area of task
acts are problems of communication, evaluation,
and control. Problems of communication involve
analyzing a situation; problems of evaluation
refer to determining members' attitudes toward
that situation; and problems of control refer to
suggestions for action within the situation. Within the area of social-emotional acts are problems
of decision, tension reduction, and reintegration. Problems of decision refer to acceptance or
rejection of proposed courses of action, whereas
problems of tension reduction and reintegration
refer to the establishment of emotional relations
within the group.

In practice, the IPA system might be used as
follows. Assume that a task group of seven
persons is meeting in a room to discuss a given
problem. Members are seated around a table
with a number in front of each person. The room
itself has one unusual feature—it is equipped
with a one-way mirror along one wall so that
observers seated behind the wall can monitor
interaction among group members. Typically,
the observers record the interaction in terms of
who speaks to whom and what types of acts are
communicated. For example, if member 2 turns
to member 5 and asks 5's opinion on an issue,
the observers score this by writing "2–5" in IPA
category 8 (asks for opinion). By following this
procedure through an entire group session
(which might run an hour or two), the observers
can obtain a complete record of interaction
within the group according to the IPA system.

**Who Talks in Groups?** The IPA system provides a means of observing who talks in groups
and what kinds of things they say. Investigators

have used it to study a wide variety of problem-solving and discussion groups involving college students, military enlisted men, patients in therapy, prison inmates, jury members, third-grade boys, and so on (Bales & Hare, 1965).

One finding that regularly emerges in these studies is that group members do not participate equally in a discussion. Some members talk more than others. Although it can vary, the most talkative person in a problem-solving group will typically initiate 40–45 percent of all communicative acts. The second-most active person will do approximately 20–30 percent of the talking. This pattern is seen clearly in Table 13-1, which summarizes initiated acts for groups ranging from three to eight members. As the size of the group increases, the most talkative person still initiates a consistently large percentage of communicative acts, while less talkative persons are crowded out almost completely (Stephan & Mishler, 1952; Bales, 1970).

Another finding from IPA studies concerns the stability of participation: The group member who initiates the most communication during the beginning minutes of interaction is very likely to continue doing so throughout the life of the group. If a problem-solving group meets for several sessions on successive days, for example, the member who ranked highest in participation during the first session is likely to rank highest during subsequent sessions. In general, the ranking of members in terms of participation is stable over time (Fisek & Ofshe, 1970; Fisek, 1974).

**Types of Acts Occurring** In addition to providing information on the number of acts, IPA provides a way to measure the types of acts that occur in groups. As we have noted, the IPA system consists of 12 categories. Of these, six pertain to task-oriented acts (categories 4–9), whereas the other six pertain to social-emotional acts (categories 1–3 and 10–12). Table 13-2 summarizes the typical, or average, pattern of acts for many problem-solving groups based on these 12 categories. As this table indicates, about two thirds of the acts in a group session are task-oriented, and about one third are social-emotional (Bales & Hare, 1965). Within the task-oriented categories, most acts fall into categories 4–6 (gives suggestion, gives opinion, gives orientation), while fewer acts fall into categories 7–9 (asks for orientation, asks for opinion, asks for suggestion). Within the social-emotional categories, most acts fall into the positive categories 1–3 (shows solidarity, shows tension release, agrees), while fewer acts fall into the negative categories 10–12 (disagrees, shows tension, shows antagonism). Intuitively, this pattern makes sense. The efforts of a problem-solving group would surely be self-defeating if there were more questions than problem-solving attempts and more negative emotional reactions than positive ones.

## Task Leaders and Social-Emotional Leaders

IPA has proved useful in documenting the emergence of role differentiation and leadership within groups. One study (Bales, 1953) investi-

▪ **TABLE 13-1**

PERCENTAGE OF TOTAL ACTS INITIATED BY EACH GROUP MEMBER AS A FUNCTION OF GROUP SIZE

| Member Number | Group Size | | | | | |
|---|---|---|---|---|---|---|
| | 3 | 4 | 5 | 6 | 7 | 8 |
| 1 | 44 | 32 | 47 | 43 | 43 | 40 |
| 2 | 33 | 29 | 22 | 19 | 15 | 17 |
| 3 | 23 | 23 | 15 | 14 | 12 | 13 |
| 4 | | 16 | 10 | 11 | 10 | 10 |
| 5 | | | 6 | 8 | 9 | 9 |
| 6 | | | | 5 | 6 | 6 |
| 7 | | | | | 5 | 4 |
| 8 | | | | | | 3 |

Note: Data are based on a total of 134,421 acts observed via the IPA system in 167 groups consisting of three to eight members.

*Source: Adapted from Bales, 1970, pp. 467–70.*

■ **TABLE 13-2**

INTERACTION PROFILE: MEAN PERCENTAGE OF IPA CATEGORY ACTS

| Type of Act (IPA) | Mean Percentage |
|---|---|
| 1. Shows solidarity | 2.97 |
| 2. Shows tension release | 8.17 |
| 3. Agrees | 10.70 |
| 4. Gives suggestion | 6.56 |
| 5. Gives opinion | 22.24 |
| 6. Gives orientation | 28.72 |
| 7. Asks for orientation | 5.89 |
| 8. Asks for opinion | 3.27 |
| 9. Asks for suggestion | 0.60 |
| 10. Disagrees | 4.73 |
| 11. Shows tension | 3.43 |
| 12. Shows antagonism | 2.41 |

*Source: Adapted from Bales and Hare, 1965.*

gated a number of groups, each consisting of five men, that met to discuss a case study problem in a laboratory setting. Group interaction was scored by observers using the IPA system. At the end of the discussion period, members filled out questionnaires in which they were asked to rate each other. Items included such questions as "Who had the best ideas in the group?" "Who did the most to guide the group discussion?" and "Which group member was the most likable?" Typically, there was high agreement among group members in their answers regarding ideas and guidance but less agreement in their answers regarding liking.

In addition, this study showed a high correlation between participation rank as measured by the IPA system and members' perceptions of one another as measured by the questionnaire. The person who initiated the most acts was perceived by others as providing the most guidance and offering the best ideas. In short, the person initiating the most acts was perceived as the

group's leader. Results of this type have been observed in other studies as well (Reynolds, 1984; Sorrentino & Field, 1986).

But this is not the entire picture. The person initiating the most acts was usually not the best-liked member and, indeed, was sometimes the most disliked member. The best-liked person was typically the second-highest initiator. Why does this occur?

In general, the highest initiator is someone who drives the group toward the attainment of its goals. A high proportion of acts initiated by this person are task oriented (IPA categories 4–6). For this reason, the high initiator is referred to as the **task leader.** This person contributes many ideas and suggestions and pushes the group toward its objectives. However, in the effort to get things done, the task specialist tends to be pushy and openly antagonistic in some instances. He or she makes the most impact on the group's opinion, but this aggressive behavior frequently creates tension. Thus, it remains for some other member, called the **social-emotional leader,** to ease the tension and soothe hurt feelings. The acts initiated by this person are likely to be acts showing tension, tension release, and solidarity (IPA categories 11, 2, and 1). The social-emotional leader is the one who exercises tact or tells a joke at just the right moment. This person helps to release tensions and maintain good spirits within the group. Not surprisingly, the social-emotional leader is often the best-liked member of the group.

Thus, in small, task-oriented groups, there are two basic leadership functions: getting things done and keeping relations pleasant. IPA results show that these will typically be performed by different members. Frequently, these persons work closely together and complement one another. For example, the task and social-emotional leaders tend to interact more and to agree more with each other than with other members of the group (Burke, 1972).

The emergence of distinct task leaders and social-emotional leaders is a common occurrence, although it does not happen invariably. In

Box 13-1

## SOCIOMETRY

Interaction Process Analysis (IPA) can be used to measure not only status differences between members but also the flow of conversation in a group. These measurements reveal the naturally occurring communication channels in a group and, therefore, provide some insight into group structure.

Communication is only one aspect of group structure, however. Other aspects include patterns of influence (who gives orders to whom) and patterns of attraction (who likes whom). Consider, for instance, the measurement of attraction in groups. One of the first empirical approaches designed to measure patterns of attraction was **sociometry** (Moreno, 1934; Hallinan, 1981). This technique—which can be used in various groups such as classes, committees, and residential units—usually involves a short questionnaire. Each person in a group is asked to name others in the group whom he or she likes according to some criterion. For example, students in a classroom may be asked whom they want to sit next to, work with, or play with during recess. In residential settings, persons may be asked whom they want to eat with or have as a roommate. Sometimes they are also asked the opposite question: Whom do you dislike, whom do you not want as a roommate? The resulting choices can then be diagrammed in a distinctive figure called a *sociogram*.

Conventionally, each person in a sociogram is represented by a small circle. Choices made by persons are represented by arrows; solid arrows indicate a positive choice or liking, whereas broken arrows indicate disliking. The distance between circles represents the closeness of a relationship. Two persons who choose one another are positioned close together (see persons 1 and 5 in Figure B-13-1), whereas two persons who reject one another are far apart (see persons 5 and 8).

Various aspects of group structure are reflected in a sociogram. For instance, person 5, who is liked by many other members, is considered a "star." Person 10, who was chosen by no one, is considered an "isolate." Persons 1, 2, 3, 4, and 5—who chose one another and form a tight subgroup—constitute a "clique." Patterns such as these affect the emotional climate in a group. In general, groups characterized by many reciprocal positive choices usually have a happy, positive climate. In contrast, groups with several isolates or several distinct cliques usually have a chilly climate.

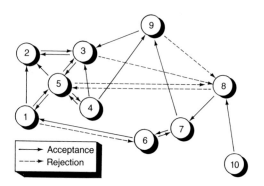

■ **FIGURE B-13-1**

A TYPICAL SOCIOGRAM

*Source: Adapted from Jahoda, Deutsch, and Cook, 1951.*

some groups, both the task and the social-emotional functions are performed by a single member, called a "great man" or "great person." Observations have shown that for problem-solving groups in laboratory settings, a single member successfully performs both of these functions in about 15 percent of the cases (Lewis, 1972). For groups in natural, nonlaboratory settings, the incidence may be even higher (Rees & Segal, 1984).

## ■ Status Characteristics and Social Interaction

So far we have considered interaction in newly formed, problem-solving groups that consist of members who are initially equal in status. Results based on the IPA technique show that despite initial equality among members, differences in status quickly emerge within a group setting. Members differ in their rate of participation, influence over group decisions, and the types of acts they contribute.

But what about interaction in newly formed groups whose members are not equal? In particular, what about groups that include members of different sexes, races, ages, and occupations? We encounter such groups every day—PTA's, student committees, neighborhood associations, juries, church groups, and so on. In fact, groups of this type are probably more common than groups consisting of equal-status members.

Individual properties such as race or occupation are referred to as status characteristics. A **status characteristic** is any property of a person around which evaluations and beliefs about that person come to be organized. In addition to race and occupation, properties such as age, sex, ethnicity, education, and physical attractiveness are status characteristics (Berger, Cohen, & Zelditch, 1972). Status characteristics involve evaluation that is based to a large degree on cultural stereotypes. In the U.S., for example,

A status hierarchy in a church. A priest, bishop, altar boy, and members of the choir participate in a service. Each person's function in the recessional, as well as his or her physical location, depends on formal status.

many consider it preferable to be male rather than female, an adult rather than a child, white rather than black, and white collar rather than blue collar. Although we may not think of ourselves as influenced by someone's age, sex, or race, we are, in fact, very sensitive to these properties. In group settings, we develop different evaluations of members having different status characteristics (Berger, Rosenholtz, & Zelditch, 1980; Meeker, 1981).

## Status Generalization

Status characteristics can significantly affect interaction in newly forming groups. Persons with high standing on various status characteristics tend to be accorded high status within such groups. They receive more respect and esteem than other members, and they are chosen more frequently as leaders. Their contributions to group problem solving are evaluated more highly, they are given more chance to participate in discussions, and they exert more influence over group decisions. The process through which differences in members' status characteristics affect group interactions is called **status generalization.** Various studies have documented status generalization in group settings (Berger, Cohen, & Zelditch, 1972; Freese & Cohen, 1973; Webster & Driskell, 1983).

As an illustration, consider how status generalization can occur in a jury. The members of a jury, who usually differ among themselves in status characteristics, must discuss a complex problem and come to a group decision. A classic study by Strodtbeck, Simon, and Hawkins (1965) demonstrated that certain status characteristics, such as occupation and sex, can have an impact on jury deliberations. Because the law did not permit the observation of actual jury delibera-

tions, researchers studied mock juries designed to be as authentic as possible. Like real juries, members were selected from the voter registration list in an urban area. The juries consisted of men and women of varying occupational status, including proprietors, clerical workers, and skilled and unskilled laborers. They listened to a tape recording of a trial, after which they were instructed to do everything a real jury does—select a foreman, deliberate on the case, and reach a verdict. Researchers recorded the jurors' deliberations. At the end of the session, jurors completed a questionnaire to indicate their impressions of each other.

From the recording of the deliberations, it was possible to determine which members talked the most. Table 13-3 shows the rates of participation by jury members as a function of their occupation and sex. The results indicated clearly that men initiated more interaction than women in the mock juries. The data also showed the impact of occupational status: The higher the occupational status, the greater was the rate of participation. This held for both males and females.

The questionnaire completed by jurors at the end of the session provided information on their perceptions of one another. Table 13-4 shows the number of votes received by the jury

■ **TABLE 13-3**

RATES OF PARTICIPATION DURING DELIBERATION BY OCCUPATION AND SEX OF JUROR

| Sex | Occupation | | | | Combined Average |
|---|---|---|---|---|---|
| | Proprietor | Clerical | Skilled | Laborer | |
| Male | 12.9 | 10.8 | 7.9 | 7.5 | 9.6 |
| Female | 9.1 | 7.8 | 4.8 | 4.6 | 6.6 |
| Combined average | 11.8 | 9.2 | 7.1 | 6.4 | 8.5 |

Note: Entries in this table are percentage rates of participation. Since there were 12 persons on a jury, the "average" juror would theoretically have a rate of participation of 8.3 percent. This can be used as a frame of reference against which to compare tabled values. (The "combined-combined" value in the table is 8.5 rather than 8.3 because 26 of 588 jurors in the study were not satisfactorily classified by occupation and were omitted from the analysis.)

Source: Adapted from Strodtbeck, Simon, and Hawkins, 1965.

■ **TABLE 13-4**

AVERAGE VOTES RECEIVED AS "HELPFUL JUROR" BY OCCUPATION AND SEX OF JUROR

| | Occupation | | | | |
| | Proprietor | Clerical | Skilled | Laborer | Combined Average |
|---|---|---|---|---|---|
| **Sex** | | | | | |
| Male | 6.8 | 4.2 | 3.9 | 2.7 | 4.3 |
| Female | 3.2 | 2.7 | 2.0 | 1.5 | 2.3 |
| Combined average | 6.0 | 3.4 | 3.5 | 2.3 | 3.6 |

*Source: Adapted from Strodtbeck, Simon, and Hawkins, 1965.*

members for being "most helpful in reaching the verdict." This measure reflected the amount of influence each member had over the group decision, as perceived by other members. The findings are very similar to those on the rates of participation. Male jurors were more often perceived as helpful than female jurors, and jurors of high occupational status were more often perceived as helpful than those of lower occupational status. These findings illustrate the status generalization effect.

Overall, this jury study revealed the impact of status generalization. Persons with higher standing in terms of sex and occupation became the group members with the higher status inside the group. They participated at a higher rate, were perceived as contributing more to the group's problem-solving efforts, and were more frequently chosen as the formal leader.

Although the findings in this study seem clear cut, the interpretation in terms of status generalization is admittedly open to criticism. A critic could argue, for example, that a person's status inside a group is not a function of his or her status outside but merely reflects the same qualities that determined that person's outside status. For example, people of high occupational status may be more intelligent, or possibly more experienced in leadership, than people of lower occupational status. One or both of these qualities could enable them to contribute heavily within a group.

To further our understanding of status generalization, several controlled studies have attempted to manipulate status characteristics experimentally. One of these studies (Moore, 1968) investigated female subjects in small, two-person groups. Both women were shown a series of large figures made up of smaller black and white rectangles. Their task was to judge which of the two colors, black or white, covered the greater area in each of the large figures. It was a difficult task because the areas were in fact approximately equal, making the figures ambiguous. The subjects, who were seated so that they could not see or talk with each other, indicated their preliminary judgments by pressing buttons on consoles in front of them. Each subject's answer was revealed to the other through a system of lights. After making their initial judgments, the subjects (who had been told that they should weigh their own answers against the answers of their partners) proceeded to make their final judgments. The subjects did not know, however, that the connections on the consoles were, in fact, controlled by the experimenter. He could manipulate how often one subject perceived that the partner disagreed with her. Between their first judgment and their final judgment, subjects were free to reverse their decisions when their partners appeared to disagree with them.

All of the subjects in this experiment were junior-college students. To manipulate status,

Jury members return to the courtroom after deliberating a case. Although formally of equal status, some jury members exercise more influence than others in determining the verdict. Characteristics such as occupation, gender, and race can affect the patterns of influence among members.

one half was told that their partner was a high school student (low-status partner), whereas the other half was led to believe that their partner was from Stanford University (high-status partner). The results show that the women who believed their partner to be of higher status changed their answers on the judgmental task more often than those who thought themselves to have higher status than their partner. In other words, group members were receptive to influence from a higher-status person but relatively unreceptive to influence from a lower-status person. The random assignment of subjects to experimental treatments eliminated the possibility that subjects differed systematically in ability on the judgmental task. Thus, receptiveness to influence depended on status expectations. These findings, like those from the mock-jury research, support the case for status generalization.

## Expectation States Theory

Why does status generalization occur? That is, why do the status characteristics of group members affect the interaction occurring within the group?

One answer to this question is provided by **expectation states theory** (Berger & Fisek, 1974; Berger et al., 1977; Skvoretz, 1981). This theory proposes that status characteristics cause members of a group to form expectations regarding one another's potential performance on the group's task. These performance expectations, which are to some degree based on cultural stereotypes, affect subsequent interaction among members. Group members are more likely to defer to and accept influence from those persons whom they expect to perform well than from those they expect to perform poorly.

In order for status characteristics such as race, sex, or age to affect behavior, they must be salient to group members. Any circumstance that fosters this will increase the probability of status generalization. Thus, status generalization is most likely to occur when group members (1) have no prior history of interaction, (2) have no information about one another except for their standing on status characteristics, and (3) have no special experience or information about the group's task.

When generalizing from status characteristics, group members behave as though the burden of proof is placed on demonstrating that characteristics are *not* relevant to the task at hand rather than showing that they are relevant. In the absence of such a demonstration, group members will treat status characteristics as relevant, even when they are not. For example, suppose a group is developing plans to take a weekend trip. Although status characteristics (such as race, sex, and occupation) may not, in fact, be relevant to performance on this task, the group's members will behave as if they are relevant. The members will be willing to ignore these status characteristics only after there has been some explicit demonstration or proof that they are not relevant to the performance in planning the trip. This may seem surprising, but the process has been tested and confirmed in several studies (Berger, Cohen, & Zelditch, 1972; Freese, 1976; Webster & Driskell, 1978).

As already noted, each member in a newly forming group possesses a variety of status characteristics. How, then, do members combine information on two or more status characteristics to form performance expectations regarding a given person? What performance expectations result if that person's standing on one status characteristic is inconsistent with his or her standing on another?

For example, consider the case of a female physician. Such a person has inconsistent status characteristics. On the one hand, she is a female, and many studies show that this is perceived as a lower-status characteristic than being male (Zimmerman & West, 1975; Berger, Rosenholtz, & Zelditch, 1980; Ridgeway 1982). On the other hand, she is a physician, and this is a high-status characteristic. How will group members respond to her? Clearly, there are several possibilities. They might ignore one of the status characteristics and focus on the other. In other words, they might respond to her only as a woman or only as a physician. Alternatively, they might combine both status characteristics to form some kind of intermediate expectation. Studies have shown that, in general, members form an intermediate expectation in these cases rather than focusing on only one or another of these status characteristics (Berger & Fisek, 1970; Zelditch, Lauderdale, & Stublarec, 1980). Thus, confronted with a female physician, group members might consider her more competent than female nonphysicians but less competent than male physicians.

## Overcoming Status Generalization

Status generalization often works to an individual's disadvantage. For example, in a mixed setting with both males and females, the females may find that they are not permitted to influence the group's decision significantly even though they are as qualified as males with respect to the problem under discussion. Verbal protests of gender equality may be to no avail (Pugh & Wahrman, 1983). Likewise, in an interracial interaction between blacks and whites, the blacks may feel that they are treated as low-status, minority members. Because irrelevant status characteristics can so easily place someone at a disadvantage, we might ask whether status generalization can be overcome or eliminated in face-to-face interaction.

Some persons have suggested that the best way to overcome status generalization is by direct methods—that is, to raise the expectations of lower-status persons regarding their own performance so that they can, in turn, force a change in other people's expectations regarding their performance on group tasks. Unfortunately, this approach does not work especially well. In one study, for example, blacks at a northern university were trained in assertiveness techniques. They then participated with whites in biracial groups. The training raised blacks' expectations for themselves, and they behaved in an assertive and confident manner. But because the whites' expectations regarding the blacks' performance were not affected by the training, what ensued was not smooth interaction but a status struggle. From the whites' viewpoint, the black members behaved inappropriately, considering their "low ability." From the blacks' viewpoint, the whites behaved inappropriately by refusing to recognize the blacks' equal ability. The whites thought the blacks seemed arrogant, whereas the blacks viewed the whites as racist bigots (Katz & Cohen, 1962; Katz, 1970).

The key to overcoming status generalization is to supply group members with information that contradicts expectations derived solely from knowledge of a status characteristic. Moreover, one must change not only the expectations held by low-status members but also those held by high-status members. Although this is not easy, it can sometimes be accomplished. Studies of black and white boys have shown that by modifying the expectations of *both*, it is possible to create status-equal interaction. For instance, in one study (Cohen & Roper, 1972), investigators taught black junior high school students how to build a radio, then showed them how to teach another pupil to build a radio. This created two specific status characteristics inconsistent with students' conception of race. They then had the blacks train white pupils to build a radio, thereby establishing the relative superiority of the blacks on this task. Finally, they informed some of the students that the skills involved in building the

radio and teaching others to build it were relevant to another, entirely different task. This task was a decision-making game called "Kill the Bull," which the boys subsequently played. Results of the study indicated that this pattern of training modified the performance expectations held by both black and white boys. The change in expectations produced a significant increase in equality between blacks and whites, as indicated by who exercised influence over the decisions when the boys played Kill the Bull. The overall conclusion is that status generalization can be overcome, provided that the expectations of both low-status and high-status persons are modified simultaneously. Similar findings appear in related studies (Cohen, Lockheed, & Lohman, 1976; Riordan & Ruggiero, 1980; Cohen, 1982).

Even when direct methods are not applicable, there are indirect methods of overcoming status generalization. One of these is to block the burden-of-proof assumption noted earlier. That is, to convince group members that a status characteristic is not relevant to the task at hand. In some settings, this can be accomplished by having a credible authority say something like, "We know from experience that race is not relevant to the task at hand. That is, some whites do well and some do poorly, and some blacks do well and some do poorly." Such methods have been shown to prevent members from forming task expectations on the basis of such status characteristics as race or sex (Webster & Driskell, 1978; Berger, Rosenholtz, & Zelditch, 1980; Sell & Freese, 1984).

# ■ Equity and Reward Distribution

So far we have considered the *sources* of status differences among group members in newly forming groups. Status differences arise from variations in those skills and motivations that affect group goal attainment; they also arise from various characteristics (age, sex, race, or occupation) that create expectations regarding performance in group contexts. In this section, we will change our focus and look at some of the *consequences* that result from status differences. In particular, we will discuss the distribution of rewards and benefits among group members.

When members contribute to the attainment of group goals, they typically receive benefits in return. This form of exchange occurs in many groups. It is especially evident in work situations, where group members contribute their labor and skill to produce the group's product and receive rewards such as money and approval in return.

Although all members obtain some benefits in exchange for their contributions, not everyone receives the same amount. In a work group, for example, higher rewards are often allotted to the high-status members—that is, higher rewards go to those persons who have higher levels of skill relevant to the group's task and who contribute more toward goal attainment (Podsakoff, 1982; Rusbult et al., 1988). The distribution of rewards within a group is a matter of concern to all members because it raises questions of justice and fairness. Most individuals care not only about their own rewards but also about the rewards their fellow members receive.

There are many criteria that members may use in judging the fairness and appropriateness of the distribution of rewards (Deutsch, 1975; Elliott & Meeker, 1986). They may follow the *equity principle* and distribute rewards in proportion to members' contributions. Alternatively, they may use the *equality principle* and distribute rewards equally among members, regardless of members' contributions. Or, they may follow the *needs principle* and distribute rewards in accordance with members' personal needs. Although there are other possible criteria, the principles of equity, equality, and need are among the most important and widely used criteria in reward distribution (Lamm & Schwinger, 1980).

A group may rely exclusively on one of these principles when allocating rewards among mem-

bers, or it may apply several of them simultaneously. These principles appear contradictory in the sense that they lead to different distributions of rewards, but what really matters is the relative importance (or weighting) accorded each principle. Not surprisingly, their relative importance varies from group to group and from situation to situation. For instance, the equity principle may be very important in work situations where persons are concerned with receiving their share of the profit. The equality principle often prevails in situations where members are concerned with solidarity and wish to avoid conflict (Leventhal, Michaels, & Sanford, 1972). There is some evidence that females favor the equality principle more than males (Leventhal & Lane, 1970). The needs principle is frequently salient in intimate relationships involving friends, lovers, and relatives. However, this principle has also been invoked in other contexts: Karl Marx, for example, advocated the adoption of the needs principle in communist societies where individuals would contribute according to their abilities and receive according to their needs.

### Equity Theory

By definition, a state of **equity** exists when members receive rewards in proportion to the contributions they make to their group. Equity is relevant to many settings. For example, in an industrial setting in the U.S., a worker normally would expect to receive better outcomes (salary, benefits) than another worker if his or her job required more input (higher skill, more hours per week, and so on). In a marriage, a wife would feel that the outcomes were inequitable if she contributed more to the relationship than her husband but received little help or love in return. As these examples show, equity judgments are made when one group member compares his or her own outcomes and inputs against the outcomes and inputs of another member. This tendency to compare one's own outcomes and inputs against those of another has been discussed by many theorists (Adams, 1965; Homans, 1974; Walster, Walster, & Berscheid,

1978; Greenberg & Cohen, 1982), and it serves as a basis for what is termed *equity theory*.

Although there is some disagreement regarding how equity should be operationalized (Alessio, 1980; Harris, Messick, & Sentis, 1981), for our purposes this comparison can be expressed in terms of the following equation:

$$\frac{\text{Person A's outcomes}}{\text{Person A's inputs}} = \frac{\text{Person B's outcomes}}{\text{Person B's inputs}}$$

This equation states that equity exists when the ratio of person A's outcomes to inputs is equal to the ratio of person B's outcomes to inputs. What matters is not merely the level of outcomes or inputs but the equality in the ratio of outcomes to inputs.

To make this concrete, consider the case of two women employed by the same industrial corporation. One of the women (person A) receives a high outcome—a salary of $50,000 a year, four weeks of paid vacation, reserved parking in the company's lot, and a fancy corner office with thick rugs and a nice view. The other woman (person B) is about the same age but receives a lesser outcome—a salary of $20,000 a year, no paid vacation, no reserved parking, and a cramped, noisy office with no windows.

Will persons A and B feel that this distribution of rewards is equitable? They may or they may not—it depends on their relative inputs. If their inputs to the company are identical, then the arrangement will almost certainly be experienced as inequitable, especially by person B. For example, if both A and B work a 40-hour week, have only high school educations, and are equally lacking in experience, there is little basis for paying person A more than person B. Person B will feel angry because the reward distribution is inequitable. Person A may also sense the inequity and feel guilty or uncomfortable.

Suppose instead that person A's inputs are much greater than B's. Let's say that person A works a 60-hour week, holds an advanced degree such as an MBA, and has twelve years of relevant experience in the industry. Suppose, also, that the work being done by person A involves a high level of stress because it entails the risk of serious failure and financial loss for the company. In this event, A not only has greater "investments" (that is, education and experience) but is also bearing greater immediate "costs" (60 hours of work a week, including stress). Person A may receive better outcomes, but she also contributes more to the company. Under these conditions, both A and B may feel that their outcomes, while not equal, are nevertheless equitable.

Although this illustrates the basic difference between an equitable and an inequitable relationship, precise calculations regarding equity can be difficult to make in everyday life. Two persons may view the same situation in different ways. They may disagree, for instance, over how to evaluate particular inputs and outcomes. If one worker holds an advanced university degree while another worker has seven years' job seniority, who is contributing the more important input? Persons may also disagree over which inputs and outcomes are to be included in the equity calculation. This issue arises, for example, with respect to pay for women in our society. In some jobs, women perform the same activities as men, but they receive less pay for their efforts; in short they do not receive equal pay for equal work. Many people feel such an arrangement is inequitable. They argue that a worker's gender should be irrelevant (not considered as an input) in the equity calculation. As these examples show, application of the equity concept can be complicated by the fact that consensus may be difficult to achieve regarding what inputs and outcomes are to be included in the calculation.

## Responses to Inequity

Inequity produces not only strong emotional reactions (anger, guilt) but also direct attempts to change the conditions that produce it. By eliminating inequity, group members can rid themselves of emotional distress. There are two distinct types of inequity: underreward and overreward. **Underreward** occurs when a person's outcomes are too small relative to his or her inputs; **overreward** occurs when a person's outcomes are too large relative to his or her inputs.

**Responses to Underreward** Persons who are underrewarded usually become dissatisfied or angry (Austin & Walster, 1974). The greater the degree of underreward, the greater will be their dissatisfaction and desire to restore equity. To illustrate, suppose that person A is underrewarded in a relationship. In this case, inequity may be expressed as follows:

$$\frac{\text{Person A's outcomes}}{\text{Person A's inputs}} < \frac{\text{Person B's outcomes}}{\text{Person B's inputs}}$$

This states that person A's ratio of outcomes to inputs is less than B's ratio. Given this situation of underreward, any of various actions could restore equity between A and B. These include (1) increasing the outcomes received by A, (2) reducing the inputs from A, (3) reducing the outcomes received by B; or (4) increasing the inputs from B.

Studies show that most underrewarded people take direct steps to reduce inequity. If the situation permits, they might attempt to reduce inequity by increasing their outcomes. This would occur, for example, if a person had the ability to reallocate rewards in the group directly (Schmitt & Marwell, 1972). A related method of reducing inequity is sometimes used by industrial employees who are paid on a piecework basis. A worker who feels underrewarded might reduce the quality of his or her effort on each piece in order to increase the total number of pieces produced per hour. This would allow the worker to increase his or her outcomes without increasing inputs (Andrews, 1967; Lawler & O'Gara, 1967).

**Responses to Overreward** What happens when a person receives more than his or her fair share in a relationship? Will he or she be content to just enjoy the benefits? Indeed, overreward is apparently less troubling to individuals than is underreward. However, a person who feels guilty and/or indebted may attempt to rectify

the inequity (Pritchard, Dunnette, & Jorgenson, 1972; Austin & Walster, 1974).

Suppose that person A is overrewarded in a relationship. In this case, inequity is expressed as follows:

$$\frac{\text{Person A's outcomes}}{\text{Person A's inputs}} > \frac{\text{Person B's outcomes}}{\text{Person B's inputs}}$$

This states that person A's ratio of outcomes to inputs is greater than person B's ratio. Given this situation, any of the following changes could restore equity: (1) reducing the outcomes received by A, (2) increasing the inputs from A, (3) increasing the outcomes received by B, or (4) reducing the inputs from B.

Research findings indicate that, in some situations, overrewarded individuals will sacrifice some of their rewards to increase those of others. Frequently, however, the extent of the redistribution will not be complete, and equity will be only partially restored (Leventhal, Weiss, & Long, 1969). At other times, overrewarded persons may prefer to restore equity by increasing their inputs. In a work situation, for example, overrewarded members may strive to produce more or better products as a means of reducing inequity (Goodman & Friedman, 1971).

This process was demonstrated in a classic study in which students were hired to work as proofreaders (Adams & Jacobson, 1964). In one condition, subjects were told that they were not really qualified for the job (due to inadequate experience and poor test scores) but that they would nevertheless be paid the same rate as professional proofreaders (30 cents a page). In a second condition, subjects were told that due to their lack of qualifications, they would be paid a reduced rate (20 cents a page). In a third condition, subjects were told that they had adequate experience and ability for the job and that they would be paid the full rate (30 cents a page). Subjects in the first condition were overrewarded, whereas those in the second and third conditions saw the pay as fair. Measures of the quality of the students' work showed that the overrewarded students caught significantly more errors than the equitably paid students. In fact, the overrewarded students were so vigilant that they often challenged the accuracy of material that was correct. These results indicate that the overrewarded students increased their inputs, thereby restoring equity. Similar findings appear in related studies (Adams & Rosenbaum, 1962; Goodman & Friedman, 1969).

**Other Responses to Inequity**  We have noted that inequity—both underreward and overreward—can be resolved by changing outcomes and/or inputs. There are other ways of coping with inequity, however. Although these methods do not rectify the conditions producing the inequity, they make it more tolerable. For instance, a person who is underrewarded might choose a different person for comparison. Instead of comparing her inputs and outcomes with those of person B, person A might compare them with person C. By adjusting her frame of reference, person A might see the situation in a different light and experience less inequity.

Another way to cope with inequity is to withdraw from the situation entirely. An individual might choose to resign from her job, for example, rather than continue to perform under severe inequity. This might lead to behavior that seems economically irrational. For instance, one study found that workers were often willing to accept an alternative position that paid less in order to remove themselves from severe inequity (Schmitt & Marwell, 1972).

Still another way to reduce inequity is through perceptual distortion. By distorting inputs and outcomes, equity can be reestablished in a psychological sense. For example, an employer might convince himself that his overworked and underpaid secretary is, in fact, being treated equitably by minimizing the secretary's inputs ("You wouldn't believe how incompetent she is!") or exaggerating her outcomes ("Work gives her a chance to socialize with all her friends"). This distortion precludes any need to reduce the secretary's work load or increase her

salary (Walster, Berscheid, & Walster, 1973; Austin & Hatfield, 1980).

Perceptual distortion is especially convenient for those who are overrewarded. If a person is receiving more than he deserves, he might simply conclude that his inputs are greater than he originally believed. This will eliminate any need to reallocate rewards. Perceptual distortion of this type has been observed in task situations (Gergen, Morse, & Bode, 1974).

## Equity in Intimate Relationships

So far, most of our examples regarding equity have involved work and business settings. However, equity theory may also be applied to intimate relationships. Initially, one might think that the concept of equity is not especially relevant in intimate relationships or that these relationships are somehow above considerations of give and take (Clark & Mills, 1979; Mills & Clark, 1982). Nevertheless, theorists such as Blau (1964) have argued that equity is relevant to personal relationships and that, indeed, people end up paired only with partners with whom they can establish equitable exchanges. Moreover, equity has a bearing on emotions in close relationships (Sprecher, 1986). In dating and marital relationships, as in work situations, people make contributions and receive outcomes. Thus, we might expect that persons will be happier and more content if they believe their intimate relationship is equitable rather than inequitable.

This hypothesis was tested in a study of 500 undergraduates involved in heterosexual dating relationships (Walster, Walster, & Traupmann, 1978). These individuals were asked to evaluate their inputs in the relationships, including personal contributions (physical attractiveness, intelligence), emotional contributions (being loving and understanding), and day-to-day contributions (taking care of housework, helping to make decisions). Individuals were also asked to evaluate the outcomes they were receiving from the relationship. Results indicated that students who believed their relationship was equitable (that is, involved a balance between their own and their partners' outcomes and inputs) were more content and happier with the relationship. Students who felt their relationship was inequitable (because they were putting more into it than they received) were more resentful and angry. The results also show that persons with equitable relationships were more physically intimate and sexually involved than persons with inequitable relationships.

Just as equity is relevant to dating relationships, it is also a significant variable in marriage. Each spouse contributes some things to their relationship, including financial support, parenting, day-to-day maintenance of the household, emotional companionship, and love. Suppose that one member feels that he or she is contributing more than is being returned. A wife, for example, may feel that she is giving far more to the family than she is receiving. When she compares herself with her husband—a guy who brings home a small paycheck, doesn't help with the housework, and refuses to spend much time with the kids—the relationship may seem inequitable. A situation like this can induce not only anger but also psychological depression.

In one study investigating the relationship between inequity and depression (Schafer & Keith, 1980), 333 married couples responded to questions about equity/inequity in the performance of five family roles (cook, housekeeper, provider, companion, and parent). The study also measured symptoms of depression, including the frequency with which respondents experienced poor appetite, lack of enthusiasm, boredom, loss of sexual interest, trouble sleeping, crying easily, feeling downhearted or blue, feeling low in energy, feeling lonely, and feeling hopeless about the future. Results indicated that psychological depression was related to marital inequity. In other words, husbands and wives who felt there was equity in the performance of family roles were less depressed than respondents who felt either underbenefited or overbenefited.

Equality in marriage can mean various things. For this couple, it means sharing the housework. What might happen to their relationship if the husband insisted on doing less around the house?

## ■ Stability of Status in Formal Groups

To this point, many of the groups considered in this chapter have been *informal groups*. In groups of this type, interaction among members is not very structured or regulated, collective tasks are often ill defined or even nonexistent, and formation may be spontaneous and short lived. Informal groups are familiar to us because most of us participate frequently as members in them.

Yet, most of us have also participated in another type of group, the *formal group*. Formal groups typically exist in the context of a larger organization or bureaucracy, and they often endure for extended periods of time (sometimes many years). Usually they deal with well-specified tasks and have an elaborate division of labor. Most important, they have a well-defined set of roles and norms, with clear lines of authority and control.

One of the most stable features of formal groups is the status structure—that is, the pattern of authority and control. If one observes interaction in a formal group—for instance, interaction among members of the sales group marketing laboratory equipment that we discussed at the beginning of this chapter—it becomes apparent that differences in status among members is quite consistent over time. The persons who yesterday exercised command and control (the vice-president and two managers) are the same persons who exercise control today.

Yet, stability of status in formal groups is the outcome of competing forces. On the one hand, there are pressures for change stemming from innovations in the group's environment and from internal contradictions within the group. On the other hand, there are pressures that delay or prevent change and that reaffirm the status quo. In the remainder of this section, we examine some forces that support the stability in the status structures of formal groups as well as some that create change.

### Endorsement of Formal Leaders

Although role systems in some formal groups are structured like networks or matrices, in many others they are structured in a hierarchy or pyramid that may involve three, four, or more levels of status. For instance, the sales group mentioned above was structured hierarchically into three levels (eight salespersons who reported to two managers, who in turn reported to one vice-president). In most cases, the high-status person at the "top" of such a pyramid is expected to perform various leadership functions for the group. These leadership functions—which

are essential if a group is to achieve its goals—include planning, organizing, and controlling the activity of group members (Stogdill, 1974).

There is, in effect, a tacit exchange between high-status members, who fulfill leadership functions, and the other group members (Hollander & Julian, 1970; Hollander, 1985). High-status members, through their leadership efforts, help a group and its members attain their collective goals. The achievement of these goals is valuable and rewarding to group members. In return for providing leadership, high-status members usually receive certain rewards, benefits, and privileges from other group members. More than this, they also receive support for their continued control over the group's activities. Specifically, they receive support for their actions (plans, decisions, orders) and for their right to continue occupying high-status positions within the group.

This support is essential if the status and leadership structure within a group is to remain stable. Without it, the very legitimacy of that group's status structure is called into question and may crumble or collapse (Walker, Thomas, & Zelditch, 1986).

**Endorsement as Support**   Support for any given member comes from two distinct sources within a group: from members having higher status and from members having equal or lower status (Dornbusch & Scott, 1975; Zelditch & Walker, 1984). Support for a formal leader from group members of even higher status is usually termed authorization, while support from members of equal or lower status is referred to as endorsement. That is, **endorsement** is an attitude held by a lower-status group member, indicating the extent to which he or she supports the leader (Hollander & Julian, 1970; Michener & Tausig, 1971). Endorsement can be measured in various ways. For example, one scale measuring endorsement includes the following items:

1. Consider the person occupying the position of leadership. How legitimate is it for that person to occupy this position?

2. How satisfied are you with the leader's use of power in arriving at group decisions?

3. How satisfied are you with the performance of the leader in directing the group?

4. To what extent do you support or oppose the leader?

5. How willing would you be to have the person serving as leader continue to head the group? (Michener & Lawler, 1975)

A lower-status group member responding favorably to most or all of these items endorses the group's high-status leader. Group members are not always unanimous in their attitude; they may endorse a leader to different degrees. Moreover, the endorsement accorded a leader fluctuates over time. A leader may enjoy high levels of endorsement on some occasions but have to face lower levels on other occasions.

With a high level of endorsement from members, a leader may feel confident when taking vigorous action. With a low level, leadership may be difficult, if not impossible, to exercise. History is full of leaders who gained and subsequently lost endorsements from their followers. One notable example is Richard Nixon, who enjoyed a high level of political support in the first years of his presidency—enough to win reelection in 1972. In the declining days of his presidency, Nixon lost the endorsement not only of the American people but of his own political party, Congress, and even some members of his cabinet. His level of endorsement dropped so low that he could no longer lead the country effectively, and he chose to resign rather than face the possibility of impeachment.

**Factors Affecting Endorsement** Why do some leaders receive high levels of endorsement from members while others receive only low levels? Leaders generally receive endorsement in proportion to the benefits they deliver to group members. Thus, the extent to which a group attains its objectives is one factor. If members perceive a group as moving toward the attainment of its goals, endorsement of leadership will remain high; if members see the group as failing,

Construction workers at a building site accept directives from the contractor. Compliance to an authority's orders depends not only on the rights that inhere in roles but also on other factors such as surveillance and peer support for the authority.

endorsement will decline. A distinction must be drawn, however, between initial failure and repeated failure on group objectives. When a group fails initially, members may actually increase their endorsement and rally around the leader in hopes of improving the situation (Hollander, Fallon, & Edwards, 1977). Repeated failure, however, usually indicates that something is fundamentally wrong. If the responsibility for failure is attributed to faulty leadership, endorsement will certainly decline, and group members may even attempt to change leaders (Julian, Hollander, & Regula, 1969; Michener & Lawler, 1975). Situations of this type sometimes arise in voluntary associations and industrial work groups. They have also been known to occur in military organizations (especially during wartime) and in professional sports (especially at the end of a losing season).

Failure to attain collective goals leads to low endorsement primarily because members infer that the leader is incompetent. Repeated failure is strong evidence that the leader is not sufficiently skilled to move the group in the direction of its goals. In addition, group members may construe other unrelated behaviors such as difficulty on high-skill tasks or failure on written tests as evidence of leader incompetence. Thus, behavior of this type by the leader will also produce low levels of endorsement (Julian, Hollander, & Regula, 1969; Suchner & Jackson, 1976).

Another determinant of endorsement is the level of consideration a leader shows toward group members. If a leader treats members equitably, this will help create a climate of high endorsement. For instance, if a leader controls the allocation of rewards and decides to share these rewards equitably, this will demonstrate concern for the welfare of other group members and induce high levels of endorsement (Michener & Lawler, 1975). A selfish leader—even one who is very competent—will suffer reduced levels of endorsement. Group members are sensitive not only to the level of rewards they receive but also to the procedures used by a leader in allocating rewards. Several studies have shown that formal leaders receive higher levels of endorsement if they allocate rewards by means of procedures viewed as fair rather than those viewed as unfair or arbitrary (Tyler & Folger, 1980; Tyler & Caine, 1981).

## Revolutionary and Conservative Coalitions

When group members agree on their relative status, and especially on that of their leader, a condition of **status consensus** is said to exist (Heslin & Dunphy, 1964). Although status consensus commonly exists at a moderate-to-high level in established groups, it may collapse under certain conditions. For example, if a group experiences unexpected difficulty in attaining its goals, some members may continue to endorse the existing leadership, whereas others may pre-

fer to change leaders. This difference of opinion would indicate a lack of consensus regarding the status order.

## Mobilization of Revolutionary Coalitions

Lack of status consensus poses serious problems for an established leader, who will perceive his or her position in the group as very precarious. Under certain conditions, a lack of consensus may lead to the formation of a **revolutionary coalition,** which is a union of some medium-status and low-status members who oppose the existing leader (Crosbie, 1975). By combining forces, members of a revolutionary coalition hope to displace the leader and his supporters.

Revolutionary coalitions are difficult to mobilize. Members must first have reason to end their allegiance to the established order and then agree to take joint action to overthrow that order. For example, before a mutiny can occur at sea, the seamen must have a serious basis for discontent with the ship's captain. They must feel that existing channels of appeal provide no remedy for their discontent, and they must be willing to face the punishment involved if the mutiny fails. Finally, they must agree who will be the new commander if the mutiny succeeds.

Nevertheless, certain factors increase the likelihood that a revolutionary coalition will form. For the most part, these are the same as those that produce low levels of endorsement. For instance, when a group fails repeatedly to achieve its goals, a revolutionary coalition may try to reverse the group's fortunes by removing the formal leader from power. If successful, the coalition will reallocate important responsibilities among group members and/or change some of the group's operational procedures (Michener & Lawler, 1971).

Inequitable treatment of group members by high-status persons may also encourage revolutionary activity. If a leader shows extreme favoritism or selfishness in allocating rewards, group members may chafe under the inequity and form a revolutionary coalition to restore equity (Ross, Thibaut, & Evenbeck, 1971; Michener & Lyons, 1972; Webster & Smith, 1978).

Similarity of interest and attitudes among lower-status members is still another important factor in the emergence of a revolutionary coalition. Sharing common interest heightens members' expectations of support from other members. This, in turn, increases the probability that individuals will join an emerging revolutionary coalition (Lawler, 1975a, 1975b).

**Reactions to Revolutionary Coalitions** Established leaders have a lot to lose by the emergence of a revolutionary coalition, and they seldom sit idly by during the upheaval. For instance, the revolt by some of Hitler's generals during World War II led only to their own death, not Hitler's. A leader will usually engage in certain counterstrategies to thwart the coalition. One of these is to threaten to punish insurgent members. This may suppress the coalition if the threats are sufficiently large and credible. A second counterstrategy is to mobilize a **conservative coalition**—a union of medium-status and low-status members who support the existing status order against revolutionaries.

A third counterstrategy available to a leader is the use of co-optation. By definition, **co-optation** is a strategic attempt to weaken the bond among potential revolutionaries by singling out one or several lower-status members for favored treatment. Co-optation might be used by a calculating leader who anticipated a future upheaval. For instance, a leader could offer several low-status members the possibility of future promotion. This would increase their investment in the existing status order, thereby making it more difficult for a revolutionary coalition to influence them or to recruit them as members (Lawler, Youngs, & Lesh, 1978; Lawler, 1983).

# ■ Authority and Obedience

In the normal, everyday operations of many groups, high-status persons exercise a substantial amount of influence over other members.

Typically, this is based on authority rather than on the direct use of threats and promises. **Authority** refers to the capacity of one group member to influence other members by invoking rights that are vested in his or her role.

Orders issued by police officers, decisions rendered by judges, directives specified by corporate executives, and exhortations made by clergymen all involve the invocation of norms. A person can exercise authority only when, by virtue of the role that he or she occupies in a group, other members accept his or her right to prescribe behavior for the issue at hand (Raven & Kruglanski, 1970). In essence, a person exercising authority activates an obligation that requires the target person(s) to comply. The more persons that can be directly or indirectly influenced in this manner, the greater is a person's authority within the group (Zelditch, 1972; Michener & Burt, 1974).

Because it involves invocation of norms, authority can be exercised only in certain contexts and relative to certain persons. If you happen to observe a military parade on the 4th of July, for example, and you hear the commanding officer issue and order to "about-face," the troops will turn around and face the opposite direction, but you yourself will not obey the order. The officer's authority does not extend to you because you are not a member of the military unit. You are outside the group's boundaries and beyond the officer's legitimate authority.

## Milgram's Study of Obedience

One issue regarding authority concerns its limits. Although obedience to authority frequently produces beneficial results, it sometimes produces negative consequences, especially when the orders are morally questionable or reprehensible. The Third Reich of Nazi Germany provided a chilling example. In complying with the dictates of Hitler's authoritarian government, German citizens committed what most people consider morally unconscionable acts—beatings, torture, confiscation of property, even murder—against millions of others.

A series of classic experiments by Milgram (1963, 1965a, 1974) explored the limits of obedience to legitimate authority. In a controlled setting, these studies created a situation in which one person (the authority) directed another person (the subject) to engage in actions that hurt a third person (the victim). Subjects in these experiments, recruited through newspaper advertisements, were adults (age 20–50) with diverse occupations (labor, blue collar, white collar, and professional). Subjects were paid a small monetary fee for participating in the study.

When a subject arrived for the experiment, he found that another person (a gentle, 47-year-old male accountant) had also responded to the newspaper advertisement. The experimenter, dressed in a lab coat, indicated that the purpose of research was to study the effects of punishment (that is, electric shock) on learning. One of the subjects was to occupy the role of teacher; the other, of learner. A random drawing was held, and the accountant was selected to be the learner. He was then taken into the adjacent room and strapped into an "electric chair." During this process, the accountant mentioned that he had some heart trouble and expressed concern that the shock might prove excessively dangerous. The experimenter replied that the shock would be painful but would not cause permanent damage.

The other subject, now in the role of teacher, was taken into a separate room and seated in front of an electric shock generator. This generator was equipped with 30 voltage levels ranging from 15 to 450 volts. The lowest voltage level was labeled "slight shock"; a higher level read "danger: severe shock"; the highest level was ominously marked "XXX." The experimenter gave the teacher a sample shock of 45 volts to show him how the generator worked. Even at this low level, the shock was painful.

The learning session was structured somewhat like a multiple-choice exam. The teacher read word pairs over an intercom system to the accountant in the adjacent room. After reciting the entire list, the teacher then read aloud the first word of a pair and four alternatives for the second word of the pair. The learner's task was to select the correct alternative response. In accordance with directives from the experimenter, the teacher was to shock the learner whenever he gave an incorrect response. The first shock was to begin at the lowest level (15 volts) and then, with each successive error, increase to the next higher voltage. Thus, the voltages were to increase from 15 to 30 to 45, on up to the 450-volt maximum.

Unhappily, the accountant proved to be a slow learner. Although getting a few answers right, he responded incorrectly on numerous trials. The required level of shock moved up quickly. Although a few subjects serving in the teacher role refused to administer high levels of shock to the learner, most were willing to proceed with the task.

When the shock level reached 75 volts, the victim grunted loudly. At 120 volts, he shouted that the shocks were becoming painful. At 150 volts, he demanded to be let out of the experiment ("Get me out of here! I won't be in the experiment anymore! I refuse to go on!"). When the shock level reached 180 volts, he cried out that he could not withstand the pain. At 270 volts, his response to the shock was an agonized scream. If a subject administering the shock expressed concern about the procedure, the experimenter calmly instructed him to persist ("You have no other choice, you must go on").

At the 300-volt level, the accountant shouted in desperation that he wanted to be released from the electric chair and would not provide any further answers to the test. The experimenter, however, instructed the subject to treat any refusal to answer as an incorrect response. At the 315-volt level, the learner gave out a violent scream. At the 330-volt level, he fell completely silent. The experimenter directed the subject to continue routinely toward the 450-volt maximum even though the learner did not respond.

The focus of this experiment by Milgram concerns the subject's compliance to the experimenter's authoritative directives. What percentage of the subjects in the teacher's role continued to administer shock up to the 450-volt maximum? That is, how effective is legitimate power?

As you may have guessed, the accountant receiving shock was a professional actor hired by the experimenter. He was not actually receiving any shocks. The shouts and screams that subjects heard from the adjacent room were from carefully crafted tape recordings. Nevertheless, the actor's performance was extremely realistic, and postexperimental interviews showed that subjects believed they were administering shocks to the learner.

The results of this study are astonishing to many people. Milgram (1963, 1974) ran several variations of the basic research design described here. Hundreds of subjects were involved, and more than 60 percent of those in the basic experimental situation continued to administer shock up to the highest possible level (450 volts). These results are shown in Figure 13-3. Despite the tortured reaction of the victim, subjects complied with the experimenter's directives.

Although they followed orders and continued to administer shocks, this situation was extremely stressful for subjects. Most were concerned about the victim's welfare. As the shock level rose, subjects grew increasingly worried and agitated. Some laughed nervously. Many requested that the experimenter check the victim to make sure he was all right. A few subjects became so distressed that they disobeyed the experimenter and refused to follow his orders. The overall level of compliance on this study, however, was extremely high, which illustrates the enormous impact of directives from a legitimate authority.

## Factors Affecting Obedience to Authority

Authorities customarily attain compliance to their directives, in part because these are often backed by the threat of coercive power. But exceptions do occur. Although most subjects in Milgram's study obeyed the experimenter's orders, some did not. Other studies have also shown that although obedience to authority is common, defiance does occur in some cases (French, Morrison, & Levinger, 1960; Michener & Burt, 1975). This raises a general question: Under what conditions will people comply with authority, and under what conditions will they refuse to comply?

Extensions of the basic study by Milgram offer some insights. Some general factors affecting compliance include (1) the amount of surveillance maintained by the authority, (2) the role a person occupies within a network of authority, and (3) the level of peer support favoring or opposing the authority.

First, obedience depends on the extent to which the subject is under surveillance by the authority issuing orders. In a series of variations on his basic experiment, Milgram (1965a, 1974) varied the physical closeness (degree of surveillance) maintained by the experimenter. In one condition, the experimenter sat just a few feet away from the subject. In another condition, after giving preliminary instructions, the experimenter departed from the laboratory and issued orders by telephone. Obedience was greatest when subjects were under close surveillance; it dropped sharply as the experimenter was physically removed from the situation. The number of obedient subjects in the face-to-face condition was almost three times as great as in the order-by-telephone condition. During the telephone conversations, some subjects specifically assured the experimenter that they were raising the shock level, when in actuality they were using only the lowest shock. This tactic permitted them to ease their conscience and, at the same time, avoid an open break with authority.

Another factor affecting obedience to authority is the subject's position in a larger chain of command. One study (Kilham & Mann, 1974) used a Milgram-like situation, in which a subject (the executant) actually pushed the buttons to administer shock, while another subject (the transmitter) simply conveyed the orders from the experimenter. Results showed that the obedience rates were approximately twice as high

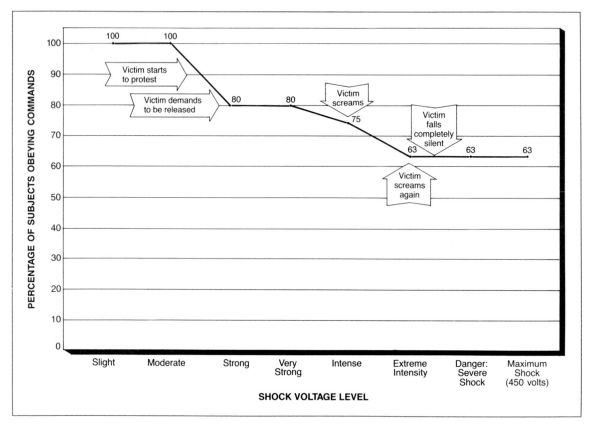

**■ FIGURE 13-3**

PERCENTAGE OF SUBJECTS WHO OBEYED
ORDERS TO ADMINISTER ELECTRIC SHOCK

Neither protests nor screams by the victim were sufficient to deter subjects from complying with orders from an authority to administer electric shock. Although more subjects were willing to administer low-voltage shocks than high-voltage shocks, as many as 63 percent administered the most severe level (450 volts).

*Source: Adapted from Milgram, 1974, Experiment 2.*

among transmitters as among executants. In other words, persons positioned closer to the authority but farther from the unhappy task of throwing the switch were more obedient.

Another variable affecting compliance to authority is pressure from peers. If peers of the subject comply with and support the authority, this increases the pressure on the subject to comply. However, if peers resist orders from the authority, the subject will also tend to disobey. To investigate the propensity to defy authority, Milgram (1965b, 1974) set up a situation involv-

ing five people: one experimenter, three teachers, and one learner. The learner (victim) and two of the teachers were confederates. The third teacher was the subject. Teacher 1 (a confederate) read aloud the list of word pairs. Teacher 2 (a confederate) indicated whether the victim's answer was correct or incorrect. Teacher 3 (the naive subject) administered punishment. As usual, the experimenter directed the subject to raise the level of shock with each wrong answer. By prearrangement, the confederates followed the experimenter's orders to the 150-volt level, at which point teacher 1 indicated that he would not participate further because of the victim's protests. Despite the authority's insistence, he got up from his chair in front of the shock generator and moved to another part of the room. The experiment continued until the 210-volt level, at which point teacher 2 refused to participate further.

Figure 13-4 shows the level of obedience by subjects in the company of two defiant peers. Note that most subjects disobeyed the experimenter at some point and that only 10 percent administered shock to the highest level (450 volts). The average maximum shock was approximately 240 volts. These findings contrast with the experiments in which subjects were alone with the authority, which showed that approximately 65 percent of subjects administered shock to the highest level and that the average maximum shock was 370 volts. The defiant activity of the peers reduced subjects' willingness to administer shock to the victim. Apparently, the peers had a liberating effect on subjects, because their defiance showed that refusal to comply with authority was a behavioral possibility. Moreover, their defiance may have intensified the moral pressure on subjects to break away from the experimenter's demands.

## ■ Summary

In any group, members occupy positions having different value or importance. As a result, each member's social status is determined by the position in the group that he or she occupies. In this chapter, we have discussed both the origins and the consequences of status differences among members.

### Status Emergence in Newly Formed Groups
In task-oriented groups, status differences among members usually develop during the course of interaction. (1) Interaction Process Analysis (IPA) is a technique for investigating interaction and communication in groups. Studies using this technique have shown not only that some group members consistently send and receive more messages than others but also that patterns of communication stabilize over time. (2) Studies using IPA have also analyzed the emergence of status differences in groups. Differentiation of leadership functions—task versus social-emotional—appears regularly in problem-solving groups.

### Status Characteristics and Social Interaction
A status characteristic is any property of a person around which beliefs about that person come to be organized. Age, race, sex, and occupation are important status characteristics. (1) When groups are composed of persons having different status characteristics, this can have a significant effect on interaction. Studies of juror interaction, for example, have shown that characteristics such as occupation and sex affect which members participate the most, exert the most influence, and hold positions of leadership. (2) This phenomenon, known as status generalization, occurs because group members base their expectations regarding one another's performance on status characteristics. Status generalization is most likely to occur when members have no prior history of interaction and no experience with the group's task. (3) To overcome or eliminate status generalization, one must change not only the expectations of low-status members but also those of high-status members. Status generalization can be eliminated only when both low-status and high-status members believe that a characteristic is not relevant to the group's task.

### Equity and Reward Distribution
Status differences in a group determine the amount of rewards that each member will receive. (1) A state of inequity exists if group members receive rewards that are not proportional to their contributions. (2) If inequity prevails within a group, members may react emotionally (with anger, or guilt) and initiate efforts to restore equity. These efforts usually entail increasing the outcomes or

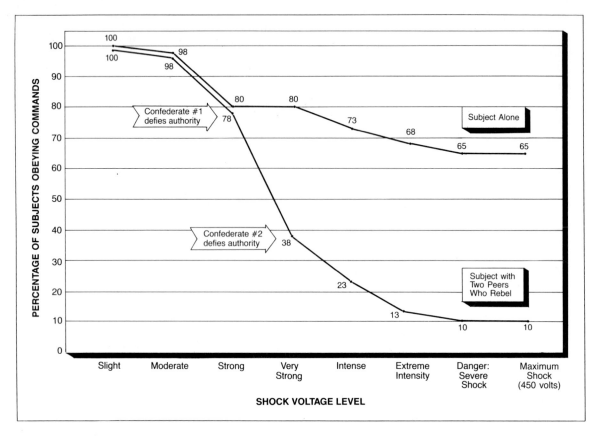

**■ FIGURE 13-4**

THE EFFECT OF REBELLIOUS PEERS ON
OBEDIENCE TO AUTHORITY

When alone, a high percentage of subjects administered the maximum shock (450 volts) to the victim. In the presence of two peers who defy orders, however, most subjects disobeyed the authority at some point, and few administered the highest level of shock.

*Source: Adapted from Milgram, 1974, Experiments 5 and 17.*

reducing the inputs of underrewarded members or, alternatively, increasing the inputs or reducing the outcomes of overrewarded members. (3) Considerations of equity apply not only to task situations but also to intimate relationships. Couples in inequitable dating relationships, for example, are less satisfied and more likely to break up than those in equitable relationships.

**Stability of Status in Formal Groups**  A group member can exercise leadership effectively only with the support of other group members. (1) In return for helping the group achieve its objectives, a leader receives endorsement as well as special rewards and privileges. Endorsement will decline if the group fails repeatedly to achieve its goal(s), if the leader is judged incompetent, or if the level of consideration the leader shows toward members is low. (2) Without status consensus, discontented members may attempt to oust an established leader by forming a revolutionary coalition. Coalitions of this type are most likely to emerge when a group fails to achieve its goals or when the leader distributes rewards inequitably among members. Revolutionary coalitions are not always successful because they may be thwarted by conservative coalitions or by a leader's use of co-optation.

**Authority and Obedience** Authority is influence based on invocation of group norms; it can be exercised only by persons occupying particular roles in certain contexts. (1) Research on obedience to authority has shown that subjects will comply with orders to administer extreme levels of electric shock to an innocent victim. (2) Obedience to authority is more likely to occur when subjects are under direct surveillance by the person issuing orders, when subjects are transmitters rather than executants of a command, and when counterpressure from defiant peers is absent.

## Key Terms

| | |
|---|---|
| AUTHORITY | 389 |
| CONSERVATIVE COALITION | 389 |
| CO-OPTATION | 389 |
| ENDORSEMENT | 368 |
| EQUITY | 383 |
| EXPECTATION STATES THEORY | 380 |
| INTERACTION PROCESS ANALYSIS (IPA) | 372 |
| OVERREWARD | 384 |
| REVOLUTIONARY COALITION | 388 |
| SOCIAL-EMOTIONAL ACT | 373 |
| SOCIAL-EMOTIONAL LEADER | 375 |
| SOCIAL STATUS | 371 |
| SOCIOMETRY | 376 |
| STATUS CHARACTERISTIC | 377 |
| STATUS CONSENSUS | 388 |
| STATUS GENERALIZATION | 378 |
| TASK ACT | 373 |
| TASK LEADER | 375 |
| UNDERREWARD | 384 |

14

chapter

# Group Performance and Leadership Effectiveness

# ■ Introduction

Many groups in society exist primarily to attain specific goals. A work group in a clothing factory strives to meet its production quotas for swimsuit apparel; an airline crew seeks to transport passengers to their specified destination on time; a police detective squad works around the clock to crack an urgent murder case; a professional ball club tries to win games and take the league championship. Not every group, of course, is concerned with producing something tangible. Some groups—such as a Wednesday-night poker club—exist for their own sake, for simple pleasure. But a large number of groups in society exist to produce something valuable or to bring about some desired end state.

## Group Performance

The term **group performance** refers to the achievement of a goal or end state through group activity. Group performance is usually considered to be a matter of degree—some groups perform very well, others less well. Performance is usually measured in terms of observable products or outcomes that result from group activity rather than the activity itself. The way in which performance is measured depends on the particular goals of the group (Gilbert, 1978). An industrial work group that manufactures clothing, for example, may be concerned with the number of garments produced, the amount of material wasted, the number of "seconds" or rejects, the speed of production, and so on. Problem-solving groups such as a police detective squad may base performance on the percentage of cases cleared by arrest, time elapsed between offenses and arrests, and percentage of charges that result in a conviction (a measure reflecting the quality of the evidence gathered by police).

Even when groups are similar in size and talent, they can differ greatly in their levels of performance. Some groups produce more than others, win competitions more frequently, achieve better solutions, work at a higher speed, commit fewer errors, and create less waste. From this basic observation stem some questions regarding group performance that are addressed in this chapter:

1. What types of tasks do groups typically confront? In what ways is group performance affected by the nature of the task(s) at hand?

2. To what extent is performance affected by a group's internal structure? How do such factors as group size, cohesiveness, reward structure, and communication channels affect the level of group performance?

3. Does the style of leadership in a group affect performance? Are some leadership styles more effective than others in fostering a high level of performance by groups?

4. Does a group's organization affect the quality of its decisions? Under what conditions do pressures for conformity lead to poor decisions?

## ■ Group Tasks

Most groups in everyday life face at least one, and sometimes several, important tasks. Whether the group is a basketball team, a jazz band, a production group on an assembly line, a crew in a space shuttle, a study commission, a jury in deliberation, or an infantry squad in combat, they all face tasks and strive to accomplish certain desired outcomes.

Just as there are many different types of groups, there are many different group tasks. The quality and quantity of a group's performance on its task depend not only on the characteristics of the group (size, cohesiveness, reward structure, communication channels, and so on) but also on the nature of the task itself. The nature of the task facing a particular group affects that group's performance directly by restricting the ways in which members can join together and interact with one another. The task also affects performance indirectly by interacting

with other group attributes. For this reason, we will begin by considering, in general terms, the types of tasks that groups face.

### Unitary Tasks and Divisible Tasks

A group task has many identifiable characteristics, including its difficulty, the rules or method for carrying it out, and the criteria that specify when it is completed. Another quality of special importance is a task's divisibility (Steiner, 1972, 1974; Shiflett, 1979). **Unitary tasks** (sometimes called nondivisible tasks) are those in which all group members perform identical activities. A typing pool at an insurance company is one example. Typically, all the persons in the pool sit at computer terminals and type insurance policies for new customers. They also retrieve and revise information on existing policies to keep the company's records current. This is a unitary task because all members in the typing pool have similar skills and perform the same activities.

In contrast, **divisible tasks** are those in which members perform different, although complementary, activities. Because divisible tasks can be broken into various subtasks, they involve a division of labor among group members. Flying a commercial aircraft is one example. The pilot, copilot, navigator, flight attendants, and baggage handlers all perform different activities between takeoff and landing. Both unitary and divisible tasks are shown schematically in Figure 14-1.

**Unitary Tasks** Unitary tasks can be further divided into three types: conjunctive, disjunctive, or additive. A **conjunctive task** is a unitary task in which the group's performance depends entirely on the performance of its weakest or slowest member. A **disjunctive task** is one in which the group's performance depends entirely on the performance of its strongest or fastest member. And an **additive task** is one in which the group's performance is equal to the sum of the performances of its members (Steiner, 1972).

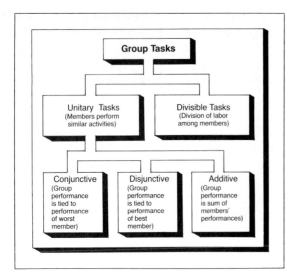

■ **FIGURE 14-1**

GROUP TASKS

Group tasks can be either unitary (that is, members perform similar activities) or divisible (members perform complementary activities). Unitary tasks can be conjunctive, disjunctive, or additive.

*Source: Adapted from Steiner, 1972.*

A conjunctive task can be illustrated by a mountain-climbing expedition that is striving to ascend a steep precipice. Although all members of the expedition may have experience, they will nevertheless differ in dexterity and strength. Because all members of the expedition are held together by a rope, the group's performance (as measured by speed and height of ascent) will be constrained by its slowest member. The expedition can move no faster than the slowest member, and, therefore, the group's performance depends entirely on this person.

A disjunctive task is one in which all the members perform the same activity, but the group's performance depends entirely on whoever happens to be its strongest or fastest member. Achievement by that member is tantamount to achievement by the entire group. As an illustration, suppose we gave the following intellective problem to a five-person group and mea-

sured group performance in terms of the speed with which the solution is found:

> Three missionaries and three cannibals are standing on one side of a wide river. The missionaries want to cross the river to the other side. One rowboat is available, but it can hold no more than two persons at a time. All of the missionaries and one of the cannibals can row the boat. The cannibals pose a danger to the missionaries, because if more cannibals than missionaries are placed on either bank of the river, those cannibals will eat the missionaries. Problem: How can all six persons cross the river alive?*

Only one member is needed to solve this problem. As soon as one finds a solution, he or she can reveal it to the other members. Thus, the group's score is the solution time of its fastest member. The task is disjunctive, because solutions by other, slower members do not affect the group's performance.

An additive task is illustrated by the insurance typing pool mentioned earlier. The number of policies typed for new customers in a day (a measure of group performance) is simply the sum of policies typed by individual members. If the eight members of the pool type 27, 41, 43, 34, 52, 40, 29, and 46, respectively, the group's overall performance is 312 policies (the sum). Another example of an additive task is a group of five men using shovels to clear a snowy driveway. Each man performs a similar activity. The group's performance can be measured in terms of the total time required to clear the driveway.

As these examples show, all unitary tasks—whether conjunctive, disjunctive, or additive—require similar skills and resources from group members. For a group to perform well on a unitary task, several conditions must be met. First, group members must have the necessary abilities and talents to perform the task. Second,

---

*The answer to the missionaries (M) and cannibals (C) problem is: (1) M1 and C1 cross; M1 returns. (2) M1 and C2 cross; C2 (who can row) returns. (3) M2 and C2 cross; C2 returns. (4) C2 and C3 cross; C2 returns. (5) M3 and C2 cross. Everyone is now on the other side of the river.

In a relay race, teams of runners face a unitary additive task. Each team member must complete a lap, and the team with the fastest total time wins.

subtasks into a group effort. Coordination becomes increasingly important as the task becomes more complex (Sorenson, 1971). One sure way to lose a football game, for instance, is to assign members to the wrong positions—make the team's weakest thrower the quarterback, the slowest runners the wide receivers, and the smallest players the linemen. Another sure route to defeat is to neglect coordination—snap the football without informing each of the players beforehand what play has been chosen.

## ■ Factors Affecting Group Performance

In everyday life, group performance and productivity are matters of considerable importance. Groups that perform well are praised and prized, while those that perform poorly are viewed as problematic. A natural question is, What factors (beyond the task itself) affect the level of a group's performance? That is, in what ways do a group's composition, culture, and structure have an impact on performance?

In this section we will address this issue by focusing on several factors that have been shown to affect group performance. These factors are group size, cohesiveness, reward structure, and communication structure.

### Group Size

How does the size of a group affect its performance and productivity? Are large groups superior to smaller ones? Large groups generally have the advantage of greater resources (information, skills, muscle), which may lead to better performance. On the other hand, large groups require extensive organization and coordination among members, which may inhibit performance.

To understand the effects of a group's size on its performance, we must take into account both group size and the type of task facing that group. For example, a five-person group can shovel the snow from a driveway faster than a

group members must have sufficient motivation —they must want to do the task. Third, group members must know what is expected of them and receive feedback indicating the quality of their performance. The absence of any of these conditions may result in poor performance.

**Divisible Tasks**  Divisible tasks pose even greater hurdles to successful performance than unitary tasks. In divisible tasks, group members must have ability, motivation, and information, and the group also must match skills with task requirements—that is, put the right member in the right job. Beyond this, the group must establish coordination among its members. It takes careful planning to blend a number of

## TYPOLOGY OF GROUP TASKS

A number of investigators have attempted to develop classificatory schemes for group tasks (Steiner, 1972; Hackman & Morris, 1975; Laughlin, 1980). Recently, McGrath (1984) proposed a typology of group tasks that is quite general and consistent with other approaches. McGrath suggests that group tasks can be divided usefully into the following eight distinct types.

(1) *Planning Tasks.* In tasks of this type, a group's major goal is to generate an action-oriented plan to achieve some objectives upon which the members have already agreed. The group must consider alternative paths to the goal, contemplate constraints imposed by resource availability, and develop a viable program of concerted action. Examples include planning a military maneuver or mapping out an advertising campaign to introduce a new product.

(2) *Creativity Tasks.* Tasks of this type require a group to generate new, original ideas. Usually, group interaction emphasizes creativity or entails brainstorming to generate new ideas, alternatives, or images. Examples include brainstorming to develop a new product concept or a new comedy routine.

(3) *Intellective Tasks.* Tasks of this type require a group to solve an intellective problem for which there is (or is believed to be) a correct answer. To solve problems of this type, group members typically must consider many alternative solutions and discard unsuitable ones. The group must strive to discover, select, or compute the right answer. Once uncovered, the answer usually seems intuitively right and compelling. Examples of intellectual tasks include arithmetic or logical reasoning problems, puzzles, cryptograms, bus routing problems, and computer programs.

(4) *Decision-Making Tasks.* Tasks of this type require a group to solve a problem for which there is no inherently right answer. Through interaction, the group must consider a variety of alternatives and reach a consensus regarding which alternative is preferred. Examples include jury deliberations, investment decisions, and decision tasks used in risky-shift and group polarization studies.

(5) *Cognitive Conflict Tasks.* In tasks of this type, group members hold varying viewpoints and strive to resolve their differences. Conflicts of this type occur not because members have different interests and goals but because they

two-person team, and eight people can probably do the job even faster. On this additive task, larger size is clearly an advantage. But can 30 chefs prepare a better soup than two? Certainly the coordination problems among 30 chefs would be enormous. And if it happens that the soup's quality is limited by the least-skilled person in the kitchen (a conjunctive task), then too many cooks might literally spoil the broth.

**Disjunctive and Conjunctive Tasks** According to a theory advanced by Steiner (1972), a group's size affects its performance in one of two ways. Performance will (1) increase with group size when the task is disjunctive and (2) decrease with group size when the task is conjunctive.

Consider the logic underlying these predictions. When a task is disjunctive, a group's performance is determined by the most competent member. Assuming that groups are composed at random, a large group has more chance than a small group of containing members of very high ability. The larger the group, the more likely it is to contain the necessary skills. Therefore, as size increases, performance should also improve. Of course, there may be an upper limit

Box 14-1

disagree regarding the use of information. That is, they disagree regarding what information is relevant to the group's goals and how the information should be weighted and combined. A resolution of these differences usually entails discussion and negotiation among members. Examples include some jury tasks and cognitive conflict tasks used in social judgment studies.

(6) *Mixed-Motive Tasks.* In tasks of this type, group members face an underlying conflict of interest with respect to conditions for reward. To resolve this conflict, group members must negotiate and bargain with one another. Examples include labor-management wage negotiations, prisoner's dilemma studies, and reward allocation tasks.

(7) *Contests/Battles.* In tasks of this type, group members compete as a unit against an external opponent or enemy. The outcome is interpreted in terms of a winner and a loser, with corresponding payoffs. Examples include competition between sports teams, battles between military combat groups, and other winner-take-all conflicts.

(8) *Performances/Psychomotor Tasks.* Tasks of this type require group members to exercise manual or psychomotor skills to bring about

desired results. Much of the heavy work of the world—lifting, connecting, extruding, digging, pushing—falls into this category. These tasks can be subclassified in a myriad of ways, including the type of material being worked on, the type of activity involved, the intended product from the activity, and many others. In general, when performing these psychomotor tasks, groups strive to meet objective or absolute standards of excellence. Examples include laying a pipeline, loading a ship to meet a deadline, or achieving high productivity on an assembly line.

Underneath the eight types of tasks proposed by McGrath, there are four general processes. These are generating ideas and alternatives (task types 1 and 2), choosing among options (types 3 and 4), negotiating resolutions to conflict (types 5 and 6), and executing manual and psychomotor tasks (types 7 and 8). Note also that four task types are basically cognitive or conceptual in nature (types 2, 3, 4, and 5), while four are behavioral or action oriented (types 6, 7, 8, and 1). Similarly, four task types are essentially cooperative in nature (types 1, 2, 3, and 8), while four involve at least some degree of conflict (types 4, 5, 6, and 7).

---

to this effect—some maximum point beyond which additional members will have relatively little or no added impact. If a group already has twenty members, for example, the twenty-first member may not add much.

When a group's task is conjunctive, the predicted effects of size are different. Group performance can be no better than that of its weakest member. Again assuming that groups are composed at random, a large group is more likely to have very weak members than a small group. These weak members will be slower or less competent on their task than other group

members. Consequently, large groups should perform less well than small ones on conjunctive tasks.

To test these predictions, one study (Frank & Anderson, 1971) investigated the effects of group size on conjunctive and disjunctive tasks by comparing the performances of groups with two, three, five, and eight members. Each group worked on a series of tasks for 15 minutes; performance was measured by the number of tasks completed successfully. The tasks were intellective in nature, requiring group members to generate ideas or images (for example, "Write three points pro and con on the issue of legalizing gambling"). These tasks were made either disjunctive or conjunctive by manipulating the instructions. Some groups received disjunctive instructions, which stated that the group was to work on various tasks in sequence. As soon as *any* member had completed the task, the group could move on to the next task. Other groups received conjunctive instructions, which stated that the group was not permitted to move to the next task until *every* member had completed the task. Thus, a group in the disjunctive condition could proceed at the speed of its most competent member, whereas a group in the conjunctive condition could move no faster than its slowest member. The observed outcomes of this study support the predictions made earlier (Steiner, 1972). On disjunctive tasks, large groups performed better than small ones. On conjunctive tasks, large groups performed worse than small ones, although this difference was not particularly marked. These findings are shown in Figure 14-2.

**Additive Tasks**    What is the effect of a group's size when it faces an additive task, which depends on the sum of the performances of all members? On tasks of this type, there often occur what are termed process losses (Steiner, 1972; Hill, 1982). By definition, a **process loss** is a negative discrepancy between a group's potential performance and its actual performance; losses of this type occur from inefficiencies introduced during interaction among group members.

Steiner (1972) suggests that on additive tasks, the amount of process loss is related to group size. Specifically, he hypothesizes that total performance increases with group size but that performance per member may decline. Thus, as group size increases, there may emerge an increasingly larger discrepancy between the group's potential performance and its actual performance. Such a discrepancy is the result of process losses.

For example, consider the task of pulling a heavy object by means of a rope. In a classic

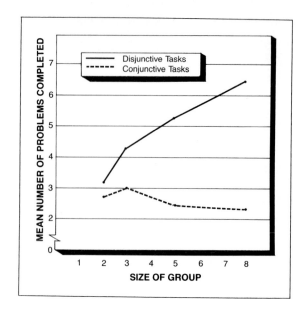

■ **FIGURE 14-2**

EFFECTS OF DISJUNCTIVE AND CONJUNCTIVE TASKS ON GROUP PRODUCTIVITY

On disjunctive tasks—tasks in which a group's performance depends on the performance of its best member —large groups were shown to perform better than small ones. However on conjunctive tasks—tasks in which a group's performance is restricted by the performance of its worst member—large groups performed slightly less well than small ones.

*Source: Adapted from Frank and Anderson, 1971.*

study published before World War I, a French agricultural engineer named Ringelmann investigated the effects of group size on the total pulling power exerted by the members of a group (Ringelmann, 1913; Moede, 1927; Kravitz & Martin, 1986). Ringelmann asked individuals to pull as hard as they could on a rope by themselves or with varying numbers of other persons (up to a group of eight). The rope was attached to heavy weights, and Ringelmann used a gauge to measure the amount of pull exerted. Results showed that although a group of eight men could pull harder than a smaller group or a single individual, the average contribution of members declined as the group's size increased (see Table 14-1). Working alone, each individual pulled at 100 percent of his capacity (that is, an average of 63 kilograms). But working as members of a two-person team, each individual pulled at only 93 percent of his capacity; and working as members of an eight-person team, each individual pulled at only 49 percent of his capacity. In other words, the larger the group, the more its actual performance fell short of its theoretic potential. Overall, these results indicate that on additive tasks, group performance increases with group size, but the efficiency per person declines.

This phenomenon—termed the *Ringelmann effect*—has been replicated in recent studies with the rope-pulling task (Ingham et al., 1974). The findings in these studies were similar to Ringelmann's, although the per-capita loss in the larger groups was not quite as dramatic as in the original study. In fact, reduction of individual performance in group settings has been found to occur on many tasks. For instance, it occurs with respect to tasks that require physical effort, such as clapping (Harkins, Latané, & Williams, 1980), pumping air (Kerr & Bruun, 1981), and cheering and shouting (Latané, Williams, & Harkins, 1979). It also occurs with respect to intellective tasks that require cognitive effort,

such as solving mazes (Jackson & Williams, 1985), reacting to proposals (Brickner, Harkins, & Ostrom, 1986), and brainstorming and vigilance (Harkins & Petty, 1982). In all these tasks, diminishing returns occurred: although larger groups produced more than smaller ones, the output per member in the larger groups was lower than that in the smaller groups.

There are several explanations for this discrepancy between a group's potential performance and its actual performance on additive tasks. The first is faulty coordination among group members. In the case of Ringelmann's rope-pull task, for instance, the possibility of poor coordination (pulling in different directions at different times) increases with group size. Although this explanation does account for some of the decline in per-capita performance, some studies have shown that the Ringelmann effect will occur even on tasks structured so that there is no possibility of loss from faulty coordination (Ingham et al., 1974; Harkins & Petty 1982).

The second explanation is what is termed **social loafing**—that is, workers tend to slack off and reduce their effort on additive tasks, which causes the group's output to fall short of its potential. This phenomenon is fairly widespread,

■ **TABLE 14-1**
ACTUAL AND POTENTIAL PULLING POWER OF GROUPS OF VARIOUS SIZES

| Number of Workers in Group | Actual Pull (kilograms) | Potential Pull (kilograms) | Ratio of Actual Pull to Potential Pull (%) |
|---|---|---|---|
| 1 | 63 | 63 | 100 |
| 2 | 118 | 126 | 93 |
| 3 | 160 | 189 | 85 |
| 4 | 194 | 252 | 77 |
| 5 | 221 | 315 | 70 |
| 6 | 238 | 378 | 63 |
| 7 | 248 | 441 | 56 |
| 8 | 248 | 504 | 49 |

*Sources: Reconstructed from Ringelmann, 1913, Moede, 1927, and Kravitz and Martin, 1986.*

and numerous studies have documented a falloff in effort by group members on additive tasks (Latané, Williams, & Harkins, 1979; Harkins & Szymanski, 1987). Research shows that one reason for the occurrence of social loafing on additive tasks like rope pulling is the inability of group members to evaluate the quality of their individual contributions. In fact, when the task is explicitly structured to make an evaluation of individual contributions possible, social loafing effects are minimized (Harkins & Jackson, 1985; Szymanski & Harkins, 1987). Other factors that minimize social loafing include high personal involvement in the task (Brickner, Harkins, & Ostrom, 1986) and high effort by other group members (Jackson & Harkins, 1985).

## Group Cohesiveness

Group cohesiveness refers to the degree to which the members of a group are attracted to that group and desire to remain in it. Although the reasons for this attraction may vary, a highly cohesive group will maintain a firm hold over its members' time, energy, and commitment (Lott & Lott, 1965; Stokes, 1983).

**Cohesiveness and Group Norms**  Is a highly cohesive group more productive than a less cohesive one? The relationship here is complex. Groups with high cohesiveness are not always more productive than those with low cohesiveness. Yet, along with other factors such as group norms and task structure, cohesiveness does have an impact on group performance.

If a group is highly cohesive and has norms that require high levels of performance, the group will be highly productive. On the other hand, if the group is highly cohesive but has norms requiring only low levels of performance, its members may spend more time socializing than producing. This hypothesis was tested by research that manipulated both group production standards and cohesiveness (Berkowitz, 1954). Results indicated that in highly cohesive groups with low production norms, performance was markedly suppressed. In general,

cohesiveness does not affect group performance directly. Instead it amplifies the effects of whatever production norms prevail within the group.

**Cohesiveness and Task Type**  Group cohesiveness interacts not only with group norms but also with task structure to affect performance (Nixon, 1976, 1977a). For instance, consider the performance of an athletic team. What we call "team sports" involves divisible tasks—a division of labor among team members. One team sport is football, which entails different squads (offensive team, defensive team, specialty team) and also requires specialization within those squads (on the offensive team there are linemen, wide receivers, a quarterback, running backs, and so on). Success in football requires individual skills as well as careful coordination of the individual performances. Thus, for team sports, group cohesiveness is positively related to successful team performance (Stogdill, 1963). This is true not only of football but also of volleyball (Vos & Brinkman, 1967), basketball (Nixon, 1977b), and baseball (Landers & Crum, 1971)—all sports that require coordination among team members. Harmonious personal relationships enhance effective teamwork, and the team's success tends to make its members friendlier toward each other.

Other sports, like singles tennis, involve unitary additive tasks. If a high school or college team is assembled to compete in tennis, its score is simply the sum of the individual members' scores. Sports organized this way (tennis, bowling, rifle shooting) do not entail a division of labor because the players perform more or less independently on identical tasks. Studies of rifle teams (McGrath, 1962) and bowling teams (Landers & Luschen, 1974) indicate a negative relationship between cohesiveness and successful team performance. In activities not requiring much coordination among members, teams that have many individual stars tend to do best. Although the stars may dislike each other, their rivalry can motivate higher levels of individual excellence. Thus, in sports involving unitary additive tasks, low levels of personal harmony (high rivalry) can actually lead to greater success in competition against other teams.

## Interdependence and Reward Structure

Task groups differ markedly in their reward structures—that is, who gets rewarded and why. In this section, we look at several common reward structures and their impact on group performance.

**Cooperation and Competition** Nicoletti and Strauss are moving their law offices across town, and they want the move completed quickly. They have hired four young men from Personpower Inc. to load their office equipment into a rented truck, drive it to the new building, and unload it there. If the men finish the job in one day or less, Nicoletti has promised them each a $25 bonus on top of the regular fee for the work. If they take more than one day, nobody gets a bonus.

In this case, a *cooperative reward structure* exists within the group of four workers (Deutsch, 1973). A cooperative structure has two main hallmarks. First, the members are functionally interdependent in the sense that they rely on one another's efforts to accomplish the overriding goal. None of the members can complete the task of moving the office equipment in one day by himself. Second, the members' interests are linked in the sense that each man can get a reward if, and only if, the others also get rewards. If one man earns a $25 bonus from Nicoletti, the other three will earn theirs.

Studies have shown that a cooperative reward structure induces open communication, mutual trust, and a readiness on the part of each person to consider the suggestions of his fellows. Group members interacting in the context of a cooperative structure are prone to trust one another and to coordinate their energies, provided they believe the group will eventually succeed in its task (Myers, 1962; Steiner, 1972).

In contrast, a *competitive reward structure* is one that pits group members against one another in an attempt to gain scarce rewards for themselves. A competitive structure involves a low level of functional interdependence and a high level of differential reward (rewards distributed unequally), with the best performers receiving the greatest rewards. For example, a competitive situation occurs in university classes when the professor decides to grade students on a curve. If only the top 10 percent of the class is allowed to receive As, successful actions by one student will obstruct goal attainment by others. Thus, classmates' interests are clearly opposing.

What type of reward structure—cooperative or competitive—brings about the best possible group performance? On the one hand, a competitive structure within a group can produce a breakdown of coordination and trust among members. We might, therefore, expect lower levels of group performance if group members are competing rather than cooperating. On the other hand, a competitive structure can heighten members' motivation and cause them to try harder in hopes of attaining superior rewards. This might lead to higher levels of performance under competition than under cooperation.

Numerous studies indicate that, in general, cooperative reward structures are superior to competitive reward structures in fostering group performance and achievement. This generalization holds especially true when (1) groups are small rather than large and (2) the group task is complex rather than simple (Johnson et al., 1981, 1982; Cotton & Cook, 1982; McGlynn, 1982).

**Functional Interdependence and Differential Rewarding** Some theorists have noted that competitive reward structures usually involve low functional interdependence among members and high differential rewarding, while cooperative reward structures usually involve high functional interdependence and low differential rewarding. Thus, they suggest that what really matters about reward structures of any type is the particular combination of underlying task interdependence and differential rewarding that is involved.

The effect of these variables on group performance was demonstrated vividly in a study (Miller & Hamblin, 1963) in which groups of three members were asked to work on a series of problems. Members were placed in separate booths connected by a system of electrical lights and switches. The group's task was to determine which one of thirteen numbers had been selected by the experimenter. Each member was privately informed of four numbers that had *not* been selected by the experimenter. These sets of four numbers were different for each member; thus, if the three members pooled their information (by the electrical signaling system), they would immediately see the correct answer. The task was completed only when all three members knew the answer. The group's performance was measured in terms of time: the faster it discovered the correct answer, the higher the score it received.

Two independent variables were manipulated in the study. One was the degree of task interdependence among group members. In the high-interdependence condition, guessing was discouraged by a substantial penalty. Group members were, therefore, forced to coordinate their efforts and share information in order to solve the problem. In the low-interdependence condition, there was no restriction on guessing, making coordination among members unnecessary. To solve the problem, individuals merely continued to guess until they hit the correct solution.

The other independent variable in the study was the degree of differential rewarding among group members. Rewards for group members were based both on their own efforts and on those of the group as a whole. Each group started with 90 points, and 1 point was subtracted for every second that elapsed before all three members had solved the problem. For example, if the group took 60 seconds to reach a solution, it scored 30 points. The group's points were

then distributed among its members. In one experimental condition (no differential rewarding), each member received one third of the group reward. In a second condition (medium differential rewarding), the member who solved the problem first received one half of the group's points, the member who finished second received one third, and the one who finished third received one sixth. In a third condition (high differential rewarding), the member who solved the problem first received two thirds of the group's points, the member who finished second received one third, and the one who finished third received none.

The results of this study are depicted in Figure 14-3. Under conditions of low interdependence (in which all three individuals could freely guess at the answer), differential rewarding did not significantly affect group performance. Under conditions of high interdependence (in which guessing was prohibited and members had to share information), higher levels of differential rewarding sharply reduced the group's level of performance.

To interpret these results, note that under conditions of high task interdependence and no differential rewarding, members realized that their best strategy for achieving higher rewards was to cooperate with each other. Thus, they shared information quickly and, as a result, had high performance scores. Under conditions of high task interdependence and high differential rewarding, however, members realized that they could achieve high rewards for themselves only by outperforming others in their group. This created a competitive atmosphere in which members tried to hinder one another. For example, a member might try to obtain information from others without yielding information in return, thereby blocking others' performance. This would engender a lack of coordination and inevitably result in poor overall group performance.

Under conditions of low task interdependence, in contrast, group members could not use a blocking strategy on one another (because

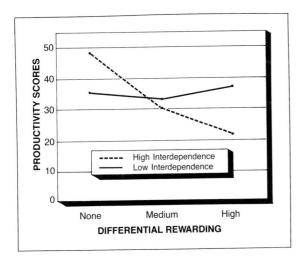

■ **FIGURE 14-3**

GROUP PRODUCTIVITY AS A FUNCTION OF
INTERDEPENDENCE AND DEGREE OF DIFFERENTIAL
REWARDING

Under low task interdependence, differential rewarding
(that is, unequal reward distribution) had little effect on
group performance. However under high task interdependence, a high level of differential rewarding induced
low group productivity as well as competition among
members, and a low level of differential rewarding
induced high productivity and cooperation.

*Source: Adapted from Miller and Hamblin, 1963.*

guessing was permitted). To increase their rewards, they were forced to raise their own performance independently. This led to moderate
group performance regardless of the degree of
differential rewarding. Related research shows
similar results (Rosenbaum et al., 1980).

## Communication Structure

We frequently hear about the frustrations of
working within a large organization like a military or industrial bureaucracy, where communication must go through prescribed channels.
Channels help to make sure that messages are
sent to the "right" person, but they also restrict
who can talk with whom. To some degree, this
holds true for smaller groups as well. Free
communication among members of a group is

also affected by the nature of a group's task, its
status patterns, and even the physical distance
between members.

Years ago a method was developed for investigating the impact of alternative communication
structures on a group's performance (Bavelas,
1950). The method involves placing the members of a group in separate cubicles. These
members are not permitted to talk to one another; they can communicate only by passing notes
through slots in the cubicle walls. When all of
these slots are opened simultaneously, each
member can communicate directly with every
other member. Closing some of the slots eliminates certain channels and restricts the flow of
information, so investigators can use this method
to study the effects of alternate communication
networks. The term **communication network**
refers to the pattern of communication opportunities within a group. Figure 14-4 displays a
variety of communication networks for a group
of five persons. While the *comcon* (or "*com*pletely
*con*nected") network allows each member to talk
freely with all the others, the other networks
restrict communication. For example, in the
*wheel* network, one person is at the hub, and all
communication must pass through him or her.
In a *chain* network, all information transmitted
from a person at one end must pass through a
number of others to reach the other end; people
located nearer the middle of the chain will,
therefore, be more central to the flow of communication than those located at the extremes.

Does the type of communication network
affect a group's performance? This question
arises partly from practical considerations. People want to know how to organize work groups,
committees, teams, and other task-oriented
groups in the most efficient way. In typical
research studies on this issue, each group is
assigned a task that requires communication
among members for a successful resolution. In
some cases these are simple intellective tasks,
such as identifying which of various symbols
(stars, triangles, circles, and so on) appear on

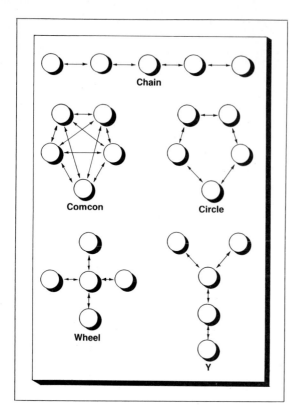

■ **FIGURE 14-4**

COMMUNICATION NETWORKS IN FIVE-PERSON
GROUPS

This figure shows different types of communication networks that can exist within five-person groups. Three of
the networks shown—the wheel, the chain, and the
Y—are centralized. That is, messages flowing from one
end of the network to the other must pass through a
central person (or hub). The other two networks—the
circle and the comcon—are decentralized, because
messages need not pass through a single hub.

*Source: Adapted from Leavitt, 1951.*

cards. In other cases these are more complex
tasks that require the members to transform
information, such as arranging words, constructing sentences, and performing arithmetic.
Whether the task is simple or complex, a group's
efficiency in problem solving is measured by the
time it takes to achieve a solution, by the number

of messages sent by members in order to achieve
a solution, or by the number of errors made
while attempting to solve a problem.

A group's problem-solving efficiency depends in part on its communication network,
because the network restricts how group members can behave when attempting to solve problems. An early study (Leavitt, 1951) compared
the relative efficiency of the wheel, chain, Y, and
circle networks in five-person groups. Results
indicate that members of groups having wheel,
chain, and Y networks adopt a centralized organization in which the person occupying the
central position (for example, the "hub" in the
wheel) is likely to emerge as highly influential.
Those occupying the peripheral positions are
unlikely to emerge as influential.

In contrast, members of groups having a
circle network tend to adopt a decentralized
organization in which all members actively communicate with and influence one another. Studies of comcon groups have also found that
members adopt a decentralized organization
(Shaw & Rothschild, 1956; Guetzkow & Dill,
1957). In general, a group's communication
network determines whether its organization is
centralized or decentralized.

This point is important, because centrality
of organization in turn affects a group's
problem-solving efficiency. This was demonstrated by a compilation of findings from numerous
experiments on communication networks (Shaw,
1964). Each study was classified as involving
either centralized networks (wheel, chain, Y) or
decentralized networks (circle, comcon) and as
entailing either simple tasks (collating information) or complex tasks (carrying out operations
on the information). The summary of results is
shown in Table 14-2. Notice that for simple
problems, centralized communication networks
led to superior performance. Groups with centralized networks solved simple problems faster,
sent fewer messages, and made fewer errors than
groups with decentralized networks. For complex problems, however, the findings were re-

■ **TABLE 14-2**

COMPARISONS OF CENTRALIZED (WHEEL, CHAIN, Y) AND DECENTRALIZED (CIRCLE, COMCON) NETWORKS' EFFICIENCY AS A FUNCTION OF TASK COMPLEXITY

| | Simple Problems[a] | Complex Problems[b] | Total |
|---|---|---|---|
| Time | | | |
| Centralized faster | 14 | 0 | 14 |
| Decentralized faster | 4 | 18 | 22 |
| Messages | | | |
| Centralized sent more | 0 | 1 | 1 |
| Decentralized sent more | 18 | 17 | 35 |
| Errors | | | |
| Centralized made more | 0 | 6 | 6 |
| Decentralized made more | 9 | 1 | 10 |
| No difference | 1 | 3 | 4 |
| Satisfaction | | | |
| Centralized higher | 1 | 1 | 2 |
| Decentralized higher | 7 | 10 | 17 |

[a]Simple problems: symbol-, letter-, number-, and color-identification tasks.
[b]Complex problems: arithmetic, word arrangement, sentence construction, and discussion problems.

*Source: Adapted from Shaw, 1964.*

versed: groups with decentralized networks were more efficient. Even though they sent more messages, they were faster and tended to make fewer errors than groups with centralized networks.

Numerous studies support the generalization that centralized networks are more efficient for simple tasks, whereas decentralized networks are more efficient for complex ones (Lawson, 1964; Morrissette, 1966). One explanation for this pattern lies in the concept of **saturation**— the degree of communication overload experienced by group members occupying the central positions within communication networks (Gilchrist, Shaw, & Walker, 1954; Shaw, 1978). The level of saturation depends not only on the communication requirements imposed by a network but also on factors such as the data manipulation and decisions required by the group's

task. In general, the greater the level of saturation, the less efficient a group is at solving problems.

As an example, consider the wheel network (see Figure 14-4). In this structure, the hub position is central. When the group is working on a simple task, communication requirements are not very demanding. The hub position does not become saturated, and the person occupying this position can work quickly and efficiently. When the group confronts a more complex problem, however, the communication requirements placed on the hub are substantial, and the person occupying that position may become saturated and overburdened. When this happens, the group's problem-solving efficiency will be low. Decentralized networks do not have this problem, because no position in the network is subject to extreme saturation.

## Box 14-2

### COMMUNICATION NETWORKS IN AN ORGANIZATION

Communication networks have a significant impact not only in laboratory groups but also in organizational settings. In one case study (Mears, 1974), a group of division representatives within an aerospace firm tried out several different communication networks. This group consisted of an administrative officer and representatives from several divisions, such as manufacturing, quality control, procurement, contracts, and engineering. At first this group was organized in a *comcon network*. This permitted virtually unrestricted communication among all members. Satisfaction was very high, but only a modest amount of work was accomplished because members wasted a great deal of time in useless discussion and debate.

Top management grew dissatisfied with the group's performance and restructured it in a *wheel network*, with the administrative officer in a position of authority at the hub. This restriction in communication reduced worker motivation and satisfaction. It also lowered productivity

because the hub position became saturated, which caused many errors in relaying complex information.

Finally, the group was again reorganized in a *modified comcon network*, which permitted each member to communicate only with persons who were directly involved in the task at hand. This reduced the communication overload on the administrative officer and also protected the time of members who did not need to be consulted. As a result, the group experienced high levels of satisfaction and performance.

As this case study shows, communication networks are not theoretical abstractions found only in research laboratories. They come into existence—either through natural evolution or by intentional design—in real-world work groups and organizations. And just as with laboratory groups, they can significantly affect the productivity and satisfaction of workers in industry.

## ■ Leadership Effectiveness

As we have seen, a group's performance depends on several factors, including the nature of the task, communication structure, and group cohesiveness. Still another factor affecting performance is the type of leadership exercised within the group.

By definition, **leadership** is the process whereby one group member (the leader) influences and coordinates the behavior of other members in pursuit of group goals (Yukl, 1981). In return for support from the others, a leader provides guidance, specialized skills, and environmental contacts that help the group attain its goals. Some of the factors that determine whether or not a person will become a leader were discussed in Chapter 13. Here, we will first consider the activities of leaders—what leaders actually do—and then look at the attributes and behaviors that determine whether or not a leader will be effective in helping a group achieve its goals.

### Activities of Leaders

A person serving as a leader fulfills certain functions necessary for successful group performance. These functions include planning, organizing, and controlling the activity of group members (Stogdill, 1974). In formal groups, where roles are organized in a status hierarchy, leadership functions are typically fulfilled by high-status members. These members have both the right and the responsibility to provide leadership for the group. In informal groups, these functions may be fulfilled by one or several persons who emerge during interaction as task leaders and social-emotional leaders.

In essence, leadership involves a tacit exchange between the leader and the other group members. By fulfilling the planning, organizing, and controlling functions in a group, the leader helps move the group toward the attainment of its goals. In return, the leader receives support for continued control, as well as special rewards and privileges. This perspective, which characterizes leadership as an exchange between the leader and group members, is termed the **transactional view of leadership** (Hollander & Julian, 1970; Homans, 1974; Hollander, 1985).

Exactly what do leaders do? That is, what behaviors do they engage in when providing leadership for a group? The behavior of leaders has been investigated in many settings, including military units, industrial work groups, training groups, and laboratory groups (Wofford, 1970; Schriesheim & Kerr, 1974). As might be expected, the behavior of leaders depends on various situational factors, such as the size of the group, the group's task, and the communication structure. Nevertheless, certain behaviors are universal to leadership, including planning, organizing, and controlling. Specifically, leaders usually do some or all of the following: (1) formulate a clear conception of the group's goals and objectives, and communicate this to group members; (2) develop specific strategies for the attainment of group goals; (3) specify role assignments and standards of performance for members; (4) establish and maintain channels of communication among members; (5) recruit new members (if needed) and train members in crucial skills; (6) interact with members personally to maintain good relations; (7) influence the task activities of group members by means of persuasion, rewards, and punishments; (8) monitor the group's progress toward its goals and take corrective

steps if the group is off track; (9) resolve conflict among members to reduce tension and maintain harmony; and (10) serve as a representative of the group to outside agencies and organizations.

In any group, some of these leadership behaviors will be more important than others. To perform successfully, leaders must have good technical and interpersonal skills as well as decision-making and problem-solving abilities.

## Leadership Style

Depending on the situation and on their personalities, leaders use different techniques to plan, organize, and control group activities. For example, some leaders adopt an authoritarian style; they make most of the decisions and issue orders to members. Other leaders adopt a democratic style; they allow the group as a whole to make important decisions and act primarily to coordinate activities. Observed differences of this type immediately raise some questions: Does leadership style have an impact on group performance? If so, which style—authoritarian or democratic—produces better group performance?

One of the earliest studies of leadership style was conducted years ago by Lewin, Lippitt, and White (1939). In this study and later work (Lippitt & White, 1952), groups of 10-year-old boys met after school for three 6-week periods. The groups performed various activities, such as painting and lettering signs, building box furniture out of wood, and carving soap. These activities were directed by adult leaders using specific leadership styles. Three leadership styles —authoritarian, democratic, and laissez-faire— were investigated in the study. The *authoritarian leader* exercised a great deal of control. He determined all policies, dictated techniques and work partners, and administered praise and criticism without explaining his reasons. Moreover, he intentionally kept the group uninformed about many matters and remained aloof from group participation. The *democratic leader* behaved quite differently. He encouraged group

A drill instructor uses the strict authoritarian leadership style to command his troops. Would the instructor's leadership

members to discuss policy issues, and he collaborated with them in making decisions. He suggested alternatives but did not give orders. His group members chose their own tasks and working companions. He was friendly and tried to be "one of the group" as much as possible. He also tried to be objective and to explain his reasons when praising and criticizing. A third type, the *laissez-faire leader*, adopted a "hands off" approach. He was friendly but extremely passive, keeping his participation to an absolute minimum. The laissez-faire leader supplied information when asked, but otherwise he left the group pretty much to itself. He made no attempt to evaluate or regulate the group's activities.

While the groups were at work, observers kept extensive records of members' behavior. The results show that leadership style had a marked influence not only on group performance but also on the interpersonal relationships of members. The lowest level of group performance was associated with the laissez-faire leadership style. The groups with a laissez-faire leader were not very goal oriented, and the members engaged in a lot of horseplay. As a result, the work was of relatively poor quality, although the boys were friendly with the laissez-faire leader.

Groups with an authoritarian leader displayed the highest level of performance in terms

style be effective in another setting, such as conducting a symphony orchestra?

of quantity of output. These groups responded submissively to their leader's directives, spent more time at work, and, as a result, produced more than the other groups. However, their members showed a high level of negativism, anger toward the leader, and destruction of property. Despite their relatively high output, the group members did not internalize the motivation to work. When the leader was called away from the room, for example, they stopped producing.

Groups with a democratic leader proved superior in some respects to those with authoritarian or laissez-faire leaders. Not only were they highly productive, they also enjoyed good inter-personal relations. Although the quantity of work completed under the democratic leader was slightly lower than that completed under the authoritarian leader, the quality and originality of the work were superior. Boys in the democratically led groups manifested less submissive and dependent behavior than in the authoritarian groups. Work motivation was also high; when the leader left the room in a democratic group, the boys continued to work. The groups were cohesive, and the boys engaged in friendly playfulness, offered mutual praise, and readily shared group property.

Many studies since this one have investigated the effects of authoritarian versus democratic leadership style on group performance. In general, findings show that members of groups under democratic leadership express more satisfaction with their group and its leaders than members of groups under authoritarian leadership. This occurs in both laboratory experiments (Shaw, 1955) and field studies of work groups within larger organizations (Likert, 1961). Although a few studies have shown that members prefer authoritarian to democratic leadership, these findings pertain to military or bureaucratic settings, where people have no expectation of sharing in decision making (Gibb, 1969).

The effects of leadership style on group performance is less clear. Neither authoritarian nor democratic leadership is reliably associated with higher levels of group performance (Anderson, 1963). Although some studies have found that a democratic style promotes higher levels of performance and effectiveness (Kahn & Katz, 1960; McGregor, 1960), other studies show the opposite effect—with authoritarian groups having higher performance levels (Shaw, 1955). Authoritarian leadership tends to be more effective, for example, among sailors on shipboard. Clearly, democratic leadership would be inefficient and potentially disastrous in a crisis at sea. Thus, the relationship between leadership style and group performance is complex.

In recent years, research on leadership effectiveness has moved in several new directions. First, leadership style has increasingly been conceptualized in terms other than democratic versus authoritarian. For instance, a leader's style might be treated as task oriented versus relationship oriented or as supportive, achievement oriented, or participative (House, 1971; Vroom & Yetton, 1973). Second, increasing emphasis has been placed on the fact that a given style of leadership will be effective in some group settings but not in others. That is, leadership style interacts with other factors in affecting group performance. We turn now to a model of leadership effectiveness that addresses both of these considerations.

## Contingency Model of Leadership Effectiveness

A leader's effectiveness in directing a group depends not only on style but also on characteristics of the situation. Someone with a specific leadership style might be effective in one situation but not in another. This basic notion underlies the **contingency model of leadership effectiveness** (Fiedler, 1978a, 1978b, 1981). The contingency model pertains primarily to groups working on divisible tasks that require coordination among members for successful performance. This model is highly general and has been applied to a variety of groups in military, educational, and industrial settings.

**The LPC Score** There are four independent variables in the contingency model. One of these is leadership style; the other three are properties of the situation in which a leader performs. The contingency model characterizes leadership style as being either relationship oriented or task oriented. To determine leadership style, a leader is first asked to recall all the people with whom he or she has ever worked in a group setting and to select the one who was most difficult to get along with. This person is designated the *least-preferred co-worker* (LPC). Next, the leader is asked to rate this person on dimensions such as pleasant-unpleasant, helpful-frustrating, cooperative-uncooperative, and efficient-inefficient. Some leaders give very low, negative ratings to their least-preferred co-worker; these are called *low-LPC leaders*. Other leaders find positive qualities even in their least-preferred co-worker; these are *high-LPC leaders*. It has been found that low-LPC leaders are oriented toward achieving successful task performance, whereas high-LPC leaders are primarily concerned with establishing congenial interpersonal relations with others (Fiedler, 1978a; Rice et al., 1982).

**Situational Factors** As already noted, a leader's effectiveness depends not only on leadership style but on various characteristics of the situation. According to the contingency model, three characteristics of the situation are crucial. In order of importance, these are (1) the leader's personal relations with other group members (good or poor), (2) the degree of structure in the group's task (structured or unstructured), and (3) the leader's position of power in the group (strong or weak).

A leader's *personal relations* with other members is assumed to be the most important factor determining the leader's influence in the group. A leader whom the other members trust, like, and respect is in a more favorable situation than one with poor rapport, and he or she will usually have the support of group members. The second factor, *task structure,* refers to how clearly defined the task requirements of a group are. Is their goal clearly spelled out? Is there a single path by which this goal can be achieved? Is there only one correct solution or decision? Is there some method of verifying the accuracy of the solution or decision? The more these things are characteristic of the task, then the greater is the task's structure and the more favorable is the situation for the leader. The third factor, *position power,* has to do with a leader's formal authority over group members and the degree to which he or she can buttress directives with rewards and punishments. A leader who wields more power is in a more favorable situation.

For analytic convenience, each of these factors is considered to be dichotomous: leader-member relations are either good or poor, the group's task is either structured or unstructured, and the leader's power is either strong or weak. Taken together, these three dichotomies produce eight possible situations. These situations are shown in Figure 14-5 and are listed according to their overall favorableness for a leader.

From a leader's standpoint, a situation is favorable if it enables him or her to exercise a great deal of influence. Thus, a situation involving good leader-member relations, a highly structured task, and strong leader power is favor-

able. At the other extreme, a situation involving poor leader-member relations, an unstructured task, and weak leader power is very unfavorable for the leader.

**Predictions** According to the contingency model, task-oriented (low-LPC) leaders are effective in situations that are either highly favorable or highly unfavorable, but they are ineffective in other situations. In contrast, relationship-oriented (high-LPC) leaders are effective in situations that are moderately favorable, but they are ineffective in the other situations. These predictions are shown in Figure 14-5.

When conditions are extremely unfavorable, a group needs strong task leadership, and it is the task-oriented (low-LPC) leader who best provides this. A low-LPC leader is willing to overlook interpersonal conflicts in order to concentrate on the group's task. A high-LPC leader attempts to smooth over interpersonal problems rather than work on the group's fundamental task.

However, when conditions for leadership are intermediate in favorableness, the high-LPC leader is the one who best provides what the group needs. Under these conditions, some situational factors are positive and others negative. Interpersonal problems may come to the surface and the relationship-oriented (high-LPC) leader will be more effective than the task-oriented (low-LPC) leader.

Finally, when conditions are highly favorable, the low-LPC leader again emerges as most effective. Under these circumstances, low-LPC leaders feel they can relax, because goal attainment is not very problematic. They turn their attention to interpersonal relations with their co-workers, because they know that the task will get done. In other words, low-LPC leaders undergo a change in behavior that causes them to be highly effective when conditions are favorable. Under these same conditions, however, relationship-oriented (high-LPC) leaders are relatively ineffective. They start looking for things

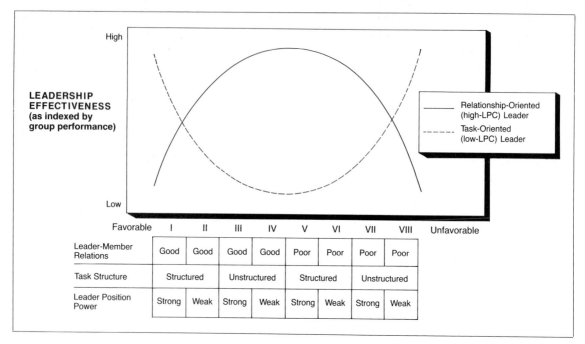

| | | Favorable | I | II | III | IV | V | VI | VII | VIII | Unfavorable |
|---|---|---|---|---|---|---|---|---|---|---|---|
| Leader-Member Relations | | | Good | Good | Good | Good | Poor | Poor | Poor | Poor | |
| Task Structure | | | Structured | | Unstructured | | Structured | | Unstructured | | |
| Leader Position Power | | | Strong | Weak | Strong | Weak | Strong | Weak | Strong | Weak | |

■ **FIGURE 14-5**

LEADERSHIP EFFECTIVENESS AS A FUNCTION OF LEADERSHIP STYLE AND SITUATIONAL FACTORS

The contingency model predicts that group performance depends on both leadership style and situational factors. According to this model, task-oriented (low-LPC) leaders are most effective in situations that are either highly favorable (that is, involving good leader-member relations, a structured task, and strong leader power) or highly unfavorable (that is, involving poor leader-member relations, an unstructured task, and weak leader power). In contrast, relationship-oriented (high-LPC) leaders are most effective in situations that are of medium favorableness.

*Source: Adapted from Fiedler, 1978.*

to do and become bossy and unconcerned with the feelings and opinions of their co-workers, which diminishes their effectiveness (Fiedler, 1978a; Larson & Rowland, 1973).

**Tests of the Contingency Model** Although there are some exceptions, research generally supports the contingency model's prediction (Strube & Garcia, 1981; Chemers, 1983; Peters, Hartke, & Pohlmann, 1985). One study (Chemers & Skryzpek, 1972; Shiflett, 1973) used West Point cadets as subjects and manipulated all three dimensions of situational favorableness. Groups of four men were assembled on the basis

of previous information regarding the relations among members. In half of the groups the leader-member relations were good, whereas in the other half they were poor. Each group performed one structured task (converting blueprints from metric units to inches) and one unstructured task (discussing an issue and making policy recommendations). In half of the groups the leader had strong power—that is, members believed that the leader would assess their performance and this would affect their standing in the Academy. In the other half the leader had only weak power—members believed he would have little effect on their standing. As predicted by the contingency model, the task-oriented (low-LPC) leaders proved to be more effective when the situation was either very unfavorable or very favorable. In contrast, the relationship-oriented (high-LPC) leaders were more effective when the situation was of intermediate favorableness. Similar findings have emerged in studies of college and elementary school students (Hardy, Sack, & Harpine, 1973; Hardy, 1975) and naval personnel (Fiedler, 1966).

# ■ Group Decision Making

Collective decisions are important outcomes resulting from interaction among group members. Examples of decisions of this type are commonplace. Family members decide where to travel together on their vacation; managers of a business firm decide which new products to develop; experts on a review committee decide which research proposals to support and which to turn down; a group of basketball coaches decides what games to schedule for the upcoming season. In every instance, the end product resulting from interaction among group members is a decision—a choice among alternative courses of action.

Deliberations by a group must follow certain steps if a good decision is to result. Information must be gathered, alternative courses of action must be formulated in detail, the potential consequences of each option must be explored, and the relative value of each option must be weighed. Although these steps cannot, by themselves, guarantee that the decisions will be successful, careful deliberations result frequently in a good outcome.

The point to note, however, is that group decision making entails certain hazards. The decision-making process can go awry for a variety of reasons. Group members may differ in the value they place on various options—a circumstance that can lead to a compromise that satisfies nobody. Then, too, conformity pressures

within the group may impel members to bypass or short-circuit the information-collection or decision-making processes. If this happens, group deliberations can produce poor or unrealistic decisions. In this section, we will look at various processes involved in group decision making.

## Groupthink

Aberrations in decision making can plague any group, even those at the highest levels of business and government. The history of U.S. foreign policy provides numerous examples. The decisions by the U.S. to invade the Cuban Bay of Pigs, to cross the 38th parallel in the Korean War, and to escalate the Vietnam War were all made by committees. The Bay of Pigs invasion, for example, was planned by a small group of top government officials immediately after President Kennedy took office in 1961. The group included the nation's best and brightest: McGeorge Bundy, Dean Rusk, Robert McNamara, Douglas Dillon, Robert Kennedy, Arthur Schlesinger, Jr., and President Kennedy himself, together with representatives of the Pentagon and the Central Intelligence Agency. The decision was made to invade Cuba in April 1961 with a small band of 1,400 Cuban exiles. The invasion was to be staged at the Bay of Pigs and assisted covertly by the U.S. Navy and Air Force and the CIA. As it turned out, the invasion was poorly conceived. The material and reserve ammunition on which the exiles were depending never arrived because Castro's air force sank the supply ships. The exiles were promptly surrounded by 20,000 well-equipped Cuban soldiers, and within three days virtually all were captured or killed. The U.S. suffered a humiliating defeat in the eyes of the world, and the Castro government became more strongly entrenched in the Caribbean.

How could it happen? How could a group of such capable and experienced men make a decision that turned out so poorly? One post-hoc analysis offered by Irving Janis (1982) suggests that high levels of group cohesiveness and conformity pressures may have produced a defective decision. In his detailed study, Janis used the term **groupthink** to refer to a mode of thinking within a group whereby pressures for unanimity overwhelm the members' motivation to appraise realistically the alternative courses of action. In their effort to preserve group cohesiveness and consensus, the members neglect critical thought, the inspection of alternatives, and the weighing of pros and cons—all the requirements of carefully reasoned decisions. A typical result of groupthink is poor, ill-conceived decisions.

Groupthink is more likely to occur under some conditions than others. Four conditions that increase the probability of groupthink are (1) a crisis situation, (2) a highly cohesive group, (3) the insulation of group members from the judgments and criticisms of qualified outsiders, and (4) a leader who actively promotes his or her own favored solution to the problem facing the group. According to Janis, the simultaneous occurrence of these conditions increases the probability that groupthink will occur.

How can the presence of groupthink be detected? Janis (1982) suggests that there are a number of symptoms, including:

1. An illusion of invulnerability shared by most or all of the members that produces excessive optimism and encourages taking extreme risks.

2. An unquestioned belief in the group's inherent morality, inclining the members to ignore the ethical consequences of their decisions.

3. Collective efforts to rationalize in order to discount warnings that might lead the members to reconsider their assumptions.

4. In the political sphere, a stereotyped view of enemy leaders as too evil to warrant genuine attempts to negotiate, or as too weak and stupid to counter whatever attempts are made to defeat their purposes.

5. Self-censorship of deviation from the apparent group consensus, with each member

inclined to minimize the importance of doubts and counterarguments.

6. A shared illusion of unanimity concerning judgments conforming to the majority view.

7. Direct pressure on any member who expresses strong arguments against any of the group's stereotypes, illusions, or commitments, making clear that dissent is contrary to what is expected of all group members.

8. The emergence of self-appointed "mindguards"—members who protect the group from adverse information that might shatter their shared complacency about the effectiveness and morality of their decisions.

Some of these symptoms were present during the decision-making processes for the Bay of Pigs invasion. For example, there was an assumed air of consensus that caused members of the group to ignore some glaring defects in their plan. Although several of Kennedy's senior advisors had strong doubts about the planning, the group atmosphere inhibited them from voicing criticism. Several members emerged as "mindguards" within the group; they suppressed opposing views by arguing that the decision to invade had already been made and that everyone should help the president instead of distracting him with dissension. Open inquiry and clearheaded exploration were discouraged. Even the contingency planning was unrealistic. For instance, if the exiles did not succeed in their prime military objective at the Bay of Pigs, they were supposed to join the anti-Castro guerrillas known to be operating in the Escambray Mountains. Apparently no one was concerned that 80 miles of impassible swamp and jungle stood between the mountains and the invasion site.

There are several ways to prevent groupthink from occurring, even in highly cohesive groups (Janis, 1982). Basically these methods increase the probability that a group will obtain all the information relevant to a decision and evaluate that information with great care. First, a group's leader should encourage dissent and call on each member to express any objections and doubts. Second, a leader should be impartial and not announce a preference for any plan at the outset. There is some research evidence that a closeminded leadership style fosters groupthink (Flowers, 1977). By describing a problem, rather than recommending a solution, a leader can foster an atmosphere of open inquiry and impartial exploration. Third, a group should establish several independent subgroups to work on the same problem, each carrying out its deliberation independently. This will prevent the premature development of consensus. Finally, after a tentative consensus has been reached, a group should hold a "second chance" meeting at which each member can express any remaining doubts before a firm decision is taken. The net result of the steps will be a better, more realistic decision.

## Polarization in Decisions Involving Risks

Groupthink is not the only anomaly that can occur in group decision making. Interaction in groups may also cause individuals to favor courses of action that are riskier than what they would choose if they made the decision alone. This was demonstrated in an early study (Stoner, 1962) in which subjects responded individually to a series of twelve problems called *choice dilemmas*. In each problem, the subjects were asked to advise a fictional character how much risk he or she should assume. The following item illustrates this task:

> Mr. A, an electrical engineer who is married and has one child, has been working for a large electronics corporation since graduating from college five years ago. He is assured of a lifetime job with a modest, although adequate, salary and liberal pension benefits upon retirement. On the other hand, it is very unlikely that his salary will increase much before he retires. While attending a convention, Mr. A is offered a job with a small, newly founded company which has a highly uncertain future. The new job would pay more to start and would offer the possibility of a share in the ownership if the company survived the competition of the larger firms.

> Imagine that you are advising Mr. A. Listed below are several probabilities or odds of the new company proving financially sound. Please check the lowest probability that you would consider acceptable to make it worthwhile for Mr. A to take the new job.

> _____ The chances are 1 in 10 that the company will prove financially sound.

> _____ The chances are 3 in 10 that the company will prove financially sound.

> _____ The chances are 5 in 10 that the company will prove financially sound.

> _____ The chances are 7 in 10 that the company will prove financially sound.

> _____ The chances are 9 in 10 that the company will prove financially sound.

> _____ Place a check here if you think Mr. A should not take the new job no matter what the probabilities. (Kogan & Wallach, 1964)

After individually offering their advice on this and other items, the participants assembled in groups of six and discussed each item until they reached a unanimous decision. Each subject was then asked to review each item and once again make an individual decision. The basic finding was that the decisions made after the group discussion were, on the average, riskier than the decisions made by individual members prior to the discussion. The term **risky-shift** is used to designate this shift toward risk following a group discussion. This phenomenon has been observed in numerous studies (Dion, Baron, & Miller, 1970; Cartwright, 1971).

Other studies, however, indicate that although a group discussion frequently produces a shift in individual opinions, this shift is not necessarily toward greater risk. In some circumstances, members shift toward a more conservative position. The term **cautious-shift** is used to designate this shift away from risk following a group discussion (Fraser, Gouge, & Billig, 1971; Knox & Safford, 1976).

A group of investors plans for a new building. In circumstances like this, group polarization may lead to ill-considered—and costly—decisions.

Risky-shift and cautious-shift are part of a more general phenomenon called group polarization. By definition, **group polarization** occurs when group members shift their opinions toward a position that is similar to, but more extreme than, their initial pregroup responses. For instance, if members initially advocated a moderately risky position prior to a group discussion, polarization would occur if they shifted toward greater risk following the discussion. On the other hand, if they initially advocated a moderately cautious position, polarization would

occur if they shifted in the direction of even greater caution after the group discussion.

Group polarization is a widespread phenomenon. It occurs not only on the choice-dilemma questionnaire but also in other contexts. It has been observed with respect to political attitudes (Paicheler & Bouchet, 1973), jury decisions (Myers & Kaplan, 1976), satisfaction with new consumer products (Johnson & Andrews, 1971), judgments of physical dimensions (Vidmar, 1974), ethical decisions (Horne & Long, 1972), perceptions of other persons

(Myers, 1975), and interpersonal bargaining and negotiating (Lamm & Sauer, 1974).

Why does group polarization occur? That is, what causes group members to shift their risk-taking responses toward an extreme position? Two basic explanations have been proposed (Myers & Lamm, 1976; Isenberg, 1986). According to one theory, group polarization results from a process of *social comparison* (Jellison & Riskind, 1970; Goethals & Zanna, 1979). This theory suggests that people often value opinions that are more extreme than those they personally advocate. They fail to adopt these ideal (extreme) positions as their own because they fear being labeled extremist or deviant. However, during a group discussion in which members compare their positions, these persons may discover that other members hold opinions closer to their ideal position than they had realized. This motivates the moderate members to adopt more extreme positions. The overall result is a polarization of opinions.

Although controversial, the social comparison theory has been supported by various studies. The major source of support stems from demonstrations that mere exposure to simple information about other group members' positions by itself can produce polarization effects (Blascovich, Ginsburg, & Veach, 1975; Baron & Roper, 1976; Myers et al., 1980).

A second theory explains group polarization as resulting from *persuasive argumentation* (Burnstein & Vinokur, 1973; Burnstein, 1982). According to this view, group polarization occurs whenever the preponderance of compelling arguments advanced during a group discussion favors a position more extreme than that held initially by the average member. Discussion within a group serves to persuade members who, because they had been unaware of the arguments, initially chose relatively moderate positions. After discussion, the moderate members shift their opinions in the direction of the most compelling, and relatively extreme, arguments. This produces group polarization.

Some research supports the persuasive argumentation theory. It has been shown, for instance, that the greater the proportion of arguments favoring a particular point of view, the greater is the shift of opinion in its direction (Ebbesen & Bowers, 1974; Madsen, 1978). Thus, subjects who are exposed to mostly risky arguments become more risk taking, whereas those who hear mostly conservative arguments become more cautious.

Overall, then, there is support for both the social comparison and the persuasive argumentation theories. Both of these processes occur in combination to produce polarization, although the effects of persuasive argumentation tend to be larger (Isenberg, 1986).

## ■ Summary

This chapter has discussed group performance and decision making. Group performance refers to the output or end state resulting from the activity of group members.

**Group Tasks** (1) Group tasks can be categorized as unitary (members perform similar activities) or divisible (members perform complementary activities). Unitary tasks can be conjunctive, disjunctive, or additive. Performance on conjunctive tasks depends on the group's slowest or weakest member; performance on disjunctive tasks depends on the fastest or strongest member; performance on additive tasks depends on the sum (or average) of members' performances.

**Factors Affecting Group Performance** Several factors determine how well a group will perform. (1) Group size affects performance. On disjunctive tasks, group performance increases directly with group size; on conjunctive tasks, it decreases with group size. On additive tasks, the Ringelmann effect occurs: Total performance increases directly with group size, but performance per member decreases. (2) Highly cohesive groups will be more productive than less cohesive groups, but only when they have norms favoring high performance. (3) The group's reward structure affects performance. Under low task interdependence, differential rewarding has little effect on group performance. Under high task interdependence, a high level of differential rewarding induces low group performance, whereas a low level of differential rewarding induces high group performance. (4) Centralized communication networks in groups lead to faster, more accurate performance than decentralized ones on simple problems. Decentralized networks lead to a superior performance on complex problems because they prevent excessive saturation.

**Leadership Effectiveness**  Leadership is an important factor affecting group performance and decision making. (1) The transactional view of leadership treats leadership as an exchange between a leader and group members. In virtually all groups, the activities of leaders include planning, organizing, and controlling. (2) Leadership style affects group performance. In general, groups led by authoritarian leaders produce large quantities of moderate-quality output, whereas those led by democratic leaders produce moderate-to-large quantities of high-quality output. (3) The contingency model of leadership effectiveness maintains that group performance is a function not only of leadership style but also of the situation in which a leader performs. According to this model, task-oriented (low-LPC) leaders are most effective both in situations that are highly favorable (that is, involving good leader-member relations, a structured task, and strong leader power) and in situations that are highly unfavorable (that is, involving poor leader-member relations, an unstructured task, and weak leader power). In contrast, relationship-oriented (high-LPC) leaders are most effective in situations that are moderately favorable. A number of studies support the contingency model.

**Group Decision Making**  Although many decisions made by groups are good ones, the process of group decision making entails potential hazards that can lead to poor or inferior choices. (1) One factor affecting decisions is groupthink—a mode of thinking that occurs when pressures for unanimity overwhelm members' motivation to realistically appraise alternative actions. Groupthink is most likely to occur in groups that are highly cohesive, facing a crisis situation, and insulated from outside criticism. Groupthink can be prevented or reduced if group leaders not only strive to obtain all information relevant to a decision but also encourage an atmosphere of impartial exploration of alternatives. (2) Group members often shift their opinions toward a more extreme position following a group discussion, a phenomenon termed group polarization. Polarization can have significant effects if it occurs for decisions involving some risk, because it will cause the group to shift its choice toward higher or lower levels of risk. Underlying group polarization are two distinct processes: social comparison and persuasive argumentation.

## Key Terms

# 15

chapter

# Intergroup Conflict

# ■ Introduction

Kanawha County in West Virginia contains the state capital of Charleston as well as a number of smaller communities. The county school board has five members, elected at large. Some years ago, the school board became embroiled in a controversy regarding the selection of 325 language-arts textbooks (English, composition, journalism, and speech). The conflict was instigated by a member of the school board, the articulate wife of a fundamentalist minister. She opposed the new books on the grounds that they were excessively liberal in viewpoint. At the April 11 meeting of the school board, she objected to the method of textbook selection. At the June 2 meeting, she again objected to the new books and observed that there was little in the texts to support a traditional, fundamentalist conception of God, the Bible, and religion. On June 23, she spoke in opposition to the books to the congregation of a local Baptist church. All these events were covered by the local news media.

On June 27, over 1,000 textbook protestors appeared at the regularly scheduled school board meeting. Nevertheless, after hours of testimony, the board voted formally to adopt the disputed books.

During the months of July and August, the textbook protestors organized their ranks and developed a strategy. There were several distinct protest groups within the movement. One of these, the Concerned Citizens of Kanawha County, was a large coalition of church congregations. A second group was the Businessmen and Professional People's Alliance for Better Textbooks, a middle-class group composed mainly of businessmen, teachers, and other professionals. A third protest group was the Christian American Parents. These groups held marches and rallies, circulated petitions, appealed to elected officials, and planned a boycott of the school system for September.

Opposing these groups were the Kanawha County School Board and the Citizens for Quality Education. Further support for the textbooks came from such liberal organizations as the American Civil Liberties Union and the National Association for the Advancement of Colored People as well as from teachers and school administrators.

On September 3, the new school term started for some 45,000 students. Protesting parents withheld their children from the schools (10,000 by some estimates) and prevented school and city buses from operating. On September 12, the school board closed the schools for a three-day cooling-off period. The controversial textbooks were removed from the classrooms pending review by a special citizen's committee.

During this period, some violence broke out. Random gunfire and sniping were reported. Vandalism of school property was commonplace. The protestors demanded that the board members and the superintendent resign and that the liberal textbooks be banned permanently. On October 28, the citizens' review committee recommended that all but 35 of the 325 books be returned to the classroom. In reaction, the county board building was dynamited and partially destroyed the night of October 30.

On November 9, the school board voted 4–1 to reintroduce most of the textbooks in the classrooms. In response, the protestors had the police arrest the school superintendent and four board members on November 15 for "contributing to the delinquency of minors." Many parents continued to withhold their children from public schools, while other students were enrolled in newly created, private Christian schools stressing a fundamentalist curriculum.

By the end of the year, the protestors appeared to have won a number of concessions from the school board. New guidelines were issued for the selection of textbooks, and a number of "alternative" elementary schools with a more traditional approach to education were planned for the following semester. Although the vehemence of the protest subsided, the anger of protestors lingered on (Page & Clelland, 1978).

## Intergroup Conflict

The Kanawha County textbook controversy is an instance of **intergroup conflict**, a situation in which groups take antagonistic actions toward one another in order to control some outcome important to them. The major participants in the textbook controversy were groups—the Concerned Citizens of Kanawha County, the Businessmen and Professional People's Alliance for Better Textbooks, and the Christian American Parents on one side and the Kanawha County School Board and the Citizens for Quality Education on the other. The textbook controversy shows many features typical of intergroup conflict. The issues at stake become more apparent as the conflict progresses. Distrust and hostility grow, and members of opposing groups develop antagonistic attitudes toward each other. Overt hostility becomes increasingly severe and destructive. Groups commit themselves to various positions, and the conflict becomes more and more difficult to resolve.

The orthodox definition of "social group" (Chapter 12) is excessively restrictive when applied to intergroup conflict. In everyday speech, we often use the term "intergroup conflict" to refer to what might be more correctly described as conflict between persons belonging to social categories, not organized groups. For instance, conflict between members of racial categories (such as blacks and Hispanics in Miami) or ethnic categories (such as Arabs and Jews) is usually considered intergroup conflict, even though the conflictants may not be members of organized groups. In such cases, the "groups" involved in conflict are not comprised solely of people who are members with well-defined role relationships or with interdependent goals. Instead, these "groups" are collections of individuals who perceive themselves as members of the same social category and are involved emotionally in this common definition of themselves. Thus, throughout this chapter, we will use the term intergroup conflict to refer both to conflict

The Kanawha County textbook controversy began with disagreement over the selection of textbooks. It escalated into a classroom boycott by students and parents, attacks on the Board of Education building, and even the creation of alternative schools.

between organized groups and to conflict between members of different social categories.

Intergroup conflict usually involves confrontation between competing beliefs and norms. Thus, a concerned mother in Kanawha County who intentionally blocks the movement of a school bus may, in the eyes of the school board, be performing an unlawful and deviant act. But, in fact, she may be conforming to a different set of norms—those of the antitextbook coalition. In most intergroup conflicts, behavior viewed as appropriate by members of one group is considered unacceptable by members of another. The conflict is rooted not merely in individual behavior but in the different goals, norms, and belief systems of opposing groups.

Although peaceful relations among groups are probably more widespread than conflicting ones, it is intergroup conflict that receives the most attention on television and in newspaper headlines. Every day in the media we encounter intergroup conflict in many forms. Street fights between teen gangs, hostilities between the Ku Klux Klan and the black community, labor strikes against management, strife between religious groups, economic competition and rivalry among ethnic groups, and long-standing family feuds—these are all examples of intergroup conflict.

In discussing intergroup conflict, this chapter will address the following questions:

1. What causes groups to shift from peaceful to hostile relations? That is, what factors cause the development and escalation of intergroup conflict?

2. What sustains the conflict? When intergroup conflict persists over a long period of time—as it often does—what mechanisms support its persistence?

3. What effect does intergroup conflict have on relationships among members within each of the groups? In other words, when a group is involved in conflict with another, what impact does this have on the group's structure and on the way its members relate to one another?

4. How can intergroup conflict be reduced or stopped before it escalates to extreme levels?

## ■ Development of Intergroup Conflict

There are several basic causes of intergroup conflict. Conflict may develop (1) because groups have directly opposing goals (that is, a real opposition of interest), (2) because one group suddenly threatens or deprives another group, provoking an aggressive reaction, or (3) because members of one group act in an ethnocentric and prejudicial way toward members of another group. These factors are not mutually exclusive; in fact, they often work simultaneously to foster intergroup conflict. We will consider each of them in turn.

## Realistic Group Conflict Theory

A number of years ago, an important study on intergroup conflict was conducted at Robbers Cave, Oklahoma, by Muzafer Sherif and his co-workers (Sherif, 1966; Sherif et al., 1961; Sherif & Sherif, 1982). The participants in this experiment were well-adjusted, academically successful, white, middle-class American boys, ages 11 and 12. These boys attended a two-week experimental summer camp and participated in camp activities, unaware that their behavior was under observation. Throughout the two-week period, the boys were organized into two groups, the Eagles and the Rattlers. The overall objective of the research was to investigate conditions that cause intergroup conflict, as well as conditions that reduce conflict.

The experiment was structured in several stages. The first stage, which lasted about a week, was intended to produce a high level of cohesiveness within each of the two groups. The boys arrived at the camp on two separate buses and settled in cabins located a considerable distance apart. Contact between the two groups was minimal.

The boys within each group engaged in numerous activities, many of which were interdependent in character and required cooperative effort for achievement. They camped out, cooked, worked on improving swimming holes, transported canoes over rough terrain to the water, and played various games. As they worked together, the boys in each group pooled their efforts, organized duties, and divided up tasks of work and play. Eventually, the boys in each unit developed a high degree of group cohesiveness.

The next stage of the experiment induced conflict between the two groups. The researchers set up several competitive situations in which one group could attain its goal only at the other's expense. The camp staff arranged a tournament of games—baseball, touch football, tug-of-war, a treasure hunt, and so on—in which prizes were awarded only to the victorious group.

The tournament started in the spirit of good sportsmanship, but as it progressed the good feeling began to fade. The "good sportsmanship" cheer that is customarily given after a game, "2-4-6-8, who do we appreciate," turned into "2-4-6-8, who do we appreci-*hate.*" Intergroup hostility intensified, and members of each group began to call their rivals "sneaks" and "cheats". After suffering a defeat in one game, the Eagles burned a banner left behind by the Rattlers. The next morning, the Rattlers seized the Eagles' flag when they arrived on the athletic field. Name-calling, threats, physical scuffling, and cabin raids became increasingly frequent. When asked by the experimenters to rate each other's characters, a large proportion of the boys in each group gave negative ratings to all the boys in the other group. When the tournament was over, members of the two groups refused to have anything to do with each other.

In later stages of the experiment, when the level of antagonism was quite high, various strategies for reducing strife were introduced. Several techniques failed, but by introducing goals that were valued by both groups and that required intergroup cooperation for attainment, the experimenters did succeed in reducing conflict (Sherif et al., 1961).

This study is a classic illustration of **realistic group conflict theory,** which is the most firmly established theory for explaining the development of intergroup conflict. This theory explains the development of intergroup conflict in terms of the goals of each group. Its central hypothesis is that groups will engage in conflictive behavior when their goals collide head-on. It was precisely under these conditions that conflict erupted between the Eagles and the Rattlers. It was also under these conditions that conflict erupted in Kanawha County regarding the content of text-books and the social values to be taught in public schools. Realistic group conflict theory has received support from Sherif's study as well as various others (Blake & Mouton, 1961b; Bobo, 1983).

The basic propositions of realistic group conflict theory are as follows: First, when groups are pursuing objectives such that one group's success necessarily results in the other's loss, by definition they have an *opposition of interest*. Second, this opposition of interest causes members of each group to experience frustration and develop antagonistic attitudes and unfavorable stereotypes regarding members of the other group. Third, as members of one group develop negative attitudes toward the other group, they become more strongly identified with and attached to their own group. Finally, as solidarity and cohesiveness within each group increases, the conflict between groups intensifies and becomes increasingly difficult to resolve.

## Aversive Events and Escalation

Sometimes a single event will trigger open conflict where none existed previously. For example, several years ago an unexpected defeat in an important high school basketball game on Long Island, New York, led to an argument between fans that quickly escalated into a serious brawl. These fans had different racial and ethnic identities—those supporting the losing home team were largely black, whereas those supporting the visitors were mostly Irish and Italian. During the fight, some persons were badly beaten, and school buses were overturned. Squads of police arrived and eventually suppressed the brawl. But the conflict continued for several weeks among groups of teenagers from each community, with occasional outbreaks of street violence.

This example shows how a single aversive event can provoke open hostilities between groups (Berkowitz, 1972; Konecni, 1979). By

definition, an **aversive event** is a situation caused by (or attributed to) an outside group that produces negative or undesirable outcomes for members of a target group. The unexpected loss of the basketball game was an aversive event for fans of the home team, and it provoked wider conflict. In general, aversive events are situations that most people would want to avoid, such as being physically or verbally attacked, being slighted or humiliated, or facing a loss of income or property.

In addition to provoking new conflicts, aversive events can also activate latent conflicts between groups. Consider a situation in which several groups have an underlying opposition of interest but have managed to avoid open hostilities over the issue. Should an aversive event occur, it may activate the latent conflict. In the Kanawha County textbook controversy, for example, a latent conflict existed between fundamentalist church groups and the liberal school board. But it took an aversive event—the adoption of the new textbooks—to provoke demonstrations and overt hostility. Once this event occurred, groups differing in ideologies and interests mobilized quickly, and the conflict escalated.

The idea that aversive events lead to intergroup conflict is one form of the psychological hypothesis that frustration leads to aggression. When group members respond to an aversive event, they often develop antagonistic attitudes and negative stereotypes. In extreme cases, a group under attack will mobilize and counterattack its adversary. This is most likely to happen when an underlying opposition of interest already exists between groups and/or when obvious differences (such as language, religion, or skin color) serve as the basis for differentiation between them.

## Social Identity Theory of Intergroup Behavior

So far we have seen how such factors as an underlying opposition of interest or an aversive event can cause intergroup conflict. Another factor in intergroup conflict is the extent to which members identify with the group to which they belong. In conjunction with aversive events or an opposition of interest, strong group identification can greatly intensify a conflict between groups.

**In-Group Identification and Ethnocentrism**  A number of years ago, William Graham Sumner (1906) noted that people tend to like their own group (the **in-group**) and to dislike competing or opposing groups (the **out-groups**). He hypothesized that persons having strong group identification are especially prone to favor the in-group and to hold negative sentiments toward out-groups. Sumner's term for this phenomenon was **ethnocentrism**—that is, the tendency to regard one's own group as the center of everything and to evaluate other groups with reference to it. In its purest form, ethnocentrism is based on a pervasive and rigid distinction between an in-group and one or more out-groups. It entails stereotyped, negative imagery and hostile attitudes regarding the out-group and stereotyped, positive imagery and favorable attitudes regarding one's in-group. Ethnocentrism can be expressed in many ways. A summary of the in-group and out-group orientations in ethnocentrism is presented in Table 15-1.

Ethnocentrism involves a generalized prejudice against the out-group and a glorification of the in-group. Thus, many specific attitudes comprising the ethnocentrism syndrome occur simultaneously in a given setting—seeing the in-group as superior and the out-group as inferior, viewing the in-group as strong and the out-group as weak, and construing the in-group as honest and peaceful and the out-group as treacherous and hostile (LeVine & Campbell, 1972; Wilder, 1981). Note, however, that not all facets of ethnocentrism appear in every intergroup conflict. In some instances, only a portion of the orientations listed in Table 15-1 occur (Brewer & Campbell, 1976; Brewer, 1986). Nor is it necessary that all members of the in-group hold ethnocentric prejudices regarding the out-

■ **TABLE 15-1**
ETHNOCENTRIC ORIENTATIONS TOWARD THE IN-GROUP AND THE OUT-GROUP

| Members' Orientations toward the In-Group | Members' Orientations toward the Out-Group |
|---|---|
| See themselves as virtuous and superior | See the out-group as contemptible, immoral, and inferior |
| See their own standards of value as universal and intrinsically true | Reject out-group values |
| See themselves as strong | See the out-group as weak |
| Maintain cooperative relations with other in-group members | Refuse to cooperate with the out-group |
| Obey authorities within the group | Disobey authorities in the out-group |
| Demonstrate a willingness to retain membership in the group | Reject membership in the out-group |
| Trust in-group members | Distrust and fear out-group members |
| Hold positive attitudes toward other in-group members | Show negative affect and hate toward out-group members |
| Take credit for in-group successes | Blame the out-group for in-group troubles |

*Source: Adapted with modifications from LeVine and Campbell, 1972.*

group. Members who identify less strongly with the in-group may hold a more objective viewpoint (Turner, 1975).

Nevertheless, ethnocentrism in one form or another often occurs in intergroup relations. Combined with aversive events or an opposition of interest, ethnocentric attitudes can create ill will and a hostile interpretation of the motives of out-group members. These attitudes devalue and demean out-group members, causing them to be seen as something less than human. In turn, this makes them ready targets for hostile action. And when out-group members fight back, they are often seen as the source of all the problems.

**Minimal Intergroup Situation** Bias favoring the in-group is very likely to come into play during situations of direct competition between groups. More remarkable, several studies have shown that even when an underlying opposition of interest is not present, the mere awareness that an out-group exists can provoke discriminatory responses by in-group members. In some cases, the simple process of social categorization —dividing people arbitrarily into groups—is sufficient to trigger intergroup discrimination.

This was demonstrated in an experimental paradigm called the *minimal intergroup situation* (Tajfel et al., 1971; Tajfel & Billig, 1974; Tajfel, 1982b). English schoolboys, ages 14–16, who knew each other well served as laboratory subjects. The boys were divided into two groups, based on their performance on a trivial task such as estimating the number of dots flashed onto a screen. Although subjects were told what group they themselves were in, they did not know who else was in their group or who was in the other group. Next, the boys were told to distribute points worth money to other subjects. They did not know the personal identities of those to whom they were awarding money—only their group membership. The boys received no directions on how to award these points; they merely were told that at the end of the experiment each of them would receive amounts of money allotted by the other subjects.

Results of these studies show that subjects awarded more money to anonymous in-group members than to anonymous out-group members. The effect was widespread; approximately 75 percent of the subjects showed a clear bias favoring the in-group. This occurred in spite of the fact that (1) such responses had no utilitarian value for the subjects themselves, because they were giving money to other people; (2) there was

no social interaction, either within a group or between groups; and (3) there was neither an opposition of interest nor any previously existing hostility between the groups.

**Social Identity Theory** Since the subjects' discriminatory responses in these studies had no direct utilitarian value, it is very hard to explain the results in terms of a realistic opposition of interest between groups. A more satisfactory explanation is found in the **social identity theory of intergroup behavior,** which was developed by Tajfel and others (Tajfel, 1981, 1982a; Tajfel & Turner, 1986). This theory begins with the assumption that persons are motivated to achieve and maintain a positive self-concept. According to this view, the self-concept has two components: a personal identity and a social identity. Any person can enhance his or her self-concept by improving the evaluation of either (or both) of these components. The social identity component depends primarily on the groups or social categories to which one belongs. The evaluation of one's own group is determined by a comparison with other, specific groups. Thus, positive social identity is based, to a large degree, on favorable comparisons that can be made between one's membership group (in-group) and some relevant out-groups.

It follows that the desire to maintain a positive self-concept creates pressures to evaluate one's own group positively relative to various out-groups. In the minimal intergroup situation, one consequence is that when people are assigned to a group, they immediately and automatically think of that group (which is the in-group for them) as better than the alternative (an out-group), and they engage in actions to support this idea (such as allocating money to members of their own group).

With respect to groups in natural, nonlaboratory settings, Tajfel and others suggest that individuals may respond to a negative or unsatisfactory social identity in any of several ways. These reactions include (1) leaving an existing group and joining some group that is more positively evaluated, (2) passing themselves off as something else so that others do not recognize them as members of a particular group (such as homosexuals passing as straight, blacks passing as white in South Africa, and so on), and (3) engaging in social action to elevate the status or welfare of that group to which they belong. This final alternative, of course, may entail overt conflict with other groups.

In natural settings, there is a much stronger basis for intergroup discrimination than in the minimal intergroup laboratory setting. Salient factors such as skin color or language make group membership easily recognizable. Members of groups based on such sharp distinctions often show strong in-group bias. Thus, as emphasized by the social identity theory of intergroup behavior, the process of social categorization, along with attendant discrimination, can be important in intergroup relations and conflict.

# ■ Persistence of Intergroup Conflict

Probably the most famous family feud in the history of the U.S. was the long-enduring conflict between the Hatfields and the McCoys. In the early days, these mountain people lived peacefully on opposite sides of a narrow river, the Hatfields in West Virginia and the McCoys in Kentucky. The feud began one day in 1873 when Floyd Hatfield drove a razorback sow and her piglets into his pigsty on the McCoy side of the river. The pigs settled in comfortably, but trouble broke out a few days later when Randolph McCoy, who was Floyd's brother-in-law, came up beside the pigsty and accused Floyd of stealing the pigs. A furious argument broke out, and the dispute eventually wound up in a backwoods court. During the trial, witness upon witness went to the stand to testify regarding the ownership of the pigs. All those witnesses named Hatfield swore that Floyd owned the pigs, while all those named McCoy pointed to Randolph as the rightful owner. When the trial finally ended, Floyd Hatfield retained possession of the animals.

This was only the beginning, however. The McCoys were very angry. What once had been familial affection toward the Hatfields quickly turned to hatred. The feud grew and deepened in intensity. Time passed and the pigs were forgotten, but the entire McCoy family joined the fight against the Hatfield clan. Several months later, Ellison Hatfield was murdered by a McCoy; in retaliation, three McCoy boys were shot by the Hatfields. Eventually, the civil authorities of West Virginia and Kentucky were drawn into the fight as they attempted to maintain order and protect human rights. According to one estimate, more than 100 persons lost their lives during the course of the feud. It was not until 1928, when Tennis Hatfield and Uncle Jim McCoy shook hands in public, that the conflict finally ended (Jones, 1948).

The Hatfield-McCoy feud is an interesting and puzzling event. It began as a simple disagreement over the ownership of some pigs and escalated into an enduring conflict that lasted 55 years and cost many lives. The feud illustrates a fundamental point about intergroup conflict: Even in the absence of outside provocation, processes internal to a conflict can cause it to escalate and persist over time. Conflicts often feed on themselves.

What processes support the persistence of intergroup conflict? It is possible to identify several processes, including (1) distorted perception regarding the intentions and character of out-group members, (2) biased evaluation of the out-group's performance, and (3) changes in the structure of the relationship between adversaries. We will consider each of these below.

### Distorted Perception of the Out-Group

In most intergroup conflicts, in-group members hold cognitive schemas (that is, stereotypes) regarding both the in-group and the

This clinic has been the scene of demonstrations both for and against abortions. Although most Americans have attitudes about abortion, only a minority act on their beliefs. People who are more certain of their attitudes, whether pro or con, are more likely to engage in such behavior.

out-group. They also observe actions by the out-group and attribute motives and characteristics to out-group members based on those actions. Yet, it is common for in-group members to have unrealistic impressions regarding the out-group. These mistaken impressions can result from erroneous stereotypes, selective perception, and attributional biases.

**Stereotypic Distortion**   Group stereotypes often exaggerate or accentuate the differences between an in-group and an out-group; that is, they make the groups seem to differ more than they really do (Eiser, 1984). First, the in-group's stereotype of the out-group typically is quite negative and unflattering, as well as unrealistic. Distorted stereotypes of out-group members result in part because in-group members have less information about the out-group than they do about their own group (Linville & Jones, 1980). Second, there is a general tendency in stereotyp-

ing to exaggerate the degree of homogeneity among out-group members. Individuals minimize the perceived differences among members of the out-group to a much greater extent than the differences among members of the in-group (Rothbart, Dawes, & Park, 1984; Quattrone, 1986). In other words, we tend to see "them" as all alike, whereas we see "us" as more diverse and heterogeneous. This effect is heightened when we anticipate future competition with the out-group (Judd & Park, 1988). Third, cognitive schemas regarding the out-group tend to be less complex—sometimes far less complex—than those regarding the in-group. One consequence of this lower schematic complexity is that in-group members often react in an extreme manner to information about the activities of the out-group.

Even though the in-group's stereotypes regarding out-group members may be overly simplistic, excessively negative, and far from realistic, they are strikingly resistant to change. Even very accurate information—such as that which can be gotten when members of hostile sides are in close contact with one another—often fails to modify them.

**Selective Perception**  How can in-group members maintain their negative stereotypes of an out-group in the face of evidence to the contrary? One answer is based on the idea of selective perception: Expectancies held by in-group members lead to "discoveries" of evidence that seem to confirm the original expectation, although no such confirmatory evidence actually exists.

One illustration of such an *illusory correlation* occurs in a study by Hamilton and Gifford (1976). Subjects in this study read a number of statements, each of which described a member of a majority group or a minority group (identified only as groups A and B) engaging in either a desirable or an undesirable behavior. The minority group was mentioned less frequently than the majority group, and undesirable behaviors were described less frequently than desirable ones. Although, in fact, no association existed between group membership and the desirability of the described behavior, subjects overestimated the correlation between the two statistically infrequent events—members of the minority group performing undesirable behaviors. In consequence, subjects rated the minority group significantly less favorably than the majority group on measures of both social and intellectual standing.

**Attributional Bias**  A somewhat similar view, based on attribution theory, suggests that unflattering stereotypes of the out-group are maintained by an *attributional bias* on the part of the in-group. If an in-group observer sees a member of the out-group engaging in undesirable behavior, the observer will ascribe it to the personal or dispositional characteristics of the out-group member. If, however, the behavior happens to be desirable or to have positive consequences for the in-group, the prejudiced in-group observer will ascribe it to situational pressures, accident, or luck. Thus, the out-group will be blamed for negative behaviors but not given credit for positive behaviors (Regan, Straus, & Fazio, 1974; Cooper & Fazio, 1986).

There are many ways that the in-group can explain away or discount positive behavior by out-group members. For example, an out-group member may be viewed as an "exceptional case," someone who differs from other out-group members ("He's really very capable—not like the rest of those stupid Arabs"). Likewise, positive behavior by an out-group member may be ascribed to situational pressures ("Admittedly that mercenary Jew offered us a good price on this business deal, but what else could he do with three of our accountants looking over his shoulder?").

One study illustrating this phenomenon was conducted in India, where Hindus and Muslims have a long history of severe conflict (Taylor & Jagge, 1974). In this study, researchers observed Hindu office clerks making attributions regarding positive or negative behaviors involving either Hindus or Muslims. The behaviors in the study included situations such as a shopkeeper either behaving generously or cheating and a person either helping or ignoring an injured individual. Results indicated that negative behavior by an out-group member (a Muslim) was attributed more to internal (personality) factors than positive behavior, whereas negative behavior by an in-group member (Hindu) was only rarely attributed to personal dispositions and was usually viewed as caused by external pressures. These attributions enabled the Hindus to maintain their prevailing stereotypes regarding Muslims. Biased perception of this kind is widespread in intergroup conflict.

## Biased Evaluation of Performance

Another common bias that feeds intergroup conflict involves an in-group's evaluation of its own performance. Members tend to overrate the performance of their own group relative to that of the out-group. Research has consistently shown that heightened attraction of a member to the in-group is positively related to the tendency to evaluate in-group performance more favorably than out-group performance (Hinkle & Schopler, 1986).

This bias was demonstrated clearly in the Robbers Cave study discussed earlier in this chapter (Sherif et al., 1961). When antagonism between the two groups of boys was at its peak, investigators arranged a bean-collecting contest. Beans were scattered on the ground, and the boys collected as many as they could in one minute. Each person stored his beans in a sack with a narrow opening, so he could not check the number of beans in it. Later, the beans gathered by each boy were projected on a screen in a large room, and all persons were asked to estimate the number in each boy's collection. The projection time was very short and prohibited counting. In reality, the same number of beans (35) was projected on the screen each time. The boys' estimates revealed a strong in-group bias. They overestimated the number of beans collected by members of their own group and underestimated the out-group's performance. This bias was more pronounced among members of the group that had won the preceding tournament of competitive sports events.

This phenomenon also has been demonstrated in other studies. For instance, research on competing industrial work groups (consisting of adult employees of diverse age and rank) reveals a clear tendency for members to judge their own group's performance as superior to that of an out-group (Blake & Mouton, 1961a, 1962b).

What causes this bias in the evaluation of in-group performance? One explanation is based on cognitive balance theory, discussed in chapter 6. An individual's cognitions are "balanced" when that person holds a high evaluation of products from a group whose members he or she likes and to which he or she personally belongs. It would obviously be difficult to hold a low or negative evaluation of in-group products

while identifying with group members. Several studies have supported this hypothesis (Ferguson & Kelley, 1964; Hinkle, 1975).

The phenomenon of bias in the evaluation of in-group performance is real and pervasive; it can produce both positive and negative consequences. On the one hand, this bias can serve as an in-group motivational device that enhances group effort, boosts group morale, avoids complacency, and so on (Worchel, Lind, & Kaufman, 1975). On the other hand, overvaluation of an in-group's performance and/or undervaluation of an out-group's abilities can engender faulty group decision making (groupthink), which in turn may aggravate conflict between groups (Janis, 1982).

## Changes in Relations between Adversaries

Once a conflict is under way, changes occur in the relationship between conflicting groups. Often there is an expansion of the issues under dispute and an increased polarization of relations. These changes may, in turn, lead to a further escalation of the conflict (Kriesberg, 1973).

**Expansion of the Issues**   When groups are actively disputing a particular matter, they often compound the conflict by introducing additional issues. Typically, the process of expansion moves from very specific issues to more general ones. The Kanawha County textbook controversy, discussed at the beginning of this chapter, provides a good illustration. The initial issue concerned which textbooks should be adopted by the school system. This quickly expanded to the larger issue of what curriculum should be taught. Soon the conflict broadened further to include the issue of who should be members of the school board and what behaviors were appropriate for board members. Opposing sides also raised a larger philosophical issue: What viewpoint—liberal or traditional—should be taught in the public schools? The conflict then expanded to include another issue: Should schools teaching the "wrong" viewpoint be permitted to operate without interference, or should they be shut down by force? Some participants went still further and raised another issue: Should legitimacy be withdrawn entirely from the public school system and invested instead in a new system of "alternate" schools?

Expansion of issues occurs for several reasons. First, as relations between conflicting groups deteriorate, latent issues that have previously been denied or ignored come to the fore. Conflicting parties feel less need to deny the repressed issues. Indeed, they may view the overt conflict as a good occasion to get even or "settle accounts" (Ikle, 1971). Second, as one group in the conflict imposes punishment on another, those actions become issues themselves. In the Kanawha County conflict, various actions by the participants—blocking the passage of school buses, arresting the school board members, and dynamiting the county board's building—were highly provocative. These actions became issues in themselves and drew new participants into the fray.

**Polarization of Relations** As a conflict expands, the relationship between adversaries often becomes polarized. There is an increase in interaction among members within each of the conflicting groups. At the same time, there is a decrease in communication between the groups; this causes the adversaries to become increasingly isolated from each other. This can only feed the conflict further because as the number of nonconflicting relations declines, opposing groups are less constrained by cross pressures and increasingly free to use coercion.

The polarization of relations between conflicting groups means that there are fewer opportunities to communicate openly about issues. Consequently, people on one side will lack accurate information about the plans and desires of the other side. Even if one group wants to de-escalate the conflict, it becomes increasingly difficult to signal that intention. Tentative efforts to reduce the conflict may, under conditions of low communication, be viewed by the other side as a trick or trap. When groups are polarized, stereotypes come heavily into play and guide behavior. Thus, polarization and the accompanying lack of communication between groups further perpetuate the conflict.

# ■ Effects of Conflict on Group Structure

So far we have discussed the effect of intergroup conflict on the relationship *between* groups. However, intergroup conflict also restructures the relationships among members *within* a given group. Once a struggle has begun, each group in the conflict undergoes changes that contribute to further escalation. In this section, we will consider the effects of conflict on (1) group cohesiveness, (2) the behavior of group leaders, and (3) the normative structure of the in-group, particularly on standards defining fairness.

### Group Cohesiveness

Social theorists have long recognized that external threats and conflict affect the internal structure of groups. Coser (1967) proposed that conflict with an outside group heightens in-group cohesiveness and reaffirms members' identification with the in-group. He also suggested that conflict can provide a "safety valve" whereby members release tension by directing aggression and hostility toward outside groups rather than toward each other.

Various studies have documented the effects hypothesized by Coser and have shown that intergroup conflict increases cohesiveness within each of the opposing groups. For instance, consider again the Robbers Cave study, in which groups of preadolescent boys at a summer camp engaged in competitive activities (Sherif et al., 1961; Sherif, 1966). As conflict between the Eagles and the Rattlers escalated, various measures of in-group cohesiveness—such as cooperativeness and friendship choice—rose to high

## GROUP MEMBERSHIP AND BIASED PERCEPTION

Group membership can bias the way in which a situation is perceived. Thus, a single event may be perceived and evaluated differently by in-group members and out-group members. This is illustrated in a classic case study of student reaction to an Ivy League football game between perennial rivals Princeton and Dartmouth (Hastorf & Cantril, 1954). It was the last game of the season for both teams. A few minutes after the opening kickoff, it was obvious that the game was going to be rough. Referees were kept busy blowing their whistles, calling penalties on both teams. Tempers flared. In the second quarter, Princeton's star back left the game with a broken nose and a mild concussion. In the third quarter, a Dartmouth player was taken off the field with a broken leg.

After the game (which Princeton won), accusations and recriminations were made by persons at both schools. What was interesting from the investigators' standpoint was that the discussions of the game on each campus, and particularly the coverage in the school newspapers, described starkly different versions of the contest. Four days after the game, the *Daily Princetonian* was quoted as saying: "This observer has never seen quite such a disgusting exhibition of so-called 'sport.' Both teams were guilty but the blame must be laid primarily on Dartmouth's doorstep. Princeton, obviously the better team, had no reason to rough up Dartmouth . . ." Meanwhile, Dartmouth stu-

dents were seeing a different game through the editorial eyes of the *Dartmouth*. That paper asserted that after Princeton's star player had been injured early in the game, the Princeton coach "instilled the old see-what-they-did-go-get-them attitude into his players. His talk got results . . . Results: one bad leg and one leg broken."

Investigators administered a questionnaire to both Dartmouth and Princeton undergraduates one week after the game. Table B-15-1 displays some of their responses. Nearly all the Princeton students judged the game as "rough and dirty"—not one respondent thought it was "clean and fair." Of them, 86 percent thought that Dartmouth's team started the rough play during the game. A plurality of Dartmouth students also felt the game was "rough and dirty," although 13 percent thought the game was "clean and fair." Although some Dartmouth students felt that their own team initiated the rough play, the majority of Dartmouth students thought both sides were to blame.

In addition to the questionnaire, the investigators arranged to show a film of the game to a group of students at each school. These viewers were asked to report any infractions they noticed. Princeton students "saw" the Dartmouth team commit more than twice as many infractions as their own team. Dartmouth students "saw" the Princeton team commit slightly more infractions than the Dartmouth team.

levels. These groups became more cohesive internally and more antagonistic externally as they participated in the win-or-lose competition.

Why does intergroup conflict lead to high levels of in-group cohesiveness? First, as conflict escalates, members endow their cause with additional significance and increase their commitment to it. This, in turn, leads to higher levels of cohesiveness. Second, a threat by an out-group can intensify hatred of a common enemy, and

this also heightens in-group cohesiveness (Holmes & Grant, 1979). In one study (Samuels, 1970) researchers varied the degree of cooperation and competition within groups separately from the presence or absence of competition between groups. The results indicated that groups facing intergroup competition were more cohesive than those not facing intergroup competition, regardless of whether members within a given group related cooperatively or

Box 15-1

This study shows the effect of group membership on the students' perceptions of the event. Perceptual distortion made one's own group appear in a favorable light and the out-group in an unfavorable one. In conflictual situations such as this, a vicious circle is established in which the in-group expects the worst from the out-group, looks for the worst, and consequently finds it.

■ TABLE B-15-1
DIFFERING OPINIONS OF DARTMOUTH AND PRINCETON STUDENTS

| Question | Percentage of Dartmouth Students | Percentage of Princeton Students |
|---|---|---|
| From your observations of what went on at the game, or from what you have heard and read about the game, do you believe the game was clean and fairly played or that it was unnecessarily rough and dirty? | | |
| Clean and fair | 13 | 0 |
| Rough and dirty | 42 | 93 |
| Rough and fair | 39 | 3 |
| Don't know | 6 | 4 |
| From what you saw in the game or the movies, or from what you have read, which team do you feel started the rough play? | | |
| Dartmouth started it | 36 | 86 |
| Princeton started it | 2 | 0 |
| Both started it | 53 | 11 |
| Neither | 6 | 1 |
| No answer | 3 | 2 |

Source: Adapted from Hastorf and Cantril, 1954.

competitively with each other. Thus, the common antagonism of group members toward an opposing group overshadowed any friction among them.

## Leadership Rivalry

The actions of group leaders are especially important under conditions of conflict. Leaders of a group plan and direct the group's strategic moves, allocate resources applicable to the conflict, and serve as spokespersons for the group to outside agencies or persons. Activities of this type can have an important impact on a group's success or failure in intergroup conflict.

When a group is embroiled in a conflict, its members may disagree as to whether the fight is worth the effort. And even if there is general agreement to pursue the conflict, there may still be disagreement regarding the best strategy to use. If action against an opposing group is not progressing favorably, rival leaders will probably emerge within the in-group. These rivals may attempt to displace the existing leaders and to change their policies. Frequently, the rivals are more militant than existing leaders. The threat of being outflanked by more militant rivals can cause established leaders to adopt an increasingly hard line and to intensify the action against the out-group. Leaders are especially prone to react in this manner when their own position is unstable or insecure (Rabbie & Bekkers, 1978). Thus, competition for leadership within groups can actually intensify conflict between groups (Kriesberg, 1973).

This was illustrated in a study of civil rights leaders in 15 U.S. cities in the mid-1960s (McWorter & Crain, 1967). At that time, civil rights organizations were trying to bring about societal changes favoring blacks and other minorities. Interviews were conducted with civil rights leaders to determine the extent to which there was rivalry for leadership within civil rights groups. Results showed that organizations with higher levels of rivalry also had greater militancy. Militancy was measured both in terms of attitudinal responses and in terms of the frequency of civil rights demonstrations conducted in the cities. Rivalry for leadership within these civil rights groups created pressure to escalate intergroup conflict.

## In-Group Normative Structure

Intergroup conflict affects not only cohesiveness and leadership but also group norms. With the onset of a conflict, a group will become more concerned with winning (or surviving) the conflict and less concerned with the rights and liberties of its individual members. Greater sanction will be placed on behavior that helps the entire group and less on what is best for individual members (Korten, 1962).

Consequently, there will be less tolerance of dissent. Enforcement of norms will stiffen and penalties will increase. When internal dissent does occur, attempts will be made to suppress it or to force the dissidents out. Pressure for conformity will increase within the social unit

A demonstrator blocking the entrance to a building is physically removed by the police. During intergroup conflict, conformity to group standards may lead to actions that violate the law.

(group, organization, or nation). In some cases, group members may turn against each other if they suspect their fellows of sympathizing with the adversary or engaging in behaviors that reduce the chances of victory. Allegedly dangerous members may find their rights reduced and their civil liberties abridged. This was demonstrated at the national level during World War II, when the U.S. government relocated many American citizens of Japanese ancestry into detention centers (Miyamoto, 1973).

Not only will the enforcement of norms intensify under conditions of conflict, but the norms themselves may change. This is particularly likely with respect to standards of equity and fairness. As a conflict intensifies, an in-group will reorder its priorities and then reallocate task assignments and resources to achieve a strategic advantage (Zaleznik, 1966). Actions that were previously considered valuable may suddenly be judged useless or even harmful. As a result, the distribution of status and rewards among members will shift in a direction that may not appear fair.

The reallocation of tasks and resources will have additional ramifications. First, there may be an unequal or disproportionate sharing of costs and hardships. If a nation becomes involved in a conventional war, for example, the persons conscripted into military combat service may pay costs greater than those producing munitions back home. Second, the group may disregard members' past contributions and seniority. The nation will favor members who can help win the war, not persons who contributed heavily during peacetime. Finally, some persons may resent the reordering of values produced by the external conflict. Certain groups within the nation may feel that the war is unjust and should not be waged. Each of these concerns may become a serious source of tension within the in-group (Leventhal, 1979).

# ■ Resolution of Intergroup Conflict

Conflicts consume—and ultimately dissipate—time, energy, and resources. Even worse, they have the capacity to expand beyond rational bounds, as illustrated by the Kanawha County textbook controversy and the Hatfield-McCoy family feud. Conflicts begin as small disagreements that grow in scope, pulling in new participants and escalating in intensity. Because intergroup conflicts are potentially very dangerous and costly, many theorists have wondered how to bring them to a halt in the early or middle stages, before they escalate beyond all control.

The solution to this problem is surprisingly complex, however. One cannot resolve intergroup conflict merely by reversing the processes that initially caused it. In a practical sense, it is often impossible to eliminate underlying opposition of interest, to prevent aversive events, or to diminish ethnocentric identification with the in-group.

Nevertheless, a number of techniques to reduce or resolve intergroup conflict have been proposed. These include (1) establishing an overriding, superordinate goal to induce collaborative action between the in-group and the out-group; (2) increasing contact and communication between the in-group and the out-group; (3) deploying group representatives to negotiate a settlement; and (4) initiating unilateral conciliatory moves in the hope that the out-group will respond in a manner that lessens hostility. Each of these approaches merits consideration, and each will be discussed in detail below.

## Superordinate Goals

One of the most effective techniques for resolving intergroup conflict is to interpose one or more superordinate goals. By definition, a **superordinate goal** is an objective held in common by all groups in a conflict that cannot be achieved by any one group without the supportive efforts of the others.

The Robbers Cave study, discussed earlier, provides a clear demonstration that superordinate goals can reduce conflict. After a high level of conflict had developed between two groups of 12-year-old boys, researchers introduced a series of superordinate goals. First, they arranged for the system that supplied water to both groups to break down so that the two groups had to work together to restore water to the camp. Next, the food delivery truck became stuck along the roadway. If the boys were to eat, they all had to work together to free the vehicle. These overriding goals induced cooperation between the groups and eventually decreased hostility (Sherif et al., 1961).

The impact of superordinate goals is not immediate, but gradual and cumulative. Results are most effective when a series of goals is introduced one after another, rather than a single goal on a one-shot basis. Under these conditions, the effects of superordinate goals are cumulative and their impact greater (Sherif et al., 1961; Blake, Shepard, & Mouton, 1964).

Superordinate goals reduce intergroup conflict for several reasons. First, they serve as a basis for restructuring the relationship between groups. By changing a hostile win-lose situation into one of the collaborative problem solving, a superordinate goal reduces friction between groups. Members of conflicting groups will consider a range of alternative actions rather than a fixed position, and they will be likely to look for points of similarity as well as differences. Second, the effort to achieve a common goal reduces the salience of the distinction between an in-group and an out-group. In other words, group boundaries may become less distinct as attention switches from group membership to the shared task at hand. Reduced salience of group boundaries has an ameliorating effect on intergroup relations (Worchel, 1986). Third, in the presence of a superordinate goal, the activities of out-group members assume greater value for in-group members and vice versa. All of these reduce hostility and conflict between groups.

While the conflict-reducing effects of superordinate goals are well known and widely appreciated, this approach to conflict reduction is not always practical. Although the experimenters were able to introduce superordinate goals in the Robbers Cave study, it is not true that outside agents are always able to introduce superordinate goals in any naturally occurring intergroup conflict. In some cases, no valued superordinate goal may be readily apparent, while in other cases the conflicting groups may be unwilling to listen to outside agents. Of course, superordinate goals are sometimes introduced by conflicting groups themselves, and occasionally they arise naturally from changes in the environments faced by conflicting groups. Still, the difficulties of introducing superordinate goals into ongoing conflicts limits the use of this approach.

## Intergroup Contact Hypothesis

Some theorists have suggested that intergroup conflict will be reduced by establishing contact and opening communication between members of opposing groups. They maintain that increased contact will eradicate stereotypes and reduce prejudice and, consequently, reduce antagonism between groups. This concept, called the **intergroup contact hypothesis,** has been proposed primarily with respect to relations between racial and ethnic groups (Cook, 1972; Amir, 1976; Stephan, 1987).

Although intergroup contact does reduce prejudice and conflict between groups in some cases, it does not do so in all cases (Brewer & Kramer, 1985). For instance, intergroup contact between blacks and whites in desegregated schools frequently fails to have positive effects on intergroup relations (Stephan, 1978; Gerard, 1983; Cook, 1984). In some instances, high levels of intergroup contact can actually increase conflict (Brewer, 1986). This raises the question, Under what conditions will intergroup contact lead to favorable outcomes? Usually the answer depends on whether the persons involved in the contact are of equal status, whether the contact is intimate rather than casual, and whether the contact has institutional and normative support.

**Equal-Status Contact** Intergroup contact is more likely to reduce prejudice when in-group and out-group members occupy positions of equal status than when they occupy positions of unequal status (Robinson & Preston, 1976; Riordan, 1978). The effect of equal-status contact was demonstrated, for instance, by a classic study conducted in the military during World War II (Mannheimer & Williams, 1949). This study showed that white soldiers changed their attitudes toward black soldiers after the two racial groups fought together in combat side by side. When asked how they felt about their company including black as well as white platoons, only 7 percent of the white soldiers from integrated units had a negative reaction. In contrast, 62 percent of the soldiers in completely segregated units indicated a negative reaction at the prospect of having black platoons in their unit.

Equal-status contact has been effective in reducing prejudice in other situations as well. It

Box 15-2

## EQUAL-STATUS CONTACT AND INTERRACIAL ATTITUDES

According to various theorists (Allport, 1954; Amir, 1969), contact between ethnic and racial groups is likely to reduce prejudice when members of different groups have equal status during the encounter. Because blacks do not enjoy the same economic position as whites in U.S. society, interracial contacts are often structured in terms of status inequality instead of status equality. Nevertheless, contacts that can be structured in terms of equal status have the potential to change interracial attitudes. One clear demonstration of this occurred in an experimental interracial summer camp for children (Clore et al., 1978).

This camp was structured to foster equal-status contact among children, ages 8–12. Half of the children were white, and half were black. Likewise, half of the counselors and administrative staff were white, and half were black. Living assignments in the camp ensured that each unit was half-white and half-black. The living situation provided an opportunity for intimate acquaintances rather than casual associations typical of many integrated social settings. All children had equal privileges and duties around the camp. Moreover, counselors provided tasks —such as firebuilding and cooking—that required cooperative efforts among the children.

Approximately 200 children attended the camp during the summer. They attended in groups of 40, with each group staying at the camp for one week. Interracial attitudes were measured in several ways. First, researchers asked the children how they felt toward persons of the opposite race; responses were recorded in terms of evaluation scales such as good-bad, clean-dirty, pleasant-unpleasant, valuable-worthless, and so on. In addition, researchers assessed the extent of interracial liking and selection of friends by means of games that required the children to indicate their interpersonal choices. For instance, in the "name game," children designated the others they knew well by circling their names on a card with a pencil. At the end of the week, children indicated the names of three other campers whose telephone numbers and addresses they wanted to have.

Results showed that a change in attitude occurred for girls although not for boys. Boys began camp sessions with neutral attitudes toward persons of the opposite race and did not have much room to change. Girls began with negative attitudes and shifted in a positive direction as a function of the interracial contact. Similarly, there was a significant increase in cross-race interpersonal choices from the beginning of the camp to the end. This increase was more pronounced for girls than for boys. Overall, the results of this study suggest that changes in interracial attitudes can be brought about by prolonged, intimate, equal-status contact across races.

---

has been shown to operate among black and white children at interracial summer camps (Clore et al., 1978) and in interracial housing situations (Hamilton & Bishop, 1976). To reduce prejudice through contact, members of different groups should ideally enter a situation on an equal footing. This establishes norms of equality from the very start of interaction and helps to ensure that members of both groups will view one another as having equal status.

**Personal versus Casual Contact** Another factor affecting reduction in prejudice through intergroup contact is whether the contact is personal rather than superficial in nature (Amir, 1976; Brown & Turner, 1981). During conflict, members of an in-group often tend to depersonalize members of the out-group; that is, out-group members are seen as different from in-group members and often are viewed stereotypically. If social contact fosters high levels of acquaintance and intimacy between members of different racial and ethnic groups, it can overcome these stereotypes and reduce prejudice. In contrast, if contact is superficial and merely involves lower levels of intimacy, it will have little effect (Segal, 1965).

There is some evidence that intergroup contact in work situations produces only limited attitude change, if any (Harding & Hogrefe, 1952). One explanation is that work situations frequently involve relatively impersonal or superficial contact, and even if a work relationship becomes more personal, it is generally confined to the job setting. In contrast, intergroup contact in housing or residential settings is potentially more intimate and personal. Thus, racially intermixed housing provides an opportunity for reducing negative racial attitudes and stereotypes. In one study (Deutsch & Collins, 1951), researchers compared segregated versus integrated occupancy patterns. Segregation was favored by approximately 70 percent of the tenants in segregated projects but only by 40 percent of the tenants in the integrated projects. Much greater intimacy between black and white housewives was found in the integrated projects. When contact between the races was more intimate, the stereotyping of blacks by whites was less common.

**Institutional Support** Intergroup contact is more likely to reduce stereotyping and create favorable attitudes if the contact is regulated by positive social norms (Williams, 1977; Cohen, 1980; Adlerfer, 1982). If the norms support openness, friendliness, and mutual respect, the contact is more likely to produce a change in attitudes. Thus, intergroup contacts sanctioned by an outside authority or by established law or custom are more likely to produce a change in attitudes.

In the absence of institutional support, members of an in-group may be reluctant to interact with outsiders because they feel it is deviant or simply inappropriate. But in the presence of institutional support, contact between

groups may be viewed as appropriate and possible. Several studies have demonstrated that institutional support increases the probability that intergroup contact will lead to positive attitude change. For instance, with respect to desegregation in elementary schools, there is evidence that students are more highly motivated and learn more in classes conducted by teachers (that is, authority figures) who support rather than oppose desegregation (Epstein, 1985).

In sum, intergroup contact tends to reduce conflict under the following conditions: (1) when there is equal-status contact between members of the in-group and the out-group, (2) when the contact is of a personal rather than a superficial nature, and (3) when the contact is based on institutional or authoritative support.

## Group Representatives

In conflicts between large groups or organizations, it may be impossible to bring together all members of the contending sides for face-to-face negotiations. Even with smaller groups, efforts to resolve conflict are frequently more effective when they involve only a few persons. For these reasons, groups frequently rely on representatives to negotiate a settlement with the opposing side. Representatives have been used to resolve a wide variety of conflicts, ranging from community arguments regarding land use and rezoning to labor-management negotiations or family disputes over inheritances.

The role of a group representative, or spokesperson, is difficult to perform effectively, because this person must deal with a range of cross pressures (Adams, 1976; Holmes, Ellard, & Lamm, 1986). Consider the pressures confronting a negotiator representing a union during a strike over wages. First, he or she faces the demands of the opposing negotiator, who represents management. Pressure to agree with management's terms may stem from the conditions prevailing in the economy, from the intrinsic logic of management's position, or from positive personal bonds with representatives of management. At the same time, the union negotiator faces pressure to meet the expectations of his or her own group. Any agreement reached with management on the wage issue must ultimately be acceptable to the union; this constraint will determine whether the union negotiator can accede to demands from the other side. Finally, the union negotiator must cope with external pressures from third parties—the community, the government, and others who demand a speedy, constructive resolution to the ongoing strike.

Although the role is complex, certain conditions increase the likelihood that a group representative will succeed in resolving intergroup conflict. If the representative treats the conflict as a problem to be solved rather than as a battle to be won, he or she is more likely to bring about a lasting resolution (Fisher & Ury, 1981; Kelman & Cohen, 1986). By adopting a problem-solving orientation, the representative will be able to think creatively about alternative resolutions or compromises. In contrast, if the representative approaches the conflict on a win-or-lose basis, adopting a rigid position or defensive reaction, he or she may doom negotiations to fail.

In practice, representatives do not always have a free hand in negotiations, and they may not be able to adopt a problem-solving orientation. Groups often constrain their representatives by giving them detailed instructions prior to negotiations. Studies show that if a group develops rigid positions and strategies rather than studying and discussing the issues broadly, the level of conflict between groups is likely to increase at the negotiation stage (Kahn & Kohls, 1972).

Another factor that may prevent a representative from adopting a problem-solving orientation is loyalty. Because representatives spend a lot of time in contact with opposing groups, their loyalty may be questioned by members of their own group. Persons in their own group may want to monitor what they are doing—that is, to make sure they are not "collaborating with the enemy" —and to limit their freedom of action. These pressures reduce the possibility of joint problem solving with the opposing side. A negotiator whose behavior is being monitored by his or her own group is more likely to adopt a competitive, tough orientation when his or her job is at risk (Bartunek, Benton, & Keyes, 1975) and when constituents control the level of monetary rewards he or she receives (Benton, 1972). Thus, the imposition of surveillance and sanctions may inhibit the representative from adopting a problem-solving orientation and ultimately preclude a realistic, lasting resolution of the impasse.

## The GRIT Strategy

Some intergroup conflicts are particularly difficult to resolve. This is likely to be the case when the conflicting groups have goals that are directly opposed and when severe distrust prevails between them. Under these conditions, superordinate goals may not be invocable, and representatives on each side are sure to be pressured by their groups to adopt a tough stance in any negotiations. The problem can be even more difficult to resolve if no third-parties can intervene and ameliorate the situation; in this case, should either of the conflicting groups wish to lessen the tensions, that group will be limited to taking steps unilaterally.

In general, resolving intergroup conflict through unilateral action is a difficult undertaking. One interesting approach to unilateral conflict reduction is a strategy called **GRIT,** which stands for "Graduated and Reciprocated Initiatives in Tension-reduction" (Osgood, 1962, 1979, 1980). The GRIT strategy pertains primarily to conflicts in which opposing sides have approximately equal power. Originally a product of the cold war between the U.S. and the Soviet Union during the 1950s and 1960s, GRIT was inspired by a fear of nuclear holocaust and a desire to find a way to reduce tension between the superpowers. These days—in an era of evergreater nuclear destructive capabilities—GRIT has a renewed relevance for the relationship between the U.S. and the Soviet Union (Granberg, 1978). However, the GRIT proposal is truly general in character; it applies not only to conflicts between nations but also to those between smaller groups.

The basic idea underlying GRIT is a sort of arms race in reverse: One side initiates deescalatory steps in the hope that they will eventually be reciprocated by the other side. GRIT is based on the assumption that each side in a conflict has an interest in reducing tension, so that resources devoted to the conflict can be reallocated to other, more productive purposes. The GRIT strategy is designed to build trust and reduce tension. By acting on the principles embodied in GRIT, one group in the conflict assumes the initiative rather than merely responding to the moves of the other.

Specifically, the principles in the GRIT strategy are as follows:

1. The group initiating the strategy issues a public statement describing its plan to reduce tension through subsequent actions.

2. The initiating group publicly announces each unilateral move in advance, and indicates that it is part of the overall strategy.

3. Each announcement of a unilateral move explicitly invites reciprocation in some form by the out-group.

4. Each unilateral move is carried out on schedule to demonstrate credibility.

5. Initiatives are continued for some time, even in the absence of reciprocation. This entails risk, but it also intensifies pressure on the out-group to reciprocate.

6. The initiatives are unambiguous and open to verification by the out-group.

7. Ideally, initiatives should be sufficiently risky that they are vulnerable to exploitation, but they must be structured so that they do not seriously impair the capacity of the in-group to retaliate against any attack that may be launched by the out-group.

8. Further moves by the initiating group are graduated to match the responses by the out-group. If the out-group responds in a friendly manner, the in-group reacts by assuming further risk. If the out-group responds in a hostile manner, the in-group responds in like fashion, but avoids excessive retaliation.

9. Ideally, unilateral moves by the initiating group should be diversified in nature. This protects against developing a large gap in defenses, and it also illustrates to the out-group that a variety of moves might be made in reciprocation. (Osgood, 1980; Lindskold, 1986)

The GRIT strategy attempts not only to change the out-group's perceptions of the initiating group but also to change its behavior through the principles of reinforcement. If the out-group reciprocates, it is rewarded with further conciliatory initiatives, but if the out-group takes aggressive action, it is punished with retaliation.

How effective is the GRIT strategy in practice? In the context of East-West relations, no one knows whether GRIT really works. Neither the U.S. nor the Soviet Union has ever fully used it, although there have been several instances in which minor conciliatory initiatives have been reciprocated. During the cold war of the 1960s, the closest approximation to a real test of GRIT was the so-called Kennedy peace offensive of 1963, in which President John F. Kennedy took various unilateral steps to lessen East-West tensions (the U.S. halted atmospheric nuclear testing and agreed to give full status to the Hungarian delegation at the United Nations). These moves did lead to various reciprocal moves (the Soviet Union stopped jamming broadcasts by the Voice of America, withdrew its objection to dispatching U.S. observers to war-torn Yemen, and announced a halt in the production of Soviet strategic bombers). Regrettably, the peace initiative stalled in late 1963 because Kennedy faced mounting criticism within the U.S. for failure to oppose communism, and he could not risk losing the next year's election (Etzioni, 1967).

Ironically, although the U.S. attempted the GRIT strategy in 1963, it is the Soviet Union that today is deploying this strategy in international relations. Under Premier Gorbachev, the Soviet Union has recently undertaken various unilateral moves to lessen world tension (it has withdrawn troops from Afghanistan, reduced troop levels in Eastern Europe, dismantled some mid-range nu-

Predicated in part on the personal relationship between the Soviet leader Mikhail Gorbachev and President Ronald Reagan, tensions between East and West began to moderate in the late 1980s. Use of a GRIT-like strategy by the Soviet Union has produced some reciprocal concessions by the U.S. and Western allies, as the U.S. has increasingly adopted a "trust but verify" policy in response.

clear weapons, and so on). Although these moves by Gorbachev have led to the perception by U.S. citizens that world tensions are declining (as well as a high approval rating in the U.S. for Gorbachev himself), they have not at this writing fostered sufficient trust of the Soviet Union to produce substantial reciprocal moves. Perhaps still more unilateral moves by the Soviet Union are required before the U.S. government will respond significantly.

Further evidence on the effectiveness of GRIT comes from research on smaller-scale conflicts and from simulation studies and experimental games. These studies indicate that some of the points in the strategy are more important than others in reducing conflict (Lindskold, 1978, 1986).

Various studies support principles 1 and 2 in the GRIT strategy (announcing intentions in advance). By issuing a public statement, the initiating group makes clear its goals and intentions. Conciliatory moves without such announcements may be incorrectly construed by the out-group as indicating weakness or passivity and may lead to exploitation (Oskamp, 1971). Studies show that repeated truthful announcements that one intends to cooperate on the next move produces greater reciprocation from the other group than does the same rate of cooperation unaccompanied by announcements (Voissem & Sistrunk, 1971; Lindskold & Finch, 1981). These findings suggest that if a group's strategy is conciliation, it should announce this fact to minimize ambiguity.

Little evidence is available regarding principle 3 (inviting reciprocation), although it is clear that the GRIT initiator must be cautious and tactful when inviting reciprocation lest the invitation be interpreted as a lure into false cooperation that will be exploited in the future. Principle 4 (carrying out initiatives as announced) has been amply supported. Carrying out initiatives as announced increases the credibility of the initiator (Ayers, Nacci, & Tedeschi, 1973; Schlenker et al., 1973), whereas failure to carry out initiatives makes the strategy ineffective in bringing about reciprocation (Gahagan & Tedeschi, 1968).

Principle 5 (continuing initiatives without reciprocation) is complex, because its effectiveness depends on the relative power positions of the two sides. Initiatives from a strong party are likely to be met with concessions by the other side, whereas initiatives from a weak party are not likely to be met with concessions by a stronger opponent (Michener et al., 1975; Lindskold & Aronoff, 1980). In general, the GRIT strategy will bring about reciprocity only when the initiatives come from a group having equal or superior power in the conflict.

Several studies have tested principle 6 (making initiatives unambiguous and open to inspection). These studies have not offered much support for this step, suggesting that it might not be crucial to the overall strategy (Pilisuk et al., 1967; Pilisuk & Skolnick, 1968). In contrast, principle 7 (maintaining one's capacity to retaliate) has received substantial support in experimental studies. If one side makes conciliatory moves but fails to maintain the capacity to retaliate, it will quickly weaken itself and create an imbalance of power. Studies have shown that cooperation drops off when a power imbalance is created (Aronoff & Tedeschi, 1968; Michener & Cohen, 1974). The GRIT strategy is viable only under conditions of approximate power equality.

The remaining points in the GRIT strategy are principle 8 (matching the response of the out-group) and principle 9 (diversifying initiatives). Although not much evidence exists with respect to principle 9, findings provide some support for principle 8, which holds that once the out-group responds, the initiating group should match the action of the out-group. Thus, if the out-group responds in a cooperative manner, the initiating group should make further concessions. If the out-group responds with hostility, the initiating group should retaliate but at moderate intensity so as not to escalate conflict. Studies regarding reciprocity show that GRIT with communication is superior to simple tit-for-tat responses (Han & Lindskold, 1983). Moreover, there is a tendency for one side in a conflict to match concessions by the other side in frequency although not in size (Pruitt & Drews, 1969; Lindskold & Collins, 1978). This suggests that it may be unrealistic to expect precise matching because of biases in perception and differences in values between the groups.

In general, available evidence supports most of the principles in the GRIT strategy. Although GRIT is not likely to be effective between groups of very unequal power, it is a promising strategy of conciliation when groups are closely matched in strength. Each principle in the GRIT strategy contributes to the reduction of tension, but the real value of GRIT lies in its unity as a strategy. Perhaps the 1990s will provide an opportunity to further test its effectiveness in international relations.

## ■ Summary

Intergroup conflict is a situation in which groups engage in antagonistic actions toward one another in order to control some outcome important to them.

**Development of Intergroup Conflict** Three major sources of intergroup conflict can be identified. (1) Often a fundamental opposition of interests underlies conflict between groups. This opposition prevents them from achieving their goals simultaneously and leads to friction, hostility, and overt conflict. (2) One group, by threatening or depriving another, may create an aversive event that turns latent antagonism into overt conflict. (3) A high level of in-group identification, accompanied by ethnocentric attitudes, may create ill will between groups and foster actions that escalate conflict. In many intergroup conflicts, two or more of these sources occur simultaneously.

**Persistence of Intergroup Conflict**  Although some conflicts between groups are resolved quickly, others extend for a long period of time. Several mechanisms support the persistence of intergroup conflict. (1) Perception of the out-group by in-group members is often biased and distorted. This distortion, caused by insufficient information regarding the out-group and excessive reliance on stereotypes, produces a biased understanding of the characteristics and intentions of out-group members. (2) In-group members tend to underestimate the performance of out-group members and overestimate the performance of in-group members. (3) Once a conflict escalates, changes occur in the relationship between the conflicting sides. The number of issues under dispute may expand, and the number of parties in the conflict may increase. Relations between adversaries will polarize, communication across groups will decline, trust will become harder to establish, and the conflict will be more difficult to resolve.

**Effects of Conflict on Group Structure**  Intergroup conflict changes the internal structure of the in-group. (1) Conflict increases the level of cohesiveness of the in-group, as members endow their cause with increased commitment and unite to face a common enemy. (2) Conflict may increase rivalry for leadership among in-group members, especially if the group appears to be losing the conflict. Rivalry can exert pressure on existing leaders to become more militant, which in turn may escalate the conflict. (3) Conflict increases the demands for conformity and often changes the normative structure of the in-group. Standards of fairness may shift as the in-group reorders its priorities to achieve a strategic advantage in the conflict.

**Resolution of Intergroup Conflict**  Intergroup conflict has the potential to escalate to severe levels, dissipating resources and imposing large losses. Several techniques have been suggested to reduce intergroup conflict before it escalates seriously. (1) One technique is to introduce superordinate goals into the conflict. Because goals of this type can be achieved only through the joint efforts of opposing sides, they promote cooperative behavior and serve as a basis for restructuring the relationship between groups. (2) Another technique is to increase intergroup contact in order to limit stereotyping and prejudice. This approach is relatively more effective in reducing conflict when contact is intimate rather than superficial and is based on equal status. (3) Another technique is to deploy group representatives to negotiate a settlement between opposing sides. This is most effective when representatives have a free hand to adopt a problem-solving orientation in the negotiations. (4) A final technique is the GRIT strategy, which is a unilateral approach to conflict reduction. Under GRIT, one side initiates de-escalatory steps in the hope that the other side will reciprocate. The GRIT approach works best when the initiating group is at least as strong as its opponent.

## Key Terms

chapter

# Life Course and Sex Roles

# ■ Introduction

"I still can't get over Liz," said Sally. "I sat next to her in almost every class for three years, and still I hardly recognized her. Put on some weight since high school, of course, and dyed her hair. But mostly it was the defeated look on her face. When she and Hank announced they were getting married they were the happiest couple ever. But that lasted long enough for a baby. Then years of underpaid jobs. She works part-time in sporting goods at Sears now. Had to take that job when her real estate work collapsed in the recession. Pity, just when she was beginning to get on her feet!"

Jim had stopped listening. How could he get excited about Sally's Central High School reunion—people he'd never met? But Sally's mind kept racing. A lot had happened in 20 years:

*John*—Still bigger than life. Football coach at the old school and assistant principal too. Must be a fantastic model for the tough kids he works with. That scholarship to Indiana was the break he needed.

*Frank*—Hard to believe he's in a mental hospital! He started okay as an engineer. Severely burned in a helicopter crash and then hooked on pain killers. Just fell apart. And we voted him most likely to succeed.

*Andrea*—Thinking about a career in politics. She didn't start college until her last kid entered school. Now she's an urban planner in the mayor's office. Couldn't stop saying how she feels like a totally new person.

*Tom*—Head Nurse at Westside Hospital's emergency ward. Quite a surprise. Last I heard he was a car salesman. Started his nursing career at 28. Got the idea when lying in the hospital for a year after a car accident.

*Julie*—Right on that one, voting her most ambitious. Finished Yale Law School, clerked for the New York Supreme Court, and just promoted to senior partner with Wine and Zysblat. Raised two kids at the same time. Having a husband who writes novels at home made life easier. Says she was lucky things were opening up for women just when she came along.

*Linda*—Too bad she quit journalism school to put her husband through med school. She was a great yearbook editor. Still, says she enjoys writing stories as a stringer for the *News*. Leaves time for family and travel.

Sally's reminiscences show how different lives can be and how unpredictable. When we think about people like Liz or Tom or Frank, change seems to be the rule. There is change throughout life for all of us. But there is continuity too. Julie's string of accomplishments is based on her continuing ambition, hard work, and competence. John is back at Central High, once a football hero, now the football coach. Though not a journalist as planned, Linda writes occasionally for a paper, and she may yet develop a serious career in journalism over the next 20 years. Even Frank had started on the predicted path to success before his tragic helicopter crash.

As we look into the future, we cannot project with any certainty what will happen to us. But people's lives have patterns that allow us to make sense of them. Each of us will experience a life characterized both by continuity and by change. This chapter will examine the **life course**—one's progression through a series of socially defined, age-linked social roles—and the important influences that shape the life course that one experiences.

Our examination of the life course will be organized around four broad questions:

1. What are the major components of the life course?
2. What are the major influences on progression through the life course? That is, what causes people's careers to follow the paths they do?
3. What are the typical courses of life for men and women in U.S. society? What happens when people depart from the typical patterns?
4. In what ways do historical events and trends modify the typical life course pattern?

## ■ Components of the Life Course

Lives are too complex to study in all their aspects. Consequently, we will start by identifying the three main components of the life course on which we will concentrate: (1) careers, (2) identity and self-esteem, and (3) stress and satisfaction. By examining these components, we will trace the continuities and changes that occur in what we do, who we are and how we feel.

### Careers

A **career** is a sequence of roles that a person enacts during his or her lifetime, each role with its own set of activities. We can view our lives as a set of intertwining careers in the worlds of school, work, family life, and so on (Abeles, Steel, & Wise, 1980). Our most important careers are in three major social domains: family and friends, education, and work. The idea of careers comes from the work world, where it refers to the sequence of jobs held. Liz's work career, for example, consisted of a sequence of jobs as waitress, checkout clerk, clothing salesperson, real estate agent, and sporting-goods salesperson.

The careers of one person differ from those of another in several ways—in the roles that make up the careers, in the order in which the roles are performed, and in the timing and duration of role-related activities. For example, one person's family career may consist of roles as infant, child, adolescent, spouse, parent, grandparent, and widow; while another's family career may include roles as stepsister and divorcée but exclude the parent role. The order of roles also may vary. "Parent before spouse" has very different consequences than "spouse before parent." In addition, the timing of career events is important. Having a first child at 36 has different life consequences than having a first child at 18. Finally, the duration of enacting a role may vary. For example, some couples end their marriages before the wedding champagne has gone flat, whereas others go on to celebrate their golden wedding anniversary.

Societies provide structured career paths that determine the options available to individuals and constrain their choices. The cultural norms, social expectations, and laws that organize life in a society make various career options more or less attractive, accessible, and necessary. In the U.S., for example, educational careers are socially structured so that virtually everyone attends kindergarten, elementary school, and at least a few years of high school. Thereafter, educational options are more diverse—night schools, technical and vocational schools, apprenticeships, community colleges, universities, and so on. But individual choice among these options is also socially constrained. The norms and expectations of our social groups strongly influence our educational careers.

Many societies, including the U.S., provide different career paths for men and women. In U.S. society, some paths open to men are not open to women, although the number of these is declining. Other paths are readily accessible to both men and women.

A person's total life course consists of intertwined careers in the worlds of work, family, and education (Elder, 1975). The shape of the life course is derived from the contents of these careers, from the way they intermesh with each other, and from the way they interweave with those of family members. Sally's classmates, Julie and Andrea, enacted similar career roles: both finished college, held full-time jobs, married, and raised children. Yet the courses of their lives were very different. Julie juggled all these roles simultaneously, helped by a husband who was able to work at home. Andrea waited until her children were attending school before continuing her education and then adding an occupational role. The different content, order, timing, and duration of intertwining careers make each person's life course unique.

## Identities and Self-Esteem

As we engage in career roles, we observe our own performances and other peoples' reactions to us. Using these observations, we construct *role identities*—conceptions of the self in specific roles. The role identities available to us depend on our location in society and on the career paths we are following. When Liz's work in real estate collapsed, she got a job in sales at Sears; she was qualified to sell sporting goods because of her prior work experience.

As we enact major roles, especially familial and occupational ones, we evaluate our performances and, thereby, gain or lose *self-esteem*—one's sense of how successful and worthy one is. Self-esteem is influenced by one's achievements; Julie has high self-esteem as a consequence of being a senior partner in a prestigious law firm. Self-esteem is also influenced by the feedback one receives from others.

Identities and self-esteem are crucial guides to behavior, as discussed in chapter 4. We, therefore, take identities and self-esteem as the second component of the life course.

## Stress and Satisfaction

Positive feelings such as satisfaction and negative feelings such as stress accompany virtually all career activities. These feelings reflect how we experience the quality of our lives; thus, stress and satisfaction are the third component of the life course.

Many events or experiences place emotional and physical demands on a person. Such events include moving, serious conflict with a parent, lover, or spouse, changing jobs, and having a child (Holmes & Rahe, 1967). At times, the demands made on a person exceed the individual's ability to cope with them; such a discrepancy is referred to as **stress** (Dohrenwend, 1961). People who are under stress often engage in behavior designed to reduce the discrepancy, for example, by resolving the conflicts. They may also become tense, anxious, worried, generally unhappy—or all of these.

These feelings vary in their intensity in response to life course events. Levels of stress, for example, change as career roles become more or less demanding (parenting roles become increasingly demanding as children enter adolescence), as different careers compete with each other (familial and occupational demands often conflict), and as unanticipated setbacks occur (such as layoffs or a business going bankrupt). Levels of satisfaction vary as career rewards change (salary increases or cuts) and as we cope more or less successfully with career demands (meeting sales quotas, passing exams) or with life events (a heart attack or skiing accident).

The extent to which particular events or transitions are stressful depends on several factors. First, the more extensive the changes associated with an event are, the greater the stress is. For example, a change in employment that requires a move to an unfamiliar city is more stressful than a new job located across town. Second, the availability of social support, in the form of advice or emotional or material aid, increases our ability to cope successfully with change. To help their members, families reallocate their resources and reorganize their activities. Thus, parents lend money to young couples, and older adults provide care for their grandchildren so that their children can work.

Personal resources and competence influence how one copes with stress. Coping successfully with earlier transitions prepares individuals for later transitions. Men who develop strong ego identities in young adulthood perceive events later in their lives as less negative (Sammon, Reznikoff, & Geisinger, 1985). Conversely, early experiences of failure reduce an individual's sense of competence (Duncan &

Morgan, 1980) and may leave that person with a sense of helplessness as he or she faces later events and transitions.

# ■ Influences on Life Course Progression

At the beginning of this chapter, we noted many events that had important impacts on the lives of Sally's classmates: loss of a job due to economic recession, a helicopter crash, a car accident, having a baby and graduating from a prestigious law school. All of these are **life events**—episodes that mark transition points in our lives. They provoke coping and readjustment (Hultsch & Plemons, 1979). For many young people, for example, a move from home to college is a life event, marking a transition from adolescence to young adulthood. Such a move initiates a period during which students work out new behavior patterns and revise their self-expectations and priorities.

This section will discuss the three major influences on life course: (1) biological aging, (2) social age grading, and (3) historical trends and events. These influences act on us through specific life events (Brim & Ryff, 1980). Some life events are carefully planned—a trip to Europe, for example. Other events, no less important, occur by chance—like meeting one's future spouse in an Amsterdam hostel (Bandura, 1982).

## Biological Aging

Throughout the life cycle we undergo biological changes in body size and structure, in the brain and central nervous system, in the endocrine system, in our susceptibility to various diseases, and in the acuity of our sight, hearing, taste, and so on. Changes are rapid and dramatic in childhood. Their pace slows considerably after adolescence, picking up again in old age. Even in the middle years, however, biological changes may have substantial impacts. The shifting hormonal levels associated with menstrual periods in women and with aging in men and women, for example, are thought by many to affect mood and behavior (Hoyenga & Hoyenga, 1979; Doering, 1980).

Biological aging is inevitable and irreversible. But it is only loosely related to chronological age. Biological puberty may come at any time between ages 8 and 17, for example, and serious decline in the functioning of body organs may begin before age 40 or after age 85. The neurons of the brain die off steadily throughout life and do not regenerate. Yet, intellectual functioning, long assumed to be determined early in life and to decline with aging, is now known to be capable of increasing over the life course. Even in old age, mental abilities can improve with opportunities for learning and practice (Baltes & Willis, 1982).

Biologically based capacities and characteristics limit what we can do. Their impacts on the life course depend, however, on the social significance we give them. How does the first appearance of gray hair affect careers, identities, and stress, for instance? For some, this biological event is a painful source of stress. It elicits dismay, sets off thoughts about mortality, and instigates desperate attempts to straighten out family relations and to make a mark in the world before it is too late. Others take gray hair as a sign to stop worrying about trying to look young, to start basing their priorities on their own values, and to demand respect for their experience. In a similar manner, the impacts of other biological changes on the life course—such as a growth spurt during adolescence or menopause in middle age—also depend on the social significance given them.

## Social Age Grading

Which members of a society are supposed to raise children, and which are supposed to be cared for by others? Who should attend school, and who should work full-time? Who should be single, and who should marry? Age is the primary criterion that every known society uses to assign people to such activities and roles (Riley,

1987). Throughout life, individuals move through a sequence of age-graded social roles. Each role consists of a set of expected behaviors, opportunities, and constraints. Movement through these roles shapes the course of life.

Each society prescribes a customary sequence of age-graded activities and roles. Table 16-1 presents a sampling of widely held age norms for transitions between roles in the domains of family, education, and work. The findings indicate that in the early 1960s, when these data were gathered, there were well-defined age norms for several role transitions. The desirable ages for some transitions may have changed somewhat in recent years, and the amount of agreement has undoubtedly changed. Preferred marriage ages may be later, for example. Whatever the specific content of current age norms, however, people continue to be aware that norms exist. As a result, age norms serve as a basis for planning, as prods for action, and as brakes against moving too fast (Neugarten & Datan, 1973).

Pressure to make the expected transitions between roles at the appropriate times means that the life course consists of a series of normative life stages. A **normative life stage** is a discrete period in the life course during which individuals are expected to perform the set of activities associated with a distinct age-related role. The fact that the order of the stages is prescribed means that people try to shape their own lives to fit socially approved career paths. In addition, people perceive deviations from expected career paths as undesirable.

Not everyone experiences major transitions in the socially approved progression. Consider the transition to adulthood; the normative order of events is leaving school, performing military service, getting a job, and getting married. Analyzing data about the high school class of 1972 (collected in 1972–80), researchers found that half of the men and women experienced a sequence that violated the "normal" path (Rindfuss, Swicegood, & Rosenfeld, 1987). Common violations included entering military service be-

■ **TABLE 16-1**
AGE NORMS FOR ROLE TRANSITIONS

| Age Norm Question | Age Range | Percentage Who Mention an Age in This Range Spontaneously | |
| | | Men (N = 50) | Women (N = 43) |
| --- | --- | --- | --- |
| Best age for a man to marry | 20–25 | 80 | 90 |
| Best age for a woman to marry | 19–24 | 85 | 90 |
| Best age for most people to finish school and go to work | 20–22 | 86 | 82 |
| When most men should be settled into a career | 24–26 | 74 | 64 |
| When most men should hold their top jobs | 45–50 | 71 | 58 |
| When most people should become grandparents | 45–50 | 84 | 79 |
| When most people should be ready to retire | 60–65 | 83 | 86 |

Note: A representative sample of middle-class American men and women, ages 40–70, shows considerable agreement. Essentially the same pattern of agreement was also found in other samples, ages 20–80.

Source: Adapted from Neugarten, Moore, and Lowe, 1964.

fore finishing school and returning to school after a period of full-time employment.

In some cases, violating the age norms associated with a transition has lasting consequences. As indicated in Table 16-1, the transition to marriage is expected to occur between the ages of 19 and 25. Research consistently finds that making this transition earlier than usual has long-term effects on marital and occupational careers. A survey of 63,000 adults allowed researchers to compare men who married as adolescents with men of a similar age who married as adults (Teti, Lamb, & Elster, 1987). Since the sample included people of all ages, the researchers could study the careers of men who married 20, 30, and 40 years earlier. Men who married as adolescents completed fewer years of education, held lower-status jobs, and earned less income. In addition, the marriages of those who married early were less stable. These effects were evident 40 years after marriage.

Insofar as individuals follow customary paths, their ages roughly locate them in normative life stages. Transitions from one life stage to another influence a person in three ways: they change the roles available for building identities,

modify the allocation of privileges and responsibilities, and alter the socialization experiences to which the person is exposed.

**Building Identities** Since our identities are built around the roles we enact, role transitions often produce identity change. The period between ages 20 and 24, for example, brings substantial, sometimes wrenching changes in self-conceptions. Those who marry, start their first job, or have their first child are likely to begin to view themselves as spouses and parents responsible for others rather than as students relatively free of responsibility or as dependent children. Much smaller changes in identity are likely for those whose roles change less, who remain single, or who pursue graduate studies.

**Allocating Privileges and Responsibilities**
The shape of the life course is further affected by normative life stages, because societies use age-graded social positions to allocate privileges and responsibilities to individuals. For example, our age largely determines whether or not we can be employed for pay or legally drive a car. The

Violating the age norms associated with a major transition, such as the transition to parenthood, may have lasting consequences. Having a baby at the age of 16 may force a young woman to leave school and may limit her to a succession of poorly paid jobs.

emphasis on teaching ideal standards of morality and performance and more emphasis on teaching ways to find realistic, workable compromises between contradictory demands. The content of socialization tends to shift from regulating biological drives in childhood to instilling broad values in adolescence and to transmitting specific, role-related norms for behavior in adulthood (Brim, 1966). The nature of socialization relationships also changes as we grow older. These relationships become less emotionally charged and less central in our lives. The power differences between us—the "socializees"—and our socialization agents also diminish as we grow older and move into higher education and occupational organizations. As a result, adults are more able to resist socialization than children (Mortimer & Simmons, 1978).

## Historical Trends and Events

Recall that Sally's classmate Julie attributed her rapid rise to senior law partner to lucky historical timing. Julie applied to Yale Law School shortly after the barriers to women had been broken, and she sought a job just when affirmative action came into vogue at the major law firms. Sally's friend Liz attributed her setback as a real estate broker to an economic recession coupled with high interest rates that crippled the housing market. As the experiences of Julie and Liz illustrate, historical trends and events are another major influence on the life course. The lives of individuals are shaped by *trends* that extend across historical periods (such as increasing equality of the sexes and improved nutrition) and by *events* that occur at particular points in history (such as recessions, wars, and earthquakes).

**Birth Cohorts**  To aid in the understanding of how historical events and trends influence the life courses of individuals, social scientists have developed the concept of cohorts (Ryder, 1965). A **birth cohort** is a group of people who were born during the same period. The period could be one year or several years, depending on the issue under study. What is most important about

specific ages granting access to privileges (holding political office) and imposing responsibilities (serving in the military) vary with each community, society, and even historical period. This variability makes it apparent that chronological age itself does not determine the allocation of privileges and responsibilities. Rather, allocation depends on the social definition of age-appropriate roles and activities.

**Shaping Socialization Experiences**  Role transitions also bring with them changes in the nature of socialization experiences. As we assume adult roles, socialization agents place less a birth cohort is that its members are all approximately the same age when they encounter particular historical events. The birth cohort of 1950, for example, was 13 years old when President Kennedy was assassinated and was in college at the height of the protests against the Vietnam War. Many of these young people were profoundly influenced by those events. Members of the birth cohort of 1970 do not remember the Kennedy assassination or Vietnam protests but were in college when the protests against racism erupted in 1987 and 1988. Some of them were influenced by those events.

A person's membership in a specific birth cohort locates that person historically in two ways. First, it points to the trends and events the person is likely to have encountered. Second, it indicates approximately where the individual was located in the sequence of normative life stages when historical events occurred. Life stage location is crucial, because historical events or trends have different impacts on individuals who are in different life stages.

To illustrate, consider the effects of a cutback in funding to support graduate studies in counseling and social work. Such a cutback would have quite different impacts on the cohort of current college seniors, as compared with the cohort of those who graduated 5 years ago. For some seniors, this historical event might mean switching professional goals or taking jobs and postponing graduate school. For the cohort that graduated 5 years ago, however, the situation is quite different, because, by now, most will have completed their graduate training. Thus, the funding cutbacks pose no particular problem for them. In fact, members of that cohort who went to graduate school in counseling and social work might even benefit because future competition in that profession will be reduced.

Of course, not all members of a cohort experience historical events in the same way. A cutback in funds would strongly affect only those members of the senior cohort who were planning careers in counseling or social work, not those planning to become lawyers or stock brokers. Thus, differential vulnerability to historical events causes differences within a single cohort as well as between cohorts.

Placement in a birth cohort also affects access to opportunities. Members of large birth cohorts, for example, are likely to be disadvantaged throughout life. They begin their education in overpopulated classrooms. They then must compete for scarce openings in professional schools and crowded job markets. And, as they age, they face reduced retirement benefits because their numbers threaten to overwhelm the social security system. Table 16-2 presents examples of how the same historical events affect members of cohorts in distinct ways. These historically different experiences mold the unique values, ideologies, personalities, and behavior patterns that characterize each cohort through the life course. Within each cohort there are differences too. For example, the Vietnam War led to a father's absence for some children but not for others.

**Cohorts and Social Change**  Due to differences in their experiences, each birth cohort ages in a unique way. Each cohort has its own set of collective experiences and opportunities. As a result, cohorts differ in their career patterns, attitudes, values, and self-concepts. As cohorts age, they succeed one another in filling the social positions in the family, as well as in political, economic, and cultural institutions. Power is transferred from members of older cohorts, with their historically based outlooks, to members of younger cohorts, with different outlooks. In this way, the succession of cohorts produces social change. It also gives rise to intergenerational conflict around issues on which successive cohorts disagree (Elder, 1975).

In this section, we have provided an overview of changes during the life course. Based on this discussion, it is useful to think of ourselves as living simultaneously in three types of time, each deriving from a different source of change. As we age biologically, we move through *developmental time* in our own biological life cycle. As we pass through the intertwined sequence of roles in our society, we move through *social time*. And as we

■ **TABLE 16-2**

HISTORY AND LIFE STAGE

| Historical Event | Cohort of 1945–50 | | Cohort of 1970–72 | |
|---|---|---|---|---|
| | *Life Stage When the Event Occurred* | *Some Life Course Implications of the Event* | *Life Stage When the Event Occurred* | *Some Life Course Implications of the Event* |
| Vietnam War (1964–73) | Young Adulthood | For draftees, disrupted family and work careers. For others, increased occupational chances and advancement. | Infancy–Toddlerhood | For some, death of father; socialization by mother; or by mother and stepfather. |
| Women's Movement (1972–78) | Young Adulthood | For women, increased work opportunities and delayed marriage and childbearing. For men, developed more egalitarian attitudes and increased competition for some jobs. | Childhood | Introduced nontraditional sex-role socialization to parents and teachers. Increased opportunities for girls and boys. |
| Recession (1980–82) | Middle Adulthood | Created financial difficulties in sending children to college. Increased wives' employment to maintain living standard. Created blue-collar unemployment. | Adolescence | Created family situations with two working parents and, thus, allowed for greater autonomy for adolescents. Caused economic hardship if one or both parents were unemployed. |

respond to the historical events that impinge on our lives, we move together with our cohort through *historical time*.

## ■ Stages in the Life Course: Age and Sex Roles

People in every society experience a fairly standard sequence of normative life stages. In this section, we will examine some typical life course patterns in U.S. society and some important variations on these patterns. We will consider experiences that distinguish males and females as well as experiences common to both sexes.

Every society expects certain role behaviors, values, and attributes of males and others of females (Riley, Johnson, & Foner, 1972). Differences in expectations for males and females are least pronounced at very young ages, increase through adolescence, and become even sharper in young adulthood. Before children are 10, for example, similar amounts of self-reliance, obedience, and leadership are usually expected of both girls and boys. By age 20, however, males and females typically face sex-differentiated expectations for these and other attributes and for choices of college major, occupation, and life style (Stoll, 1978). Expectations remain strongly differentiated by sex throughout adulthood. With retirement and old age, expectations for males and females become more similar again.

The four postchildhood stages are shown in Table 16-3. Each stage is labeled according to the major social task or challenge that characterizes it. The labels point to the fact that these stages are socially derived rather than the direct products of biological or cognitive development. In discussing each stage, we will focus on the three major components of the life course we have identified—careers, identity and self-esteem, and stress and satisfaction. We will also note the role transitions inherent in moving through these stages and will consider how people cope with the problems posed by these transitions.

## Stage I: Achieving Independence

Most college students are in the stage of achieving independence. This is a period of transition from lives centered psychologically and economically around parents to lives in which we stand on our own. This stage challenges us to disengage from parents and to take responsibility for ourselves. For both men and women, achieving independence means taking actions that are competent, persevering, and task oriented. In addition, women—but not men—

tend to view the expression of warmth, kindness, and empathy as signs of independence (Johnson et al., 1975). In contrast, there is evidence that as college men become more competent and task oriented, they see themselves as increasingly less warm, less open, and less interested in others (Mortimer, Finch, & Kumka, 1982).

Several major social transitions are typically associated with this life stage: leaving one's home, leaving school, entering the work force, getting married, and establishing an independent household. A century ago, these transitions were usually spread over many years, occurring one at a time in a fixed order. More recently, these transitions have been compressed into the age range 17–25 for most American youth. Individuals are freer now to choose when to make each transition; but they are constrained in these choices by the requirements of getting a formal education and preparing for an occupation. Individuals who make transitions too early or too late pay a price. Being out of step leads to lost income and lost occupational and marital opportunities (Hogan, 1981), as noted earlier.

**Careers** During this stage, individuals explore the fit between their personal abilities and interests on the one hand and available work options on the other. Following high school, many youths join the work force in unskilled or semiskilled jobs (as sales and stock clerks, workers in restaurants and fast-food chains, construction workers, and so on). Others, especially minority group

■ TABLE 16-3
POSTCHILDHOOD LIFE STAGES

| Stage | Major Challenge | Conventional Labels | Age Range |
|-------|-----------------|---------------------|-----------|
| I | Achieving independence | Youth, Late adolescence | 16–23 |
| II | Balancing family and work commitments | Young adulthood | 18–40 |
| III | Performing adult roles | Adulthood, Maturity, Middle age | 35–70 |
| IV | Coping with loss | Late maturity, Old age | 60–90 |

Note: The overlap in ages between the stages shows that age is only a rough indicator of life stage.

Socially important role transitions are often marked by rites of passage—public ceremonies that affirm the individual's new status. These ceremonies signify both to the individual and to others that he or she now has a new identity and that new behaviors, rights, and duties are now appropriate.

members in large cities, are frustrated in their search for jobs. They find only sporadic employment during this life stage. Still others enter military service. About one third of all youths attend college, and about one half of these youths work as well (Sweet & Bumpass, 1987; U.S. Bureau of the Census, 1987). As a result, most youths try out a variety of jobs during this life stage, acquiring skills and preferences that eventually lead to more permanent employment.

College students benefit from institutional support in their movement toward independence. College provides a partially protected environment in which one can develop self-reliance with the support of peers. College also instills in many students the motivation to pursue long-range, socially approved goals—such as studying eight years to earn an advanced degree —even if this is not personally attractive. Such motivation is very helpful in the struggle for occupational success (Becker, 1964). The aca-

demic interests of students often reflect the sex typing of subject areas (physics and engineering for males, humanities and social work for females). Sex-differentiated preferences for subjects arise in high school, when boys or girls see particular subjects as useful in their future occupational choices (Hilton & Berglund, 1974).

One ultimate goal for most men and women is to establish a long-term, intimate relationship with a member of the opposite sex. In U.S. society, dating is the mechanism by which potential partners get to know each other. As discussed in chapter 11, the development of intimacy involves increasing trust and self-disclosure over time. The conventions of dating encourage men to take the initiative and women to be more passive and dependent. If followed, this pattern fosters a sense of autonomy among men, but it inhibits the achievement of independence among women.

**Identities and Self-Esteem** A central challenge of this life stage is to solidify a personal identity—to develop a firm sense of continuity and direction in one's life (Erikson, 1968). Because men and women face different adult role expectations, they tend to build their identities on different bases. National surveys reveal that males tend to construct their identities around their anticipated or actual occupational roles. By anchoring their identities in their own vocational choices, young men can gain a concrete sense of self-direction (Lowenthal, Thurnher, & Chiriboga, 1975).

Most men do not perceive their occupational career as dependent on future family roles. In contrast, most females in recent decades have tended to construct their identities around their anticipated roles as wives and mothers. Young women's identity commitments may be more tentative and ambiguous than men's because the exact nature of their future roles depends so much on when and whom they will marry. It can be adaptive for young women to hold back from firm identity commitments and to maintain flexibility regarding goals. Young women, thus, avoid

Money, power, excitement—yes! Women? Try to find one. Despite progress toward equality, men continue to far outnumber women in better-paid and higher-powered jobs. Women who build their identities around their work are therefore at a disadvantage.

prematurely narrowing the range of potentially compatible marital partners and retain the capacity to adjust to their future husbands' goals (Angrist & Almquist, 1975; Sales, 1978). Nonetheless, an increasing number of young women, especially the daughters of happily employed mothers, have recently begun to anchor their identities in their own vocational aspirations (Zuckerman, 1981).

Studies of self-evaluation during this life stage show that self-esteem may drop after high school as youths struggle to achieve independence. Illustrating this drop, a sample of male college undergraduates from Michigan rated themselves as less competent, successful, active, and strong in their senior year than they did in their freshman year (Mortimer, Finch, & Kumka, 1982). Self-esteem rises later when people suc-

cessfully adapt to family and work roles. For the Michigan sample, self-ratings rose over the ten years following college graduation.

**Stress and Satisfaction**   Both men and women experience this period of achieving independence as relatively stressful. Women are considerably more likely than men to feel frightened, to feel that life is hard, and to feel financially insecure (Campbell, Converse, & Rodgers, 1976). These sentiments may well reflect the fact that young women have less control than men over the important directions their lives are taking. Once they set their life directions by marrying, however, the stress women feel is usually reduced. On the other hand, some women seriously pursue occupational goals at this time; they may experience social disapproval.

## SEX STEREOTYPES AND SEX DIFFERENCES: MYTH OR REALITY?

We are all taught that some behaviors are more appropriate for males and that others are more appropriate for females. Not surprisingly, many people come to believe that males and females have different personality traits. What are the traits believed to be typical of each sex? Do females and males really differ on these traits, or are these sex stereotypes a myth?

The best-known series of studies on sex stereotyping of personality characteristics began by asking male and female college students to list all the ways they thought men and women differed psychologically (Broverman et al., 1972). The researchers then listed each trait that the students mentioned at least twice on a scale that looked like this:

| Not at all aggressive | | | | | | Very aggressive |
|---|---|---|---|---|---|---|
| 1 | 2 | 3 | 4 | 5 | 6 | 7 |

Various groups of adults, ages 17 to 60, then indicated how much they thought each item characterized adult men and adult women. In numerous studies, people cited consistent differences between males and females for twenty traits. To see how aware you are of these stereotypes, indicate with a check on Table B-16-1 whether you think a trait is usually seen as more

typical of men or of women. Note also whether you consider each trait as desirable or undesirable for an adult to have.

Broverman and her associates (1972) found that both men and women agreed on the stereotypes and on the desirability of each trait. The first five traits listed were seen as more typical of men, the next ten as more typical of women, and the last five as more typical of men. That is, men were seen as more independent, aggressive, and ambitious, and women were seen as more passive, emotional, and easily influenced. In general, men were perceived as stronger and more confident than women, and women were perceived as weaker and more expressive than men. Recent studies have found that these stereotypes persist (Deaux & Lewis, 1983).

The researchers also found that most traits stereotyped as masculine were evaluated as desirable, whereas most traits stereotyped as feminine were evaluated as undesirable. In other words, traits associated with men were usually considered to be better. Did your evaluations of trait desirability favor the male stereotyped traits? If not, you may fit a trend among the educated toward valuing some of the traditionally feminine traits (such as emotionality) more positively and some of the traditionally masculine traits (such as ambitiousness) more nega-

Box 16-1

tively (Pleck, 1976; Der-Karabetian & Smith, 1977). This trend means that even if sex stereotypes persist, women may be evaluated less negatively than before.

Do men and women actually differ in the ways suggested by these stereotypes? The picture is still far from clear. There appear to be some real differences between males and females in social behavior. But the stereotypes exaggerate both the number and size of the differences. Here is a summary of some major conclusions reached in critical surveys of the research.

### Aggression

Males are indeed more physically aggressive than females. Starting with more hitting and shoving at age 2 or 3 and progressing through more physical fights, violent crime, and spouse beating, males exhibit more physical aggression through the life course (Eagly, 1987). This difference appears in many different cultures, suggesting that biological factors may be involved. Note, however, that even though males are more likely to initiate physical aggression, females are no less aggressive than males when they are directly provoked and when aggression is socially acceptable (Maccoby & Jacklin, 1974; Frieze et al., 1978).

### Emotionality

The stereotype holds that women are more fearful, anxious, and easily upset than men. The majority of more than 30 studies on this topic based on people's descriptions of their own and others' traits supports this stereotype. Women and girls tend to describe themselves as more fearful and anxious, and other raters also tend to describe women as more anxious than men. However, studies in which researchers observed how people actually behaved in fear-arousing situations and that measured physiological signs of emotionality (pulse, heartbeat, and respiration), have revealed no consistent differences between males and females. Taken together, the evidence on emotionality is inconclusive: women may score higher than men on self-reports of emotionality because women are given more freedom and encouragement to express their feelings in our society (Frieze et al., 1978; Tavris & Offir, 1984).

### Dependency

Included in the stereotype that women are more dependent is the idea that women rely more on others for protection and help and are more easily influenced and persuaded. The evidence offers only weak support for this stereotype (Becker, 1986). Young girls and boys do not differ in clinging to their parents, resisting separation, or wandering about freely. Studies of adult behavior indicate that women are more likely than men to accept suggestions, comply with requests, and be persuaded in face-to-face interaction. These gender differences may be due to the fact that the presence of others makes gender role more salient, so that in face-to-face interaction women yield more readily than men (Eagly, 1987). Also, women often hold low-status positions or are assumed to have lower status if no other status indications are available; persons of lower status are expected to yield to influence attempts by higher-status persons.

### Sociability

Are women more interested in people, friendlier, and more capable of establishing interpersonal relationships than men? Recent evidence suggests that neither men nor women consistently seek more social contact. The number of friends people have depends mainly on opportunities to meet and spend time with others, not on gender. Thus, college women and men—

having equal social opportunities—have about the same number of friends. The same holds for unmarried women and men. Young married women, whose social contacts are typically more confined, tend to have fewer friends. Studies have revealed differences in the quality of male and female friendships, however. Same-sex friendships tend to be more intimate and spontaneous for women than for men. Women are more likely to talk about their feelings and concerns; men, to engage in activities such as sports (Fischer, 1980; Caldwell & Peplau, 1982; Rubin, 1983).

Research on other social behaviors and on intellectual abilities have revealed few differences between males and females (Hyde & Linn, 1986), except those that emerge after age 10 (such as females' greater verbal ability and males' greater visual-spatial ability). Future research may reveal additional differences, however. The causes of most of the differences we observe in everyday behavior of men and women are clear; they are the product of socialization, of responses to social expectations, and of the channeling of opportunities in the family, school, and workplace.

Some research suggests that the lives of successful career women usually include at least one strong, nontraditional figure—a working mother, egalitarian father, or supportive teacher or relative—who provides encouragement and helps the woman to withstand this added source of stress (Huston-Stein & Higgins-Trenk, 1978).

Both overall life satisfaction and satisfaction with specific aspects of life tend to be lower for women and men during this life stage than during most later stages (Lowenthal et al., 1975; Campbell et al., 1976; Gould, 1978). Individuals in this stage are unsure of their objectives, their

abilities, and their futures. Aspirations for occupational and marital success may be high, but individuals worry about translating their dreams into realities. As people make firm career commitments and solidify their identities—and thereby move into the next stage—life satisfaction increases.

### Stage II: Balancing Family and Work Commitments

During their 20s and early 30s, most men and women hold worker, spouse, and parent roles. The central challenge of this stage is to

# Box 16-1

Continued

■ **TABLE B-16-1**

STEREOTYPICAL BELIEFS ABOUT MEN AND WOMEN

| Trait | Most Typical of | | Desirable | | Trait | Most Typical of | | Desirable | |
|---|---|---|---|---|---|---|---|---|---|
| | Men | Women | Yes | No | | Men | Women | Yes | No |
| Independent | ___ | ___ | ___ | ___ | Aware of others' feelings | ___ | ___ | ___ | ___ |
| Aggressive | ___ | ___ | ___ | ___ | | | | | |
| Ambitious | ___ | ___ | ___ | ___ | Submissive | ___ | ___ | ___ | ___ |
| Strong | ___ | ___ | ___ | ___ | Strong need for security | ___ | ___ | ___ | ___ |
| Blunt | ___ | ___ | ___ | ___ | | | | | |
| Passive | ___ | ___ | ___ | ___ | Feelings easily hurt | ___ | ___ | ___ | ___ |
| Emotional | ___ | ___ | ___ | ___ | Self-confident | ___ | ___ | ___ | ___ |
| Easily influenced | ___ | ___ | ___ | ___ | Adventurous | ___ | ___ | ___ | ___ |
| Talkative | ___ | ___ | ___ | ___ | Acts as a leader | ___ | ___ | ___ | ___ |
| Tactful | ___ | ___ | ___ | ___ | Makes decisions easily | ___ | ___ | ___ | ___ |
| Excitable in minor crises | ___ | ___ | ___ | ___ | Likes math and science | ___ | ___ | ___ | ___ |

establish oneself firmly in these roles—to forgo other options and to commit one's energy, time, and self-definition to a particular job and to a particular mate. Priorities must also be set between work and family roles. Women usually give first priority to their roles as wife and mother; men, to their work roles.

**Careers** Both familial and occupational roles are developed during this stage. During the past century, the family careers of over 90 percent of Americans have included marriage. Young adults are marrying later now than in recent decades

(see Table 16-4). In 1980, half of the women had married by age 22.1, and half the men had married by age 24.4. Men and women were marrying at about the same ages in 1980 as they were in the early 1900s. What is not yet clear is whether the proportion who will never marry is increasing.

Obtaining a college education usually delays marriage about two years or more for women and one year for men (Marini, 1978). The expanding premarital period of adulthood is being filled increasingly by the less permanent commitment of cohabitation. In recent years, probably

■ **TABLE 16-4**

PERCENT OF MEN AND WOMEN MARRIED BY VARIOUS AGES AT THREE DIFFERENT PERIODS

| | **Women** | | | | **Men** | | |
|---|---|---|---|---|---|---|---|
| **Age** | *1960* | *1970* | *1980* | **Age** | *1960* | *1970* | *1980* |
| 18 | 24 | 18 | 12 | 18 | 5 | 5 | 3 |
| 22 | 75 | 66 | 52 | 22 | 48 | 48 | 31 |
| 25 | 87 | 86 | 72 | 25 | 72 | 73 | 57 |
| 29 | 91 | 92 | 85 | 29 | 84 | 86 | 76 |

*Source: Adapted from Cherlin, 1981.*

one third of young adults have lived with a member of the opposite sex for a period of six months or more before marriage (Cherlin, 1981). Though the proportion who marry eventually may not be changing, remaining single for an extended period has now become a viable life style (Bernard, 1981).

Divorce rates reveal that many young adults have difficulties establishing firm family commitments. If recent trends continue, about half of those who marry this year will eventually divorce, most within the first ten years of marriage. Recent analyses indicate a strong cohort effect on divorce rates; the rate among persons who married between 1960 and 1964 is much lower than the rate among persons who married between 1975 and 1979 (Morgan & Rindfuss, 1985). But people who get divorced typically remarry, and they tend to do so within three years. As a result, most divorces and remarriages occur before people reach age 35 (Cherlin, 1981).

The most striking transition of this stage is the transition to parenthood. For about 13 percent of couples, the first child arrives within seven months after marriage. For an additional 26 percent of married couples, the first child arrives between eight and twenty-three months after marriage. The probability of having a child slowly declines as the length of marriage increases beyond two years (Sweet & Bumpass, 1987).

Newly married couples need to develop priorities, a shared life style, and commitments to each other. When the first child arrives, it brings the challenge of developing a parental role and a division of labor regarding child care. If the first child arrives shortly after marriage, couples are forced to make both transitions—to marriage and to parenthood—at virtually the same time.

For men, parental and occupational roles are relatively independent. Most men work full-time throughout this stage, barring problems of unemployment or disability. A study comparing parents and nonparents of similar age and marital duration found no differences in career orientation or job characteristics between men who had children and men who did not (Waite, Haggstrom, & Kanouse, 1986). For many women, parental and occupational roles are intertwined. Most women work full-time until they have their first child, stay home to care for their young child(ren), and then return to work as the youngest child grows out of infancy. Statistics for women's employment in 1980 reflect this pattern. Over 89 percent of childless women ages 25–34 (single or married) were in the labor force as compared to only 41 percent of women with children under age 3. Regular jobs outside the home were held by 51 percent of those with children ages 3–5 and 62 percent of women with children ages 6–17. Many women with young children choose to stay home. But women are

Some parents are able to blend work roles and family roles by working at home. As further advances occur in telecommunication, more men and women may choose this option.

also constrained to stay home because their housework load typically doubles with the birth of a child and because child care is both difficult to arrange and expensive.

With the arrival of a first child comes the need to develop a division of home-related work between husbands and wives. Studies consistently find that women spend more time performing housework than men. According to one survey, women spend almost three hours per day on housework as compared to about one hour by men (Bielby & Bielby, 1988). Working mothers spend 3.5 hours per day on child care as compared to 1.75 hours by working fathers. Furthermore, there are differences in the kinds of tasks performed. Typically, women are responsible for

maintenance activities—shopping, cooking, and cleaning—while men are responsible for repairs (Schooler et al., 1984). Thus, "women's work" must be done every day; "men's work" can be done intermittently.

There is considerable interest in the conditions that influence how much time men spend on housework. One influence is gender-role attitudes; nontraditional attitudes, particularly of the husband, are positively related to men's participation in housework (Seccombe, 1986). Employment schedules are also important; when husband and wife work different shifts, men are more involved in child care (Presser, 1988). A third influence is the family life cycle; although wives consistently perform more housework than husbands, the gap is much bigger when there are young children at home, as illustrated in Figure 16-1 (Rexroat & Shehan, 1987).

**Identities and Self-Esteem**  Men committed to a particular line of work become increasingly caught up in their occupational careers during this stage. Men build their identities around their performance at work and around the future advancement they anticipate (Maines & Hardesty, 1987). They are likely to think of themselves as "rising junior executives," "budding craftsmen," "maturing scholars," or "salesmen on the move." Success in their occupational careers brings a rise in self-esteem during this stage. In contrast, men who fail to establish themselves in a work career by the end of this stage experience confusion about their identities and have relatively low self-esteem (VanMaanen, 1976).

Most women become increasingly committed to their families during this stage. Even if they are employed, married women tend to build their identities primarily around their relationships to their husbands and children. They are likely to think of themselves as wives and mothers, as persons who try to provide pleasant, supportive homes for their children and husbands. Women who enter business and the professions often do construct firm identities around their occupational careers. They are sometimes hindered in this, however, by the treatment they receive as a minority in the upper reaches of large organizations. Women are more often excluded from informal, male peer networks and find it harder to be taken seriously than men (Kanter, 1976).

It is evidently more difficult to build a sense of self-esteem on work as a homemaker than on work outside the home (Mackie, 1983). In fact, when we talk about "work," we are usually referring to activities for which people get paid. We devalue housework because it is unpaid labor (Daniels, 1987). There are no clear criteria for judging the quality of homemaking and no raises or promotions to signify a job well done. In contrast, jobs outside the home usually provide supportive social contacts, and the regular paycheck indicates that other people value one's work. Thus, even low-prestige jobs can bolster self-esteem. The effects of outside work on self-esteem can be seen clearly in a survey of working-class women. Three out of four housewives felt incompetent at running their homes, whereas over half the employed wives said they were "extremely good at their jobs," and not one said she felt incompetent (Ferree, 1976). A study of educated middle-class women also showed the effects of work on self-esteem: housewives said they felt "worthless" almost twice as often as employed wives (Shaver & Freedman, 1976).

**Stress and Satisfaction**  Marriage brings a dramatic drop in feelings of stress for most women but an increase in stress for men. Recently married husbands are more likely than their wives to feel rushed, to feel life is hard, and to worry about paying their bills (Campbell et al., 1976).

The birth of a first child increases stress to the highest level in the life course for both husbands and wives. Strain between spouses increases. One study followed couples who were having their first child from late pregnancy until nine months after the baby was born (Belsky, Lang, & Rovine, 1985). Marital quality declined significantly over the period, particularly as reported by wives. The level of stress associated with the transition to motherhood is influenced by the degree to which the birth disrupts a wife's relationship with her husband (Stemp, Turner, & Noh, 1986). On the other hand, the transition is less stressful if a new mother's social network includes other parents (McCannell, 1988) and if she perceives network members as concerned and caring. We can see that intimate relationships are an important buffer during stressful transitions.

Levels of satisfaction are only partly determined by stress; hence, patterns of satisfaction differ somewhat across the life course. Figure 16-2 portrays satisfaction levels for men and

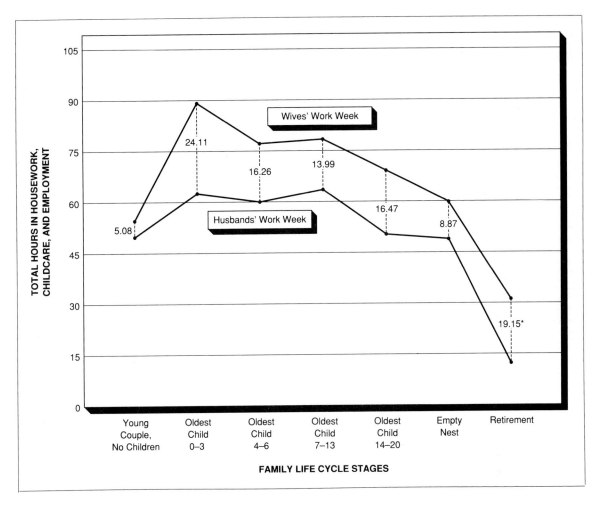

■ **FIGURE 16-1**

TOTAL HOURS (PER WEEK) SPENT BY WORKING SPOUSES IN HOUSEWORK, CHILDCARE, AND EMPLOYMENT

*The retirement stage includes spouses who are not employed.

*Source: Rexroat and Shehan, 1987, Figure 1.*

women in different life cycle stages. Young married women report higher levels of satisfaction with their lives than any other group. Men are also very happy during the childless married years. The arrival of children reduces satisfaction, especially among women. When young children are present, there is less satisfaction with standards of living, with savings, with housing, and with the marriage relationship. It is the divorced and separated men and women who are the least satisfied, however.

In dual-earner families, satisfaction is influenced by time spent together; couples who spend more time talking, eating meals, and having fun together report higher levels of satisfaction with their marriages (Kingston & Nock, 1987). On the other hand, couples in which the wife works more than 40 hours per week outside the home

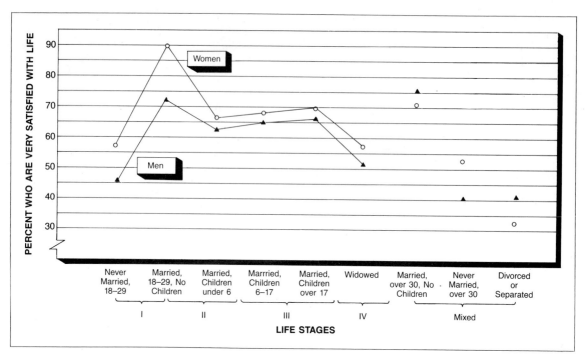

**■ FIGURE 16-2**

SATISFACTION OVER THE LIFE COURSE

A national sample of Americans responded to the question How satisfied are you with your life as a whole these days? Responses were scaled from 1 (completely dissatisfied) to 7 (completely satisfied). This chart shows the percentages of men and women who indicated that they were very satisfied (by choosing 6 or 7 on the scale). The sample has been divided into life cycle categories based on age, marital status, and age of youngest child. The first six categories, arranged in sequence from the left, represent the standard life course pattern, beginning with the young unmarried status, through marriage and parenthood, to widowhood. The curves connecting the responses of these groups show the trends of satisfaction across the life course. The final three categories represent people who diverge from the common pattern. The chart also shows the match between the life cycle categories and the life stages.

*Source: Adapted from Campbell et al., 1976, p. 398.*

are characterized by higher levels of marital instability (Booth et al., 1984). Work and family demands are obviously intertwined and must be balanced if satisfaction is to be kept high and stress kept low.

Although most of the available research has focused on white Americans, a few recent studies have analyzed data from blacks. With regard to global satisfaction, married black men and women are more satisfied than nonmarried ones (Zollar & Williams, 1987). Further, older black adults report higher levels of satisfaction than

younger ones, but this is not a consequence of having children (Broman, 1988b). The same patterns are found in data from white samples. Among blacks, employed persons and men who report doing most of the housework are less satisfied with their family life as compared to whites. (Broman, 1988a).

## Stage III: Performing Adult Roles

What is the major challenge around which people of your parents' age organize their lives? For most adults, the major challenge from their late 30s until into their 60s is to put their lives to a useful purpose—to make a meaningful social contribution (Erikson, 1968). Of course, few people talk about making "meaningful social contributions." Instead, people try to be good workers, parents, and spouses; that is, they try to meet high standards for performance in the adult roles to which they are committed.

**Careers** Most men and childless career women spend the first part of this stage working their way up the occupational ladder. Devoting themselves primarily to their jobs, they seek the increased responsibility, respect, and financial rewards available in their occupations.

The occupational experiences of women and men are likely to be different. More than half of all employed women are in clerical (secretarial, typing, data entry), retail, and service occupations. Men are much more likely to be managers or craftsmen (carpenters, plumbers, electricians) or to be involved in the operation of machinery (Reskin & Hartmann, 1986). Although almost equal percentages of working women and working men are in the professions, women are frequently in the lower-status professions, such as teaching and nursing. These occupational differences produce differences in the prestige and income that men and women derive from employment.

At some point in their 40s or 50s, many workers recognize that their occupational life has reached a plateau that may extend to retirement. Taking stock of where they are, these workers often become less concerned with their own achievements and more interested in promoting others (Lowenthal et al., 1975; Gould, 1978). Thus, lawyers, machinists, managers, or researchers who earlier enjoyed the guidance of mentors may now take pleasure in guiding younger associates themselves. They may also concern themselves more with the social values of their occupations—with fairness and safety in the work setting or with the politics and ethics of their professions.

Workers who are inclined toward greater self-determination may respond to the occupational plateau by pursuing new experiences and challenges, often with the objective of increasing their income. Some seek jobs similar to their current ones in firms or institutions where the path to further advancement is still open (such as leaving one automobile company for another or a local government job for one in Washington). Others return to school or study on their own in order to launch new careers (such as switching from teaching to stockbrokering or from banking to public law). Starting one's own business, moonlighting at a second job, and turning a hobby into a money-making venture are other self-determining responses at this life stage. Career switching in middle adulthood has become easier in the U.S. because there is an increasing cultural emphasis on self-realization and a growing tolerance for varied occupational alternatives for both men and women (Sarason, 1977).

Most employed mothers devote themselves primarily to their families when their children are young. This causes them to fall behind men in their work careers. Thus, women who seek jobs in this stage, after several years at home, are often at a disadvantage; they have no established record of reliable employment, nor have they developed or maintained marketable skills. Consequently, they must often settle for jobs that are below their educational level and less than fulfilling (Sewell, Hauser, & Wolf, 1980). As a result, these women may pursue occupational advancement and personal achievement most intensively during their 40s and 50s, after their children become teenagers. Paradoxically, this may be just when their husbands are beginning to feel less driven or to perceive restricted opportunities for further advancement.

Men change little in their parental role during this lengthy stage. For women, however, the departure of their youngest child from the home may mark a major transition. At this point the "basic tasks of womanhood, as defined in American society, those of bearing and rearing children, are completed" (Lopata, 1971, p. 41). Women who have devoted themselves exclusively

## MALE DOMINANCE IN COMMUNICATION

Asymmetric patterns of verbal and nonverbal communication often reflect status differences between people. Is it, therefore, surprising to find that asymmetry pervades communication between males and females? Numerous studies offer subtle but convincing evidence of a widespread tendency for males to dominate females in cross-sex communication (Thorne, Kamerae, & Henley, 1983). Do any of the following patterns appear in your own interactions?

Regardless of status, women are more likely than men to be addressed by first name rather than by title and last name. This use of lower-status address forms for women has been observed in such work settings as hospitals, universities, and news rooms. Men tend to signal dominance through freer staring, pointing, unreciprocated touch, and walking slightly ahead of the women they are with. Women are more likely to avert or lower their eyes, react passively to touch, and move out of a man's way when passing him on the sidewalk (Henley, 1977; LaFrance & Mayo, 1978; Leffler, Gillespie, & Conaty, 1982).

Joking and laughing also reveal asymmetry. Males tell jokes much more often—a sign of high status; but women laugh harder at the jokes—a sign of submissiveness (Coser, 1960).

Women also tend to be less assertive than men when introducing new topics into a conversation. Women are much more likely to ask a question ("Did you see the report on . . . ?") rather than simply to declare their interest in the topic ("There was a report on . . ."). Women also reveal their lower power by checking more frequently whether their conversational partner is still listening. They do this by adding unnecessary questions onto their assertions ("Isn't it?"), and by inserting "You know" into their remarks (Fishman, 1978, 1980).

Paralinguistic behavior also suggests that women have less interpersonal power. Women use less intrusive responses than men to indicate attention or agreement during conversation. Women prefer head nods and "M-hmn," rather than the more assertive "yeah" or "right." In conversations between men and women, men interrupt more, but women do not protest; and women hesitate longer after interruptions before starting to talk again (Argyle, Lalljee, & Cook, 1968; Zimmerman & West, 1975). Women are more encouraging and supportive listeners; but, contrary to cultural stereotypes, men are generally more talkative (LaFrance & Mayo, 1978).

Why do men use higher status, more domi-

---

to child rearing may feel useless when their children no longer need them, and they may become deeply depressed (Bart, 1975). But because most women today are involved in non-family activities and careers by this transition point, they react to the "empty nest" in a more positive way (Radloff, 1980). Though some feel depressed initially, they tend to adjust quickly and to experience a sense of relief at their reduced family responsibilities. Many women feel more free, self-directed, and open to oppor-

tunity after their children leave home than ever before (Lowenthal et al., 1975; Rubin, 1979).

During this stage, another type of family role may become important: the role of sibling. Brothers and sisters may provide emotional support and direct services throughout one's life (Goetting, 1986). In mid-life, the need to care for elderly parents may lead to increased sibling interaction and cooperation. A study of 50 pairs of sisters noted distinct styles of participation in parental care, ranging from routinely providing

Box 16-2

nant styles of communication than women? One explanation points to gender roles. In Western cultures, men are socialized from early childhood to be dominant and women to be submissive. As a result, men and women use speech styles that reflect the personalities and self-images they have acquired through socialization (Lakoff, 1979). A study of 91 college students compared the influence of sex and of gender identity on speech style. The gender identity of each student was assessed; later, 30 men and 26 women participated in two-person conversations. The results showed that the more masculine the student's gender role, regardless of sex, the greater the likelihood that the person interrupted and engaged in overlapping conversation (Drass, 1986).

A second explanation points to differences in power rather than differences in identity. It proposes that the different communication styles are situational adaptations to the gender-linked distribution of power in society. Because women are granted lower status and power than men in most situations, women lack the power needed to control communication through dominant styles. Instead, women adopt verbal and nonverbal styles that enable low-power individuals to gain some control over communica-

tion (Fishman, 1980; O'Barr & Atkins, 1980).

A study of the influence of power on communication style focused on couples who were living together. The power of each person relative to the partner was measured by an eight-item scale. Couples were selected so that in half of them power was shared equally, while in the other half one had greater power. There were five male-female couples, five male-male couples, and five female-female couples. The results showed that interruptions and back-channel feedback ("M-hmn," "yeah") were linked to power regardless of sex; the more powerful person interrupts more and gives less feedback. Amount of time the person talked was linked to both sex and power; males and women with greater relative power talked more (Kollock, Blumstein, & Schwartz, 1985).

These findings come during a period of social inequality between man and women. Future research will reveal whether this dominance will persist. Both explanations suggest that if and when women achieve equality, signs of male dominance in communication styles will disappear.

care to providing no assistance at all (Matthews & Rossner, 1988). The results of interviews with the women suggest that birth order and geographic proximity are important influences on style; oldest siblings and those who live closest to the parent(s) usually provide the routine care.

**Identities and Self-Esteem** It is widely assumed that men continue to anchor their identities in their occupational roles throughout this mid-life stage. The level of men's self-esteem is

thought to depend primarily on their relative occupational success. People commonly gauge their success by whether they receive promotions and raises early, late, or on time compared to others in similar occupational roles (Levinson, 1978; Lowenthal et al., 1975). However, the results of two surveys of national samples of men (and women) challenge these assumptions (Pleck, 1985). On measures of role involvement, including items such as "My main satisfaction in life comes from my work," employed men and women reported higher levels of involvement in *family* roles than in work roles. In addition, satisfaction with family was more closely associated with overall well-being than satisfaction with work.

Women anchor their identities primarily in their family roles. Therefore, loss of the maternal role when children leave home might weaken a woman's sense of identity temporarily and undermine her self-esteem. But the ensuing period of freedom is usually accompanied by an expanding sense of competence, maturity, and self-assurance (Harris, Ellicott, & Holmes, 1986). Menopause was also once thought to weaken women's self-assurance and self-esteem. Perhaps this was true when menopause signified the end of childbearing—the loss of what was considered women's most important capacity. But menopause no longer has this social meaning in the U.S., because women end childbearing earlier today and cultivate alternative roles around which to build identities. There is even cross-cultural evidence that following menopause women become more assertive, dominant, and autonomous; whereas men at this age become more nurturant and affiliative (Guttman, 1977; Freedman, 1979).

Being employed increases the self-esteem of married women (Kessler & McCrae, 1982; Baruch & Barnett, 1986). In particular, women who complete college and work as professionals have a firmer sense of identity and higher self-esteem. This was illustrated in a study (Birnbaum, 1975) that compared the sense of identity and self-esteem of a group of married female professionals, single female professionals, and housewives (see Table 16-5). Single female professionals resembled married female professionals on both dimensions, although the group of singles felt as lonely and unattractive as did housewives. The responses of each group suggest that employment for women contributes to a firmer identity and sense of competence and that the combination of employment and marriage combats feelings of loneliness and being unattractive to men.

**Stress and Satisfaction**  For both married men and married women, this stage is a period of declining psychological stress. For parents, the following types of stress are greatest when children are under 6, and decline steadily as children grow older: feeling tied down, feeling life is hard, worrying about having a nervous breakdown, and—especially—worrying about finances (Campbell et al., 1976).

The drop in stress levels at this stage fits with a general theory about "life cycle squeeze" as a determinant of stress and satisfaction (Wilensky, 1961). The term **life cycle squeeze** refers to the fact that in certain life stages, financial and family burdens are typically great while job status and rewards are low. This produces an unfavorable balance between aspirations and resources. Life cycle squeeze is relatively low among young singles and childless couples. It increases sharply for couples with preschool children, producing the increased stress and dissatisfaction noted above. During this stage, life cycle squeeze eases: workers' earnings and status rise toward their peak, while family and financial burdens lessen. This leads to decreased stress and increased life satisfaction. However, if people are divorced or separated, they may experience the imbalance of resources and aspirations that characterizes life cycle squeeze. Surveys reveal that life satisfaction, stress, and happiness generally rise and fall together with the experience of life cycle squeeze (Estes & Wilensky, 1978).

Although life cycle squeeze eases as the mid-life stage progresses, two other sources of stress gradually increase: physical illness and the death of parents or close friends. People ages 45–64 reported three times more anxiety about physical illness than people ages 21–34 in one national survey (Gurin, Veroff, & Feld, 1960). In a more recent study, women's reports of deteriorating health (involving their eyesight, hearing, teeth, hair, and so on) increased greatly after age 45 (Rossi, 1980). The death of a parent or close friend can be stressful because it increases one's

■ **TABLE 16-5**

EFFECTS OF MARRIAGE AND EMPLOYMENT ON IDENTITY AND SELF-ESTEEM OF EDUCATED MIDDLE-AGED WOMEN

| | Percentage of Housewives (N = 29) | Percentage of Married Professionals (N = 25) | Percentage of Single Professionals (N = 27) |
|---|---|---|---|
| **Identity** | | | |
| Feelings of uncertainty about who you are and what you want: | | | |
|   Hardly ever | 34 | 64 | 58 |
|   Fairly often | 66 | 36 | 42 |
| Feeling lonely: | | | |
|   Hardly ever | 28 | 72 | 27 |
|   Sometimes to often | 72 | 28 | 73 |
| **Self-Esteem** | | | |
| Self-evaluated competence in five areas (domestic, social, child care, cultural, intellectual): | | | |
|   Poor to average | 31 | 4 | 15 |
|   Average to good | 55 | 42 | 31 |
|   Good to very good | 14 | 54 | 54 |
| Feels "not very" attractive to men: | 61 | 12 | 58 |

Note: Housewives who were interviewed had graduated with honors from the University of Michigan 15–25 years earlier. The employed single or married professionals held Ph.D.'s or M.D.'s, and were matched with housewives of similar age.

Source: Adapted from Birnbaum, 1975.

own sense of mortality, deprives one of important roles as child or friend, may increase fears about health, and may impose added financial burdens. But the disruption of day-to-day living resulting from parental death is seldom severe. This is because elderly parents rarely play a central role in the lives of their adult children in our society (Kalish, 1976).

## Stage IV: Coping with Loss

Most of us will enter the final life stage during our 60s. Retirement or the onset of a major physical disability are the key markers of the transition into this stage. The central challenge of this stage for older adults is to cope with a series of practically unavoidable losses: loss of one's occupational role through retirement,

of significant relationships through death, and of health, energy, income, and independence. Despite the severity of these losses, most older people say they are quite satisfied with life and that they are coping fairly well (Harris et al., 1982).

Most people who reach age 65 in the U.S. today will spend as many years in this final stage of life as they spent in childhood and adolescence. Men who reach 65 can expect to live another 14 years; women, another 18.

Through many of these years, older people can actively enjoy a wide range of activities because most remain reasonably healthy at least until age 75. Yet the aged, as a group, are often mistakenly believed to be narrow-minded, unteachable, not very bright, uninterested in sex,

## THE MID-LIFE CRISIS: HOW NORMAL? HOW NECESSARY?

It is widely believed that some time between ages 35 and 44 most of us—especially men—will pass through a period of severe inner turmoil. If we open ourselves to the inevitable questions and repressed self-doubts this period raises, we will emerge with renewed vigor and authenticity to face the second half of our lives. If we fail to confront our inner worries and dissatisfactions and to reassess profoundly our past commitments and behavior, we will lose vitality and remain constricted in our actions during our remaining years. This **mid-life crisis** is the critical stage of transition into middle age, when people become aware that time is running short and that unless they make changes in their personal relations, in their work, and in themselves, it will be too late.

One factor contributing to this crisis is the *aspiration-achievement gap.* As we begin our occupational careers, in stages I and II, we often have high aspirations, hoping that we will rise to the top of our chosen field. In mid-life, after 20 years of hard work, many realize that they have not achieved and, more importantly, will not

achieve such high status. This realization can trigger distress and a reevaluation of one's commitments. Note that anyone with high aspirations may experience this gap, whether male or female.

The mid-life crisis has been discussed in social science research (Gould, 1978; Levinson, 1978), popularized in the best-seller *Passages* (Sheehy, 1976), and discussed on countless talk shows. When a Wisconsin psychiatrist set up a "mid-life crisis hotline," his phone rang off the hook. Despite the widespread acceptance of this concept, many remain skeptical (Brim, 1976; Bush & Simmons, 1981). The existence and nature of the mid-life crisis have been established primarily through in-depth interviews and discussions with highly educated men. This research may be biased, however, because the interviews have been carried out by researchers who assumed that adulthood consists of a series of discrete stages and transitions.

The major concerns that have been raised regarding the mid-life crisis revolve around two questions: (1) Is a mid-life crisis a normal, virtu-

and not good at getting things done (Harris et al., 1975). Thus, in addition to coping with losses, older people must often cope with **ageism**—prejudice and discrimination against the elderly based on negative beliefs about aging.

**Careers**  Most older people maintain strong primary relationships. These include ties to their adult children. As children grow into adulthood, parents gradually give up responsibility for their offsprings' personal care, work, and financial status; the relationship evolves toward status equality (Blieszner & Mancini, 1987). Parents and children usually continue to have strong emotional ties, communicate regularly, and may

provide various kinds of help to each other. Parents are selective in providing aid, giving it to the offspring who needs it—for example, temporary financial support in the event of unemployment (Aldous, 1987).

More than half the elderly live with their spouses, and marriage continues to be their most important social tie. But the marital relationship often becomes more egalitarian during this stage. Women typically continue the shift toward greater assertiveness and independence begun in the preceding life stage, and men continue their shift toward greater nurturance and expressiveness (Guttman, 1977). Retirement adds to equality by reducing differences between spouses'

Box 16-3

ally universal part of aging for men? and (2) Is it necessary for a vital, healthy later adulthood?

In exploring these questions, researchers began by developing an inventory of the problems that men undergoing a mid-life crisis presumably experience according to the literature (Costa & McCrae, 1980). They identified types of problems (such as inner turmoil, change in time perspective, increased awareness of the repressed parts of self, and marital and job dissatisfaction). Members of a first sample ($N = 233$) reported whether they were experiencing these problems by completing a questionnaire. In order to determine whether the problems presumably typical of the mid-life crisis became more frequent or intense between the ages of 35 and 45, the researchers compared various age groups between 33 and 79. To their surprise, there were no age differences at all in the number or intensity of problems reported. Replicating this study with a second sample ($N = 315$), the researchers again found no differences between age groups.

What do these findings mean? Apparently a mid-life crisis is not a universal, normal part of aging associated with any one period in adulthood. But is a mid-life crisis necessary for later health and vitality? Some additional findings suggest that the answer is no. Men who reported many of the problems presumably associated with the mid-life crisis also scored high on a neuroticism questionnaire. These same men also scored high on neuroticism 10 years earlier. This strongly suggests that the current crises of these men were "simply one more incident in a life-long history of maladjustment" (Costa & McCrae, 1980, p. 84).

Results of these studies suggest that men (and women) do confront problems with their work and family commitments, identities, and self-evaluations through adulthood. But, contrary to popular dramatization, people do not struggle intensely with all of these problems at any single adult stage. In fact, a struggle of crisis proportions is neither common in the general population nor a sign of future vitality.

activities. The happiest marital relationships among the elderly are those characterized by relative equality, mutual emotional support, and flexible sharing of household tasks (Sinnot, 1977).

Death of one's spouse is the most severe trauma the elderly confront. Women are typically widowed because they have longer life expectancies than their husbands. Half of married women lose their husbands by age 70. When husbands outlive their wives, they usually become widowers only after age 85. Death of a spouse causes many types of loss. It severs the deepest of emotional bonds, takes away the main companion in day-to-day activities, frustrates the

fulfillment of sexual needs, removes the key significant other for affirming one's identity, and—especially for women—produces economic loss.

How well a widow copes with her husbands' death depends on several factors. First, the death of a spouse is more stressful if it is sudden and unexpected, if other family members or friends die within weeks of the spouse's death, or if the surviving spouse is in poor health prior to the death (Sanders, 1988). Further, a woman whose identity and life style were built on her marital role will experience greater disorganization when her husband dies (Lopata, 1988). Finally, women with strong social support networks cope

Despite the loss associated with old age, most elderly people say they are quite satisfied with life. As long as they stay healthy and independent, most elderly people maintain their social involvements, activities, and self-esteem.

with bereavement more successfully. Continuing interaction with married sisters seems to be a particularly important contributor to a widow's well-being (O'Bryant, 1988).

Although there are a few tragic exceptions, older Americans—married or not—are rarely abandoned by their families. In fact, more than half see at least one of their adult children almost every day, and the vast majority have contact with a child or sibling every week (Shanas, 1979; Harris et al., 1982). The image of the elderly as isolated in institutions such as nursing homes is inaccurate. Only about one in four is likely to spend *any* part of this life stage in an institution (Tobin, 1980). Very few older people live with their adult children; this reflects a preference for independence on both sides,

not the lack of an emotional bond. When the need arises, adult children provide the overwhelming proportion of personal care (Steere, 1981).

Occupations, the second main career line, come to an end for most older people with retirement between ages 60–65. Ideally, retirement should be a gradual process, loosely linked to aging, because occupational abilities and inclinations diminish only gradually. People who have control over giving up their occupational careers (for example, top management and the self-employed) do, indeed, withdraw more gradually (Hochschild, 1975). Retirement brings many losses; income, prestige, one's sense of competence and usefulness, and one's social contacts may all decline. Retired persons may need to fill free time, develop new everyday routines, and—if married—adjust to spending more time with their spouses. Men experience substantial role discontinuity upon retiring. Women tend to experience less discontinuity because they retain at least their homemaking responsibilities.

The main fear of retirees is that they will be cut off from social participation. Does this happen? As a general rule, people continue the level of social involvement they developed in their preretirement years. Those who were constantly busy and involved with people find new outlets in voluntary activities, hobbies, and social visits. Individuals who were uninvolved remain so. Few withdraw further (Palmore, 1981).

Reduced financial resources do, however, disrupt social participation, especially among retirees from the working class (Robson, 1982). Retirees who must worry about finances hesitate to spend money traveling to visit friends and entertaining them. Nor can they buy "proper" clothes or tickets for social events. Working-class couples also have more difficulty with the increased presence of retired husbands in the home. Because they are accustomed to exclusive control over the household, working-class wives tend to experience this as an irritating invasion. Middle-class wives are more likely to welcome the added companionship (Kerckhoff, 1966).

**Identities and Self-Esteem** Retirement, widowhood, and declining health deprive people of many of the central roles and relationships around which their identities have been built. Considering these losses, identity change in this stage is less than one might expect (Atchley, 1980). The relative stability of identities is due primarily to the fact that older people continue to think of themselves in terms of their former roles. Though retired, individuals still think of themselves as nurses, accountants, or musicians, for example. Though widowed, they remain "John's wife" or "Sarah's husband" in their own eyes. Interactions with siblings often reinforce continuing identification with roles occupied

earlier in life—for example, through reminiscences (Goetting, 1986). Identities are also preserved because many personal qualities remain stable. Individuals are likely to see themselves as unchanged in their honesty, outspokenness, religiousness, and so on.

Neither widowhood nor retirement alone do serious damage to self-esteem (Atchley, 1980). Several aspects of aging are, however, associated with self-esteem loss. Whatever deprives older people of their independence and control over their own lives—such as ill health or falling into poverty—weakens self-esteem. Moving into an institution or family residence where one becomes highly dependent also undermines self-esteem if caretakers make all the decisions for an older person. These well-meaning actions communicate the assumption that the older person is mentally and physically incompetent.

**Stress and Satisfaction** Fear of death is not a major source of stress for older people. Most accept its inevitability. The three main sources of stress are perceived financial difficulties, failing health, and reduced social activity (Atchley, 1980; Palmore, 1981). Anxiety over financial difficulties is particularly high among widows. Financial anxiety also plagues retired persons whose incomes are fixed in periods of rapid inflation. Stress from failing health rises throughout this life stage, especially after age 75. Older people become anxious and depressed when they experience reduced energy levels, lack of motivation, memory loss, a slowdown in their ability to process information, and chronic or acute diseases. Depression induced by ill health is often misdiagnosed as mental deterioration and confusion. As a result, depressed older people often fail to receive the psychotherapeutic treatment that could restore them to alertness and satisfactory functioning.

A widely publicized theory of adjustment to aging, **disengagement theory,** holds that withdrawal from social commitments is inevitable with aging and that withdrawal promotes satisfaction with life because it frees up time and energy for introspection (Cumming & Henry, 1961). The evidence, however, contradicts this theory. First, disengagement is not inevitable. Older people whose social and occupational opportunities permit usually remain engaged. For instance, elderly carpenters may continue their cabinetmaking, and emeritus professors may continue their research and writing. Second, life satisfaction is more often associated with engagement than with disengagement. With few exceptions, older people who are more involved in social activities and relationships report higher levels of satisfaction (Palmore, 1981; Hanson, 1986).

# ■ Historical Variations

Throughout this chapter we have based our description of stages in the life course on recent findings and on projected future trends. But unique historical events—wars, depressions, medical breakthroughs—change life courses. And historical trends—fluctuating birth and divorce rates, rising education levels, varying patterns of women's work—also influence the life courses of individuals born in particular historical periods.

No one can predict with confidence the future changes that will result from historical trends and events. What can be done is to examine how major events and trends have influenced life courses in the past. Two examples will be treated: the historical trend toward greater involvement of women in the occupational world and the effects of military service during the Vietnam War on the lives of men. The goals of this section are (1) to emphasize the fact that historical trends continually influence the typical life course and (2) to illustrate how to analyze the links between historical events and the life course.

## Women's Work: Sex-Role Attitudes and Behavior

At their graduation exercise in 1955, the women of Smith College were launched into the adult world with the following advice from Adlai

Stevenson (senator, U.N. ambassador, and presidential candidate): A women's place in politics is to "influence man and boy" through the "humble role of housewife." Misguided as these remarks may sound today, they probably struck a responsive chord in most of the audience; for during the 1950s, it was widely accepted that "there is some work that is men's and some that is women's, and they shouldn't be doing each other's" (Thornton & Freedman, 1979).

**Sex-Role Attitudes** In the past three decades, attitudes toward women's roles in the world outside the family have changed dramatically. The historical trend in attitudes has been away from the traditional division of labor (paid occupations for men and homemaking for women) to a more egalitarian view. But much of this change did not occur until well into the 1960s.

Consider some of the following statements of sex-role attitudes. Do you agree with them?

1. It is better if the man is the achiever outside the home and the woman takes care of home and family.
2. Most of the important decisions in life should be made by the man of the house.
3. If her husband can support the family, a woman should not work for pay.
4. It is wrong for a woman to be very active in clubs, politics, and other outside activities before her children are grown.
5. Preschool children are likely to suffer if their mother works.

These are typical of attitude statements included in one or more large-scale surveys of adult women during the 1960s and 1970s. In the 1960s, about two thirds or more of women surveyed agreed with these statements. However, by the late 1970s, three of these attitudes had become minority views (2, 3, and 4), and statements (1) and (5) were endorsed only by about half the women. This shift from traditional to egalitarian sex-role attitudes has been quite strong among women (Thornton & Freedman, 1979; Spitze & Huber, 1980).

**Work Force Participation** This historical trend is not limited to attitudes. Women's actual participation in the work force has been on the increase for almost a century. Figure 16-3 shows the percentage of women employed outside the home since 1951. The proportion of married women who are employed has been growing steadily, with a slight acceleration of growth between 1970 and 1980. Among young single women, the employment level, which was already very high in 1951, has remained high. The proportion of women who work during pregnancy and who return to work while their child is still an infant has also grown steadily over this time period (Sweet & Bumpass, 1987).

Why have women joined the work force in ever-greater numbers throughout the twentieth century? Has the spread of egalitarian attitudes been an important source of influence? Probably not. The idea that wives and mothers should not work, except in cases of extreme need, was widely held until the 1940s. Yet women's employment increased steadily between 1900 and 1940. The change in sex-role attitudes occurred largely in the 1970s, yet women's employment rose rapidly during the two decades preceding these attitude changes. Therefore, it seems likely that sex-role attitude changes have not been a cause of the increased employment of women but a response to it—an acceptance of what more and more women are in fact doing.

What, then, are the causes? Perhaps most convincing is the argument that the types of industries and occupations that demand female labor are the ones that have expanded most rapidly in this century. Light industries like electronics, pharmaceuticals, and food processing have grown rapidly, for example, and service jobs in education, health services, and secretarial and clerical work have multiplied. Many of these occupations were so strongly segregated by sex that men were reluctant to enter them (Oppenheimer, 1970). In addition, male labor has been scarce during much of the century due to rapidly expanding industry and commerce. The majority of the slack was taken up by a large pool of unemployed married women. These women could be pulled into the work force at a lower wage because they were often supplementing their family income.

The changes noted in the preceding paragraph led to increased job opportunities for women. Other factors influenced women's desire to work outside the home. One of these was continuing inflation and rising interest rates; in many families, two incomes became necessary to make ends meet. Other factors that may have promoted the increased employment of women include rising divorce rates, falling birth rates, rising education levels, and the invention of labor-saving devices for the home. None of these factors alone can explain the continuing rise in the employment of women over the whole century. However, at one time or another each of these factors probably strengthened the historic trend, along with changes in sex-role attitudes.

The specific changes in women's work behavior demonstrate that the timing of a person's birth in history greatly influences the course of his or her life. Whether you will join the work force depends in part on historical trends during your lifetime. So does the likelihood that you will get a college education, marry, have children, divorce, die young or old, and so on.

## Effects of Historical Events: The Vietnam War

An important recent historical event was the Vietnam War. It affected the lives of millions of Americans, some directly and others indirectly. One group whose lives were directly touched by the war were the hundreds of thousands of men and women who served in the military during the period 1964–73, when U.S. combat troops fought in Vietnam.

There is growing literature on the effect of military service in Vietnam, much of it focused on the association between exposure to traumatic experiences during the war and subsequent psychological and social difficulties. To assess the impact of a historical event such as this, one must compare those who were directly involved in the war with similar persons who were not. In this case, we want to compare men who served in

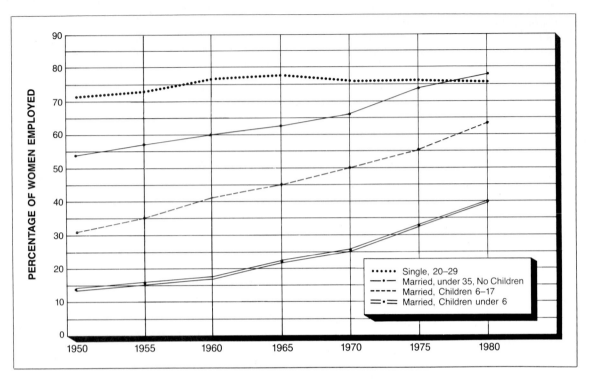

■ **FIGURE 16-3**

WOMEN'S EMPLOYMENT: 1951–1980

The percentage of married women who are employed has risen steadily since 1950. Young single women have maintained virtually the same high level of employment throughout this period. Among married women, in each year, women with young children were least likely and childless married women were most likely to be employed.

*Source: Adapted from Sweet, 1985.*

Vietnam with men who are similar in age, race, and education but did not serve. It is possible, of course, that serving in the military has certain effects regardless of whether the person is involved in combat. We can take this into account by studying a group of veterans who were in the military service at the same time but did not serve in Vietnam. Comparing these three groups —nonveterans, non-Vietnam veterans, and Vietnam veterans—is a comparison *within* a birth cohort.

Laufer and Gallops (1985) have carried out a within-cohort analysis of this type. Their data are based on a sample of 1,259 men who were draft eligible (ages 18–26) during the war. The data were collected in 1977 and 1979, providing information on the life experiences of the veterans for a period of up to twelve years following their discharge from the military. What was the effect of military service and combat experience on familial careers? Table 16-6 summarizes the results. Looking first at the likelihood of marriage, veterans—especially those who served in Vietnam—were more likely to have married by the time of the survey. Also, veterans married for the first time at somewhat younger ages, with 67 percent of those who served in Vietnam married by age 25 compared with 56 percent of the nonveterans. Additional analyses of the data indicate that the veterans married quickly after

■ **TABLE 16-6**

EFFECTS OF MILITARY SERVICE AND COMBAT EXPERIENCE ON MARITAL PATTERNS

| | | Military Experience | |
| --- | --- | --- | --- |
| Marital Patterns | Nonveterans (N = 590) | Non-Vietnam Veterans (N = 341) | Vietnam Veterans (N = 326) |
| Rates of marriage | 70% | 81% | 84% |
| Percentage married by 25 | 56% | 58% | 67% |
| Rates of divorce | 24% | 27% | 20% |

Source: Adapted from Laufer and Gallups, 1985.

they left the service. Thus, at the same time they were experiencing the transition back to civilian life and trying to settle into an occupational role, they also made the transition to marriage. Such a pile-up of life events and transitions can be especially stressful.

What happened to these marriages? We might expect that because of continuing combat-related stress the veterans' marriages were more likely to end quickly in divorce. The non-Vietnam veterans did experience the highest divorce rate; at the same time, those who served in Vietnam had the lowest divorce rate. However, when those who served in Vietnam were grouped according to their amount of combat experience, the results indicate an increasing likelihood of divorce as exposure to combat increased. Additional analyses indicated that Vietnam veterans with combat experience reported the largest number of symptoms of stress in the year prior to the interview and were most likely to report drinking heavily during the two years before the survey.

From this study, we draw two general conclusions. First, the more exposed individuals are to a historic event, the more it will impact on their lives. For example, the more directly a man was involved in the war, the greater it affected the likelihood of divorce. Second, historical events influence individuals by changing everyday interactions, socialization experiences, and/or life chances. Men who served in the military in Vietnam seem to have rushed into marriage, creating a pile-up of stressful life events.

■ **Summary**

This chapter discusses the life course and sex roles in U.S. society.

**Components of the Life Course** This chapter focuses on three components of the life course. (1) The life course consists of careers—sequences of roles and associated activities. The principal careers involve work, family, and friends. (2) As we engage in career roles, we develop role identities; and evaluations of our performance contribute to self-esteem. (3) The emotional reactions we have to career and life events include feelings of stress and satisfaction.

**Influences on Life Course Progression** There are three major influences on progression through the life course. (1) The biological growth and decline of the body and brain set limits on what we can do. The impacts of biological developments on the life course, however, depend on the social meanings we give them. (2) Each society has a customary, normative sequence of age-graded roles and activities. This normative sequence largely determines the bases for building identities, the responsibilities and privileges, and the socialization experiences available to individuals of different ages. (3) Historical trends and events modify an individual's life course. The impact of a historical event depends on a person's life stage when the event occurs.

**Stages in the Life Course: Age and Sex Roles**  Four broad life stages characterize the life course beyond childhood. (1) Achieving independence (ages 16–23) entails numerous crucial transitions in education, work, and family life. Stress is high, and satisfaction with life is relatively low during this stage. Young men tend to emphasize jobs in building their identities; young women emphasize family ties. (2) Balancing family and work commitments (ages 18–40) through marriage and work careers is the norm. The patterns and timing of family and work careers have undergone major changes in recent decades. For most men, identity and self-esteem are tied to occupational success; for many women, family remains central. Satisfaction rises with marriage, but the birth of a first child increases stress and reduces satisfaction. An important source of stress is unequal division between men and women of household labor. (3) Performing adult roles competently (ages 35–70) leads to an occupational plateau for most men, who then seek other challenges. Most women turn more to occupational advancement or to community involvement as family demands recede. Self-esteem is supported by occupational achievement for both women and men. Stress eases during this stage because resources grow faster than needs. (4) Coping with loss (ages 60–90) is the key challenge for the elderly. Most cope well with retirement and even with the death of a spouse unless or until they experience financial difficulties, failing health, or reduced social participation. As long as the elderly retain independence, their self-esteem remains stable and their identities change little.

**Historical Variations**  The historical timing of one's birth influences the life course through all stages. (1) Over the past 30 years, women's participation in the work force has increased dramatically and attitudes towards women's employment have become much more favorable. The likelihood that women will experience pressures and opportunities to work outside the home is now greater at every life stage. (2) Military service in Vietnam, particularly combat experience, has had long-term effects on the family careers of men. Those who served in Vietnam married at younger ages, and those with combat experience were more likely to experience divorce and other difficulties a decade after the war.

## Key Terms

17

chapter

# Social Structure and Personality

# ■ Introduction

Fred is 38, married, the father of two children, and sells pacemakers and artificial joints to hospitals. He travels two or three days a week and works at home the rest of the time in his $150,000 house in the suburbs. He earns $65,000 a year. Because his income is based entirely on commission, Fred worries about his sales falling off; but, on the whole, he is satisfied with his life. His values are conservative, and he votes for Republican candidates.

Larry is also 38 and has a wife and two children. He runs a service station and works six days a week, from early morning until 6:00 or 7:00 P.M. Larry and his family live in a small, three-bedroom house. Last year, he made about $22,000. He worries a lot about money and has been very tense the past year. He has liberal values and usually votes for Democratic candidates.

Marie is 39. She is head nurse in a hospital pediatric ward. Although she enjoys her patients, she hates all the paperwork and the personnel problems. Some of her values are conservative, whereas others are liberal; she considers herself an independent.

Fred, Larry, and Marie are three very different people. They have different occupations, which produce differences in income and life style. They differ in their values—in what they believe is important—and in the amount of stress they feel.

Where do these differences come from? Often they are the result of one's location in society. Every person occupies a **position**—a designated location in a social system (Biddle & Thomas, 1966). The ordered and persisting relationships among these positions in a social system make up the **social structure** (House, 1981a).

This chapter is concerned with the impact of social structure on the individual. There are three ways in which social structure influences a person's life. First, every person occupies one or more positions in the social structure. Each position carries a set of expectations about the behavior of the occupant of that position, called a **role** (Rommetveit, 1955). Role expectations are anticipations of how people will behave based on knowledge of their position. Through socialization and personal experience, each of us knows the role expectations associated with our positions (Heiss, 1981). For example, Fred enacts several roles, including salesman, husband, and father. The expectations associated with these roles are a major influence on Fred's behavior.

A second influence on the individual are **social networks**—the sets of relationships associated with the various positions a person occupies. Each of us is woven into several networks, including those involving co-workers, family, and friends. Two examples of social networks are depicted in Figure 17-1. Both networks center on a single individual, Mary. The network on the left depicts the pattern of relationships among Mary and her friends that evolved out of their shared experiences. For example, the ties between Mary and Margie and John developed because they took classes together, whereas Mary got to know Kathy at work. Mary described this network when asked to name the people to whom she is closest, excluding relatives. When asked to include relatives, Mary described the network on the right of Figure 17-1. The most striking difference is the greater number of direct ties between the persons in the network on the right. This network is *dense*—most of the persons in it know each other independently of their ties to Mary (Milardo, 1988). In both of these social networks, a tie between Mary and another person reflects a **primary relationship,** one that is personal, emotionally involving, and of long duration. Such relationships have a substantial effect on one's behavior and self-image (Cooley, 1902).

A third way that social structure influences the individual is through **status**—the social ranking of a person's position. In every society, some positions are accorded greater prestige than others. Differences in ranking indicate a person's relative standing—his or her status—in the social structure. Each of us occupies several positions of differing status. In the U.S., occupational status is especially influential. It is the major determinant of income, which has a substantial effect on one's life style. One of the obvious differences between Fred and Larry, for instance, is their annual income.

This chapter will focus on how social structure influences the individual—via roles, social networks, and status. It considers the impact of social structure on four areas: achievement, values, physical and mental health, and a person's sense of belonging. Specifically, it considers the following four questions:

1. What are the main influences on the status a person attains?

2. How does social position influence a person's values?

3. How does social location influence a person's physical and mental health?

4. How does position influence a person's sense of belonging in society or the lack thereof?

# ■ Status Attainment

An individual's relative standing in the social structure—that is, status—is perhaps the single most important influence on his or her life. Status determines access to resources—to money and influence over others. In the U.S., one's occupation is the main determinant of status. This section will consider the nature of occupational status, the determinants of the status that particular individuals attain or achieve, and the impact of social networks on the attainment of status.

## Occupational Status

Occupational status is a key component of social standing and a major determinant of income and life style. Fred is a sales representative for a company that makes artificial hip and

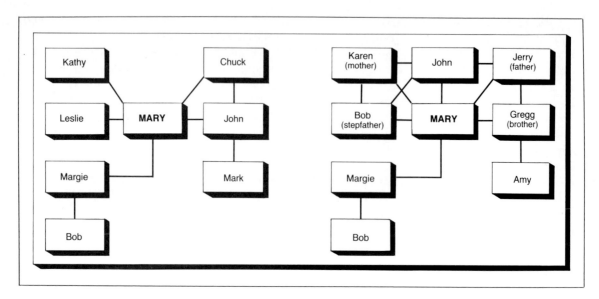

■ **FIGURE 17-1**

SOCIAL NETWORKS

These two networks are focused on Mary. When asked to name her best friends, excluding relatives, Mary described the network on the left. John is her boyfriend, whereas Mark is John's roommate. Mary met Chuck through John, and since then she and Chuck have become friends. Mary's three best women friends are Kathy (a co-worker), Leslie (whose parents are friends of Mary's parents) and Margie (a classmate). Margie has recently married Bob. This is a loosely knit network, because few of Mary's friends are close to each other. The network on the right represents the people to whom Mary feels closest, including her relatives. Notice that only Margie and John remain from the friendship network. Mary's intimates now include Gregg (her brother) and Amy (Gregg's girlfriend). They also include her mother (Karen), stepfather (Bob), and her father (Jerry). This network is dense, because most of its members are close to Mary and to each other.

elbow joints, pacemakers, and other medical equipment. These items are in great demand, and few companies make them. Fred sells a single pacemaker for $3,000 and keeps half of the money as his commission. He only needs to be on the road two or three days a week to earn $65,000 each year. He has a beautiful suburban home and two cars. Larry, by contrast, owns a service station. He works from morning until night pumping gas and repairing cars. His station is in a good location, but his overhead is high; he only earned $22,000 last year, and he worries that this year will be worse. Larry and his family live in a smaller, older house and have a six-year-old car.

The benefits Fred and Larry receive from their occupational statuses are clearly different.

First, Fred earns four times as much money as Larry. This determines the quality of housing, clothing, and medical care his family receives. Fred also has much greater control over his own time. Within limits, he can choose which days he works and how much; this, in turn, affects the time he can spend with family and friends. Larry doesn't have much free time. Finally, Fred receives a great deal of respect from the people he works with. He controls a scarce resource, so doctors and hospital personnel generally treat him well. Larry, however, deals with people who are usually preoccupied or angry because their cars are not running properly.

In addition to these tangible benefits, occupational status is associated with prestige. Several surveys in the U.S. have found that there is

Uniforms are a very efficient way of communicating one's role. We can tell at a glance what role this woman is enacting, and as a result, we know how to behave toward her.

widespread agreement about the prestige ranking of specific occupations. In these studies, respondents typically are given a list of occupations and asked to rate each occupation in terms of its "general standing" or "social standing." The average rating is often used as a measure of relative prestige. The prestige scores for the U.S., shown in Table 17-1, were taken from the Standard International Occupational Prestige Scale, which ranges from 10–90. Surprisingly, there is substantial agreement across diverse societies in the average ranking of occupations. Even adults in China give rankings that are similar to those displayed in Table 17-1 (Lin & Xie, 1988).

The social structure of the U.S. can be viewed as consisting of several groups or social classes. A *social class* consists of persons who share a common status in society. There are various views regarding the nature of social classes in the U.S. One view of social class emphasizes occupational prestige, in conjunction with income and education, in defining class boundaries. This approach ordinarily classifies people into upper-upper, lower-upper, upper-middle, lower-middle, working, and lower classes (Coleman & Neugarten, 1971). A very different approach emphasizes the control, or lack thereof, an individual has over his or her work and co-workers as the main determinant of class standing (Wright et al., 1982).

## Intergenerational Mobility

When a person moves from an occupation lower in prestige and income to one higher in prestige and income, he or she is said to be engaged in **upward mobility.** To what extent is upward mobility realistically possible in the U.S.? On the one hand, we have the Horatio Alger rags-to-riches imagery in our culture. Supposedly, anyone who is determined and works hard can achieve economic success. This imagery is fueled by stories about the astonishing success of the man who invented the transistor, the woman who founded Mary Kay cosmetics, the man who borrowed from his friends to establish Motown Records, and so on. On the other hand, some argue that the U.S. is not an open society, that our eventual occupational and economic achievements are fixed at birth by our parents' social class, our ethnicity, and our gender. To be sure, every city includes families who have been wealthy for generations and families who have been poor for just as long. This suggests that the U.S. is characterized by *castes*—groups whose members are prevented from changing their social status.

These two views of upward mobility in U.S. society are concerned with *intergenerational mobility*—the extent of change in social status from one generation to the next. To measure intergenerational mobility, we need to compare individuals' social status with that of their parents. If

■ TABLE 17-1

OCCUPATIONAL PRESTIGE IN THE U.S.

| Occupation | Prestige |
|---|---|
| Physician | 78 |
| College or university professor | 78 |
| Lawyer | 72 |
| Dentist | 70 |
| Airplane pilot | 66 |
| Electronic engineer | 65 |
| Sales representative | 61 |
| Clergy | 60 |
| Elementary school teacher | 57 |
| Social worker | 56 |
| Office manager | 55 |
| Registered nurse | 54 |
| Legal secretary | 53 |
| Computer programmer | 51 |
| Radio/TV announcer | 50 |
| Airline stewardess | 50 |
| Athlete | 49 |
| Dental hygienist | 44 |
| Insurance salesperson | 44 |
| Auto mechanic | 43 |
| Farmer | 40 |
| Salesclerk | 38 |
| Carpenter | 37 |
| Hairdresser | 35 |
| Mail carrier | 33 |
| Streetcar operator | 29 |
| Waiter | 23 |
| Gas station attendant | 22 |
| Garbage collector | 13 |

Source: Adapted from Treiman, 1977.

the rags-to-riches image is accurate, we should find that a substantial number of adults attain a social status significantly higher than their parents. If the caste-society image is correct, we should find little or no upward mobility, though there may be considerable movement to an occupation similar in prestige.

What are the influences on upward (intergenerational) mobility in U.S. society? In this section we will consider three factors: the impact of socioeconomic background, gender, and occupational segregation.

**Socioeconomic Background** Occupational attainment in U.S. society rests heavily on educational achievement. In order to be a doctor, dental assistant, computer programmer, lawyer, or business executive, one needs to complete the required education. In order to be a registered nurse, Marie (whom we met at the beginning of this chapter) had to complete nursing school. Fred, our medical-equipment salesman, completed a bachelor's degree in business. Thus, movement to an occupation with higher prestige often requires additional education.

Beyond education, what other factors influence occupational attainment? To answer this question effectively we need to trace the occupational careers of individuals over their life course. Such longitudinal data are available in a study that began by surveying large samples of high school students (Sewell & Hauser, 1980). In 1957, all high school seniors in Wisconsin were surveyed regarding their plans after high school. From this population, a random sample of 10,317 were selected for more intensive analysis. In 1964, researchers obtained information from their parents about the students' post-high-school education, military service, marital status, and current occupation. Subsequently, they obtained information about the students' earnings and about the colleges/universities they attended. In 1975, 97 percent of the original sample were located, and most were interviewed by telephone. The interview focused on post-high-school education, work history, and family characteristics. Data from this study enabled researchers to trace the impact of characteristics of high school seniors on subsequent education, occupation, earnings, and work experience.

Figure 17-2 presents a diagram of the relationships found among the variables studied. The arrows indicate causal impacts. Variables are arranged from left to right, reflecting the order in which the variables affect the person through

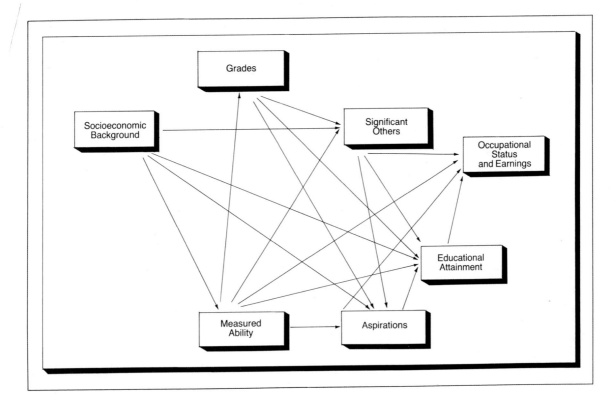

■ **FIGURE 17-2**

THE DETERMINANTS OF OCCUPATIONAL STATUS ATTAINMENT

This figure summarizes the influences that determine educational and occupational status over the life course. Socioeconomic background (parents' education, occupation, and income) influences ability, aspirations, and educational attainment. Ability influences grades, which, in turn, affect encouragement from significant others and aspirations for educational attainment. Occupational status is affected by education and also by ability, aspirations, and significant others.

*Source: Adapted from Sewell and Hauser, 1980.*

time. For example, socioeconomic background influences ability, which, in turn, influences educational attainment. Specifically, these results indicate that children from more affluent homes have greater ability, higher aspirations, and receive more education. Children with higher ability get better grades, which reward them for their academic work and reinforce their aspirations. Children who do well are also encouraged by significant others, such as teachers and relatives, which also contributes to high aspirations. These children are likely to choose courses that will prepare them for college. They are likely to spend more time on academic pursuits and less time on dating and social activities (Jessor et al., 1983). As a result, they are likely to continue their education beyond high school and perhaps beyond college. Finally, high ability, encouragement from significant others, and high educational attainment lead to greater occupational status and earnings.

Note that socioeconomic background and

grades have an indirect effect on occupational status and a direct influence on educational attainment. This does not mean that parental socioeconomic status and an individual's grades are unrelated to occupational status. Rather, it indicates that status and grades influence occupational attainment through other variables—like aspirations—that have a direct impact on occupational attainment (Sewell & Hauser, 1975).

In the research summarized in Figure 17-2, the family characteristics studied were mothers' and fathers' socioeconomic standings—education, occupation, and income. How is it that variables such as your father's education and your mother's income influenced your educational attainment? Parents often use their resources to create a home environment that facilitates doing well in school (Teachman, 1987). Thus, they provide such aids as a quiet place to study and encyclopedias. In addition, they may provide cultural enrichment activities such as attending concerts and arts events (DiMaggio & Mohr, 1985). These have been shown to influence educational attainment. It is estimated that family membership is the most important single influence on how much education one completes.

The experiences of Fred and Larry clearly reflect the importance of these processes. Fred's parents were upper-middle class; they sent him to preschool at age 4 and encouraged him to learn to read. Larry's parents were working class; they encouraged him to get out and play, not to waste time reading. Fred did well in school; his grades were always high. Larry struggled with his schoolwork, especially math. By eighth grade, Fred had an excellent record, and his teachers gave him lots of encouragement; Larry's teachers, on the other hand, didn't pay much attention to him. Fred worked hard in high school, got good grades, and, with the support of his teachers and family, went to a university. After finishing high school, Larry went into the army, where he learned vehicle mechanics. When Fred finished college, he got a job in a medical equipment firm. Ten years after he graduated from high school, Fred was selling $200,000 worth of equipment a year and earning 20-percent commissions. After he finished his military service, Larry went to work in a gas station. Ten years after Larry graduated from high school, he was earning $16,000 a year working as a mechanic.

Studies like the one described earlier indicate that there is upward mobility in U.S. society and that socioeconomic background does not fix one's occupational attainment and earnings. Through greater education, many persons achieve an occupational status and income substantially greater than would be expected based solely on their background. Thus, the U.S. is not a caste society. At the same time, socioeconomic background is not irrelevant to one's educational and occupational attainment. Not everyone has the ability to be a doctor, lawyer, or engineer. Opportunities for upward mobility are not unlimited.

**Gender** Is the process of status attainment different for men and women? According to the data obtained on Wisconsin high school students, the determinants of occupational status, as depicted in Figure 17-2, are the same for both men and women; although the size of some of the relationships varies.

Most striking were the findings with regard to occupational status. Using a prestige scale ranging from 0–100, the first jobs held by women were, on the average, 6 points higher on the scale than the first jobs held by men. That is, women, in general, started out in higher-prestige jobs than men. Women's first jobs were concentrated within a narrow range of prestige, whereas the first jobs held by men varied greatly in prestige scores (Sewell, Hauser, & Wolf, 1980). Table 17-1 reveals how this occurred. The first jobs women held included elementary school teacher, social worker, registered nurse, airline stewardess, dental hygienist, and salesclerk. The prestige scores of these jobs ranged from 57–38, respectively. In contrast, men's first jobs ranged from physician (78) to garbage collector (13).

When the researchers looked at 1975 occupations, they found men had gained an average of 9 points in status (in the 18-year period since graduation from high school). Women, on the other hand, had actually lost status; the average prestige of current occupations for women was 2 points lower than the average prestige of their first jobs. Men experience upward mobility because they work continuously. In addition, they are in occupations with possibilities of promotion and advancement. Women's work careers are often interrupted by marriage, by moving (due to spouses being transferred), and by raising children; when they return to work, they often take up the same job. Thus, women are often unable to build up enough continuous experience to gain promotion. Advancement is also more limited in occupations held largely by women. The top positions in schools, social work, airlines, and sales are more often held by men than women. So the occupational status achieved by men and women differs over the course of their careers.

These differences are evident in the lives of Fred, Larry, and Marie, who were introduced at the beginning of this chapter. After college, Fred began in sales, and his income increased substantially every year. If he wanted, he could move up in the company to regional sales manager, national sales manager, and perhaps vice president of sales. Larry has moved from gas station attendant (prestige score of 22) to owner of a service station (prestige score of 50). Like Fred, Marie went to college and earned a bachelor's degree. Her first job involved working on a surgical unit in a large hospital (prestige score of 54). As head nurse in pediatrics, she works days now, gets weekends off, and earns more—but her occupation is basically unchanged. She could move up to director of nursing, but she is unlikely to strive for this because the added responsibility isn't balanced by added pay.

**Occupational Segregation** In the preceding section, we saw that the influence of factors such as socioeconomic background and ability on occupational attainment is similar for men and women. At the same time, working men and women are not proportionately distributed across occupational categories. Refer to Table 17-1. As you look at each occupation in the list, what gender comes to mind? Chances are that when you think of engineers, auto mechanics, and carpenters, you picture males performing these jobs. In 1980, of those employed in these occupations, 96 percent, 99 percent, and 99 percent, respectively, were men (Reskin & Hartmann, 1986). Similarly, when you think of registered nurses or dental hygienists you probably picture females in these roles. In 1980, of those employed in these occupations, 96 percent and 98 percent were women. Clearly, many occupations consist overwhelmingly of either men or women; there is substantial occupational segregation by gender.

Experience with occupational segregation begins in adolescence. Data from a sample of 3,101 tenth-grade and eleventh-grade students in suburban high schools provide concrete evidence (Greenberger & Steinberg, 1983). Adolescents' first jobs are segregated by sex, and girls earn a lower hourly wage than boys in their first jobs. These differences reflect differential opportunity; employers hire primarily girls or primarily boys for a particular job—such as newspaper carrier or fast-food employee—and they pay boys more. The other contributor to occupational segregation is differential socialization, which includes the observation that men and women are unequally distributed across jobs, reinforced by mass media portrayals.

Adults often experience sex segregation in the workplace. In a survey of 290 organizations with more than 50,000 employees, the results indicate that men and women rarely perform similar work in a single organization; when they do, they usually have different job titles (Bielby & Baron, 1986). There is little evidence that employers' practices in these regards is a rational response to differences between men and women. Jobs held by men in one organization are held by women in other organizations. Moreover, differences in work performed or in job titles often result in substantial differences in pay. In 1987, the median earnings of a woman employed full-time was $10,810, while the median for men was $19,859 (U.S. Bureau of the Census, 1989).

## Social Networks

We have seen that socioeconomic background, ability, educational attainment, and earlier jobs influence occupational attainment over the life course. In part, this is because differences in experiences create differences in an individual's aspirations and abilities to cope with the occupational world. Varied experiences also move people into different social networks. This exposes them to varied social contacts that have an important effect on their upward mobility. This section will consider the ways in which position in social networks promotes entry into specific jobs.

Networks provide channels for the flow of information, including information about employment opportunities. What types of networks are likely to provide information on finding new jobs? One might think it is networks characterized by strong ties, such as families or peer groups. Surprisingly, employment opportunities are often found through networks characterized by *weak ties*—relationships involving infrequent interaction and little closeness or emotional depth (Marsden & Campbell, 1984). Weak ties are more likely to characterize relationships involving members of two different groups. Those to whom our ties are weak are involved in different groups and activities. Consequently, they will be exposed to information that is different from the information we already have. For this reason, new information is more likely to come from a weak tie than a strong one.

In one study (Granovetter, 1973), a random sample of persons who had recently changed jobs was asked how they had found out about their new job. Those who had heard about their job through another person were asked how often they had seen that contact. If the person saw the contact at least twice a week, this was considered a strong tie; if the person saw the contact less often, it was considered a weak tie. Of those who found jobs through contacts, only 17 percent were obtained through people who were considered to be strong ties. The remainder found jobs through people they saw less than twice a week. The contacts were often friends from school, former co-workers, or former employers—people with whom they had little recent interaction. It was often a chance meeting or a reintroduction by a mutual friend that led to the individual learning about the job. Thus, people were more likely to hear about jobs from those to whom they were weakly tied.

We noted earlier that women are less likely than men to experience upward occupational mobility during their careers. Might this occur because men and women differ in their access to networks that carry job information? Our ties to networks grow out of the activities we share with others. The organizations we belong to are a major setting for such activities (Feld, 1981). The larger an organization is, the larger the potential number of weak rather than strong ties is. If men belong to larger organizations than women, it follows that they should have more weak ties and, hence, better access to information useful in finding jobs.

To examine this possibility, a sample of 1,799 adults were asked the name and size of each organization to which they belonged (Miller-McPherson & Smith-Lovin, 1982). On the average, men belonged to organizations such as business and professional groups and labor unions, whereas women were more likely to belong to smaller charitable, church, neighborhood, and community groups. Thus, men were likely to develop a larger number of weak ties. Moreover, job-related contacts are more likely to develop in business, professional, and union groups. The study found that males had an average of 170 job-related potential contacts, whereas females had an average of less than 35.

Apparently, men are in networks that allow greater access to information about and opportunities for advancement.

Moreover, just as occupations are segregated by gender, so are the organizations to which people belong. Using data on 815 voluntary organizations in ten communities, researchers found that 50 percent of them were exclusively female and 20 percent were exclusively male (McPherson & Smith-Lovin, 1986). The typical woman member was likely to have face-to-face contact with 29 others, only 4 of them males. The typical male was likely to have contact with 37 others, only 8 of them females. Thus, the gender segregation of these organizations, in conjunction with the significance of weak ties in finding jobs, further contributes to occupational segregation.

Although finding a job is influenced by social networks, the status of the job one finds is influenced by the contact's status. You are more likely to get a management job at the phone company, for example, if your friend's father is a vice president than if he is a lineman. In a study of males ages 21–64 in the Albany, New York, area, 57 percent of the men reported using contacts to get their first job. The higher the contact's occupational status was, the greater the prestige of the position the job seeker obtained (Lin, Ensel, & Vaughn, 1981).

## ■ Individual Values

Last year, Fred, Larry, and Marie were each approached by a labor union organizer. Fred, the sales representative, was approached by a member of Retail Clerks International. The organizer explained that under a union contract, Fred would spend fewer days on the road and would be entitled to a travel allowance from his employer. Larry was approached by a representative of the Teamsters. The organizer sympathized with the problems of independent service station owners and urged Larry to let the Team-sters represent his interests in dealing with his supplier. Marie was approached by the president of United Health Care Workers; she was promised higher wages and greater respect from physicians if she would join.

Fred flatly rejected the invitation, believing that a union contract would limit his freedom and perhaps reduce his income. Larry's reaction was mixed. On the one hand, he felt he was at the mercy of "big oil." On the other hand, he was also a self-employed businessman; like Fred, he didn't want to join a labor organization that might limit his ability to determine his prices and the pace at which he worked. Marie reacted very favorably to her invitation and began to attend union meetings "to see what they were like." She felt that a union might lead to higher pay and might force the hospital to give her more freedom in determining the pace at which she worked.

In making their decisions, Fred, Larry, and Marie used their personal **values**—enduring beliefs that certain patterns of behavior or end states are preferable to others (Rokeach, 1973). All three were concerned with protecting or enhancing their freedom and their wealth, or income. These values provided criteria for making decisions. Thus, each person weighed the potential effect of joining a union on freedom and income. Fred felt the effect on both of these would be negative. Larry was sure union membership would limit his freedom but uncertain about its effect on his income. Marie perceived a potential gain in both freedom and income, so she decided to explore union membership.

Each of us has his or her own values. Each of us believes that particular goals and modes of behavior are more desirable than others. Although our values are general, they influence many specific attitudes, behaviors, and choices. For example, values are related to our attitudes toward public policy. Thus, the importance we place on personal property and on social equality is related to our attitudes toward paying higher taxes in order to help the poor (Tetlock,

1986). Those who place greater value on property will oppose higher taxes, whereas those who place greater importance on equality will favor increasing taxes to help the poor. Those who feel these values are equally important will find it hard to decide.

How do value systems arise? To some extent, they are influenced by our location in the social structure. This section will examine three aspects of social position that affect individual values: occupational role, education, and social class.

## Occupational Role

Because we spend up to half of our waking hours at work, it is not surprising that our work influences our values. But occupational experiences vary tremendously. To determine their effect on our values, it is necessary to identify the basic differences between occupations. Three important characteristics have been suggested (Kohn, 1969). The first is closeness of supervision—the extent to which a worker is under the direct surveillance and control of a supervisor. As a traveling salesman, Fred is rarely under close supervision, whereas Marie's work is supervised by the director of nursing and various physicians. The second occupational characteristic is routinization of work—the extent to which tasks are repetitive and predictable. Much of Larry's work is quite routine—pumping gas, tuning engines, relining brakes. But Larry's work is not highly predictable. From one day to the next, he never knows what kind of auto breakdown he will encounter or what unusual request some customer might make. The third characteristic is substantive complexity of the work—how complicated the work tasks are. Working with people is usually more complex than working with data or things. Marie's occupation as a nurse is especially complex because she must constantly cope with the problems posed by doctors, patients, and families.

All three of these characteristics were measured in several studies of employed men to determine the impacts of occupational role on

Workers on an assembly line often experience alienation. Assembly-line jobs are monotonous, do not allow workers to exercise initiative, and give them no influence over working conditions.

values and personality (Kohn, Schooler, & Associates, 1983). Results of these studies show a relationship between particular occupational characteristics and particular values. Men whose jobs were less closely supervised, less routine, and more complex placed especially high value on responsibility, good sense, and curiosity. Men whose work was closely supervised, routine, and not complex were more likely to value conformity. Thus, the occupational conditions that encourage self-direction—less supervision and nonroutine and complex tasks—are associated with valuing individual qualities that facilitate adjustment and success in a self-directed environment, such as responsibility, curiosity, and good sense. Occupational conditions that encourage adherence to a prescribed routine—close supervision and routine and simple tasks such as bolting bumpers on new cars—are associated with qualities that facilitate success in that environment, such as neatness and obedience. This pattern has emerged in studies of employed men and women (Miller et al., 1979) and in studies conducted in several countries, including the U.S., Japan, and Poland (Slomczynski, Miller, & Kohn, 1981).

Early studies of the relationships between workers' values and their occupational conditions revealed that workers exposed to particular conditions tended to hold particular values. However, these studies were unable to determine with certainty whether adjustment to occupational conditions actually *caused* people to value particular qualities. Perhaps men who value curiosity and desire responsibility select occupations that allow them to exercise these traits (Kohn & Schooler, 1973). In attempting to identify the causal order, researchers compared the men's values and occupational conditions in 1974 with their values and occupational conditions ten years earlier (Kohn & Schooler, 1982). What they found indicated causal impacts in *both* directions between values and occupational conditions. In other words, men who valued self-direction highly in 1964 were more likely to be in work roles that were more complex, less routine, and less closely supervised ten years later. Thus, values influenced job selection. At the same time, men who were in occupations that allowed or required self-direction in 1964 tended to place greater value on responsibility, curiosity, and good sense in 1974. Thus, their earlier job conditions influenced their later values.

## Education

Are differences in education also related to differences in an individual's values? The research by Kohn and his colleagues, described in the preceding section, demonstrated that men in jobs that are not closely supervised, nonroutine, and substantively complex value self-direction; whereas men in jobs with the opposite characteristics value conformity. Education is associated with the value one places on these characteristics; the more education one has, the greater value he or she places on self-direction.

Substantively complex occupations involve working independently with people, things, or data. Such work requires intellectual flexibility—the ability to evaluate information or situations and to solve problems. These abilities should be related to educational attainment, so education should be related to intellectual flexibility. Analyses of data from a sample of 3,101 men indicate that as education increases, so does intellectual flexibility (Kohn & Schooler, 1973). Thus, education influences both the value placed on self-direction and the abilities needed for success in substantively complex occupations.

In fact, it is possible to identify variations in self-direction among students. In one study, researchers assessed the complexity of a student's course work, of his or her most recent term paper or project, and of his or her extracurricular activities (Miller, Kohn, & Schooler, 1986). The sample included students from seventh grade through senior year of college. The results showed that the greater the substantive complexity of the student's work was, the greater the value he or she placed on self-direction. Thus, the exercise of self-direction in school or at work increases the value one places on self-direction.

## Social Class

How important is each of the following to you: a comfortable life, a sense of accomplishment, an exciting life, family security, salvation? What order of importance would you place them in? Evidence suggests that there is a systematic difference in the way people of different social classes order a number of values—including the five listed above. In one study (Rokeach, 1973), persons who were of high status gave high ratings to a sense of accomplishment and family security, whereas persons low in education and income gave high ratings to a comfortable life and salvation. At the same time, there was no relationship between social class and the importance of an exciting life, which was given a low rating by most respondents.

These differences probably reflect the social conditions experienced by rich and poor in the U.S. Persons with low incomes perhaps aspire to a comfortable life because they don't have it, or at least can't count on having it. Their economic situation may also make religion an important source of comfort and, thus, lead to a concern for salvation. High-status persons already have a comfortable life; for them, maintaining the family is important because it contributes to their social standing in the community. Those with high status also have the time and resources available to make individual accomplishment possible.

The five values mentioned above are *terminal values*—enduring beliefs about desirable end states. Two other terminal values are freedom and equality. A survey of 1,397 adults revealed systematic differences in the average rankings of freedom and equality by whites and blacks (Rokeach, 1973). Each respondent was asked to rank a list of eighteen values in order of importance. Whereas white respondents ranked freedom as more important (5.6) than equality (9.6), blacks ranked them about equally important (5.0 and 4.6, respectively). On the whole, whites in the U.S. have more education and income than blacks. Many blacks believe that discrimination or racism prevents them from achieving higher levels of education and income. They rank equality higher than whites because they view equality as essential to improving their status in American society.

# ■ Social Influences on Health

Most of us attribute diseases to biological rather than social factors. But the transmission of disease obviously depends on people's interactions,

and our physical susceptibility to disease is influenced by our life styles. This is quite true, for instance, with a disease such as AIDS. Likewise, our mental health is influenced by our relationships with relatives, friends, lovers, professors, supervisors, and so on. Thus, social position affects both physical and mental health. This section examines the impact of occupation, gender, and marital roles on physical health. It also considers the influence of gender, marital and work roles, social networks, and social class on mental health.

## Physical Health

**Occupational Roles**   What do the physician addicted to demerol, the executive with an ulcer, the coal miner with black lung, and the factory worker with a heart condition all have in common? The answer is a health problem that may be due largely to one's occupational role.

Occupational roles affect physical health in two ways. First, some occupations directly expose workers to health hazards. For example, miners who are exposed to coal dust may suffer damage to lung tissue, a condition known as black lung. Workers whose jobs involve direct contact with asbestos are more likely to contract lung cancer. Workers who manufacture polyvinyl chloride, used in making plumbing fixtures, are particularly susceptible to cancer of the liver (Epstein, 1976).

Second, many occupational roles expose individuals to stresses that affect physical health indirectly. Each of the roles we play carries a set of obligations or duties. We experience **role overload** when the demands of our roles exceed the amount of time, energy, and other resources we have to meet them (Goode, 1960).

Many physicians are subject to role overload. They are expected to carefully diagnose each patient's condition, to make sure each is treated correctly, to react professionally to emergencies, to keep up with advances in their areas of specialization, and to be active in the medical society and on hospital boards. At the same time, physicians are expected to fulfill their roles as family members, to contribute to housework, to travel with their children when they are out of school. Many physicians seek relief from role overload through narcotics, such as demerol, to which they may become addicted (Winick, 1964). Physicians and nurses are more likely to use narcotics than other professionals because the drugs are readily available to them. Availability alone is not enough to explain addiction, however; pharmacists also have access to narcotics but rarely use them. Pharmacists, however, are less likely to experience role overload. It is the combination of role overload and availability that promotes drug use by medical professionals.

The most widely studied relationship between job characteristics and physical health is the impact of occupational stress on coronary heart disease. As work load increases—including perceived demands on one's time, number of hours worked, and feelings of responsibility—so does the incidence of coronary heart disease (House, 1974). One acute source of overload is an important deadline, especially when failure to meet the deadline may have negative consequences. Heart attacks are associated with high levels of serum cholesterol in the blood. Several studies report that the level of serum cholesterol rises among persons under high work-related stress (Sales, 1969). This suggests one tangible link between role demands and physical health.

Recent research suggests that an important aspect of jobs that are associated with an increased risk of heart attack is a lack of control over work pace and task demands (Karasek et al., 1988). Occupations associated with the highest risk include cooks, waiters, assembly line operators and gas station attendants. These jobs are characterized by high demand—heavy work load and rapid pace—over which a worker has little or no control. Cashiers and waiters are four to five times more likely to have a heart attack than foresters or civil engineers.

People are not necessarily at the mercy of occupational and other role demands. There are three common *coping strategies* for dealing with

role overload. First, people set priorities, deciding on the relative importance of obligations to family, work or school, friends, and other activities. They establish a queue—an order in which they will complete tasks or occupy various roles (Schwartz, 1978). Once established, the queue takes on moral and psychological significance. We have all experienced the feeling that we "have to" do something even though a much more attractive opportunity is offered. Thus, we turn down an invitation to a party because we have to study, visit parents or other relatives, or do some volunteer task for a club or neighborhood organization. Although limiting, queues reduce the pressures associated with role overload.

A second strategy is to renegotiate the role expectations causing the overload (Handel, 1979). A student with two important exams on the same day may be able to arrange to take one of them earlier or later. An overworked doctor may substantially reduce the number of patients she is seeing; alternatively, she may add a nurse clinician or a partner to her practice to handle part of the heavy patient load.

A third strategy is to exit from one or more of the roles (Goode, 1960). A college student who finds that he is unable to meet the school's performance standards may quit school and go to work. A heart attack victim will assume a slower pace, and may even resign from some of his roles.

**Gender Roles** Who is more likely to experience coronary heart disease, cirrhosis of the liver, or lung cancer—men or women? If you picked men, you are right. Men are two to six times more likely than women to die from these conditions. Although there is evidence that genetics and hormones play a role, traditional role expectations for males and females in our society are another significant factor (Waldron, 1976).

Earlier we mentioned that role overload is associated with coronary heart disease. Professionals such as physicians, lawyers, accountants, and so on are especially vulnerable to overload, and the persons holding these positions are primarily male. Other studies have shown that heart attacks are correlated with certain personality traits known as *coronary prone behavior patterns* (Jenkins, Rosenman, & Zyzanski, 1974). People who exhibit this behavior pattern are work oriented, aggressive, competitive, and impatient. Males are much more likely to be characterized by this behavior pattern than females.

Men are more prone than women to have cirrhosis of the liver, because they are four times more likely than women to be heavy drinkers (Cahalan, 1970). Moreover, men are more likely than women to contract lung cancer and emphysema (until recently, men were much more likely to smoke cigarettes). They are also more likely to die in auto accidents, both because of higher rates of driving under the influence of alcohol and because of poor driving habits (Waldron, 1976).

**Marital Roles**  A study compared single, married, divorced, separated, and widowed persons on a variety of health-related measures (Verbrugge, 1979). Divorced and separated persons had the highest rates of illness and disability, followed by widowed and single persons. Married persons were healthiest. Other data indicated that rates of hospital residence were highest for singles and lowest for married persons. Mortality rates at a given age were also correlated with marital status (Kobrin & Hendershot, 1977). A study of 20,000 deaths revealed that singles had the highest death rate in any age group. Following singles were single heads of households (unmarried people with others dependent on them), followed by married persons without children and married persons with children. Among married persons, the death rate of males and females is virtually the same (Waldron, 1976).

Why is it that being married and having children protects people against illness and accidents? The most likely explanation is that married persons are less likely to engage in behaviors that expose them to illness and accidents (Verbrugge, 1979). They probably eat and sleep better than unmarried persons. They are perhaps less likely to smoke and drink. They may

take fewer risks, reducing the likelihood that they will be involved in accidents. Finally, they may be more likely to seek medical care when ill (Verbrugge, 1979). Thus, being married is associated with a life style that reduces the risks of illness and death.

## Mental Health

At the beginning of this chapter we introduced Larry, who owns a service station, works long hours, and earns about $22,000 a year. He has two children, a large mortgage on his home, and has trouble making ends meet. He comes home from work every day exhausted. He worries about the economy and whether or not there will be another energy crisis leading to inadequate supplies of gasoline or an oil glut leading to gasoline price wars. Either one would ruin his business, because more than one half of his income is from gasoline sales.

Like many Americans, Larry finds that his life situation is very demanding. His customers expect him to do high-quality repair work at low prices, his wife expects him to support the family and spend time with her, and his children want more toys than he can afford. At times, the demands made on him by others exceed his ability to cope with them, causing psychological *stress* (Dohrenwend, 1961). People who are under stress often become tense and anxious, are troubled by a poor appetite, or experience insomnia. A widely used questionnaire designed to measure stress-related symptoms is reproduced in Box 17-1. Many of the items on this scale measure behavior, thoughts, and feelings associated with depression.

Stress is often temporary. A move from one apartment to another, for example, is stressful during the weeks all the arrangements are being made and during the move itself. As one becomes settled, however, the demands decline. At the same time, the ability to respond to the demands of moving may increase as one learns how to cope with packing, disconnecting the old phone and getting a new one, and so on.

On the other hand, stress may be continuous (as it is for Larry, who constantly experienc-

Family members play an important role in helping us to cope with stressful events, such as the death of a relative or close friend. They are an important source of emotional support, and may help by temporarily taking over some of our role responsibilities.

es tension and anxiety due to his economic worries). Particularly debilitating is a condition called **pileup of demands,** in which stressful life events (such as serious illness) and normative life transitions (such as the birth of a child) are added to the continuing stress from prior stressors (Lavee, McCubbin, & Olson, 1987).

Chronic stress from a pileup of demands may lead to physical illness. Excellent evidence of the link between stress and illness comes from a longitudinal study of two samples of adults (140 and 190) employed in a large company (Maddi, Bartone, & Puccetti, 1987). Each person's level of stress was assessed by a carefully designed measure of stressful life events. One or two years later, each person completed a questionnaire regarding illness that included both mild (such as influenza) and serious (such as heart attack)

conditions. There were substantial associations between the level of stress experienced initially and reported illness one or two years later.

Chronic stress may also lead to impaired psychological functioning. Neuroses, schizophrenia, and affective disorders such as depression are among the mental illnesses associated with severe stress. The experience of stress and impaired psychological functioning varies by gender, marital, and work roles, by membership in social networks, and by social class.

**Gender, Marital, and Work Roles** Adult women in the U.S. have somewhat poorer mental health than men. On stress scales such as the one shown in Box 17-1, women attain significantly higher scores than men (Warheit et al., 1976). Women also have higher rates of two types of

# HOW DO YOU RESPOND TO STRESS?

Stress is a discrepancy between the demands on a person and his or her ability to successfully respond to those demands. Individuals under stress often manifest a variety of physical and psychological symptoms. One of the most common measures of stress was developed by Langner (1963). His original scale consists of the 22 questions reproduced below. As you read each question, circle the appropriate response.

| Question | Response Categories |
|---|---|
| **1.** I feel weak all over much of the time. | 1. yes<br>2. no |
| **2.** I have had periods of days, weeks, or months when I couldn't take care of things because I couldn't "get going." | 1. yes<br>2. no |
| **3.** In general, would you say that most of the time you are in high (very good) spirits, good spirits, low spirits, or very low spirits? | 1. high<br>2. good<br>3. low<br>4. very low |
| **4.** Every so often I suddenly feel hot all over. | 1. yes<br>2. no |
| **5.** Have you ever been bothered by your heart beating hard? Would you say: often, sometimes, or never? | 1. often<br>2. sometimes<br>3. never |
| **6.** Would you say your appetite is poor, fair, good, or too good? | 1. poor<br>2. fair<br>3. good<br>4. too good |
| **7.** I have periods of such great restlessness that I cannot sit long in a chair (cannot sit still very long). | 1. yes<br>2. no |
| **8.** Are you the worrying type (a worrier)? | 1. yes<br>2. no |
| **9.** Have you ever been bothered by shortness of breath when you were *not* exercising or working hard? Would you say: often, sometimes, or never? | 1. often<br>2. sometimes<br>3. never |
| **10.** Are you ever bothered by nervousness (irritable, fidgety, tense)? Would you say: often, sometimes, or never? | 1. often<br>2. sometimes<br>3. never |
| **11.** Have you ever had any fainting spells (lost consciousness)? Would you say: never, a few times, or more than a few times? | 1. never<br>2. a few times<br>3. more than a few times |
| **12.** Do you ever have any trouble in getting to sleep or staying asleep? Would you say: often, sometimes, or never? | 1. often<br>2. sometimes<br>3. never |

Box 17-1

| Question | Response Categories |
|---|---|
| **13.** I am bothered by acid (sour) stomach several times a week. | 1. yes<br>2. no |
| **14.** My memory seems to be all right (good). | 1. yes<br>2. no |
| **15.** Have you ever been bothered by "cold sweats"? Would you say: often, sometimes, or never? | 1. often<br>2. sometimes<br>3. never |
| **16.** Do your hands ever tremble enough to bother you? Would you say: often, sometimes, or never? | 1. often<br>2. sometimes<br>3. never |
| **17.** There seems to be a fullness (clogging) in my head or nose much of the time. | 1. yes<br>2. no |
| **18.** I have personal worries that get me down physically (make me physically ill). | 1. yes<br>2. no |
| **19.** Do you feel somewhat apart even among friends (apart, isolated, alone)? | 1. yes<br>2. no |
| **20.** Nothing ever turns out for me the way I want it to (turns out, happens, comes about, that is, my wishes aren't fulfilled). | 1. yes<br>2. no |
| **21.** Are you ever troubled with headaches or pains in the head? Would you say: often, sometimes, or never? | 1. often<br>2. sometimes<br>3. never |
| **22.** You sometimes can't help wondering if anything is worthwhile anymore. | 1. yes<br>2. no |

These questions measure response to stress. To determine your stress score, give yourself one point if you circled 3 or 4 on item 3, if you answered 3 on item 11, and if you answered 2 on item 14. For each of the other items, give yourself 1 point if you circled a 1.

Langner collected data using this and other measures from 1,660 adults living in Manhattan. He found that adults who scored 0–3 on this scale appeared to be "normal" on various other indices of mental health. Those who scored 4–6 on the scale also showed some psychological impairment on other indices. Adults who scored 7 or more were very likely to be diagnosed as severely impaired. Thus, according to Langner, as response to stress increases, mental health deteriorates.

*Source: Adapted from Langner, 1963.*

mental illness: neuroses—which involve high levels of anxiety—and depression (Dohrenwend & Dohrenwend, 1976).

Can differences in the social roles that men and women typically enact account for the higher level of psychological impairment among women? The majority of people in these studies were married. Traditional marital roles require men to provide financially for the family and women to take care of the home and to provide emotional support for their husbands. Thus, traditional marriages may reduce stress for men because they provide a supportive environment (Gerstel, Riessman, & Rosenfield, 1985). For women, however, being married may increase stress. Women in a traditional marriage may be subordinate to their husbands and, thus, have less control over their lives. A lack of control is, in itself, stressful, as discussed earlier. Until recently, married women also tended to remain in the home, isolated from social networks outside the family that could provide support. The responsibility of married women to care for husbands and children may also reduce their ability to respond adaptively to stress—for example, to take care of themselves when they are ill (Gove & Hughes, 1979). Figure 17-3 presents some results from a study that asked questions about people's ability to take care of themselves; note that married women, and especially those with children, were shown to be less able to take care of themselves than married men.

Is greater social isolation and the presence of children associated with levels of depression among married women? This hypothesis was tested, using data from interviews with more than 2,300 men and women (Pearlin & Johnson, 1977). The study measured three potential sources of stress, including economic hardship, social isolation, and parental responsibility. Economic hardship was assessed by asking each person how often he or she lacked enough money for food, clothing, and medical care. Social isolation was measured by the person's membership in voluntary associations and number of "really good friends" and by the length of

time the person had lived in his or her neighborhood. Parental responsibility was measured by the number of children in the home. The study also measured symptoms associated with depression. The results indicated that economic strain was the major variable associated with depression scores. Neither social isolation nor having children was related to depression. Both married men and women were less depressed than singles under equivalent conditions of hardship, social isolation, and parental responsibility. Thus, marriage appears to help both men and women cope with stress.

Although marriage appears to benefit both men and women, the conflicting demands of family and work roles may cause stress, as discussed in chapter 16. If severe, this stress may lead to impaired psychological functioning. A survey of 1,800 adults (Gove & Geerken, 1977) examined the effects of women's employment on couples with children. Four variables were measured: feeling "incessant demands" (overworked); the desire to be alone; loneliness; and the frequency with which respondents experienced 14 psychiatric symptoms over the past two weeks. Married men scored the lowest on each of these four measures, unemployed wives scored the highest, and employed wives scored in between these two groups. The presence of children was associated with feeling more demands and the desire to be alone for both men and women. Higher scores on these two measures were associated with psychiatric symptoms. Thus, for married women, employment is associated with greater psychological well-being, whereas having children is associated with increased stress and reduced psychological well-being.

Thus, the relationships between work, marital and parental roles, and psychological well-being are complex. Men typically have higher levels of well-being than women. Employment of wives appears to improve their mental health. The demands of child care appear to challenge the psychological well-being of both parents, especially working mothers.

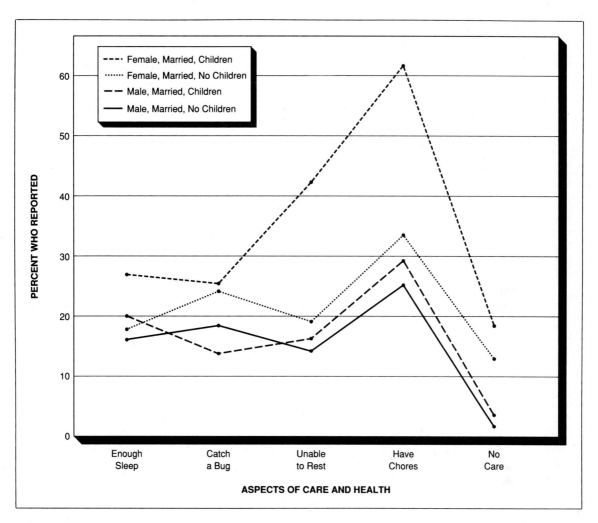

▪ **FIGURE 17-3**

ROLE ATTRIBUTES AND TAKING CARE OF ONESELF

The roles we occupy influence whether we can take care of ourselves. In a study of the impact of marital and parental roles, respondents were asked five questions: (1) Do you get enough sleep? (2) If a family member has a bug, do you catch it? (3) When you are really sick, do you get a good rest? (4) When you are really sick, are there chores you have to do? and (5) When you are really sick, does someone take care of you?

The graph displays the percentage of people who did not get enough sleep, reported catching bugs from family members, were unable to rest, had chores to do, and had no one to take care of them when they were sick. Married men, with and without children, were shown to be more likely to take care of themselves and to be taken care of when they were sick. Married women, especially those with children, were less likely to take care of themselves.

*Source: Adapted from Gove and Hughes, 1979.*

**Social Networks** To this point, we have reviewed some of the evidence showing that our relationships with others—that is, our membership in social networks—can be major sources of stress. At the same time, social networks can serve as an important resource in coping with stress. Networks provide us with continuing social support and with help during stressful events. They also influence the way we define the sources of the stress we experience.

First, a network of close friends and kin eases the impact of stressful events by providing various types of support (House, 1981b; Cooke et al., 1988). One type is *emotional support*—letting us know that they care for and are concerned about us. A second type is *esteem support*—providing us with positive feedback about our abilities and worth as a person. A poor grade, for example, is less stressful if your friends let you know they think you are a good student. *Informational support* from others prepares us to avoid problems or to handle them when they arise. Advice from friends on how to handle job interviews, for example, improves our ability to cope with this situation. Finally, network members provide each other with *instrumental support*—money, labor, and time.

Research has documented the impact of supportive relationships on an individual's ability to cope with stress. A longitudinal study of a representative sample of 900 adults focused on the relationship between social network membership and physical health (Seeman, Seeman, & Sayles, 1985). Persons who reported in the initial interview that they had instrumental support available were in better physical health one year later. (These were persons who, when ill, had others who would call on them, express concern, and offer help.) Another longitudinal study assessed the impact of family support on mental health (Aldwin & Revenson, 1987). The sample consisted of 245 men and 248 women from randomly selected families in an urban area. The availability of support from one's family at the time of the initial survey was associated with better psychological adjustment one year later. Other research indicates that individuals with fam-

ily support are more likely to cope with stressful events by using active strategies rather than avoidance or withdrawal strategies (Holohan & Moos, 1987). Finally, a longitudinal study of the relationship between coping strategies and mental health found that people who used active strategies at the time of the initial survey reported fewer psychological symptoms at the time of the second survey (Aldwin & Revenson, 1987).

A second way social networks reduce stress is by helping people cope with stressful events or crises when they occur. When members of a group are all subjected to similar stressors, the group may develop coping strategies. A study of interns and residents in a single hospital found that they were subjected to long hours of demanding work in often poor facilities (Mizrahi, 1984). These physicians coped with stress by minimizing the time spent with each patient, by limiting interaction with patients to "relevant topics," and by treating patients as nonpersons, for example, by focusing exclusively on their illnesses. These strategies were passed on from experienced members to new ones.

Third, network members influence how we react to stress. When we experience anxiety or tension for a prolonged period, we often discuss it with friends or family in an attempt to define the problem (Emerson & Messinger, 1977). The reactions of others often influence how we handle our problem. A student experiencing chronic tension, for example, might talk it over with a close friend. A friend with strong religious beliefs might suggest that the student be "born again" and find renewed purpose in life. Another friend might say that a ski trip to Colorado is the necessary tonic.

**Social Class** The lower a person's socioeconomic status is, the greater the amount of stress is reported (Mirowsky & Ross, 1986). Socioeconomic status is jointly determined by a person's education, occupation, and income. Does each of these components contribute to stress independently? Or is stress the result of only one or two of these components?

An analysis of data from surveys of eight,

quite diverse samples (Kessler, 1982) shows a consistent pattern: low education, low occupational attainment, and low income contribute separately to stress. The relative importance of these three components as sources of stress is not the same for men as for women, however. For men, income appears most important; for women (employed or not) education appears to be the most important component. Occupational attainment is the least important determinant of stress for both sexes.

Further analyses have sought to identify the causes of the negative relationship between social status and stress. Are lower-class persons exposed to greater stress, or are they simply less able to cope effectively with stressful events? The answer is both (Kessler & Cleary, 1980). On the one hand, lower-class persons are more likely to experience economic hardship—not having enough money to provide adequate food, clothing, or medical care (Pearlin & Radabaugh, 1976). They also experience higher rates of a variety of physical illnesses (Syme & Berkman, 1976). Both economic hardship and illness increase the stress an individual experiences. At the same time, persons who are low in income, education, and occupational attainment lack the resources that would enable them to cope with these stresses effectively. Low income reduces their ability to cope with illness. In addition, low-status persons are less likely to have a sense of control over their environment and less access to political power or influence. For this reason, they are less likely to attempt to change stressful conditions or events.

If stress increases as socioeconomic status decreases, we would expect persons lower in status to have poorer mental health. Research over the past 30 years has consistently confirmed this expectation; there is a strong correlation between social class and serious mental disorders (Eaton, 1980). This correlation has been found in studies conducted in numerous countries (Dohrenwend & Dohrenwend, 1974). In general, persons in the lowest socioeconomic class have the highest rates of mental illness.

A study of first admissions to mental hospitals in Maryland (Eaton, 1980) analyzed the relationship between class (measured by education) and schizophrenia. Individuals with only an elementary school education were ten times as likely to be diagnosed as schizophrenic upon admission as those with a college education. This finding pertains to persons who reach mental hospitals. What of the rate of schizophrenia among those who remain in the community? In community surveys—which include psychiatric assessments of persons who are not institutionalized—the same concentration of schizophrenia in the lowest class is found. With respect to the less serious conditions such as neuroses, the results are less consistent. In some studies the rate of neurosis is highest among lower-class persons, whereas other studies show no relation to social class (Dohrenwend & Dohrenwend, 1974). In the case of the affective psychoses (manic, depressive, and manic-depressive disorders) there appears to be no relationship between the incidence of illness and social class (Eaton, 1980).

There are three major explanations for why membership in the lower class might cause mental disorders (Eaton, 1980). The first suggests that lower-class infants are more likely to have suffered damage during pregnancy and that this damage may contribute to later psychiatric illness (Mednick & Schulsinger, 1969). Although no direct evidence supports this explanation, some important indirect evidence is provided by the higher rate of infant mortality in the lower class. By extension, it is argued that conditions that are not fatal but that could cause later psychological disorders are more common in this class. A study of first admissions to mental hospitals revealed that the negative relationship between class and illness was strongest for disorders with an organic cause, which are often due to hereditary factors (Rushing & Ortega, 1979). These findings are consistent with the hypothesis that lower-class infants are more often physically damaged during pregnancy.

The second explanation contends that personality characteristics vary by social class and that socialization patterns typical of lower-class families contribute to mental illness. These patterns may reduce people's ability to deal successfully with stress. As we have seen, the values taught to children by parents differ by social class (Kohn, 1969, 1977). The values for conformity and obedience to authority that lower-class persons tend to acquire may reduce their flexibility and make them less able to respond adaptively to new situations. There is no direct evidence, however, that these differences in socialization produce differential rates of mental illness.

Finally, it has been proposed that the different rates of mental disorder are caused by differences in stress. Both low socioeconomic class (Warheit et al., 1976) and economic hardship (Pearlin & Radabaugh, 1976; Pittman & Lloyd, 1988) are positively associated with stress. Perhaps the highest rates of schizophrenia and personality disorders among the lower class are a consequence of the greater stress associated with poverty, low education, low occupational status, and other deprivational aspects of lower-class life. Evidence for such an association is based on a survey of 1,660 adults living in Manhattan (Langner & Michael, 1963). Results of this study, illustrated in Figure 17-4, show a strong association between social class, stress, and the incidence of mental disorder.

## ■ Alienation

Jim dragged himself out of bed and headed for the shower. As the water poured over him he thought: "Thursday . . . another 10-hour shift . . . if the line doesn't shut down. I'll bolt 500 bumpers . . . sick of car frames . . . I'd rather do almost anything else . . . if only I'd finished high school . . . damn the money! . . . Let 'em take the job and shove it . . . but what else pays a guy who quit school $11.28 an hour?"

Jim is experiencing **alienation**—the sense that one is uninvolved in the social world or lacks control over it. Many types of alienation have been identified (Seeman, 1975). Two of the most important are self-estrangement and powerless-

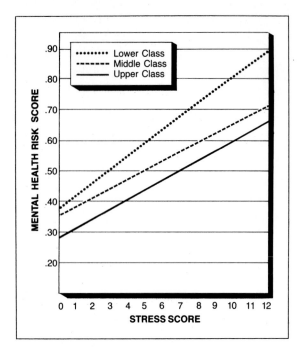

### ■ FIGURE 17-4
SOCIAL CLASS, STRESS, AND MENTAL HEALTH RISK

The midtown Manhattan study investigated the relationship between social class, stress, and mental health. Respondents completed an early version of the Langner scale (see Box 17-1) and were questioned in detail about their lives. Based on the interviews, two psychiatrists independently rated the mental health of each person; these ratings were averaged and transformed into mental health risk scores, which ranged from .00–1.0. The risk score for the average person in the study was .50; the higher the score was above .50, the greater the risk was. The results show that both social class and the stress score are related to the mental health risk score: As socioeconomic status declines and stress increases, a person's risk of mental health impairment increases dramatically.

*Source: Adapted from Langner and Michael, 1963, Figure 14-2.*

ness. We will consider both of these in detail as well as four additional types: normlessness, meaninglessness, social isolation, and cultural estrangement.

### Self-Estrangement

Jim's hatred for his job reflects **self-estrangement**—the awareness that one is engaging in activities that are not rewarding in themselves. Work is an important part of an individual's waking hours. When work is meaningless, the individual perceives the self as devoting time and energy to something unrewarding —that is, something "alien."

What makes a job intrinsically rewarding? Perhaps the most important feature is autonomy. Work that requires an individual to use judgment, exercise initiative, and surmount obstacles contributes to self-respect and a sense of mastery. A second feature is variety in the tasks that the person performs. Jim has no autonomy; his job does not allow him to exercise judgment or initiative. It also has no variety; it is monotonous and boring.

Four features of industrial technology tend to produce self-estrangement. First, self-estrangement will be higher if a worker has no connection with the finished product itself. Second, it will be higher if the worker has no control over company policies. Third, it will be higher if the worker has little influence over the conditions of employment, over which days, which hours, or how long he or she works. Finally, it will be higher if the worker has no control over the work process—for example, the speed with which the tasks must be performed (Blauner, 1964).

These features are especially characteristic of assembly-line work, in which each person performs the same highly specialized task many times a day. Thus, workers on assembly lines should be more likely to experience self-estrangement than other workers. A study testing this hypothesis (Blauner, 1964) compared assembly-line workers in textile and automobile plants with skilled printers and technicians in the chemical industry. As expected, assembly-line workers were more alienated than skilled workers, who had jobs that were more varied and involved the exercise of judgment and initiative.

It has also been argued that work in bureaucratic organizations—like large insurance companies or government agencies—may produce self-estrangement. In many bureaucratic organizations, workers have little or no control over the work process and do not participate in organizational decision making. Thus, workers at the lowest levels of such organizations should experience self-estrangement or dissatisfaction with their work. Conversely, workers who are involved in decision making should be less alienated. A survey of 8,000 employees in 100 companies located in the U.S. and Japan found that workers involved in participatory decision making structures had higher commitment to their work (Lincoln & Kalleberg, 1985). Such workers were willing to work harder and were proud to be employed by and wanted to remain with the company.

According to the theory developed years ago by Karl Marx (1964), whether a person will experience self-estrangement is determined by his or her relation to the means of production. The most alienated employees are hypothesized to be those who have no autonomy, who do not have the freedom to solve nonroutine problems, and who have no subordinates. Marx referred to such workers as the *proletariat*. In contemporary society, assembly-line workers, salesclerks, file clerks, and laborers are all in occupations that have these characteristics. A recent survey of 1,499 working adults found that 46 percent were in jobs of this type (Wright et al., 1982). Several studies have found that men whose jobs are characterized by lack of autonomy and complexity typically have high scores on measures of self-estrangement (Kohn, 1976) and low scores on measures of job involvement (Lorence & Mortimer, 1985).

Some have suggested that the widespread alienation of assembly-line workers results in poorer quality and lower productivity in the automobile and other manufacturing industries in the U.S. However, self-estrangement does not necessarily lead to dissatisfaction with work. Satisfaction is influenced not only by the presence or absence of intrinsic gratification but also by extrinsic factors like pay, hours, and job security.

There is some evidence that characteristics of the work environment influence psychological well-being. One researcher assessed the common environment of office workers by averaging the

The grafitti gracing subway trains in some American cities are responses by urban youth to alienation. Spray-painted messages reflect the lack of control over their lives that disadvantaged youths feel.

ratings of all of those employed in each of 37 branch offices; each worker's own ratings were used as a measure of his or her immediate environment (Repetti, 1987). Workers who rated their branch more positively (on interpersonal climate and support and respect from co-workers) reported lower levels of anxiety and depression. Aggregate ratings by the workers of the environment in the branch were also related to anxiety and depression scores.

## Powerlessness

Consider the fact that vandalism is widespread in certain sections of large cities, that many middle-class and upper-class adults do not vote in presidential elections, and that some people on welfare make no effort to find a job. These facts all have something in common. They reflect, at least in part, people's sense of **powerlessness**—the sense of having little or no control over events.

Powerlessness is a generalized orientation toward the social world. People who feel powerless believe they have no influence on political affairs and world events; this is different from feeling a lack of control over events in day-to-day life. A typical measure of powerlessness is reproduced in Table 17-2. Most people's scores on measures of powerlessness are quite stable over a period of many years (Neal & Groat, 1974). There is some evidence that a sense of powerlessness develops during childhood (Seeman, 1975). Interestingly, a sense of powerlessness is not associated with social class—that is, income, occupation, or education.

Statements that measure powerlessness, in-

■ **TABLE 17-2**

A MEASURE OF POWERLESSNESS

|  | Strongly Agree | Agree | Disagree | Strongly Disagree |
|---|---|---|---|---|
| **1.** People like me can change the course of world events if we make ourselves heard. | SA | A | D | SD |
| **2.** I think each of us can do a great deal to improve world opinion of the United States. | SA | A | D | SD |
| **3.** There's very little that persons like myself can do to improve world opinion of the United States. | SA | A | D | SD |
| **4.** The average citizen can have an influence on government decisions. | SA | A | D | SD |
| **5.** This world is run by the few people in power, and there is not much the little guy can do about it. | SA | A | D | SD |
| **6.** It is only wishful thinking to believe that one can really influence what happens in society at large. | SA | A | D | SD |
| **7.** A lasting world peace can be achieved by those of us who work toward it. | SA | A | D | SD |
| **8.** More and more, I feel helpless in the face of what's happening in the world today. | SA | A | D | SD |

Note: Agreement with items 3, 5, 6, and 8 indicates a sense of powerlessness, as does disagreement with statements 1, 2, 4, and 7. How powerless do *you* feel?

Source: Adapted from Zeller, Neal, and Groat, 1980.

cluding "people like me have no say" and "politicians don't care what I think," were included in several surveys between 1952 and 1980. Analysis of patterns of agreement with these items shows that powerlessness or political alienation declined from 1952 to 1960, rose steadily from 1960 to 1976, and then declined (Rahn & Mason, 1987). The increase in the 1960s and 1970s was associated with increased concern about such political and social issues as civil rights for blacks, the Vietnam War, and Watergate. Thus, fluctuations in powerlessness reflect, at least in part, events in the larger society.

Although a sense of powerlessness is found in all classes, upper and lower classes may have different means of expressing it. Whereas middle-class and upper-class persons may be more likely to stay home on election day or to feel apathetic about political affairs or organizations that influence public policy, lower-class persons may be more likely to have a hostile attitude toward city officials and to vandalize city buses, subway trains, and businesses in their neighborhoods. In other words, how an individual expresses frustration over lack of influence on the world may depend on his or her social position.

### Other Forms of Alienation

What other ways might people feel estranged from society? Four other varieties of alienation include normlessness, meaninglessness, social isolation, and cultural estrangement.

**Normlessness** Conformity to social norms is a major feature of social life. Not everyone feels that conformity will help them get ahead, however. People characterized by **normlessness** believe that socially disapproved behavior is necessary to achieve their goals (Neal & Groat, 1974).

Like powerlessness, normlessness is measured by using an attitude scale. The most widely used measure of normlessness is Srole's (1956) *anomia scale*. One item on this scale states, "Nowadays, a person has to live pretty much for today and let tomorrow take care of itself." Persons low in education, income, or occupational prestige attain higher scores on scales measuring normlessness. The association of normlessness with low status probably reflects the fact that persons of low status often lack socially approved means of achieving their goals (Merton, 1957). They may, in fact, have to use disapproved methods, such as theft, to earn a living. The relationship between social class and normlessness may also reflect the difference between upper and lower classes in the availability of education and jobs.

**Meaninglessness**  Have you ever felt that things have become so complicated in the world today that you don't understand what is going on? If so, you have experienced **meaninglessness**—a sense that what is going on around you is incomprehensible. Meaninglessness refers to the absence of a definition of the situation—to the lack of a set of meanings that the individual can use as guidelines for behavior. Meanings arise out of interaction with others. Thus, meaninglessness can result from a lack of group ties and the absence of the familial and other roles that reflect such ties.

**Social Isolation**  Closely related to meaninglessness is **social isolation**—the lack of involvement in meaningful relationships. Critics of American society frequently argue that social isolation is the primary source of alienation in the U.S. Is isolation from family and community common? According to research, the answer is no. Most adults are embedded in networks of primary relationships that include family and close friends (Shulman, 1975; Wellman, 1979). Even weak ties (short-term, nonintimate relationships) appear to be a source of social support.

At the same time, some people are isolated and have virtually no ties. For example, some cities have sizable populations of chronically mentally ill, former mental patients who have been released to the community. Many of these persons were released from mental hospitals in the late 1970s in an attempt to reintegrate them into society. Often they have no friends or family to help them, and many are unable to find employment. Also, large metropolitan areas like New York City have hundreds of homeless persons who live on the streets, in railroad and bus terminals, and in subway tunnels. These people are truly alienated. They are cut off from familial and work roles and have few ties to others.

**Cultural Estrangement**  A final variety of alienation is **cultural estrangement**—the rejection of the basic values and life styles available in society. Whereas self-estrangement is alienation from one's self, cultural estrangement is alienation from society. Cultural estrangement was the basis for the counterculture movement of the late 1960s and early 1970s and for the founding of various "utopian" communities. Such a rejection promotes either a lack of commitment to established cultural values and a withdrawal from organized social life or militant attempts to change the society. The former leads to deviant behavior, whereas the latter is a source of social protest and social movements. The consequences of cultural estrangement will be examined in more detail in the next two chapters.

# ■ Summary

This chapter considers the impact of social structure on four areas of an individual's life: achievement, values, physical and mental health, and sense of belonging in society. Social structure influences the individual through the expectations associated with his or her roles, the social networks to which he or she belongs, and the status associated with his or her positions.

**Status Attainment**  An individual's status determines access to resources—to money, life style, and influence over others. Three generalizations can be made about status in the U.S.: (1) An individual's status is closely tied to his or her occupation. (2) Occupational attainment is influenced directly by the individual's educational level and ability and indirectly by his or her socioeconomic background. Among women, occupational status and income is limited by gender segregation. (3) Information about job opportunities is often obtained through social networks, especially those characterized by weak ties.

**Individual Values**   Three aspects of an individual's position in society influence his or her values. (1) Particular values are reliably associated with certain occupational role characteristics. Men and women whose jobs are closely supervised, routine, and not complex value conformity; whereas those whose jobs are less closely supervised, less routine, and more complex value self-direction. (2) A formal education influences values. Higher education is associated with greater value on self-direction and greater intellectual flexibility. (3) Social-class standing influences the relative importance of such values as a sense of accomplishment, a comfortable life, freedom, and equality.

**Social Influences on Health**   (1) Physical health is influenced by occupation, gender, and marital roles. Occupational roles determine the health hazards individuals are exposed to and whether they experience role overload. The traditional role expectations for males and females make males more vulnerable than females to illnesses such as coronary heart disease. Marriage seems to protect both men and women from illness and premature death.

(2) Mental health is also influenced by social factors. Women have somewhat poorer mental health than men, although marriage is associated with reduced stress for both men and women. The presence of children may lead to reductions in psychological well-being for both women and men; working outside the home is associated with increased well-being for women. Social networks are an important resource in coping with stress; they provide a person with emotional, esteem, informational and instrumental support. Lower-class persons report greater stress and experience a higher incidence of mental illness.

**Alienation**   There are several types of alienation. (1) Self-estrangement is associated with occupational roles that do not allow workers a sense of autonomy, such as assembly-line jobs. (2) Powerlessness, which seems to develop in childhood, is a generalized sense that one has little or no control over the world. (3) Other forms of alienation include normlessness, meaninglessness, social isolation, and cultural estrangement.

## Key Terms

| | |
|---|---|
| ALIENATION | 514 |
| CULTURAL ESTRANGEMENT | 518 |
| MEANINGLESSNESS | 518 |
| NORMLESSNESS | 517 |
| PILEUP OF DEMANDS | 507 |
| POSITION | 492 |
| POWERLESSNESS | 516 |
| PRIMARY RELATIONSHIP | 493 |
| ROLE | 493 |
| ROLE OVERLOAD | 504 |
| SELF-ESTRANGEMENT | 514 |
| SOCIAL ISOLATION | 518 |
| SOCIAL NETWORK | 493 |
| SOCIAL STRUCTURE | 492 |
| STATUS | 493 |
| UPWARD MOBILITY | 495 |
| VALUES | 501 |

18

chapter

# Deviant Behavior and Social Reaction

# ■ Introduction

Virginia and Susan wandered through the department store, stopping briefly to look at blouses and then going to the jewelry counter. Each looked at several bracelets and necklaces. Susan kept returning to a 24-karat gold bracelet with several jade stones, priced at $139.50. Finally, she picked it up, glanced quickly around her, and dropped the bracelet into her shopping bag.

The only other shopper in the vicinity, a well-dressed man in his 40s, saw Susan take the bracelet. He too looked around the store, spotted a security guard, and walked toward him. Virginia stammered, "I, uh, I don't think we should do this."

Susan replied, "Oh, it's okay. Nothing will happen." Susan walked quickly out of the store. Moments later, Virginia followed her. As Susan entered the mall, the security guard stepped up to her, took her by the elbow, and said, "Come with me, please."

Shoplifting episodes like this one occur dozens of times every day in the U.S. Shoplifting is one of many types of **deviant behavior**—behavior that violates the norms that apply in a given situation. Other types of deviance include such behaviors as "knavery, skulduggery, cheating, unfairness, crime, sneakiness, malingering, cutting corners, immorality, dishonesty, betrayal, graft, corruption, wickedness, and sin." (Cohen, 1966).

There are two major reasons why social psychologists study deviant behavior, one theoretical and one practical. First, social norms and conformity are the basic means by which the orderly social interaction necessary to maintain society is attained. By studying nonconformity, we learn about the processes that produce social order. For example, we might conclude that Susan took the bracelet because there were no store employees nearby, suggesting the importance of surveillance in maintaining order. Second, social psychologists study deviant behavior to better understand its causes. Deviant behaviors such as alcoholism, drug addiction, and crime are perceived as serious threats to society. Once we understand its causes, we may be able to develop better programs that reduce or eliminate deviance or that help people change their deviant behavior.

Four fundamental questions will be addressed in this chapter.

1. What are the causes of deviant behavior?
2. How important for deviant behavior is the reaction of observers? That is, does someone have to react to behavior in particular ways in order for it to be considered deviant?
3. Why do some people engage in deviance regularly? Why do they adopt a life style that involves participation in deviant activities?
4. What determines how authorities and agents of social control deal with incidents of deviance? Is their reaction influenced by the deviant's gender, social status, or other characteristics of the situation?

## ■ The Violation of Norms

When we read or hear that someone is accused of murder or embezzling money from a bank, we often ask why? In Susan's case we would ask, "Why did she take that bracelet?" In this section we will consider first the meaning of norms and then look at several theories about the causes of deviant behavior. These include anomie, control, and differential association theories.

### Norms

Most people would regard Susan's behavior in the department store as deviant because it violated social norms. Specifically, she violated laws that define taking merchandise from stores without paying for it as a criminal act. Thus, deviance is a social construction; whether a behavior is deviant or not depends on the norms or expectations for behavior in the situation in which it occurs.

In any situation, our behavior is governed by norms derived from several sources (Suttles, 1968). First, there are purely "local" and group norms. Thus, roommates and families develop norms about what personal topics can and cannot be discussed. Second, there are subcultural norms that apply to large numbers of persons who share some characteristic. For example, there are racial or ethnic group norms governing the behavior of blacks or persons of Slavic descent that do not apply to other Americans. Third, there are societal norms such as those requiring certain types of dress or those limiting sexual activity to certain relationships and situations. Thus, the norms that govern our daily behavior have a variety of origins, including family, friends, socioeconomic or religious or ethnic subcultures, and the society in general.

The repercussions of deviant behavior depend on which type of norm an individual violates. Violations of local norms may be of concern only to a certain group. Failing to do the dishes when it is your turn may result in a scolding from your roommate, although your friends may not care about that deviance. Subcultural norms are often held in common by most of those with whom we interact, whether friends, family members, or co-workers. Violations of these norms may affect most of one's day-to-day interactions. Violations of societal norms may subject a person to action by formal agencies of control such as the police or the courts. Earlier in the book we discussed the violation of local norms (chapter 9) and group norms (chapter 12). In this chapter we will focus on the violation of societal norms and reactions to norm violations.

### Anomie Theory

The **anomie theory** of deviance, developed by Robert Merton (1957), suggests that deviance arises when people who strive to achieve culturally valued goals such as wealth find they do not have any legitimate way to attain these goals. These people then break the rules, often in an attempt to attain these goals illegitimately.

Most Americans are socialized to strive for economic success. But some people do not have access to legitimate employment, so they seek wealth via alternative, sometimes illegal, means.

Every society provides its members with goals to aspire to. If the members of a society value religion, they are likely to socialize their youths and adults to aspire to salvation. If the members value power, they will teach people to seek positions in which they can dominate others. Merton argued that U.S. culture extols wealth as the appropriate goal for most members of society. In every society, there are also norms that define acceptable ways of striving for goals, called **legitimate means.** In the U.S., legitimate means for attaining wealth include learning and acquiring roles that serve as routes to success (such as student or apprentice), working

hard at a job to earn money, starting a business, and making wise investments.

A person socialized into U.S. society will most likely desire material wealth and will strive to succeed in a desirable occupation—to become a teacher, nurse, business executive, or doctor. The legitimate means of attaining these goals are to obtain a formal education and to climb the ladder of occupational prestige. The person who has access to these means—who can afford to go to college and has the accepted skin color, ethnic background, and gender—can attain these socially desirable goals.

What about those who do not have access to the legitimate means? As Americans, these people will desire material wealth like everyone else, but at the same time they will be blocked in their strivings. Because of the way society is structured, certain members will be denied access to legitimate means. Government decisions regarding budgeting and building or closing schools will determine the availability of education to individuals. Similarly, certain members of society will be denied access to jobs. Not only individual characteristics, such as lack of education, but also social factors, such as the profitability of making steel in Ohio, determine who will be unemployed.

A person who strives to attain a legitimate goal but is denied access to legitimate means will experience *anomie*—a state that reduces commitment to norms and the pursuit of goals. There are four ways a person may respond to anomie; each is a distinct type of deviance. First, an individual may reject the goals, give up trying to achieve success, but continue to conform to social norms; this adaptation is termed *ritualism*. The poorly paid stock clerk who never misses a day of work in 45 years is a ritualist. He is deviant because he has given up the struggle for success. Second, the individual might reject both the goals and the means, withdrawing from active participation in society by *retreatism*. This may take the form of drinking, drug use, withdrawal into mental illness, or other kinds of escape. Third, one might remain committed to the goals but turn to disapproved or illegal ways of achieving success; this adaptation is termed *innovation*. Earning a living as a burglar, fence, or loan shark is an innovative means of attaining wealth. Finally, one might attempt to overthrow the existing system and create different goals and means through *rebellion*. Examples include the type of countercultural behavior seen in the U.S. in the late 1960s and early 1970s.

Within this theory, shoplifting is a form of innovation. Like other types of economic crime, it represents a rejection of the normatively prescribed means (buying what you want) while continuing to strive for the goal (possessing merchandise). According to anomie theory, Susan, the shoplifter, has been socialized to desire wealth but does not have access to a well-paying job due to her poor education. As a result, she steals what she wants because she cannot pay for it.

Why is Susan an innovator? Why didn't she accept her lack of wealth and become a ritualist? The adaptation of the individual depends partly on the relative strength of socialization to the goals and to the means. Given that Susan does not have access to legitimate means to success, she will become an innovator if her socialization to the goals is stronger than her socialization to the means. If her socialization to the means is stronger, however, she will become a ritualist. Withdrawal is the likely outcome when socialization to both goals and means is weak.

Another influence on an individual's adaptation is access to deviant roles. Utilizing a means of goal achievement—whether legitimate or illegitimate—requires access to two structures (Cloward, 1959). The first is a **learning structure**—an environment in which an individual can learn the information and skills required. A shoplifter needs to learn how to quickly conceal objects, to spot plainclothes detectives, and so forth. The second is an **opportunity structure**—an environment in which an individual has opportunities to play a role, which usually requires the assistance of those in complementary roles. Anomie theory assumes that anyone can be an innovator—through shoplifting, prostitution, or professional theft. But not everyone has access to the special knowledge and skills needed to succeed as a prostitute (Heyl, 1977) or a professional thief (Sutherland, 1937). Just as access to legitimate means to achieve goals is limited, so is access to illegitimate means. Only those who have both the learning and opportunity structures necessary to become a shoplifter, prostitute, or embezzler can utilize these alternative routes to success (Coleman, 1987).

The opportunities for deviance available to a person depend on age, sex, kinship, ethnicity, and social class (Cloward, 1959). All these characteristics, with the possible exception of class, are beyond the individual's control. Thus, prostitution in our society primarily involves young, physically attractive persons. People who do not have access to the necessary learning or opportunity structures are double failures; they can succeed neither through legitimate nor illegitimate means. Double failure often produces retreatism. Drug addicts, alcoholics, and mentally ill persons may be losers in both the conventional and criminal worlds.

Anomie theory directs our attention to the importance of education and employment in attaining wealth. Because lower-class persons are more frequently excluded from quality education and jobs, the theory predicts that they will commit more crimes. Data collected by police departments and the FBI in the 1950s and 1960s confirmed this prediction, showing that a disproportionate number of those arrested for crimes were poor, minority males. This led some to conclude that crime and social class are inversely related—that the highest crime rates are found in the lower social strata (Cloward, 1959).

However, there is a class bias built into official statistics on crime. For example, not all illegitimate activities are included in these statistics. While data on burglary, robbery, and larceny are compiled by police departments, data on income tax evasion, price-fixing, and stock swindles are not. Police and FBI statistics are much more likely to include crimes committed by the lower classes than the kinds of economic white-collar crimes committed by members of the middle and upper classes.

Another bias in official statistics is the fact that many crimes are never reported. In the past, store employees rarely turned shoplifters over to the police. It has been suggested that they were unlikely to call the police if the person was middle or upper class. If in fact police were more likely to be notified when the suspect was from a lower socioeconomic class, then this, too, would contribute to a relationship between class and crime.

In response to these limitations in official statistics, researchers began to gather information using self-report measures of crime. Such studies ask persons—frequently high school or college students—whether they have engaged in various behaviors in the recent past. Typically, the questionnaire asks whether they have taken things that did not belong to them, damaged, destroyed, or mistreated others' property, smoked marijuana, or committed a variety of other crimes (See Box 18-1). Contrary to the results of research using official statistics, these studies have found little or no relationship between social class and the commission of crime (Tittle & Villemez, 1977).

How can we reconcile these differences? Are lower-class people more likely to commit crimes as anomie theory predicts and official statistics seem to verify? Or is there no relationship as self-report studies indicate? These questions assume that self-reports of behavior and official statistics measure the same thing—that is, they measure crime—and that studies where the sample consists of adolescents measure delinquency. This assumption may be false (Hindelang, Hirschi, & Weis, 1979). Arrest statistics compiled by police departments include seven crimes: homicide, sexual assault, robbery, aggravated assault, burglary, larceny/theft, and auto theft. Self-report measures ask about assault, theft, auto theft, receiving stolen property, use of drugs such as marijuana, cocaine, and tran-

quilizers, and academic cheating. Thus, official statistics include serious and relatively infrequent crimes, whereas self-report measures include less serious, much more common offenses. Both sets of results appear to be valid. Lower-class persons are more likely to commit the serious, officially tabulated crimes, but no more likely than middle-class persons to commit less serious violations of the law (Elliott & Ageton, 1980; Thornberry & Farnsworth, 1982).

## Control Theory

If you were asked why you don't shoplift compact discs or tapes from stores, you might reply, "Because my parents (or lover or friends) would kill me if they found out." According to **control theory,** social ties influence our tendency to engage in deviant behavior. We often conform to social norms because we are sensitive to the wishes and expectations of others. This sensitivity creates a bond between the individual and other persons; the stronger the bond is, the less likely the individual is to engage in deviant behavior.

According to Hirschi (1969), there are four components of this social bond. The first is *attachment*—ties of affection and respect to others. Attachment to parents is especially important, because they are the primary socializing agents of a child; a strong attachment leads the child to internalize social norms. The second component is *commitment* to long-term educational and occupational goals. Someone who aspires to go to law school is unlikely to commit a crime, because a criminal record would be an obstacle to a career in law. The third component is *involvement*. People who are involved in sports, scouts, church groups, and other conventional activities simply have less time to engage in deviance. The fourth component is *belief*—a respect for the law and persons in positions of authority.

We can apply control theory to the shoplifting incident described at the beginning of this chapter. Susan does not feel attached to law-abiding adults; therefore, she was not concerned about their reactions to her behavior. Nor did she seem deterred by commitment when she said, "Nothing will happen." Susan's deviant act seems to reflect the absence of a strong bond with conventional society.

The relationship between delinquency and the four components of the social bond has been the focus of numerous studies. Several studies have found a relationship between a lack of attachment and delinquency; young people from homes characterized by a lack of parental supervision, communication, and support report more delinquent behavior (Hirschi, 1969; Jensen, 1972; Hundleby & Mercer, 1987). Attachment to school, measured by grades, is also associated with delinquency; boys and girls who do well in school are less likely to be delinquent. Regarding commitment to long-term goals, research indicates that youths who are committed to educational and career goals are less likely to engage in property crimes such as robbery and theft (Johnson, 1979; Shover et al., 1979). Findings relevant to the third component, involvement, are mixed. Whereas involvement in studying and homework is negatively associated with reported delinquency, participation in athletics, hobbies, and work is unrelated to reported delinquency. Involvement in religion, as reflected in frequent church attendance and rating religion as important in one's life, is associated with reduced delinquency (Sloane & Potrin, 1986). Finally, evidence suggests that a person's beliefs have a less important impact on delinquent behavior than do other influences (Jensen, 1972).

Control theory asserts that attachment to parents leads to reduced delinquency. Implicitly, the theory assumes that parents do not encourage delinquent behavior. While this assumption may be valid in most instances, there are exceptions. Studies suggest that some parents encourage some delinquent behaviors. Adolescent drinking is associated with parental alcohol consumption; parents who are heavy drinkers are more likely than nondrinking parents to have adolescents who are heavy drinkers (Barnes, Farrell, & Cairns, 1986). In this instance, parental attachment leads to increased delinquency.

## Differential Association Theory

Are all types of deviance explained by the absence of a social bond? Perhaps not. Sometimes people deviate from one set of norms because they are being influenced by a contradictory set of norms. U.S. society is composed of many groups with different values, norms, and behavior patterns. With respect to many behaviors there is no single, societywide norm. An adolescent's use of marijuana may deviate from her parents' norms, for example, but may conform to her friends' norms. Hence, the deviance involved in marijuana use reflects a conflict between the norms of two groups rather than an insensitivity to the expectations of others. In fact, the use of marijuana may reflect a high degree of sensitivity to the expectations of one's peers.

This view of deviance is the basis of **differential association theory,** which was developed by Edwin Sutherland who argued that even though the law provides a uniform standard for deviance, one group may define that behavior as deviant, whereas another defines it as desirable. Shoplifting, for example, is legally defined as a crime. Some groups believe it is wrong because (1) it leads to increased prices, which hurts everyone; (2) it violates the moral principle against stealing; and (3) it constitutes lawbreaking. Other groups, in contrast, believe shoplifting is acceptable because (1) businesses deserve to have things taken because they overcharge; (2) the loss is covered by insurance; and (3) the shoplifter won't be caught. Susan's comment "It's okay. Nothing will happen." reflects the latter belief.

Beliefs and attitudes about behaviors are learned through associations with others, usually in primary group settings. People learn motives, drives, and techniques of engaging in specific

## HOW DEVIANT ARE YOU?

Researchers frequently use self-report to study the incidence of deviant behavior. By asking people direct questions, researchers hope to avoid the biases found in official statistics that suggest that crime is concentrated in the lower class. Table B-18-1 is a typical questionnaire on deviance. Take a few minutes and fill it out.

Most people have done at least a few things that others would consider wrong. For each item in Table B-18-1, circle the number of times you have engaged in the activity in the past two years.

Answers to these questions provided by a sample of college sophomores, juniors, and

■ TABLE B-18-1

| Behavior | Number of Times | | | | | |
|---|---|---|---|---|---|---|
| 1. Taken an item from a store without paying for it. | 0 | 1 | 2 | 3 | 4 | 5+ |
| 2. Taken things without permission from someone else's room or home that did not belong to you. | 0 | 1 | 2 | 3 | 4 | 5+ |
| 3. Bought, kept, or used something that you knew had been stolen from someone else. | 0 | 1 | 2 | 3 | 4 | 5+ |
| 4. Damaged, destroyed, or mistreated property on purpose. | 0 | 1 | 2 | 3 | 4 | 5+ |
| 5. Beaten up or hurt someone on purpose. | 0 | 1 | 2 | 3 | 4 | 5+ |
| 6. Threatened to beat up or hurt someone unless they did what you wanted. | 0 | 1 | 2 | 3 | 4 | 5+ |
| 7. Smoked or used marijuana. | 0 | 1 | 2 | 3 | 4 | 5+ |
| 8. Used cocaine. | 0 | 1 | 2 | 3 | 4 | 5+ |
| 9. Used drugs such as LSD, pep pills ("uppers," "speed"), tranquilizers ("downers"), or sleeping pills without a doctor's prescription. | 0 | 1 | 2 | 3 | 4 | 5+ |
| 10. Cheated on an exam or quiz in class. | 0 | 1 | 2 | 3 | 4 | 5+ |
| 11. Worked with other students on homework or a take-home project when you weren't supposed to. | 0 | 1 | 2 | 3 | 4 | 5+ |

Source: Adapted from Tittle and Villemez, 1977.

Box 18-1

seniors (97 males and 196 females) are reproduced in Table B-18-2. The sample was obtained from a survey of two large undergraduate classes at the University of Wisconsin (Zimmerman & DeLamater, 1983). These responses indicate that students commit some crimes frequently, such as taking things without permission, vandalism, and the use of marijuana. Compare your own responses to the questionnaire with theirs.

■ **TABLE B-18-2**

| Behavior | | Number of Times | | | | | |
|---|---|---|---|---|---|---|---|
| | | 0 | 1 | 2 | 3 | 4 | 5+ |
| **1.** Taken something from a store. | males | 81% | 8 | 4 | 3 | 0 | 3 |
| | females | 85 | 7 | 4 | 2 | 0 | 3 |
| **2.** Taken something from someone's room or home. | males | 56 | 12 | 12 | 10 | 2 | 7 |
| | females | 54 | 17 | 14 | 5 | 2 | 8 |
| **3.** Bought or used stolen property. | males | 61 | 20 | 15 | 2 | 1 | 1 |
| | females | 85 | 6 | 4 | 1 | 1 | 4 |
| **4.** Vandalized property. | males | 58 | 16 | 13 | 6 | 1 | 5 |
| | females | 77 | 14 | 6 | 1 | 0 | 2 |
| **5.** Beaten up someone. | males | 86 | 8 | 2 | 2 | 1 | 1 |
| | females | 89 | 8 | 3 | 0 | 0 | 0 |
| **6.** Threatened someone. | males | 79 | 7 | 8 | 2 | 1 | 2 |
| | females | 91 | 1 | 4 | 0 | 1 | 3 |
| **7.** Used marijuana. | males | 36 | 10 | 9 | 6 | 8 | 29 |
| | females | 34 | 16 | 8 | 12 | 9 | 20 |
| **8.** Used cocaine. | males | 69 | 5 | 2 | 2 | 2 | 19 |
| | females | 75 | 5 | 3 | 4 | 3 | 11 |
| **9.** Used other drugs. | males | 72 | 5 | 4 | 0 | 0 | 19 |
| | females | 66 | 8 | 6 | 6 | 1 | 15 |
| **10.** Cheated in class. | males | 55 | 25 | 8 | 3 | 4 | 5 |
| | females | 58 | 18 | 12 | 7 | 5 | 1 |
| **11.** Worked with other students when you weren't supposed to. | males | 65 | 12 | 10 | 4 | 3 | 5 |
| | females | 59 | 19 | 13 | 4 | 2 | 3 |

Source: Zimmerman and DeLamater, 1983.

behaviors. What they learn depends on whom they interact with—that is, on their differential associations. Whether someone engages in a specific behavior depends on how frequently one is exposed to attitudes and beliefs that are favorable toward that behavior.

The principle of differential association states that a "person becomes delinquent because of an excess of definitions favorable to violation of the law over definitions unfavorable to violation of the law" (Sutherland & Cressey, 1978). Studies designed to test this principle typically ask individuals questions about their attitudes toward a specific behavior and about their participation in that behavior. One study revealed that the number of definitions favorable to delinquency accurately predicted which young males reported delinquent behavior (Matsueda, 1982). The larger the number of definitions that were endorsed by a youth, the larger the number of delinquent acts he reported having committed in the preceding year.

Another study involving several hundred college students found systematic differences in the beliefs of those who had engaged in shoplifting and those who had not (Kraut, 1976). Those who had taken merchandise without paying for it believed that they would not be caught or that they would not be severely punished even if they were caught. On a measure of self-image, the ratings they gave themselves were similar to the ratings they gave in describing a "typical shoplifter." Students who had not taken things defined the risk of getting caught as high and believed the consequences would be serious. They were especially concerned about informal sanctions and reactions of friends and family. They rated themselves as quite different from the typical shoplifter.

The theory of differential association does not specify the process by which people learn criminal or deviant behavior. For this reason, Burgess and Akers (1966) developed a modified theory of differential association. Unlike the original theory that emphasizes the influence of interactions with others, this newer version em-phasizes the influence of positive and negative reinforcement on the acquisition of behavior. Much of this reinforcement comes from friends and associates. Thus, if beliefs are learned through interaction with others, then people whose attitudes are favorable toward a behavior should have friends who also have favorable attitudes toward that behavior. Alternatively, people whose attitudes are opposed to the activity should have friends who share those negative views.

A survey of 3,056 high school students was conducted to test these hypotheses (Akers et al., 1979). In particular, it assessed the relationship between differential association, reinforcement, and adolescents' drinking behavior and marijuana use. Differential association was measured by three questions: How many of your (1) best friends, (2) friends you spend the most time with, and (3) friends you have known longest, smoke marijuana and/or drink? The survey also assessed students' definitions of drug and alcohol laws and laws in general. Both social reinforcement (whether adolescent expected praise or punishment for use from parents and peers) and nonsocial reinforcement (whether the effects of substance use were positive or negative) were measured. Findings of this survey showed that differential association was closely related to the use of alcohol and/or drugs—that is, the larger the number was of friends who drank and/or smoked marijuana, the more likely the student was to drink and/or smoke marijuana. Reinforcement was also related to behavior; those who used a substance reported that it had positive effects. The students' definitions were also related to whom they associated with; if their friends drank and/or used marijuana, they were more likely to have positive attitudes toward the behavior and negative attitudes toward laws defining that behavior as criminal. Finally, students' attitudes were consistent with their behavior; those who opposed marijuana use and supported the marijuana laws were much less likely to use that substance.

Survey data collected at one point in time

often cannot be used to test hypotheses about cause-and-effect relationships. However, survey data collected from the same people at two or more times can be. Stein and colleagues (1987) analyzed data from 654 young people who were surveyed three times: in junior high school, four years later (in senior high), and four years later. The measures included peer drug use, adult drug use, and community approval for drug use. The results showed that adolescents who believed that both peers and adults were using drugs were more likely to become drug users. Thus, association with persons who use alcohol and drugs, especially in primary relationships, is one cause of substance use by adolescents.

Because drinking, smoking marijuana, and using other drugs often occur in group settings, it is not surprising that these behaviors are influenced by the norms of the group. Even more interesting is the fact that the attitudes and behavior of our friends influence very private behaviors. For example, studies have found that whether and how frequently single persons engage in various sexual activities reflect differential association. One survey of college seniors found that whether individuals engaged in premarital sex was related to how many of their best friends had engaged in that behavior (Schultz et al., 1977). Results of the survey are illustrated in Figure 18-1. If none of their best friends were sexually experienced, only 15–35 percent of the students reported engaging in sexual intercourse; if all five of their best friends were sexually experienced, more than 80 percent of the respondents reported they were also.

Of course, we don't simply conform to the norms of whatever group we happen to be in. Our behavior is also influenced by our own beliefs. For example, researchers in this survey found that the more a person endorsed "conventional religious values," the less likely he or she was to have engaged in sexual intercourse, regardless of how many friends had done so. The least religious students were 18 percent more likely to be sexually experienced than the most religious students.

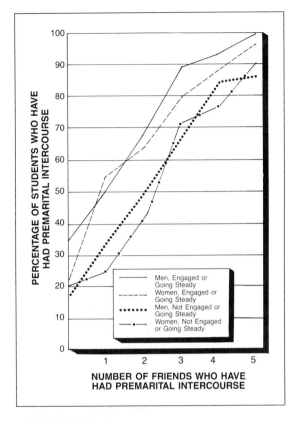

■ **FIGURE 18-1**

PERCEPTION OF FRIENDS' PARTICIPATION IN PREMARITAL SEX AND INDIVIDUAL PREMARITAL BEHAVIOR

More than 1,800 college seniors were asked whether they had engaged in premarital intercourse as well as how many of their five best friends of the same sex had engaged in premarital intercourse. Note that as the number of best friends who were sexually experienced increased, the likelihood that the student was sexually experienced increased dramatically. For each additional friend who was experienced, the number of sexually active persons increased 12 percent. Note that going steady or being engaged was also associated with an increased likelihood of premarital intercourse.

*Source: Adapted from Schulz et al., 1977.*

Since each person usually associates with several groups, the consistency or inconsistency in definitions across groups is an important influence on behavior (Krohn, 1986). *Network multiplexity* refers to the degree to which individ-

## THE POWER OF SUGGESTION

Rape, robbery, murder, and other types of deviant behavior receive a substantial amount of coverage in newspapers and on radio and television. One function of publicizing deviance is to remind us of norms—to tell us what we should not do (Erikson, 1964). But is this the only consequence? Could the publicity given particular deviant activities increase the frequency with which they occur? In some cases, the answer appears to be yes.

A study of the relationship between the publicity given suicides and suicide rates suggests that the two are positively correlated (Phillips, 1974). This study identified every time a suicide was publicized in three major U.S. daily papers from 1947 to 1968. Next, the researcher calculated the number of expected suicides for the following month by averaging the suicide rates for that same month from the year before and the year after. For example, researchers noted that the suicide of a Ku Klux Klan leader on November 1, 1965, was widely publicized. They then obtained the expected numbers of suicides (1,652) by averaging the total number of suicides for November 1964 (1,639) and November 1966 (1,665). In fact, there were 1,710 suicides in November 1965; the difference between the observed and expected rates (58) could be due to suggestion via the mass media.

Results of this study showed that suicides increased in the month following reports of a suicide in major daily papers. Moreover, the more publicity a story was given—as measured by the number of days the story was on the front page—the larger the rise in suicides was. If a story was published locally—in Chicago but not in New York, for example—the rise in suicides occurred only in the area where it was publicized.

Why should such publicity lead other persons to kill themselves? There must be some factor that predisposes a small number of persons to take their own lives following a publicized suicide. That predisposing factor may be *anomie*. According to this theory, suicide is a form of retreatism, of withdrawal from the struggle for success. Persons who don't have access to legitimate means are looking for some way to adapt to their situation. Publicity given to a suicide may suggest a solution to their problem.

When we think of suicide, we think of shooting oneself, taking an overdose of a drug, or jumping off a building. We distinguish suicide from accidents when we presume the person didn't intend to harm himself or herself. But the critical difference is the person's intent, not the event itself. Some apparent accidents may be suicides. For example, when a car hits a

uals who interact in one context also interact in other contexts. When you interact with the same people at church, at school, on the athletic field, and at parties, multiplexity is high. When you interact with different people in each of these settings, multiplexity is low. When multiplexity is high, definitions of an activity will be consistent across groups; when it is low, definitions may be inconsistent across groups. Thus, associations should have the greatest impact on attitudes and

beliefs when multiplexity is high. A survey of 1,435 high school students measured the extent to which individuals interacted with parents and with the same peers in each of several activities (Krohn, Massey, & Zielinski, 1988). Students who participated jointly with parents and peers in various activities were less likely to smoke.

The anomie, control, and differential association perspectives are not incompatible. Anomie theory suggests that culturally valued goals

Box 18-2

bridge abutment well away from the pavement on a clear day with no evidence of mechanical malfunction, this may be suicide.

If some auto accidents are in fact suicides, we should observe the increase in motor vehicle accidents following newspaper stories about a suicide. In fact, data from newspapers and motor vehicle deaths in San Francisco and Los Angeles verify this hypothesis (Phillips, 1979). Statistics show a marked increase in the number of deaths in automobile accidents two and three days after a suicide is publicized—especially accidents involving one vehicle. In the Detroit metropolitan area, an analysis of motor vehicle fatalities between 1973 and 1976 revealed an average increase in fatalities of 35–40 percent the third day after a suicide story appeared in the daily papers (Bollen & Phillips, 1981); again, the more publicity, the greater the increase. Finally, if the person whose suicide is publicized was young, deaths of young drivers increase; whereas if the person killing himself was older, the increase in fatalities involves more older drivers.

Does an increase in suicide follow any publicized suicide, or are some suicides more likely to be imitated than others? Stack (1987) studied instances in which celebrities killed themselves. Each was classified according to whether the person was an entertainer, political

figure, artist, member of the economic elite, or villain. The results showed that when entertainers and politicians took their own lives, there was an increase in suicide rates. Suicides by artists, members of the elite, and villains were not followed by an increase. Moreover, the findings suggest that the effect of publicized suicide is gender and race specific. Suicide by a male celebrity was followed by an increase in the number of males who killed themselves but not in the number of females who took their lives, and vice versa. Similarly, an increase in suicides by whites followed a publicized case involving a white celebrity, while rates among blacks were unaffected. The facts that the effects of publicized suicide are age, gender, and race specific are all consistent with the concept of imitation.

A very different form of deviance, the hijacking of aircraft, also appears to be influenced by publicity. An analysis of all instances of air piracy in the U.S. between 1968 and 1972 (Holden 1986) found that successful hijackings were followed by an increase in hijacking attempts. There was no increase following unsuccessful hijacking attempts.

Thus, media reports of some types of deviance may suggest behavioral responses to some problem, suggestions that are acted on by some persons.

---

and the opportunities available to achieve these goals are major influences on behavior. Opportunities to learn and occupy particular roles are influenced by age, social class, gender, race, and ethnic background. According to control theory, we are also influenced by our attachments to others and our commitment to attaining success. Our position in the social structure and our attachments to parents and peers determine our differential associations—the kinds of groups to

which we belong. Within these groups, we learn definitions favorable to particular behaviors, and we learn that we face sanctions when we choose behaviors group members define as deviant.

## ■ Reactions to Norm Violations

When we think of murder, robbery, or rape, we think of cases we have read about or heard of through radio or television. We frequently refer to police and FBI statistics as measures of the number of crimes that have occurred in our city or county. Our knowledge of alcohol or drug abuse and homosexual behavior depends on knowing or hearing about persons who engage in these behaviors. All of these instances of deviance share another important characteristic as well. In every case, the behavior was discovered by someone who called it to the attention of others.

Does it matter that these instances involve both an action (by a person) and a reaction (by a victim or an observer)? Isn't an act just as deviant whether others find out about it or not? Let's go back to Susan's theft of the bracelet. Suppose Susan had left the store without being stopped by the security guard. In this case, she and Virginia would have known she had taken the bracelet, but she would not have faced sanctions from others. She would not have experienced the embarrassment of being confronted by a store detective and accused of deviance. Moreover, she would have had a beautiful bracelet. But the fact is that she was stopped by the guard. She will be questioned, the police will be called, and she may be arrested. Thus, the consequences for committing a deviant act are quite different when certain reactions follow. This reasoning is the basis of **labeling theory**—the view that reactions to a norm violation are a critical element in deviance. Only after an act is discovered and labeled "deviant" is the act recognized as such. If the same act is not discovered and labeled, it is not deviant (Becker, 1963).

If deviance depends on the reactions of others to an act rather than on the act itself, the key social psychological question becomes Why do particular audiences choose to label an act deviant (whereas other audiences may not)? Labeling theory is an attempt to understand how and why acts are labeled deviant. In the case of the stolen bracelet, labeling analysts would not be concerned with Susan's behavior. Rather, they would be interested in Virginia's response to Susan's act and the reactions of the male customer and security guard. Only if an observer

challenges Susan's behavior or alerts a store employee does the act of taking the bracelet become deviant. In this case, the action taken by the security guard clearly defined her act as a violation of norms.

### Reactions to Rule Breaking

Labeling theorists refer to behavior that violates norms as **rule breaking** to emphasize that the act by itself is not deviant. Most rule violations are "secret" in the sense that no one other than the actor (and on occasion the actor's accomplices) is aware of them. Many cases of theft and tax evasion, many violations of drug laws, and some burglaries are never detected; these activities can be carried out by a single person. Other acts, such as robberies, assaults, and various sexual activities, involve other people who will know about them, but may not label the act deviant.

How will members of an audience respond to a rule violation? It depends on the circumstances, but various studies suggest that very often people *ignore* it. When wives of men hospitalized for psychiatric treatment were asked how they reacted to their husbands' bizarre behavior, for example, they generally replied that they had not considered their husbands ill or in need of help (Yarrow et al., 1955). People react to isolated episodes of unusual behavior in one of four ways. A common response is *denial*, in which the person simply does not recognize that a rule violation occurred. In one study, denial was typically the first response of women to their husband's excessive drinking (Jackson, 1954). A second response is *normalization*, in which the observer recognizes that the act occurred but defines it as normal or common. Thus, wives often reacted to excessive drinking as normal, assuming that many men drink a lot. Third, the person may excuse the act, recognizing it as a rule violation but attributing its occurrence to situational or transient factors; such a reaction is known as *attenuation*. Thus, some of the wives of men who were later hospitalized believed that the episodes of bizarre behavior were caused by

The reactions of others to rule-breaking behavior depend on the characteristics of the actor. The dress and grooming of this shoplifter will influence whether her behavior is reported by the man observing it.

unusually high levels of stress or by physical illness. Finally, people may respond to the rule violation by *balancing* it, recognizing it as a violation but deemphasizing its significance due to the actor's good qualities.

The man who had witnessed Susan's behavior could not deny it. He was looking directly at her. He might have normalized the act, believing that Susan intended to go to the cashier's counter and pay for the bracelet before she left the store. He could have excused the act, noting that both girls' clothes were worn, suggesting that they were not well off. Finally, he could have

balanced the fact that Susan was stealing against the fact that both girls were young; perhaps they reminded him of his daughters. In this case, he might have felt that the theft was not serious enough to have them apprehended, questioned, and perhaps charged with a crime.

In fact, the man did not react in these ways; he looked around, spotted a security guard, and reported the act. In doing so, he labeled the actor. Labeling involves a redefinition of the actor's social status; the man placed Susan into a category of "shoplifter" or "thief." The security guard, in turn, probably defined Susan as a "typical shoplifter." Although labeling is triggered by a behavior, it results in a redefinition or typing of the actor. As we shall see, this has a major impact on people's perceptions of and behavior toward the actor.

## Determinants of the Reaction

What determines how an observer reacts to rule breaking? Reactions depend on three aspects of the rule violation, including the nature of the actor, the audience, and the situation (Goode, 1978).

**Actor Characteristics**   Reaction to a rule violation often depends on who performs the behavior violating the rule. First, people are more tolerant of rule breaking by family members than by strangers. The research cited above revealed extraordinary tolerance of spouses for bizarre, disruptive, and even physically abusive behavior. Many of us probably know of a family who is attempting to care for a member whose behavior creates problems for them. Second, people are more tolerant of rule violations by persons who make positive contributions in other ways. In small groups, tolerance is greater for persons who contribute to the achievement of group goals (Hollander & Julian, 1970). We seem to tolerate deviance when we are dependent on the person committing the act, perhaps because if we punish the actor it will be costly for us. Third, we are less tolerant if the person has a history of rule breaking (Whitt & Meile, 1985).

Does gender affect reactions to behavior? An ingenious field experiment suggests that gender does not affect an audience's response to shoplifting. With the cooperation of store employees, shoplifting events were staged in the presence of customers who could see the event. The experiment was conducted in a small grocery store, a large supermarket in a shopping mall, and a large discount department store. Three aspects of the situation were varied: the gender of the shoplifter, the appearance of the shoplifter, and the gender of the observer. Neither the shoplifter's nor the customer's gender had an effect on the frequency with which the customer reported the apparent theft. The appearance of the shoplifter, however, had a substantial effect. If the man or woman who took an item was wearing soiled, patched clothing and had unkempt hair, the customer was much more likely to report the theft than if the shoplifter was neatly dressed and well groomed. Perhaps the customers balanced the theft against what they presumed were the good qualities of the well-dressed shoplifter (Steffensmeier & Terry, 1973).

More generally, analysts have suggested that we are less likely to label women than men for violations of the criminal law (Haskell & Yablonsky, 1983). Research indicates that women are less likely to be kept in jail between arraignment and trial and receive more lenient sentences than men. One explanation for this differential treatment is that women are subject to greater informal control, by family members and friends, and, so, are treated more leniently in the courts. A recent study of the influences on pretrial release and sentence severity found that both men and women with families received more lenient treatment; the effect was stronger for women (Daly, 1987).

**Audience Characteristics** The reaction to a violation of rules also depends on who witnesses it. Because groups vary in their norms, audiences vary in their expectations. People enjoying a city park on a warm day will react quite differently to a nude man walking through the park than will a group of nudists in a nudist park. Recognizing this variation in reaction, people who contemplate breaking the rules often make sure no one is around who will punish them—as when smoking marijuana, drinking in public, or jaywalking.

Citizens react very differently to suspected criminals than do officials who routinely deal with suspects. A study of various officials working in a court-affiliated unit who evaluate suspected murderers following arrest found that these officials had a stereotyped image of the type of person who commits murder, called the "normal primitive" (Swigert & Farrell, 1977). When lower-class male members of ethnic minorities committed murder, these officials believed that it was in response to a threat on their masculinity. For example, they would be more likely to assume that an Italian had killed another man in response to verbal insults and taunts. This labeling based on a stereotype had important consequences. Suspects who fit this image were less likely to be defended by a private attorney, more likely to be denied bail, more likely to plead guilty, and more likely to be convicted on more severe charges.

Consider the example of a student with a drinking problem seeking help at a university counseling center. The treatment will depend on how counselors view student "troubles." One study found that the staff of a university clinic believed that students' problems could be classified into one of the following categories: problems in studying, choosing a career, achieving sexual intimacy, or handling personal finances; conflict with family or friends; and stress arising from sociopolitical activities. When a student came to the clinic because of excessive drinking, the therapist decided which of these categories applied to this particular person's troubles— that is, which type of problem might cause the student to drink excessively. How the problem was defined in turn determined what the therapist did to try to help the student (Kahne & Schwartz, 1978).

Because audiences vary in their standards, conditions defined as deviant by one audience

Observers differ in their reaction to rule-breaking behavior. While some people consider Oliver North a criminal who acted for personal gain, others consider him a patriot and a hero. A jury found him guilty of three violations of the law, including shredding and altering documents to cover up events in the Iran-Contra case.

may be considered normal by another. A woman in rural Wisconsin claimed that the Virgin Mary appeared and spoke with her on eight occasions. To many, these experiences are hallucinations, a symptom of serious mental illness. But to tens of thousands, these were genuine religious experiences, and thousands made pilgrimages to the woman's farm (Scheff, 1967).

**Situational Characteristics** Whether a behavior is construed as normal or labeled deviant also depends on the definition of the situation in which the behavior occurs. Marijuana and alcohol use, for example, are much more acceptable at a party than at work (Orcutt, 1975). Various

sexual activities expected between married persons in the privacy of their home would elicit condemnation if performed in a public park.

Consider so-called "gang" violence. In some major cities, incidents in which teenage gangs assault each other are common. News media, police, and other outsiders often refer to such incidents as "gang wars." These events often occur in the neighborhoods where the gang members live. How do their parents, relatives, and friends react to such incidents? According to a study of one Chicano community, it depends on the situation (Horowitz, 1987). Young men are expected to protect their families, women, and masculinity. When violence results from a challenge to honor, the community generally tolerates it. On the other hand, if the violence disrupts a community affair, such as a dance or wedding, it is not tolerated.

We often rely on the behavior of others to help us define situations. Our reaction to a rule violation may be influenced by the reactions of other members of the audience. The influence of the reactions of others is demonstrated in a field experiment of intrusions into waiting lines (Milgram et al., 1986). Members of the research team intruded into 129 waiting lines, with an average length of six persons. One or two confederates approached the line and stepped in between the third and fourth persons. In some cases, other confederates served as buffers; they occupied the fourth and fifth positions and did not react to the intrusion. When the buffers were present, others in the line were much less likely to react verbally or nonverbally to the intrusion.

## Consequences of Labeling

Assume that an audience defines an act as deviant. What are the consequences for the actor and the audience? We will consider four possible outcomes.

**Institutionalization of Deviance** In some cases, individuals who label a behavior deviant may decide that it is in their own interest for the person to continue the behavior. They may, in fact, reward that person for the deviant behavior. If you learn that a good friend is selling drugs, you may decide to use this person as a source and purchase drugs from him. Over time, your expectations will change; you will come to expect him to sell drugs. If your drug-selling friend decides to stop dealing, you may then treat him as a rule breaker. The process by which members of a group come to expect and support deviance by another member over time is called **institutionalization of deviance** (Dentler & Erikson, 1959).

Consider prostitution, which has been a fixture in most societies throughout history. Despite attempts of religious and political leaders to stamp it out, why does the impersonal exchange of sexual gratification for money persist? Institutionalization is partly the answer. Every society specifies the legitimate means of achieving sexual gratification; in U.S. society, adult, heterosexual, voluntary relationships are the appropriate means. Not everyone has access to these means, however, and prostitution provides an alternative (Davis, 1976). For instance, houses of prostitution developed rapidly on the West Coast of the U.S. in the 1880s, where there was a tremendous surplus of Chinese men relative to the number of Chinese women. These brothels became widely known and were patronized by large numbers of men (Light, 1977). Brothels continue to operate legally in parts of Nevada, perhaps because there is an excess of men relative to women in those counties.

**Backtracking** Even when an audience reacts favorably to a rule violation, the actor may decide to discontinue the behavior. This second consequence of labeling is called *backtracking*. It may occur after the actor learns that others label his or her act deviant. Even though some audiences react favorably, the actor may wish to avoid the reaction of those who would not react favorably and the resulting punishment. Many teenagers try substances like marijuana once or twice. Although their friends may encourage its continued use, some youths backtrack because they want to avoid their parents' negative reactions.

**Effective Social Control** An audience that reacts negatively to rule breaking and attempts to punish the actor or threatens to do so may force the actor to give up further involvement in the activity. This third consequence of labeling is known as *effective social control*. This reaction is common among friends or family members who often threaten to end their association with an actor who continues to engage in deviance. Similarly, they may threaten to break off their relationship if the person does not seek professional help. In these instances, the satisfaction of the actor's needs is contingent on changing his or her behavior. Members of the audience may also insist that the actor renounce aspects of his or her life that they see as contributing to future deviance (Sagarin, 1975). If excessive drinking is due to job-related stresses, for example, family members may demand that the person find a different type of employment.

**Unanticipated Deviance**   Still another possibility is that the individual may engage in further or unanticipated deviance. Note the use of the term "unanticipated." Negative reactions by members of an audience are intended to terminate rule breaking activity. However, such reactions may, in fact, produce further deviance. This occurs when the audience's response sets in motion a process that leads the actor to greater involvement in deviance. This process and its outcomes are the focus of the next section.

# ▪ Labeling and Secondary Deviance

Labeling a person deviant may set in motion a process that has important effects on the individual. The process of societal reaction produces changes in the behavior of others toward the individual and may lead to corresponding changes in his or her self-image. A frequent consequence of the process is involvement in secondary deviance and a deviant subculture. In this section, we will consider this process in detail.

## Societal Reaction

Earlier in this chapter, we mentioned that labeling is a process of redefining a person. By categorizing a person as a particular kind of deviant, we place that person in a stigmatized social status. The deviant (addict, pimp, thief) is defined as undesirable, not fully acceptable in conventional society, and frequently treated as inferior. Two important consequences of this stigmatized status follow. First, it leads to changes in the behavior of others toward the person. Second, the person gradually comes to perceive himself or herself as deviant.

**Changes in the Behavior of Others**   When we learn that someone is an alcoholic, homosexual, or mentally ill, our perceptions and behavior toward that person change. For example, if we learn that someone has a drinking problem, we may respond to his or her request for a drink with "Do you think you should?" or "Do you really need it?" to convey our own objections. We may avoid jokes about drinking in the person's presence, or we may stop inviting him or her to parties or dinners.

A more severe behavior reaction involves withdrawal from the stigmatized person (Kitsuse, 1964). For instance, the labeled shoplifter, alcoholic, or homosexual may be fired from his or her job. Behavioral withdrawal may occur because of hostility toward the deviant (Kitsuse, 1964), or it may reflect a sincere desire to help the person. For example, the employer who fires an alcoholic may do so because he dislikes alcoholics or because he believes that relief from work obligations will reduce stress that may be causing the drinking problem.

Paradoxically, our reaction to deviance may produce additional rule breaking by the labeled person. We expect people who are psychologically disturbed to be irritable or unpredictable, so we avoid them in order to avoid an unpleasant interaction. The other person may sense that he is being avoided and respond with anger or distrust. His anger may cause co-workers to talk about him behind his back; he may respond with suspicion and become paranoid. When members of an audience behave toward a person according to a label and cause the person to respond in ways that confirm the label, they have produced a **self-fulfilling prophecy** (Merton, 1957). Lemert (1962) documents a case in which just such a sequence led to a man's hospitalization for paranoia.

**Self-Perception of the Deviant**   Another consequence of stigmatized social status is that it changes the deviant's self-image. A person labeled deviant often incorporates the label into his or her identity. This redefinition of oneself is due partly to feedback from others, to the extent that they treat the person as a deviant. The new self-image may be reinforced by the individual's own behavior. Repeated participation in shoplifting, for example, may lead Susan to define herself as a thief.

Redefinition is facilitated by the institutions that deal with specific types of deviant persons. Such institutions pressure persons to acknowledge that they are deviant. Admitting that one is a thief will often lead police and prosecutors to go easy on a shoplifter, especially if it is a first offense. Failure to acknowledge this may lead to a substantial prison sentence. Many social programs also help shape the deviant's self-image. Admitting that one is mentally ill is often a prerequisite for psychiatric treatment (Goffman, 1959). Mental health professionals often believe that a patient cannot be helped until the individual recognizes his or her problem. A number of agencies that serve the blind view getting a blind person to admit and accept his or her limitations as a major part of their job. These agencies try to convince the blind that their disability imposes serious obstacles to their independence—obstacles seen as insurmountable by a majority of people (Scott, 1969).

Thus, there are numerous pressures on the deviant to accept a stigmatized identity. Acceptance of a stigmatized identity has important effects on self-perception. Everyone has beliefs about what people think of specific types of deviant persons. Accepting a label such as "thief," "drunk," or "crazy" leads a person to expect that others will stigmatize and reject him or her, which, in turn, produces self-rejection. Self-rejection makes subsequent deviance more likely (Kaplan, Martin, & Johnson, 1986). In a study of junior high school students, data were collected three times, each one year apart. Self-rejection (feeling that one is no good, a failure, rejected by parents and teachers) was related to more favorable dispositions (definitions) toward deviance and an increased likelihood of associating with deviant persons one year later. A high disposition and associations with deviant peers

were related to increased deviance—theft, gang violence, drug use, and truancy—one year later (Kaplan, Johnson, & Bailey, 1987). Figure 18-2 summarizes these relationships. Delinquent behavior, in turn, is associated with reduced self-esteem (McCarthy & Hoge, 1984).

In short, labeling may set in motion a cycle in which changes in behavior produce changes in other people's behavior, which, in turn, changes the deviant's self-image and subsequent behavior.

While most attention has been given to situations in which others label the person, some persons become committed to deviance in the absence of such labeling. For example, some persons voluntarily seek psychiatric treatment; some of these cases reflect *self-labeling* (Thoits, 1985). Since individuals know that others view certain behaviors as symptoms of mental illness, if they observe themselves engaging repeatedly in those behaviors, they may label themselves mentally ill.

## Secondary Deviance

A frequent outcome of the societal reaction process is **secondary deviance,** in which a person engages increasingly in deviant behavior as an adjustment to others' reactions (Lemert, 1951). Usually the individual becomes openly and actively involved in the deviant role, adopting the clothes, speech, and mannerisms associated with it. For example, initially a person with a drinking problem may drink only at night and on weekends to prevent his or her drinking from interfering with work. Once the person adopts the role of "heavy drinker" or "alcoholic," however, he or she may drink continuously. For male homosexuals, "coming out" constitutes an act of self-labeling and a public commitment to homosexuality (Dank, 1971). Prior to coming out, the homosexual tries to limit his homosexual involvement, whereas after coming out he engages in sexual activity with males more openly and frequently.

As an individual becomes openly and regularly involved in deviance, he or she may increas-

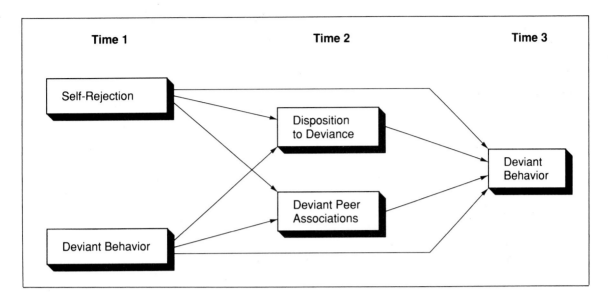

▪ **FIGURE 18-2**

THE RELATIONSHIP OF SELF-REJECTION TO DEVIANT BEHAVIOR

A person who engages in deviant behavior anticipates that others will reject him or her, which, in turn, can lead to self-rejection. A longitudinal study collected data from junior high school students three times, each one year apart. At time 1, reported participation in deviance was positively related to self-rejection (feeling one is no good, a failure, rejected by parents and teachers). Self-rejection at time 1 was associated with more favorable dispositions (attitudes) toward deviance and an increased likelihood of associating with other deviants one year later (at time 2). Favorable dispositions and deviant associations at time 2, as well as deviance and self-rejection at time 1, were related to increased deviance—theft, gang violence, drug use, and truancy—at time 3.

*Source: Kaplan, Johnson, and Bailey, 1987, Figure 2.*

ingly associate with others who routinely engage in the same or related activity. In this way the individual becomes a member of a **deviant subculture**—a group of people whose norms encourage participation in the deviance and who regard positively those who engage in it. Subcultures provide not only acceptance but also the opportunity to enact deviant roles. Through a deviant subculture the would-be drug dealer or prostitute can gain access to customers more readily.

Subcultural groups are an attractive alternative for deviants for two reasons. First, these people are often forced out of straight relationships and groups through others' reactions. As family and friends progressively break off relationships with them, they are compelled to seek

acceptance elsewhere. Second, membership in subcultural groups may result from deviants' desire to associate with people who are similar and who can provide them with feelings of social acceptance and self-worth (Cohen, 1966). Deviants are no different from others in their need for such interpersonal rewards.

Deviant subcultures help persons cope with the stigma associated with deviant status. We have already noted that deviants are often treated with disrespect and sanctioned by others for their activity. Such treatment threatens self-esteem and produces fear of additional sanctions. Subcultures help the deviant cope with these feelings. They provide a *vocabulary of motives*—beliefs that explain and justify the individual's participation in the behavior.

The norms and belief systems of subcultures support a positive self-conception. A study of a group of black male heroin addicts (Finestone, 1964) found that these men valued being "cool" and earning a living through a "hustle"—such as gambling or running a "stable" of prostitutes. These "cats," in turn, disdained the "square" who worked hard at a repetitive, boring, legitimate job. Similarly, many people think that nudists are exhibitionists who take off their clothes in order to get sexual kicks. Nudists consider themselves morally respectable and hold several beliefs designed to enhance that claim: (1) nudity and sexuality are unrelated, (2) there is nothing shameful about the human body, (3) nudity promotes a feeling of freedom and natural pleasure, and (4) nude exposure to the sun promotes physical, mental, and spiritual well-being. There are also specific norms—"no staring," "no sex talk," and "no body contact" —designed to sustain these general beliefs (Weinberg, 1976). The contrast between the stigmatized image of the deviant and the deviant's self-image is illustrated in Figure 18-3. The belief systems of deviant subcultures provide the social support the person needs to maintain a positive self-image.

Joining a deviant subculture may not only stabilize participation in one form of deviance, it may also lead to involvement in additional forms of deviant behavior. Black heroin users, for example, were encouraged by group norms and values to engage in a variety of other illegal, money-making activities. Similarly, many prostitutes become drug users through participation in a subculture.

# ■ Formal Social Controls

So far, this chapter has been concerned with **informal social control**—the reactions of family, friends, and acquaintances to rule violations by individuals. Informal controls are probably the major influence on an individual's behavior.

In modern societies, however, there are often elaborate systems set up specifically to process deviants. Collectively, these are called **formal social controls**—agencies given responsibility for dealing with violations of rules or laws. Typically the rules enforced are written, and, in some cases, punishments may also be specified. The most prominent system of formal social control in our society is the criminal justice system, which includes police, courts, jails, and prisons. A second system of formal social control is the juvenile justice system, which includes juvenile officers, social workers, and probation officers, courts, and treatment or detention facilities. A third system of formal social control deals with mental illness. It includes mental health professionals, commitment procedures, and institutions for the mentally ill and mentally impaired.

## Formal Labeling and the Creation of Deviance

Most of us think of formal agencies as reactive—as simply processing individuals who have already committed crimes, who are mentally retarded, or in need of psychiatric treatment. But these agencies do much more than take care of persons already known to be deviant. It can be argued that the function of formal social control agencies is to select members of society and identify or certify them as deviant (Erikson, 1964).

**Functions of Labeling** Of what value is labeling people "criminals," "delinquents," or "mentally ill"? There are three functions of labeling persons deviant: (1) to provide concrete examples of undesirable behavior, (2) to provide scapegoats for the release of tensions, and (3) to unify the group or society.

First, the public identification of deviance provides concrete examples of how we should not behave (Cohen, 1966). When someone is actually apprehended and sanctioned for deviance, the norms of society are made starkly clear.

■ **FIGURE 18-3**

IMAGES OF THE DEVIANT

We often have very negative images of many types of deviant persons. For example, many people view marijuana users as dirty, unkempt drop-outs, something like the person on the left. Because these images are widely shared, persons who engage in some form of deviant behavior are usually aware that others look down on them. To counter this stigma, deviants attempt to create a positive self-image, which is reinforced by members of deviant subcultures. The user views himself as clean, cool, and in touch, something like the person on the right. It is easier to view oneself as "normal" when others support that view.

For instance, the arrest of someone for shoplifting dramatizes the possible consequences of taking things that do not belong to us. According to the **deterrence hypothesis,** the arrest and punishment of some individuals for violations of the law deters other persons from committing the same violations. To what extent does general deterrence really affect people's behavior? Most analysts agree that the objective possibility of arrest and punishment does not deter people from breaking the law. Rather, conformity is based on people's perceptions of the likelihood

of punishment and the severity of punishment. Thus, youths who perceive a higher probability that they will be caught and that the punishment will be severe are less likely to engage in delinquent behavior (Jensen, Erickson, & Gibbs, 1978). Similarly, a study of theft of company property by employees found that those who perceived greater certainty and severity of organizational sanctions for theft were less likely to have taken property (Hollinger & Clark, 1983).

Perceived certainty of sanctions generally has a much greater effect on persons who have low levels of moral commitment (Silberman, 1976). People whose morals define a behavior as wrong are not as affected by the threat of punishment. For example, a person's moral beliefs are a more important influence on whether adults use marijuana than the fear of legal sanctions (Meier & Johnson, 1977). Adults who believe that the use of marijuana is wrong do not use it, regardless of their perception of the likelihood that they will be sanctioned for its use. Thus, by itself, deterrence is not a primary influence on whether or not people break the law. The perceived threat of legal sanctions deters mainly those who are not morally opposed to an activity.

A second function of public identification of deviants is to provide a scapegoat for the release of tension. Many people face threats to the stability and security of their daily lives. Some fear the possibility that they will be victimized by aggressive behavior or the criminal activity of others. The existence of such threats arouses tension. Persons identified publicly as deviants provide a focus for these fears and insecurities. In other words, the publicly identified deviant becomes the concrete threat that we can deal with decisively.

This scapegoating process is illustrated by the Puritans, who came to New England during the 1600s in order to establish a community based on a specific Christian theology. As time passed, groups within the community periodically challenged the ministers' claims that they were the sole interpreters of the theology. In addition, they faced the threat of Indian attacks and the problems of daily survival in a harsh environment. In 1692, when a group of young women began to behave in such bizarre ways as screaming, convulsing, crawling on all fours, and barking like·dogs, the community focused attention on these women. The physician defined them as "witches," representatives of Satan, and the entire community banded together in search of others who were under the "devil's influence." The community imprisoned many persons suspected of sorcery and sent 22 persons to death.

Public identification of a person as deviant involves mass media coverage, such as the intense publicity given to House Speaker Jim Wright's financial dealings in 1989. The final act was Speaker Wright's tearful resignation. One important function of such publicity is to provide concrete examples of how not to behave.

Thus, the witch hunt provided a scapegoat, an outlet for people's fears and anxieties (Erikson, 1966).

A third function of public identification of deviants is to increase the cohesion and solidarity of society. Nothing unites the members of a group like a common enemy (Cohen, 1966). Deviants, in this context, are "internal enemies," persons whose behavior threatens the morale and efficiency of a group. Should the solidarity of the group be threatened, it can be restored by identifying one member as deviant and imposing appropriate sanctions. Suppose you are given the case study of a boy with a history of delinquency who is to be sentenced for a minor crime. You are asked to discuss the case with three other persons and decide what should be done. One member argues for extreme discipline, while you and the other two favor leniency. Suddenly, an expert in criminal justice who has been sitting quietly in the corner announces that your group should not be allowed to reach a decision. How might you deal with this threat to the group's existence? The reasoning above suggests that the person who took the extreme position will be identified as the cause of the group's poor performance and that other members will try to exclude him from future group meetings. A laboratory study used exactly this setup, contrasting the reaction of threatened groups to the person taking the extreme position with the reaction of nonthreatened groups. In the former condition, the person taking the extreme position was more likely to be stigmatized and rejected (Lauderdale, 1976).

Thus, controlled amounts of deviant behavior serve important functions. If deviance is useful, we might expect control agencies to "create" deviance when the functions it serves are needed. In fact, the number of persons who are publicly identified as deviant seems to reflect the levels of stress and integration in society (Scott, 1976). When integration declines, there is an increased probability of deviance. Eventually, the level and severity of deviance may reach a point where citizens will demand a "crackdown," and social control agencies will increase their activity, increasing the number of publicly identified deviants. Thus, in turn, will increase solidarity and lower stress, leading to an increase in the amount of informal control and a reduction in deviance.

**The Process of Labeling** Labeling by formal agencies is not a simple, one-step procedure. The processing of rule breakers usually involves

a sequence of decisions. At each step, someone has to decide whether to terminate the process or to pass the rule breaker on to the next step. Figure 18-4 portrays the sequence of steps involved in processing criminal defendants.

Each of the controlling agents—police officers, prosecutors, and judges—has to make many decisions every day. Like anyone else, they develop cognitive categories and rules that simplify their decision making. A very common police-citizen encounter occurs when an officer stops a motorist who has been drinking. What determines whether a driver who has been drinking is labeled a "drunk driver"? Research suggests that police officers develop a series of informal guidelines that they use in deciding whether to arrest the motorist. Officers on the street have to rely on a variety of subjective data, since the breathalyzer or blood or urine test may only be available at the police station. The decision to arrest also depends on situational circumstances. In one study of 195 police encounters with persons who had been drinking, arrests were more likely if the encounter occurred downtown and if the citizen was disrespectful (Lundman, 1974).

Prosecutors also develop informal rules that govern their decisions. For example, in one large midwestern city, taking an object worth less than $100 is a misdemeanor, and conviction normally results in a fine. Theft of a more valuable object is a felony and results in a prison sentence. Because felony theft cases require much more time and effort, the prosecutor charged virtually all persons arrested for shoplifting with misdemeanors, even if they took jewelry worth hundreds of dollars.

In many jurisdictions, probation officers are asked to prepare a presentencing report and to recommend a sentence for the convicted person. Research indicates that these officers have a set of typologies or *cognitive schema* into which they sort persons (Lurigio & Carroll, 1985). Semi-structured interviews with probation officers in one community identified ten schemas, including burglar, addict, gang member, welfare fraud,

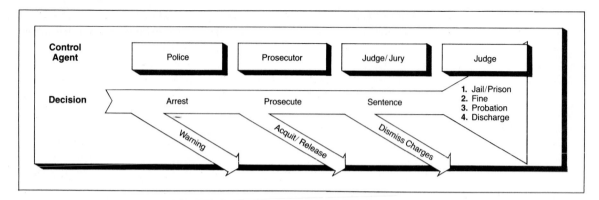

■ **FIGURE 18-4**

FORMAL SOCIAL CONTROL: PROCESSING CRIMINAL DEFENDANTS

Formal social control often involves several control agents, each of whom makes one or more decisions. The first step in the criminal justice system is an encounter with a law enforcement officer. If you are arrested, the case is passed to a prosecutor, who decides whether to prosecute. If your case goes to court, the judge or jury decides whether you are guilty. Finally, the judge renders a sentence. These decision makers are influenced by their own personal attitudes, attributional processes, role expectations, and the attitudes of others regarding their decisions. Much research is devoted to the social psychological aspects of decision making in the criminal justice system.

and conman. Each schema was associated with beliefs about the motive for the crime and the appropriate treatment and prognosis. When officers were asked to evaluate sample cases, those fitting a schema were evaluated more quickly and confidently. More experienced officers were more likely to use similar schemas. At the same time, probation officers vary in the emphasis they place on particular items of information about the convicted person (Drass & Spencer, 1987).

**Biases in Social Control** Not all persons who violate the rules are necessarily labeled. Most social control agencies process only some of the persons who violate the rules. In the study of police encounters with drunken persons, only 31 percent were arrested (Lundman, 1974). In some cases, control agents may be influenced by the demeanor of the rule breaker, by the agent's schema, or by where the violation occurs. This leads one to ask whether systematic biases exist in the social control system.

It has been suggested that control agents are more likely to label those people who have the least power to resist certification as being deviant (Quinney, 1970). This hypothesis predicts that people from the lower class and members of racial and ethnic minorities are more likely to be certified as deviant than upper-class, middle-class, and white persons. This hypothesis offers a radically different explanation for the correlation between crime and social class. Earlier in this chapter, we suggested that crime rates are higher for lower-class persons because they do not have access to nondeviant means of economic and social success. Here we are suggesting that crime rates are higher among lower-class persons because they are more likely to be arrested, prosecuted, and found guilty, even though the underlying rate of deviant activity may not vary as a function of social class.

Does social class or race influence how an individual is treated by control agents? One way to answer this question is by studying police-citizen encounters through the "ride-along" method, in which trained observers ride in squad cars and systematically record data about police-citizen encounters. In the largest study of this kind, observers rode with some officers on all shifts every day for seven weeks. Data were collected in Boston, Washington, and Chicago and included 5,713 encounters. There was no evidence that blacks were more likely to be arrested than whites. Rather, arrests were more likely when a third party demanded an arrest, when the evidence was strong, and when the crime was serious (Black, 1980). A study of how police officers managed violent encounters between citizens found that arrest was more likely if the incident involved white persons, two men (instead of one or two women), and if the person acted abusively toward the officer (Smith, 1987).

What about decisions by prosecutors? Do they entail discrimination based on race or class? Prosecutors are generally motivated to maximize the ratio of convictions to trials. This may be one criterion citizens use in evaluating the performance of a district attorney. Prosecutors develop beliefs about which cases are "strong"—those likely to result in conviction. A study of a random sample of 980 defendants charged with felonies found that prosecutors are more likely to prosecute cases involving serious crimes where the evidence is strong and the defendant has a serious prior record. Race was not generally influential (Myers & Hagan, 1979).

Does the social class of an arrested person influence how he or she is treated by the courts? Research on this question offers no straightforward answer. For instance, a study of the handling of juvenile cases by the courts in Denver and Memphis found no evidence of class or race bias in the processing of cases. In most instances, the seriousness of the offense and prior record were the major determinants of the severity of the sentence (Cohen & Kluegel, 1978). A longitudinal study of 9,945 boys born in Philadelphia in 1945 followed them from their tenth to their eighteenth birthdays (Thornberry & Christenson, 1984). During these years, 3,475 were officially labeled delinquent. Analyses found that

the most important influence on the disposition of a charge was the disposition imposed for prior offense(s). A study of all delinquency cases in the state of Florida between 1979 and 1982 in which there were two or more events involving the same person reached the same conclusion (Henretta, Frazier, & Bishop, 1986).

A common practice in adult criminal cases is *plea bargaining,* in which a prosecutor and a defendant's lawyer negotiate a plea to avoid the time and expense of a trial. A single action frequently violates several laws. Thus, if a driver who has been drinking runs a red light and hits a pedestrian who later dies, that incident involves at least three crimes: drunken driving, failure to obey a signal, and vehicular manslaughter. These offenses vary in seriousness and, thus, in their associated sentences. The prosecutor may offer not to indict the driver for manslaughter if a plea of guilty is entered to a drunken driving charge; the attorney may accept the offer, provided the prosecutor also recommends a suspended sentence.

Are the members of certain groups more likely to be tried or to get bigger reductions in sentences? An analysis of charge reduction or plea bargaining in a sample of 1,435 criminal defendants found that women and whites received slightly more favorable reductions than men and blacks (Bernstein et al., 1977). Another study of 1,213 men charged with felonies found that the characteristics of an offense—especially the seriousness of the crime and the strength of the evidence—were most important in determining the disposition. Neither age, ethnicity, nor employment status were shown to be related to the outcome of the case (Bernstein, Kelly, & Doyle, 1977).

Among persons convicted, do we find a class or racial bias in the length of sentences given? One study focused on the sentences received by 10,488 persons in three southern states: North Carolina, South Carolina, and Florida (Chiricos & Waldo, 1975). Researchers examined sentences for 17 different offenses and found no relationship between socioeconomic status or race

Whether or not a police officer gives a citizen a traffic ticket will depend partly on the demeanor of the citizen. Officers are more likely to ticket or arrest hostile, argumentative persons than polite and submissive ones.

and sentence length. Once again, the individual's prior record was the principal variable related to sentence length. A study of a random sample of 16,798 felons convicted between 1976 and 1982 in Georgia looked at racial differences in sentencing (Myers & Talarico, 1986). In general, the seriousness of a crime was the principal influence on the sentence length.

Thus, research on decisions by social control agents does not find a systematic bias against members of particular groups. Cases involving serious deviance and strong evidence are more likely to be prosecuted and to lead to punishment.

## Long-Term Effects of Formal Labeling

How long does the official label of deviant stick to a person? Can it be shaken? In contrast to the setting in which a person is formally certified as deviant, there is no formal ceremony terminating one's deviant status (Erikson, 1964). People are simply released from prison or a mental hospital, or the final day of probation passes—with no fanfare. Does this mean that deviant status in our society tends to be for life?

Some argue that ex-convicts, ex-patients, and others who have been labeled as deviant face continuing pressures from family and friends that could prevent them from readjusting to normal life; it constitutes a reminder of their former stigmatized status. However, two studies of former psychiatric patients (Sampson et al., 1964; Greenley, 1979) found no evidence of continuing stigmatization by members of their families.

Another area in which former prison inmates and mental patients might face discrimination is employment. This may occur because others continue to perceive these persons as deviant and expect them to behave in ways consistent with that label. In one study (Schwartz & Skolnick, 1964), researchers prepared four versions of a job application to be shown to prospective employers. All four applications were largely identical; all applicants had a succession of short-term jobs. The only variable was their legal record. In one condition, the applicant had been convicted of assault; in the second, he had been charged and acquitted. In the third condition, he had been acquitted, and the application included a letter from the trial judge verifying the acquittal. In the fourth (control) condition, he had no legal record. Each of the four versions was shown to 25 employers. The employers' responses are summarized in Table 18-1. Compared to the control condition, employers were less likely to respond favorably when the person had been arrested. Employers were about equally likely to respond negatively whether the man had been acquitted or convicted. When the letter from the judge was included, the employer's interest was higher.

A related concern is the impact of mental illness on occupational careers (Huffine & Clausen, 1979). A study of psychiatrically disturbed persons compared the income and employment status of those who had been treated (labeled) with the income and status of those who had not been treated. Treatment was negatively associated with both income and employment status (Link, 1982). The impact seemed to depend partially on whether occupational competence was developed before or after the onset of the illness. Men who had no history of competent work performance had more difficulty obtaining employment following hospitalization, whereas men who had a history of occupational competence usually kept their jobs, even during periods when their work performance was seriously affected.

A recent study of the long-term impact of being labeled mentally ill suggests that it is not the label by itself that has impact but the label combined with changes in self-perception (Link, 1987). The study compared samples of residents and clinic patients from the same area within New York City. Three samples involved people who had been labeled: first-treatment contact patients, repeat-treatment contact patients, and formerly treated community residents. The other two groups were untreated "cases" (people with symptoms) and a sample of residents. All participants completed a scale that measured the belief that mental patients are stigmatized and discriminated against. High scores on the measure were associated with reduced income and unemployment in the labeled groups but not in the unlabeled ones.

The long-term effects of formal labeling may be limited, because persons who have been labeled in the past engage in various tactics to prevent others from learning about their stigma. These tactics include selective concealment of past labeling, preventive disclosure to close friends, and various deception strategies (Miall, 1986).

■ **TABLE 18-1**

EFFECTS OF LEGAL RECORD ON EMPLOYMENT OPPORTUNITIES

| Percentage of Employers Who Responded: | Applicant's Legal Record | | | | |
|---|---|---|---|---|---|
| | Convicted (N = 25) | Acquitted without Letter (N = 25) | Acquitted with Letter (N = 25) | Control (N = 25) | Total |
| Positively | 4 | 12 | 24 | 36 | 19 |
| Negatively | 96 | 88 | 76 | 64 | 81 |

Source: Adapted from Schwartz and Skolnick, 1964.

# ■ Summary

Deviant behavior is any act that violates the social norms that apply in a given situation.

**The Violation of Norms** (1) Norms are local, subcultural, or societal in origin. The repercussions of deviant behavior depend on which type of norm an individual violates. (2) Anomie theory asserts that deviance occurs when persons do not have legitimate means available for attaining cultural success goals. Possible responses to anomie include ritualism, retreatism, innovation, and rebellion. (3) Control theory states that deviance occurs when an individual is not responsive to the expectations of others. This responsiveness, or social bond, includes attachment to others, commitment to long-term goals, involvement in conventional activities, and a respect for law and authorities. (4) Differential association theory emphasizes the importance of learning through interaction with others. Individuals often learn the motives and actions that constitute deviant behavior just as they learn socially approved behavior.

**Reactions to Norm Violations** Deviant behavior involves not only acts that violate social norms but also the societal reactions to these acts. (1) There are numerous possible responses to rule breaking. Very often we ignore it. At other times, we deny that the act occurred, define the act as normal, excuse the perpetrator or recognize the act but deemphasize its significance. Only after an act is discovered and labeled "deviant" is it recognized as such. (2) Our reaction to rule breaking depends on the characteristics of the actor, the audience, and the situation. People often have a stereotyped image of deviant persons; these stereotypes influence how audiences react to rule violations. (3) The consequences of rule breaking depend on the reactions of the audience and the response of the rule breaker. If members of the audience reward the person, the deviance may become institutionalized. Alternatively, the person may decide to avoid further deviance, in spite of others' encouragement. If the person is punished, he or she may give up the behavior or respond with additional rule violations.

**Labeling and Secondary Deviance** The process of labeling has two important consequences. (1) It leads members of an audience to change their perceptions of and behavior toward the actor. If they withdraw from the stigmatized person, they may create a self-fulfilling prophecy and elicit the behavior they expected from the actor. (2) Labeling often causes the actor to change his or her self-image and come to define the self as deviant. This, in turn, may lead to secondary deviance—an open and active involvement in a life style based on deviance. Such life styles are often embedded in deviant subcultures.

**Formal Social Controls** Every society gives certain agents the authority to respond to deviant behavior. (1) In U.S. society, the major formal social control agents are the criminal justice and mental health systems. These agencies select persons and identify them as deviant through a sequence of decisions. Within the criminal justice system, the sequence includes the decisions to arrest, prosecute, and sentence the person. Various factors influence each step in decision making, including the strength of the evidence, the seriousness of the rule violation, and the individual's prior record. (2) Contrary to popular belief, people do not systematically stigmatize former deviants. Most families do not continue to stigmatize relatives following their release from mental hospitals, and most employers do not stigmatize ex-patients and ex-convicts who have established competent work records.

## Key Terms

| | |
|---|---|
| ANOMIE THEORY | 523 |
| CONTROL THEORY | 526 |
| DETERRENCE HYPOTHESIS | 543 |
| DEVIANT BEHAVIOR | 522 |
| DEVIANT SUBCULTURE | 541 |
| DIFFERENTIAL ASSOCIATION THEORY | 527 |
| FORMAL SOCIAL CONTROL | 542 |
| INFORMAL SOCIAL CONTROL | 542 |
| INSTITUTIONALIZATION OF DEVIANCE | 538 |
| LABELING THEORY | 534 |
| LEARNING STRUCTURE | 525 |
| LEGITIMATE MEANS | 524 |
| OPPORTUNITY STRUCTURE | 525 |
| RULE BREAKING | 534 |
| SECONDARY DEVIANCE | 540 |
| SELF-FULFILLING PROPHECY | 539 |

# 19

chapter

# Collective Behavior and Social Movements

# ■ Introduction

—Following a victory by the home football team, hundreds of excited spectators pour onto the field and tear down the goalposts.

—Rumors that a major auto manufacturer is bankrupt set off waves of panicky selling on Wall Street, completely disrupting the stock market.

—In the wake of the shooting of a black teenager by white police officers, thousands of blacks march through the streets to City Hall, and a series of speakers demand changes in police practices.

Events such as these occur daily and sometimes receive national media coverage. In part because they are so common, they have been of interest to social scientists since the turn of the century.

**Collective behavior** refers to "emergent and extrainstitutional" behavior (Lofland, 1981). By *emergent,* we mean subject to norms created by the participants themselves. By *extrainstitutional,* we mean that the norms involved are not derived from and may even be opposed to those of society (Turner & Killian, 1972).

Collective behavior has three dimensions: the underlying goals, the degree of organization among participants, and the duration. First, the goal of collective behavior may be expressive (to release tension or publicly display emotions such as patriotism or hostility) or it may be instrumental (to achieve some concrete outcome such as an escape from danger or a change in the distribution of power). Many collective behaviors have expressive as well as instrumental aspects. Second, collective behavior can vary from being unorganized (such as a spontaneous event with no formal leaders) to being highly organized (such as a planned program of activities with formal leaders). Third, the duration of collective behavior can vary greatly from lasting only a few hours (such as victory celebrations) to lasting several days (such as racial disturbances that have occurred in large cities) or even years (such as the Pentecostal movement within various religions).

The diagram in Figure 19-1 compares various forms of collective behavior in terms of these three dimensions. Most research focuses on either relatively short, unorganized events—often referred to as crowds—or on long-term, relatively organized social movements.

The first part of this chapter will be concerned with collective behavior such as crowds, while the second part will discuss social movements. Specifically, this chapter will address the following questions:

1. What social processes occur within a crowd situation?

2. What causes collective behavior? That is, what are the underlying causes, and what precipitates particular collective actions?

3. What influences the behavior of members of a crowd?

4. How do social movements develop? What are the processes by which they define issues and attract members?

5. How do movement organizations mobilize supporters? How is their operation affected by processes inside and outside of the organization?

6. How do social movements affect the larger society?

# ■ Collective Behavior

## Crowds

A **crowd** refers to a substantial number of persons, usually strangers, who engage in behavior recognized as unusual by participants and observers. Crowd events are characterized by a unanimity of feeling (Turner & Killian, 1972). Examples include a victory celebration, a mass looting of retail stores during a blackout, a stampede of people waiting to be admitted to a theater, and epidemics in which physical symptoms or behavior spread rapidly.

Crowd events sometimes result in injuries and death. On December 3, 1979, several thousand people gathered outside the Riverfront Coliseum in Cincinnati to attend a rock concert. When the doors opened, persons in the middle and at the edges of the crowd pushed forward. As a result, at least 25 persons near the doors were pushed and fell into a pile (Johnson, 1987). There were 11 deaths and numerous injuries. Such tragedies and the desire to prevent them have led many people to study what happens within crowds.

**Le Bon** One of the first social scientists to analyze crowds was Gustave Le Bon (1895). Le Bon proposed the crowd mind, or "psychological crowd," to explain people's behavior in such settings. In 1895, he wrote: "The sentiments and ideas of all the persons in the gathering take one and the same direction, and their conscious personality vanishes. A collective mind is formed." According to Le Bon, this "mental unity of the crowd" determines behavior. The best evidence for the existence of the crowd mind is that members of the aggregation "feel, think, and act in a manner quite different from that in which each individual of them would feel, think, and act were he in a state of isolation." In the Cincinnati incident, each person had the same idea: to get inside the coliseum quickly in order to get a good seat. Most people did not consider the effects of their pushing on others.

Le Bon identified three mechanisms that bring about the crowd mind. First, a crowd provides anonymity. When an individual is among others, often strangers, there is a decline in the sense of responsibility for one's own behavior. This makes it easier for the person to act on impulse and to engage in behavior that violates the norms of society—such as looting, vandalism, or assault. Second, contagion occurs; behavior or ideas expressed by one person spread quickly through a crowd. This process gives direction to the crowd's activity; it is the mechanism through which uniformity of behavior is produced. In the concert incident, the fact that some people were pushing forward undoubtedly led others to do the same. The third

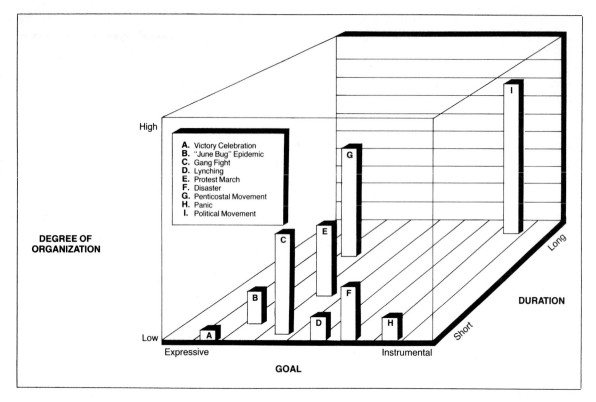

**DEGREE OF ORGANIZATION**

High

A. Victory Celebration
B. "June Bug" Epidemic
C. Gang Fight
D. Lynching
E. Protest March
F. Disaster
G. Penticostal Movement
H. Panic
I. Political Movement

Low

Expressive          Instrumental

**GOAL**

Long

**DURATION**

Short

■ **FIGURE 19-1**

FORMS OF COLLECTIVE BEHAVIOR

Collective behaviors differ on three dimensions: goal, degree of organization, and duration. The goal may be expressive (as in the celebration of an athletic victory), or it may be instrumental (as in the panic selling of stock in response to a rumor that a company is going bankrupt). Some collective behaviors have almost no organization (as in the aimless wandering of hundreds of people after a natural disaster), whereas others have a high degree of organization (such as a Pentecostal religious movement). Finally, some collective behaviors involve only one event (such as a lynching), whereas others involve numerous activities over an extended period of time (such as a political movement).

mechanism is suggestibility. Le Bon believed that individuals in crowds lose conscious awareness of their behavior and that unconscious aspects of their personalities come to the fore. This accounts for the seemingly irrational or bizarre character of some crowd behavior.

Le Bon's analysis of crowd behavior has had considerable influence on subsequent research. Some of the concepts used by contemporary social psychologists have their roots in his writings.

**Deindividuation**  Le Bon's view of anonymity is similar to the contemporary concept of **deindividuation**—a reduction in awareness of one's individuality and feelings of personal responsibility.

In one study of deindividuation (Festinger et al., 1952), students participated in a group discussion under one of two conditions. In one condition (public identity), participants, wearing their own clothing, were seated in a well-lit room. In the other condition (deindividuation), each person put on a gray coat, and the discussion was conducted in a dimly lit room. Each group was asked to make hostile remarks about their parents. Such remarks were much more frequent in the deindividuation condition. In another study (Zimbardo, 1969), subjects were asked to administer electric shocks to another person (a confederate). In one condition, they wore hoods, were forbidden to use names, and were placed in a dimly lit room. In the other condition, subjects wore large name tags in a well-lit room. Not surprisingly, deindividuated subjects administered electric shocks for longer periods to the target person. Of course, the equipment was wired so that the target person did not receive the shocks.

Thus, it appears that an increase in anonymity (deindividuation) produces an increase in antisocial behavior. This evidence, however, comes from experiments in which participants performed a particular behavior in small groups. Thus, the extent to which it applies to crowd settings is not clear. In a crowd, activities are initiated by one member and then spread to others.

**Contagion**   A second process identified by Le Bon is **contagion**—the rapid spread through a group of visible and often unusual symptoms or behavior. In any interaction, one person acts and witnesses the reaction of others to that behavior. Usually, others respond in different but related ways. In a crowd, however, others respond by engaging in the *same* behavior. If one person shouts a hostile message, others will do likewise. Their reaction reinforces that person's behavior, making it more likely that it will be repeated. This *circular reaction* is a fundamental mechanism of collective behavior (Blumer, 1969).

One illustration of contagion is the "June bug" epidemic, which occurred in a clothing manufacturing plant in the southern U.S. in 1962. Within one week, 62 employees received medical attention for what they said were insect bites. Typical symptoms reported by the affected persons included feeling faint, nausea, severe

Behavioral contagion often occurs in a crowd. One person acts, liberating others from the restraints that have prevented them from performing the act. Behavioral contagion can produce something trivial like a "wave" at a football game—or something serious like this attack on a spectator at a soccer match.

pain, or feeling disoriented. Almost all of those affected were white women who worked the same shift in the same area of the plant. A thorough investigation could not identify any insect or chemical that could have produced these symptoms.

The "June bug" epidemic is an example of *hysterical contagion* (Kerckhoff, Back, & Miller, 1965). We are all frequently exposed to models of hysterical behavior, such as people expressing fears of insect bites. We generally ignore these behaviors completely or define persons who display them as deviant. However, when two other conditions are met, contagion may occur: (1) the persons involved must be experiencing tension, and (2) the behavior must be relevant to their situation. It appears that both of these conditions were present in the clothing plant. Pressures to produce were great, and many of the employees were working overtime. At the same time, there was the possibility of layoffs at the end of the peak period. The behavior was also relevant because insects are common in clothing plants.

An analysis of friendship patterns revealed that the sociometric structure of workers in the plant affected the spread of the "epidemic." People who were considered isolates—those who had no friends at the plant—were most likely to be affected at the beginning of the epidemic. As the contagion spread, however, it increasingly affected friends of those who had already been affected. These results suggest that contagion initially occurs among those most susceptible to tension, the isolates; as it spreads, ties to social networks become increasingly important in influencing who is affected (Kerckhoff, Back, & Miller, 1965).

Closely related to hysterical contagion is *behavioral contagion,* in which the behavior of one person reduces constraints that prevent others from performing the same behavior (Wheeler, 1966). For example, many people in a crowd may have thought of rushing onto a playing field but restrained themselves for fear of being arrested. Seeing others run onto the field without interference from security police shows observers that they will not be punished for that behavior.

Although contagion is useful in understanding why persons engage in the same behavior, it does not explain why some people in a crowd engage in various other behaviors and why others do not act at all (Turner & Killian, 1972). By itself, contagion is not a complete explanation of crowd behavior.

**Convergence**  So far we have assumed that a crowd is composed of dissimilar persons and that its initial makeup depends simply on who happens to be in the vicinity. Next, conditions such as deindividuation reduce the effectiveness of self-control. Finally, through contagion, some

behavior spreads through the crowd. But suppose that persons in the crowd share certain qualities that predispose them to act in some ways and not others (Milgram & Toch, 1969).

When this happens, similar behavior may reflect the convergence of those with similar predispositions, not reduction in self-control and contagion. For example, electronic media often rapidly spread word of an event. The news of John Lennon's death in December of 1980 was transmitted nationally within hours. While millions heard the news, only a few hundred converged at the apartment building where he lived in New York City. These were people for whom his death was especially meaningful. Their behavior—singing his songs, lighting candles, and various other memorial rituals at the scene of his murder—reflected their shared sense of loss.

**Emergent Norms**  Deindividuation, contagion, and convergence provide only a partial explanation of collective behavior. Most crowds include some people who know each other; in fact, many participants arrive and remain with family and friends or acquaintances (McPhail & Wohlstein, 1983). In addition, crowds involve various types of persons engaging in a variety of related but different actions. A much more general perspective on crowd activity is provided by the emergent norm theory, derived from the symbolic interactionist perspective.

According to *emergent norm theory,* collective behavior occurs when people find themselves in an undefined or unanticipated situation (Turner & Killian, 1972). The situation may be novel, so that there are no cultural norms or directives for action. For instance, in the late 1980s, several incidents occurred in the U.S. in which a person with a gun opened fire in a school. These incidents were completely unexpected, and there were no behavioral guidelines for students, teachers, or administrators. Alternatively, the social structure may be temporarily disrupted by a natural disaster, such as a tornado, or by an event, such as a citywide strike by police officers.

People do not always panic and run when a disaster occurs. When it is clear that escape is not possible, as it was when a section of this plane ripped off in flight, people will remain in their seats, stationary if not calm.

through informal and often novel channels that cannot be validated—exerts a major influence on the emerging definition of the situation. *Milling*—the movement of persons within a setting and the consequent exchange of information between crowd members—is the major method through which information is transmitted.

As noted earlier, a crowd usually consists of a variety of people with initially differing definitions of the situation. Thus, diverse action tendencies are present in any crowd situation (Turner & Killian, 1972). Each person will have some sense of the likelihood that others will support his or her own disposition (Johnson & Feinberg, 1977). Someone will initiate an act, perhaps in the belief that others will support him or her. Once a person initiates an act, the support of those nearby will determine whether that person persists in attempts to influence others. If enough people reinforce that person's position or behavior, a consensus will emerge. The definition of the situation that results from interaction in an initially ambiguous situation is termed an **emergent norm.** The emergent norm is usually not completely novel; it involves a modification or transformation of preexisting norms (Killian, 1984).

Once a definition of the situation develops, people are able to act. In a crowd, behaviors consistent with the emergent norm are encouraged, whereas behaviors inconsistent with the norm are discouraged. Thus, there are normative limits on the behavior of crowd participants. Crowds celebrating a football championship do not engage in looting; conversely, looters usually do not congregate in bars for several hours and drink alcoholic beverages.

The emergent norm model incorporates all of the ideas discussed earlier in this section. It emphasizes the situational character of collective behavior without assuming that participants behave in the same way or that they are all strangers. In fact, the pressure to conform to the emergent norm may be more effective when participants know each other.

Another possibility is that there may be conflicting definitions of how people should behave. In order to act in these situations, those present must develop a shared definition of the situation and the associated behavioral norms.

In circumstances such as these, people are unable to find out what is going on or what they should do. Because of the need for information, conventional barriers to communication break down. Strangers talk to each other or to members of groups they usually avoid. In addition, the usual standards of judgment and morality may be suspended. **Rumor**—communication

Box 19-1

## REACTIONS TO DISASTERS

A **disaster** is an event that produces widespread physical damage or destruction of property accompanied by social disruption (Quarantelli & Dynes, 1977). Many disasters occur with no warning. Sometimes, however, there may be prior warning, such as increasingly strong tremors prior to an earthquake or warnings by authorities prior to an enemy attack. There are three main types of reactions to the threat of disaster (Perry & Pugh, 1978). Denial is the most common when the likelihood of the disaster is perceived as small or the warning is ambiguous. Extreme emotion, such as terror and hysteria, is likely if the threat is accurately perceived but not imminent. Effective adaptation is most likely if there are repeated, accurate warnings that include information about what people can do to enhance their survival. Research shows that reactions to threat are more appropriate and effective when information is communicated through official channels; inappropriate and ineffective behavior is more likely when information is transmitted through unofficial, interpersonal channels, such as rumor.

Little is known about how people behave at the time a disaster hits. What data there are indicate that most people do not panic (Perry & Pugh, 1978). Panic in response to a disaster occurs only under certain conditions. First, the situation must be defined as dangerous. Panic will not occur if most people deny that there is any threat to themselves. Second, the danger must be perceived as escapable by some action. If there is a fire in a building, for example, at least some of those present must believe there are doors or windows through which they can escape. People often remain calm when it is obvious that there is no possibility of escape, for example, in an emergency on an airplane at high altitudes. Thus, when a section of the roof of a commercial jetliner disintegrated at 26,000 feet, the passengers, while scared, remained seated until the plane landed about one hour later. Third, at least some of those present must believe that escape routes are inadequate or will be cut off. Under these conditions, people are influenced by the behavior of others. If there is an emergency during a concert, and the emcee calmly announces the procedures to be followed, everyone may leave in an orderly fashion. On the other hand, if several people suddenly run toward the exits, others may attempt to do the same. This clearly illustrates the emergent norm model. An experiment (Kelley et al., 1965) in which all three conditions were simulated revealed that people were more likely to successfully escape when the threat was low rather than high and when the group was small (three or four) rather than large (five or six).

What happens after a disaster depends, in part, on the ability of existing emergency organizations to respond, including police and fire departments, hospitals, utility companies, and civil defense agencies (Quarantelli & Dynes, 1977). Such organizations often have drills to prepare their personnel to respond to emergencies rapidly and efficiently. At the scene of a disaster, an *emergency social system* emerges (Perry & Pugh, 1978). At first, residents of the impact area may work together in informal groups. They will take responsibility for the recovery of victims, removal of debris, and the initial assessment of damages. A tremendous sense of community and high morale often develop. Gradually, the emergency organizations move in to supplement or supplant the emergency social system.

## Underlying Causes of Collective Behavior

Having considered the internal dynamics of crowds, we turn now to the causes of collective behavior. In some instances, collective behavior is simply a response to a precipitating event, like a natural disaster, an athletic victory, or an assassination. Other types of collective behavior —demonstrations, boycotts, lynchings, lootings, and epidemics—frequently involve not only a precipitating event but also more basic causes that can be traced to underlying conditions in the larger society. Three such conditions are strain, relative deprivation, and grievances.

**Strain** Society may be viewed as normally in a state of equilibrium, maintaining a balance between the emphasis on achieving society's goals and the provision of means to achieve them— education and jobs (Merton, 1957). At times, however, social change may disrupt this equilibrium, so that one aspect of society is no longer in balance with other aspects. Advances in technology, for example, demand changes in occupational structure. Machines and robots have increasingly replaced many blue-collar workers in automobile plants. This has produced high unemployment in cities, like Detroit, that depend heavily on the auto industry. Such change produces strains in society affecting some individuals more than others. Although those who are affected may not recognize the source (for example, automation), they experience discontent or frustration, which could result in an outbreak of collective behavior.

Historically, economic issues have frequently been at the heart of collective protest (Rude, 1964). "Food riots" to protest the lack of sufficient food, attacks on factories and businesses to prevent mechanization, and sabotage to disable machinery and other property are frequently economically motivated. These activities were common in preindustrial England and France. More recently during the 1980s, we have seen farmers in the U.S. dump milk and slaughter beef cattle rather than sell them at depressed prices. Such protests seek to maintain or improve an area's standard of living by reducing supplies and, thus, keeping prices high.

Evidence that economic issues are related to strain comes from a study of support for radical political proposals (Plutzer, 1987). Telephone interviews were conducted with 912 adults living in the largest U.S. cities and with 458 unemployed men and women. Each was asked whether he or she supported several radical proposals. These included having the government limit personal income or corporate profits and doing away with capitalism, if necessary, in order to limit unemployment. Results showed that economic insecurity—concern about one's economic future—was associated with support for the proposals.

**Relative Deprivation** In the eighteenth century, the revolt against the feudal socioeconomic structure occurred first in France. Yet France had already lost many feudal characteristics by the time the French Revolution began in 1789. The French peasant was free to travel, to buy and sell goods, and to contract services. In Germany, however, the feudal social structure was still intact. Thus, based on objective conditions, one would have expected a revolution to occur in Germany before it did in France. Why didn't it? One analyst (de Tocqueville, 1856/1955) argued that the decline of medieval institutions in France caused peasants to become obsessed with the ownership of land. The improvement in their objective situation created subjective expectations for further improvement. Peasant participation in the French Revolution was motivated by the desire to fulfill subjective expectations— to obtain land—rather than by a desire to eliminate oppressive conditions.

This basic insight on the causes of revolutions was expanded into a more systematic view. According to the **J-curve theory** (Davies, 1962, 1971) the "state of mind" of citizens determines whether there is political stability or revolution.

On the basis of external conditions, individuals develop expectations regarding the satisfaction of their needs. Expectations may be derived from one's own past experience or from a comparison with the experiences of other groups. Under certain conditions, persons expect continuing improvement in the satisfaction of their needs. As long as these expectations are met, people are content and political stability results. But if the gap between expectations and reality becomes too great, people become frustrated and engage in protest and rebellious activity.

Revolutions usually occur when the level of actual satisfaction declines following a period of rising expectations and their relative satisfaction (Davies, 1971). These relationships are summarized in Figure 19-2. Note the J shape of the actual need satisfaction curve; as satisfaction declines, an intolerable gap between expected need satisfaction and actual need satisfaction is created. The J-curve theory applies well to several major revolutions, including the Russian Revolution of 1917, the Cairo riots of 1952, and the Hungarian uprising of 1956.

The gap between expected and actual need satisfaction is called **relative deprivation.** As relative deprivation becomes greater, the likelihood of protests and social movements increases. As noted earlier, expectations regarding need satisfaction may be based on one's own past experience or on a comparison with the current experiences of other groups. The French peasant in prerevolutionary times experienced deprivation because, even though his freedom had increased dramatically, he had not gained the freedom to own land like the aristocracy. Recent research confirms that it is the feeling that members of one's own group are deprived relative to members of other groups that is associated with collective protest (Begley & Alker, 1982; Guimond & Dube-Simard, 1983).

A common form of collective protest in the contemporary U.S. is the strike, in which union members collectively withhold their labor in the hope of improving their economic position. Can strikes be explained by relative deprivation? The

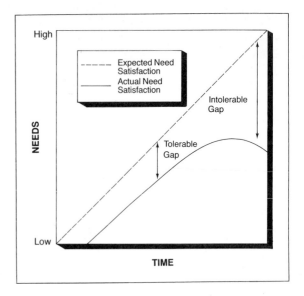

■ **FIGURE 19-2**

THE J-CURVE MODEL

One theory of the causes of revolutions is the J-curve theory. According to this model, revolutions occur when there is an intolerable gap between people's expectations of need satisfaction and the actual level of satisfaction they experience. In response to improved economic and social conditions, people expect continuing improvement in the satisfaction of their needs. As long as they experience satisfaction, there is political stability, even if there is a gap between expected and actual satisfaction (called *relative deprivation*). If the level of actual satisfaction declines, the gap gets bigger; at some point it becomes intolerable, and revolution occurs.

*Source: Adapted from Davies, 1962.*

answer is yes, under certain conditions. First, union members base their expectations regarding contract provisions on their own past experience, not the experience of other groups. Strikes are more likely when economic conditions are good and there is a large gap between what workers expect and what management offers— that is, high relative deprivation (Snyder, 1975). Furthermore, strikes are only effective when labor unions are institutionalized—that is, when there is ongoing labor-management accommodation, when union membership is large and stable, and when organized labor has influence in the national political system (Rubin, 1986).

There have been numerous attempts to study the relationship between collective behavior and relative deprivation. Many researchers have attempted to measure people's subjective evaluation of their level of need satisfaction using questionnaires. Results indicate that while relative deprivation is associated with collective behavior, it is not the only determinant of participation (Marx & Wood, 1975).

**Grievances**  In any society, certain resources will be highly valued but scarce; these resources include income or property, skills of certain types, and power and influence over others. Because of their scarcity, such resources are unequally distributed. Some groups have more access to a given resource than others. When one group has a *grievance*—is discontent with the existing distribution of resources—collective behavior may occur in order to change that distribution (Oberschall, 1973). Attempts to change the existing arrangement frequently elicit responses by other groups designed to preserve the status quo.

Generally there are three types of collective action (Tilly et al., 1975). *Competitive* action involves conflict between communal groups, usually on a local scale. Gang fights in cities, long-standing feuds between ethnic or kin groups, and similar phenomena are concerned with dominance. The high rates of lynching of blacks in the South in the 1890s is another example. From 1865 to 1890, blacks enjoyed substantial gains in political influence. By 1890, however, the white middle and upper class were attempting to regain political control. Between 1890 and 1900, several state legislatures discussed laws that would have taken the vote away from blacks. During these years, the number of blacks lynched in Alabama, Georgia, Louisiana, Mississippi, and South Carolina reached a peak (Wasserman, 1977). Both disenfranchisement and lynchings can be seen as a reassertion of white dominance over blacks.

Civil disorders or riots, which occur periodically in American cities, often involve members of minority groups. Looting and vandalism are common, with businesses whose owners are perceived as racist as the likely targets.

A second type of collective action, called *reactive*, involves a conflict between a local group and agents of a national political system. Tax rebellions, draft resistance movements, and protests of governmental policy are reactive. Such behavior is a response to attempts by the state to enforce its rules (regarding military service, for example), or extend its control (such as imposing a new tax). Thus, such events represent resistance to the centralization of authority.

A third type of collective action, called *proactive*, involves demands for material resources, rights, or power. Unlike reactive behavior, it is an attempt to influence rather than resist authority. Strikes by workers, demonstrations in favor of equal rights or equal opportunity, and various nonviolent protest activities are all proactive. Most proactive situations involve broad coalitions rather than one or two locally based groups.

The three underlying conditions discussed in this section differ in their emphasis. The strain model emphasizes the individual's emotional state in explaining collective behavior, whereas the relative deprivation viewpoint emphasizes the person's subjective assessment of need satisfaction. The grievance model suggests that collective behavior results from rational attempts to redistribute resources in society (Zurcher & Snow, 1981).

## Precipitating Events

Conditions of strain, relative deprivation, and grievances may be present in a society over extended periods of time. By contrast, incidents of collective behavior are often sporadic. Frequently, there are warning signals that a group is frustrated or dissatisfied. Members of the dissatisfied group or third parties may attempt to convince those in power to make changes (Oberschall, 1973). If changes are not made, members will increasingly perceive legitimate channels as ineffective, leading to marches, protests, or other activities. Eventually, an incident may occur that adversely affects members of the group and symbolizes the problem, triggering collective behavior by group members; such an incident is termed a **precipitating event.**

An incident is more likely to trigger collective behavior if it occurs in an area accessible to many members of the affected group at a time when social controls are weak (Turner & Killian, 1972). It is also more likely to become a precipitating event if it occurs in a location that has special significance to group members (Oberschall, 1973). An event that occurs in such a place may produce a stronger reaction than would the same incident in a less meaningful location.

On Monday, January 18, 1989, a black motorcyclist in Miami was being pursued by police officers for a traffic violation and was fatally wounded by a gunshot fired by an officer. Within minutes of the shooting, a crowd of 100 blacks gathered at the scene and began throwing rocks and bottles at police. As word of the shooting spread, more people became involved, and crowds began looting businesses and setting fire to stores and autos. The riot spread into two neighborhoods—Overtown and Liberty City—and continued for 36 hours. Initially, police limited their action to protecting life and property; as the riot continued, officers began to sweep areas and make arrests. By Wednesday evening, when relative calm had returned, 700 officers had been deployed, and more than 325 people had been arrested.

This precipitating event had several significant characteristics. It occurred on the evening of the national holiday commemorating the birthday of Martin Luther King, Jr. It involved a police officer killing a black person under ambiguous circumstances; for many blacks, such killings symbolize white racism. Finally, it occurred in a neighborhood where there have been protests and riots in the past.

## Empirical Studies of Crowds

Because they are unpredictable, hostile crowd events such as the Miami riot are difficult to study empirically. Nevertheless, extensive and sophisticated research has been conducted on racial disturbances, such as those that occurred in many U.S. cities between 1965 and 1969. These studies support many of the theories presented earlier in this chapter. In the first nine months of 1967, there were 164 racial disturbances. In response, President Lyndon Johnson appointed the National Advisory Commission on Civil Disorders to study the causes of these incidents. In its report (1968), the Commission concluded that the racial disturbances were caused by the underlying social and economic conditions affecting blacks in our society. It pointed to the high rates of unemployment, poverty, poor health and sanitation conditions in black ghettos, exploitation of blacks by retail merchants, and the experience of racial discrimination, all of which produced a sense of deprivation and frustration among blacks.

The Commission studied 24 disorders in 23 cities in depth. It concluded that

> disorder was generated out of an increasingly disturbed social atmosphere, in which typically a series of tension-heightening incidents over a period of weeks or months became linked in the minds of many in the Negro community with a reservoir of underlying grievances. At some point in the mounting tension, a further incident—in itself often routine or trivial—became the breaking point and the tension spilled over into violence. Violence usually occurred almost immediately following the occurrence of the final precipitating incident, and then escalated rapidly. Disorder generally began with rock and bottle throwing and window breaking. Once store windows were broken, looting usually followed. (National Advisory Commission on Civil Disorder, 1968, p. 6)

The precipitating incident frequently involved contacts between police officers and blacks. In Tampa, Florida, a disturbance in 1967 began after a policeman shot a fleeing robbery suspect. A rumor quickly spread that the black suspect was surrendering when the officer shot him. In other cities, disorder was triggered by incidents involving police attempts to disperse a crowd in a shopping district or to arrest patrons of a tavern selling alcoholic beverages after the legal closing time. To many blacks, police officers symbolize white society and are, therefore, a readily available target for grievances and frustration. When a police officer arrests or injures a black under ambiguous circumstances, it provides a concrete focus for discontent.

**Severity of Disturbances** Of particular concern to researchers was the intensity or severity of these disturbances. Numerous studies were based on the assumption that severity was determined by the degree of deprivation experienced by blacks.

In one study of racial disorders in 42 American cities during 1967 (Morgan & Clark, 1973), a "racial disorder" was defined as four or more persons engaging in behavior involving personal injury, property damage, or civil disobedience. A distinction was drawn between the intensity of a disorder and the extent of participation in it. Four empirical measures of intensity were employed: duration in days, number of injuries, estimated property damage, and a militancy index. Measures of extent included the number of participants, number of police, and number of arrests. The data showed that measures of intensity and extent were highly correlated, so they were combined into an overall index of severity.

Researchers hypothesized that the more widespread the grievances were among blacks, the more persons supported disorders and, hence, the more severe the disorders were. A high level of grievance may result from relative deprivation. Deprivation was measured by comparing the situation of blacks with that of whites on several dimensions, including housing and employment. Results showed that the severity of disturbances was positively associated with housing inequality and a city's population size. This suggests that greater relative deprivation in housing produces greater dissatisfaction and that the larger the black population was, the greater the number of potential participants was (Morgan & Clark, 1973).

Although underlying frustrations may account for the severity of disturbances, they do not explain why disorders occur in some cities and not others (Spilerman, 1976). Presumably, blacks in most U.S. cities were experiencing the same types of deprivation in the 1960s. Yet, disturbances occurred in only 170 of the 673 cities whose 1960 population exceeded 25,000. Thus, there must be factors other than grievance level that differentiate those cities where disturbances occur.

One possibility is that disturbances occurred in cities where the deprivation experienced by blacks was greatest. This hypothesis was tested in an analysis of 322 incidents that occurred in 1967 and 1968 (Spilerman, 1976). The study measured both absolute and relative levels of deprivation. The absolute level of deprivation was measured by the unemployment rate, the average income, and the average education of nonwhites in each city. Relative deprivation was measured by the differences between white and nonwhite unemployment rates, average income, average education, and average occupational status. To measure the severity of disorders, the study used the composite riot severity scale reproduced in Table 19-1. This scale distinguishes four degrees of severity based on the amount of personal injury, property damage, crowd size, and number of arrests.

Both the severity and frequency of disturbances were associated with the size of the nonwhite population of a city, rather than a city's overall population. But, in sharp contrast to earlier research, the level of deprivation was not associated with the severity of disorders. These results suggest that black protests were not due to local community conditions but to general features of the society such as increased black consciousness, heightened racial awareness, and greater identification with other blacks due to the civil rights movement. Television may have contributed to the disorders of the late 1960s by providing role models, as blacks in one city witnessed and subsequently copied the actions of those in other cities. Blacks engaged in vandalism, looting, and other collective behavior served as a model for blacks experiencing deprivation.

Thus, although deprivation was an underlying cause of black protest, disorders were facilitated by the civil rights movement and increased black consciousness. The magnitude of deprivation was not related to the location or severity of the disturbances. Rather, the severity was associated with the number of people experiencing deprivations—that is, the number of people available to participate in the disorders.

**Ambient Temperature and Violence** It is often suggested that high temperatures are a contributing factor in large-scale racial disturbances. In its report, the National Advisory Commission (1968) noted that 60 percent of the 164 racial disorders that occurred in U.S. cities

■ **TABLE 19-1**
RIOT SEVERITY SCALE

| | |
|---|---|
| 0 | *Low intensity*—rock and bottle throwing, some fighting, little property damage. Crowd size <125; arrests <15; injuries <8. |
| 1 | Rock and bottle throwing, fighting, looting, serious property damage, some arson. Crowd size 75–250; arrests 10–30; injuries 5–15. |
| 2 | Substantial violence, looting, arson, and property destruction. Crowd size 200–500; arrests 25–75; injuries 10–40. |
| 3 | *High intensity*—major violence, bloodshed, and destruction. Crowd size >400; arrests >65; injuries >35. |

*Source: Adapted from Spilerman (1976).*

in 1967 took place in July during hot weather. Of the 24 serious disturbances studied in detail, in most instances, the temperature during the day on which violence first erupted was quite high.

One study of the relationship between temperature and collective violence focused on 102 incidents that occurred between 1967 and 1971 (Baron & Ransberger, 1978). Results showed a strong relationship between temperature and the occurrence of violence. The incidents of collective violence were much more likely to have begun on days when the temperature was high (between 71 and 90 degrees). This relationship is depicted in Figure 19-3.

As the figure indicates there were few riots on days when the temperature was greater than 91 degrees. Does this reflect the fact it is too hot on such days, or the fact that there are very few days when the temperature is that high? In order to answer this question, researchers estimated the probability of a disturbance controlling for the number of days in each temperature range; the results show a direct relationship (Carlsmith & Anderson, 1979). In other words, the higher the temperature is, the more likely a disturbance is to occur. One interpretation of this relationship is that in high-density neighborhoods with

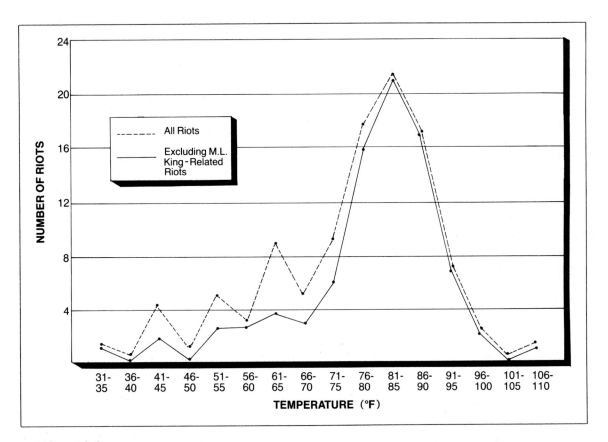

■ **FIGURE 19-3**

AMBIENT TEMPERATURE AND COLLECTIVE VIOLENCE

There is a strong relationship between mean temperature and the occurrence of collective disturbances. An analysis of 102 incidents between 1967 and 1971 generated this graph. As the temperature increased, the frequency of riots also increased. As the temperature increases, so does the number of people who are outside. Large numbers of people on the streets facilitate the development of a crowd.

*Source: Adapted from Baron and Ransberger, 1978.*

little air-conditioning, the number of people on the streets increases with the temperature. Large street crowds facilitate the transmission of rumors, increase the likelihood of supportive responses by others to acts initiated by a single individual, and increase the number affected by the process of circular reaction.

If temperature is associated with the occurrence of collective violence, an obvious question is whether high temperature is related to other types of violent behavior. One readily available

measure of behavior is the rate of violent crime —murder, sexual assault, and assault. One recent study analyzed the relationship between rates of violent crime and average temperature in data from 260 cities for the year 1980 (Anderson, 1987). As expected, the higher the average temperature was, the higher the rates of violent crime in a city were. Another study analyzed data from Dayton, Ohio, for a two-year period (Rotton & Frey, 1985). The researchers looked at daily variations in the number of reports to police of assaults and family disturbances. The number of each was positively associated with the temperature. These results suggest that people are more irritable in hot weather and, thus, more likely to engage in aggressive or violent behavior.

**Selection of Targets** Looting during a civil disturbance does not occur in a random fashion. During racial disorders, such as the riot in Miami in January 1989, property damage is almost always confined to retail stores. Residences, public buildings such as schools, and medical facilities such as clinics and hospitals are usually unaffected. In addition, the looting and vandalism of businesses is often selective. Some stores are cleaned out, whereas others in the same block are untouched.

According to one survey (Berk & Aldrich, 1972), the reason for this discrepancy is that businesses with higher average prices of goods sold (that is, more attractive merchandise) were more likely to be attacked. A second factor was familiarity with the interior of the store; the larger the percentage of black customers was, the more likely the store was to be looted. Retaliation was also a factor; stores whose owners refused to cash checks and give credit to blacks were more likely to be attacked. White ownership by itself was the least important factor.

Thus, the selection of targets during a riot reflects the desire of participants to obtain expensive consumer goods and to retaliate against antiblack owners. This may reflect the operation of social control within the crowd; emergent norms may define some buildings and types of stores as appropriate targets and others as inappropriate targets. These norms are probably enforced by members of the crowd itself (Oberschall, 1973).

**Social Control and Collective Behavior** Social control agents such as police officers strongly influence the course of a collective incident. In some cases, the mere appearance of authorities at the scene of an incident will set off collective action. Police-citizen encounters are frequently

the precipitating factor in racial disturbances, such as the one in Miami in January 1989.

The importance of control agents is especially clear in protest situations. Protesters usually enter a situation with (1) beliefs about the efficacy of violence and (2) norms regarding the use of violence (Kritzer, 1977). If participants' norms do not oppose violence, and if they believe violence may be effective, they are predisposed to choose violent tactics. Similarly, control agents enter a situation with (1) beliefs about what tactics the protesters are likely to use and (2) informal norms regarding violence. If the police anticipate violence, they will prepare for violence by bringing specially trained personnel and special equipment. Based on these beliefs and expectations, either the control agents or the protesters may initiate violence. Violence by one group is likely to produce a violent response from the other. Data from 126 protest events support this view (Kritzer, 1977).

In some instances, the response of authorities determines the severity and duration of disorders (Speigel, 1969). In any disturbance, there are two critical points at which undercontrol or overcontrol can cause a protest to escalate. First, there is the authorities' response to the initial, or *street confrontation*, phase. Undercontrol by police in reaction to the initial disorder may be interpreted by protesters as "an invitation to act"; it suggests that illegal behavior will not be punished. Overcontrol at this point, such as an unnecessary show of force or large numbers of arrests, may arouse moral indignation, which may attract new participants and increase violence. In the Miami disorders of January 1989, it appears that the police initially maintained a low profile, perhaps hoping that the crowds would quickly dissipate. The second critical point in a disturbance is the response to widespread disorder and looting during the later *Roman Holiday* phase. As it progresses, participants gradually become physically exhausted. Undercontrol by authorities may facilitate the collapse of the protest, whereas overcontrol may result in incidents that fuel hostility, draw in new participants, and move the disturbance into the final *siege* phase.

The response of authorities to one incident may also affect the severity of subsequent disorders in the same city (Spilerman, 1976). One study investigating incidents of collective violence in France, Germany, and Italy between 1830 and 1930 (Tilly et al., 1975) found that episodes involving violence were often preceded by nonviolent, collective action. Moreover, a substantial amount of the violence consisted of the forcible *reaction* of authorities (often military or police forces employed by the government) to the nonviolent protests of citizens. Thus, violence was not necessarily associated with attempts to influence authority; if anything, it was associated with reactions to such attempts by the agents of authority. A survey of the death rate associated with political violence in 49 countries in the decade from 1968 to 1977 found that the death rate was higher in countries with moderate scores on an index of regime repressiveness (Muller, 1985).

## Smelser's Theory of Collective Behavior

The most ambitious theory of collective behavior was proposed by Neil Smelser (1963). Because it is very comprehensive, it can be used to summarize our discussion. According to Smelser, there are six determinants of collective behavior that occur sequentially. Each stage depends on those that precede it, and each adds its own ingredient to the final product. To illustrate this sequence, Smelser describes the process of making cars from iron. Included in this process are mining, smelting, shaping, combining steel with other parts, assembling the car, delivering it to the dealer, and selling it. These stages must occur in this order. If the order varied, one would not get the final product. In addition, at each stage the possible outcome is progressively narrowed, until a car is the only possible product.

The determinants of collective behavior are as follows. First, certain social conditions, called *structural conduciveness,* are necessary in order

for certain types of collective behavior to occur. A financial panic can only occur in a money economy. A protest march can only occur in a political system that permits such activity. The second determinant, called *structural strain,* involves ambiguous situations, relative deprivation, and intergroup conflict. The third determinant is the growth and spread of a *generalized belief* among participants. This identifies the cause of the strain and specifies appropriate actions for those afflicted by it.

Together, conduciveness, strain, and generalized beliefs constitute the underlying causes of collective behavior. The fourth condition, a *precipitating factor,* is necessary to produce a collective episode. Once the episode begins, it is necessary to *mobilize participants* of the affected group. A collective episode provides a concrete focus for behavior. In response, *social controls* are mobilized; authorities act in ways designed to interrupt, deflect, or inhibit the collective activity. This, in turn, determines the intensity and direction of the episode.

It is the occurrence of each of these determinants in order and the particular social conditions, strain, belief, precipitating factor, mobilization, and control effects that determine the character of any particular incident. A change in any one of these will produce a change in the character of the incident.

There is widespread agreement among critics that structural conduciveness plays a major role in influencing collective behavior. Moreover, there is agreement that specific events do serve to crystalize discontent and trigger collective action and that the responses of authorities play a major role in determining the course of the incident. However, with respect to strain as the underlying cause, like other analysts, Smelser's theory does not specify how much is necessary. Nor does it offer a precise statement of the relationship between strain and collective behavior. There is less agreement that a generalized belief is necessary; incidents such as victory celebrations and responses to disasters do not seem to require a generalized belief. Finally, it is very difficult to identify good empirical measures for each determinant. For this reason, the predictive validity of Smelser's theory is not yet fully known (Marx & Wood, 1975).

# ■ Social Movements

The difference between collective behavior and social movements is one of degree. Both involve extrainstitutional behavior—behavior that is not consistent with the norms of society. Both are caused by social conditions that generate strain, frustration, or grievances. Their differences lie in degree of organization. Crowd incidents are relatively unorganized; they occur spontaneously, with no widely recognized leaders and no specific goals.

A **social movement** is collective activity that expresses a high level of concern about some issue (Zurcher & Snow, 1981). Participants are people who feel strongly enough about an issue to take action. Persons involved in a movement do a tremendous variety of things—talk to family or friends, sign petitions, participate in demonstrations, campaign in particular elections, or donate time and money to their cause. In this sense, a social movement involves a very broad segment of society. Within the movement, an organization may emerge—a group of persons with defined roles who engage in sustained activity to promote or resist social change (Turner & Killian, 1972).

In this part of the chapter, we will first discuss the development of a social movement. Then we will consider the movement organization and some of the influences on how it operates. Finally, we will discuss the consequences of social movements.

## The Development of a Movement

**Preconditions** By itself, strain or frustration cannot create a social movement. For a movement to occur, people must perceive their discontent as the result of controllable forces external to themselves (Ferree & Miller, 1985). If they

attribute their discontent to such internal forces as their own failings or bad luck, they will not be predisposed to attempt to change their environment. In addition, people must believe they have a right to the satisfaction of their unmet expectations (Oberschall, 1973). These attributions are often the result of interaction with others in similar circumstances. The moral principles used to legitimate their demand may be taken from the culture or from a specific ideology or philosophy. Thus, at the core of any social movement are beliefs rooted in the larger society.

A current social movement taking place in the U.S. involves abortion. In the late 1960s, an organized social movement developed that pressed for change in the laws that restricted the availability of abortion. This movement culminated in the Supreme Court decision *Roe v. Wade* on January 22, 1973, which held that the state cannot interfere in an abortion decision by a woman and her physician during the first three months of pregnancy. The increasingly widespread availability of abortion created strain for others in our society. Many people view a fetus as a person and, thus, define abortion as murder. These people, drawing primarily on conservative Christian theology, gradually organized a movement in the mid-1970s to obtain legislation that would sharply restrict a woman's right to choose abortion.

In addition to perceiving unmet needs, people in a social movement must believe that the satisfaction of their needs cannot be achieved through established channels. Frequently, this perception is based on the failure of prior attempts to bring about change through those channels. At first, the antiabortion movement emphasized lobbying and attempts to influence elections. As these activities failed to produce legislative or judicial action, the movement increasingly adopted more aggressive tactics.

Given a large population experiencing deprivation, there are four structural conditions that increase the probability a social movement will emerge (Morrison, 1971). The first is proximity and interaction among members of that popula-

This demonstration is part of the social movement aimed at reducing the availability of pornography. Organizers of such events hope to gain coverage by the mass media, for reports of movement activities may attract additional supporters.

tion. This facilitates the emergence of shared definitions and beliefs. Second, people experiencing the same deprivation must be similar in social status; otherwise, an individual is likely to attribute the problem to himself or herself rather than the social environment. Third, there must be a social hierarchy with "visible power differences between the strata" (Morrison, 1971). This increases the likelihood that the problem will be attributed to the social structure. Finally, the existence of considerable collective activity creates the perception that concerted effort may help.

**Ideology**  As affected individuals interact, an ideology or generalized belief emerges. **Ideology** is a conception of reality that emphasizes certain values and justifies a movement (Turner & Killian, 1972; Zurcher & Snow, 1981).

The antiabortion, or pro-life, ideology rests on several assumptions. First, each conception is an act of God; abortion, thus, violates God's will. Second, the fetus is an individual who has a constitutional right to life. Third, every human life is unique and should be valued by every other human being. Pro-life forces view the current status of abortion as temporary, a departure from the past when it was morally unacceptable. Persons and programs (such as sex education) are evaluated in terms of whether they support or undermine these beliefs. Any person or group who favors continued legal abortion is defined as immoral. In recent elections in many communities, a candidate's position on abortion has been a major political issue. Pro-life activists believe that by opposing people and programs that encourage abortion, they will bring about a sharp decline in its availability and redefine it as illegitimate.

Such an ideology fulfills a variety of functions (Turner & Killian, 1972). First, it provides a way of identifying people and events and a set of beliefs regarding appropriate behavior toward them. Ideology is usually oversimplified, because it emphasizes one or a small number of values at the expense of others. A second function of ideology is that it gives a movement a temporal perspective. It provides a history (what caused the present undesirable situation) and a conception of the future (what goals can be attained by the movement). Third, it defines group interests and gives preference to them. Finally, it creates villains; it identifies certain persons or aspects of society as responsible for the discontent. This latter function is essential, because it provides the rationale for activity designed to produce change (Oberschall, 1973).

**Recruitment**   The development and continuing existence of any movement depends on *recruitment*—the process of attracting supporters. Some people are attracted to a movement because they share some distinctive attributes (Zurcher & Snow, 1981). In many instances, these are persons who experience the discontent or grievances at the base of the movement. A study comparing people who participated in the movement to prevent the reopening of Pennsylvania's Three Mile Island nuclear power plant with a group of nonparticipants found that the activists had opposed commercial and military uses of nuclear energy before the accident and that the accident served to substantially increase their discontent (Walsh & Warland, 1983). There are limitations, however, to this "grass roots" view of recruitment (Turner & Killian, 1972). Many studies have found that supporters of a movement are not the most deprived or frustrated. Also, the goals of a movement may not be

aimed at removing the sources of the discontent. The content of the ideology reflects several influences, not just a desire to eliminate a particular source of frustration. Once developed, people may be attracted by the ideology who do not share the discontent.

Recruitment depends on two catalysts: the ideology and existing social networks (Zurcher & Snow, 1981). The content of the ideology is what attracts supporters. The ideology spreads, at least in part, through existing social networks. Supporters communicate the ideology to their friends, families, and co-workers in the course of their ongoing relationships.

The patterns of recruitment into religious movements document the importance of friendship and kinship ties (Stark & Bainbridge, 1980). Adherents to the Mormon religion, for example, are prone to proselytize. They establish friendship ties with nonmembers, then gradually introduce their beliefs to their friends. Likewise, members of a doomsday cult who believed that the earth would soon be destroyed were, in many cases, relatives of other members. Those with kinship ties were less likely to leave the cult.

In the past decade, many persons have been "born again,"—that is, become Pentecostals. One study of this phenomenon compared 150 converts to Catholic Pentecostalism with a control group of non-Pentecostal Catholics who were similar in age, social class, and gender (Heirich, 1977). The major difference between the two groups was their social networks. Converts reported that they had been introduced to Pentecostalism by a "trusted person," often a teacher, priest, or nun. Non-Pentecostal Catholics, however, had not been introduced to Pentecostalism by such a person. Converts were more likely to have received positive or neutral reactions to their initial participation from family and friends. Following their introduction to the Pentecostal movement, converts also tended to seek out other converts and spend less time with friends who were not part of the movement.

Sometimes entire groups are recruited all at

Social movements vary greatly in their impact. The National Farm Labor Movement of 1946–55 was generally unsuccessful, whereas the United Farm Workers led by Cesar Chavez is today achieving major benefits for agricultural workers. One difference is widespread support for the UFW by other labor and political organizations.

once (Oberschall, 1973). The civil rights movement in the South in the 1950s is one example. Because of their religious views, black ministers were predisposed to support a movement whose ideology emphasized freedom and equality. These ministers recruited their entire congregations and communicated the ideology to other ministers. As a consequence, the movement spread rapidly. More recently, the pro-life movement has grown by recruiting entire congregations of Catholics and Mormons. Such bloc recruitment is much more efficient than recruiting individuals (Jenkins, 1983).

An alternative mechanism of recruitment is the mass media. On October 3, 1970, an estimated 15,000 to 20,000 persons participated in

Reverend Carl McIntire's March for Victory in Washington. McIntire, a fundamentalist pastor who supported the Vietnam War, communicated his views in a weekly radio program carried by 600 stations and a weekly newsletter. Interviews with 201 march demonstrators revealed that most had come from outside the Washington area, and the overwhelming majority learned of the march through McIntire's radio program or newsletter (Lin, 1974–75).

Thus, the media play an important role in social movements. In the McIntire case, prior political beliefs appear to have led persons to seek specific information about the organizational program and the march. Media reports convey a movement's ideology and attract members by providing role models or by providing information about the time and place of activities. It is no accident that movement groups devote considerable effort in getting television camera crews to cover their activities.

## Social Movement Organizations

If a movement is to have any impact, there must be some degree of organization in order to exert continuing pressure for (or resistance to) change. A group of persons with defined roles engaged in sustained activity that reflects a movement's ideology is called a **movement organization.** It develops through the mobilization of people and resources. Once developed, the organization is influenced by its environment and by its own internal processes.

**Resource Mobilization**  Having attracted supporters, a movement must induce some of them to become committed members (Zurcher & Snow, 1981). Commitment involves the creation of links between an individual's interests and those of the movement so that the individual will be willing to contribute actively to the achievement of movement goals. Committed members are necessary if the movement is to become active and self-sustaining. **Mobilization** is the process through which individuals surrender personal resources and commit them to the pursuit of group goals (Oberschall, 1978). Resources can be many things: money or other material goods, time and energy, leadership or other skills, or moral or political authority. From the individual's viewpoint, mobilization involves a rational decision about investing one's resources. The person weighs the costs and benefits of various actions; if the potential rewards seem to outweigh the potential risks, a particular course of action is undertaken.

Leadership is obviously essential to an organization. Taking a leadership role in a movement organization may be risky but potentially very rewarding (Oberschall, 1973). If the movement is successful, leaders may attain substantial pres-

tige, visibility, a permanent, well-paid position with the organization, and opportunities to interact with powerful, high-status members of society. Leaders of movements are often persons with substantial education (such as lawyers, writers, professors, and students) and at least moderate status in society. They are frequently persons whose skills cannot be confiscated by authorities, who can expect social support, and who will be dealt with leniently if arrested. Thus, their risk-reward ratio is favorable for involvement.

For a movement to succeed, others must also be induced to work actively in the organization (Zurcher & Snow, 1981). One basis of commitment is moral—anchoring an individual's world view in the movement's ideology. Members who are attracted primarily by the content of ideology tend to see their own interests as furthered by the achievement of movement goals. Many women become involved in proabortion organizations because preserving freedom of choice for all women will benefit them. At the same time, movement adherents clarify, extend, and even reinterpret the ideology as they attempt to persuade others to commit resources to the movement (Snow et al., 1986). A second basis of commitment is a sense of belonging. This is facilitated by collective rituals in which members participate. One advantage of recruiting preexisting groups, such as church congregations, is that this sense of belonging is already developed. A third basis of commitment is instrumental. If the organization has enough resources at its disposal, it can provide utilitarian rewards for committed members. These rewards may be distributed equally among members or selectively to members who make a particular contribution (Oliver, 1980).

Depending on its overall strategy, a movement organization can use moral, affective, or instrumental rewards as bases for building commitment. These are sufficient for most organizations to induce members to contribute time, materials, and other resources. Still other organizations demand that members commit themselves to exclusive participation. They require that members renounce other roles and commitments and undergo **conversion**—the process through which a movement's ideology becomes the individual's fundamental perspective. This degree of commitment is required by some religious movements and by "utopian" communities. Conversion involves persuasion and "consciousness raising"; it is an attempt to change the individual's world view. Conversion is usually accomplished during a period of intensive interaction with other movement members. It is, thus, a very labor-intensive mobilization strategy (Ferree & Miller, 1985).

Movement efforts to mobilize resources vary according to what potential supporters are willing to contribute. If there is considerable latent support for a movement's goals but supporters are unwilling to make large contributions, mobilization efforts may focus on monetary donations. Another low-cost means of translating weak support into a resource is through the use of initiative or referendum ballots, asking supporters to vote in favor of a movement's goal, such as clean water or a nuclear free zone (Ennis & Schreuer, 1987). Mobilization efforts also vary according to the resources being sought. An organization seeking volunteers for high-risk activities, such as those involving the risk of arrest, must use powerful inducements. A study of volunteers for a project to register voters in rural Mississippi in 1964 found that several characteristics distinguished participants from volunteers who decided not to go (McAdam, 1986). Participants had intense ideological commitment, previous experience with activism, and ties to other activists (see Figure 19-4).

**Organization-Environment Relations**  The social environment is another influence on the development of a movement organization. First, changes in the environment can increase or decrease the size of the movement and the probability of its success. This is illustrated in a study comparing the evolution of the National

Farm Labor Union movement in 1946–55 and the United Farm Workers movement in 1965–72 (Jenkins & Perrow, 1977). Both movements had very similar ideologies and organizational structures. However, whereas the NFLU was generally unsuccessful, the UFW achieved major benefits for many farm workers. The difference was the amount of resource support from the environment due to a change in the political climate in the U.S. An analysis of newspaper articles published during each period indicates that government, liberal political organizations, and organized labor gave much greater support to the UFW. Thus, challenges to the established order are more likely to succeed when there is sustained support from the environment and when organized opposition is absent.

Second, the growth and effectiveness of a movement organization depends, in part, on whether there are other organizations with similar goals. The National Association for the Advancement of Colored People was the principal black civil rights organization prior to 1960. Its goal was racial integration, and it relied heavily on a combination of litigation and lobbying. From 1960 to 1973, other black organizations developed with more radical goals, means, or both. The Congress of Racial Equality and the Student Nonviolent Coordinating Committee favored direct action, and both later excluded whites from membership. Membership in the NAACP remained stable, while the other organizations grew rapidly. The competition between these organizations for resources forced the NAACP to shift from its goal of integration to economic advances. The NAACP also shifted from relying on members for financial support to relying on donations from foundations and corporations (Marger, 1984). Paradoxically, the more radical programs of CORE and SNCC made it easier for more moderate organizations like the NAACP to attract financial support from whites, especially after the riots of 1967 and 1968 (Haines, 1984).

Under some circumstances, organizations

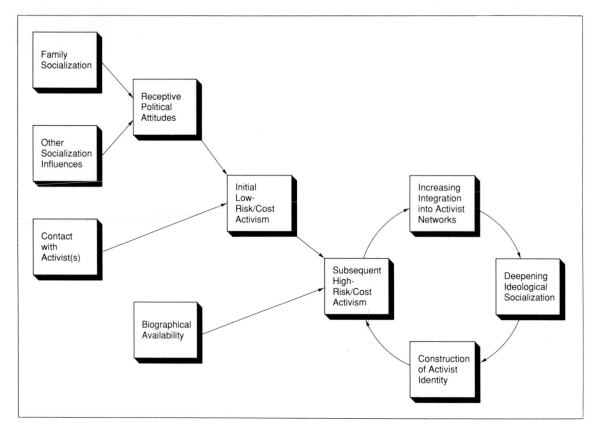

■ **FIGURE 19-4**

INFLUENCES ON RECRUITMENT TO HIGH-RISK ACTIVISM

Many social movement activities are low risk—involving little or no cost to participants. Such activities include donating money, distributing leaflets, and attending rallies. Sometimes, however, movement leaders decide to engage in activities that are high risk—involving the risk of injury or death. Political activism in the rural South in 1964 is an example of a high-risk activity. A study of the characteristics associated with the willingness to participate in this activity compared those who went with those who initially volunteered but decided not to go. Analyses suggested that a sequence of experiences, beginning with particular socialization experiences and including contact with other activists and prior activism, were associated with being available for and participating in the high-risk activity. Once involved, participation facilitated greater involvement.

Source: McAdam, 1986, Figure 1.

with similar goals will form coalitions. Coalitions are possible if organizations appear to have similar values and clearly defined common interests (Ferree & Miller, 1985). A study of cooperation and competition between movement organizations focused on thirteen organizations and six coalitions involved in proabortion activity between 1966 and 1983 (Staggenborg, 1986). Coalitions were likely under conditions of exception-al opportunity or threat. On the other hand, abortion coalitions often dissolved due to conflicting ideologies, for example, differences with respect to the use of civil disobedience, or conflicts between the maintenance needs of the organizations.

A third, important aspect of the environment is countermobilization. The development of an organization that seeks change of some kind is likely to stimulate those who would be adversely affected by that change to mobilize as well. For instance, an analysis of the resistance by corporate managers to the union movement in the 1920s identified a wide range of tactics that were used to counter union efforts to mobilize workers (Griffin, Wallace, & Rubin, 1986). Managers tried to diminish the common interests of the workers by hiring immigrants and mechanizing factories. They attempted to reduce the power of the union movement by boycotting union goods and harassing other firms that recognized unions. Often these tactics were successful.

The woman's suffrage movement in the early 1900s and the Equal Rights Amendment movement in the 1970s both provoked counter movements. The National Association Opposed to Women Suffrage defended the homemaker's role and life style and attempted to safeguard the material privileges enjoyed by well-to-do women with an "educational campaign" (Marshall, 1986). The anti-ERA movement in the 1970s used more overtly political strategies than the NAOWS, including rallies, marches, and parades. The leader of the anti-ERA movement, Phyllis Schlafly, had considerable political experience working in the National Federation of Republican Women, and she used her experience and contacts (Marshall, 1985).

**Internal Processes**  Other influences on the development of a movement organization are internal processes. These include factionalization, professionalization, and radicalization.

*Factionalization* is the emergence of subgroups within an organization. It is more likely to occur in a democratic movement organization. The Clamshell Alliance, an antinuclear protest group, was based on the ideology that every person was equal; this was an important element in mobilizing people to protest the activities of utilities and government agencies. As a result, every decision had to be unanimous; one dissenter could prevent the group from acting. Over a period of five years, sharp disa-

greements developed over strategy, leading to factionalism and reduced effectiveness (Downey, 1986).

Factionalization may, in turn, produce *schismogenesis,* which occurs when a faction splits off or leaves a movement organization. Schismogenesis is particularly likely when the social base is heterogeneous and when organizational authority is based on doctrine. When the support base is heterogeneous, there is a tendency for persons to seek out others within the movement who are similar and form factions. Similarly, when a movement is based on doctrine, whether religious or political, there is a tendency for disputes over doctrinal issues to result in schisms.

Another internal process involves increasing reliance on paid professionals. Over time, as an organization develops, it is subject to increasing routinization. The organization develops an administrative structure. Roles are increasingly filled by secretaries, accountants, and lawyers, rather than by volunteers committed to the movement. This is likely to produce an increasingly conservative stance among the movement's leadership, which may alienate members who are more radical in orientation.

A recent study analyzes the effects of these changes in the black protest movement from 1953 to 1980 (Jenkins & Eckert, 1986). The data indicate that as the flow of financial resources into movement organizations grew, there was increasing *professionalization* of organization staff. Furthermore, this process did not generate further growth and may have contributed to later decline. At the same time, professionalization does not appear to have produced changes in movement goals and tactics.

There are several other reasons why social movements often become more conservative over time (Myers, 1971). First, the leader-follower relationship is characterized by increasing distance and formality. Second, there is increased bureaucracy—including a role hierarchy, rules and procedures, and membership criteria. Third, there is a tendency for goal

## PROTEST AGAINST PORNOGRAPHY: THE VAGARIES OF COLLECTIVE ACTION

The underlying causes of collective behavior are present more or less continuously, yet protests and other collective acts occur only occasionally. Consider the example of pornography. Many communities have adult, or X-rated, bookstores and theaters that show X-rated movies. Many of these businesses have been operating for years. There are undoubtedly persons in these communities who find the goods and services sold by these businesses offensive. Yet public attention and protests focus on these businesses only once or twice a year (or less often) and for only a few days at a time.

One measure of the level of public concern about such businesses is the number of newspaper articles published per month or per year. Generally, collective action aimed at the availability of pornography is reported in local newspapers. Madison, Wisconsin, is a community of 170,000 people. There are two daily newspapers, one more liberal in editorial policy and one more conservative in policy. Thus, it is likely that collective action concerning pornography would be reported in at least one of these papers.

A case study of the antipornography movement in Madison included a count and an analysis of newspaper articles reporting local collective action (Duesterhoeft, 1987). The results of the count for the years 1972–86 are displayed in Figure B-19-1. The study identified two distinct groups that engaged in collective action, one comprised of Christian women and one of feminists. In general, these two groups had been unable to form a coalition due to ideological differences. The Christian group defined as pornographic any depiction of sex outside of heterosexual, monogamous marriage. The feminists defined as pornographic any depiction that was degrading to women or children. The Christian group did not distinguish between acceptable portrayals ("erotica") and pornography, whereas the feminist group did. Because of these differences, the two groups engaged in separate activities; most newspaper articles during the period of the study reported action by one group or the other but not of the two in coalition.

A look at Figure B-19-1 indicates that in 9 of the 16 years, there were fewer than ten newspaper articles. These were years in which there was little protest activity. There were two periods in which numerous articles were published: 1980–81 and 1984–85. In the first of these periods, there were three small protests in response to the showing of notorious films (*Caligula* and *Windows*) and a single, large protest march that focused on pornography and sexual assault.

The only instance of sustained protest activity during the 16-year period occurred in 1984–85. In August of 1984, local interest developed in passing a county ordinance that would define the sale or exhibition of pornography as a violation of civil rights, similar to legislation passed in Minneapolis. Protests, news conferences, forums, and other activities occurred regularly from August 1984 through 1985. The protests were aimed at the sale of magazines such as *Playboy* and *Penthouse* at both adult book stores and other businesses such as convenience stores. Thus, the introduction, debate, and ultimately action on the proposed ordinance increased public interest in the availability of pornography. Both the Christian and the feminist groups took advantage of this interest and engaged in a series of actions.

This case study illustrates a number of

Box 19-2

principles about collective behavior and social movements. First, while underlying causes are always present, collective action occurs sporadically. Second, collective action is often a response to a precipitating incident, such as the showing of a notorious film. Third, in order to sustain collective action over time, there must be support from the community—for example, people willing to attend speeches and demonstrations. Finally, when there is more than one organization, differences in ideology may prevent the formation of a coalition and, thus, limit the effectiveness of the protest.

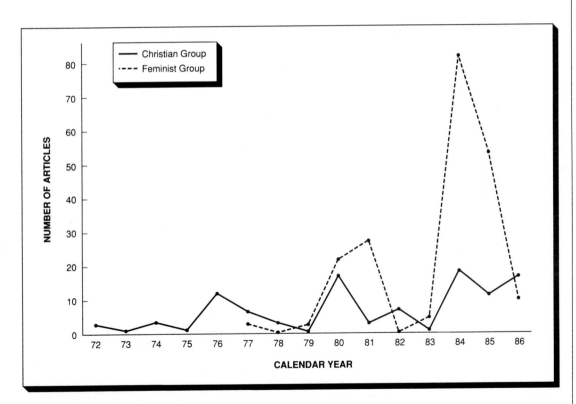

■ **FIGURE B-19-1**

displacement to occur; leaders become increasingly concerned with maintaining the organization and less concerned with the goals of the broader movement.

Increasing conservatism is not inevitable. An organization committed to democracy (not necessarily equality) may adopt practices that allow members to retain control. These include a system of open elections, an open decision-making process, and regular communication between leaders and members (Nyden, 1985). A study of the United Steelworkers of America documented one case in which a bureaucratized union organization was turned into a democratic one by a reform group that emerged from within.

Much more rare is a movement that undergoes increased *radicalization,* like the Committee of 100 in Great Britain (Myers, 1971). The Committee was formed to protest the increasing reliance of the British government on nuclear weapons for national defense. From the beginning, the group was committed to the use of civil disobedience to achieve its goals, and this commitment led to greater radicalization rather than conservatism. In general, groups committed to illegal tactics tend to attract members who are more radical. This commitment also produces extreme demands on the time and energy of leaders, which leads to a predominance of younger leaders who tend to have more energy and fewer competing commitments. Organizations of this type are less likely to achieve their goals, because the use of illegal tactics tends to alienate the larger society and reduce the amount of support received. Such tactics may even elicit violence by control agents (Kritzer, 1977).

### The Consequences of Social Movements

Once a movement has established an ideology, attracted supporters, and developed an organization that embodies its ideals, its ultimate success or failure depends on the reaction of authorities outside the movement organization. Those in positions of power—whether political,

economic, or religious—can use various strategies in dealing with a social movement (Oberschall, 1973).

One obvious strategy is to restrict a movement through the exercise of social control. For example, authorities can control access to social roles; they determine the educational, occupational, or religious roles available. A student protesting racism on campus may be expelled from school by administrators; a Catholic priest challenging his vow of celibacy may be removed from his parish by the bishop. Beyond this, authorities can manipulate material benefits and punishments, which may affect the risk-reward assessments of movement members. Authorities can also use physical power in the form of police or troops. Of course, in most cases, this method results in violence.

A second general strategy is conciliation. Authorities may open up channels of communication and negotiate with movement leaders in an attempt to resolve grievances. Conciliation is more likely when a movement is relatively powerful and has the support of large numbers of influential members of society. It is also more likely if both sides are highly organized, united internally, and have strong leadership.

By responding to conciliation from authorities, a movement may achieve at least some of its objectives. As a result of negotiation, there may be changes in the distribution of resources or in power relationships in society. A movement that succeeds in this sense may itself become institutionalized. The changes it brings may ensure a continuing flow of resources that perpetuate the movement organization. This is what happened to the labor, civil rights, and woman's movements in the U.S.

### ■ Summary

This chapter discusses both collective behavior and social movements.

**Collective Behavior** (1) Many instances of collective behavior involve crowds. Four processes that contribute to crowd behavior are deindividuation, contagion, convergence, and normative emergence. (2) There are three underlying causes of collective behavior: strain, relative deprivation, and grievances. (3) Collective behavior is often triggered by a precipitating event that adversely affects those experiencing strain, deprivation, or grievances. (4) Empirical studies of crowds suggest that the severity of a disturbance is influenced by the extent of relative deprivation and the number of potential participants. Once it begins, the course of a collective incident and the likelihood of future disorders are influenced by the behavior of police and other social control agents. (5) According to a general theory proposed by Smelser, there are six determinants of collective behavior. They occur in sequence and determine the outcome of any collective incident.

**Social Movements** A social movement is collective activity that expresses a high level of concern about some issue. (1) The development of a social movement rests on several factors. First, people must experience strain or deprivation, believe that they have a right to the satisfaction of their unmet needs, and believe that satisfaction cannot be achieved through established channels. Second, as participants interact, an ideology must emerge that justifies collective activity. Third, to sustain the movement, additional people must be recruited by spreading the ideology, often through existing social networks. (2) A movement organization is a group of persons engaged in sustained activity that reflects the movement ideology. Development of a movement organization depends on resource mobilization—getting individuals to commit personal resources to the group. The organization's effectiveness depends in part on its external environment, on whether there are cooperating or competing organizations, and on whether countermobilization occurs. Over time, organizations may experience such internal processes

as factionalization, professionalization, or radicalization. (3) The ultimate success or failure of a social movement depends on the reactions of authorities outside the movement—whether those in positions of power decide to exercise control or to negotiate and attempt to ameliorate the sources of strain or deprivation.

## Key Terms

# Glossary

**Access display** A signal (verbal or nonverbal) from one person indicating to another that further social interaction is permissible.

**Accounts** Explanations people offer after they have performed acts that threaten their social identities. Accounts take two forms—excuses that minimize one's responsibility and justifications that redefine acts in a more socially acceptable manner.

**Achievement motive** A conscious or unconscious desire to reach high standards of excellence.

**Activation (of an attitude)** The bringing of an attitude from memory into conscious awareness (which may then influence behavior).

**Active minority** In groups, a subgroup of members that adopts a distinct viewpoint on some important issue(s) and that tries to persuade the majority to change its position.

**Actor-observer difference** The bias in attribution whereby actors tend to see their own behavior as due to characteristics of the external situation, whereas observers tend to attribute actors' behavior to the actors' internal, personal characteristics.

**Additive model** A theoretical model for combining information about individual traits in order to predict how favorable the overall impression people form of another will be. This model *adds* together the favorability values of all the single traits people associate with the person. *See also* Averaging model; Weighted averaging model

**Additive task** A type of unitary group task in which the group's performance is equal to the sum (or average) of the performances of its members.

**Ageism** Prejudice and discrimination against the elderly based on negative beliefs about aging.

**Aggression** Behavior intended to harm another.

**Alienation** The sense that one is uninvolved in the social world or lacks control over it.

**Aligning actions** Actions people use to define their apparently questionable conduct as actually in line with cultural norms, thereby repairing social identities, restoring meaning to situations, and reestablishing smooth interaction.

**Altercasting** Tactics we use to impose roles and identities on others that produce outcomes to our advantage.

**Altruism** Voluntary, self-sacrificing behavior intended to benefit another person with no expectation of external reward.

**Anomie theory** The theory that deviant behavior arises when people striving to attain culturally valued goals find that they do not have access to the legitimate means of achieving these goals.

**Anticipatory socialization** The process of learning how to perform a role that is not yet assumed. Unlike explicit training, anticipatory socialization is not intentionally designed as role preparation by socialization agents.

**Archival research** A research method that involves acquisition and analysis (or reanalysis) of existing information that was collected by others.

**Attachment** A warm, close relationship with an adult that provides an infant with a sense of security and stimulation.

**Attitude** A predisposition to respond to a particular object in a generally favorable or unfavorable way.

**Attitudinal similarity** The sharing by two people of beliefs, opinions, likes, and dislikes.

**Attractiveness stereotype** The belief that "what is beautiful is good"; the assumption that an attractive person possesses other desirable qualities.

**Attribution** The process by which we link behavior to its causes—to the intentions, dispositions, and events that explain why people act as they do.

**Authority** The capacity to influence the behavior of group members by invoking rights that are vested in one's role.

**Availables** Those persons with whom we come into contact and who constitute the pool of potential friends and lovers.

**Averaging model** A theoretical model for combining information about individual traits in order to predict how favorable the overall impression people form of another will be. This model *averages* the favorability values of all the single traits people associate with the person. *See also* Additive model; Weighted averaging model

**Aversive event** In intergroup relations, a situation or event caused by (or attributed to) an outside group that produces negative or undesirable outcomes for members of the target group.

582

**Back channel feedback** The small vocal and visual comments a listener makes while a speaker is talking, without taking over the speaking turn. This includes responses such as "Yeah," "Huh?," head nods, brief smiles, and completions of the speaker's words. Back channel feedback is crucial to coordinate conversation smoothly.

**Balance theory** A theory concerning the determinants of consistency in three-element cognitive systems.

**Balanced state** The state in which the sentiment relations among three cognitive elements are all positive, or in which one is positive and the other two are negative.

**Bargaining** A process of interaction in which two (or more) persons with different preferences make a sequence of concessions in an attempt to reach an agreement that is mutually acceptable.

**Bargaining range** In bargaining, the difference between the limit of person A and the limit of person B; the set of possible outcomes that might result from bargaining.

**Bilateral threat** In bargaining, a situation where both bargainers can issue threats and inflict punishments on one another.

**Birth cohort** A group of people who were born during the same period of one or several years and who are therefore all exposed to particular historical events at approximately the same age.

**Body language** Communication through the silent motion of body parts—scowls, smiles, nods, gazes, gestures, leg movements, postural shifts, caresses, slaps, and so on. Because body language entails movement, it is known as kinesics.

**Bystander effect** The tendency for bystanders in an emergency to help less often and less quickly as the number of bystanders present increases.

**Career** A sequence of roles that a person enacts during his or her lifetime, each role with its own set of activities. People's most important careers are in the domains of family and friends, education, and work.

**Case study** An intensive investigation of one incident or event that entails the use of multiple sources of information about the event in question.

**Catharsis** The reduction of aggressive arousal by means of performing aggressive acts. The catharsis hypothesis states that we can purge ourselves of hostile emotions by intensely experiencing these emotions while performing aggression.

**Cautious-shift** In group decision making, the tendency for decisions made in groups after discussion to be more cautious (less risky) than decisions made by individual members prior to discussion.

**Cognition** An element of cognitive structure. Cognitions include attitudes, beliefs, and perceptions of behavior.

**Cognitive dissonance** A state of psychological tension induced by dissonant relationships between cognitive elements.

**Cognitive processes** The mental activities of an individual, including perception, memory, reasoning, problem solving, and decision making.

**Cognitive structure** Any form of organization among a person's cognitions (concepts and beliefs).

**Cognitive theory** A theoretical perspective based on the premise that an individual's mental activities (perception, memory, and reasoning) are important determinants of behavior.

**Collective behavior** Emergent and extrainstitutional behavior that is often spontaneous and subject to norms created by the participants.

**Communication** The process through which people transmit information about their ideas and feelings to one another.

**Communication network** The pattern of communication opportunities in groups; typical communication networks include the comcon, the wheel, the circle, the chain, and so on.

**Communicator credibility** In persuasion, the extent to which the communicator is perceived by the target audience as a believable source of information.

**Comparison level** A standard used to evaluate the outcomes of a relationship, based on the average of the person's experience in past relevant relationships.

**Comparison level for alternatives** A standard specifying the lowest level of outcomes a person will accept in light of available alternatives. The level of profit available to an individual in his or her best alternative relationship.

**Compliance** In social influence, adherence by the target to the source's requests or demands. Compliance may occur either with or without concomitant change in attitudes.

**Concession magnitude** In bargaining, the average size of a bargainer's concessions.

**Concept** An idea that specifies in what way various objects or stimuli are similar or related to each other.

**Conditioning** A process of learning in which, if a person performs a particular response and if this response is then reinforced, the response is strengthened.

**Conformity** Adherence by an individual to group norms, so that behavior lies within the range of tolerable behavior.

**Conjunctive task** A type of unitary group task in which the group's performance depends entirely on that of its weakest or slowest member.

**Connotative meaning**  The personal associations and emotional responses that an individual associates with a word or an object.

**Consensus**  In person perception, information used to make attributions about a behavior by referring to whether all or only a few people perform that behavior.

**Conservative coalition**  In groups, a union of medium- and low-status members who support the existing leadership and status order against revolutionaries.

**Consistency**  In person perception, information used to make attributions about a behavior by referring to whether the actor behaves the same way at different times and in different settings.

**Contagion**  The rapid spread through a group of visible and often unusual symptoms or behavior.

**Content analysis**  A research method that involves a systematic scrutiny of documents or messages to identify specific characteristics and then making inferences based on their occurrence.

**Contingency model of leadership effectiveness**  A theory of leadership effectiveness (proposed by Fiedler) which maintains that group performance is a function of the interaction between a leader's style (task-oriented or relationship-oriented) and various situational factors such as the leader's personal relations with members, the degree of task structure, and the leader's position power.

**Control theory**  The theory that social ties influence our tendency to engage in deviant behavior.

**Controllability-leakage hierarchy**  The ranking that exists among communication channels, with the order in terms of controllability being the direct opposite of the order in terms of leakage.

**Convenience sample**  A sample of respondents or subjects selected on the basis of ready availability.

**Conversion**  The process through which the ideology of a social movement becomes the individual's fundamental perspective.

**Cooling-out**  A response to repeated or glaring failures that gently persuades an offender to accept a less desirable, though still reasonable, alternative identity.

**Cooperative principle**  The assumption conversationalists ordinarily make that a speaker is behaving cooperatively by trying to be (1) informative, (2) truthful, (3) relevant to the aims of the ongoing conversation, and (4) clear.

**Co-optation**  In groups, a strategy by which an existing leader singles out one or several lower-status members for favored treatment; this is done to weaken the bonds among lower-status persons who might otherwise form revolutionary coalitions.

**Correspondence**  The degree to which the action, context, target, and time in a measure of attitude is the same as those in a measure of behavior.

**Correspondent inference**  In person perception, a correspondent inference occurs when the inferred disposition (attributed by the observer to the actor) corresponds directly to the nature of the observed act (by the actor). Correspondence in inference is usually treated as a matter of degree.

**Crowd**  A substantial number of persons, usually strangers, who engage in behavior recognized as unusual by participants and observers.

**Cultural estrangement**  The rejection of the basic values and lifestyles available in the society.

**Death instinct**  In Freud's theory, the innate urge we carry within us to destroy. This instinct constantly generates hostile impulses that demand release and is the basis for aggression.

**Definition of the situation**  In symbolic interactionist theory, a person's interpretation or construal of a situation and the objects in it. An agreement among people who are interacting about who they are, what actions are appropriate, and what their behaviors mean.

**Deindividuation**  The temporary loss of self-awareness and of the sense of personal responsibility; it may be brought on by such situational conditions as anonymity, an undifferentiated crowd, darkness, and consciousness-altering drugs.

**Denotative meaning**  The literal, explicit properties associated with a word as defined in a dictionary. In contrast to the connotative meaning of a word, which varies depending on who is defining it, the denotative meaning is shared by most people.

**Dependent variable**  In an experiment, the variable that is measured to determine whether it is effected by the manipulation of one or more other variables (i.e., independent variables).

**Deterrence hypothesis**  The view that the arrest and punishment of some individuals for violation of laws deters other persons from committing the same violations.

**Deviant behavior**  Behavior that violates the norms that apply in a given situation.

**Deviant subculture**  A group of people whose norms encourage participation in a specific form of deviance and who regard those who engage in it positively.

**Diffusion of responsibility**  The process of accepting less personal responsibility to act because responsibility is shared with others. Diffusion of responsibility among bystanders in an emergency is one reason why they sometimes fail to help.

**Disaster**  An event that produces widespread physical damage or destruction of property, accompanied by social disruption.

**Disclaimers**  Verbal assertions people use before acting in ways they anticipate will be disruptive, in order to ward off any negative implications the actions may have for their social identities.

**Discrepant message**  In persuasion, a message advocating a position that is different from what the target believes.

**Disengagement theory**  A theory of adjustment to aging that holds that withdrawal from social commitments is inevitable with aging, and that withdrawal promotes satisfaction with life because it frees up time and energy for introspection.

**Disinhibition**  In aggression, the loosing of ordinary internal controls against socially disapproved behavior.

**Disjunctive task**  A type of unitary group task in which the group's performance depends entirely on that of its strongest or fastest member.

**Displacement**  The release of pent-up anger in the form of aggression against a target other than the original cause of anger.

**Display rules**  Cultural-specific norms for modifying facial expressions of emotion to make them fit with the social situation.

**Dissonance effect**  The greater the reward or incentive for engaging in counterattitudinal behavior, the less the resulting attitude change.

**Distinctiveness**  In person perception, information used to make attributions about a behavior by referring to whether the actor behaves differently toward a particular object than toward other objects.

**Divisible task**  A group task in which members perform different, although complementary, activities.

**Dyadic withdrawal**  The state of an increased reliance on one person for gratification and a decreased reliance on others.

**Elaborated code**  A relatively abstract speech style attuned to the characteristics of the particular listener. It allows for subtle differences in meaning by employing qualifications and extended perspectives on time and events. The code is used in relationships that tend toward flexible, personal social control, such as in the middle class.

**Embarrassment**  The feeling people experience when interaction is interrupted because the identity they have claimed in an encounter is discredited.

**Emergent norm**  The definition of a situation that results from interaction in an initially ambiguous situation.

**Emotion work**  An individual's attempt to change the intensity or quality of emotions in order to bring them into line with the requirements of the occasion. Emotion work may be used to evoke feelings that are not present but should be (psyching oneself up), or to suppress feelings that are present but should not be (calming oneself down).

**Empathy**  An emotional response to the distress of others as if we ourselves were in that person's situation, feeling pleasure at another's pleasure or pain at another's pain.

**Endorsement**  An attitude held by a group member indicating the extent to which he or she supports the group's leader.

**Equity**  A state of affairs that prevails in a dyad or group when people receive rewards in proportion to the contributions they make toward attainment of group goals.

**Equitable relationship**  A relationship in which the outcomes received by each person are equivalent.

**Ethnocentrism**  In intergroup relations, the tendency to take one's own group as the center of everything and to evaluate other groups with reference to it.

**Evaluation apprehension**  A state of concern about what others expect of us and how others will evaluate our behavior. Evaluation apprehension inhibits helping when bystanders fear others will view their intervention as foolish. It promotes helping when bystanders believe others expect them to help.

**Exchange**  An interaction in which person A gives person B something that B values in exchange for B giving A something that A values.

**Expectation states theory**  A theory that proposes that status characteristics cause group members to form expectations regarding one another's potential performance on the group's task; these performance expectations affect subsequent interaction among members.

**Experiment**  A research method designed to investigate cause-and-effect relationships between one variable (the independent variable) and another (the dependent variable). In an experiment, the investigator manipulates the independent variable, randomly assigns subjects to various levels of that variable, and measures the dependent variable.

**External validity**  The extent to which an investigator can generalize the results of one study to other populations, settings, or times.

**Extraneous variable**  A variable that is not explicitly included in a research hypothesis but has a causal impact on the dependent variable.

**Extrinsically motivated behavior**  A behavior that results from the motivation to obtain a reward (food or praise) or avoid a punishment (spanking or criticism) controlled by someone else.

**Feeling rules**  Social rules that dictate what an individual with a particular public identity ought to feel in given situation.

**Field study** An investigation that involves the collection of data about on-going activity in everyday settings.

**Focus of attention bias** The tendency to overestimate the causal impact of whomever or whatever our attention is focused on.

**Formal social control** Agencies that are given responsibility for dealing with violations of rules or laws.

**Frame** The type of social occasion people recognize themselves to be in as they interact. Each frame is governed by a set of stable, widely known rules or conventions that indicate what roles are operative and what behaviors are appropriate.

**Front and back regions** Settings used to manage appearances. Front regions are settings where people carry out performances and exert efforts to maintain appropriate appearances. Back regions, are settings inaccessible to outsiders where people allow themselves to violate appearances while they prepare, rehearse, and rehash performances.

**Frustration** The blocking of goal-directed activity. According to the frustration-aggression hypothesis, frustration leads to aggression.

**Frustration-aggression hypothesis** The hypothesis that asserts that every frustration leads to some form of aggression and every aggressive act is due to some prior frustration.

**Fundamental attribution error** The tendency to underestimate the importance of situational influences and to overestimate personal, dispositional factors as causes of behavior.

**Gender role** The behavioral expectations associated with gender.

**Generalized other** A conception of the attitudes and expectations held in common by the members of the organized groups with whom one interacts.

**Goal consensus** In groups, a state of affairs in which a substantial proportion of members agrees on the group's goal and accepts the desirability of trying to achieve it.

**GRIT** A strategy for reducing intergroup conflict whereby one side initiates de-escalatory steps in the hope that they will eventually be reciprocated by the other side; GRIT is an acronym for Graduated and Reciprocal Initiatives in Tension-reduction.

**Group** A set of persons who relate to one another as parts of a system.

**Group cohesiveness** The degree to which individuals are attracted to a group and desire to remain in the group.

**Group goal** A desirable outcome that group members strive collectively to accomplish or bring about.

**Group performance** The achievement of a goal or end state through group activity; the output or end-state resulting from group activity.

**Group polarization** In group decision making, the tendency for group members to shift their opinions toward a position that is similar to, but more extreme than, the positions they held prior to group discussion; both the risky-shift and the cautious-shift are instances of group polarization.

**Groupthink** A mode of thinking within a cohesive group whereby pressures for unanimity overwhelm the members' motivation to appraise realistically alternative courses of action.

**Halo effect** The tendency to perceive personalities as clusters of either good or bad traits based on the knowledge of only one or a few of the traits.

**Helping** Any behavior intended to benefit another person. Also referred to as prosocial behavior, because helping has positive social consequences and is approved by prevailing social standards.

**Hypothesis** A statement that a specific behavior or event is caused by some other event or social process.

**Identity** The categories people use to specify who they are, to locate themselves relative to other people.

**Identity degradation** A response to repeated or glaring failures that destroys the offender's current identity and transforms him or her into a "lower" social type.

**Ideology** In the study of social movements, a conception of reality that emphasizes certain values and justifies the movement.

**Imbalanced state** The state in which the relations among three cognitive elements are all negative, or in which two are positive and one is negative.

**Imitation** A process of learning in which the learner watches another person's response and observes whether that person receives reinforcement.

**Implicit personality theory** A set of assumptions people make about how personality traits are related —which ones go together and which do not. Theories of this type are implicit because people do not subject them to explicit examination nor do they ordinarily know their content.

**Impression management** The intentional use of tactics to manipulate the images others form of us.

**Incentive effect** The greater the reward or incentive for engaging in counterattitudinal behavior, the greater the resulting attitude change.

**Independent variable** In an experiment, the variable that the investigator manipulates to study the effects on one or more other (dependent) variables.

**Informal social control** The reactions of family, friends, and acquaintances to rule violations by individuals.

**Informational influence** In groups, a form of influence that occurs when a group member accepts information from others as valid evidence about reality. Influence of this type is particularly likely to occur in situations of uncertainty, or where there are no external or "objective" standards of reference.

**Informed consent** Voluntary consent by an individual to participate in a research project based upon information received about what their participation will entail.

**Ingratiation** The deliberate use of deception to increase a target person's liking for us in hopes of gaining tangible benefits that the target person controls. Techniques such as flattery, expressing agreement with the target person's attitudes, and exaggerating one's admirable qualities may be used.

**In-group** In intergroup relations, one's own membership group; contrasted with the out-group.

**Institutionalization of deviance** The process by which members of a group come to expect and support deviance by another member over time.

**Instrumental conditioning** The process through which an individual learns a behavior in response to a stimulus to obtain a reward or avoid a punishment.

**Integrative proposal** In bargaining, a proposition or alternative that reconciles bargainers' divergent interests by providing high benefits to both of them.

**Interaction process analysis (IPA)** A coding system of 12 categories used to measure and analyze communication patterns and processes in groups.

**Intergroup conflict** A situation in which groups having opposing interests take antagonistic actions toward one another in order to control some outcome important to them.

**Intergroup contact hypothesis** A theoretical perspective holding that, in intergroup relations, increased interpersonal contact between groups will reduce stereotypes and prejudice, and consequently reduce antagonism between groups.

**Internalization** The process through which initially external behavioral standards become internal and subsequently guide an individual's behavior.

**Internal validity** The extent to which the findings of a study are free from contamination by extraneous variables.

**Interpersonal attraction** A positive attitude held by one person toward another person.

**Interpersonal spacing (proxemics)** Nonverbal communication involving the ways people position themselves at varying distances and angles from others. Because interpersonal spacing refers to the proximity of people, it is known as proxemics.

**Interview** A research method that involves gathering information by asking people a series of questions.

**Intrinsically motivated behavior** A behavior that results from the motivation to achieve an internal state that an individual finds rewarding.

**Labeling theory** The view that reactions by others to a norm violation are an essential element in deviance.

**Leadership** In groups, the process whereby one member influences and coordinates the behavior of other members in pursuit of group goals. The enactment of several functions necessary for successful group performance; these functions include planning, organizing, and controlling the activities of group members.

**Learning structure** An environment in which the individual can learn the information and skills required to enact a role.

**Legitimate means** Those ways of striving to achieve goals which are defined as acceptable by social norms.

**Level of aspiration** In bargaining, the highest price that a person realistically hopes to get for his product or service. In group decision making, the level of difficulty of the goal(s) that group members agree to pursue.

**Life course** An individual's progression through a series of socially defined, age-linked social roles.

**Life cycle squeeze** Periods in the life course when financial and family burdens are great while job status and rewards are low, producing an unfavorable balance between aspirations and resources. Life cycle squeeze tends to be greatest for couples with preschool children.

**Life event** An episode marking a transition point in the life course that provokes coping and readjustment.

**Likert scale** A technique for measuring attitudes that asks a respondent to indicate the extent to which he or she agrees with each of a series of statements about an object.

**Linguistic relativity hypothesis** The hypothesis proposed by Whorf that language shapes ideas and guides the individual's mental activity. The strong form of the hypothesis holds that people cannot perceive distinctions absent from their language. The weak form holds that languages facilitate thinking about events and objects that are easily codable or symbolized in them.

**Linguistic universals** Features that are common to all languages such as nouns, verbs, and terms for expressing distance and time.

**Loneliness** An unpleasant, subjective experience that results from the lack of satisfying social relationships in either quantity or quality.

**Mass media** Those channels of communication (TV, radio, newspapers) that enable a source to reach and influence a large audience.

**Matching hypothesis** The hypothesis that each person looks for someone to date who is of approximately the same level of desirability.

**Meaning** In symbolic interactionist theory, the implications that stem from an object or thing when it is judged in terms of a person's plan of action.

**Meaninglessness** The sense that what is going on around the person is incomprehensible; the absence of a definition of the situation.

**Media campaign** A systematic attempt by an influencing source to use the mass media to change attitudes and beliefs of a target audience.

**Mere exposure effect** Repeated exposure to the same stimulus which produces a positive attitude toward it.

**Methodology** A set of systematic procedures used to conduct empirical research. Usually these procedures pertain to how data will be collected and analyzed.

**Mid-life crisis** A critical stage of transition into middle age, when people become aware that time is running short and that unless they make changes in their personal relations, in their work, and in themselves, it will be too late.

**Mobilization** The process through which individuals surrender personal resources and commit them to the pursuit of group or organizational goals.

**Moral development** The process through which children become capable of making moral judgments.

**Movement organization** A group of persons with defined roles who engage in sustained activity that reflects the ideology of a social movement.

**Norm** In groups, a standard or rule that specifies how members are expected to behave under given circumstances. Expectations concerning which behaviors are acceptable and which are unacceptable for specific persons in specific situations.

**Norm of homogamy** A social norm requiring that friends, lovers, and spouses be characterized by similarity in age, race, religion and socioeconomic status.

**Normative influence** In groups, a form of influence that occurs when a member conforms to group norms in order to receive the rewards and/or avoid the punishments that are contingent on adherence to these norms.

**Normative life stage** A discrete period in the life course during which individuals are expected to perform the set of activities associated with a distinct age-related role.

**Normlessness** The belief that socially disapproved behavior is necessary to achieve one's goals.

**Observational learning** The acquisition of behavior based on observation of another person's behavior and of its consequences for that person.

**Opportunity structure** An environment in which an individual has opportunities to enact a role, which usually requires the assistance of those in complementary roles.

**Outcomes** The rewards expected from an interpersonal relationship minus the expected costs.

**Out-group** In intergroup relations, a group of persons different from one's own group.

**Overreward** In equity theory, a condition of inequity in which a group member receives more rewards than would be justified on the basis of his or her contribution to the group.

**Paralanguage** All the vocal aspects of speech other than words, including vocal pitch, speed of speaking, pauses, sighs, laughter, and so on.

**Participant observation** A research method in which investigators not only make systematic observations but also play an active role in the ongoing events.

**Passionate love** A state of intense absorption and physiological arousal toward another person.

**Personal norm** A feeling of moral obligation to perform specific actions that stem from individuals' internalized systems of values.

**Persuasion** An effort by a source to change the beliefs or attitudes of a target person through the use of information or argument.

**Pileup of demands** The condition in which stressful life events (such as serious illness) and normative life transitions (such as the birth of a child) are added to the continuing stress from prior stressors.

**Population** A set of people whose attitudes, behavior, or characteristics are of interest to the researcher.

**Position** A designated location in a social system.

**Powerlessness** The sense of having little or no control over events.

**Precipitating event** An incident that adversely affects members of a group and symbolizes their discontent, triggering collective behavior by group members.

**Prejudice** A strong like or dislike for members of a specific group.

**Primacy effect** The tendency when forming an impression to be most influenced by the earliest information received. The primacy effect accounts for the fact that first impressions are especially powerful. *See also* Recency effect

**Primary group** In contrast with **Secondary Group,** a small group in which there exist strong emotional ties among members.

**Primary relationship** An interpersonal relationship that is personal, emotionally involving, and of long duration.

**Principle of consistency** In cognitive theory, a principle maintaining that if a person holds several ideas that are incongruous or inconsistent with each other, he or she will experience discomfort or conflict and will subsequently change one or more of the ideas to render them consistent.

**Principle of covariation** A principle that attributes behavior to the potential cause that is present when the behavior occurs and absent when the behavior fails to occur.

**Principle of determinism** A principle that discoverable causes exist for all events in a science's domain of interest.

**Process loss** A negative discrepancy between the group's potential performance and its actual performance; losses of this type stem from inefficiencies introduced during interaction among group members.

**Promise** An influence technique that is a communication taking the general form: "If you do X (which I want), then I will do Y (which you want)."

**Punishment** A painful or discomforting stimulus that reduces the frequency with which the behavior occurs.

**Questionnaire** A research method in which individuals read through a series of printed questions and record their own answers.

**Random assignment** In an experiment, the assignment of subjects to experimental conditions on the basis of chance.

**Range of tolerable behavior** With respect to group norms, that portion or segment of the behavior dimension that group members approve and evaluate positively; the range of behavior that group members find acceptable.

**Realistic group conflict theory** A theoretical perspective regarding intergroup conflict that explains the development and the resolution of conflict in terms of the goals of each group; its central hypothesis is that groups will engage in conflictive behavior when their goals involve opposition of interest.

**Recency effect** The tendency when forming an impression to be most influenced by the latest information received. Although the direct opposite effect (primacy) is more common, the recency effect occurs when there is reason for the perceiver to attend especially to more recent information. *See also* Primacy effect

**Reciprocity norm** A widely accepted social norm stating that people should help those who help them and avoid hurting those who help them.

**Reinforcement theory** A theoretical perspective based on the premise that social behavior is governed by external events, especially rewards and punishments.

**Reinforcer** In reinforcement theory, a pleasant or unpleasant stimulus that follows a behavior and that makes the response more or less likely.

**Relative deprivation** A gap between the expected level and the actual level of satisfaction of the individual's needs, in which the level expected by the individual exceeds the level of need satisfaction experienced.

**Reliability** The degree to which a measurement instrument produces the same results each time it is employed under a set of specified conditions.

**Response rate** In a survey, the percentage of people contacted who complete the survey.

**Restricted code** A concrete and egocentric speech style that is direct, rooted in the here and now, lacking qualifications, and emotionally expressive. The code is used in relationships that tend toward rigid, positional social control, such as in the working class.

**Return potential model** A formal model of a social norm. This model treats norms as having two dimensions—the behavior dimension and the evaluation dimension. A norm is a function that specifies what level of evaluation will correspond to what behaviors.

**Revolutionary coalition** In groups, a union of medium- and low-status members who oppose the existing leadership and wish to overturn it.

**Risk-benefit analysis** A technique that weighs the potential risks to research subjects against anticipated benefits to subjects and the importance of the knowledge that may result from the research.

**Risky-shift** In group decision making, the tendency for decisions made in groups after discussion to be riskier than decisions made by individual members prior to discussion.

**Role** A set of functions performed by a person on behalf of a group of which he or she is a member. The set of expectations governing the behavior of an occupant of a specific position within a social structure.

**Role discontinuity** The discontinuity that results when the values and identities associated with a new role contradict those of earlier roles.

**Role identity** An individual's concept of self in specific social roles.

**Role overload**  The condition in which the demands placed on a person by his/her roles exceed the amount of time, energy, and other resources available to meet those demands.

**Role taking**  In symbolic interactionist theory, the process of imaginatively occupying the position of another person and viewing the situation and the self from that person's perspective; the process of imagining the other's attitudes and anticipating that person's responses.

**Role theory**  A theoretical perspective based on the premise that a substantial portion of observable, day-to-day social behavior is simply persons carrying out role expectations.

**Romantic love ideal**  Five beliefs regarding love, including belief (1) in true love, (2) in love at first sight, (3) that love conquers all, (4) that there is one and only one true love for each person, and (5) that one should follow his or her heart.

**Rule breaking**  Behavior that violates social norms.

**Rumor**  Communication via informal and often novel channels that cannot be validated.

**Salience**  The relative importance of a specific role identity to the individual's self-concept. The salience hierarchy refers to the ordering of an individual's role identities according to their importance.

**Sample**  A subset of a population chosen for investigation.

**Saturation**  In communication networks, the degree of overload experienced by members occupying central positions.

**Schema**  A specific cognitive structure that organizes the processing of complex information about other persons, groups, and situations. Our schemas guide what we perceive in our environment, how we organize information in memory, and what inferences and judgments we make about people and things.

**Secondary deviance**  Deviant behavior employed by a person as a means of defense or adjustment to the problems created by others' reactions to rule breaking by him or her.

**Secondary group**  In contrast with **primary group,** a group whose members have few emotional ties with one another and relate in terms of limited roles. Interaction in secondary groups is formal, impersonal, and nonspontaneous.

**Self**  The individual viewed as both the active source and the passive object of reflexive behavior.

**Self-awareness**  A state in which we take the self as the object of our attention and focus on our own appearance, actions, and thoughts.

**Self-concept**  An individual's thoughts and feelings about himself or herself.

**Self-disclosure**  The process of revealing personal information (aspects of our feelings and behaviors) to others. Self-disclosure is sometimes used as an impression management tactic.

**Self-esteem**  The evaluative component of the self-concept. The positive and negative evaluations people have of themselves.

**Self-estrangement**  The awareness that one is engaging in activities that are not rewarding in themselves.

**Self-fulfilling prophecy**  When persons behave toward another person according to a label (impression), and cause the person to respond in ways that confirm the label.

**Self-presentation**  All conscious and unconscious attempts by people to control the images of self they project in social interaction.

**Self-reinforcement**  An individual's use of internalized standards to judge his own behavior and reward the self.

**Self-schema**  The organized structure of information that people have about themselves; the primary influence on the processing of information about the self.

**Self-serving bias**  In attribution, the tendency for people to take personal credit for acts that yield positive outcomes, and to deflect blame for bad outcomes, attributing them to external causes.

**Semantic differential scale**  A technique for measuring attitudes that asks a respondent to rate an object on each of a series of bipolar adjective scales, that is, scales whose ends are two adjectives having opposite meanings.

**Sentiment relations**  In balance theory, a positive or negative relationship between two cognitive elements.

**Sentiments**  Socially significant feelings such as grief, love, jealousy, or indignation that arise out of enduring social relationships. Each sentiment is a pattern of sensations, emotions, actions, and cultural beliefs appropriate to a social relationship.

**Shaping**  The learning process in which an agent initially reinforces any behavior that remotely resembles the desired response and subsequently requires increasing correspondence between the learner's behavior and the desired response before providing reinforcement.

**Significant other**  A person whose views and attitudes are very important and worthy of consideration. The reflected views of a significant other have great influence on the individual's self-concept and self-regulation.

**Simple random sample**  A sample of individuals selected from a population in such a way that everyone is equally likely to be selected.

**Situated self**  The subset of self-concepts that constitutes the self people recognize in a particular situation. Selected from a person's various identities, qualities, and self-evaluations, the situated self depends on the demands of the situation.

**Situated social identity**  A person's conception of who he or she is in relation to other people in a specific situation, adopted to facilitate smooth social interaction.

**Situational constraint**  An influence on behavior due to the likelihood that other persons will learn about that behavior and respond positively or negatively to it.

**Small group**  Any group of persons sufficiently limited in size that its members are able to interact directly or face-to-face.

**Socialization**  The process through which individuals learn skills, knowledge, values, motives, and roles appropriate to their positions in a group or society.

**Social-emotional act**  In group interaction, emotional reactions, both positive and negative, directed by one member toward other members.

**Social-emotional leader**  In groups, a person who strives to keep emotional relationships pleasant among members; a person who initiates acts that ease the tension and soothe hurt feelings.

**Social exchange theory**  A theoretical perspective, based on the principle of reinforcement, that assumes that people choose from available options those actions that maximize rewards and minimize costs.

**Social identity theory of intergroup behavior**  A theory of intergroup relations based on the premise that people spontaneously categorize the social world into various groups (especially, in-groups and out-groups) and experience high self-esteem to the extent that the in-groups to which they belong have more status than the out-groups.

**Social influence**  An interaction process in which one person's behavior causes another person to change an opinion or to perform an action that he or she would not otherwise do.

**Social isolation**  The lack of involvement in meaningful relationships.

**Social learning theory**  A theoretical perspective based on the premise that a person acquires new responses through the application of reinforcement.

**Social loafing**  The tendency by group members to slack off and reduce their effort on additive tasks, which causes the group's output to fall short of its potential.

**Social movement**  Collective activity that expresses a high level of concern about some issue; the activity may include participation in discussions, petition drives, demonstrations, or election campaigns.

**Social network**  The sets of relationships associated with the social positions a person occupies.

**Social perception**  The process through which we construct an understanding of the social world out of the data we obtain through our senses. More narrowly defined, the process through which we use available information to form impressions of people.

**Social psychology**  The field that systematically studies the nature and causes of human social behavior.

**Social responsibility norm**  A widely accepted social norm stating that individuals should help people who are dependent on them.

**Social status**  In groups, the evaluation by members of a social position in terms of its prestige, importance, or value to the group. Also, the evaluation or ranking of the person occupying that position.

**Social structure**  The ordered and persisting relationships among the positions in a social system.

**Sociolinguistic competence**  Knowledge of the implicit rules for generating socially appropriate sentences that make sense because they fit with the listeners' cultural and social knowledge.

**Sociometry**  An empirical research method, based on questionnaire nominations, used to measure patterns of interpersonal attraction in groups.

**Source**  In social influence, the person who intentionally engages in some behavior (persuasion, threat, promise) to cause another person to behave in a manner different from what he or she would otherwise do.

**Speech act**  The smallest unit of verbal social behavior intended to express a purpose such as to warn, inform, question, invite, and so on.

**Spoken language**  A socially acquired system of sound patterns with meanings agreed on by the members of a group.

**Status**  In groups, a member's relative standing vis-a-vis others. The social evaluation or ranking assigned to a position.

**Status characteristic**  Any property of a person around which evaluations and beliefs about that person come to be organized; properties such as race, occupation, age, sex ethnicity, education, and so on.

**Status consensus**  A state of affairs in which group members agree on their relative status and especially on that of their leader.

**Status generalization**  A process through which differences in members' status characteristics lead to different performance expectations and hence affect patterns of interaction in groups.

**Stereotype**  A fixed set of characteristics that is attributed to all the members of a group. A simplistic and rigid perception of members of one group that is widely shared by others.

**Stigma** Personal characteristics that others view as insurmountable handicaps preventing competent or morally trustworthy behavior.

**Stratified sample** In a survey, a sampling design whereby researchers subdivide the population into groups according to characteristics known or thought to be important, select a random sample of groups, and then draw a sample of units within each selected group.

**Stress** The condition in which the demands made on the person exceed the individual's ability to cope with them.

**Subjective expected value (SEV)** With respect to threats, the product of a threat's credibility times its magnitude; with respect to promises, the product of a promise's credibility times its magnitude.

**Subjective norm** An individual's perception of others' beliefs about whether or not a behavior is appropriate in a specific situation.

**Summons-answer sequence** The most common verbal method for initiating a conversation in which one person summons the other as with a question or greeting, and the other indicates availability for conversation by responding. This sequence establishes the mutual obligation to speak and to listen that produces conversational turn taking.

**Superordinate goal** In intergroup conflict, an objective held in common by all conflicting groups that cannot be achieved by any group without the supportive efforts of the others.

**Survey** A research method that involves gathering information by asking questions of individuals, either through face-to-face interview or through questionnaire.

**Symbol** A form used to represent ideas, feelings, thoughts, intentions, or any other object. Symbols represent our experiences in a way that others can perceive with their sensory organs—through sounds, gestures, pictures, and so on.

**Symbolic interactionist theory** A theoretical perspective based on the premise that human nature and social order are products of communication among people.

**Target** In social influence, the person who is impacted by a social influence attempt from the source.

**Task act** In group interaction, instrumental behaviors that move the group toward realization of its goal.

**Task leader** In groups, a member who pushes the group toward attainment of its goals; a person who contributes many ideas and suggestions to the group.

**Theoretical perspective** A theory that makes broad assumptions about human nature and offers general explanations for a wide range of diverse behaviors.

**Theory** A set of interrelated propositions that organizes and explains a set of observed facts; a network of hypotheses that may be used as a basis for prediction.

**Theory of cognitive dissonance** A theory concerning the sources and effects of inconsistency in cognitive systems with two or more elements.

**Theory of reasoned action** The theory that behavior is determined by behavioral intention, which in turn is determined by both attitude and subjective norm.

**Theory of speech accommodation** The theory that people express or reject intimacy with others by adjusting their speech behavior (accent, vocabulary, or language) during interaction. They make their own speech behavior more similar to their partner's to express liking, and more dissimilar to reject intimacy.

**Threat** An influence technique that is a communication taking the general form: "If you don't do X (which I want), then I will do Y (which you don't want)."

**Threat-counterthreat spiral** In bargaining, a situation in which bargainers stand firm in their positions and issue increasingly stronger threats to inflict more and more damage on one another.

**Transactional view of leadership** A theoretical perspective that characterizes leadership in groups as an exchange between the leader and other group members.

**Trust** The belief that a person is both honest and benevolent.

**Underreward** In equity theory, a condition of inequity in which a group member receives fewer rewards than would be justified on the basis of his or her contribution to the group.

**Unitary task** A group task in which all members perform identical activities.

**Unit relations** In balance theory, the extent of perceived association between two cognitive elements.

**Upward mobility** Movement from an occupation that is lower in prestige and income to one that is higher in prestige and income.

**Validity** The degree to which a measurement instrument measures the concept the investigator intends to measure.

**Values** Enduring beliefs that certain patterns of behavior or end states are preferable to others.

**Vocabularies of motive** Sets of explanations of unsuitable behavior regarded as appropriate by particular groups in specific situations.

**Weighted averaging model** A theoretical model for combining information about individual traits in order to predict how favorable the overall impression people form of another will be. This model *averages* the favorability values of all the single traits people associate with the person after first *weighting* each trait according to criteria such as its credibility, negativity, and consistency with prior impressions. *See also* Additive model; Averaging model

# References

Abeles, R. P., Steel, L., & Wise, L. (1980). Patterns and implications of life-course organization: Studies from project TALENT. In P. Baltes & O. Brim, Jr. (Eds.), *Life-span development and behavior* (Vol. 3). New York: Academic Press.

Abramson, L. Y., Seligman, M. E. P., & Teasdale, J. (1978). Learned helplessness in humans: Critique and reformulation. *Journal of Abnormal Psychology, 87*, 49–74.

Acock, A., & Scott, W. (1980). A model for predicting behavior: The effect of attitude and social class on high and low visibility political participation. *Social Psychology Quarterly, 43*, 59–72.

Adams, J. S. (1963). Toward an understanding of inequity. *Journal of Abnormal and Social Psychology, 67*, 422–436.

Adams, J. S. (1965). Inequity in social exchange. In L. Berkowitz (Ed.), *Advances in experimental social psychology* (Vol. 2). New York: Academic Press.

Adams, J. S. (1976). The structure and dynamics of behavior in organization boundary roles. In M. D. Dunnette (Ed.), *Handbook of industrial and organizational psychology*. Chicago: Rand McNally.

Adams, J. S., & Jacobsen, P. R. (1964). Effects of wage inequities on work quality. *Journal of Abnormal and Social Psychology, 69*, 19–25.

Adams, J. S., & Rosenbaum, W. B. (1962). The relationship of worker productivity to cognitive dissonance about wage inequities. *Journal of Applied Psychology, 46*, 161–164.

Adler, R. P., Friedlander, B. Z., Lesser, G. S., Meringoff, L., Robertson, T. S., Rossiter, J. R., & Ward, S. (1977). *Research on the effects of television advertising on children*. Washington, DC: Government Printing Office.

Adlerfer, C. P. (1982). Problems in changing white males' behavior and beliefs concerning race relations. In P. Goodman & Associates (Eds.), *Change in organizations* (pp. 122–165). San Francisco: Jossey-Bass.

Ainsworth, M. (1979). Infant-mother attachment. *American Psychologist, 34*, 932–937.

Ajzen, I. (1982). On behaving in accordance with one's attitudes. In M. Zanna, E. Higgins, & C. Herman (Eds.), *Consistency in social behavior: The Ontario symposium* (Vol. 2). Hillsdale, NJ: Erlbaum.

Ajzen, I., & Fishbein, M. (1977). Attitude-behavior relations: A theoretical analysis and review of research. *Psychological Bulletin, 84*, 888–918.

Ajzen, I., & Fishbein, M. (1980). *Understanding attitudes and predicting social behavior*. Englewood Cliffs, NJ: Prentice-Hall.

Ajzen, I., & Holmes, W. H. (1976). Uniqueness of behavioral effects in causal attribution. *Journal of Personality, 44*, 98–108.

Ajzen, I., Timko, C., & White, J. (1982). Self-monitoring and the attitude behavior relation. *Journal of Personality and Social Psychology, 42*, 426–435.

Akers, R., Krohn, M., Lanza-Kaduce, L., & Radosevich, M. (1979). Social learning and deviant behavior: A specific test of a general theory. *American Sociological Review, 44*, 636–655.

Aldous, J. (1987). New views on the family life of the elderly and the near-elderly. *Journal of Marriage and the Family, 49*, 227–234.

Aldwin, C., & Revenson, T. (1987). Does coping help? A reexamination of the relationship between coping and mental health. *Journal of Personality and Social Psychology, 53*, 337–348.

Alessio, J. C. (1980). Another folly for equity theory. *Social Psychology Quarterly, 43*, 336–340.

Alexander, C. N., & Lauderdale, P. (1977). Situated identities and social influence. *Sociometry, 40*, 225–233.

Alexander, C. N., Jr., & Rudd, J. (1984). Predicting behaviors from situated identities. *Social Psychology Quarterly 47*, 172–177.

Alexander, C. N., & Wiley, M. G. (1981). Situated activity and identity formation. In M. Rosenberg & R. Turner (Eds.), *Social psychology: Sociological perspectives*. New York: Basic Books.

Allen, H. (1972). Bystander intervention and helping on the subway. In L. Bickman & T. Henchy (Eds.), *Beyond the laboratory: Field research in social psychology*. New York: McGraw-Hill.

Allen, V. L., & Levine, J. M. (1969). Consensus and conformity. *Journal of Experimental Social Psychology, 5*, 389–399.

Allen, V. L., & Levine, J. M. (1971). Social support and conformity: The role of independent assessment of reality. *Journal of Experimental Social Psychology, 7*, 48–58.

Allport, F. H. (1924). *Social psychology*. Cambridge, MA: Houghton Mifflin.

Allport, G. W. (1935). Attitudes. In C. Murchison (Ed.), *Handbook of social psychology* (pp. 798–844). Worcester, MA: Clark University Press.

Allport, G. W. (1954). *The nature of prejudice*. Reading, MA: Addison-Wesley.

Allport, G. W. (1961). *Pattern and growth in personality*. New York: Holt, Rinehart and Winston.

Allport, G. W. (1985). The historical background of social psychology. In G. Lindzey & E. Aronson (Eds.), *Handbook of social psychology: Vol. I* (3rd ed.). New York: Random House.

Altman, I., & Taylor, D. A. (1973). *Social penetration: The development of interpersonal relationships*. New York: Holt, Rinehart and Winston.

Alvarez, R. (1968). Informal reactions to deviance in a simulated work organization: A laboratory study. *American Sociological Review, 33*, 895–912.

Alwin, D. F. (1986). Religion and parental child-rearing orientations: Evidence of Catholic-Protestant convergence. *American Journal of Sociology, 92*, 412–440.

Amir, Y. (1969). Contact hypothesis in ethnic relations. *Psychological Bulletin, 71,* 319–342.

Amir, Y. (1976). The role of intergroup contact in change of prejudice and ethnic relations. In P. A. Katz (Ed.), *Towards the elimination of racism* (pp. 245–308). New York: Pergamon.

Anderson, A. B. (1975). Combined effects of interpersonal attraction and goal-path clarity on the cohesiveness of task oriented groups. *Journal of Personality and Social Psychology, 31,* 68–75.

Anderson, B. A., & Silver, B. D. (1987). The validity of survey responses: Insights from interviews of married couples in a survey of Soviet emigrants. *Social Forces, 66,* 537–554.

Anderson, C. A. (1987). Temperature and aggression: Effects on quarterly, yearly, and city rates of violent and nonviolent crime. *Journal of Personality and Social Psychology, 52,* 1161–1173.

Anderson, N. H. (1968). Likeableness ratings of 555 personality trait words. *Journal of Personality and Social Psychology, 9,* 272–279.

Anderson, N. H. (1981). *Foundations of information integration theory.* New York: Academic Press.

Anderson, R. C. (1963). Learning in discussions: A resume of the authoritarian-democratic studies. In W. W. Charters, Jr. & N. L. Gage (Eds.), *Readings in the social psychology of education.* Boston: Allyn and Bacon.

Anderson, R., Manoogian, S. T., & Reznick, J. (1976). The undermining and enhancing of intrinsic motivation in preschool children. *Journal of Personality and Social Psychology, 34,* 915–922.

Andrews, I. R. (1967). Wage inequity and job performance: An experimental study. *Journal of Applied Psychology, 51,* 39–45.

Angrist, S. S., & Almquist, E. (1975). *Careers and contingencies: How college women juggle with gender.* New York: Dunellen.

Appleton, W. (1981). *Fathers and daughters.* New York: Doubleday.

Archer, D., & Akert, R. (1977). Words and everything else: Verbal and nonverbal cues in social interpretation. *Journal of Personality and Social Psychology, 35,* 443–449.

Archer, R. L. (1980). Self-Disclosure. In D. M. Wegner & R. R. Valacher (Eds.), *The self in social psychology.* New York: Oxford University Press.

Argyle, M., Lalljee, M., & Cook, M. (1968). The effects of visibility of interaction in a dyad. *Human Relations, 21,* 3–17.

Arkin, R. M. (1980). Self presentation. In D. M. Wegner & R. R. Vallacher (Eds.), *The self in social psychology.* New York: Oxford University Press.

Aronfreed, J., & Reber, A. (1965). Internalized behavior suppression and the timing of social punishment. *Journal of Personality and Social Psychology, 1,* 3–16.

Aronoff, D., & Tedeschi, J. T. (1968). Original stakes and behavior in the prisoner's dilemma game. *Psychonomic Science, 12,* 79–80.

Aronson, E., Brewer, M., & Carlsmith, J. M. (1985). Experimentation in social psychology. In G. Lindzey & E. Aronson (Eds.), *Handbook of social psychology: Vol. 1.* (3rd ed.). New York: Random House.

Aronson, E., Turner, J. A., & Carlsmith, J. M. (1963). Communicator credibility and communication discrepancy as determinants of opinion change. *Journal of Abnormal and Social Psychology, 67,* 31–36.

Asch, S. E. (1946). Forming impressions of personality. *Journal of Abnormal and Social Psychology, 41,* 258–290.

Asch, S. E. (1951). Effects of group pressure upon the modification and distortion of judgments. In H. Guetzkow (Ed.), *Groups, leadership, and men.* Pittsburgh: Carnegie Press.

Asch, S. E. (1955). Opinions and social pressure. *Scientific American, 193,* 31–35.

Asch, S. E. (1957). An experimental investigation of group influence. *Symposium on Preventive and Social Psychiatry.* Walter Reed Army Institute of Research, Washington, DC: Government Printing Office.

Asch, S. E. & Zukier, H. (1984). Thinking about persons. *Journal of Personality and Social Psychology, 46,* 1230–1240.

Asendorpf, J. B. (1987). Videotape reconstruction of emotions and cognitions related to shyness. *Journal of Personality and Social Psychology, 53,* 542–549.

Ashmore, R. D. (1981). Sex stereotypes and implicit personality theory. In D. L. Hamilton (Ed.), *Cognitive processes in stereotyping and intergroup behavior.* Hillsdale, NJ: Erlbaum.

Atchley, R. C. (1980). *The social forces in later life.* Belmont, CA: Wadsworth.

Atkin, C. K. (1981). Mass media information campaign effectiveness. In R. E. Rice & W. J. Paisley (Eds.), *Public communication campaigns.* Beverly Hills, CA: Sage.

Austin, J. L. (1962). *How to do things with words.* Cambridge, MA: Harvard University Press.

Austin, W. G., & Hatfield, E. (1980). Equity theory, power, and social justice. In G. M. Kula (Ed.), *Justice and social interaction.* Bern, Switzerland: Hans Huber.

Austin, W., & Walster, E. (1974). Participants' reactions to "equity with the world." *Journal of Experimental Social Psychology, 10,* 528–548.

Ayers, L., Nacci, P., & Tedeschi, J. T. (1973). Attraction and reaction to noncontingent promises. *Bulletin of the Psychonomic Society, 1* (1B), 75–77.

Bacharach, S. B., & Lawler, E. J. (1981). *Bargaining: Power, Tactics, and Outcomes.* San Francisco: Jossey-Bass.

Bachman, J. G. (1970). *Youth in transition* (Vol. 2). *The impact of family background and intelligence on tenth-grade boys.* Ann Arbor, MI: Institute for Social Research.

Backman, C. (1981). Attraction in interpersonal relationships. In M. Rosenberg & R. Turner (Eds.), *Social psychology: Sociological perspectives* (pp. 235–268). New York: Basic Books.

Backman, C., & Secord, P. (1959). The effect of perceived liking on interpersonal attraction. *Human Relations, 12,* 379–384.

Backman, C., & Secord, P. (1962). Liking, selective interaction, and misperception in congruent interpersonal relations. *Sociometry, 25,* 321–325.

Backman, C. W., & Secord, P. (1968). The self and role selection. In C. Gordon & K. J. Gergen (Eds.), *The self in social interaction.* New York: Wiley.

Bagozzi, R. P. (1981). Attitudes, intentions and behavior: A test of some key hypotheses. *Journal of Personality and Social Psychology, 41,* 607–627.

Bales, R. F. (1950). *Interaction process analysis.* Reading, MA: Addison-Wesley.

Bales, R. F. (1953). The equilibrium problem in small groups. In T. Parsons, R. F. Bales, & E. A. Shils (Eds.), *Working papers in the theory of action.* New York: Free Press.

Bales, R. F. (1970). *Personality and interpersonal behavior.* New York: Holt, Rinehart and Winston.

Bales, R. F., & Hare, A. P. (1965). Diagnostic use of the interaction profile. *Journal of Social Psychology, 67,* 239–258.

Ball, D. W. (1976). Failure in sports. *American Sociological Review, 41,* 726–739.

Baltes, P., & Willis, S. (1982). Plasticity and enhancement of intellectual functioning in old age: Penn State's adult development and enrichment project. In F. Craik & S. Trehub (Eds.), *Aging and cognitive processes.* New York: Plenum.

Bandura, A. (1965). Influences of models' reinforcement contingencies on the acquisition of imitative responses. *Journal of Personality and Social Psychology, 1,* 589–595.

Bandura, A. (1969). Social-learning theory of identificatory processes. In D. Goslin (Ed.), *Handbook of socialization theory and research.* Chicago: Rand McNally.

Bandura, A. (1973). *Aggression: A social learning analysis.* Englewood Cliffs, NJ: Prentice-Hall.

Bandura, A. (1977). *Social learning theory.* Englewood Cliffs, NJ: Prentice-Hall.

Bandura, A. (1978). The self system in reciprocal determinism. *American Psychologist, 33,* 344–358.

Bandura, A. (1982). The psychology of chance encounters and life paths. *American Psychologist, 37,* 747–755.

Bandura, A. (1982). Self-efficacy mechanism in human agency. *American Psychologist, 37,* 122–147.

Bandura, A. (1982). The self and the mechanisms of agency. In J. Suls (Ed.), *Psychological perspectives on the self* (Vol. 1). Hillsdale, NJ: Erlbaum.

Barker, R. G., Dembo, T., & Lewin, K. (1941). Frustration and regression: An experiment with young children. *University of Iowa Studies in Child Welfare, 18,* 1–34.

Barnes, E. J. (1972). The black community as a source of positive self-concept for black children: A theoretical perspective. In R. L. Jones (Ed.), *Black psychology.* New York: Harper and Row.

Barnes, G., Farrell, M., & Cairns, A. (1986). Parental socialization factors and adolescent drinking behavior. *Journal of Marriage and the Family, 48,* 27–36.

Baron, R. A. (1971). Reducing the influence of an aggressive model: The restraining effects of discrepant modeling cues. *Journal of Personality and Social Psychology, 20,* 240–245.

Baron, R. A. (1977). *Human aggression.* New York: Plenum.

Baron, R. A., & Kepner, C. R. (1970). Model's behavior and attraction toward the model as determinants of adult aggressive behavior. *Journal of Personality and Social Psychology, 14,* 335–344.

Baron, R., & Ransberger, V. (1978). Ambient temperature and the occurrence of collective violence: The "long, hot summer" revisited. *Journal of Personality and Social Psychology, 36,* 351–360.

Baron, R. S., & Roper, G. (1976). A reaffirmation of a social comparison view of choice shifts, averaging, and extremity effects in autokinetic situations. *Journal of Personality and Social Psychology, 33,* 521–530.

Bart, P. B. (1975). The loneliness of the long-distance mother. In J. Freeman (Ed.), *Women: A feminist perspective.* Palo Alto, CA: Mayfield.

Bar-Tal, D. (1976). *Prosocial behavior: Theory and research.* New York: Halsted.

Barton, A. H. (1969). *Communities in disaster: A sociological analysis of collective stress situations.* New York: Doubleday.

Bartos, O. J. (1974). *Process and outcome in negotiation.* New York: Columbia University Press.

Bartunek, J. M., Benton, A. A., & Keys, C. B. (1975). Third party intervention and the bargaining behavior of group representatives. *Journal of Conflict Resolution, 19,* 532–557.

Baruch, G., & Barnett, R. (1986). Role quality, multiple-role involvement, and psychological well-being in mid-life women. *Journal of Personality and Social Psychology, 51,* 578–585.

Batson, C. D., & Coke, J. S. (1981). Empathy: A source of altruistic motivation for helping? In J. P. Rushton & R. M. Sorrentino (Eds.), *Altruism and helping behavior.* Hillsdale, NJ: Erlbaum.

Batson, C. D., O'Quin, K., Fultz, J., Vanderplas, M., & Isen, A. M. (1983). Influence of self-reported distress and empathy on egoistic versus altruistic motivation to help. *Journal of Personality and Social Psychology, 45,* 706–718.

Bauer, R. (1964). The obstinate audience: The influence process from the point of view of social communication. *American Psychologist, 19,* 319–328.

Baum, A., Riess, M., & O'Hara, J. (1974). Architectural variants of reaction to spatial invasion. *Environment and Behavior, 6,* 91–100.

Baumeister, R. F. (1982). A self-presentational view of social phenomena. *Psychological Bulletin, 91,* 3–26.

Baumeister, R. F., & Hutton, D. G. (1987). A self-presentational perspective on group processes. In B. Mullen & G. R. Goethals (Eds.), *Theories of group behavior.* New York: Springer-Verlag.

Baumrind, D. (1964). Some thoughts on ethics of research: After reading Milgram's "Behavioral study of obedience." *American Psychologist, 19,* 421–423.

Baumrind, D. (1980). New directions in socialization research. *American Psychologist, 35,* 639–652.

Bavelas, A. (1950). Communication patterns in task-oriented groups. *Journal of the Acoustical Society of America, 22,* 725–730.

Beaman, A. L., Klentz, B., Diener, E., & Svanum, S. (1979). Objective self awareness and transgression in children: A field study. *Journal of Personality and Social Psychology, 37,* 1835–1846.

Beck, S. B., Ward-Hull, C. I., & McLear, P. M. (1976). Variables related to women's somatic preferences of the male and female body. *Journal of Personality and Social Psychology, 34,* 1200–1210.

Becker, B. J. (1986). Influence again: Another look at studies of gender differences in social influence. In J. S. Hyde & M. C. Linn (Eds.), *The psychology of gender: Advances through meta-analysis.* Baltimore, MD: Johns Hopkins University Press.

Becker, H. S. (1963). *Outsiders: Studies in the sociology of deviance.* New York: Free Press.

Becker, H. S. (1964). What do they really learn at college? *Trans-Action, 1,* 14–17.

Beebe, L. M., & Giles, H. (1984). Speech accommodation theories: A discussion in terms of second language learning. *International Journal of the Sociology of Language, 46,* 5–32.

Begley, T., & Alker, H. (1982). Anti-busing protest: Attitudes and actions. *Social Psychology Quarterly, 45,* 187–197.

Bell, R. (1979). Parent, child and reciprocal influences. *American Psychologist, 34,* 821–826.

Belsky, J., Lang, M., & Rovine, M. (1985). Stability and change in marriage across the transition to parenthood: A second study. *Journal of Marriage and the Family, 47,* 855–865.

Belsky, J., Steinberg, L. D., & Walker, A. (1982). The ecology of day care. In M. E. Lamb (Ed.), *Nontraditional families: Parenting and child development.* Hillsdale, NJ: Erlbaum.

Bem, D. J. (1970). *Beliefs, attitudes and human affairs.* Belmont, CA: Brooks/Cole.

Bem, D. J. & Allen, A. (1974). On predicting some of the people some of the time: The search for cross-situational consistency in behavior. *Psychological Review, 81,* 506–520.

Benassi, M. (1982). Effect of order of presentation, primacy and physical attractiveness on attributions of ability. *Journal of Personality and Social Psychology, 43,* 48–58.

Benedict, R. (1938). Continuities and discontinuities in cultural conditioning. *Psychiatry, 1,* 161–167.

Benton, A. A. (1972). Accountability and negotiations between group representatives. *Proceedings of the 80th Annual Convention of the American Psychological Association, 7.*

Benton, A. A., Kelley, H. H., & Liebling, B. (1972). Effects of extremity of offers and concession rate on the outcomes of bargaining. *Journal of Personality and Social Psychology, 24,* 73–83.

Berger, J., Cohen, B. P., & Zelditch, M., Jr. (1972). Status characteristics and social interaction. *American Sociological Review, 37,* 241–255.

Berger, J., & Fisek, M. H. (1970). Consistent and inconsistent status characteristics and the determination of power and prestige orders. *Sociometry, 33,* 287–304.

Berger, J., & Fisek, M. H. (1974). A generalization of the theory of status characteristics and expectation states. In J. Berger, T. L. Conner, & M. H. Fisek (Eds.), *Expectation states theory.* Cambridge, MA: Winthrop.

Berger, J., Fisek, M. H., Norman, R. Z., & Zelditch, M., Jr. (1977). *Status characteristics and social interaction: An expectation states approach.* New York: Elsevier.

Berger, J., Rosenholtz, S. J., & Zelditch, M., Jr. (1980). Status organizing processes. *Annual review of sociology* (Vol. 6, pp. 479–508). Palo Alto, CA: Annual Reviews.

Berk, R. A., & Aldrich, H. (1972). Patterns of vandalism during civil disorders as an indicator of selection of targets. *American Sociological Review, 37,* 533–547.

Berkowitz, L. (1954). Group standards, cohesiveness, and productivity. *Human Relations, 7,* 509–519.

Berkowitz, L. (1972). Frustrations, comparisons, and other sources of emotion arousal as contributors to social unrest. *Journal of Social Issues, 28,* 77–91.

Berkowitz, L. (1978). Whatever happened to the frustration-aggression hypothesis? *American Behavioral Scientist, 21,* 691–708.

Berkowitz, L., Klanderman, S. B., & Harris, R. (1964). Effects of experimenter awareness and sex of subject and experimenter on reactions to dependency relationships. *Sociometry, 27,* 327–337.

Berkowitz, L., & LePage, A. (1967). Weapons as aggression-eliciting stimuli. *Journal of Personality and Social Psychology, 7,* 202–207.

Berkowitz, M. W., Mueller, C. W., Schnell, S. V., & Pudberg, M. T. (1986). Moral reasoning and judgments of aggression. *Journal of Personality and Social Psychology, 51,* 885–891.

Bernard, J. S. (1981). *The female world.* New York: Free Press.

Bernstein, B. (1974). *Class, codes and control I* (rev. ed.). London: Routledge & Kegan Paul.

Bernstein, B. (1975). *Class, codes and control III.* London: Routledge & Kegan Paul.

Bernstein, I., Kelly, W., & Doyle, P. (1977). Societal reaction to deviants: The case of criminal defendants. *American Sociological Review, 42,* 743–755.

Bernstein, I., Kick, E., Leung, J., & Schulz, B. (1977). Charge reduction: An intermediary stage in the process of labelling criminal defendants. *Social Forces, 56,* 362–384.

Bernstein, W. M., Stephan, W. G., & Davis, M. H. (1979). Explaining attributions for achievement: A path analytic approach. *Journal of Personality and Social Psychology, 37,* 1810–1821.

Berscheid, E. (1966). Opinion change and communicator-communicatee similarity and dissimilarity. *Journal of Personality and Social Psychology, 4,* 670–680.

Berscheid, E., Dion, K., Walster (Hatfield), E., & Walster, G. (1971). Physical attractiveness and dating choice: A test of the matching hypothesis. *Journal of Experimental Social Psychology, 7,* 173–189.

Berscheid, E., & Walster (Hatfield), E. (1974a). A little bit about love. In T. Huston (Ed.), *Foundations of interpersonal attraction* (pp. 355–381). New York: Academic Press.

Berscheid, E., & Walster (Hatfield), E. (1974b). Physical attractiveness. In L. Berkowitz (Ed.), *Advances in Experimental Social Psychology.* New York: Academic Press.

Berscheid, E., & Walster (Hatfield), E. (1978). *Interpersonal attraction* (2nd ed.). Reading, MA: Addison-Wesley.

Bertenthal, B. I., & Fischer, K. (1978). Development of self-recognition in the infant. *Developmental Psychology, 14*, 44–50.

Bickman, L. (1971). The effect of another bystander's ability to help on bystander intervention in an emergency. *Journal of Experimental Social Psychology, 7*, 367–379.

Biddle, B. J. (1979). *Role theory: Expectations, identities, and behaviors.* New York: Academic Press.

Biddle, B. J. (1986). Recent developments in role theory. In A. Inkeles, J. Coleman, & N. Smelser (Eds.), *Annual review of sociology* (Vol. 12, pp. 67–92). Palo Alto, CA: Annual Reviews.

Biddle, B. J., & Thomas, E. (Eds.). (1966). *Role theory: Concepts and research.* New York: Wiley.

Bielby, D., & Bielby, W. (1988). She works hard for the money: Household responsibilities and the allocation of work effort. *American Journal of Sociology, 93*, 1031–1059.

Bielby, W., & Baron, J. (1986). Men and women at work: Sex segregation and statistical discrimination. *American Journal of Sociology, 91*, 759–799.

Birdwhistell, R. L. (1970). *Kinesics in context: Essays on body motion communications.* Philadelphia: University of Pennsylvania Press.

Birnbaum, J. A. (1975). Life patterns and self-esteem in gifted family-oriented and career committed women. In M. Mednick, S. Schwartz, & L. Hoffman (Eds.), *Women and achievement: Social and motivational analyses.* New York: Halsted.

Black, D. (1980). *The manners and customs of the police.* New York: Academic Press.

Blake, R. R., & Mouton, J. S. (1961a). Reactions to intergroup competition under win-lose conditions. *Management Science, 7*, 420–435.

Blake, R. R., & Mouton, J. S. (1961b). Comprehension of own and of outgroup positions under intergroup competition. *Journal of Conflict Resolution, 5*, 304–310.

Blake, R. R., & Mouton, J. S. (1962). Overevaluation of own group's product in intergroup competition. *Journal of Abnormal and Social Psychology, 64*, 237–238.

Blake, R. R., Shepard, H. A., & Mouton, J. S. (1964). *Managing intergroup conflict in industry.* Houston: Gulf.

Blanck, P. D., & Rosenthal, R. (1982). Developing strategies for decoding "leaky" messages. In R. S. Feldman (Ed.), *Development of nonverbal behavior in children.* New York: Springer-Verlag.

Blascovich, J., Ginsburg, G. P., & Veach, T. L. (1975). A pluralistic explanation of choice shifts on the risk dimension. *Journal of Personality and Social Psychology, 31*, 422–429.

Blau, P. (1964). *Exchange and power in social life.* New York: Wiley.

Blauner, R. (1964). *Alienation and freedom.* Chicago: University of Chicago Press.

Blieszner, R., & Mancini, J. (1987). Enduring ties: Older adults' parental role and responsibilities. *Family Relations, 36*, 176–180.

Blom, J. P., & Gumperz, J. J. (1972). Social meaning and linguistic structure: Code-switching in Norway. In J. J. Gumperz & D. Hymes (Eds.), *Directions in sociolinguistics.* New York: Holt, Rinehart and Winston.

Blumenthal, M., Kahn, R. L., Andrews, F. M., & Head, K. B. (1972). *Justifying violence: Attitudes of American men.* Ann Arbor, MI: Institute for Social Research.

Blumer, H. (1962). Society and symbolic interactionism. In A. M. Rose (Ed.), *Human behavior and social processes.* Boston: Houghton Mifflin.

Blumer, H. (1969). Elementary collective groupings. In A. M. Lee (Ed.), *Principles of sociology* (3rd ed.). New York: Barnes and Noble.

Blumer, H. (1969). *Symbolic interactionism: Perspective and method.* Englewood Cliffs, NJ: Prentice-Hall.

Blumstein, P. W. (1974). The honoring of accounts. *American Sociological Review, 39*, 551–566.

Blumstein, P. W. (1975). Identity bargaining and self-conception. *Social Forces, 53*, 476–485.

Bobo, L. (1983). Whites' opposition to busing: Symbolic racism or realistic group conflict? *Journal of Personality and Social Psychology, 45*, 1196–1210.

Bochner, S., & Insko, C. A. (1966). Communicator discrepancy, source credibility, and opinion change. *Journal of Personality and Social Psychology, 4*, 614–621.

Bodenhausen, G. V., & Lichtenstein, M. (1987). Social stereotypes and information processing strategies: The impact of task complexity. *Journal of Personality and Social Psychology, 52*, 871–880.

Bodenhausen, G., & Wyer, R., Jr. (1985). Effects of stereotypes on decision making and information-processing strategies. *Journal of Personality and Social Psychology, 48*, 267–282.

Bollen, K. A., & Phillips, D. P. (1981). Suicidal motor vehicle fatalities in Detroit: A replication. *American Journal of Sociology, 87*, 404–412.

Booth, A., Johnson, D., White, L., & Edwards, J. (1984). Women, outside employment and marital instability. *American Journal of Sociology, 90*, 567–583.

Bord, R. J. (1976). The impact of imputed deviant identities in structuring evaluations and reactions. *Sociometry, 39*, 108–116.

Borkenau, P., & Ostendorf, F. (1987). Fact and fiction in implicit personality theory. *Journal of Personality, 55*, 415–443.

Boucher, J. D., & Ekman, P. (1975). Facial areas of emotional information. *Journal of Communication, 25*, 21–29.

Boulding, K. E. (1981). *Ecodynamics: A new theory of societal evolution.* Beverly Hills, CA: Sage.

Bourhis, R. Y., Giles, H., Leyens, J. P., & Tajfel, H. (1979). Psycholinguistic distinctiveness: Language diversity in Belgium. In H. Giles & R. N. St. Clair (Eds.), *Language and social psychology.* Oxford: Blackwell.

Bowlby, J. (1965). Maternal care and mental health. In J. Bowlby (Ed.), *Child care and the growth of love.* London: Penguin.

Bradley, G. W. (1978). Self-serving biases in the attribution process: A reexamination of the fact or fiction question. *Journal of Personality and Social Psychology, 36*, 56–71.

Braine, M. (1963). The ontogeny of English phrase structure: The first phrase. *Language, 39*, 1–13.

Brehm, J. W. (1956). Postdecision changes in the desirability of alternatives. *Journal of Abnormal and Social Psychology, 52*, 384–389.

Brehm, J. (1972). *Responses to loss of freedom: A theory of psychological reactance.* Morristown, NJ: General Learning Press.

Brehm, J. W., & Cohen, A. (1962). *Explorations in cognitive dissonance.* New York: Wiley.

Brenner, M. W. (1976). *Memory and interpersonal relations.* Unpublished doctoral dissertation, Univ. of Michigan.

Brewer, M. B. (1986). The role of ethnocentrism in intergroup conflict. In S. Worchel & W. G. Austin (Eds.), *Psychology of intergroup relations* (2nd ed.) (pp. 88–102). Chicago: Nelson-Hall Publishers.

Brewer, M. B., & Campbell, D. T. (1976). *Ethnocentrism and intergroup attitudes: East African evidence.* New York: Halsted.

Brewer, M. B. & Kramer, R. M. (1985). The psychology of intergroup attitudes and behavior. *Annual Review of Psychology, 36,* 219–243.

Brickman, P., Rabinowitz, V. C., Karuza, J., Coates, D., Cohn, E., & Kidder, L. (1982). Models of helping and coping. *American Psychologist, 37,* 368–384.

Brickner, M. A., Harkins, S. G., & Ostrom, T. M. (1986). Effects of personal involvement: Thought provoking implications for social loafing. *Journal of Personality and Social Psychology, 51,* 763–769.

Brim, O. G., Jr. (1966). Socialization through the life-cycle. In O. Brim, Jr., and S. Wheeler (Eds.), *Socialization after childhood,* New York: Wiley.

Brim, O. G., Jr., (1976). Theories of the male mid-life crisis. *Counseling Psychologist, 6,* 2–9.

Brim, O. G., Jr., & Ryff, C. (1980). On the properties of life events. In P. Baltes & O. Brim, Jr. (Eds.), *Life-span development and behavior* (Vol. 3). New York: Academic Press.

Broman, C. (1988a). Household work and family life satisfaction of blacks. *Journal of Marriage and the Family, 50,* 743–748.

Broman, C. (1988b). Satisfaction among blacks: The significance of marriage and parenthood. *Journal of Marriage and the Family, 50,* 45–51.

Broverman, I., Vogel, S., Broverman, D., Clarkson, F., & Rosenkrantz, P. (1972). Sex-role stereotypes: A current appraisal. *Journal of Social Issues, 28* (2), 59–78.

Brown, P., & Elliott, R. (1965). Control of aggression in a nursery school class. *Journal of Experimental Child Psychology, 2,* 103–107.

Brown, R. (1964). The acquisition of language. In D. Rioch & E. Weinstein (Eds.), *Disorders of communication* (Vol. 42). Proceedings of the Association for Research in Nervous and Mental Disease. Baltimore: Williams and Wilkins.

Brown, R. (1965). *Social psychology.* Glencoe, IL: Free Press.

Brown, R. (1986). *Social psychology: The second edition.* New York: Free Press.

Brown, R., & Belugi, U. (1964). Three processes in the child's acquisition of syntax. *Harvard Educational Review, 34,* 133–151.

Brown, R., & Fraser, C. (1963). The acquisition of syntax. In C. Cofer & B. Musgrave (Eds.), *Verbal behavior and learning.* New York: McGraw-Hill.

Brown, R. J., & Turner, J. C. (1981). Interpersonal and intergroup behavior. In J. Turner & H. Giles (Eds.), *Intergroup Behavior* (pp. 33–65). Chicago: University of Chicago Press.

Bruner, J. S. (1958). Social psychology and perception. In E. E. Maccoby, T. M. Newcomb, & E. L. Hartley (Eds.), *Readings in social psychology.* New York: Holt, Rinehart and Winston.

Bruner, J. (1964). The course of cognitive growth. *American Psychologist, 19,* 1–15.

Bryan, J. H., & Davenport, M. (1968). *Donations to the needy: Correlates of financial contributions to the destitute* (Research Bulletin No. 68–1). Princeton, NJ: Educational Testing Service.

Bugenthal, D. E. (1974). Interpretations of naturally occurring discrepancies between words and intonation: Modes of inconsistency resolution. *Journal of Personality and Social Psychology, 30,* 125–133.

Burgess, R. L., & Akers, K. L. (1966). A differential association-reinforcement theory of criminal behavior. *Social Problems, 14,* 128–147.

Burke, P. J. (1972). Leadership role differentiation. In C. G. McClintock (Ed.), *Experimental social psychology.* New York: Holt, Rinehart and Winston.

Burke, P. J., & Reitzes, D. (1981). The link between identity and role performance. *Social Psychology Quarterly, 44,* 83–92.

Burnstein, E. (1982). Persuasion as argument processing. In H. Brandstatter, J. H. Davis, & G. Stocher-Kreichgauer (Eds.), *Contemporary problems in group decision-making* (pp. 103–124). New York: Academic Press.

Burnstein, E., & Vinokur, A. (1973). Testing two classes of theories about group-induced shifts in individual choice. *Journal of Experimental Social Psychology, 9,* 123–137.

Bush, D., & Simmons, R. (1981). Socialization processes over the life course. In M. Rosenberg & R. Turner (Eds.), *Social psychology: Sociological perspectives.* New York: Basic Books.

Byrne, D. (1961a). The influence of propinquity and opportunities for interaction on classroom relationships. *Human Relations, 14,* 63–69.

Byrne, D. (1961b). Interpersonal attraction and attitude similarity. *Journal of Abnormal and Social Psychology, 62,* 713–715.

Byrne, D. (1971). *The attraction paradigm.* New York: Academic Press.

Byrne, D., & Clore, G. L. (1970). A reinforcement model of evaluative responses. *Personality: An International Journal, 1,* 103–128.

Byrne, D., Ervin, C. & Lamberth, J. (1970). Continuity between the experimental study of attraction and real-life computer dating. *Journal of Personality and Social Psychology, 16,* 157–165.

Byrne, D., & Nelson, D. (1965). Attraction as a linear function of proportion of positive reinforcements. *Journal of Personality and Social Psychology, 1,* 659–663.

Cahill, S. (1987). Children and civility: ceremonial deviance and the acquisition of ritual competence. *Social Psychology Quarterly, 50,* 312–321.

Calahan, D. (1970). *Problem drinkers.* San Francisco: Jossey-Bass.

Caldwell, M. A., & Peplau, L. (1982). Sex differences in same-sex relationships. *Sex roles, 8*, 721–732.

Callero, P. (1985). Role-identity salience. *Social Psychology Quarterly, 48*, 203–215.

Campbell, A., Converse, P., & Rodgers, W. (1976). *The quality of American life.* New York: Russell Sage Foundation.

Campbell, D. T. (1967). Stereotypes in the perception of group differences. *American Psychologist, 22*, 817–829.

Campbell, D. T., & Stanley, J. C. (1963). *Experimental and quasi-experimental designs for research.* Chicago: Rand McNally.

Cantor, N., & Kihlstrom, J. F. (Eds.). (1981). *Personality, cognition, and social interaction.* Hillsdale, NJ: Erlbaum.

Cantor, N., & Mischel, W. (1979). Prototypes in person perception. In L. Berkowitz (Ed.), *Advances in experimental social psychology* (Vol. 12). New York: Academic Press.

Cantor, J., Alfonso, H., & Zillmann, D. (1976). The persuasive effectiveness of the peer appeal and a communicator's first hand experience. *Communication Research, 3*, 293–310.

Caplan, F. (1973). *The first twelve months of life.* New York: Grosset and Dunlap.

Carey, M. (1978). The role of gaze in the initiation of conversation. *Social Psychology, 41*, 269–271.

Carlsmith, J. M., & Anderson, C. (1979). Ambient temperature and the occurrence of collective violence. *Journal of Personality and Social Psychology, 37*, 337–344.

Carlsmith, J. M., Ellsworth, P., & Whiteside, J. (1968). Guilt, confession, and compliance. Cited in Freedman, J. L., Transgression, compliance and guilt. In J. R. Macaulay and L. Berkowitz (Eds.), *Altruism and helping behavior.* New York: Academic Press.

Carlston, D. E., & Shovar, N. (1983). Effects of performance attributions on others' perceptions of the attributor. *Journal of Personality and Social Psychology, 44*, 515–525.

Carroll, L. (1981). *Through the looking glass.* New York: Bantam. (Original publication 1871)

Cartwright, D. (1968). The nature of group cohesiveness. In D. Cartwright and A. Zander (Eds.), *Group dynamics* (3rd ed.). New York: Harper and Row.

Cartwright, D. (1971). Risk taking by individuals and groups: An assessment of research employing choice dilemmas. *Journal of Personality and Social Psychology, 20*, 361–378.

Cartwright, D. S., & Robertson, R. J. (1961). Membership in cliques and achievement. *American Journal of Sociology, 66*, 441–445.

Cartwright, D., & Zander, A. (1968). Motivational processes in groups: Introduction. In D. Cartwright & A. Zander (Eds.), *Group dynamics: Research and theory* (3rd ed.). New York: Harper and Row.

Centers, R. (1975). Attitude similarity-dissimilarity as a correlate of heterosexual attraction and love. *Journal of Marriage and the Family, 37*, 305–312.

Chaffee, S. (1981). Mass media in political campaigns: An expanding role. In R. E. Rice & W. J. Paisley (Eds.), *Public communication campaigns.* Beverly Hills, CA: Sage.

Chaiken, S. (1979). Communicator physical attractiveness and persuasion. *Journal of Personality and Social Psychology, 37*, 1387–1397.

Chaiken, S. (1980). Heuristic versus systematic information processing and the use of source versus message cues in persuasion. *Journal of Personality and Social Psychology, 39*, 752–766.

Chaiken, S., & Yates, S. (1985). Affective-cognitive consistency and thought-induced polarization. *Journal of Personality and Social Psychology, 49*, 1470–1481.

Chemers, M. M., & Skrzypek, G. J. (1972). An experimental test of the contingency model of leadership effectiveness. *Journal of Personality and Social Psychology, 24*, 172–177.

Chemers, M. M. (1983). Leadership theory and research: A systems-process integration. In P. B. Paulus (Ed.), *Basic group processes* (pp. 9–39). New York: Springer-Verlag.

Cherlin, A. J. (1981). *Marriage, divorce, remarriage.* Cambridge, MA: Harvard University Press.

Chertkoff, J. M., & Conley, M. (1967). Opening offer and frequency of concession as bargaining strategies. *Journal of Personality and Social Psychology, 7*, 181–185.

Cherulnik, P. D. (1983). *Behavioral research: Assessing the validity of research findings in psychology.* New York: Harper and Row.

Chesterfield, Earl of (P. D. Stanhope). 1901 (original publ. 1774). In W. M. Dunne (Ed.), *Letters to his son.* New York: Wiley.

Chiricos, T., & Waldo, G. (1975). Socioeconomic status and criminal sentencing: An empirical assessment of a conflict proposition. *American Sociological Review, 40*, 753–772.

Christenson, P. G. (1982). Children's perceptions of TV commercials and products: The effects of PSAs. *Communication Research, 9*, 491–524.

Cialdini, R., & Baumann, D. (1981). Littering: A new unobtrusive measure of attitudes. *Social Psychology Quarterly, 44*, 254–259.

Cialdini, R. B., Borden, R., Thorne, A., Walker, M., & Freeman, S. (1976). Basking in reflected glory: Three (football) field studies. *Journal of Personality and Social Psychology, 34*, 366–375.

Cialdini, R. B., Schaller, M., Houlihan, D., Arps, K., Fultz, J., & Beaman, A. (1987). Empathy-based helping: Is it selflessly or selfishly motivated? *Journal of Personality and Social Psychology, 52*, 749–758.

Cicourel, A. V. (1972). Basic and normative rules in the negotiation of status and role. In D. Sudnow (Ed.), *Studies of social interaction.* New York: Free Press.

Clark, E. V. (1976). From gesture to word: On the natural history of deixis in language acquisition. In J. S. Bruner & A. Gartner (Eds.), *Human growth and development.* Oxford: Clarendon Press.

Clark, H. H., & Clark, E. V. (1977). *Psychology and language.* New York: Harcourt Brace Jovanovich.

Clark, K. B. (1963). *Prejudice and your child* (2nd ed.). New York: Beacon Press.

Clark, M. S., & Mills, J. (1979). Interpersonal attraction in exchange and communal relationships. *Journal of Personality and Social Psychology, 37*, 12–24.

Clark, R. D., III, & Word, L. E. (1972). Why don't bystanders help? Because of ambiguity? *Journal of Personality and Social Psychology, 24*, 392–400.

Clarke-Stewart, K. A. (1978). Popular primers for parents. *American Psychologist, 33*, 359–369.

Clausen, J. (1966). Family structure, socialization and personality. In M. Hoffman & L. Hoffman (Eds.), *Review of*

*Child Development Research* (Vol. 2). New York: Russell Sage Foundation.

Clausen, J. A. (1968). *Socialization and society.* Boston: Little, Brown.

Clore, G. L., Bray, R. M., Itkin, S. M., & Murphy, P. (1978). Interracial attitudes and behavior at a summer camp. *Journal of Personality and Social Psychology, 36,* 107–116.

Cloward, R. (1959). Illegitimate means, anomie and deviant behavior. *American Sociological Review, 24,* 164–176.

Code of Federal Regulations (1983). Part 46: *Protection of human subjects* (rev. March 8, 1983). Washington, DC: Office for Protection from Research Risks, National Institutes of Health.

Cohen, A. (1966). *Deviance and control.* Englewood Cliffs, NJ: Prentice-Hall.

Cohen, E. (1980). Design and redesign of the desegregated school: Problems of status, power, and conflict. In W. G. Stephan & J. Feagin (Eds.), *School desegregation* (pp. 251–280). New York: Academic Press.

Cohen, E. G. (1982). Expectation states and interracial interaction in school settings. In R. H. Turner (Ed.), *Annual review of sociology* (Vol. 8, pp. 209–35). Palo Alto, CA: Annual Reviews.

Cohen, E. G., Lockheed, M. E., & Lohman, M. R. (1976). The center for interracial cooperation: A field experiment. *Sociology of Education, 49,* 47–58.

Cohen, E. G., & Roper, S. (1972). Modification of interracial interaction disability: An application of status characteristic theory. *American Sociological Review, 37,* 643–657.

Cohen, L., & Kluegel, J. (1978). Determinants of juvenile court dispositions: Ascriptive and achieved factors in two metropolitan courts. *American Sociological Review, 43,* 162–176.

Cohen, R. (1982). In the end, this son is the father. *Washington Post Syndicate.*

Cohn, N. B., & Strassberg, D. S. (1983). Self-disclosure reciprocity among preadolescents. *Personality and Social Psychology Bulletin, 9,* 97–102.

Cohn, R. M. (1978). The effect of employment status change on self-attitudes. *Social Psychology, 41,* 81–93.

Coleman, J. W. (1987). Toward an integrated theory of white-collar crime. *American Journal of Sociology, 93,* 406–439.

Coleman, R. P., & Neugarten, B. (1971). *Social status in the city.* San Francisco: Jossey-Bass.

Collett, P. (1971). On training Englishmen in the nonverbal behavior of Arabs: An experiment in inter-cultural communication. *International Journal of Psychology, 6,* 209–215.

Comstock, G. (1984). Media influences on aggression. In A. Goldstein (Ed.), *Prevention and control of aggression: Principles, practices and research.* New York: Pergamon.

Comstock, G. S., Chaffee, S., Katzman, N., McCombs, M., & Roberts, D. (1978). *Television and human behavior.* New York: Columbia University Press.

Condon, W. S., & Ogston, W. D. (1967). A segmentation of behavior. *Journal of Psychiatric Research, 5,* 221–235.

Condry, J. (1977). Enemies of exploration: Self-initiated versus other initiated learning. *Journal of Personality and Social Psychology, 35,* 459–477.

Cook, K. (Ed.). (1987). *Social exchange theory.* Newbury Park, CA: Sage.

Cook, S. W. (1972). Motives in a conceptual analysis of attitude-related behavior. In J. Brigham & T. Weissbach (Eds.), *Racial attitudes in America: Analyses and findings of social psychology.* New York: Harper and Row.

Cook, S. W. (1984). The 1954 social science statement and school desegregation: A reply to Gerard. *American Psychologist, 39,* 819–832.

Cooke, B., Rossmann, M., McCubbin, H., & Patterson, J. (1988). Examining the definition and measurement of social support: A resource for individuals and families. *Family Relations, 37,* 211–216.

Cooley, C. H. (1902). *Human nature and the social order.* New York: Scribner.

Cooley, C. H. (1908). A study of the early use of self-words by a child. *Psychological Review, 15,* 339–357.

Cooper, H. M. (1979). Statistically combining independent studies: A meta-analysis of sex differences in conformity research. *Journal of Personality and Social Psychology, 37,* 131–146.

Cooper, J., & Fazio, R. H. (1986). The formation and persistence of attitudes that support intergroup conflict. In S. Worchel & W. G. Austin (Eds.), *Psychology of intergroup relations* (2nd ed.) (pp. 183–195). Chicago: Nelson-Hall Publishers.

Coopersmith, S. (1967). *The antecedents of self-esteem.* San Francisco: W. H. Freeman.

Coser, L. A. (1967). *Continuities in the study of social conflict.* New York: Free Press.

Coser, R. L. (1960). Laughter among colleagues. *Psychiatry, 23,* 81–95.

Costa, P. T., Jr., & McCrae, R. (1980). Still stable after all these years: Personality as a key to some issues in adulthood and old age. In P. Baltes & O. G. Brim, Jr. (Eds.), *Life-span development and behavior* (Vol. 3). New York: Academic Press.

Cota, A. A., & Dion, K. L. (1986). Salience of gender and sex composition of ad hoc groups: An experimental test of distinctiveness theory. *Journal of Personality and Social Psychology, 50,* 770–776.

Cotton, J. L., & Cook, M. S. (1982). Meta-analyses and the effects of various reward systems: Some different conclusions from Johnson et al. *Psychological Bulletin, 92,* 176–183.

Cozby, P. C. (1973). Self-disclosure: A literature review. *Psychological Bulletin, 79,* 73–91.

Crocker, J., Thompson, L. L., McGraw, K. M., & Ingerman, C. (1987). Downward comparison, prejudice, and evaluations of others: Effects of self-esteem and threat. *Journal of Personality and Social Psychology, 52,* 907–916.

Crosbie, P. V. (Ed.). (1975). *Interaction in small groups.* New York: Macmillan.

Cumming, E., & Henry, W. (1961). *Growing old: The process of disengagement.* New York: Basic Books.

Cunningham, J. D. (1981). Self-disclosure intimacy: Sex, sex-of-target, cross-national and "generational" differences. *Personality and Social Psychology Bulletin, 7,* 314–319.

Cunningham, M. R. (1986). Measuring the physical in physical attractiveness: Quasi-experiments on the sociobiology of female facial beauty. *Journal of Personality and Social Psychology, 50,* 925–935.

Cunningham, M., Steinberg, J., & Grev, R. (1980). Wanting to and having to help: Separate motivations for positive mood and guilt induced helping. *Journal of Personality and Social Psychology, 38,* 181–192.

Cutrona, C. E. (1982). Transition to college: Loneliness and the process of social adjustment. In L. A. Peplau & D. Perlman (Eds.), *Loneliness: A resource book of current theory, research and therapy.* New York: Wiley.

Dabbs, J. M., Jr., & Leventhal, H. (1966). Effects of varying the recommendations in a fear-arousing communication. *Journal of Personality and Social Psychology, 4,* 525–531.

Daher, D. M., & Banikiotes, P. G. (1976). Interpersonal attraction and rewarding aspects of disclosure content and level. *Journal of Personality and Social Psychology, 33,* 492–496.

Daly, K. (1987). Discrimination in the criminal courts: Family, gender and the problem of equal treatment. *Social Forces, 66,* 152–175.

Daniels, A. (1987). Invisible work. *Social Problems, 34,* 403–415.

Dank, B. (1971). Coming out in the gay world. *Psychiatry, 34,* 180–197.

Darley, J. M., & Batson, C. D. (1973). From Jerusalem to Jericho: A study of situational and dispositional variables in helping behavior. *Journal of Personality and Social Psychology, 27,* 100–108.

Darley, J. M., & Fazio, R. H. (1980). Expectancy confirmation processes arising in the social interaction sequence. *American Psychologist, 35,* 867–881.

Darley, J. M., & Latane, B. (1968). Bystander intervention in emergencies: Diffusion of responsibility. *Journal of Personality and Social Psychology, 8,* 377–383.

Darley, J. M., Teger, A. I., & Lewis, L. D. (1973). Do groups always inhibit individuals' response to potential emergencies? *Journal of Personality and Social Psychology, 26,* 395–399.

Davidson, A. R., & Jaccard, J. (1979). Variables that moderate the attitude-behavior relation: Results of a longitudinal survey. *Journal of Personality and Social Psychology, 37,* 1364–1376.

Davidson, A. R., Yantis, S., Norwood, M., & Montano, D. (1985). Amount of information about the attitude object and attitude-behavior consistency. *Journal of Personality and Social Psychology, 49,* 1184–1198.

Davidson, L. R., & Duberman, L. (1982). Friendship: Communication and interaction patterns in same sex dyads. *Sex Roles, 8,* 809–822.

Davies, J. C. (1962). Toward a theory of revolution. *American Sociological Review, 27,* 5–19.

Davies, J. C. (Ed.). (1971). *When men revolt—and why.* New York: Free Press.

Davis, D., & Perkowitz, W. T. (1979). Consequences of responsiveness in dyadic interaction: Effects of probability of response and proportion of content-related responses on interpersonal attraction. *Journal of Personality and Social Psychology, 37,* 534–550.

Davis, F. (1961). Deviance disavowal: The management of strained interaction by the visibly handicapped. *Social Problems, 9,* 120–132.

Davis, J. D. (1976). Self-disclosure in an acquaintance exercise: Responsibility for level of intimacy. *Journal of Personality and Social Psychology, 33,* 787–792.

Davis, K. (1947). Final note on a case of extreme isolation. *American Journal of Sociology, 52,* 432–437.

Davis, K. (1976). Sexual behavior. In R. Merton & R. Nisbet (Eds.), *Contemporary social problems* (4th ed.). New York: Harcourt Brace Jovanovich.

Davis, K. E. (1985, February). Near and dear: Friendship and love. *Psychology Today, 22,* 22–30.

Davis, M. H. (1983). Empathic concern and the muscular dystrophy telethon. *Personality and Social Psychology Bulletin, 9,* 223–229.

Davis, M. H., & Franzoi, S. L. (1986). Adolescent loneliness, self-disclosure, and private self-consciousness: A longitudinal investigation. *Journal of Personality and Social Psychology, 51,* 595–608.

Deaux, K., & Lewis, L. (1983). Components of gender stereotypes. *Psychological Documents, 13,* 25 (No. 2583).

Deci, E. (1975). *Intrinsic motivation.* New York: Plenum.

Deci, E., Nezlek, J., & Sherman, L. (1981). Characteristics of the rewarder and intrinsic motivation of the rewardee. *Journal of Personality and Social Psychology, 40,* 1–10.

de Jong-Gierveld, J. (1987). Developing and testing a model of loneliness. *Journal of Personality and Social Psychology, 53,* 119–128.

DeLamater, J., & MacCorquodale, P. (1979). *Premarital sexuality: Attitudes, relationships, behavior.* Madison, WI: University of Wisconsin Press.

DeLamater, J., & McKinney, K. (1982). Response-effects of question content. In W. Dijkstra & J. Van der Zouwen (Eds.), *Response behavior in the survey interview.* London: Academic Press.

DeMartini, J. R. (1983). Social movement participation: Political socialization, generational consciousness, and lasting effects. *Youth and Society, 15,* 195–223.

Dembroski, T. M., Lasater, T. M., & Ramires, A. (1978). Communicator similarity, fear-arousing communications, and compliance with health care recommendations. *Journal of Applied Social Psychology, 8,* 254–269.

Demerath, N. J., III, Marwell, G. & Aiken, M. T. (1971). *Dynamics of idealism: White activists in a black movement.* San Francisco: Jossey-Bass.

Demo, D. H. (1985). The measurement of self-esteem: Refining our methods. *Journal of Personality and Social Psychology, 48,* 1490–1502.

Dentler, R., & Erikson, K. (1959). The functions of deviance in groups. *Social Problems, 7,* 98–107.

Denzin, N. (1977). *Childhood socialization: Studies in the development of language, social behavior, and identity.* San Francisco: Jossey-Bass.

Denzin, N. (1983). *On understanding emotion.* San Francisco: Jossey-Bass.

DePaulo, B. M., & Fisher, J. D. (1980). The costs of asking for help. *Basic and Applied Social Psychology, 1,* 23–35.

DePaulo, B. M., Lanier, K., & Davis, T. (1983). Detecting the deceit of the motivated liar. *Journal of Personality and Social Psychology, 45,* 1096–1103.

DePaulo, B. M., Lassiter, G. D., & Stone, J. T. (1982). Attentional determinants of success at determining deception and truth. *Personality and Social Psychology Bulletin, 8,* 273–279.

DePaulo, B. M., Nadler, A., & Fisher, J. D. (1983). *New directions in helping: Help seeking* (Vol. 2). New York: Academic Press.

DePaulo, B. M., & Rosenthal, R. (1979). Ambivalence, discrepancy, and deception in nonverbal communication. In R. Rosenthal (Ed.), *Skill in nonverbal communication.* Cambridge, MA: Oelgeschlager, Gunn and Hain.

DePaulo, B. M., Rosenthal, R., Eisenstat, R. A., Rogers, P. L., & Finkelstein, S. (1978). Decoding discrepant nonverbal cues. *Journal of Personality and Social Psychology, 36,* 313–323.

DePaulo, B. M., Stone, J. I., & Lassiter, G. D. (1985). Deceiving and detecting deceit. In B. R. Schlenker (Ed.), *The self and social life.* New York: McGraw-Hill.

Der-Karabetian, A., & Smith, A. (1977). Sex-role stereotyping in the United States: Is it changing? *Sex Roles, 3,* 193–198.

Derlega, V. J., & Chaiken, A. L. (1975). *Sharing intimacy: What we reveal to others and why.* Englewood Cliffs, NJ: Prentice-Hall.

Derlega, V. J., Durham, B., Gockel, B., & Sholis, D. (1981). Appropriate self-disclosure. In G. J. Chelune (Ed.), *Self-disclosure: Origins, patterns, and implications of openness in interpersonal relationships.* San Francisco: Jossey-Bass.

Derlega, V. J., & Grzelak, J. (1979). Appropriateness of self-disclosure. In G. J. Chelune (Ed.), *Self-Disclosure.* San Francisco: Jossey-Bass.

Dervin, B. (1981). Mass communicating: Changing conceptions of the audience. In R. E. Rice & W. J. Paisley (Eds.), *Public communication campaigns.* Beverly Hills, CA: Sage.

Deschamps, J. C. (1972–1973). Attribution of responsibility for failure (or success) and social categorization. *Bulletin de Psychologie, 26,* 794–806.

de Tocqueville, A. (1955). *The old regime and the French Revolution* (Stuart Gilbert, Trans.). Garden City, NY: Doubleday. (Original work published 1856)

Deutsch, M. (1973). *The resolution of conflict: Constructive and destructive processes.* New Haven, CT: Yale University Press.

Deutsch, M. (1975). Equity, equality, and need: What determines which value will be used as the basis of distributive justice? *Journal of Social Issues, 31,* 137–149.

Deutsch, M., & Collins, M. E. (1951). *Interracial housing: A psychological evaluation of a social experiment.* Minneapolis: University of Minnesota Press.

Deutsch, M., & Gerard, H. B. (1955). A study of normative and informational social influences upon individual judgment. *Journal of Abnormal and Social Psychology, 51,* 629–636.

Deutsch, M., & Krauss, R. M. (1960). The effect of threats upon interpersonal bargaining. *Journal of Abnormal and Social Psychology, 61,* 181–189.

Deutsch, M., & Krauss, R. M. (1962). Studies of interpersonal bargaining. *Journal of Conflict Resolution, 6,* 52–76.

Dickoff, H. (1961). *Reactions to evaluations by others as a function of self-evaluation and the interaction context.* Unpublished doctoral dissertation, Raleigh, NC: Duke University.

Diener, E. (1980). Deindividuation: The absence of self-awareness and self-regulation in group members. In P.

B. Paulus (Ed.), *The psychology of group influence.* Hillsdale, NJ: Erlbaum.

Diener, E., & Wallbom, M. (1976). Effects of self-awareness on antinormative behavior. *Journal of Research in Personality, 10,* 107–111.

DiMaggio, P., & Mohr, J. (1985). Cultural capital, educational attainment, and marital selection. *American Journal of Sociology, 90,* 1231–1257.

Dion, K. L. (1979). Intergroup conflict and intergroup cohesiveness. In W. G. Austin & S. Worchel (Eds.), *The social psychology of intergroup relations.* Monterey, CA: Brooks/Cole.

Dion, K., Baron, R., & Miller, N. (1970). Why do groups make riskier decisions than individuals? In L. Berkowitz (Ed.), *Advances in experimental social psychology* (Vol. 5). New York: Academic Press.

Dion, K., Berscheid, E., & Walster (Hatfield), E. (1972). What is beautiful is good. *Journal of Personality and Social Psychology, 24,* 285–290.

Dipboye, R. L. (1977). Alternative approaches to deindividuation. *Psychological Bulletin, 84,* 1057–1075.

Doering, C. H. (1980). The endocrine system. In O. Brim, Jr. & J. Kogan (Eds.), *Constancy and change in human development.* Cambridge, MA: Harvard University Press.

Dohrenwend, B. P. (1961). The social psychological nature of stress: A framework for causal inquiry. *Journal of Abnormal and Social Psychology, 62,* 294–302.

Dohrenwend, B. P., & Dohrenwend, B. S. (1974). Social and cultural influences on psychopathology. *Annual Review of Psychology, 25,* 417–452.

Dohrenwend, B. P., & Dohrenwend, B. S. (1976). Sex differences and psychiatric disorders. *American Journal of Sociology, 81,* 1447–1454.

Dollard, J., Doob, J., Miller, N., Mowrer, O., & Sears, R. (1939). *Frustration and aggression.* New Haven, CT: Yale University Press.

Dollinger, S., & Thelen, M. (1978). Overjustification and children's intrinsic motivation: Comparative effects of four rewards. *Journal of Personality and Social Psychology, 36,* 1259–1269.

Dornbusch, S. M., & Scott, W. R. (1975). *Evaluation and the exercise of authority.* San Francisco: Jossey-Bass.

Dovidio, J. F., & Ellyson, S. L. (1982). Decoding visual dominance: Attributions of power based on relative percentages of looking while speaking and looking while listening. *Social Psychology Quarterly, 45,* 106–113.

Downey, G. L. (1986). Ideology and the Clamshell identity: Organizational dilemmas in the anti-nuclear power movement. *Social Problems, 33,* 357–373.

Drabeck, T. E., & Stephenson, J. S. (1978). When disaster strikes. *Journal of Applied Social Psychology, 1,* 187–203.

Drake, G. F. (1980). The social role of slang. In H. Giles, W. P. Robinson, & P. Smith (Eds.), *Language: Social psychological perspectives.* New York: Pergamon.

Drass, K. A. (1986). The effect of gender identity on conversation. *Social Psychology Quarterly, 49,* 294–301.

Drass, K., & Spencer, J. W. (1987). Accounting for presentencing recommendations: Typologies and probation officers' theory of office. *Social Problems, 34,* 277–293.

Dreben, E. K., Fiske, S. T., & Hastie, R. (1979). The

independence of evaluative and item information: Impression and recall order effects in behavior-based impression formation. *Journal of Personality and Social Psychology, 37,* 1758–1768.

Driver, E. D. (1969). Self-conceptions in India and the United States: A cross-cultural validation of the twenty statements test. *The Sociological Quarterly, 10,* 341–354.

Druckman, D. (1977). *Negotiations: A social psychological perspective.* Beverly Hills, CA: Sage-Halsted.

Duesterhoeft, D. (1987). *An unholy alliance? A case study comparison of religious and feminist anti-pornography activists.* Unpublished master's thesis, University of Wisconsin.

Duncan, S., Jr., & Fiske, D. W. (1977). *Face-to-face interaction: Research methods and theory.* Hillsdale, NJ: Erlbaum.

Dunphy, D. (1972). *The primary group.* New York: Appleton-Century-Crofts.

Dutton, D., & Aron, A. (1974). Some evidence for heightened sexual attraction under conditions of high anxiety. *Journal of Personality and Social Psychology, 30,* 510–517.

Dutton, D., & Lake, R. (1973). Threat of own prejudice and reverse discrimination in interracial situations. *Journal of Personality and Social Psychology, 28,* 94–100.

Dyck, R. J., & Rule, B. G. (1978). Effect on retaliation of causal attribution concerning attack. *Journal of Personality and Social Psychology, 36,* 521–529.

Eagly, A. (1987). *Sex differences in social behavior: A social-role interpretation.* Hillsdale, NJ: Erlbaum.

Eagly, A. H., & Carli, L. L. (1981). Sex of researchers and sex-typed communications as determinants of sex differences in influenceability. *Psychological Bulletin, 90,* 1–20.

Eagly, A. H., & Chaiken, S. (1975). An attribution analysis of the effect of communicator characteristics on opinion change: The case of communicator attractiveness. *Journal of Personality and Social Psychology, 32,* 136–144.

Eagly, A., & Steffen, V. J. (1984). Gender stereotypes stem from distribution of women and men into social roles. *Journal of Personality and Social Psychology, 46,* 735–754.

Eagly, A. H., & Warren, R. (1976). Intelligence, comprehension, and opinion change. *Journal of Personality, 44,* 226–242.

Eagly, A. H., Wood, W., & Chaiken, S. (1978). Causal inferences about communicators and their effect on attitude change. *Journal of Personality and Social Psychology, 36,* 424–435.

Eaton, W. (1980). *The sociology of mental disorders.* New York: Praeger Publishers.

Ebbesen, E. B., & Bowers, R. J. (1974). Proportion of risky to conservative arguments in a group discussion and choice shift. *Journal of Personality and Social Psychology, 29,* 316–327.

Edelmann, R. J. (1985). Social embarrassment: An analysis of the process. *Journal of Social and Personal Relationships, 2,* 195–213.

Edwards, A. D. (1976). *Language in culture and class: The sociology of language and education.* London: Heinemann Educational.

Efran, M. G., & Cheyne, J. A. (1974). Affective concomitants of the invasion of shared space: Behavioral, physiological and verbal indicators. *Journal of Personality and Social Psychology, 29,* 219–226.

Ehrlich, H., & Graeven, D. B. (1971). Reciprocal self-disclosure in a dyad. *Journal of Experimental Social Psychology, 7,* 389–400.

Eibl-Eibesfeldt, I. (1979). Universals in human expressive behavior. In A. Wolfgang (Ed.), *Nonverbal behavior: Applications and cultural implications.* New York: Academic Press.

Eiser, J. R. (Ed.). (1984). *Attitudinal judgment.* New York: Springer.

Ekman, P. (1972). Universals and cultural differences in facial expression of emotion. In J. K. Cole (Ed.), *Nebraska symposium on motivation, 1971.* Lincoln: Nebraska University Press.

Ekman, P., & Friesen, W. V. (1969). Nonverbal leakage and clues to deception. *Psychiatry, 32,* 88–106.

Ekman, P., & Friesen, W. V. (1975). *Unmasking the face.* Englewood Cliffs, NJ: Prentice-Hall.

Ekman, P., Friesen, W. V., & Tomkins, S. S. (1971). Facial affect scoring technique (FAST): A first validity study. *Semiotica, 3,* 37–58.

Ekman, P., Friesen, W., et al. (1987). Universals and cultural differences in the judgments of facial expressions of emotion. *Journal of Personality and Social Psychology, 53,* 712–717.

Elder, G. H., Jr. (1975). Age differentiation and the life course. In A. Inkeles, J. Coleman, & N. Smelser (Eds.), *Annual review of sociology* (Vol. 1). Palo Alto, CA: Annual Reviews.

Elder, G., & Bowerman, C. (1963). Family structure and childbearing patterns: The effect of family size and sex composition. *American Sociological Review, 28,* 891–905.

Elkin, F., & Handel, G. (1978). *The child and society* (3rd ed.). New York: Random House.

Elkin, R. A., & Leippe, M. (1986). Physiological arousal, dissonance, and attitude change: Evidence for a dissonance-arousal link and a "Don't remind me" effect. *Journal of Personality and Social Psychology, 51,* 55–65.

Elkind, D. (1961). The child's conception of his religious denomination: I. The Jewish child. *Journal of Genetic Psychology, 99,* 209–225.

Elkind, D. (1962). The child's conception of his religious denomination: II. The Catholic child. *Journal of Genetic Psychology, 101,* 183–193.

Elkind, D. (1963). The child's conception of his religious denomination: III. The Protestant child. *Journal of Genetic Psychology, 103,* 291–304.

Elliott, D. S., & Ageton, S. S. (1980). Reconciling race and class differences in self-reported and official estimates of delinquency. *American Sociological Review, 45,* 95–110.

Elliott, G. C., & Meeker, B. F. (1986). Achieving fairness in the face of competing concerns: The different effects of individual and group characteristics. *Journal of Personality and Social Psychology, 50,* 754–760.

Ellsworth, P. C., Carlsmith, J. M., & Henson, A. (1972). The stare as a stimulus to flight in human subjects. *Journal of Personality and Social Psychology, 21,* 302–311.

Emerson, R. M. (1966). Mount Everest: A case study of

communication feedback and sustained group goal-striving. *Sociometry, 29,* 213–227.

Emerson, R. M. (1981). Social exchange theory. In M. Rosenburg & R. H. Turner (Eds.), *Social psychology: Sociological perspectives.* New York: Basic Books.

Emerson, R. M. & Messinger, S. (1977). The micro-politics of trouble. *Social Problems, 25,* 121–134.

Emmons, R. A., & Diener, E. (1986). Situation selection as a moderator of response consistency and stability. *Journal of Personality and Social Psychology, 51,* 1013–1019.

Emmons, R. A., Diener, E., & Larsen, R. J. (1986). Choice and avoidance of everyday situations and affect congruence: Two models of reciprocal interactionism. *Journal of Personality and Social Psychology, 51,* 815–826.

Emswiller, T., Deaux, K., & Willis, J. E. (1971). Similarity, sex, and requests for small favors. *Journal of Applied Social Psychology, 1,* 284–291.

Ennis, J. G., & Schreuer, R. (1987). Mobilizing weak support for social movements: The role of grievance, efficiency and cost. *Social Forces, 66,* 390–409.

Epstein, J. L. (1985). After the bus arrives: Resegregation in desegregated schools. *Journal of Social Issues, 41,* 23–44.

Epstein, S. (1976). The political and economic basis of cancer. *Technology Review, 78,* 34–43.

Erikson, E. H. (1968). *Identity: Youth and crisis.* New York: Norton.

Erikson, K. (1964). Notes on the sociology of deviance. In H. Becker (Ed.), *The other side.* New York: Free Press.

Erikson, K. (1966). *The wayward Puritans.* New York: Wiley.

Eron, L. D. (1980). Prescription for the reduction of aggression. *American Psychologist, 35,* 244–252.

Eron, L. D. (1982). Parent-child interaction, television violence, and aggression of children. *American Psychologist, 37,* 197–211.

Estes, R. L. & Wilensky, H. (1978). Life cycle squeeze and the morale curve. *Social Forces, 56,* 277–292.

Ettinger, R. F., Marino, C. J., Endler, N. S., Geller, S. H., & Natziuk, T. (1971). Effects of agreement and correctness on relative competence and conformity. *Journal of Personality and Social Psychology, 19,* 204–212.

Etzioni, A. (1967). The Kennedy experiment. *The Western Political Quarterly, 20,* 361–380.

Evans, N. J., & Jarvis, P. A. (1980). Group cohesion, a review and reevaluation. *Small Group Behavior, 11,* 359–370.

Faley, T., & Tedeschi, J. T. (1971). Status and reactions to threats. *Journal of Personality and Social Psychology, 17,* 192–199.

Farhar-Pilgrim, B., & Shoemaker, F. F. (1981). Campaigns to affect energy behavior. In R. E. Rice & W. J. Paisley (Eds.), *Public communication campaigns.* Beverly Hills, CA: Sage.

Farina, A., Gliha, D., Boudreau, L. A., Allen, J. G., & Sherman, M. (1971). Mental illness and the impact of believing others know it. *Journal of Abnormal Psychology, 77,* 1–5.

Fazio, R., Powell, M., & Herr, P. (1983). Toward a process model of the attitude-behavior relation: Accessing one's attitude upon mere observation of the attitude object. *Journal of Personality and Social Psychology, 44,* 723–735.

Fazio, R. H., Sanbonmatsu, D. M., Powell, M. C., & Kardes, F. R. (1986). On the automatic activation of attitudes. *Journal of Personality and Social Psychology, 50,* 229–238.

Fazio, R. H., & Williams, C. J. (1986). Attitude accessibility as a moderator of the attitude-perception and attitude-behavior relations: An investigation of the 1984 presidential election. *Journal of Personality and Social Psychology, 51,* 505–514.

Fazio, R. H., & Zanna, M. (1981). Direct experience and attitude-behavior consistency. In L. Berkowitz (Ed.), *Advances in experimental social psychology, 14*: 161–202. New York: Academic Press.

Feather, N. T. (1967). A structural balance approach to the analysis of communication effects. In L. Berkowitz (Ed.), *Advances in experimental social psychology* (Vol. 3). New York: Academic Press.

Feather, N. T., & Armstrong, D. J. (1967). Effects of variations in source attitude, receiver attitude and communication stand on reactions to source and contents of communications. *Journal of Personality, 35,* 435–455.

Feld, S. (1981). The focused organization of social ties. *American Journal of Sociology, 86,* 1015–1035.

Felipe, N. J., & Sommer, R. (1966). Invasions of personal space. *Social Problems, 14,* 206–214.

Felson, R. B. (1981). Ambiguity and bias in the self-concept. *Social Psychology Quarterly, 44,* 64–69.

Felson, R. B. (1985). Reflected appraisal and the development of self. *Social Psychology Quarterly, 48,* 71–78.

Felson, R. B., & Reed, M. (1986). The effects of parents on the self-appraisals of children. *Social Psychology Quarterly, 49,* 302–308.

Ferguson, C. K., & Kelley, H. H. (1964). Significant factors in overevaluation of own group's product. *Journal of Abnormal and Social Psychology, 69,* 223–228.

Ferree, M. M. (1976). Working-class jobs: Housework and paid work as sources of satisfaction. *Social Problems, 23,* 431–441.

Ferree, M. M., & Miller, F. D. (1985). Mobilization and meaning: Toward an integration of social psychological and resource perspectives on social movements. *Sociological Inquiry, 55,* 38–55.

Festinger, L. (1954). A theory of social comparison processes. *Human Relations, 7,* 117–140.

Festinger, L. A. (1957). *A theory of cognitive dissonance.* Stanford, CA: Stanford University Press.

Festinger, L., & Carlsmith, J. (1959). Cognitive consequences of forced compliance. *Journal of Abnormal and Social Psychology, 58,* 203–210.

Festinger, L., Pepitone, A., & Newcomb, T. (1952). Some consequences of de-individuation in a group. *Journal of Abnormal and Social Psychology, 47,* 382–389.

Festinger, L., Riecken, H., & Schachter, S. (1956). *When prophecy fails.* Minneapolis: University of Minnesota Press.

Festinger, L., Schachter, S., & Back, K. W. (1950). *Social pressures in informal groups.* New York: Harper and Row.

Festinger, L. (1954). A theory of social comparison processes. *Human Relations, 7,* 117–140.

Fiedler, F. E. (1966). The effect of leadership and cultural heterogeneity on group performance: A test of the contingency model. *Journal of Experimental Social Psychology, 2,* 237–264.

Fiedler, F. E. (1978a). Recent developments in research on the contingency model. In L. Berkowitz (Ed.), *Group processes*. New York: Academic Press.

Fiedler, F. E. (1978b). The contingency model and the dynamics of the leadership process. In L. Berkowitz (Ed.), *Advances in experimental social psychology* (Vol. 11). New York: Academic Press.

Fiedler, F. E. (1981). Leadership effectiveness. *American Behavioral Scientist, 24,* 619–632.

Finestone, H. (1964). Cats, kicks and color. In H. Becker (Ed.), *The other side*. New York: Free Press.

Fink, E. L., Kaplowitz, S. A., & Bauer, C. L. (1983). Positional discrepancy, psychological discrepancy, and attitude change: Experimental tests of some mathematical models. *Communication Monographs, 50,* 413–430.

Fischer, C. S. (1980). *Friendship, gender and the life-cycle*. Unpublished paper, Institute of Urban and Regional Development, University of California, Berkeley.

Fischer, C. S. (1984). *The urban experience* (2nd ed.). San Diego: Harcourt Brace Jovanovich.

Fisek, M. H. (1974). A model for the evolution of status structures in task-oriented discussion groups. In J. Berger, T. L. Conner, & M. H. Fisek (Eds.), *Expectation states theory*. Cambridge, MA: Winthrop.

Fisek, M. H., & Ofshe, R. (1970). The process of status evolution. *Sociometry, 33,* 327–346.

Fishbein, M. (1980). A theory of reasoned action: Some applications and implications. In H. Howe & M. Page (Eds.), *Nebraska Symposium on Motivation* (Vol. 27). Lincoln: University of Nebraska Press.

Fishbein, M., & Ajzen, I. (1975). *Belief, attitude, intention and behavior*. Reading, MA: Addison-Wesley.

Fisher, J., Nadler, D., & Whitcher-Alagna, S. (1982). Recipient reactions to aid. *Psychological Bulletin, 91,* 33–54.

Fisher, R., & Ury, W. (1981). *Getting to yes: Negotiating agreement without giving in*. Boston: Houghton Mifflin.

Fishman, P. M. (1978). Interaction: The work women do. *Social Problems, 25,* 397–406.

Fishman, P. M. (1980). Conversational insecurity. In H. Giles & W. P. Robinson (Eds.), *Language: Social psychological perspectives*. New York: Pergamon.

Fiske, S. T., & Taylor, S. E. (1984). *Social cognition*. Reading, MA: Addison-Wesley.

Fitz, D. (1976). A renewed look at Miller's conflict theory of aggression displacement. *Journal of Personality and Social Psychology, 33,* 725–732.

Flavell, J., Shipstead, S., & Croft, K. (1978). *What young children think you see when their eyes are closed*. Unpublished report, Stanford University.

Flowers, M. L. (1977). A laboratory test of some implications of Janis' groupthink hypothesis. *Journal of Personality and Social Psychology, 35,* 888–896.

Ford, M. R., & Lowery, C. R. (1986). Gender differences in moral reasoning: A comparison of justice and care orientations. *Journal of Personality and Social Psychology, 50,* 777–783.

Form, W. H., & Nosow, S. (1958). *Community in disaster*. New York: Harper and Row.

Forsyth, D. R., Berger, R. E., & Mitchell, T. (1981). The effects of self-serving vs. other-serving claims of respon-

sibility on attraction and attribution in groups. *Social Psychology Quarterly, 44,* 59–64.

Frank, F., & Anderson, L. R. (1971). Effects of task and group size upon group productivity and member satisfaction. *Sociometry, 34,* 135–149.

Franks, D., & Marolla, J. (1976). Efficacious action and social approval as interacting dimensions of self-esteem. *Sociometry, 39,* 324–341.

Fraser, C., Belugi, U., & Brown, R. (1963). Control of grammar in imitation, comprehension and production. *Journal of Verbal Learning and Verbal Behavior, 2,* 121–135.

Fraser, C., Gouge, C., & Billig, M. (1971). Risky shifts, cautious shifts, and group polarization. *European Journal of Social Psychology, 1,* 7–30.

Fredericks, A., & Dossett, D. (1983). Attitude-behavior relations: A comparison of the Fishbein-Ajzen and the Bentler-Speckart models. *Journal of Personality and Social Psychology, 45,* 501–512.

Freedman, D. G. (1979). *Human sociobiology*. New York: Free Press.

Freese, L. (1976). The generalization of specific performance expectations. *Sociometry, 39,* 194–200.

Freese, L., & Cohen, B. P. (1973). Eliminating status generalization. *Sociometry, 36,* 177–193.

French, J. R. P., Jr., Morrison, H. W., & Levinger, G. (1960). Coercive power and forces affecting conformity. *Journal of Abnormal and Social Psychology, 61,* 93–101.

Freud, S. (1905). Fragment of an analysis of a case of hysteria. *Collected Papers* (Vol. 3). New York: Basic Books.

Freud, S. (1930). *Civilization and its discontents*. London: Hogarth Press.

Freud, S. (1950). Why war? In J. Strachey (Ed.), *Collected papers* (Vol. 5). London: Hogarth Press.

Frey, D. (1978). Reactions to success and failure in public and private conditions. *Journal of Experimental Social Psychology, 14,* 172–179.

Frey, D. (1982). Different levels of cognitive dissonance, information seeking and information avoidance. *Journal of Personality and Social Psychology, 43,* 1175–1183.

Frey, K. S., & Ruble, D. N. (1985). What children say when the teacher is not around: Conflicting goals in social comparison and performance assessment in the classroom. *Journal of Personality and Social Psychology, 48,* 550–562.

Frieze, I. H., Parsons, J., Johnson, P., Ruble, D., & Zellman, G. (1978). *Women and sex roles: A social psychological perspective*. New York: Norton.

Frieze, I., & Weiner, B. (1971). Cue utilization and attributional judgments for success and failure. *Journal of Personality, 39,* 591–605.

Frodi, A. (1975). The effect of exposure to weapons on aggressive behavior from a cross-cultural perspective. *International Journal of Psychology, 10,* 283–292.

Gaertner, S., & Bickman, L. (1971). A nonreactive indicator of racial discrimination: The wrong number technique. *Journal of Personality and Social Psychology, 20,* 218–222.

Gahagan, J. P., & Tedeschi, J. T. (1968). Strategy and the

credibility of promises in the Prisoner's Dilemma game. *Journal of Conflict Resolution, 12,* 224–234.

Gardner, R. A., & Gardner, B. T. (1980). Comparative psychology and language acquisition. In K. Salzinger & F. Denmark (Eds.), *Psychology: The state of the art.* Annals of the New York Academy of Science.

Garfinkel, H. (1956). Conditions of successful degradation ceremonies. *American Sociological Review, 61,* 420–424.

Garfinkel, I., & McLanahan, S. S. (1986). *Single mothers and their children: A new American dilemma.* Washington, DC: The Urban Institute.

Gecas, V. (1972). Parental behavior and dimensions of adolescent self-evaluation. *Sociometry, 34,* 466–482.

Gecas, V. (1979). The influence of social class on socialization. In W. Burr, R. Hill, F. Nye & I. Reiss (Eds.), *Contemporary theories about the family* (Vol. 1). New York: Free Press.

Gecas, V. (1981). Contexts of socialization. In M. Rosenburg & R. Turner (Eds.), *Social psychology: Sociological perspectives.* New York: Basic Books.

Gecas, V., & Nye, F. (1974). Sex and class differences in parent-child interaction: A test of Kohn's hypothesis. *Journal of Marriage and the Family, 36,* 742–749.

Gecas, V., & Schwalbe, M. (1983). Beyond the looking-glass self: Social structure and efficacy-based self-esteem. *Social Psychology Quarterly, 46,* 77–88.

Geen, R. (1968). Effects of frustration, attack, and prior training in aggressiveness upon aggressive behavior. *Journal of Personality and Social Psychology, 9,* 316–321.

Geen, R. G. (1978). Some effects of observing violence upon the behavior of the observer. In B. A. Maher (Ed.), *Progress in experimental personality research* (Vol. 8). New York: Academic Press.

Geen, R. G., & Quanty, M. G. (1977). The catharsis of aggression: An analysis of a hypothesis. In L. Berkowitz (Ed.), *Advances in experimental social psychology* (Vol. 10). New York: Academic Press.

Geen, R. G., Stonner, L., & Shope, G. L. (1975). The facilitation of aggression by aggression: A study in response inhibition and disinhibition. *Journal of Personality and Social Psychology, 31,* 721–726.

Gelfand, D. M., & Hartmann, D. P. (1982). Response consequences and attributions: Two contributors to prosocial behavior. In N. Eisenberg (Ed.), *The development of prosocial behavior.* New York: Academic Press.

Geller, V. (1977). *The role of visual access in impression management and impression formation.* Unpublished doctoral dissertation, Columbia University, NY.

Gelles, R. (1980). A profile of violence toward children in the United States. In G. Gerbner, C. Ross, & E. Zigler (Eds.), *Child abuse.* New York: Oxford University Press.

Gerard, H. B. (1983). School desegregation: The social science role. *American Psychologist, 38,* 869–877.

Gerard, H. B. Wilhelmy, R. A., & Conolley, E. S. (1968). Conformity and group size. *Journal of Personality and Social Psychology, 8,* 79–82.

Gergen, K. (1971). *The self-concept.* New York: Holt, Rinehart and Winston.

Gergen, K. J., Ellsworth, P., Maslach, C., & Siepel, M. (1975).

Obligation, donor resources and reactions to aid in three cultures. *Journal of Personality and Social Psychology, 31,* 390–400.

Gergen, K. J., Gergen, M. M., & Meter, K. (1972). Individual orientation to prosocial behavior. *Journal of Social Issues, 28,* 105–130.

Gergen, K. J., Morse, S. J., & Bode, K. A. (1974). Overpaid or overworked? Cognitive and behavioral reactions to inequitable rewards. *Journal of Applied Social Psychology, 4,* 259–274.

Gerstal, N., Riessman, C., & Rosenfield, S. (1985). Explaining the symptomatology of separated and divorced women and men: The role of material conditions and social networks. *Social Forces, 64,* 84–101.

Gesell, A., & Ilg, F. (1943). *Infant and child in the culture of today.* New York: Harper and Row.

Gibb, C. A. (1969). Leadership. In G. Lindzey & E. Aronson (Eds.), *Handbook of social psychology* (Vol. 4) (2nd ed.). Reading, MA: Addison-Wesley.

Gibbons, F. X., Smith, T. W., Ingram, R. E., Pearce, K., Brehm, S. S., & Schroeder, D. J. (1985). Self-awareness and self-confrontation: Effects of self-focussed attention on members of a clinical population. *Journal of Personality and Social Psychology, 48,* 662–675.

Gifford, R. (1982). Projected interpersonal distance and orientation choices: Personality, sex, and social situation. *Social Psychology Quarterly, 45,* 145–152.

Gilbert, G. M. (1951). Stereotype persistence and change among college students. *Journal of Abnormal and Social Psychology, 46,* 245–254.

Gilbert, T. F. (1978). *Human competence.* New York: McGraw-Hill.

Gilchrist, J. C., Shaw, M. E., & Walker, L. C. (1954). Some effects of unequal distribution of information in a wheel group structure. *Journal of Abnormal and Social Psychology, 49,* 554–556.

Giles, H. (1980). Accommodation theory: Some new directions. In S. deSilva (Ed.), *Aspects of linguistic behavior.* York, England: York University Press.

Giles, H., Hewstone, M., & St. Clair, R. (1981). Speech as an independent and dependent variable of social situations: An introduction and new theoretical framework. In H. Giles & R. St. Clair (Eds.), *The social psychological significance of speech.* Hillsdale, NJ: Erlbaum.

Gilligan, C. (1982). *In a different voice.* Cambridge, MA: Harvard University Press.

Glaser, B. G., & Strauss, A. (1971). *Status passage: A formal theory.* Chicago: Aldine.

Glass, J., Bengston, V. L., & Dunham, C. (1986). Attitude similarity in three-generation families: Status inheritance or reciprocal influence? *American Sociological Review, 51,* 685–698.

Goethals, G. R., & Zanna, M. P. (1979). The role of social comparison in choice shift. *Journal of Personality and Social Psychology, 37,* 1469–1476.

Goetting, A. (1986). The developmental tasks of siblingship over the life cycle. *Journal of Marriage and the Family, 48,* 703–714.

Goffman, E. (1952). Cooling the mark out: Some adaptations to failure. *Psychiatry, 15,* 451–463.

Goffman, E. (1959a). The moral career of the mental patient. *Psychiatry, 22,* 125–169.

Goffman, E. (1959b). *The presentation of self in everyday life.* Garden City, NJ: Doubleday/Anchor.

Goffman, E. (1963a). *Behavior in public places.* New York: Free Press.

Goffman, E. (1963b). *Stigma: Notes on the management of spoiled identity.* Englewood Cliffs, NJ: Spectrum/Prentice-Hall.

Goffman, E. (1967). *Interaction ritual.* Chicago: Aldine.

Goffman, E. (1974). *Frame analysis.* New York: Harper and Row.

Goffman, E. (1983). Felicity's condition. *American Journal of Sociology, 89,* 1–53.

Gonzales, M., Davis, J., Loney, G., Lukens, C., & Junghans, C. (1983). Interactional approach to interpersonal attraction. *Journal of Personality and Social Psychology, 44,* 1192–1197.

Goldberg, C. (1974). Sex roles, task competence, and conformity. *Journal of Psychology, 86,* 157–164.

Goldberg, C. (1975). Conformity to majority type as a function of task and acceptance of sex-related stereotypes. *Journal of Psychology, 89,* 25–37.

Gonos, G. (1977). "Situation" vs. "frame": The "interactionist" and the "structuralist" analysis of everyday life. *American Sociological Review, 42,* 854–867.

Goode, E. (1978). *Deviant behavior: An interactionist approach.* Englewood Cliffs, NJ: Prentice-Hall.

Goode, W. J. (1960). A theory of role strain. *American Sociological Review, 25,* 483–496.

Goodman, P. S., & Friedman, A. (1969). An examination of quantity and quality of performance under conditions of overpayment in piece rate. *Organizational Behavior and Human Performance, 4,* 365–374.

Goodman, P. S., & Friedman, A. (1971). An examination of Adams' theory of inequity. *Administrative Science Quarterly, 16,* 271–288.

Goodwin, C. (1987). Forgetfulness as an interactive resource. *Social Psychology Quarterly, 50,* 115–131.

Gordon, C. (1968). Self-conceptions: Configurations of content. In C. Gordon & K. J. Gergen (Eds.), *The self in social interaction, I: Classic and contemporary perspectives.* New York: Wiley.

Gordon, C. (1976). Development of evaluated role-identities. In A. Inkeles, J. Coleman, & N. Smelser (Eds.), *Annual review of sociology* (Vol. 2, pp. 405–433). Palo Alto, CA: Annual Reviews.

Gordon, S. L. (1981). The sociology of sentiments and emotion. In M. Rosenberg & R. H. Turner (Eds.), *Social psychology: Sociological perspectives.* New York: Basic Books.

Gorfein, D. S. (1964). The effects of a nonunanimous majority on attitude change. *Journal of Social Psychology, 63,* 333–338.

Gould, R. L. (1978). *Transformations.* New York: Simon and Schuster.

Gouldner, A. (1960). The norm of reciprocity: A preliminary statement. *American Sociological Review, 25,* 161–178.

Gouran, D. S., & Andrews, P. H. (1984). Determinants of punitive responses to socially proscribed behavior. *Small Group Behavior, 15,* 525–543.

Gove, W., & Geerken, M. (1977). The effect of children and employment on the mental health of married men and women. *Social Forces, 56,* 66–76.

Gove, W., & Hughes, M. (1979). Possible causes of the apparent sex differences in physical health: An empirical investigation. *American Sociological Review, 44,* 126–146.

Gove, W., Hughes, M., & Geerken, M. R. (1980). Playing dumb: A form of impression management with undesirable effects. *Social Psychology Quarterly, 43,* 89–102.

Granberg, D. (1978). GRIT in the final quarter: Reversing the arms race through unilateral initiatives. *Bulletin of Peace Proposals, 9,* 210–221.

Granovetter, M. S. (1973). The strength of weak ties. *American Journal of Sociology, 78,* 1360–1380.

Grant, P. R., & Holmes, J. G. (1981). The integration of implicit personality schemas and stereotype images. *Social Psychology Quarterly, 44,* 107–115.

Grasmick, H. G., & Bryjak, G. J. (1980). The deterrent effect of perceived severity of punishment. *Social Forces, 59,* 471–491.

Grayshon, M. C. (1980). Social grammar, social psychology, and linguistics. In H. Giles, W. P. Robinson, & P. M. Smith (Eds.), *Language: Social psychological perspectives.* New York: Pergamon.

Green, J. A. (1972). Attitudinal and situational determinants of intended behavior toward blacks. *Journal of Personality and Social Psychology, 22,* 13–17.

Greenbaum, P., & Rosenfeld, H. (1978). Patterns of avoidance in response to interpersonal staring and proximity: Effects of bystanders on drivers at a traffic intersection. *Journal of Personality and Social Psychology, 36,* 575–587.

Greenberg, J. H. (1966). *Language universals.* The Hague: Mouton.

Greenberg, J., & Cohen, R. L. (1982). *Equity and justice in social behavior.* New York: Academic Press.

Greenberg, M. (1980). A theory of indebtedness. In K. J. Gergen, M. S. Greenberg, & R. H. Willis (Eds.), *Social exchange: Advances in theory and research.* New York: Plenum.

Greenberg, M., & Frisch, D. (1972). Effect of intentionality on willingness to reciprocate a favor. *Journal of Experimental Social Psychology, 8,* 99–111.

Greenberger, E., & Steinberg, L. (1983). Sex differences in early labor force participation. *Social Forces, 62,* 467–486.

Greenberger, E., & Steinberg, L. (1986). *When teenagers work: The psychological and social costs of adolescent employment.* New York: Basic Books.

Greenfield, L. (1972). Spanish and English usage self-ratings in various situational contexts. In J. A. Fishman (Ed.), *Advances in the sociology of language II.* The Hague: Mouton.

Greenley, J. (1979). Familial expectations, post-hospital adjustment and the societal reaction perspective on mental illness. *Journal of Health and Social Behavior, 20,* 217–227.

Greenwald, A. G., & Pratkanis, A. (1984). The self. In R. S. Wyer & T. K. Srull (Eds.), *Handbook of social cognition.* Hillsdale, NJ: Erlbaum.

Grice, P. H. (1975). Logic and conversation. In P. Cole, & J.

L. Morgan (Eds.), *Syntax and semantics* (Vol. 3): *Speech acts*. New York: Academic Press.

Griffin, L., Wallace, M., & Rubin, B. (1986). Capitalist resistance to the organization of labor before the New Deal: Why? How? Success? *American Sociological Review, 51,* 147–167.

Grimshaw, A. D. (1973). On language in society (Part 1). *Contemporary Sociology, 2,* 575–585.

Grimshaw, A. D. (1981). Talk and social control. In M. Rosenberg & R. H. Turner (Eds.), *Social psychology: Sociological perspectives.* New York: Basic Books.

Grimshaw, A. D. (1987). Disambiguating discourse: Members' skill and analysts' problem. *Social Psychology Quarterly, 50,* 186–204.

Gross, E., & Stone, G. P. (1970). Embarrassment and the analysis of role requirements. In G. P. Stone & H. A. Farberman (Eds.), *Social psychology through symbolic interaction.* Waltham, MA: Ginn-Blaisdell.

Guetzkow, H., & Dill, W. R. (1957). Factors in the organizational development of task-oriented groups. *Sociometry, 20,* 175–204.

Guimond, S., & Dube-Simard, L. (1983). Relative deprivation theory and the Quebec nationalist movement: The cognition-emotion distinction and the personal-group deprivation issue. *Journal of Personality and Social Psychology, 44,* 526–535.

Gumperz, J. J. (1976). The sociolinguistic significance of conversational code-switching. In J. Cook-Gumperz & J. J. Gumperz (Eds.), *Papers on language and context,* Berkeley, CA: University of California Language Behavior Research Laboratory.

Gurin, G., Veroff, J., & Feld, S. (1960). *Americans view their mental health.* New York: Basic Books.

Guttman, D. (1977). The cross-cultural perspective: Notes toward a comparative psychology of aging. In J. Birren & K. Schaie (Eds.), *Handbook of the psychology of aging.* New York: Van Nostrand Reinhold.

Guttman, L. (1944). A basis for scaling qualitative data. *American Sociological Review, 9,* 139–150.

Haan, N. (1978). Two moralities in action contexts: Relationships to thought, ego regulation, and development. *Journal of Personality and Social Psychology, 36,* 286–305.

Haan, N. (1986). Systematic variability in the quality of moral action, as defined in two paradigms. *Journal of Personality and Social Psychology, 50,* 1271–1284.

Haan, N., Smith, M., & Block, J. (1968). Moral reasoning of young adults: Political-social behavior, family background, and personality correlates. *Journal of Personality and Social Psychology, 10,* 183–201.

Haas, R. G. (1981). Effects of source characteristics on cognitive responses and persuasion. In R. E. Petty, T. M. Ostrom, & T. C. Brock (Eds.), *Cognitive responses in persuasion.* Hillsdale, NJ: Erlbaum.

Hacker, H. M. (1981). Blabbermouths and clams: Sex differences in self-disclosure in same-sex and cross-sex friendship dyads. *Psychology of Women Quarterly, 5,* 385–401.

Hackman, J. R., & Morris, C. G. (1975). Group tasks, group interaction process, and group performance effective-

ness: A review and proposed integration. In L. Berkowitz (Ed.), *Advances in experimental social psychology* (Vol. 8). New York: Academic Press.

Haines, H. H. (1984). Black radicalization and the funding of civil-rights: 1957–1970. *Social Problems, 32,* 31–43.

Halberstam, D. (1979). *The powers that be.* New York: Alfred Knopf.

Hall, E. T. (1966). *The hidden dimension.* Garden City, NJ: Doubleday.

Hallinan, M. T. (1981). Recent advances in sociometry. In S. R. Asher & J. M. Gottman (Eds.), *The development of children's friendships* (pp. 91–115). Cambridge, England: Cambridge University Press.

Hamilton, D. L. (1979). A cognitive-attributional analysis of stereotyping. In L. Berkowitz (Ed.), *Advances in experimental social psychology* (Vol. 12). New York: Academic Press.

Hamilton, D. L. (1981). Cognitive representations of persons. In E. T. Higgins, C. P. Herman, & M. P. Zanna (Eds.), *Social cognition: The Ontario symposium* (Vol. 1). Hillsdale, NJ: Erlbaum.

Hamilton, D. L., & Bishop, G. D. (1976). Attitudinal and behavioral effects of initial integration of white suburban neighborhoods. *Journal of Social Issues, 32,* 47–56.

Hamilton, D. L., & Gifford, R. K. (1976). Illusory correlation in interpersonal perception: A cognitive basis of stereotypic judgments. *Journal of Experimental Social Psychology, 12,* 392–407.

Hamilton, D. L., & Zanna, M. P. (1972). Differential weighting of favorable and unfavorable attributes in impressions of personality. *Journal of Experimental Research in Personality, 6,* 204–212.

Han, G., & Lindskold, S. (1983). A comparison of strategies with and without communication in a prisoner's dilemma game. Unpublished manuscript, Ohio University, OH.

Handel, W. (1979). Normative expectations and the emergence of meaning as solutions to problems: Convergence of structural and interactionist views. *American Journal of Sociology, 84,* 855–881.

Hanni, R. (1980). What is planned during speech pauses? In H. Giles, W. P. Robinson, & P. M. Smith (Eds.), *Language: Social psychological perspectives.* New York: Pergamon.

Hanson, R. (1986). Relational competence, relationships, and adjustment in old age. *Journal of Personality and Social Psychology, 50,* 1050–1058.

Harding, J., & Hogrefe, R. (1952). Attitudes of white department store employees toward Negro coworkers. *Journal of Social Issues, 8,* 18–28.

Harding, J., Proshansky, H., Kutner, B., & Chein, I. (1969). Prejudice and ethnic relations. In G. Lindzey & E. Aronson (Eds.), *The handbook of social psychology* (2nd ed.), (Vol. 5). Reading, MA: Addison-Wesley.

Hardy, R. C. (1975). A test of poor leader-member relations cells of the contingency model on elementary school children. *Child Development, 45,* 958–964.

Hardy, R. C., Sack, S., & Harpine, F. (1973). An experimental test of the contingency model on small classroom groups. *Journal of Psychology, 85,* 3–16.

Harkins, S., & Jackson, J. (1985). The role of evaluation in eliminating social loafing. *Personality and Social Psychology Bulletin, 11,* 457–465.

Harkins, S., Latane, B., & Williams, K. (1980). Social loafing: Allocating effort or taking it easy? *Journal of Experimental Social Psychology, 16,* 457–465.

Harkins, S. G., & Petty, R. E. (1981a) Effects of source magnification of cognitive effort on attitudes: An information processing view. *Journal of Personality and Social Psychology, 40,* 401–413.

Harkins, S. G., & Petty, R. E. (1981b). The multiple source effect in persuasion: The effects of distraction. *Personality and Social Psychology Bulletin, 4,* 627–635.

Harkins, S., & Petty, R. (1982). Effects of task difficulty and task uniqueness on social loafing. *Journal of Personality and Social Psychology, 43,* 1214–1229.

Harkins, S. G., & Petty, R. E. (1983). Social context effects in persuasion: The effects of multiple sources and multiple targets. In P. B. Paulus (Ed.), *Basic group processes.* New York: Springer-Verlag.

Harkins, S. G., & Petty, R. E. (1987). Information utility and the multiple source effect. *Journal of Personality and Social Psychology, 52,* 260–268.

Harkins, S. G., & Szymanski, K. (1987). Social loafing and social facilitation. In C. Hendrick (Ed.), *Group processes and intergroup relations* (pp. 167–188). Newbury Park, CA: Sage.

Harnett, D. L., & Vincelette, J. P. (1978). Strategic influences on bargaining effectiveness. In H. Sauermann (Ed.), *Contributions to experimental economics* (Vol. 7). Tubingen: Mohr.

Harper, R. G., Wiens, A. N., & Matarazzo, J. D. (1978). *Nonverbal communication: The state of the art.* New York: Wiley.

Harris, L., et al. (1975). *The myth and the reality of aging in America.* Washington, DC: National Council on Aging.

Harris, L., et al. (1982). *Aging in the 80's.* Washington, DC: National Council on Aging.

Harris, M. B. (1974). Mediators between frustration and aggression in a field experiment. *Journal of Experimental Social Psychology, 10,* 561–571.

Harris, M. B., Benson, S. M., & Hall, C. (1975). The effects of confession on altruism. *Journal of Social Psychology, 96,* 187–192.

Harris, R., Ellicott, A., & Holmes, D. (1986). The timing of psychosocial transitions and changes in women's lives: An examination of women 45 to 60. *Journal of Personality and Social Psychology, 51,* 409–416.

Harris, R. J., Messick, D. M., & Sentis, K. P. (1981). Proportionality, linearity, and parameter constancy: Messick and Sentis reconsidered. *Journal of Experimental Social Psychology, 17,* 210–225.

Harrison, A. (1977). Mere exposure. In L. Berkowitz (Ed.), *Advances in experimental social psychology* (Vol. 10). New York: Academic Press.

Harrison, A., & Saeed, L. (1977). Let's make a deal: An analysis of revelations and stipulations in lonely hearts advertisements. *Journal of Personality and Social Psychology, 35,* 257–264.

Hart, R. J. (1978). Crime and punishment in the Army. *Journal of Personality and Social Psychology, 36,* 1456–1471.

Harvey, J. H., Yarkin, K. L., Lightner, J. M., & Tolin, J. P. (1980). Unsolicited interpretation and recall of interpersonal events. *Journal of Personality and Social Psychology, 38,* 551–568.

Haskell, M. R., & Yablonsky, L. (1983). *Criminology: Crime and Criminality* (3rd ed.). Boston: Houghton Mifflin.

Hastorf, A. H., & Isen A. M. (Eds.). (1982). *Cognitive social psychology.* New York: Elsevier North-Holland.

Hastorf, A. H., & Cantril, H. (1954). They saw a game: A case study. *Journal of Abnormal and Social Psychology, 49,* 129–134.

Hastorf, A. H., Wildfogel, J., & Cassman, T. (1979). Acknowledgment of a handicap as a tactic in social interaction. *Journal of Personality and Social Psychology, 31,* 1790–1797.

Hatfield, E (1982). What do women and men want from love and sex? In E. R. Allgeier & N. B. McCormick (Eds.), *Changing boundaries: Gender roles and sexual behavior.* Palo Alto, CA: Mayfield.

Hatfield, E., & Walster, G. W. (1983). *A new look at love* (2nd ed.). Reading, MA: Addison-Wesley.

Hayduk, L. A. (1978). Personal space: An evaluation and orienting review. *Psychological Bulletin, 85,* 117–134.

Heider, E. R., & Olivier, D. (1972). The structure of the color space in naming and memory of two languages. *Cognitive Psychology, 3,* 337–354.

Heider, F. (1944). Social perception and phenomenal causality. *Psychological Review, 51,* 258–374.

Heider, F. (1958). *The psychology of interpersonal relations.* New York: Wiley.

Heirich, M. (1977). Change of heart: A test of some widely held theories about religious conversion. *American Journal of Sociology, 83,* 653–680.

Heiss, J. (1981). Social roles. In M. Rosenberg & R. Turner (Eds.), *Social psychology: Sociological perspectives.* New York: Basic Books.

Heiss, J., & Owens, S. (1972). Self-evaluations of blacks and whites. *American Journal of Sociology, 78,* 360–370.

Henley, N. M. (1977). *Body politics: Power, sex and nonverbal communication.* Englewood Cliffs, NJ: Prentice-Hall.

Henretta, J. C., Frazier, C., & Bishop, D. (1986). The effect of prior case outcome on juvenile justice decision-making. *Social Forces, 65,* 554–562.

Hensley, V., & Duval, S. (1976). Some perceptual determinants of perceived similarity, liking, and correctness. *Journal of Personality and Social Psychology, 34,* 159–168.

Hepburn, C., & Locksley, A. (1983). Subjective awareness of stereotyping: Do we know when our judgments are prejudiced? *Social Psychology Quarterly, 45,* 311–318.

Herek, G. (1987). Can functions be measured? A new perspective on the functional approach to attitudes. *Social Psychology Quarterly, 50,* 285–303.

Heritage, J., & Greatbatch, D. (1986). Generating applause: A study of rhetoric and response at party political conferences. *American Journal of Sociology, 92,* 110–157.

Heslin, R., & Dunphy, D. (1964). Three dimensions of member satisfaction in small groups. *Human Relations, 17,* 99–112.

Hetherington, E. M., Cox, M., & Cox, R. (1982). Effects of

divorce on parents and children. In M. E. Lamb (Ed.), *Nontraditional families: Parenting and child development*. Hillsdale, NJ: Erlbaum.

Hewitt, J. P. (1988). *Self and society* (4th ed.). Boston, MA: Allyn and Bacon.

Hewitt, J. P., & Stokes, R. (1975). Disclaimers. *American Sociological Review, 40*, 1–11.

Hewstone, M., & Jaspars, J. (1987). Covariation and causal attribution: A logical model of the intuitive analysis of variance. *Journal of Personality and Social Psychology, 53*, 663–672.

Heyl, B. S. (1977). The Madam as teacher: The training of house prostitutes. *Social Problems, 24*, 545–555.

Higbee, K. L. (1969). Fifteen years of fear arousal: Research on threat appeals, 1953–1968. *Psychological Bulletin, 72*, 426–444.

Higgins, E. T., & Bryant, S. L. (1982). Consensus information and the fundamental attribution error: The role of development and in-group versus out-group knowledge. *Journal of Personality and Social Psychology, 47*, 422–435.

Higgins, E. T., King, G. A., & Mavin, G. H. (1982). Individual construct accessibility and subjective impressions and recall. *Journal of Personality and Social Psychology, 43*, 35–47.

Hill, C., Rubin, Z., & Peplau, L. (1976). Breakups before marriage: The end of 103 affairs. *Journal of Social Issues, 32*(1), 147–168.

Hill, C., & Stull, D. (1981). Sex differences in effects of social and value similarity in same-sex friendship. *Social Psychology, 41*, 488–502.

Hill, G. W. (1982). Group versus individual performance. Are N + 1 heads better than one? *Psychological Bulletin, 91*, 517–539.

Hilton, T. L., & Berglund, G. (1974). Sex differences in mathematics achievement—a longitudinal study. *Journal of Educational Research, 67*, 231–237.

Hindelang, M., Hirschi, T., & Weis, J. (1979). Correlates of delinquency: The illusion of discrepancy between self-report and official measures. *American Sociological Review, 44*, 995–1014.

Hinkle, S. (1975). *Cognitive consistency effects on attitudes toward ingroup and outgroups products*. Unpublished doctoral dissertation, University of North Carolina, NC.

Hinkle, S., & Schopler, J. (1986). Bias in the evaluation of in-group and out-group performance. In S. Worchel & W. G. Austin (Eds.), *Psychology of intergroup relations* (2nd ed.) (pp. 196–212). Chicago: Nelson-Hall Publishers.

Hirschi, T. (1969). *Causes of Delinquency*. Berkeley, CA: University of California Press.

Hochschild, A. R. (1975). Disengagement theory: A critique and a proposal. *American Sociological Review, 40*, 553–569.

Hochschild, A. R. (1983). *The managed heart: Commercialization of human feeling*. Berkeley, CA: University of California Press.

Hoelter, J. W. (1983). The effects of role evaluation and commitment on identity salience. *Social Psychology Quarterly, 46*, 140–147.

Hoelter, J. W. (1984). Relative effects of significant others on self-evaluation. *Social Psychology Quarterly, 47*, 255–262.

Hoelter, J. W. (1986). The relationship between specific and global evaluations of self: A comparison of several models. *Social Psychology Quarterly, 49*, 129–141.

Hoffman, C., Lau, I., & Johnson, D. (1986). The linguistic relativity of person cognition: An English-Chinese comparison. *Journal of Personality and Social Psychology, 51*, 1097–1105.

Hoffman, M. L. (1977). Empathy, its development and prosocial implications. In C. B. Keasey (Ed.), *Nebraska symposium on motivation* (Vol. 25). Lincoln: University of Nebraska Press.

Hoffman, M. L. (1981a). Is altruism part of human nature? *Journal of Personality and Social Psychology, 40*, 121–137.

Hoffman, M. L. (1981b). The development of empathy. In J. P. Rushton & R. M. Sorrentino (Eds.), *Altruism and helping behavior*. Hillsdale, NJ: Erlbaum.

Hogan, D. P. (1981). *Transitions and social change*. New York: Academic Press.

Holden, R. T. (1986). The contagiousness of aircraft hijacking. *American Journal of Sociology, 91*, 874–904.

Hollander, E. P. (1975). Independence, conformity and civil liberties: Some implications from social psychological research. *Journal of Social Issues, 31*, 55–67.

Hollander, E. P. (1985). Leadership and power. In G. Lindzey & E. Aronson (Eds.), *Handbook of social psychology* (3rd ed.) (Vol. 2, pp. 485–537). New York: Random House.

Hollander, E. P., Fallon, B. J., & Edwards, M. T. (1977). Some aspects of influence and acceptability for appointed and elected group leaders. *Journal of Psychology, 95*, 289–296.

Hollander, E. P., & Julian, J. W. (1970). Studies in leader legitimacy, influence and innovation. In L. Berkowitz (Ed.), *Advances in experimental social psychology* (Vol. 5, pp. 34–67). New York: Academic Press.

Hollander, E. P., & Julian, J. W. (1978). Studies in leader legitimacy, influence, and innovation. In L. Berkowitz (Ed.), *Group processes*. New York: Academic Press.

Hollinger, R. C., & Clark, J. P. (1983). Deterrence in the workplace: Perceived certainty, perceived severity and employee theft. *Social Forces, 62*, 398–418.

Holmes, D. S. (1971). Compensation for ego threat: Two experiments. *Journal of Personality and Social Psychology, 18*, 234–237.

Holmes, J. G., & Grant, P. (1979). Ethnocentric reactions to social threats. In L. H. Strickland (Ed.), *Social psychology: East-west perspectives*. Oxford, England: Pergamon.

Holmes, J. G., Ellard, J. H., & Lamm, H. (1986). Boundary roles and intergroup conflict. In S. Worchel & W. G. Austin (Eds.), *Psychology of intergroup relations* (2nd ed.) (pp. 343–363). Chicago: Nelson-Hall Publishers.

Holmes, T. H., & Rahe, R. (1967). The social readjustment rating scale. *Journal of Psychosomatic Research, 11*, 213–218.

Holohan, C., & Moos, R. (1987). Personal and contextual determinants of coping strategies. *Journal of Personality and Social Psychology, 52*, 946–955.

Holsti, O. (1968). Content analysis. In G. Lindzey & E.

Aronson (Eds.), *Handbook of social psychology* (2nd ed.) (Vol. II). Reading, MA: Addison-Wesley.

Holtgraves, T. (1986). Language structure in social interaction: Perceptions of direct and indirect speech acts and interactants who use them. *Journal of Personality and Social Psychology, 51*, 305–314.

Homans, G. C. (1961). *Social behavior: Its elementary forms.* New York: Harcourt Brace Jovanovich.

Homans, G. C. (1974). *Social behavior: Its elementary forms* (2nd ed.). New York: Harcourt Brace Jovanovich.

Hood, W. R., & Sherif, M. (1962). Verbal report and judgment of an unstructured stimulus. *Journal of Psychology, 54*, 121–130.

Hopmann, P. T., & Smith, T. C. (1977). An application of the Richardson process model: Soviet-American interactions in the test ban negotiations. *Journal of Conflict Resolution, 21*, 701–726.

Horai, J., Naccari, N., & Fatoullah, E. (1974). The effects of expertise and physical attractiveness upon opinion agreement and liking. *Sociometry, 37*, 601–606.

Horai, J., & Tedeschi, J. T. (1969). The effects of threat credibility and magnitude of punishment upon compliance. *Journal of Personality and Social Psychology, 12*, 164–169.

Horne, W. C., & Long, G. (1972). Effect of group discussion on universalistic-particularistic orientation. *Journal of Experimental Social Psychology, 8*, 236–246.

Hornstein, G. (1985). Intimacy in conversational style as a function of the degree of closeness between members of a dyad. *Journal of Personality and Social Psychology, 49*, 671–681.

Hornstein, H. A. (1978). Promotive tension and prosocial behavior: A Lewinian analysis. In L. Wispe (Ed.), *Altruism, sympathy and helping.* New York: Academic Press.

Horowitz, R. (1987). Community tolerance of gang violence. *Social Problems, 34*, 437–450.

House, J. S. (1974). Occupational stress and coronary heart disease: A review and theoretical integration. *Journal of Health and Social Behavior, 15*, 17–21.

House, J. S. (1981a). Social structure and personality. In M. Rosenburg & R. Turner (Eds.), *Social psychology: Sociological perspectives* (pp. 525–561). New York: Basic Books.

House, J. S. (1981b). *Work stress and social support.* Reading, MA: Addison-Wesley.

House, R. J. (1971). A path-goal theory of leader effectiveness. *Administrative Science Quarterly, 16*, 321–338.

Hoyenga, K. B., & Hoyenga, K. (1979). *The question of sex differences: Psychological, cultural and biological issues.* Boston: Little, Brown.

Huesmann, L. R. (1982). Television violence and aggressive behavior. In D. Pearl, L. Bouthilet, & J. Lazar (Eds.), *Television and behavior: Ten years of scientific progress and implications for the eighties* (Vol. 2). Washington, DC: Government Printing Office.

Huffine, C., & Clausen, J. (1979). Madness and work: Short- and long-term effects of mental illness on occupational careers. *Social Forces, 57*, 1049–1062.

Hull, C. L. (1943). *Principles of behavior.* New York: Appleton-Century-Crofts.

Hultsch, D., & Plemons, J. (1979). Life events and life span development. In P. Baltes & O. Brim, Jr. (Eds.), *Life-span development and behavior* (Vol. 2). New York: Academic Press.

Humphreys, L. (1970). *Tearoom trade: Impersonal sex in public places.* Chicago: Aldine.

Hundleby, J. D., & Mercer, G. W. (1987). Family and friends as social environments and their relationship to young adolescents' use of alcohol, tobacco and marijuana. *Journal of Marriage and the Family, 49*, 151–164.

Huston, A. C., & Wright, J. C. (1982). Effects of communication media on children. In C. B. Kopp & J. B. Krakow (Eds.), *The child: Development in a social context.* Boston: Addison-Wesley.

Huston-Stein, A., & Higgins-Trenk, A. (1978). Development of females from childhood through adulthood: Career and feminine role orientations. In P. Baltes (Ed.), *Life-span development and behavior* (Vol. 1). New York: Academic Press.

Hyde, J., & Linn, M. (Eds.). (1986). *The psychology of gender: Advances through meta-analysis.* Baltimore, MD: Johns Hopkins University Press.

Hymes, D. (1974). *Foundations in sociolinguistics.* London: Tavistock.

Ikle, F. C. (1971). *Every war must end.* New York: Columbia University Press.

Ingham, A. G., Levinger, G., Graves, J., & Peckham, V. (1974). The Ringelmann effect: Studies of group size and group performance. *Journal of Experimental Social Psychology, 10*, 371–384.

Insko, C. A., Arkoff, A., & Insko, V. M. (1965). Effects of high and low fear-arousing communications upon opinions toward smoking. *Journal of Experimental Social Psychology, 40*, 256–266.

Insko, C. A., Drenan, S., Solomon, M. R., Smith, R., & Wade, T. J. (1983). Conformity as a function of the consistency of positive self-evaluation with being liked and being right. *Journal of Experimental Social Psychology, 19*, 341–358.

Insko, C. A., Smith, R. H., Alicke, M. S., Wade, J., & Taylor, S. (1985). Conformity and group size: The concern with being right and the concern with being liked. *Personality and Social Psychology Bulletin, 11*, 41–50.

Isen, A. M. (1970). Success, failure, attention, and reaction to others: The warm glow of success. *Journal of Personality and Social Psychology, 15*, 294–301.

Isen, A. M., Clark, M., & Schwartz, M. F. (1976). Duration of the effect of good mood on helping: Footprints on the sands of time. *Journal of Personality and Social Psychology, 34*, 385–393.

Isen, A. M., & Levin, P. F. (1972). Effect of feeling good on helping. Cookies and kindness. *Journal of Personality and Social Psychology, 21*, 384–388.

Isenberg, D. J. (1986). Group polarization: A critical review and meta-analysis. *Journal of Personality and Social Psychology, 50*, 1141–1151.

Jaccard, J. (1981). Toward theories of persuasion and belief change. *Journal of Personality and Social Psychology, 40*, 260–269.

Jackson, J. (1954). The adjustment of the family to the crisis of alcoholism. *Quarterly Journal of Studies on Alcohol, 15,* 564–586.

Jackson, J. (1965). Structural characteristics of norms. In I. D. Steiner & M. Fishbein (Eds.), *Current studies in social psychology.* New York: Holt, Rinehart and Winston.

Jackson, J. M., & Harkins, S. G. (1985). Equity in effort: An explanation of the social loafing effect. *Journal of Personality and Social Psychology, 49,* 1199–1206.

Jackson, J. M., & Williams, K. D. (1985). Social loafing on difficult tasks: Working collectively can improve performance. *Journal of Personality and Social Psychology, 49,* 937–942.

Jacques, J. M., & Chason, K. (1977). Self-esteem and low status groups: A changing scene? *Sociological Quarterly, 18,* 399–412.

James, W. (1890). *Principles of psychology.* New York: Holt, Rinehart and Winston.

Janis, I. L. (1982). *Groupthink* (2nd ed.). Boston: Houghton Mifflin.

Jellison, J. M., & Riskind, J. (1970). A social comparison of abilities interpretation of risk-taking behavior. *Journal of Personality and Social Psychology, 15,* 375–390.

Jenkins, C., Rosenman, R., & Zyzanski, S. (1974). Prediction of clinical coronary heart disease by a test for coronary prone behavior pattern. *New England Journal of Medicine, 290,* 1271–1275.

Jenkins, J. C. (1983). Resource mobilization theory and the study of social movements. In R. H. Turner & J. F. Short (Eds.), *Annual review of sociology* (Vol. 9, pp. 527–553). Palo Alto, CA: Annual Reviews.

Jenkins, J. C., & Eckert, C. (1986). Channeling black insurgency: Elite patronage and professional social movement organizations in the development of the black movement. *American Sociological Review, 51,* 812–829.

Jenkins, J. C., & Perrow, C. (1977). Insurgency of the powerless: Farm worker movements 1946–1972. *American Sociological Review, 42,* 249–268.

Jensen, G. (1972). Parents, peers and delinquent action: A test of the differential association perspective. *American Journal of Sociology, 78,* 562–575.

Jensen, G., Erickson, M., & Gibbs, J. (1978). Perceived risk of punishment and self-reported delinquency. *Social Forces, 57,* 57–78.

Jessor, R., Costa, F., Jessor, L., & Donovan, J. (1983). Time of first intercourse: A prospective study. *Journal of Personality and Social Psychology, 44,* 608–626.

Johnson, D. L., & Andrews, I. R. (1971). The risky-shift hypothesis tested with consumer products as stimuli. *Journal of Personality and Social Psychology, 20,* 382–385.

Johnson, D. W., Maruyama, G., Johnson, R., Nelson, D., & Skon, L. (1981). Effects of cooperative, competitive, and individualistic goal structures on achievement: A meta-analysis. *Psychological Bulletin, 89,* 47–62.

Johnson, D. W., Maruyama, G., & Johnson, R. T. (1982). Separating ideology from currently available data: A reply to Cotton and Cook and McGlynn. *Psychological Bulletin, 92,* 186–192.

Johnson, K. J., Lund, D. A., & Dimond, M. F. (1986). Stress, self-esteem, and coping during bereavement among the elderly. *Social Psychology Quarterly, 49,* 273–279.

Johnson, M., & Leslie, L. (1982). Couple involvement and network structure: A test of the dyadic withdrawal hypothesis. *Social Psychology Quarterly, 45,* 34–43.

Johnson, M. M., Stockard, J., Acker, J., & Naffziger, G. (1975). Expressiveness reevaluated. *School Review, 83,* 617–644.

Johnson, N. R. (1987). Panic at "The Who Concert Stampede": An empirical assessment. *Social Problems, 34,* 362–373.

Johnson, N., & Feinberg, W. (1977). A computer simulation of the emergence of consensus in crowds. *American Sociological Review, 52,* 505–521.

Johnson, R. E. (1979). *Juvenile delinquency and its origins.* New York: Cambridge University Press.

Johnson, W. T., & DeLamater, J. D. (1976). Response effects in sex surveys. *Public Opinion Quarterly, 40,* 165–181.

Johnson-George, C., & Swap, W. (1982). Measurement of specific interpersonal trust: Construction and validation of a scale to assess trust in a specific other. *Journal of Personality and Social Psychology, 43,* 1306–1317.

Jones, E. E. (1964). *Ingratiation.* New York: Appleton-Century-Crofts.

Jones, E. E. (1979). The rocky road from acts to dispositions. *American Psychologist, 34,* 107–117.

Jones, E. E. (1985). Major developments in social psychology during the past five decades. In G. Lindzey & E. Aronson (Eds.), *Handbook of social psychology* (3rd ed.) (Vol. I). New York: Random House.

Jones, E. E., & Davis, K. E. (1965). From acts to dispositions. In L. Berkowitz (Ed.), *Advances in experimental social psychology* (Vol. 2). New York: Academic Press.

Jones, E. E., Davis, K. E., & Gergen, K. J. (1961). Role playing variations and their informational value for person perception. *Journal of Abnormal and Social Psychology, 63,* 302–310.

Jones, E. E., Farina, A., Hastorf, A. H., Markus, H., Miller, D. T., & Scott, R. A. (1984). *Social stigma.* New York: W. H. Freeman.

Jones, E. E., Gergen, K. J., Gumpert, P., & Thibaut, J. (1965). Some conditions affecting the use of ingratiation to influence performance evaluation. *Journal of Personality and Social Psychology, 1,* 613–626.

Jones, E. E., & Goethals, G. R. (1971). *Order effects in impression formation: Attribution context and the nature of the entity.* Morristown, NJ: General Learning Press.

Jones, E. E., & Harris, V. A. (1967). The attribution of attitudes. *Journal of Experimental Social Psychology, 3,* 1–24.

Jones, E. E., & McGillis, D. (1976). Correspondent inferences and the attribution cube: A comparative reappraisal. In J. H. Harvey, W. J. Ickes, & R. F. Kidd (Eds.), *New directions in attribution research* (Vol. 1). Hillsdale, NJ: Erlbaum.

Jones, E. E., & Nisbett, R. (1972). The actor and observer: Divergent perceptions of the causes of behavior. In E. E. Jones et al. (Eds.), *Attribution: Perceiving the causes of behavior.* Morristown, NJ: General Learning Press.

Jones, E. E., & Pittman, T. S. (1982). Toward a general theory of strategic self-presentation. In J. Suls (Ed.), *Psychological perspectives on the self* (Vol. 1). Hillsdale, NJ: Erlbaum.

Jones, E. E., Rock, L., Shaver, K. G., Goethals, G. R., &

Ward, L. M. (1968). Pattern of performance and ability attribution: An unexpected primacy effect. *Journal of Personality and Social Psychology, 10,* 317–340.

Jones, E. E., & Wortman, C. (1973). *Ingratiation: An attributional approach.* Morristown, NJ: General Learning Press.

Jones, V. C. (1948). *The Hatfields and the McCoys.* Chapel Hill: University of North Carolina Press.

Jones, W. H., Briggs, S. R., & Smith, T. G. (1986). Shyness: Conceptualization and measurement. *Journal of Personality and Social Psychology, 51,* 629–639.

Joreskog, K. G., & Sorbom, D. (1979). *Advances in factor analysis and structural equation models.* Cambridge, MA: Abt Books.

Joseph, N., & Alex, N. (1972). The uniform: A sociological perspective. *American Journal of Sociology, 77,* 719–730.

Jourard, S. M. (1971). *Self-disclosure.* New York: Wiley.

Judd, C. M., & Park, B. (1988). Out-group homogeneity: Judgments of variability at the individual and group levels. *Journal of Personality and Social Psychology, 54,* 778–788.

Julian, J. W., Hollander, E. P., & Regula, C. R. (1969). Endorsement of group spokesman as a function of his source of authority, competence, and success. *Journal of Personality and Social Psychology, 11,* 42–49.

Jussim, L., Coleman, L., & Nassau, S. (1987). The influence of self-esteem on perceptions of performance and feedback. *Social Psychology Quarterly, 50,* 95–99.

Kahn, A. S., & Kohls, J. W. (1972). Determinants of toughness in dyadic bargaining. *Sociometry, 35,* 305–315.

Kahn, R. L., & Katz, D. (1960). Leadership practices in relation to productivity and morale. In D. Cartwright & A. Zander (Eds.), *Group dynamics* (2nd ed.). Evanston, IL: Row, Peterson and Co.

Kahne, M., & Schwartz, C. (1978). Negotiating trouble: The social construction and management of trouble in a psychiatric context. *Social Problems, 25,* 461–475.

Kalish, R. A. (1976). Death and dying in a social context. In R. Binstock & E. Shanas (Eds.), *Handbook of aging and the social sciences.* New York: Van Nostrand Reinhold.

Kandel, D. (1978). Similarity in real-life adolescent friendship pairs. *Journal of Personality and Social Psychology, 36,* 306–312.

Kanter, R. M. (1976). *Men and women of the corporation.* New York: Basic Books.

Kaplan, H. B., Johnson, R., & Bailey, C. A. (1987). Deviant peers and deviant behavior: Further elaboration of a model. *Social Psychology Quarterly, 50,* 277–284.

Kaplan, H. B., Martin, S. S., & Johnson, R. J. (1986). Self-rejection and the explanation of deviance: Specification of the structure among latent constructs. *American Journal of Sociology, 92,* 384–411.

Kaplan, M. F. (1987). The influencing process in group decisionmaking. In C. Hendrick (Ed.), *Group processes.* Newbury Park, CA: Sage.

Kaplan, M. F., & Miller, C. E. (1987). Group decision making and normative versus informational influence: Effects of type of issue and assigned decision rule. *Journal of Personality and Social Psychology, 53,* 306–313.

Karabenick, S. A. (1983). Sex-relevance of content and influenceability. *Personality and Social Psychology Bulletin, 9,* 243–252.

Karasek, R., Theorell, T., Schwartz, J., Schnall, P., Pieper, C., & Michela, J. (1988). Job characteristics in relation to the prevalence of myocardial infarction in the US Health Examination Survey (HES) and the Health and Nutrition Examination Survey (HANES). *American Journal of Public Health, 78,* 910–918.

Karlins, M., & Abelson, H. I. (1970). *How opinions and attitudes are changed* (2nd ed.). New York: Springer.

Karlins, M., Coffman, T. L., & Walters, G. (1969). On the fading of social stereotypes: Studies on three generations of college students. *Journal of Personality and Social Psychology, 13,* 1–16.

Karp, D. A., & Yoels, W. C. (1986). *Sociology and everyday life.* Itasca, IL: F. E. Peacock Publishers.

Katz, D. (1960). The functional approach to the study of attitudes. *Public Opinion Quarterly, 24,* 163–204.

Katz, D., & Braly, K. (1933). Racial stereotypes in one hundred college students. *Journal of Abnormal and Social Psychology, 28,* 280–290.

Katz, I. (1970). Experimental study in Negro-white relationships. In L. Berkowitz (Ed.), *Advances in experimental social psychology* (Vol. 5). New York: Academic Press.

Katz, I. (1981). *Stigma: A social psychological analysis.* Hillsdale, NJ: Erlbaum.

Katz, I., & Cohen, M. (1962). The effects of training Negroes upon cooperative problem solving in biracial teams. *Journal of Abnormal and Social Psychology, 64,* 319–325.

Katz, I., Wackenhut, J., & Glass, D. C. (1986). An ambivalence-amplification theory of behavior toward the stigmatized. In S. Worchel & W. G. Austin (Eds.), *Psychology of intergroup relations* (2nd ed.) (pp. 103–117). Chicago: Nelson-Hall Publishers.

Kauffman, D. R., & Steiner, I. D. (1968). Some variables affecting the use of conformity as an ingratiation technique. *Journal of Experimental Social Psychology, 4,* 400–414.

Kelley, H. H. (1950). The warm-cold variable in first impressions. *Journal of Personality, 18,* 431–439.

Kelley, H. H. (1967). Attribution theory in social psychology. In D. Levine (Ed.), *Nebraska symposium on motivation 1967.* Lincoln: University of Nebraska Press.

Kelley, H. H. (1973). The process of causal attribution. *American Psychologist, 28,* 107–128.

Kelley, H. H., Berscheid, E., Christensen, A., Harvey, H. H., Huston, T., Levinger, G., McClintock, E., Peplau, L. H., & Peterson, D. R. (1983). *Close relationships.* New York: W. H. Freeman.

Kelley, H. H., Condry, J. C., Jr., Dahlke, A. E., & Hill, A. H. (1965). Collective behavior in a simulated panic situation. *Journal of Experimental Social Psychology, 1,* 20–54.

Kelley, H. H., & Michela, J. (1980). Attribution theory and research. *Annual Review of Psychology, 31,* 457–501.

Kelley, H. H., & Thibaut, J. W. (1978). *Interpersonal relations: A theory of interdependence.* New York: Wiley.

Kelman, H. C. (1974). Attitudes are alive and well and gainfully employed in the sphere of action. *American Psychologist, 29,* 310–324.

Kelman, H. C., & Cohen, S. P. (1986). Resolution of international conflict: An interactional approach. In S.

Worchel & W. G. Austin (Eds.), *Psychology of intergroup relations* (2nd ed.) (pp. 323–342). Chicago: Nelson-Hall Publishers.

Kemper, T. D. (1973). The fundamental dimensions of social relationship: A theoretical statement. *Acta Sociologica, 16,* 41–57.

Kemper, T. D. (1978). *A social interactional theory of emotions.* New York: Wiley.

Kendon, A. (1970). Movement coordination in social interaction. Some examples described. *Acta Psychologica, 32,* 100–125.

Kendon, A., Harris, R. M., & Key, M. R. (1975). *Organization of behavior in face-to-face interaction.* The Hague: Mouton.

Kenney, D., & La Voie, L. (1982). Reciprocity of interpersonal attraction: A confirmed hypothesis. *Social Psychology Quarterly, 45,* 54–58.

Kent, G., Davis, J., & Shapiro, D. (1978). Resources required in the construction and reconstruction of conversations. *Journal of Personality and Social Psychology, 36,* 13–22.

Kerckhoff, A. C. (1966). Family patterns and morale in retirement. In I. Simpson & J. McKinney (Eds.), *Social aspects of aging.* Durham, NC: Duke University Press.

Kerckhoff, A. C. (1974). The social context of interpersonal attraction. In T. Huston (Ed.), *Foundations of interpersonal attraction* (pp. 61–78). New York: Academic Press.

Kerckhoff, A. C., Back, K. W., & Miller, N. (1965). Sociometric patterns in hysterical contagion. *Sociometry, 28,* 2–15.

Kerr, N., & Bruun, S. (1981). Ringelmann revisited: Alternative explanations for the social loafing effect. *Personality and Social Psychology Bulletin, 7,* 224–231.

Kessler, R. C. (1982). A disaggregation of the relationship between socioeconomic status and psychological distress. *American Sociological Review, 47,* 752–764.

Kessler, R. C., & Cleary, P. (1980). Social class and psychological distress. *American Sociological Review, 45,* 463–478.

Kessler, R. C., & McCrae, J., Jr. (1982). The effects of wives' employment on mental health of married men and women. *American Sociological Review, 47,* 216–226.

Kiesler, C. A., & Corbin, L. H. (1965). Commitment, attraction, and conformity. *Journal of Personality and Social Psychology, 2,* 890–895.

Kiesler, C. A., & Kiesler, S. B. (1969). *Conformity.* Reading, MA: Addison-Wesley.

Kiesler, C. A., Zanna, M., & deSalvo, J. (1966). Deviation and conformity: Opinion change as a function of commitment, attraction, and the presence of a deviate. *Journal of Personality and Social Psychology, 3,* 458–467.

Kilham, W., & Mann, L. (1974). Level of destructive obedience as a function of transmitter and executant roles in the Milgram obedience paradigm. *Journal of Personality and Social Psychology, 29,* 696–702.

Killian, L. (1984). Organization, rationality and spontaneity in the civil rights movement. *American Sociological Review, 49,* 770–783.

Kimberly, J. C. (1970). The emergence and stabilization of stratification in simple and complex social systems. *Sociological Inquiry, 40,* 73–101.

Kimberly, J. C. (1984). Cognitive balance, inequality, and consensus: Interrelations among fundamental processes in groups. In E. J. Lawler (Ed.), *Advances in group processes* (Vol. 1). Greenwich, CT: JAI Press.

Kingston, P., & Nock, S. (1987). Time together among dual-earner couples. *American Sociological Review, 52,* 391–400.

Kiparsky, P. (1976). Historical linguistics and the origin of language. *Annals of the New York Academy of Sciences, 280,* 97–103.

Kirk, R. E. (1982). *Experimental design* (2nd ed.). Belmont, CA: Brooks/Cole.

Kitsuse, J. (1964). Societal reaction to deviant behavior: Problems of theory and method. In H. Becker (Ed.), *The other side.* New York: Free Press.

Kleck, R. E. (1968). Physical stigma and nonverbal cues emitted in face-to-face interaction. *Human Relations, 21,* 19–28.

Kleck, R. E., & Strenta, A. (1980). Perceptions of the impact of negatively valued physical characteristics on social interaction. *Journal of Personality and Social Psychology, 39,* 861–873.

Kleinke, C. L., & Kahn, M. L. (1980). Perceptions of self-disclosers: Effects of sex and physical attractiveness. *Journal of Personality, 48,* 190–205.

Knox, R. E., & Safford, R. K. (1976). Group caution at the race track. *Journal of Experimental Social Psychology, 12,* 317–324.

Kobrin, F., & Hendershot, G. (1977). Do family ties reduce mortality? Evidence from the United States, 1966–68. *Journal of Marriage and the Family, 39,* 737–745.

Koestner, R., & Wheeler, L. (1988). Self-presentation in personal advertisements: The influence of implicit notions of attraction and role expectations. *Journal of Social and Personal Relationships, 5,* 149–160.

Koestner, R., Zuckerman, M., & Koestner, J. (1987). Praise, involvement, and intrinsic motivation. *Journal of Personality and Social Psychology, 53,* 383–390.

Kogan, N., & Wallach, M. A. (1964). *Risk taking: A study in cognition and personality.* New York: Holt, Rinehart and Winston.

Kohlberg, L. (1969). Stage and sequence: The cognitive-developmental approach to socialization. In D. Goslin (Ed.), *Handbook of socialization theory and research.* Chicago: Rand McNally.

Kohn, M. (1969). *Class and conformity: A study in values.* Homewood, IL: Dorsey Press.

Kohn, M. (1976). Occupational structure and alienation. *American Journal of Sociology, 82,* 111–130.

Kohn, M. (1977). Reassessment, 1977. *Class and conformity: A study in values* (2nd ed.). Chicago: University of Chicago Press.

Kohn, M., & Schooler, C. (1973). Occupational experience and psychological functioning: An assessment of reciprocal effects. *American Sociological Review, 38,* 97–118.

Kohn, M., & Schooler, C. (1982). Job conditions and personality: A longitudinal assessment of their reciprocal effects. *American Journal of Sociology, 87,* 1257–1286.

Kohn, M., Schooler, C., with the collaboration of J. Miller, K. Miller, S. Schoenbach, & R. Schoenberg. (1983). *Work and personality: An inquiry into the impact of social stratification.* Norwood, NJ: Ablex Publishing.

Kollock, P., Blumstein, P., & Schwartz, P. (1985). Sex and

power in interaction: Conversational privileges and duties. *American Sociological Review, 50,* 34–46.

Konecni, V. J. (1979). The role of aversive events in the development of intergroup conflict. In W. G. Austin & S. Worchel (Eds.), *The social psychology of intergroup relations.* Monterey, CA: Brooks/Cole.

Korten, D. C. (1962). Situational determinants of leadership structure. *Journal of Conflict Resolution, 6,* 222–235.

Kothandapani, V. (1971). Validation of feeling, belief and intention to act as three components of attitude and their contribution to prediction of contraceptive behavior. *Journal of Personality and Social Psychology, 19,* 321–333.

Kraus, S. (Ed.). (1962). *The great debates.* Bloomington: Indiana University Press.

Krauss, R. M. (1981). Impression formation, impression management, and nonverbal behaviors. In E. T. Higgins, C. P. Herman, & M. P. Zanna (Eds.), *Social Cognition: The Ontario Symposium.* Hillsdale, NJ: Erlbaum.

Krauss, R. M., Geller, V., & Olson, C. T. (1976). *Modalities and cues in the detection of deception.* Paper presented at the meeting of the American Psychological Association, Washington, DC.

Kraut, R. E. (1976). Deterrent and definitional influences on shoplifting. *Social Problems, 23,* 358–368.

Kraut, R. E. (1978). Verbal and nonverbal cues in the perception of lying. *Journal of Personality and Social Psychology, 36,* 380–391.

Kraut, R. E., Lewis, S. H., & Swezey, L. W. (1982). Listener responsiveness and the coordination of conversation. *Journal of Personality and Social Psychology, 43,* 718–731.

Kraut, R. E., & Poe, D. (1980). Behavior roots of person perception: The deception judgments of customs inspectors and laymen. *Journal of Personality and Social Psychology, 39,* 784–798.

Kravitz, D. A., & Martin, B. (1986). Ringelmann rediscovered: The original article. *Journal of Personality and Social Psychology, 50,* 936–941.

Krebs, R. (1967). *Some relations between moral judgment, attention and resistance to temptation.* Unpublished doctoral dissertation, University of Chicago.

Krebs, D. (1975). Empathy and altruism. *Journal of Personality and Social Psychology, 32,* 1134–1146.

Krebs, D. L. (1982). Psychological approaches to altruism: An evaluation. *Ethics, 92,* 147–158.

Kremer, J. F., & Stephens, L. (1983). Attributions and arousal as mediators of mitigation's effect on retaliation. *Journal of Personality and Social Psychology, 45,* 335–343.

Kriesberg, L. (1973). *The sociology of social conflicts.* Englewood Cliffs, NJ: Prentice-Hall.

Kritzer, H. (1977). Political protest and political violence: A nonrecursive causal model. *Social Forces, 55,* 630–640.

Krohn, M. D. (1986). The web of conformity: A network approach to the explanation of delinquent behavior. *Social Problems, 33,* 581–593.

Krohn, M. D., Massey, J. L., & Zielinski, M. (1988). Role overlap, network multiplexity and adolescent deviant behavior. *Social Psychology Quarterly, 51,* 346–356.

Kuhlman, C. E., Miller, M. H., & Gungor, E. (1973).

Interpersonal conflict resolution: The effects of language and meaning. In L. Rappoport & D. A. Summers (Eds.), *Human judgment and social interaction.* New York: Holt, Rinehart and Winston.

Kuhn, D., Langer, J., Kohlberg, L., & Haan, N. (1977). The development of formal operations in logical and moral judgment. *Genetic Psychology Monographs, 95,* 97–188.

Kuhn, M. H., & McPartland, T. (1954). An empirical investigation of self-attitudes. *American Sociological Review, 19,* 68–76.

Kulick, J. A., & Brown, R. (1979). Frustration, attribution of blame, and aggression. *Journal of Experimental Social Psychology, 15,* 183–194.

Kurtines, M. M. (1986). Moral behavior as rule governed behavior: Person and situation effects on moral decision-making. *Journal of Personality and Social Psychology, 50,* 784–791.

Labov, W. (1972a). *Language in the inner city. Studies in the Black English vernacular.* Philadelphia: University of Pennsylvania Press.

Labov, W. (1972b). *Sociolinguistic patterns.* Philadelphia: University of Pennsylvania Press.

LaFrance, M., & Mayo, C. (1978). *Moving bodies: Nonverbal communication in social relationships.* Monterey, CA: Brooks/Cole.

Lakoff, R. T. (1979). Women's language. In O. Buturff & E. L. Epstein (Eds.), *Women's language and style.* Akron, OH: University of Akron.

Lamb, M. E. (1979). Paternal influence and the father's role: A personal perspective. *American Psychologist, 34,* 938–943.

Lamb, M. E. (1982). Maternal employment and child development: A review. In M. E. Lamb (Ed.), *Nontraditional families: Parenting and child development.* Hillsdale, NJ: Erlbaum.

Lamm, H., & Sauer, C. (1974). Discussion-induced shift toward higher demands in negotiation. *European Journal of Social Psychology, 4,* 85–88.

Lamm, H., & Schwinger, T. (1980). Norms concerning distributive justice: Are needs taken into consideration in allocation decisions? *Social Psychology Quarterly, 43,* 425–429.

Landers, D. M., & Crum, T. F. (1971). The effects of team success and formal structure on inter-personal relations and cohesiveness of baseball teams. *International Journal of Sports Psychology, 2,* 88–96.

Landers, D. M., & Luschen, G. (1974). Team performance outcome and the cohesiveness of competitive coacting groups. *International Review of Sport Sociology, 2,* 57–69.

Landy, D., & Sigall, H. (1974). Beauty is talent: Task evaluation as a function of the performer's physical attractiveness. *Journal of Personality and Social Psychology, 29,* 299–304.

Langner, T. S. (1963). A twenty-two item screening score of psychiatric symptoms indicating impairment. *Journal of Health and Human Behavior, 3,* 269–276.

Langner, T. S., & Michael, S. (1963). *Life stress and mental health: The midtown Manhattan study.* New York: Free Press.

Lantz, H., Keyes, J., & Schultz, M. (1975). The American family in the preindustrial period: From base lines in history to change. *American Sociological Review, 40,* 21–36.

Lantz, H., Schultz, M., & O'Hara, M. (1977). The changing American family from the preindustrial to the industrial period: A final report. *American Sociological Review, 42,* 406–421.

LaPiere, R. (1934). Attitudes versus actions. *Social Forces, 13,* 230–237.

Larson, L. L., & Rowland, K. (1973). Leadership style, stress, and behavior in task performance. *Organizational Behavior and Human Performance, 9,* 407–421.

Larzelere, R., & Huston, T. (1980). The dyadic trust scale: Toward understanding interpersonal trust in close relationships. *Journal of Marriage and the Family, 42,* 595–604.

Latane, B. (1981). The psychology of social impact. *American Psychologist, 36,* 343–356.

Latane, B., & Darley, J. M. (1970). *The unresponsive bystander: Why doesn't he help?* New York: Appleton-Century-Crofts.

Latane, B., Nida, S. A., & Wilson, D. W. (1981). The effects of group size on helping behavior. In J. P. Rushon & R. M. Sorrentino (Eds.), *Altruism and helping behavior: Social, personality and developmental perspectives.* Hillsdale, NJ: Erlbaum.

Latane, B., & Rodin, J. (1969). A lady in distress: Inhibiting effects of friends and strangers on bystander intervention. *Journal of Experimental Social Psychology, 5,* 189–202.

Latane, B., Williams, K., & Harkins, S. (1979). Many hands make light the work: The causes and consequences of social loafing. *Journal of Personality and Social Psychology, 37,* 822–832.

Lau, R. R., & Russell, D. (1980). Attributions in the sports pages. *Journal of Personality and Social Psychology, 39,* 29–38.

Lauderdale, P. (1976). Deviance and moral boundaries. *American Sociological Review, 41,* 660–676.

Laufer, R., & Gallups, M. (1985). Life course effects of Vietnam combat and abusive violence: Marital patterns. *Journal of Marriage and the Family, 47,* 839–853.

Laughlin, P. R. (1980). Social combination processes of cooperative, problem-solving groups as verbal intellective tasks. In M. Fishbein (Ed.), *Progress in social psychology* (Vol. 1). Hillsdale, NJ: Erlbaum.

Lavee, Y., McCubbin, H., & Olson, D. (1987). The effect of stressful life events and transitions on family functioning and well-being. *Journal of Marriage and the Family, 49,* 857–873.

Lawler, E. E., & O'Gara, P. W. (1967). Effects of inequity produced by underpayment on work output, work quality, and attitudes toward work. *Journal of Applied Psychology, 51,* 403–410.

Lawler, E. J. (1975a). An experimental study of factors affecting the mobilization of revolutionary coalitions. *Sociometry, 38,* 163–179.

Lawler, E. J. (1975b). The impact of status differences on coalitional agreements: An experimental study. *Journal of Conflict Resolution, 19,* 271–285.

Lawler, E. J. (1983). Cooptation and threats as "divide and rule" tactics. *Social Psychology Quarterly, 46,* 89–98.

Lawler, E. J. (1986). Bilateral deterrence and conflict spiral. In E. J. Lawler (Ed.), *Advances in group processes* (Vol. 3). Greenwich, CT: JAI Press.

Lawler, E. J., Ford, R. S., & Blegen, M. A. (1988). Coercive capability in conflict: A test of bilateral deterrence versus conflict spiral theory. *Social Psychology Quarterly, 51,* 93–107.

Lawler, E. J., Youngs, J. A., Jr., Lesh, M. D. (1978). Cooptation and coalition mobilization. *Journal of Applied Social Psychology, 8,* 199–214.

Lawson, E. B. (1964). Reinforced and non-reinforced four-man communication nets. *Psychological Reports, 14,* 287–296.

Leary, M. R., Wheeler, D. S., & Jenkins, T. B. (1986). Aspects of identity and behavioral preference: Studies of occupational and recreational choice. *Social Psychology Quarterly, 49,* 11–18.

Leavitt, H. J. (1951). Some effects of certain communication patterns on group performance. *Journal of Abnormal and Social Psychology, 46,* 38–50.

Le Bon, G. (1895). *Psychologie des Foules* [The crowd]. London: Unwin (1903).

Ledvinka, J. (1971). Race of interviewer and the language elaboration of Black interviewees. *Journal of Social Issues, 27,* 185–197.

Leffler, A., Gillespie, D. L., & Conaty, J. C. (1982). The effects of status differentiation on nonverbal behavior. *Social Psychology Quarterly, 45,* 153–161.

Leippe, M. R., & Elkin, R. A. (1987). When motives clash: Issue involvement and response involvement as determinants of persuasion. *Journal of Personality and Social Psychology, 52,* 269–278.

Lemert, E. (1951). *Social Pathology.* New York: McGraw-Hill.

Lemert, E. (1962). Paranoia and the dynamics of exclusion. *Sociometry, 25,* 2–20.

Lemon, N. (1973). *Attitudes and their measurement.* New York: Wiley.

Lemyre, L., & Smith, P. M. (1985). Intergroup discrimination and self-esteem in the minimal intergroup paradigm. *Journal of Personality and Social Psychology, 49,* 660–670.

Lepper, M., Greene, D., & Nisbett, R. (1973). Undermining children's intrinsic interest with extrinsic reward: A test of the "overjustification" hypothesis. *Journal of Personality and Social Psychology, 28,* 129–137.

Leventhal, G. S. (1979). Effects of external conflict on resource allocation and fairness within groups and organizations. In W. G. Austin & S. Worchel (Eds.), *The social psychology of intergroup relations.* Monterey, CA: Brooks/Cole.

Leventhal, G. S., & Lane, D. W. (1970). Sex, age, and equity behavior. *Journal of Personality and Social Psychology, 15,* 312–316.

Leventhal, G. S., Michaels, J. W., & Sanford, C. (1972). Inequity and interpersonal conflict: Reward allocation and secrecy about reward as methods of preventing conflict. *Journal of Personality and Social Psychology, 23,* 88–102.

Leventhal, G. S., Weiss, T., & Long, G. (1969). Equity,

reciprocity, and reallocating the rewards in the dyad. *Journal of Personality and Social Psychology, 13,* 300–305.

Leventhal, H. (1970). Findings and theory in the study of fear communications. In L. Berkowitz (Ed.), *Advances in experimental social psychology* (Vol. 5). New York: Academic Press.

Leventhal, H. (1980). Toward a comprehensive theory of emotion. In L. Berkowitz (Ed.), *Advances in experimental social psychology* (Vol. 13). New York: Academic Press.

Leventhal, H., & Singer, R. P. (1966). Affect arousal and positioning of recommendations in persuasive communications. *Journal of Personality and Social Psychology, 4,* 137–146.

Levin, P. E., & Isen, A. M. (1975). Something you can still get for a dime: Further studies on the effects of feeling good on helping. *Sociometry, 38,* 141–147.

Levine, J. M. (1980). Reaction to opinion deviance in small groups. In Paul B. Paulus (Ed.), *Psychology of group influence.* Hillsdale, NJ: Erlbaum.

Levine, J. M., Saxe, L., & Harris, H. J. (1976). Reaction to attitudinal deviance: Impact of deviate's direction and distance of movement. *Sociometry, 39,* 97–107.

Levine, J. M., Saxe, L., & Ranelli, C. J. (1975). Extreme dissent, conformity reduction, and the bases of social influence. *Social Behavior and Personality, 3,* 117–126.

Levine, J. M., & Russo, E. M. (1987). Majority and minority influence. In C. Hendrick (Ed.), *Group processes.* Newbury Park, CA: Sage.

LeVine, R. A., & Campbell, D. T. (1972). *Ethnocentrism: Theories of conflict, ethnic attitudes and group behavior.* New York: Wiley.

Levinger, G. (1974). A three-level approach to attraction: Toward an understanding of pair relatedness. In T. Huston (Ed.), *Foundations of interpersonal attraction* (pp. 100–120). New York: Academic Press.

Levinger, G. (1976). A social psychological perspective on marital dissolution. *Journal of Social Issues, 32*(1), 21–47.

Levinson, D. (1978). *The seasons of a man's life.* New York: Alfred Knopf.

Levinson, R., Powell, B., & Steelman, L. C. (1986). Social location, significant others and body image among adolescents. *Social Psychology Quarterly, 49,* 330–337.

Levitin, T. E. (1975). Deviants are active participants in the labelling process: The visibly handicapped. *Social Problems, 22,* 548–557.

Lewicki, P. A. (1982). Social psychology as viewed by its practitioners: Survey of SESP members' opinions. *Personality and Social Psychology Bulletin, 8,* 409–416.

Lewin, K., Lippitt, R., & White, R. K. (1939). Patterns of aggressive behavior in experimentally created "social climates." *Journal of Social Psychology, 10,* 271–299.

Lewis, G. H. (1972). Role differentiation. *American Sociological Review, 37,* 424–434.

Lewis, M., & Brookes-Gunn, J. (1979). Toward a theory of social cognition: The development of self. In I. Uzgiris (Ed.), *Social interaction and communication during infancy. New directions for child development* (Vol. 4). San Francisco: Jossey-Bass.

Lewis, S. A., Langan, C. J., & Hollander, E. P. (1972).

Expectation of future interaction and the choice of less desirable alternatives in conformity. *Sociometry, 35,* 440–447.

Lieberman, P. (1975). *On the origins of human language: An introduction to the evolution of human speech.* New York: Macmillan.

Lieberman, S. (1965). The effect of changes of roles on the attitudes of role occupants. In H. Proshansky & B. Seidenberg (Eds.), *Basic studies in social psychology.* New York: Holt, Rinehart and Winston.

Liebert, R. M., Smith, W. P., Hill, J. H., & Kieffer, M. (1968). The effects of information and magnitude of initial offer on interpersonal negotiation. *Journal of Experimental Social Psychology, 4,* 431–441.

Liebhart, E. H. (1972). Empathy and emergency helping: The effects of personality, self-concern and acquaintance. *Journal of Experimental Social Psychology, 8,* 404–411.

Light, I. (1977). The ethnic vice industry, 1880–1944. *American Sociological Review, 42,* 464–479.

Likert, R. (1932). A technique for the measurement of attitudes. *Archives of Psychology* (Whole No. 142).

Likert, R. (1961). *New patterns of management.* New York: McGraw-Hill.

Lin, N. (1974–1975). The McIntire march: A study of recruitment and commitment. *Public Opinion Quarterly, 38,* 562–573.

Lin, N., Ensel, W., & Vaughn, J. (1981). Social resources and strength of ties: Structural factors in occupational status attainment. *American Sociological Review, 46,* 393–405.

Lin, N., & Xie W. (1988). Occupational prestige in urban China. *American Journal of Sociology, 93,* 793–832.

Lincoln, J., & Kalleberg, A. (1985). Work organization and workforce commitment: A study of plants and employees in the United States and Japan. *American Sociological Review, 50,* 738–760.

Linder, D. E., Cooper, J., & Jones, E. (1967). Decision freedom as a determinant of the role of incentive magnitude in attitude change. *Journal of Personality and Social Psychology, 6,* 245–254.

Lindskold, S. (1978). Trust development, the GRIT proposal, and the effects of conciliatory acts on conflict and cooperation. *Psychological Bulletin, 85,* 772–793.

Lindskold, S. (1986). GRIT: Reducing distrust through carefully introduced conciliation. In S. Worchel & W. G. Austin (Eds.), *Psychology of intergroup relations* (2nd ed.) (pp. 305–323). Chicago: Nelson-Hall Publishers.

Lindskold, S., & Aronoff, J. R. (1980). Conciliatory strategies and relative power. *Journal of Experimental Social Psychology, 16,* 187–198.

Lindskold, S., & Collins, M. G. (1978). Inducing cooperation by groups and individuals: Applying Osgood's GRIT strategy. *Journal of Conflict Resolution, 22,* 679–690.

Lindskold, S., Cullen, P., Gahagen, J., & Tedeschi, J. T. (1970). Developmental aspects of reaction to positive inducements. *Developmental Psychology, 3,* 277–284.

Lindskold, S., & Finch, M. L. (1981). Styles of announcing conciliation. *Journal of Conflict Resolution, 25,* 145–155.

Lindskold, S., & Tedeschi, J. T. (1971). Reward power and attraction in interpersonal conflict. *Psychonomic Science, 22,* 211–213.

Link, B. (1982). Mental patient status, work and income: An examination of the effects of a psychiatric label. *American Sociological Review, 47,* 202–215.

Link, B. G. (1987). Understanding labelling effects in the area of mental disorders: An assessment of the effects of expectations of rejection. *American Sociological Review, 52,* 96–112.

Linville, P. W., & Jones, E. E. (1980). Polarized appraisals of out-group members. *Journal of Personality and Social Psychology, 38,* 689–703.

Lippitt, R., & White, R. K. (1952). An experimental study of leadership and group life. In G. E. Swanson, T. M. Newcomb, & E. L. Hartley (Eds.), *Readings in social psychology.* New York: Holt.

Lippmann, W. (1922). *Public opinion.* New York: Harcourt Brace Jovanovich.

Liska, A. (1984). A critical examination of the causal structure of the Fishbein-Ajzen attitude-behavior model. *Social Psychology Quarterly, 47,* 61–74.

Littlepage, G., & Pineault, T. (1978). Verbal, facial, and paralinguistic cues to the detection of truth and lying. *Personality and Social Psychology Bulletin, 4,* 461–464.

Lofland, J. (1981). Collective behavior: The elementary forms. In M. Rosenberg & R. Turner (Eds.), *Social psychology: Sociological perspectives* (pp. 411–446). New York: Basic Books.

Lohr, J. M. & Staats, A. (1973). Attitude conditioning in Sino-Tibetan languages. *Journal of Personality and Social Psychology, 26,* 196–200.

Lopata, H. Z. (1971). *Occupation: Housewife.* New York: Oxford University Press.

Lopata, H. (1988). Support systems of American urban widowhood. *Journal of Social Issues, 44,* 113–128.

Lorence, J., & Mortimer, J. (1985). Job involvement through the life course: A panel study of three age groups. *American Sociological Review, 50,* 618–638.

Lord, C. G., Lepper, M. R., & Mackie, D. (1984). Attitude prototypes as determinants of attitude-behavior consistency. *Journal of Personality and Social Psychology, 46,* 1254–1266.

Lorenz, K. (1966). *On aggression.* New York: Harcourt Brace Jovanovich.

Lorenz, K. (1974). *Civilized man's eight deadly sins.* New York: Harcourt Brace Jovanovich.

Lott, A. J., & Lott, B. E. (1965). Group cohesiveness as interpersonal attraction: A review of relationships with antecedent and consequent variables. *Psychological Bulletin, 64,* 259–309.

Lott, A., & Lott, B. (1974). The role of reward in the formation of positive interpersonal attitudes. In T. Huston (Ed.), *Foundations of interpersonal attraction* (pp. 171–192). New York: Academic Press.

Lowenthal, M. F., Thurnher, M., & Chiriboga, D. (1975). *Four stages of life: A comparative study of women and men facing transitions.* San Francisco: Jossey-Bass.

Luchins, A. S. (1957). Experimental attempts to minimize the impact of first impressions. In C. I. Hovland (Ed.), *The order of presentation in persuasion.* New Haven, CT: Yale University Press.

Luckenbill, D. F. (1982). Compliance under threat of severe punishment. *Social Forces, 60,* 811–825.

Lundman, R. J. (1974). Routine police arrest practices: A commonweal perspective. *Social Problems, 22,* 127–141.

Lurigis, A. J., & Carroll, J. S. (1985). Probation officers' schemata of offenders: Content, development and impact on treatment decisions. *Journal of Personality and Social Psychology, 48,* 1112–1126.

Lynch, J. C., Jr., & Cohen, J. L. (1978). The use of subjective expected utility theory as an aid to understanding variables that influence helping behavior. *Journal of Personality and Social Psychology, 36,* 1138–1151.

Lytton, H., Conway, D., & Sauve, R. (1977). The impact of twinship on parental-child interaction. *Journal of Personality and Social Psychology, 35,* 97–107.

Maass, A., & Clark, R. D., III. (1984). Hidden impact of minorities: Fifteen years of minority influence research. *Psychological Bulletin, 95,* 428–450.

Maass, A., West, S. G., & Cialdini, R. B. (1987). Minority influence and conversion. In C. Hendrick (Ed.), *Group processes.* Newbury Park, CA: Sage.

Macaulay, J. R., & Berkowitz, L. (Eds.) (1970). *Altruism and helping behavior.* New York: Academic Press.

Maccoby, E., & Jacklin, C. (1974). *The psychology of sex differences.* Stanford, CA: Stanford University Press.

Mackie, M. (1983). The domestication of self: Gender comparisons of self-imagery and self-esteem. *Social Psychology Quarterly, 46,* 343–350.

Mackie, D. M., & Goethals, G. R. (1987). Individual and group goals. In C. Hendrick (Ed.), *Group processes.* Newbury Park, CA: Sage.

Maddi, S., Bartone, P., & Puccetti, M. (1987). Stressful events are indeed a factor in physical illness: Reply to Schroeder and Costa (1984). *Journal of Personality and Social Psychology, 52,* 833–843.

Maddux, J. E., & Rogers, R. W. (1980). Effects of source expertness, physical attractiveness, and supporting arguments on persuasion: A case of brains over beauty. *Journal of Personality and Social Psychology, 39,* 235–244.

Maddux, J. E., & Rogers, R. W. (1983). Protection motivation and self-efficacy: A revised theory of fear appeals and attitude change. *Journal of Experimental Social Psychology, 19,* 469–479.

Madsen, D. B. (1978). Issue importance and choice shifts: A persuasive arguments approach. *Journal of Personality and Social Psychology, 36,* 1118–1127.

Maines, D., & Hardesty, M. (1987). Temporality and gender: Young adults' career and family plans. *Social Forces, 66,* 102–120.

Malinowski, C. I., & Smith, C. P. (1985). Moral reasoning and moral conduct: An investigation prompted by Kohlberg's theory. *Journal of Personality and Social Psychology, 49,* 1016–1027.

Mannheim, B. F. (1966). Reference groups, membership groups and the self-image. *Sociometry, 29,* 265–279.

Mannheimer, D., & Williams, R. M., Jr. (1949). A note on Negro troops in combat. In S. A. Stouffer, E. A. Suchman, L. C. DeVinney, S. A. Star, R. M. Williams, Jr. (Eds.), *The American soldier* (Vol. 1). Princeton, NJ: Princeton University Press.

Marger, M. N. (1984). Social movement organizations and

response to environmental change: The NAACP, 1960–1973. *Social Problems, 32,* 16–30.

Marini, M. M. (1978). Sex differences in educational attainment and age at marriage. *American Sociological Review, 43,* 483–507.

Markus, H. (1977). Self-schemata and processing information about the self. *Journal of Personality and Social Psychology, 35,* 63–78.

Markus, H., & Zajonc, R. B. (1985). The cognitive perspective in social psychology. In G. Lindzey & E. Aronson (Eds.), *Handbook of social psychology* (3rd ed.) (Vol. I). New York: Random House.

Marsden, P., & Campbell, K. (1984). Measuring tie strength. *Social Forces, 63,* 482–501.

Marsh, H. W., Barnes, J., & Hocevar, D. (1985). Self-other agreement on multidimensional self-concept ratings: Factor analysis and multitrait-multimethod analysis. *Journal of Personality and Social Psychology, 49,* 1360–1377.

Marshall, S. E. (1985). Ladies against women: Mobilization dilemmas of antifeminist movement. *Social Problems, 32,* 348–362.

Marshall, S. E. (1986). In defense of separate spheres: Class and status politics in the antisuffrage movement. *Social Forces, 65,* 327–351.

Martin, C. L. (1987). A ratio measure of sex stereotyping. *Journal of Personality and Social Psychology, 52,* 489–499.

Marwell, G., Aiken, M. T., & Demerath, N. J., III. (1987). The persistence of political attitudes among 1960s civil rights activists. *Public Opinion Quarterly, 51,* 383–399.

Marwell, G., McKinney, K., Sprecher, S., Smith, S., & DeLamater, J. (1982). Legitimizing factors in the initiation of heterosexual relationships. Paper presented at the International Conference on Personal Relationships, Madison WI.

Marx, G. T., & Wood, J. L. (1975). Strands of theory and research in collective behavior. In A. Inkeles, J. Coleman, & N. Smelser (Eds.), *Annual review of sociology* (Vol. 1, pp. 363–428). Palo Alto, CA: Annual Reviews.

Marx, K. (1964). In T. B. Bottomore (Ed. and Trans.), *Early writings.* New York: McGraw-Hill.

Matsueda, R. (1982). Testing control theory and differential association: A causal modeling approach. *American Sociological Review, 47,* 489–504.

Matthews, S. H. (1977). *The social world of old women.* Beverly Hills, CA: Sage.

Matthews, S., & Rosner, T. (1988). Shared filial responsibility: The family as primary care giver. *Journal of Marriage and the Family, 50,* 185–195.

Maynard, D. W. (1978). Placement of topic changes in conversation. *Semiotica, 30* (3/4) 263–290.

Maynard, D. W. (1983). Social order and plea bargaining in the court. *Sociological Quarterly, 24,* 215–233.

McAdam, D. (1986). Recruitment to high-risk activism: The case of Freedom Summer. *American Journal of Sociology, 92,* 64–90.

McArdle, J. B. (1972). Positive and negative communications and subsequent attitude and behavior change in alcoholics. Unpublished doctoral dissertation, University of Illinois.

McArthur, L. Z. (1972). The how and what of why: Some determinants and consequences of causal attribution. *Journal of Personality and Social Psychology, 22,* 171–193.

McArthur, L. Z., & Post, D. L. (1977). Figural emphasis and person perception. *Journal of Experimental Social Psychology, 13,* 520–535.

McCall, G. J., & Simmons, J. L. (1978). *Identities and interactions.* New York: Free Press.

McCannell, K. (1988). Social networks and the transition to motherhood. In R. Milardo (Ed.), *Families and social networks.* Newbury Park, CA: Sage.

McCarthy, J. D., & Hoge, D. R. (1984). The dynamics of self-esteem and delinquency. *American Journal of Sociology, 90,* 396–410.

McCauley, C., Stitt, C. L., & Segal, M. (1980). Stereotyping: From prejudice to prediction. *Psychological Bulletin, 87,* 195–208.

McClelland, D. (1958). Risk-taking in children with high and low need for achievement. In J. Atkinson (Ed.), *Motives in fantasy, action and society.* Princeton, NJ: Van Nostrand.

McClelland, D. (1961). *The achieving society.* Princeton, NJ: Van Nostrand.

McClelland, D., & Winter, D. (1969). *Motivating economic achievement.* New York: Free Press.

McCombs, M. E., & Eyal, C. H. (1980). Spending on mass media. *Journal of Communication, 30,* 153–158.

McCullers, J. C., Fabes, R. A., & Moran, J. D., III. (1987). Does intrinsic motivation theory explain the adverse effects of rewards on immediate task performance? *Journal of Personality and Social Psychology, 52,* 1027–1033.

McFarland, C., & Ross, M. (1982). The impact of causal attributions on affective reactions to success and failure. *Journal of Personality and Social Psychology, 43,* 937–946.

McGlynn, R. P. (1982). A comment on the meta-analysis of goal structures. *Psychological Bulletin, 92,* 184–185.

McGrath, J. E. (1962). The influence of positive interpersonal relations on adjustment and effectiveness in rifle teams. *Journal of Abnormal and Social Psychology, 65,* 365–375.

McGrath, J. E. (1984). *Groups: Interaction and performance.* Englewood Cliff, NJ: Prentice-Hall.

McGregor, D. M. (1960). *The human side of enterprise.* New York: McGraw-Hill.

McGuire, W. J. (1964). Inducing resistance to persuasion: Some contemporary approaches. In L. Berkowitz (Ed.), *Advances in experimental social psychology* (Vol. 1). New York: Academic Press.

McGuire, W. J. (1972). Attitude change: The information-processing paradigm. In C. G. McClintock (Ed.), *Experimental social psychology.* New York: Holt, Rinehart and Winston.

McGuire, W. J. (1973). The yin and yang of progress in social psychology: Seven koan. *Journal of Personality and Social Psychology, 26,* 446–456.

McGuire, W. J. (1985). Attitude and attitude change. In G. Lindzey & E. Aronson (Eds.), *The handbook of social psychology* (3rd ed.) (Vol. II). New York: Random House.

McGuire, W. J., & McGuire, C. (1982). Significant others in self-space: Sex differences and developmental trends in the social self. In J. Suls (Ed.), *Psychological perspectives on the self* (Vol. 1). Hillsdale, NJ: Erlbaum.

McGuire, W. J., & McGuire, C. (1986). Differences in conceptualizing self versus conceptualizing other people as manifested in contrasting verb types used in natural speech. *Journal of Personality and Social Psychology, 51,* 1135–1143.

McGuire, W. J., & Padawer-Singer, A. (1976). Trait salience in the spontaneous self-concept. *Journal of Personality and Social Psychology, 33,* 743–754.

McGuire, W. J., & Papageorgis, D. (1961). The relative efficacy of various types of prior belief-defense in producing immunity against persuasion. *Journal of Abnormal and Social Psychology, 62,* 327–337.

McLeod, J. M., Price, K. O., & Harburg, E. (1966). Socialization, liking and yielding of opinions in imbalanced situations. *Sociometry, 29,* 197–212.

McNamara, E. F., Kurth, T., & Hansen, D. (1981). Communication efforts to prevent wildfires. In R. E. Rice & W. J. Paisley (Eds.), *Public communication campaigns.* Beverly Hills, CA: Sage.

McPhail, C., & Wohlstein, R. T. (1983). Individual and collective behavior within gatherings, demonstrations and riots. In R. H. Turner & J. F. Short (Eds.), *Annual review of sociology* (Vol. 9, pp. 579–600). Palo Alto, CA: Annual Reviews.

McPherson, J., & Smith-Lovin, L. (1986). Sex segregation in voluntary associations. *American Sociological Review, 51,* 61–79.

McWorter, G. A., & Crain, R. L. (1967). Subcommunity gladiatorial competition: Civil rights leadership as a competitive process. *Social Forces, 46,* 8–21.

Mead, G. H. (1934). *Mind, self, and society.* Chicago: University of Chicago Press.

Mears, P. (1974). Structuring communication in a working group. *Journal of Communication, 24,* 71–79.

Meddin, J. (1979). Chimpanzees, symbols and the reflective self. *Social Psychology Quarterly, 42,* 99–100.

Mednick, S. A., & Schulsinger, F. (1969). Factors related to breakdown in children at high risk for schizophrenia. In M. Rolf & D. Rocks (Eds.), *Life history studies in psychopathology.* Minneapolis: University of Minnesota Press.

Meeker, B. F. (1981). Expectation states and interpersonal behavior. In M. Rosenberg & R. H. Turner (Eds.), *Social psychology: Sociological perspectives.* New York: Basic Books.

Mehrabian, A. (1972). *Nonverbal communication.* New York: Aldine-Atherton.

Mehrabian, A., & Ksionzky, S. (1970). Models for affiliative and conformity behavior. *Psychological Bulletin, 74,* 110–126.

Meier, R. F., & Johnson, W. T. (1977). Deterrence as social control: The legal and extralegal production of conformity. *American Sociological Review, 42,* 292–304.

Mendelsohn, H. (1973). Some reasons why information campaigns can succeed. *Public Opinion Quarterly, 37,* 50–61.

Merton, R. K. (1948). The self-fulfilling prophecy. *Antioch Review, 8,* 193–210.

Merton, R. (1957). *Social theory and social structure.* Glencoe, IL: Free Press.

Meyrowitz, J. (1985). *No sense of place: The impact of electronic media on social behavior.* New York: Oxford University Press.

Miall, C. E. (1986). The stigma of involuntary childlessness. *Social Problems, 33,* 268–282.

Michaels, J. W., Edwards, J. N., & Acock, A. C. (1984). Satisfaction in intimate relationships as a function of inequality, inequity and outcomes. *Social Psychology Quarterly, 47,* 347–357.

Michener, H. A., & Burt, M. R. (1974). Legitimacy as a base of social influence. In J. T. Tedeschi (Ed.), *Perspectives on social power.* Chicago: Aldine-Atherton.

Michener, H. A., & Burt, M. R. (1975). Components of "authority" as determinants of compliance. *Journal of Personality and Social Psychology, 31,* 605–614.

Michener, H. A., & Burt, M. R. (1975). Use of social influence under varying conditions of legitimacy. *Journal of Personality and Social Psychology, 32,* 398–407.

Michener, H. A., & Cohen, E. D. (1973). Effects of punishment magnitude in the bilateral threat situation: Evidence for the deterrence hypothesis. *Journal of Personality and Social Psychology, 26,* 427–438.

Michener, H. A., & Lawler, E. J. (1971). Revolutionary coalition strength and collective failure as determinants of status reallocation. *Journal of Experimental Social Psychology, 7,* 448–460.

Michener, H. A., & Lawler, E. J. (1975). Endorsement of formal leaders: An integrative model. *Journal of Personality and Social Psychology, 31,* 216–223.

Michener, H. A., & Lyons, M. (1972). Perceived support and upward mobility as determinants of revolutionary coalitional behavior. *Journal of Experimental Social Psychology, 8,* 180–195.

Michener, H. A., Plazewski, J. G., & Vaske, J. J. (1979). Ingratiation tactics channeled by target values and threat capability. *Journal of Personality, 47,* 36–56.

Michener, H. A., & Tausig, M. (1971). Usurpation and perceived support as determinants of the endorsement accorded formal leaders. *Journal of Personality and Social Psychology, 18,* 364–372.

Michener, H. A., Vaske, J. J., Schleifer, S. L., Plazewski, J. G., & Chapman, L. J. (1975). Factors affecting concession rate and threat usage in bilateral conflict. *Sociometry, 38,* 62–80.

Milardo, R. M. (1982). Friendship networks in developing relationships: Converging and diverging social environments. *Social Psychology Quarterly, 45,* 162–172.

Milardo, R. (1988). Families and social networks: An overview of theory and methodology. In R. Milardo (Ed.), *Families and social networks.* Newbury Park, CA: Sage.

Milardo, R., Johnson, M., & Huston, T. (1983). Developing close relationships: Changing patterns of interaction between pair members and social networks. *Journal of Personality and Social Psychology, 44,* 964–976.

Milavsky, J. R., Kessler, R., Stipp, H., & Rubens, W. (1983). *Television and aggression: A panel study.* New York: Academic Press.

Miles, R. H. (1977). Role-set configuration as a predictor of role conflict and ambiguity in complex organizations. *Sociometry, 40,* 21–34.

Milgram, S. (1963). Behavioral study of obedience. *Journal of Abnormal and Social Psychology, 67,* 371–378.

Milgram, S. (1965a). Some conditions of obedience and disobedience to authority. *Human Relations, 18,* 57–76.

Milgram, S. (1965b). Liberating effects of group pressure. *Journal of Personality and Social Psychology, 1,* 127–134.

Milgram, S. (1972). The lost letter technique. In L. Bickman & T. Henchy (Eds.), *Beyond the laboratory: Field research in social psychology.* New York: McGraw-Hill.

Milgram, S. (1974). *Obedience to authority.* New York: Harper and Row.

Milgram, S., Liberty, H. J., Toledo, R., & Wackenhut, J. (1986). Response to intrusion into waiting lines. *Journal of Personality and Social Psychology, 51,* 683–689.

Milgram, S., & Toch, H. (1969). Collective behavior: Crowds and social movements. In G. Lindzey & E. Aronson (Eds.), *The handbook of social psychology* (2nd ed.) (Vol. IV, pp. 507–610). Reading, MA: Addison-Wesley.

Miller, A. G. (1976). Constraint and target effects on the attribution of attitudes. *Journal of Experimental Social Psychology, 12,* 325–339.

Miller, A. G. (Ed.). (1982). *In the eye of the beholder: Contemporary issues in stereotyping.* New York: Praeger Publishers.

Miller, G. H. (1981). *Language and speech.* San Francisco: W. H. Freeman.

Miller, G. H., & McNeill, D. (1969). Psycholinguistics. In G. Lindzey & E. Aronson (Eds.), *The handbook of social psychology* (2nd ed.) (Vol. 3). Reading, MA: Addison-Wesley.

Miller, J., Schooler, C., Kohn, M., & Miller, K. (1979). Women and work: The psychological effects of occupational conditions. *American Journal of Sociology, 85,* 66–94.

Miller, L. K., & Hamblin, R. L. (1963). Interdependence, differential rewarding, and productivity. *American Sociological Review, 43,* 193–204.

Miller, K., Kohn, M., & Schooler, C. (1986). Educational self-direction and personality. *American Sociological Review, 51,* 372–390.

Miller, L. C., Berg, J., & Archer, R. (1983). Openers: Individuals who elicit intimate self-disclosure. *Journal of Personality and Social Psychology, 44,* 1234–1244.

Miller, R. S. (1986). Embarrassment: Causes and consequences. In W. Jones, J. Cheek, & S. Briggs (Eds.), *Shyness: Perspectives on research and treatment.* New York: Plenum.

Miller, R. S. (1987). Empathic embarrassment: Situational and personal determinants of reactions to the embarrassment of another. *Journal of Personality and Social Psychology, 53,* 1061–1069.

Miller-McPherson, J., & Smith-Lovin, L. (1982). Women and weak ties: Differences by sex in the size of voluntary organizations. *American Journal of Sociology, 87,* 883–904.

Mills, C. W. (1970). Situated actions and vocabularies of motive. In G. P. Stone & H. A. Farberman (Eds.), *Social psychology through symbolic interaction.* Waltham, MA: Ginn-Blaisdell.

Mills, J., & Clark, M. S. (1982). Exchange and communal relationships. In L. Wheeler (Ed.), *Review of personality and social psychology* (Vol. III). Beverly Hills, CA: Sage.

Mills, T. M. (1967). *The sociology of small groups.* Englewood Cliff, NJ: Prentice-Hall.

Minnigerode, F., & Lee, J. A. (1978). Young adults' percep-

tions of sex roles across the lifespan. *Sex Roles, 4,* 563–569.

Mirowsky, J., & Ross, C. (1986). Social patterns of distress. In A. Inkeles, J. Coleman, & N. Smelser (Eds.), *Annual review of sociology* (Vol. 12, pp. 23–45). Palo Alto, CA: Annual Reviews.

Mischel, W., & Liebert, R. (1966). Effects of discrepancies between deserved and imposed reward criteria on their acquisition and transmission. *Journal of Personality and Social Psychology, 3,* 45–53.

Miyamoto, S. F. (1973). The forced evacuation of the Japanese minority during World War II. *Journal of Social Issues, 29,* 11–31.

Miyamoto, S. F., & Dornbusch, S. (1956). A test of interactionist hypotheses of self-conception. *American Journal of Sociology, 61,* 399–403.

Mizrahi, T. (1984). Coping with patients: Subcultural adjustments to the conditions of work among internists-in-training. *Social Problems, 32,* 156–166.

Modigliani, A. (1971). Embarrassment, face-work, and eye contact: Testing a theory of embarrassment. *Journal of Personality and Social Psychology, 17,* 15–24.

Moede, W. (1927). Die Richtlinien der Leistungs-Psychologie. [Guidelines for a psychology of achievement.] *Industrielle Psychotechnik, 4,* 193–209.

Money, J., & Ehrhardt, A. (1972). *Man and woman. Boy and girl.* Baltimore, MD: Johns Hopkins University Press.

Moore, J. C., Jr. (1968). Status and influence in small group interaction. *Sociometry, 31,* 47–63.

Moran, G. (1966). Dyadic attraction and orientational consensus. *Journal of Personality and Social Psychology, 4,* 94–99.

Moray, N. (1959). Attention in dichotic listening: Affective cues and the influence of instructions. *Quarterly Journal of Experimental Psychology, 12,* 56–60.

Moreno, J. L. (1934). *Who shall survive?* Washington, DC: Nervous and Mental Disease Publishing Co.

Morgan, S., & Rindfuss, R. (1985). Marital disruption: Structural and temporal dimensions. *American Journal of Sociology, 90,* 1055–1077.

Morgan, W., Alwin, D., & Griffin, L. (1979). Social origins, parental values and the transmission of inequality. *American Journal of Sociology, 85,* 156–166.

Morgan, W. R., & Clark, T. (1973). Causes of racial disorders: A grievance-level explanation. *American Sociological Review, 38,* 611–624.

Morrione, T. J. (1975). Symbolic interactionism and social action theory. *Sociology and Social Research, 59,* 201–218.

Morris, W. N., & Miller, R. S. (1975). The effect of consensus-breaking and consensus-preempting partners on reduction of conformity. *Journal of Experimental Social Psychology, 11,* 215–223.

Morrissette, J. O. (1966). Group performance as a function of task difficulty and size and structure of groups, II. *Journal of Personality and Social Psychology, 3,* 357–359.

Morrison, D. E. (1971). Some notes toward a theory on relative deprivation, social movements and social change. *American Behavioral Scientist, 14,* 675–690.

Morse, S., & Gergen, K. (1970). Social comparison, self-consistency, and the concept of self. *Journal of Personality and Social Psychology, 16,* 148–156.

Mortimer, J. T., Finch, M., & Kumka, D. (1982). Persistence and change in development: The multidimensional self-concept. In P. Baltes & O. Brim, Jr. (Eds.), *Life span development and behavior* (Vol. 4). New York: Academic Press.

Mortimer, J. T., & Simmons, R. (1978). Adult socialization. In R. Turner, J. Coleman, & R. Fox (Eds.), *Annual review of sociology* (Vol. 4). Palo Alto, CA: Annual Reviews.

Moscovici, S. (1980). Toward a theory of conversion behavior. In L. Berkowitz (Ed.), *Advances in experimental social psychology* (Vol. 13). New York: Academic Press.

Moscovici, S. (1985). Innovation and minority influence. In S. Moscovici, G. Mugny, & E. Van Avermaet (Eds.), *Perspectives on minority influence* (pp. 9–51). Cambridge, England: Cambridge University Press.

Moscovici, S., & Lage, E. (1976). Studies in social influence III: Majority versus minority influence in a group. *European Journal of Social Psychology, 6,* 149–174.

Moscovici, S., & Mugny, G. (1983). Minority influence. In P. B. Paulus (Ed.), *Basic group processes* (pp. 41–64). New York: Springer-Verlag.

Moss, H., & Kagan, J. (1961). Stability of achievement and recognition seeking behavior from early childhood through adulthood. *Journal of Abnormal and Social Psychology, 62,* 504–513.

Mugny, G. (1982). *The power of minorities.* New York: Academic Press.

Mugny, G. (1984). The influence of minorities: Ten years later. In H. Tajfel (Ed.), *The social dimension: European developments in social psychology* (Vol. 2, pp. 498–517). Cambridge, England: Cambridge University Press.

Muir, D., & Weinstein, E. (1962). The social debt: An investigation of lower-class and middle-class norms of social obligation. *American Sociological Review, 27,* 532–539.

Mulder, M., & Stermerding, A. (1963). Threat, attraction to the group, and need for strong leadership: A laboratory experiment in a natural setting. *Human Relations, 16,* 317–334.

Muller, E. (1985). Income inequality, regime repressiveness and political violence. *American Sociological Review, 50,* 47–61.

Murray, J. P., & Kippax, S. (1979). From the early window to the late night show: International trends in the study of television's impact on children and adults. In L. Berkowitz (Ed.), *Advances in experimental social psychology* (Vol. 12). New York: Academic Press.

Murstein, B. (1976). *Who will marry whom?* New York: Springer.

Murstein, B. (1980). Mate selection in the 1970s. *Journal of Marriage and the Family, 42,* 777–792.

Myers, A. (1962). Team competition, success, and the adjustment of group members. *Journal of Abnormal and Social Psychology, 65,* 325–332.

Myers, D. G. (1975). Discussion-induced attitude polarization. *Human Relations, 28,* 699–714.

Myers, D. G., & Kaplan, M. F. (1976). Group-induced polarization in simulated juries. *Personality and Social Psychology Bulletin, 2,* 63–66.

Myers, D. G., & Lamm, H. (1976). The group polarization phenomenon. *Psychological Bulletin, 83,* 602–627.

Myers, D. G., Bruggink, J. B., Kersting, R. C., & Schlosser, B. A. (1980). Does learning others' opinions change one's opinion? *Personality and Social Psychology Bulletin, 6,* 253–260.

Myers, F. E. (1971). Civil disobedience and organization change: The British Committee of 100. *Political Science Quarterly, 86,* 92–112.

Myers, M. A., & Hagan, J. (1979). Private and public trouble: Prosecutors and the allocation of court resources. *Social Problems, 26,* 439–451.

Myers, M. A., & Talarico, S. M. (1986). The social contexts of racial discrimination in sentencing. *Social Problems, 33,* 236–251.

Nadler, A., & Fisher, J. D. (1984a). Effects of donor-recipient relationships on recipients reactions to aid. In E. Staub, D. Bar-Tal, J. Karylowski, & J. Reykowski (Eds.), *Development and maintenance of prosocial behavior: International perspectives on positive morality.* New York: Plenum.

Nadler, A., & Fisher, J. D. (1984b). The role of threat to self-esteem and perceived control in recipient reaction to aid. In L. Berkowitz (Ed.), *Advances in experimental social psychology* (Vol. 17). New York: Academic Press.

Nadler, A., & Mayseless, O. (1983). Recipient self-esteem and reactions to help. In J. D. Fisher, A. Nadler, & B. M. DePaulo (Eds.), *New directions in helping* (Vol. 1). New York: Academic Press.

Neal, A. G., & Groat, H. (1974). Social class correlates of stability and change in levels of alienation: A longitudinal study. *Sociological Quarterly, 15,* 548–558.

Nemeth, C. J. (1986). Differential contributions of majority and minority influence. *Psychological Bulletin, 93,* 23–32.

Neugarten, B. L., & Datan, N. (1973). Sociological perspectives on the life cycle. In P. Baltes & K. Schaie (Eds.), *Life-span developmental psychology: Personality and social processes.* New York: Academic Press.

Newcomb, T. M. (1943). *Personality and social change.* New York: Dryden.

Newcomb, T. M. (1961). *The acquaintance process.* New York: Holt, Rinehart and Winston.

Newcomb, T. M. (1968). Interpersonal balance. In R. P. Abelson et al. (Eds.), *Theories of cognitive consistency: A sourcebook.* Chicago: Rand McNally.

Newcomb, T. M. (1971). Dyadic balance as a source of clues about interpersonal attraction. In B. Murstein (Ed.), *Theories of attraction and love.* New York: Springer.

Newspaper Advertising Bureau. (1980). *Mass media in the family setting: Social patterns in media availability and use by parents.* New York: Newspaper Advertising Bureau.

*Newsweek.* Gold in the streets. (1978, January 9). *Newsweek* pp. 56–57.

Newtson, D. (1974). Dispositional inference from effects of actions: Effects chosen and effects forgone. *Journal of Experimental Social Psychology, 10,* 487–496.

Nimmo, D. D., & Sanders, K. R. (Eds.). (1981). *Handbook of political communication.* Beverly Hills, CA: Sage.

Nisbett, R. E., Caputo, C., Legant, P., & Maracek, J. (1973). Behavior as seen by the actor and as seen by the observer. *Journal of Personality and Social Psychology, 27,* 154–164.

Nixon, H. L., II. (1976). *Sport and social organization.* Indianapolis, IN: Bobbs-Merrill.

Nixon, H. L. II. (1977a). "Cohesiveness" and team success: A theoretical reformulation. *Review of Sport and Leisure, 2,* 36–57.

Nixon, H. L., II. (1977b). Reinforcement effects of sports teams success on cohesiveness-related factors. *International Review of Sport Sociology, 4,* 17–38.

Nizer, L. (1973). *The implosion conspiracy.* New York: Doubleday.

Norman, R. (1975). Affective-cognitive consistency, attitudes, conformity, and behavior. *Journal of Personality and Social Psychology, 32,* 83–91.

Nyden, P. W. (1985). Democratizing organizations: A case study of a union reform movement. *American Journal of Sociology, 90,* 1179–1203.

O'Barr, W., & Atkins, B. (1980). Women's language or powerless language. In S. McConnell-Ginet, R. Borker, & N. Furman (Eds.), *Women and language in literature and society.* New York: Praeger Publishers.

O'Bryant, S. (1988). Sibling support and older widows' well-being. *Journal of Marriage and the Family, 50,* 173–183.

Oberschall, A. (1973). *Social conflict and social movements.* Englewood Cliffs, NJ: Prentice-Hall.

Oberschall, A. (1978). Theories of social conflict. In R. Turner, J. Coleman, & R. Fox (Eds.), *Annual review of sociology* (Vol. 4, pp. 291–315). Palo Alto, CA: Annual Reviews.

Oliver, P. (1980). Rewards and punishments as selective incentives for collective action: Theoretical investigations. *American Journal of Sociology, 85,* 1356–1375.

Olver, R. (1961). Developmental study of cognitive equivalence. Unpublished doctoral dissertation, Radcliffe College.

Oppenheimer, V. K. (1970). The female labor force in the United States. *Population Monograph Series* (No. 5). Berkeley, CA: Institute of International Studies.

Orcutt, J. D. (1973). Societal reaction and the response to deviation in small groups. *Social Forces, 52,* 259–267.

Orcutt, J. (1975). Deviance as a situated phenomenon: Variations in the social interpretation of marijuana and alcohol use. *Social Problems, 22,* 346–356.

Orne, M. T. (1969). Demand characteristics and the concept of quasi-controls. In R. Rosenthal & R. Rosnow (Eds.), *Artifact in behavior research.* New York: Academic Press.

Orwell, G. (1949). *1984.* New York: Harcourt Brace Jovanovich.

Osgood, C. E. (1962). *An alternative to war or surrender.* Urbana: University of Illinois Press.

Osgood, C. E. (1979). GRIT for MBFR: A proposal for unfreezing force-level postures in Europe. *Peace Research Review, 8,* 77–92.

Osgood, C. E. (1980, May). The GRIT strategy. *Bulletin of the Atomic Scientists,* 58–60.

Osgood, C. E., Suci, G., & Tannenbaum, P. (1957). *The measurement of meaning.* Urbana: University of Illinois Press.

Oskamp, S. (1971). Effects of programmed strategies on cooperation in the Prisoner's Dilemma and other mixed-motive games. *Journal of Conflict Resolution, 15,* 225–259.

Page, A. L., & Clelland, D. A. (1978). The Kanawha County textbook controversy: A study of the politics of lifestyle concern. *Social Forces, 57,* 265–281.

Paicheler, G., & Bouchet, J. (1973). Attitude polarization, familiarization, and group process. *European Journal of Social Psychology, 3,* 83–90.

Palmore, E. (1981). *Social patterns in normal aging.* Durham, NC: Duke University Press.

Papastamou, S., & Mugny, G. (1985). Rigidity and minority influence: The influence of the social in social influence. In S. Moscovici, G. Mugny, & E. Van Avermaet (Eds.), *Perspectives on minority influence* (pp. 113–136). Cambridge, England: Cambridge University Press.

Parke, R. (1969). Effectiveness of punishment as an interaction of intensity, timing, agent nurturance and cognitive structuring. *Child Development, 40,* 213–235.

Parke, R. (1970). The role of punishment in the socialization process. In R. Hoppe, G. Milton, & E. Simmel (Eds.), *Early experiences and the processes of socialization.* New York: Academic Press.

Parke, R. D., Berkowitz, L., Leyens, J. P., West, S., & Sebastian, R. J. (1977). Some effect of violent and nonviolent movies on the behavior of juvenile delinquents. In L. Berkowitz (Ed.), *Advances in experimental social psychology* (Vol. 10). New York: Academic Press.

Patterson, M. L., Mullens, S., & Romano, J. (1971). Compensatory reactions to spatial intrusion. *Sociometry, 34,* 114–121.

Patterson, R. J., & Neufeld, R. W. J. (1987). Clear danger: Situational determinants of the appraisal of threat. *Psychological Bulletin, 101,* 404–416.

Patterson, T. E. (1980). *The mass media election: How Americans choose their president.* New York: Praeger Publishers.

Pearlin, L., & Johnson, J. (1977). Marital status, life-strains and depression. *American Sociological Review, 42,* 704–715.

Pearlin, L., & Kohn, M. (1966). Social class, occupation and parental values: A cross-national study. *American Sociological Review, 31,* 466–479.

Pearlin, L., & Radabaugh, C. (1976). Economic strains and the coping functions of alcohol. *American Journal of Sociology, 82,* 652–663.

Pennebaker, J. W. (1980). Self-perception of emotion and internal sensation. In D. W. Wegner & R. R. Vallacher (Eds.), *The self in social psychology.* New York: Oxford University Press.

Pennebaker, J., Dyer, M., Caulkins, R., Litowitz, D., Ackerman, P., Anderson, D., & McGraw, K. (1979). Don't the girls get prettier at closing time: A country and western application to psychology. *Personality and Social Psychology Bulletin, 5,* 122–125.

Peplau, L., Rubin, Z., & Hill, C. (1977). Sexual intimacy in dating relationships. *Journal of Social Issues, 33*(2), 86–109.

Perlman, D. (1988). Loneliness: A life-span family perspective. In R. Milardo (Ed.), *Families and social networks.* Newbury Park, CA: Sage.

Perry, J. B., & Pugh, M. D. (1978). *Collective behavior: Response to social stress.* St. Paul: West Publishing Company.

Peters, L. H., Hartke, D. D., & Pohlmann, J. T. (1985). Fiedler's contingency theory of leadership: An application of the meta-analysis procedures of Schmidt and Hunter. *Psychological Bulletin, 97,* 274–285.

Petrunik, M., & Shearing, C. D. (1983). Fragile facades: Stuttering and the strategic manipulation of awareness. *Social Problems, 31,* 125–138.

Petty, R. E., & Cacioppo, J. T. (1979). Issue involvement can increase or decrease persuasion by enhancing message-relevant cognitive responses. *Journal of Personality and Social Psychology, 37,* 1915–1926.

Petty, R. E., & Cacioppo, J. T. (1981). *Attitudes and persuasion: Classic and contemporary approaches.* Dubuque, IA: Wm. C. Brown.

Petty, R. E., & Cacioppo, J. T. (1986a). *Communication and persuasion: Central and peripheral routes to attitude change.* New York: Springer-Verlag.

Petty, R. E., & Cacioppo, J. T. (1986b). The elaboration likelihood model of persuasion. In L. Berkowitz (Ed.), *Advances in experimental social psychology* (Vol. 19). New York: Academic Press.

Petty, R. E., Cacioppo, J. T., & Goldman, R. (1981). Personal involvement as a determinant of argument-based persuasion. *Journal of Personality and Social Psychology, 41,* 847–855.

Petty, R. E., Cacioppo, J. T., & Heesacker, M. (1981). Effects of rhetorical questions on persuasion: A cognitive response analysis. *Journal of Personality and Social Psychology, 40,* 432–440.

Phelps, E. S. (1975). *Altruism, morality and economic theory.* Chicago: Russell Sage Foundation.

Phillips, D. P. (1974). The influence of suggestion on suicide: Substantive and theoretical implications of the Werther effect. *American Sociological Review, 39,* 340–354.

Phillips, D. P. (1979). Suicide, motor vehicle fatalities and the mass media: Evidence toward a theory of suggestion. *American Sociological Review, 84,* 1150–1174.

Piaget, J. (1954). *The construction of reality in the child.* New York: Basic Books.

Piaget, J. (1965). *The moral judgment of the child.* New York: Free Press.

Piliavin, I. M., & Briar, S. (1964). Police encounters with juveniles. *American Journal of Sociology, 70,* 206–214.

Piliavin, I. M., Rodin, J., Piliavin, J. A. (1969). Good Samaritanism: An underground phenomenon? *Journal of Personality and Social Psychology, 13,* 289–299.

Piliavin, J. A., Dovidio, J. F., Gaertner, S. L., & Clark, R. D., III. (1981). *Emergency intervention.* New York: Academic Press.

Pilisuk, M., & Skolnick, P. (1968). Inducing trust: A test of the Osgood Proposal. *Journal of Personality and Social Psychology, 8,* 121–133.

Pilisuk, M., Winter, J. A., Chapman, R., & Haas, N. (1967). Honesty, deceit, and timing in the display of intentions. *Behavioral Science, 12,* 205–215.

Pittman, J., & Lloyd, S. (1988). Quality of family life, social support and stress. *Journal of Marriage and the Family, 50,* 53–67.

Plastic surgery now commonplace. (1979, July 15). *Wisconsin State Journal.* Madison, WI.

Pleck, J. H. (1976). The male sex role: Definitions, problems and sources of change. *Journal of Social Issues, 32,* 155–164.

Pleck, J. H. (1985). *Working wives/Working husbands.* Beverly Hills, CA: Sage.

Plutzer, E. (1987). Determinants of leftist radical belief in the United States: A test of competing theories. *Social Forces, 65,* 1002–1017.

Podsakoff, P. M. (1982). Determinants of a supervisor's use of rewards and punishments: A literature review and suggestions for further research. *Organizational Behavior and Human Performance, 29,* 58–83.

Poyatos, F. (1983). *New perspectives in nonverbal communication: Studies in cultural anthropology, social psychology, linguistics, literature and semantics.* Oxford: Pergamon.

Premack, D., & Premack, A. (1984). *The mind of an ape.* New York: Norton.

Prentice-Dunn, S., & Rogers, R. W. (1980). Effects of deindividuating situational cues and aggressive models on subjective deindividuation and aggression. *Journal of Personality and Social Psychology, 39,* 104–113.

Presser, H. (1988). Shift work and child care among young dual-earner American parents. *Journal of Marriage and the Family, 50,* 133–148.

Price, K. O., Harburg, E., & Newcomb, T. M. (1966). Psychological balance in situations of negative interpersonal attitudes. *Journal of Personality and Social Psychology, 3,* 265–270.

Priest, R. T., & Sawyer, J. (1967). Proximity and peership: Bases of balance in interpersonal attraction. *American Journal of Sociology, 72,* 633–649.

Pritchard, R., Dunnette, M., & Jorgenson, D. (1972). Effects of perception of equity and inequity on worker performance and satisfaction. *Journal of Applied Psychology, 56,* 75–94.

Pruitt, D. G. (1968). Reciprocity and credit building in a laboratory dyad. *Journal of Personality and Social Psychology, 8,* 143–147.

Pruitt, D. G. (1981). *Negotiation behavior.* New York: Academic Press.

Pruitt, D. G., & Carnevale, P. J. D. (1980). The development of integrative agreements in social conflict. In V. J. Derlega & J. Grzelak (Eds.), *Living with other people: Theories and research on cooperation and helping behavior.* New York: Academic Press.

Pruitt, D. G., & Drews, J. L. (1969). The effects of time pressure, time elapsed, and the opponent's concession rate on behavior in negotiation. *Journal of Experimental Social Psychology, 5,* 43–60.

Pruitt, D. G., & Insko, C. A. (1980). Extension of the Kelley attribution model: The role of comparison-object consensus, target-object consensus, distinctiveness, and consistency. *Journal of Personality and Social Psychology, 39,* 39–58.

Pruitt, D. G., & Rubin, J. Z. (1985). *Social conflict: Escalation, impasse, and resolution.* Reading, MA: Addison-Wesley.

Pugh, M. D., & Wahrman, R. (1983). Neutralizing sexism in mixed-sex groups: Do women have to be better than men? *American Journal of Sociology, 80,* 746–762.

Quarantelli, E. L., & Dynes, R. R. (1977). Response to social crisis and disaster. In A. Inkeles, J. Coleman, & N. Smelser (Eds.), *Annual review of sociology* (Vol. 3, pp. 23–49). Palo Alto, CA: Annual Reviews.

Quattrone, G. A. (1986). On the perceptions of a group's variability. In S. Worchel & W. G. Austin (Eds.), *Psychology of intergroup relations* (2nd ed.) (pp. 25–48). Chicago: Nelson-Hall Publishers.

Quinney, R. (1970). *The social reality of crime.* Boston: Little, Brown.

Rabbie, J. M., & Bekkers, F. (1978). Threatened leadership and intergroup competition. *European Journal of Social Psychology, 8,* 9–20.

Rabow, J., Neuman, C. A., & Hernandez, A. (1987). Cognitive consistency in attitudes, social support and consumption of alcohol: Additive and interactive effects. *Social Psychology Quarterly, 50,* 56–63.

Radloff, L. S. (1980). Depression and the empty nest. *Sex Roles, 6,* 775–781.

Rahn, J., & Mason, W. (1987). Political alienation, cohort size, and the Easterlin hypothesis. *American Sociological Review, 52,* 155–169.

Raven, B. H., & Rietsema, J. (1957). The effects of varied clarity of group goal and group path upon the individual and his relation to the group. *Human Relations, 10,* 29–44.

Raven, B. H., & Kruglanski, A. W. (1970). Conflict and power. In P. Swingle (Ed.), *The structure of conflict.* New York: Academic Press.

Ray, M. (1973). Marketing communication and the hierarchy of effects. In P. Clarke (Ed.), *New models for communication research.* Beverly Hills, CA: Sage.

Rees, C. R., & Segal, M. W. (1984). Role differentiation in groups: The relations between instrumental and expressive leadership. *Small Group Behavior, 15,* 109–123.

Regan, D. T., & Fazio, R. (1977). On the consistency between attitudes and behavior: Look to the method of attitude formation. *Journal of Experimental Social Psychology, 35,* 21–30.

Regan, D. T., Straus, E., & Fazio, R. (1974). Liking and the attribution process. *Journal of Experimental Social Psychology, 10,* 385–397.

Reis, H. T., Senchak, M., & Solomon, B. (1985). Sex differences in the intimacy of social interaction: Further examination of potential explanations. *Journal of Personality and Social Psychology, 48,* 1204–1217.

Rempel, J. K., Holmes, J. G., & Zanna, M. P. (1985). Trust in a close relationship. *Journal of Personality and Social Psychology, 49,* 95–112.

Repetti, R. (1987). Individual and common components of the social environment at work and psychological well-being. *Journal of Personality and Social Psychology, 52,* 710–720.

*Report of the National Advisory Commission on Civil Disorders.* (1968). New York: Bantam Books.

Reskin, B., & Hartmann, H. (Eds.). (1986). *Women's work, men's work: Sex segregation on the job.* Washington, DC: National Academy Press.

Rexroat, C., & Shehan, C. (1987). The family life cycle and spouses' time in housework. *Journal of Marriage and the Family, 49,* 737–750.

Reynolds, P. D. (1984). Leaders never quit: Talking, silence, and influence in interpersonal groups. *Small Group Behavior, 15,* 404–413.

Rhine, R. J., & Severance, L. J. (1970). Ego-involvement, discrepancy, source credibility, and attitude change. *Journal of Personality and Social Psychology, 16,* 175–190.

Rice, R. W., Marwick, N. J., Chemers, M. M., & Bentley, J. C. (1982). Task performance and satisfaction: Least preferred coworker (LPC) as a moderator. *Personality and Social Psychology Bulletin, 8,* 534–541.

Ridgeway, C. L. (1982). Status in groups: The importance of motivation. *American Sociological Review, 47,* 76–88.

Ridgeway, C. (1987). Nonverbal behavior, dominance, and the basis of status in task groups. *American Sociological Review, 52,* 683–694.

Riggio, R. E., & Friedman, H. S. (1983). Individual differences and cues to deception. *Journal of Personality and Social Psychology, 45,* 899–915.

Rigney, J. (1962). A developmental study of cognitive equivalence transformation and their use in the acquisition and processing of information. Unpublished honors thesis, Radcliffe College, Cambridge, MA.

Riley, M. (1987). On the significance of age in sociology. *American Sociological Review, 52,* 1–14.

Riley, M. W., Johnson, M., & Foner, A. (1972) (Eds.). *Aging and society* (Vol. 3): *A sociology of age stratification.* New York: Russell Sage Foundation.

Rindfuss, R., Swicegood, C. G., & Rosenfeld, R. (1987). Disorder in the life course: How common and does it matter? *American Sociological Review, 52,* 785–801.

Ring, K., & Kelley, H. H. (1963). A comparison of augmentation and reduction as modes of influence. *Journal of Abnormal and Social Psychology, 66,* 95–102.

Ringelmann, M. (1913). Recherches sur les moteurs animes: Travail de l'homme. [Research on animate sources of power: The work of man.] *Annales de l'Institut National Agronomique, 2e serie-tome XII,* 1–40.

Riordan, C. (1978). Equal-status interracial contact: A review and revision of the concept. *International Journal of Intercultural Relations, 1,* 161–185.

Riordan, C. A., Marlin, N. A., & Kellogg, R. T. (1983). The effectiveness of accounts following transgression. *Social Psychology Quarterly, 46,* 213–219.

Riordan, C., & Ruggiero, J. A. (1980). Producing equal-status interracial interaction: A replication. *Social Psychology Quarterly, 43,* 131–136.

Robinson, J. W., Jr., & Preston, J. D. (1976). Equal-status contact and the modification of racial prejudice: A reexamination of the contact hypothesis. *Social Forces, 54,* 911–924.

Robinson, W. P., & Rackstraw, S. J. (1972). *A question of answers.* London: Routledge & Kegan Paul.

Robson, P. (1982). Patterns of mobility and activity among the elderly. In E. Warnes (Ed.), *Geographical perspectives on the elderly.* New York: Wiley.

Roethlisberger, F. J., & Dickson, W. J. (1939). *Management and the worker.* Cambridge, MA: Harvard University Press.

Rogers, M., Miller, N., Mayer, F. S., & Duvall, S. (1982). Personal responsibility and salience of the request for help: Determinants of the relation between negative affect and helping behavior. *Journal of Personality and Social Psychology, 43,* 956–970.

Rogers, R. W., & Mewborn, C. R. (1976). Fear appeals and

attitude change: Effects of a threat's noxiousness, probability of occurrence, and the efficacy of coping responses. *Journal of Personality and Social Psychology, 34,* 54–61.

Rogers, T. B. (1977). Self-reference in memory: Recognition of personality items. *Journal of Research in Personality, 11,* 295–305.

Rokeach, M. (1973). *The nature of human values.* New York: Free Press.

Rommetveit, R. (1955). *Social norms and roles.* Minneapolis: University of Minnesota Press.

Ronis, D. L., & Lipinski, E. R. (1985). Value and uncertainty as weighting factors in impression formation. *Journal of Experimental Social Psychology, 21,* 47–60.

Roopnarine, J. L. (1985). Changes in peer-directed behaviors following preschool experience. *Journal of Personality and Social Psychology, 48,* 740–745.

Rosch, E. (1978). Principles of categorization. In E. Rosch & B. B. Lloyd (Eds.), *Cognition and categorization.* Hillsdale, NJ: Erlbaum.

Rosen, B., & D'Andrade, R. (1959). The psychological origins of achievement motivation. *Sociometry, 22,* 185–218.

Rosen, S. (1984). Some paradoxical status implications of helping and being helped. In E. Staub et al (Eds.), *Development and maintenance of prosocial behavior: International perspectives on positive morality.* New York: Plenum.

Rosenbaum, M. E. (1986). The repulsion hypothesis: On the nondevelopment of relationships. *Journal of Personality and Social Psychology, 51,* 1156–1166.

Rosenbaum, M. E., Moore, D. L., Cotton, J. L., Cook, M. S., Hieser, R. A., Shovar, M. N., & Gray, M. J. (1980). Group productivity and process: Pure and mixed reward structures and task interdependence. *Journal of Personality and Social Psychology, 39,* 626–642.

Rosenberg, L. A. (1961). Group size, prior experience, and conformity. *Journal of Abnormal and Social Psychology, 63,* 436–437.

Rosenberg, M. (1965). *Society and the adolescent self-image.* Princeton, NJ: Princeton University Press.

Rosenberg, M. (1973). Which significant others? *American Behavioral Scientist, 16,* 829–860.

Rosenberg, M. (1979). *Conceiving the self.* New York: Basic Books.

Rosenberg, M. (1981). The self-concept: Social product and social force. In M. Rosenberg & R. H. Turner (Eds.), *Social psychology: Sociological perspectives.* New York: Basic Books.

Rosenberg, M. J., & Abelson, R. (1960). An analysis of cognitive balancing. In C. Hovland & M. Rosenberg (Eds.), *Attitude organization and change,* New Haven, CT: Yale University Press.

Rosenberg, M., & Pearlin, L. (1978). Social class and self-esteem among children and adults. *American Journal of Sociology, 84,* 53–77.

Rosenberg, M., & Simmons, R. (1972). *Black and white self-esteem: The urban school child.* Washington, DC: American Sociological Association.

Rosenberg, S. V. (1977). New approaches to the analysis of personal constructs in person perception. In A. W. Landfield (Ed.), *Nebraska symposium on motivation 1976.* Lincoln: University of Nebraska Press.

Rosenberg, S. V., Nelson, C., & Vivekananthan, P. S. (1968). A multidimensional approach to the structure of personality impressions. *Journal of Personality and Social Psychology, 9,* 283–294.

Rosenberg, S. V., & Sedlak, A. (1972). Structural representations in implicit personality theory. In L. Berkowitz (Ed.), *Advances in experimental social psychology* (Vol. 6). New York: Academic Press.

Rosenfeld, H. (1978). Conversational control functions of nonverbal behavior. In A. Seigman & S. Feldstein (Eds.), *Nonverbal behavior and communication.* Hillsdale, NJ: Erlbaum.

Rosenhan, D. L. (1973). On being sane in insane places. *Science, 179,* 250–258.

Rosenhan, D. L., Salovey, P., & Hargis, K. (1981). The joys of helping: Focus of attention mediates the impact of positive affect on altruism. *Journal of Personality and Social Psychology, 40,* 899–905.

Rosenhan, D. L., Salovey, P., Karylowski, J., & Hargis, K. (1981). Emotion and altruism. In J. P. Rushton & R. M. Sorrentino (Eds.), *Altruism and helping behavior.* Hillsdale, NJ: Erlbaum.

Rosenthal, R. (1966). *Experimenter effects in behavioral research.* New York: Appleton-Century-Crofts.

Rosenthal, R., & Rubin, D. B. (1978). Interpersonal expectancy effects: The first 345 studies. *The Behavioral and Brain Sciences, 3,* 377–386.

Rosow, I. (1974). *Socialization to old age.* Berkeley: University of California Press.

Ross, A. S. (1971). Effect of increased responsibility on bystander intervention: The presence of children. *Journal of Personality and Social Psychology, 19,* 306–310.

Ross, L. (1977). The intuitive psychologist and his shortcomings: Distortion in the attribution process. In L. Berkowitz (Ed.), *Advances in experimental social psychology* (Vol. 10). New York: Academic Press.

Ross, L., Amabile, T., & Steinmetz, J. (1977). Social roles, social control and biases in social perception processes. *Journal of Personality and Social Psychology, 35,* 485–494.

Ross, M., & Fletcher, G. (1985). Attribution and social perception. In G. Lindzey & E. Aronson (Eds.), *The handbook of social psychology* (3rd ed.). Reading, MA: Addison-Wesley.

Ross, M., & Lumsden, H. (1982). Attributions of responsibility in sports settings: It's not how you play the game but whether you win or lose. In H. Hiebsch, H. Brandstatter, & H. H. Kelley (Eds.), *Social psychology.* East Berlin: VEB Deutscher Verlag der Wissenschaften.

Ross, M., & Sicoly, F. (1979). Egocentric biases in availability and attribution. *Journal of Personality and Social Psychology, 37,* 322–336.

Ross, M., Thibaut, J., & Evenbeck, S. (1971). Some determinants of the intensity of social protests. *Journal of Experimental Social Psychology, 7,* 401–418.

Rossi, A. S. (1980). Parenthood in the middle years. In P. Baltes & O. Brim, Jr. (Eds.), *Life-span development and behavior* (Vol. 3). New York: Academic Press.

Rossiter, J. R. (1981). Research on television advertising's general impact on children: American and Australian findings. In J. F. Esserman (Ed.), *Television advertising and children: Issues, research and findings.* New York: Child Research Service.

Roth, D. L., Snyder, C. R., & Pace, L. M. (1986). Dimensions of favorable self-presentation. *Journal of Personality and Social Psychology, 51,* 867–874.

Rothbart, M., Dawes, R., & Park, B. (1984). Stereotypes and sampling biases in intergroup perception. In J. R. Eiser (Ed.), *Attitudinal judgment* (pp. 109–134). New York: Springer.

Rothbart, M., Fulero, S., Jensen, C., Howard, J., & Birrell, B. (1978). From individual to group impressions: Availability heuristics in stereotype formation. *Journal of Experimental Social Psychology, 14,* 237–255.

Rotton, J., & Frey, J. (1985). Air pollution, weather and violent crimes: Concomitant time-series analyses of archival data. *Journal of Personality and Social Psychology, 49,* 1207–1220.

Rozin, P., McKinan, L., & Nemeroff, C. (1986). Operation of the laws of sympathetic magic in disgust and other domains. *Journal of Personality and Social Psychology, 50,* 703–712.

Ruback, R. (1987). Deserted (and nondeserted) aisles: Territorial intrusion can produce persistence, not flight. *Social Psychology Quarterly, 50,* 270–276.

Rubin, B. A. (1986). Class struggle American style: Unions, strikes and wages. *American Sociological Review, 51,* 618–631.

Rubin, J. (1962). Bilingualism in Paraguay. *Anthropological Linguistics, 4,* 52–68.

Rubin, J. Z., & Brown, B. (1975). *The social psychology of bargaining and negotiation.* New York: Academic Press.

Rubin, J. Z., & Lewecki, R. J. (1973). A three-factor experimental analysis of promises and threats. *Journal of Applied Social Psychology, 3,* 240–257.

Rubin, L. (1979). *Women of a certain age: The midlife search for self.* New York: Harper and Row.

Rubin, L. (1983). *Intimate strangers: Men and women together.* New York: Harper and Row.

Rubin, Z. (1970). Measurement of romantic love. *Journal of Personality and Social Psychology, 16,* 265–273.

Rubin, Z. (1974). From liking to loving: Patterns of attraction in dating relationships. In T. Huston (Ed.), *Foundations of interpersonal attraction* (pp. 383–402). New York: Academic Press.

Rubin, Z., Hill, C., Peplau, L., & Dunkel-Sheker, C. (1980). Self-disclosure in dating couples: Sex roles and the ethic of openness. *Journal of Marriage and the Family, 42,* 305–317.

Rubinstein, E. A. (1983). Television and behavior: Research conclusions of the 1982 NIMH report and their policy implications. *American Psychologist, 38,* 820–825.

Rude, G. (1964). *The crowd in history.* New York: Wiley.

Rule, B. F., Ferguson, T. J., & Nesdale, A. R. (1980). Emotional arousal, anger and aggression: The misattribution issue. In P. Pliner, K. Blankstein, & T. Speigel (Eds.), *Advances in communication and affect.* Hillsdale, NJ: Erlbaum.

Rusbult, C. (1983). A longitudinal test of the investment model: The development (and deterioration) of satisfaction and commitment in heterosexual involvements. *Journal of Personality and Social Psychology, 45,* 101–117.

Rusbult, C. E., Johnson, D. J., & Morrow, G. D. (1986). Predicting satisfaction and commitment in adult romantic involvements: An assessment of the generalizability of the investment model. *Social Psychology Quarterly, 49,* 81–89.

Rusbult, C. E., Lowry, D., Hubbard, M. L., Maravankin, O. J., & Neises, M. (1988). Impact of employee mobility and employee performance on the allocation of rewards under conditions of constraint. *Journal of Personality and Social Psychology, 54,* 605–615.

Rusbult, C. E., Zembrodt, I. M., & Gunn, L. K. (1982). Exit, voice, loyalty and neglect: Responses to dissatisfaction in romantic involvement. *Journal of Personality and Social Psychology, 43,* 1230–1242.

Rushing, W. A., & Ortega, S. (1979). Socioeconomic status and mental disorders: New evidence and a sociomedical formulation. *American Journal of Sociology, 84,* 1175–1200.

Ryder, N. B. (1965). The cohort as a concept in the study of social change. *American Sociological Review, 30,* 843–861.

Sacks, H., Schegloff, E., & Jefferson, G. (1978). A simplest systematics for the organization of turn-taking in conversations. In J. Schenkein (Ed.), *Studies in the organization of conversational interaction.* New York: Academic Press.

Saegert, S. C., Swap, W., & Zajonc, R. B. (1973). Exposure, context and interpersonal attraction. *Journal of Personality and Social Psychology, 25,* 234–242.

Sagarin, E. (1975). *Deviants and deviance.* New York: Praeger Publishers.

Sakurai, M. M. (1975). Small group cohesiveness and detrimental conformity. *Sociometry, 38,* 340–357.

Sales, E. (1978). Women's adult development. In I. Frieze, J. Parsons, P. Johnson, D. Ruble, & G. Zellman (Eds.), *Women and sex roles: A social psychological perspective.* New York: Norton.

Sales, S. (1969). Organizational roles as a risk factor in coronary heart disease. *Administrative Science Quarterly, 14,* 325–336.

Saltzer, E. B. (1981). Cognitive moderation of the relationship between behavioral intentions and behavior. *Journal of Personality and Social Psychology, 41,* 260–271.

Sammon, S., Reznikoff, M., & Geisinger, K. (1985). Psychosocial development and stressful life events among religious professionals. *Journal of Personality and Social Psychology, 48,* 676–687.

Sample, J., & Warland, R. (1973). Attitude and the prediction of behavior. *Social Forces, 51,* 292–304.

Sampson, E. E., & Brandon, A. C. (1964). The effects of role and opinion deviation on small group behavior. *Sociometry, 27,* 261–281.

Sampson, H., Messinger, S., Towne, R., Russ, D., Livson, F., Bowers, M., Cohen, L., & Dorst, K. (1964). The mental hospital and marital family ties. In H. Becker (Ed.), *The other side.* New York: Free Press.

Samuels, F. (1970). The intra- and inter-competitive group. *Sociological Quarterly, 11,* 390–396.

Sanders, C. (1988). Risk factors in bereavement outcome. *Journal of Social Issues, 44,* 97–111.

Santee, R., & Jackson, S. (1978). Similarity and positivity of self-description as determinants of estimated appraisal and attraction. *Social Psychology, 41*, 162–165.

Santee, R. T., & Jackson, S. (1979). Commitment to self-identification: A socio-psychological approach to personality. *Human Relations, 32*, 141–158.

Sarason, S. B. (1977). *Work, aging and social change: Professionals and the one life–one career imperative.* New York: Free Press.

Sarbin, T., & Rosenberg, B. (1955). Contributions to role-taking theory IV: A method for obtaining a qualitative estimate of the self. *Journal of Social Psychology, 42*, 71–81.

Sawyer, A. (1973). The effects of repetition of refutational and supportive advertising appeals. *Journal of Marketing Research, 10*, 23–33.

Schachter, S. (1964). The interaction of cognitive and physiological determinants of emotional state. In L. Berkowitz (Ed.), *Advances in experimental social psychology* (Vol, 1). New York: Academic Press.

Schachter, S., & Singer, J. (1962). Cognitive, social and physiological determinants of emotional state. *Psychological Review, 69*, 379–399.

Schaefer, R. B., & Keith, P. M. (1980). Equity and depression among married couples. *Social Psychology Quarterly, 43*, 430–435.

Schatzman, L., & Strauss, A. (1955). Social class and modes of communication. *American Journal of Sociology, 60*, 329–338.

Scheff, T. (1966). *Being mentally ill.* Chicago: Aldine.

Scheff, T. J. (1967). Introduction. In T. J. Scheff (Ed.), *Mental illness and social processes* (pp. 1–18). New York: Harper and Row.

Schegloff, E. (1968). Sequencing in conversational openings. *American Anthropologist, 70*, 1075–1095.

Scheier, M. F., & Carver, C. (1981). Public and private aspects of the self. In L. Wheeler (Ed.), *Review of personality and social psychology* (Vol. 2). Beverly Hills, CA: Sage.

Scheier, M. F., & Carver, C. (1983). Two sides of the self: One for you and one for me. In J. Suls & A. G. Greenwald (Eds.), *Psychological perspectives on the self* (Vol. 2). Hillsdale, NJ: Erlbaum.

Scherer, S. E. (1974). Proxemic behavior of primary school children as a function of their socioeconomic class and subculture. *Journal of Personality and Social Psychology, 29*, 800–805.

Scherer, K. R. (1979). Nonlinguistic indicators of emotion and psychopathology. In C. E. Izard (Ed.), *Emotions in personality and psychopathology.* New York: Plenum.

Schiffenbauer, A., & Schiavo, R. S. (1976). Physical distance and attraction: An intensification effect. *Journal of Experimental Social Psychology, 12*, 274–282.

Schiffrin, D. (1977). Opening encounters. *American Sociological Review, 42*, 679–691.

Schifter, D. E., & Ajzen, I. (1985). Intention, perceived control and weight loss: An application of the theory of planned behavior. *Journal of Personality and Social Psychology, 45*, 843–851.

Schlenker, B. R. (1975). Self-presentation. Managing the impression of consistency when reality interferes with self-enhancement. *Journal of Personality and Social Psychology, 32*, 1030–1037.

Schlenker, B. R. (1980). *Impression management: The self-concept, social identity, and interpersonal relations.* Belmont, CA: Brooks/Cole.

Schlenker, B. R., Helm, B., & Tedeschi, J. T. (1973). The effects of personality and situational variables on behavioral trust. *Journal of Personality and Social Psychology, 25*, 419–427.

Schmitt, D. R., & Marwell, G. (1972). Withdrawal and reward reallocation as responses to inequity. *Journal of Experimental Social Psychology, 8*, 207–221.

Schneider, D. J. (1973). Implicit personality theory: A review. *Psychological Bulletin, 79*, 294–309.

Schooler, C., Miller, J., Miller, K., & Richtand, C. (1984). Work for the household: Its nature and consequences for husbands and wives. *American Journal of Sociology, 90*, 97–124.

Schrauger, J. S., & Schoeneman, T. (1979). Symbolic interactionist view of self-concept: Through the looking glass darkly. *Psychological Bulletin, 86*, 549–573.

Schriesheim, C., & Kerr, S. (1974). Psychometric properties of the Ohio State leadership scales. *Psychological Bulletin, 81*, 756–765.

Schulz, B., Bohrnstedt, G., Borgatta, E., & Evans, R. (1977). Explaining premarital sexual intercourse among college students: A causal model. *Social Forces, 56*, 148–165.

Schuman, H., & Johnson, M. (1976). Attitudes and behavior. In A. Inkeles, J. Coleman, & N. Smelser (Eds.), *Annual review of sociology* (Vol. 2, pp. 161–203). Palo Alto, CA: Annual Reviews.

Schutte, N. S., Kendrick, D. T., & Sadalla, E. K. (1985). The search for predictable settings: Situational prototypes, constraint, and behavioral variation. *Journal of Personality and Social Psychology, 49*, 121–128.

Schutte, J., & Light, J. (1978). The relative importance of proximity and status for friendship choices in social hierarchies. *Social Psychology, 41*, 260–264.

Schwartz, B. (1978). Queues, priorities, and social process. *Social Psychology, 41*, 3–12.

Schwartz, R., & Skolnick, J. (1964). Two studies of legal stigma. In H. Becker (Ed.), *The other side.* New York: Free Press.

Schwartz, S. H. (1977). Normative influences on altruism. In L. Berkowitz (Ed.), *Advances in experimental social psychology* (Vol. 10). New York: Academic Press.

Schwartz, S. (1978). Temporal instability as a moderator of the attitude-behavior relationship. *Journal of Personality and Social Psychology, 36*, 715–724.

Schwartz, S., & Ames, R. (1977). Positive and negative referent others as sources of influence: A case of helping. *Sociometry, 40*, 12–20.

Schwartz, S. H., & Clausen, G. T. (1970). Responsibility, norms, and helping in an emergency. *Journal of Personality and Social Psychology, 16*, 299–310.

Schwartz, S., Feldman, K., Brown, M., & Heingartner, A. (1969). Some personality correlates of conduct in two situations of moral conflict. *Journal of Personality and Social Psychology, 37*, 41–57.

Schwartz, S. H., & Fleishman, J. (1978). Personal norms and the mediation of legitimacy effects on helping. *Social Psychology, 41*, 306–315.

Schwartz, S. H., & Gottlieb, A. (1976). Bystander reactions to a violent theft: Crime in Jerusalem. *Journal of Personality and Social Psychology, 34*, 1188–1199.

Schwartz, S. H., & Gottlieb, A. (1980). Bystander anonymity and reactions to emergencies. *Journal of Personality and Social Psychology, 39,* 418–430.

Schwartz, S. H., & Howard, J. A. (1980). Explanations of the moderating effect of responsibility denial on personal norm-behavior relationship. *Social Psychology Quarterly, 43,* 441–446.

Schwartz, S. H., & Howard, J. A. (1981). A normative decision-making model of altruism. In J. P. Rushton & R. M. Sorrentino (Eds.), *Altruism and helping behavior.* Hillsdale, NJ: Erlbaum.

Schwartz, S. H., & Howard, J. A. (1984). Internalized values as motivators of altruism. In E. Staub, D. Bar-Tal, J. Karylowski, & J. Reykowski (Eds.), *The development and maintenance of prosocial behavior: International perspectives on positive morality.* New York: Plenum.

Scott, M., & Lyman, S. (1968). Accounts. *American Sociological Review, 33,* 46–62.

Scott, R. (1969). *The making of blind men: A study of adult socialization.* New York: Russell Sage Foundation.

Scott, R. (1976). Deviance, sanctions and social integration in small-scale societies. *Social Forces, 54,* 604–620.

Scotton, C. M. (1983). The negotiation of identities in conversation. *International Journal of the Sociology of Language, 44,* 115–136.

Searle, J. R. (1979). *Expression and meaning: Studies in the theory of speech acts.* Cambridge, England: Cambridge University Press.

Sears, D. O., & Freedman, J. L. (1967). Selective exposure to information: A critical review. *Public Opinion Quarterly, 31,* 194–213.

Sears, D. O., & Whitney, R. E. (1973). Political persuasion. In I. deS. Pool, W. Schramm, et al. (Eds.), *Handbook of communication.* Chicago: Rand McNally.

Sears, R. R., Maccoby, E., & Levin, H. (1957). *Patterns of child rearing.* New York: Harper and Row.

Sears, R. R., Whiting, J., Nowlis, V., & Sears, P. (1953). Some child-rearing antecedents of aggression and dependency in young children. *Genetic Psychology Monograph, 47,* 135–234.

Sebeok, T. A., & Umiker-Sebeok, J. (1980). *Speaking of apes.* New York: Plenum.

Seccombe, K. (1986). The effects of occupational conditions upon the division of household labor. *Journal of Marriage and the Family, 48,* 839–848.

Seedman, A. A., & Hellman, P. (1975). *Chief.* New York: Avon Books.

Seeman, M. (1975). Alienation studies. In A. Inkeles, J. Coleman, & N. Smelser (Eds.), *Annual review of sociology* (Vol. 1, pp. 91–123). Palo Alto, CA: Annual Reviews.

Seeman, M., Seeman, T., & Sayles, M. (1985). Social networks and health status: A longitudinal analysis. *Social Psychology Quarterly, 48,* 237–248.

Segal, B. E. (1965). Contact, compliance and distance among Jewish and non-Jewish undergraduates. *Social Problems, 13,* 66–74.

Sell, J., & Freese, L. (1984). The process of eliminating status generalization. *Social Forces, 63,* 538–554.

Serbin, L., & O'Leary, K. (1975). How nursery schools teach girls to shut up. *Psychology Today, 9,* 56–58.

Serpe, R. T. (1987). Stability and change in self: A structural symbolic interactionist explanation. *Social Psychology Quarterly, 50,* 44–55.

Sewell, W. H., & Hauser, R. M. (1975). *Education, occupation and earnings: Achievement in the early career.* New York: Academic Press.

Sewell, W. H., & Hauser, R. M. (1980). The Wisconsin longitudinal study of social and psychological factors in aspirations and achievements. *Research in Sociology of Education and Socialization, 1,* 59–99.

Sewell, W. H., Hauser, R., & Wolf, W. (1980). Sex, schooling, and occupational status. *American Journal of Sociology, 86,* 551–583.

Shanas, E. (1979). The family as a support system in old age. *Gerontologist, 19,* 169–174.

Shaver, P., & Freedman, J. (1976, August). Your pursuit of happiness. *Psychology Today, 10,* 26–32.

Shaw, M. E. (1955). A comparison of two types of leadership in various communication nets. *Journal of Abnormal and Social Psychology, 50,* 127–134.

Shaw, M. E. (1964). Communication networks. In L. Berkowitz (Ed.), *Advances in experimental social psychology* (Vol. 1). New York: Academic Press.

Shaw, M. E. (1978). Communication networks fourteen years later. In L. Berkowitz (Ed.), *Group processes.* New York: Academic Press.

Shaw, M., & Costanzo, P. (1982). *Theories of social psychology* (2nd ed.). New York: McGraw-Hill.

Shaw, M. E., & Rothschild, G. H. (1956). Some effects of prolonged experience in communication nets. *Journal of Applied Psychology, 40,* 281–286.

Shaw, M. E., & Shaw, L. M. (1962). Some effects of sociometric grouping upon learning in a second grade classroom. *Journal of Social Psychology, 57,* 453–458.

Sherif, M. (1936). *The psychology of social norms.* New York: Harper and Row.

Sherif, M. (1966). *In common predicament.* Boston: Houghton Mifflin.

Sherif, M., Harvey, O. J., White, B. J., Hood, W. R., & Sherif, C. W. (1961). *Intergroup cooperation and competition: The Robbers Cave experiment.* Norman, OK: University Book Exchange.

Sherif, M., & Sherif, C. (1964). *Exploration into conformity and deviation of adolescents.* New York: Harper and Row.

Sherif, M., & Sherif, C. (1967). Group processes and collective interaction in delinquent activities. *Journal of Research in Crime and Delinquency, 4,* 43–62.

Sherif, M., & Sherif, C. W. (1982). Production of intergroup conflict and its resolution—Robbers Cave experiment. In J. W. Reich (Ed.), *Experimenting in society: Issues and examples in applied social psychology.* Glenview, IL: Scott, Foresman & Co.

Sherman, S. J. (1970). Effects of choice and incentive on attitude change in a discrepant behavior situation. *Journal of Personality and Social Psychology, 15,* 245–252.

Sherman, S. J., Judd, C. M., & Park, B. (1989). Social cognition. *Annual Review of Psychology, 40,* 281–326.

Sherwood, J. J. (1965). Self-identity and referent others. *Sociometry, 28,* 66–81.

Shibutani, T. (1961). *Society and Personality.* Englewood Cliffs, NJ: Prentice-Hall.

Shiflett, S. C. (1973). The contingency model of leadership

effectiveness: Some implications of its statistical and methodological properties. *Behavioral Science, 18,* 429–440.

Shiflett, S. (1979). Toward a general model of small group productivity. *Psychological Bulletin, 86,* 67–79.

Shover, N., Novland, S., James, J., & Thornton, W. (1979). Gender roles and delinquency. *Social Forces, 58,* 162–175.

Shrum, W., & Creek, N. A., Jr. (1987). Social structure during the school years: Onset of the degrouping process. *American Sociological Review, 52,* 218–223.

Shulman, N. (1975). Life-cycle variations in patterns of close relationships. *Journal of Marriage and the Family, 37,* 813–821.

Shweder, R. A. (1977). Likeness and likelihood in everyday thought: Magical thinking in judgments about personality. *Current Anthropology, 18,* 637–658.

Sigall, H., & Landy, D. (1973). Radiating beauty: The effects of having a physically attractive partner on person perception. *Journal of Personality and Social Psychology, 28,* 218–224.

Sigall, H., & Page, R. (1971). Current stereotypes: A little fading, a little faking. *Journal of Personality and Social Psychology, 18,* 247–255.

Silberman, M. (1976). Toward a theory of criminal deterrence. *American Sociological Review, 41,* 442–461.

Simmons, R. G., Brown, L., Bush, D., & Blyth, D. (1978). Self-esteem and achievement of black and white adolescents. *Social Problems, 26,* 86–96.

Simpson, J. A. (1987). The dissolution of romantic relationships: Factors involved in emotional stability and emotional distress. *Journal of Personality and Social Psychology, 53,* 683–692.

Simpson, R. L., & Simpson, I. H. (1959). The psychiatric attendant: Development of an occupational self-image in a low status occupation. *American Sociological Review, 24,* 389–392.

Singer, J. L., & Singer, D. G. (1981). *Television, imagination and aggression: A study of preschoolers.* Hillsdale, NJ: Erlbaum.

Singer, J. L., & Singer, D. G. (1983). Psychologists look at television: Cognitive, developmental, personality, and social policy implications. *American Psychologist, 38,* 826–834.

Sinnott, J. D. (1977). Sex-role inconstancy, biology, and successful aging. *Gerontologist, 17,* 459–463.

Sistrunk, F., & McDavid, J. W. (1971). Sex variable in conforming behavior. *Journal of Personality and Social Psychology, 17,* 200–207.

Sivacek, J., & Crano, W. (1982). Vested interest as a moderator of attitude-behavior consistency. *Journal of Personality and Social Psychology, 43,* 210–221.

Skinner, B. F. (1953). *Science and human behavior.* New York: MacMillan.

Skinner, B. F. (1957). *Verbal behavior.* New York: Appleton-Century-Crofts.

Skinner, B. F. (1971). *Beyond freedom and dignity.* New York: Alfred Knopf.

Skvoretz, J. (1981). Extending expectation states theory: Comparative status models of participation in n-person groups. *Social Forces, 59,* 752–770.

Slater, P. (1963). On social regression. *American Sociological Review, 28,* 339–364.

Sloane, D. M., & Potrin, R. H. (1986). Religion and delinquency: Cutting through the maze. *Social Forces, 65,* 87–105.

Slomczynski, K. M., Miller, J., & Kohn, M. (1981). Stratification, work and values: A Polish-United States comparison. *American Sociological Review, 46,* 720–744.

Small, K. H., & Peterson, J. (1981). The divergent perceptions of actors and observers. *Journal of Social Psychology, 113,* 123–132.

Smelser, N. (1963). *Theory of collective behavior.* New York: Free Press.

Smith, D. A. (1987). Police response to interpersonal violence: Defining the parameters of legal control. *Social Forces, 65,* 767–782.

Snow, D. A., Rochford, E., Jr., Worden, S., & Benford, R. (1986). Frame alignment processes, micromobilization and movement participation. *American Sociological Review, 51,* 464–481.

Smith, W. P., & Anderson, A. (1975). Threats, communication, and bargaining. *Journal of Personality and Social Psychology, 32,* 76–82.

Smith, W. P., & Leginski, W. A. (1970). Magnitude and precision of punitive power in bargaining strategy. *Journal of Experimental Social Psychology, 6,* 57–76.

Snodgrass, S. E., & Rosenthal, R. (1982). Teacher suspiciousness of experimenter's intent and the mediation of teacher expectancy effects. *Basic and Applied Social Psychology, 3,* 219–230.

Snyder, C. R., Lassegard, M. A., & Ford, C. E. (1986). Distancing after group success and failure: Basking in reflected glory and cutting off reflected failure. *Journal of Personality and Social Psychology, 51,* 382–388.

Snyder, D. (1975). Institutional setting and industrial conflict: Comparative analysis of France, Italy and the United States. *American Sociological Review, 40,* 259–278.

Snyder, M. (1979). Self-monitoring. In L. Berkowitz (Ed.), *Advances in experimental social psychology* (Vol. 12). New York: Academic Press.

Snyder, M. (1981). On the self-perpetuating nature of social stereotypes. In D. L. Hamilton (Ed.), *Cognitive processes in stereotyping and intergroup behavior,* Hillsdale, NJ: Erlbaum.

Snyder, M., Berscheid, E., & Glick, P. (1985). Focusing on the exterior and the interior: Two investigations of the initiation of personal relationships. *Journal of Personality and Social Psychology, 48,* 1427–1439.

Snyder, M., & Tanke, E. (1976). Behavior and attitude: Some people are more consistent than others. *Journal of Personality, 44,* 501–517.

Snyder, M., Tanke, E., & Berscheid, E. (1977). Social perception and interpersonal behavior: On the self-fulfilling nature of social stereotypes. *Journal of Personality and Social Psychology, 35,* 656–666.

Solomon, D. S. (1982). Mass media campaigns for health promotion. *Prevention in Human Services, 2,* 115–123.

Solomon, S., & Saxe, L. (1977). What is intelligent, as well as attractive, is good. *Personality and Social Psychology Bulletin, 3,* 670–673.

Sommer, R. (1969). *Personal space.* Englewood Cliffs, NJ: Prentice-Hall.

Sorenson, J. R. (1971). Task demands, group interaction and group performance. *Sociometry, 34,* 483–495.

Sorrentino, R. M., & Field, N. (1986). Emergent leadership over time: The functional value of positive motivation. *Journal of Personality and Social Psychology, 50,* 1091–1099.

Spencer, J. W. (1987). Self-work in social interaction: Negotiating role-identities. *Social Psychology Quarterly, 50,* 131–142.

Spiegel, J. P. (1969). Hostility, aggression and violence. In A. Grimshaw (Ed.), *Patterns in American racial violence.* Chicago: Aldine.

Spilerman, S. (1976). Structural characteristics of cities and severity of racial disorders. *American Sociological Review, 41,* 771–793.

Spitz, R. (1945). Hospitalism. *The Psychoanalytic Study of the Child, 1,* 53–72.

Spitz, R. (1946). Hospitalism: A follow-up report. *The Psychoanalytic Study of the Child, 2,* 113–117.

Spitze, G. D., & Huber, J. (1980). Changing attitudes toward women's nonfamily roles: 1938 to 1978. *Sociology of Work and Occupations, 7,* 317–335.

Sprecher, S. (1986). The relation between inequity and emotions in close relationships. *Social Psychology Quarterly, 49,* 309–321.

Srole, L. (1956). Social integration and certain corollaries. *American Sociological Review, 21,* 709–716.

Staats, A. W., & Staats, C. (1958). Attitudes established by classical conditioning. *Journal of Abnormal and Social Psychology, 57,* 37–40.

Stack, S. (1987). Celebrities and suicide: A taxonomy and analysis, 1948–1983. *American Sociological Review, 52,* 401–412.

Stafford, M. C., Gray, L. N., Menke, B. A., & Ward, D. A. (1986). Modelling the deterrent effects of punishment. *Social Psychology Quarterly, 49,* 338–347.

Staggenborg, S. (1986). Coalition work in the pro-choice movement: Organizational and environmental opportunities and obstacles. *Social Problems, 33,* 374–390.

Stang, D. J. (1972). Conformity, ability, and self-esteem. *Representative Research in Social Psychology, 3,* 97–103.

Stark, R., & Bainbridge, W. S. (1980). Networks of faith: Interpersonal bonds and recruitment in cults and sects. *American Journal of Sociology, 85,* 1376–1395.

Stech, F., & McClintock, C. G. (1981). Effects of communication timing on duopoly bargaining outcomes. *Journal of Personality and Social Psychology, 40,* 664–674.

Steere, G. H. (1981). The family and the elderly. In F. Berghorn, D. Schafer, et al. (Eds.), *The dynamics of aging.* Boulder, CO: Westview Press.

Steffensmeier, D. J., & Terry, R. M. (1973). Deviance and respectability: An observational study of reactions to shoplifting. *Social Forces, 51,* 417–426.

Stein, J. A., Newcomb, M. D., & Bentler, P. M. (1987). An eight-year study of multiple influences on drug use and drug use consequences. *Journal of Personality and Social Psychology, 53,* 1094–1105.

Steiner, I. D. (1972). *Group process and productivity.* New York: Academic Press.

Steiner, I. D. (1974). *Task-performing groups.* Morristown, NJ: General Learning Press.

Steiner, I. D., & Rogers, E. (1963). Alternative responses to dissonance. *Journal of Abnormal and Social Psychology, 66,* 128–136.

Stemp, P., Turner, R., & Noh, S. (1986). Psychological distress in the postpartum period: The significance of social support. *Journal of Marriage and the Family, 48,* 271–277.

Stephan, F. F., & Mishler, E. G. (1952). The distribution of participation in small groups: An exponential approximation. *American Sociological Review, 17,* 598–608.

Stephan, W. G. (1978). School desegregation: An evaluation of predictions made in Brown v. Board of Education. *Psychological Bulletin, 85,* 217–238.

Stephan, W. G. (1987). The contact hypothesis in intergroup relations. In C. Hendrick (Ed.), *Group processes and intergroup relations* (pp. 13–40). Beverly Hills, CA: Sage.

Stephenson, W. (1953). *The study of behavior.* Chicago: University of Chicago Press.

Sternberg, R. J. (1985). Implicit theories of intelligence, creativity, and wisdom. *Journal of Personality and Social Psychology, 49,* 607–627.

Sternthal, B., Dholakia, R., & Leavitt, C. (1978). The persuasive effect of source credibility: A test of cognitive response analysis. *Journal of Consumer Research, 4,* 252–260.

Stewart, R. H. (1965). Effect of continuous responding on the order effect in personality impression formation. *Journal of Personality and Social Psychology, 1,* 161–165

Stiles, W., Orth, J., Scherwitz, L., Hennrikus, D., & Vallbona, C. (1984). Role behaviors in routine medical interviews with hypertensive patients: A repertoire of verbal exchanges. *Social Psychology Quarterly, 47,* 244–254.

Stogdill, R. M. (1963). *Team achievement under high motivation.* (Monograph No. R-113). Ohio State University, Bureau of Business Research.

Stogdill, R. M. (1974). *Handbook of leadership: A survey of theory and research.* New York: Free Press.

Stokes, J. P. (1983). Components of group cohesion: Intermember attraction, instrumental value, and risk-taking. *Small Group Behavior, 14,* 163–173.

Stokes, J. P. (1985). The relation of social network and individual difference variables to loneliness. *Journal of Personality and Social Psychology, 48,* 981–990.

Stokes, J., & Levin, I. (1986). Gender differences in predicting loneliness from social network characteristics. *Journal of Personality and Social Psychology, 51,* 1069–1074.

Stoll, C. S. (1978). *Female and male; Socialization, social roles, and social structure.* Dubuque, IA: W. C. Brown.

Stone, G. P. (1962). Appearances and the self. In A. Rose (Ed.), *Human behavior and social processes.* Boston: Houghton Mifflin.

Stoner, J. A. F. (1961). *A comparison of individual and group decisions involving risk.* Unpublished master's thesis, Massachusetts Institute of Technology. Cited in Marquis, D. G. (1962). Individual responsibility and group decisions involving risk. *Industrial Management Review, 3,* 8–23.

Storms, M. D. (1973). Videotape and attribution process:

Reversing actors' and observers' points of view. *Journal of Personality and Social Psychology, 27,* 165–175.

Streufert, S., & Streufert, S. C. (1969). Effects of conceptual structure, failure, and success on attribution of causality and interpersonal attitudes. *Journal of Personality and Social Psychology, 11,* 138–147.

Streufert, S., Streufert, S. C., & Castore, C. H. (1969). Complexity, increasing failure, and decision making. *Journal of Experimental Research in Personality, 3,* 293–300.

Stricker, L. J., Jacobs, P. I., & Kogan, N. (1974). Trait interrelations in implicit personality theories and questionnaire data. *Journal of Personality and Social Psychology, 30,* 198–207.

Strodtbeck, F. L., Simon, R. J., Hawkins, C. (1965). Social status in jury deliberations. In I. D. Steiner & M. Fishbein (Eds.), *Current studies in social psychology.* New York: Holt, Rinehart and Winston.

Stroebe, W., & Diehl, M. (1981). Conformity and counterattitudinal behavior: The effect of social support on attitude change. *Journal of Personality and Social Psychology, 41,* 876–889.

Stroebe, W., Thompson, V., Insko, C., & Reisman, S. R. (1970). Balance and differentiation in the evaluation of linked attitude objects. *Journal of Personality and Social Psychology, 16,* 38–47.

Strube, M. J., & Garcia, J. E. (1981). A meta-analytic investigation of Fiedler's contingency model of leadership effectiveness. *Psychological Bulletin, 90,* 307–321.

Stryker, S. (1980). *Symbolic interactionism: A social structural version.* Menlo Park, CA: Benjamin/Cummings.

Stryker, S. (1987). The vitalization of symbolic interactionism. *Social Psychology Quarterly, 50,* 83–94.

Stryker, S., & Gottlieb, A. (1981). Attribution theory and symbolic interactionism: A comparison. In J. H. Hawes, W. Ickes, & R. F. Kidd (Eds.), *New directions in attribution theory* (Vol. 3). Hillsdale, NJ: Erlbaum.

Stryker, S., & Serpe, R. (1981). Commitment, identity salience and role behavior: Theory and research example. In W. Ickes & E. Knowles (Eds.), *Personality, roles and social behavior.* New York: Springer-Verlag.

Stryker, S., & Serpe, R. (1982). Towards a theory of family influence in the socialization of children. In A. Kerckhoff (Ed.), *Research in sociology of education and socialization* (Vol. 4). Greenwich, CT: JAI Press.

Suchner, R. W., & Jackson, D. (1976). Responsibility and status: A causal or only a spurious relationship? *Sociometry, 39,* 243–256.

Sudman, S., & Bradburn, N. M. (1974). *Response effects in surveys.* Chicago: Aldine.

Suls, J. M., & Miller, R. (Eds.). (1977). *Social comparison processes: Theoretical and empirical perspectives.* New York: Wiley.

Sumner, W. G. (1906). *Folkways.* New York and Boston: Ginn.

Surra, C. A. (1985). Courtship types: Variations in interdependence between partners and social networks. *Journal of Personality and Social Psychology, 49,* 357–375.

Sussman, N. M., & Rosenfeld, H. M. (1982). Influence of culture, language and sex on conversational distance. *Journal of Personality and Social Psychology, 42,* 66–74.

Sutherland, E. H. (1937). *The professional thief.* Chicago: University of Chicago Press.

Sutherland, E., & Cressey, D. (1978). *Principles of criminology* (10th ed.). New York: J.B. Lippincott.

Suttles, G. (1968). *The social order of the slum.* Chicago: University of Chicago Press.

Swadesh, M. (1971). J. Sherzer (Ed.), *The origin and diversification of language.* Chicago: Aldine.

Swann, W. B., Jr. (1987). Identity negotiation: Where two roads meet. *Journal of Personality and Social Psychology, 53,* 1038–1051.

Swann, W. B., Jr., Griffin, J., Jr., Predmore, S., & Gaines, B. (1987). The cognitive-affective crossfire: When self-consistency confronts self-enhancement. *Journal of Personality and Social Psychology, 52,* 881–889.

Swanson, D. L. (Ed.). (1979). The uses and gratifications approach to mass communications research. *Communication Research, 3,* 3–111.

Sweet, J. A., & Bumpass, L. (1987). *American families and households.* New York: Russell Sage Foundation.

Swigert, V., & Farrell, R. (1977). Normal homicides and the law. *American Sociological Review, 42,* 16–32.

Syme, S. L., & Berkman, L. (1976). Social class, susceptibility and sickness. *American Journal of Epidemiology, 104,* 1–8.

Szymanski, K., & Harkins, S. G. (1987). Social loafing and self-evaluation with a social standard. *Journal of Personality and Social Psychology, 53,* 891–897.

Tajfel, H. (1981). *Human groups and social categories: Studies in social psychology.* Cambridge, England: Cambridge University Press.

Tajfel, H. (1982a). *Social identity and intergroup relations.* Cambridge, England: Cambridge University Press.

Tajfel, H. (1982b). Social psychology of intergroup relations. *Annual Review of Psychology, 33,* 1–39.

Tajfel, H., Billig, M. G., Bundy, R. P., & Flament, C. (1971). Social categorization and intergroup behavior. *European Journal of Social Psychology, 1,* 149–178.

Tajfel, H., & Billig, M. (1974). Familiarity and categorization in intergroup behavior. *Journal of Experimental Social Psychology, 10,* 159–170.

Tajfel, H., & Turner, J. (1986). The social identity theory of intergroup behavior. In S. Worchel & W. G. Austin (Eds.), *Psychology of intergroup relations* (2nd ed.) (pp. 7–24). Chicago: Nelson-Hall Publishers.

Tanford, S., & Penrod, S. (1984). Social influence model: A formal integration of research on majority and minority influence processes. *Psychological Bulletin, 95,* 189–225.

Tavris, C., & Offir, C. (1984). *The longest war: Sex differences in perspective* (2nd ed.). New York: Harcourt Brace Jovanovich.

Taylor, D. M., & Jaggi, V. (1974). Ethnocentrism and causal attribution in a South Indian context. *Journal of Cross-Cultural Psychology, 5,* 162–171.

Taylor, D. M., & Royer, L. (1980). Group processes affecting anticipated language choice in intergroup relations. In H. Giles, W. P. Robinson & P. M. Smith (Eds.), *Language: Social psychological perspectives.* New York: Pergamon.

Taylor, S. E. (1981). A categorization approach to stereotyping. In D. L. Hamilton (Ed.), *Cognitive processes in stereotyping and intergroup behavior*. Hillsdale, NJ: Erlbaum.

Taylor, S. E., & Crocker, J. (1981). Schematic bases of social information processing. In E. T. Higgins, C. P. Herman, & M. P. Zanna (Eds.), *Social cognition: The Ontario symposium* (Vol. 1). Hillsdale, NJ: Erlbaum.

Taylor, S. E., & Fiske, S. T. (1978). Salience, attention and attribution: Top of the head phenomena. In L. Berkowitz (Ed.), *Advances in experimental social psychology* (Vol. 11). New York: Academic Press.

Taylor, S. E., Fiske, S. T., Etcoff, N. L., & Ruderman, A. J. (1978). The categorical and contextual bases of person memory and stereotyping. *Journal of Personality and Social Psychology, 36,* 778–793.

Taylor, S. P. (1967). Aggressive behavior and physiological arousal as a function of provocation and the tendency to inhibit aggression. *Journal of Personality, 35,* 297–310.

Teachman, J. (1987). Family background, educational resources and educational attainment. *American Sociological Review, 52,* 548–557.

Tedeschi, J. T. (Ed.). (1981). *Impression management theory and social psychological research*. New York: Academic Press.

Tedeschi, J. T., Bonoma, T. V., & Schlenker, B. R. (1972). Influence, decision, and compliance. In J. T. Tedeschi (Ed.), *The social influence processes*. Chicago: Aldine.

Tedeschi, J. T., Schlenker, B. R., & Lindskold, S. (1972). The exercise of power and influence: The source of influence. In J. T. Tedeschi (Ed.), *The social influence processes*. Chicago: Aldine.

Terrace, H. (1984). *Language in apes*. New York: Academic Press.

Tesser, A., & Campbell, J. (1983). Self-definition and self-evaluation maintenance. In J. Suls & A. G. Greenwald (Eds.), *Psychological perspectives on the self* (Vol. 2). Hillsdale, NJ: Erlbaum.

Tessler, R. C., & Schwartz, S. H. (1972). Help-seeking, self-esteem, and achievement motivation: An attributional analysis. *Journal of Personality and Social Psychology, 21,* 318–326.

Teti, D., Lamb, M., & Elster, A. (1987). Long-range socioeconomic and marital consequences of adolescent marriage in three cohorts of adult males. *Journal of Marriage and the Family, 49,* 499–506.

Tetlock, P. E. (1980). Explaining teacher evaluations of pupil performance: A self-presentation interpretation. *Social Psychology Quarterly, 43,* 282–290.

Tetlock, P. E. (1981). The influence of self-presentational goals on attributional reports. *Social Psychology Quarterly, 44,* 300–311.

Tetlock, P. (1986). A value pluralism model of ideological reasoning. *Journal of Personality and Social Psychology, 50,* 819–827.

Tetlock, P. E., & Manstead, A. S. R. (1985). Impression management versus intrapsychic explanations in social psychology: A useful dichotomy? *Psychological Review, 92,* 59–77.

Thakerar, J. N., Giles, H., & Cheshire, J. (1982). Psychological and linguistic parameters of speech accommodation theory. In C. Fraser & K. R. Scherer (Eds.), *Advances in the social psychology of language*. Cambridge, England: Cambridge University Press.

Thibaut, J., & Kelley, H. (1959). *The social psychology of groups*. New York: Wiley.

Thoits, P. A. (1985). Self-labelling processes in mental illness: The role of emotional deviance. *American Journal of Sociology, 91,* 221–249.

Thomas, W. I., & Znaniecki, F. (1918). *The Polish peasant in Europe and America*. (Vol. 1). Boston: Badger.

Thompson, W. C., Cowan, C. L., & Rosenhan, D. (1980). Focus of attention mediates the impact of negative affect on altruism. *Journal of Personality and Social Psychology, 38,* 291–300.

Thorlundsson, T. (1987). Bernstein's sociolinguistics: An empirical test in Iceland. *Social Forces, 65,* 695–718.

Thornberry, T. D., & Christenson, R. L. (1984). Juvenile justice decision-making as a longitudinal process. *Social Forces, 63,* 433–444.

Thornberry, T., & Farnsworth, M. (1982). Social correlates of criminal involvement: Further evidence on the relationship between social status and criminal behavior. *American Sociological Review, 47,* 505–518.

Thorndike, E. L. (1920). A constant error in psychological ratings. *Journal of Applied Psychology, 4,* 25–29.

Thorne, B., Kramerae, C., & Henley, H. (1983). (Eds.). *Language, gender and society*. Rowley, MA: Newbury.

Thornton, A. (1984). Changing attitudes toward separation and divorce: Causes and consequences. *American Journal of Sociology, 90,* 856–872.

Thornton, A., & Freedman, D. (1979). Changes in sex-role attitudes of women, 1962–1977: Evidence from a panel study. *American Sociological Review, 44,* 831–842.

Thornton, R., & Nardi, P. (1975). The dynamics of role acquisition. *American Journal of Sociology, 80,* 870–885.

Tilly, C., Tilly, L., & Tilly, R. (1975). *The rebellious century, 1830–1930*. Cambridge, MA: Harvard University Press.

Tittle, C., & Villemez, W. (1977). Social class and criminality. *Social Forces, 56,* 474–502.

Tobin, S. (1980). Institutionalization of the aged. In N. Datan & N. Lohman (Eds.), *Transitions of aging*. New York: Academic Press.

Toch, H. (1969). *Violent men: An inquiry into the psychology of violence*. Chicago: Aldine.

Toi, M., & Batson, C. D. (1982). More evidence that empathy is a source of altruism. *Journal of Personality and Social Psychology, 43,* 289–292.

Touhey, J. (1979). Sex-role stereotyping and individual differences in liking for the physically attractive. *Social Psychology Quarterly, 42,* 285–289.

Treiman, D. (1977). *Occupational prestige in comparative perspective*. New York: Academic Press.

Triandis, H. C. (1980). Values, attitudes and interpersonal behavior. In H. Howe & M. Page (Eds.), *Nebraska Symposium on Motivation* (Vol. 27). Lincoln: University of Nebraska Press.

Trivers, R. L. (1983). The evolution of cooperation. In D. L. Bridgeman (Ed.), *The nature of prosocial behavior*. New York: Academic Press.

Trowbridge, N. T. (1972). Self-concept and socioeconomic

status in elementary school children. *American Educational Research Journal, 9*, 525–537.

Turner, G. J. (1974). Social class and children's language of control at ages five and seven. In B. Bernstein (Ed.), *Class, codes and control II* (rev. ed.). London: Routledge & Kegan Paul.

Turner, J. C. (1975). Social comparison and social identity: Some prospects for intergroup behaviour. *European Journal of Social Psychology, 5*, 5–34.

Turner, R. H. (1962). Role-taking: Process vs. conformity. In A. Rose (Ed.), *Human behavior and social process*. Boston: Houghton Mifflin.

Turner, R. H. (1978). The role and the person. *American Journal of Sociology, 84*, 1–23.

Turner, R. H., & Killian, L. M. (1972). *Collective behavior* (2nd ed.). Englewood Cliffs, NJ: Prentice-Hall.

Turner, R. H., & Shosid, N. (1976). Ambiguity and interchangeability in role attribution. *American Sociological Review, 41*, 993–1006.

Tyler, T. R., & Caine, A. (1981). The influence of outcomes and procedures on satisfaction with formal leaders. *Journal of Personality and Social Psychology, 41*, 642–655.

Tyler, T. R., & Folger, R. (1980). Distributional and procedural aspects of satisfaction with citizen-police encounters. *Basic and Applied Social Psychology, 1*, 281–292.

Tyler, T. R., & Sears, D. (1977). Coming to like obnoxious people when we must live with them. *Journal of Personality and Social Psychology, 35*, 200–211.

Ubell, E. (1985, February 17). Can changing your looks change your life? *Parade Magazine.*

Ungar, S. (1980). The effects of certainty of self-perceptions on self-presentation behaviors: A test of the strength of self-enhancement motives. *Social Psychology Quarterly, 43*, 165–172.

U.S. Bureau of the Census. (1987). *School enrollment—Social and economic characteristics of students: Oct. 1987* (Current Population Reports, Series P-20, No. 413). Washington, DC: Government Printing Office.

U.S. Bureau of the Census. (1989). *Money income of households, families and persons in the United States: 1987* (Current Population Reports, Series P-60, No. 162). Washington, DC: Government Printing Office.

Valle, V. A., & Frieze, I. H. (1976). Stability of causal attributions as a mediator in changing expectations for success. *Journal of Personality and Social Psychology, 35*, 579–589.

Van Gennep, A. (1960, 1908). *The rites of passage.* Chicago: University of Chicago Press.

Van Maanen, J. (1976). Breaking in: Socialization to work. In R. Dubin (Ed.), *Handbook of work, organization and society.* Chicago: Rand McNally.

Van Zelst, R. H. (1952). Sociometrically selected work teams increase production. *Personnel Psychology, 5*, 175–186.

Verbrugge, L. (1979). Marital status and health. *Journal of Marriage and the Family, 41*, 267–285.

Verplanck, W. S. (1955). The control of the content of conversation: Reinforcement of statements of opinion. *Journal of Abnormal and Social Psychology, 51*, 668–676.

Vidmar, N. (1974). Effects of group discussion on category width judgments. *Journal of Personality and Social Psychology, 29*, 187–195.

Voissem, N. H., & Sistrunk, F. (1971). Communication schedule and cooperative game behavior. *Journal of Personality and Social Psychology, 19*, 160–167.

von Baeyer, C. L., Sherk, D. L., & Zanna, M. P. (1981). Impression management in the job interview: When the female applicant meets the male (chauvinist) interviewer. *Personality and Social Psychology Bulletin, 7*, 45–51.

Vos, K., & Brinkman, W. (1967). Success and cohesion in sports groups. *Sociologische Gids, 14*, 30–40.

Vroom, V. H., & Yetton, P. W. (1973). *Leadership and decision making.* Pittsburgh: University of Pittsburgh Press.

Vygotsky, L. S. (1962). *Thought and language.* Cambridge, MA: MIT Press.

Wahrman, R. (1977). Status, deviance, sanctions, and group discussion. *Small Group Behavior, 8*, 147–168.

Waite, L., Haggstrom, G., & Kanouse, D. (1986). The effects of parenthood on the career orientations and job characteristics of young adults. *Social Forces, 65*, 43–73.

Waldron, I. (1976). Why do women live longer than men? *Social Science and Medicine, 10*, 349–362.

Walker, H. A., Thomas, G. M., & Zelditch, M., Jr. (1986). Legitimation, endorsement, and stability. *Social Forces, 64*, 620–643.

Walker, L. L., & Heyns, R. W. (1962). *An anatomy for conformity.* Englewood Cliffs, NJ: Prentice Hall.

Wall, J. A., Jr. (1977). Intergroup bargaining: Effects of opposing constituent's stance, opposing representative's bargaining, and representatives' locus of control. *Journal of Conflict Resolution, 21*, 459–474.

Wallin, P. (1950). Cultural contradictions and sex roles: A repeat study. *American Sociological Review, 15*, 288–293.

Walsh, E. J., & Taylor, M. (1982). Occupational correlates of multidimensional self-esteem: Comparisons among garbage collectors, bartenders, professors and other workers. *Sociology and Social Research, 66*, 252–258.

Walsh, E. J., & Warland, R. H. (1983). Social movement involvement in the wake of a nuclear accident: Activists and free riders in the TMI area. *American Sociological Review, 48*, 764–780.

Walster, E., Aronson, E., & Abrahams, D. (1966). On increasing the persuasiveness of a low prestige communicator. *Journal of Experimental Social Psychology, 2*, 325–342.

Walster (Hatfield), E., Aronson, V., Abrahams, D., & Rottman, L. (1966). The importance of physical attractiveness in dating behavior. *Journal of Personality and Social Psychology, 4*, 508–516.

Walster (Hatfield), E., Berschied, E., & Walster, G. W. (1973). New directions in equity research. *Journal of Personality and Social Psychology, 25*, 151–176.

Walster (Hatfield), E., Walster, G. W., & Berscheid, E. (1978). *Equity: Theory and research.* Boston: Allyn and Bacon.

Walster, E., Walster, G. W., & Traupmann, J. (1978). Equity and premarital sex. *Journal of Personality and Social Psychology, 36,* 82–92.

Walters, J., & Walters, L. (1980). Parent-child relationships: A review, 1970–1979. *Journal of Marriage and the Family, 42,* 807–822.

Warheit, G., Arey, S., Bell, R., & Holzer, C., III (1976). Sex, marital status and mental health: A reappraisal. *Social Forces, 55,* 459–470.

Warner, L. C., & DeFleur, M. (1969). Attitude as an interactional concept: Social constraint and social distance as intervening variables between attitudes and action. *American Sociological Review, 34,* 153–169.

Warwick, C. E. (1964). Relationship of scholastic aspiration and group cohesiveness to the academic achievement of male freshman at Cornell University. *Human Relations, 17,* 155–168.

Wasserman, I. (1977). Southern violence and the political process. *American Sociological Review, 42,* 359–362.

Waters, H. F., & Malamud, P. (1975, March 10). Drop that gun, Captain Video. *Newsweek, 85*(10), 81–82.

Watson, D. (1982). The actor and the observer: How are their perceptions of causality divergent? *Psychological Bulletin, 92,* 682–700.

Watson, R. I. (1973). Investigation into deindividuation using a cross-cultural survey technique. *Journal of Personality and Social Psychology, 25,* 342–345.

Webb, E. J., Campbell, D. J., Schwartz, R. D., & Sechrest, L. (1966). *Unobtrusive measures: Nonreactive research in the social sciences.* Chicago: Rand McNally.

Weber, R., & Crocker, J. (1983). Cognitive processes in the revision of stereotypic beliefs. *Journal of Personality and Social Psychology, 45,* 961–977.

Webster, M., Jr., & Driskell, J. E., Jr. (1978). Status generalization: A review and some new data. *American Sociological Review, 43,* 220–236.

Webster, M., Jr., & Driskell, J. E., Jr. (1983). Processes of status generalization. In H. H. Blumberg, A. P. Hare, V. Kent, & M. Davies (Eds.), *Small groups and social interaction* (Vol. 1, pp. 57–67). London: Wiley.

Webster, M., & Smith, L. F. (1978). Justice and revolutionary coalitions: A test of two theories. *American Journal of Sociology, 84,* 267–292.

Weigel, R. H., & Newman, L. (1976). Increasing attitude-behavior correspondence by broadening the scope of the behavioral measure. *Journal of Personality and Social Psychology, 33,* 793–802.

Weinberg, M. (1976). The nudist management of respectability. In M. Weinberg (Ed.), *Sex research: Studies from the Kinsey Institute.* New York: Oxford University Press.

Weiner, B. (1985). An attributional theory of achievement motivation and emotion. *Psychological Review, 92,* 548–573.

Weiner, B. (1986). *An attributional theory of motivation and emotion.* New York: Springer-Verlag.

Weiner, B., Amirkhan, J., Folkes, V. S., & Verette, J. A. (1987). An attributional analysis of excuse giving: Studies of a naive theory of emotion. *Journal of Personality and Social Psychology, 52,* 316–324.

Weiner, B., Frieze, I., Kukla, A., Reed, L., Rest, B., & Rosenbaum, R. M. (1971). *Perceiving the causes of success and failure.* Morristown, NJ: General Learning Press.

Weiner, B., Heckhausen, H., Meyer, W. U., & Cook, R. E. (1972). Causal ascriptions and achievement behavior: A conceptual analysis of effort and reanalysis of locus of control. *Journal of Personality and Social Psychology, 21,* 239–248.

Weinstein, E. A., & Deutschberger, P. (1963). Some dimensions of altercasting. *Sociometry, 26,* 454–466.

Weiss, R. S. (1973). *Loneliness: The experience of emotional and social isolation.* Cambridge, MA: MIT Press.

Weitzman, L. J., Eifler, D., Hokada, E., & Ross, K. (1972). Sex role socialization in picture books for pre-school children. *American Journal of Sociology, 77,* 1125–1150.

Wellman, B. (1979). The community question: The intimate networks of east Yorkers. *American Journal of Sociology, 84,* 1201–1231.

Werner, C., & Parmelee, P. (1979). Similarity of activity preferences among friends: Those who play together stay together. *Social Psychology Quarterly, 42,* 62–66.

Whalen, M. R., & Zimmerman, D. (1987). Sequential and institutional contexts in calls for help. *Social Psychology Quarterly, 50,* 172–185.

Wheeler, L. (1966). Toward a theory of behavioral contagion. *Psychological Review, 73,* 179–192.

White, C., Bushnell, N., & Regnemer, J. (1978). Moral development in Bahamian school children: A 3-year examination of Kohlberg's stages of moral development. *Developmental Psychology, 14,* 58–65.

White, G. (1980). Physical attractiveness and courtship progress. *Journal of Personality and Social Psychology, 39,* 660–668.

White, J. W., & Gruber, K. J. (1982). Instigative aggression as a function of past experience and target characteristics. *Journal of Personality and Social Psychology, 42,* 1069–1075.

Whitt, H. P., & Meile, R. L. (1985). Alignment, magnification and snowballing: Processes in the definition of symptoms of mental illness. *Social Forces, 63,* 682–697.

Whorf, B. L. (1956). In J. B. Carroll (Ed.), *Language, thought and reality.* Cambridge, MA: MIT Press.

Wicker, A. W. (1969). Attitudes versus actions: The relationship of verbal and overt behavioral responses to attitude objects. *Journal of Social Issues, 25,* 41–78.

Wicklund, R. A. (1975). Objective self-awareness. In L. Berkowitz (Ed.), *Advances in experimental social psychology* (Vol. 8). New York: Academic Press.

Wicklund, R. A., & Brehm, J. (1976). *Perspectives on cognitive dissonance.* Hillsdale, NJ: Erlbaum.

Wicklund, R. A., & Frey, D. (1980). Self-awareness theory: When the self makes a difference. In D. M. Wegner & R. R. Vallacher (Eds.), *The self in social psychology,* New York: Oxford University Press.

Wiggins, J. A., Dill, F., & Schwartz, R. D. (1965). On "status-liability." *Sociometry, 28,* 197–209.

Wiggins, J. S., Wiggins, N., & Conger, J. C. (1968). Correlates of heterosexual somatic preference. *Journal of Personality and Social Psychology, 10,* 82–90.

Wilder, D. A. (1981). Perceiving persons as a group: Categorization and intergroup relations. In D. L. Hamilton

(Ed.), *Cognitive processes in stereotyping and intergroup behavior.* Hillsdale, NJ: Erlbaum.

Wilensky, H. L. (1961). Life cycle, work situation, and participation in formal associations. In R. Kleemeier (Ed.), *Aging and leisure.* New York: Oxford University Press.

Wiley, M. G. (1973). Sex roles in games. *Sociometry, 36,* 526–541.

Wilke, H., & Lanzetta, J. T. (1970). The obligation to help: The effects of amount of prior help on subsequent helping behavior. *Journal of Experimental Social Psychology, 6,* 488–493.

Williams, R. M., Jr. (1977). *Mutual accommodation: Ethnic conflict and cooperation.* Minneapolis: University of Minnesota Press.

Winick, C. (1964). Physician narcotic addicts. In H. Becker (Ed.), *The other side* (pp. 261–279). New York: Free Press.

Winch, R. (1958). *Mate selection: A study of complementary needs.* New York: Harper and Row.

Winnubst, J. A. M., & Ter Heine, E. J. H. (1985). German developments in role theory 1958–1980. *Sociology, 19,* 598–608.

Winterbottom, M. (1958). The relation of need for achievement to learning experiences in independence and mastery. In J. Atkinson (Ed.), *Motives in fantasy, action and society.* Princeton, NJ: Van Nostrand.

Wofford, J. (1970). Factor analysis of managerial behavior. *Journal of Applied Psychology, 54,* 169–173.

Wolf, S. (1985). Manifest and latent influence of majorities and minorities. *Journal of Personality and Social Psychology, 48,* 899–908.

Won-Doornink, M. J. (1979). On getting to know you: The association between the stage of a relationship and the reciprocity of self-disclosure. *Journal of Experimental Social Psychology, 15,* 229–241.

Won-Doornink, M. J. (1985). Self-disclosure and reciprocity in conversation: A cross-national study. *Social Psychology Quarterly, 48,* 97–107.

Worchel, S. (1986). The role of cooperation in reducing intergroup conflict. In S. Worchel & W. G. Austin (Eds.), *Psychology of intergroup relations* (2nd ed.) (pp. 288–304). Chicago: Nelson-Hall Publishers.

Worchel, S., Lind, E., & Kaufman, K. (1975). Evaluations of group products as a function of expectations of group longevity, outcome of competition, and publicity of evaluations. *Journal of Personality and Social Psychology, 31,* 1089–1097.

Wortman, C., Adesman, P., Herman, E., & Greenberg, R. (1976). Self-disclosure: An attributional perspective. *Journal of Personality and Social Psychology, 33,* 184–191.

Wright, E. O., Costello, C., Hachen, D., & Sprague, J. (1982). The American class structure. *American Sociological Review, 47,* 709–726.

Wrightsman, L. S. (1969). Wallace supporters and adherence to law and order. *Journal of Personality and Social Psychology, 13,* 17–22.

Wyer, R. S., Jr. (1966). Effects of incentive to perform well, group attraction, and group acceptance on conformity in a judgmental task. *Journal of Personality and Social Psychology, 4,* 21–26.

Wyer, R. S., Jr., & Srull, T. K. (1984). *Handbook of social cognition* (Vols. 1–3). Hillsdale, NJ: Erlbaum.

Wylie, R. C. (1979). *The self-concept: Theory and research on selected topics.* (rev. ed.) (Vol. 2). Lincoln: University of Nebraska Press.

Yancey, W. L., Rigsby, L., & McCarthy, J. (1972). Social position and self-evaluation: The relative importance of race. *American Journal of Sociology, 78,* 338–359.

Yarrow, M., Schwartz, C., Murphy, H., & Deasy, L. (1955). The psychological meaning of mental illness in the family. *Journal of Social Issues, 11,* 12–24.

Youngs, G. A., Jr. (1986). Patterns of threat and punishment reciprocity in a conflict setting. *Journal of Personality and Social Psychology, 51,* 541–546.

Yukl, G. A. (1974a). Effects of situational variables and opponent concessions on a bargainer's perception, aspirations, and concessions. *Journal of Personality and Social Psychology, 29,* 227–236.

Yukl, G. A. (1974b). Effects of opponent's initial offer, concession magnitude, and concession frequency on bargaining behavior. *Journal of Personality and Social Psychology, 30,* 323–335.

Yukl, G. (1981). *Leadership in organizations.* Englewood Cliffs, NJ: Prentice-Hall.

Zajonc, R. B. (1968). The attitudinal effects of mere exposure. *Journal of Personality and Social Psychology, 9,* Pt. 2, 1–27.

Zaleznik, A. (1966). *Human dilemmas of leadership.* New York: Harper and Row.

Zander, A. (1971). *Motives and goals in groups.* New York: Academic Press.

Zander, A. (1977). *Groups at work.* San Francisco: Jossey-Bass.

Zanna, M., & Fazio, R. (1982). The attitude-behavior relation: Moving toward a third generation of research. In M. Zanna, E. Higgins, & C. Herman (Eds.), *Consistency in Social Behavior: The Ontario Symposium,* (Vol. 2, pp. 283–301). Hillsdale, NJ: Erlbaum.

Zanna, M. P., & Hamilton, D. L. (1977). Further evidence for meaning change in impression formation. *Journal of Experimental Social Psychology, 13,* 224–238.

Zelditch, M., Jr. (1972). Authority and performance expectations in bureaucratic organizations. In C. G. McClintock (Ed.), *Experimental social psychology.* New York: Holt, Rinehart and Winston.

Zelditch, M., Jr., Lauderdale, P., & Stublarec, S. (1980). How are inconsistencies between status and disability resolved? *Social Forces, 58,* 1025–1043.

Zelditch, M., Jr., & Walker, H. A. (1984). Legitimacy and the stability of authority. In E. J. Lawler (Ed.), *Advances in group processes,* (Vol. 1). Greenwich, CT: JAI Press.

Zeller, R. A., Neal, A., & Groat, H. (1980). On the reliability and stability of alienation measures: A longitudinal analysis. *Social Forces, 58,* 1195–1204.

Zelnik, M., & Kantner, J. (1981). Sexual and contraceptive experience of young unmarried women in the United States, 1976 and 1971. In F. F. Furstenberg, Jr., R.

Lincoln, & J. Mencken (Eds.), *Teenage sexuality, pregnancy and childbearing.* Philadelphia: University of Pennsylvania Press.

Zillman, D. (1978). Attribution and misattribution of excitatory reactions. In J. Harvey, W. Ickes, & R. F. Kidd (Eds.), *New directions in attribution research* (Vol. 2). Hillsdale, NJ: Erlbaum.

Zillman, D. (1979). *Hostility and aggression.* Hillsdale, NJ: Erlbaum.

Zillmann, D. (1982). Transfer of excitation in emotional behavior. In J. T. Cacioppo & R. E. Petty (Eds.), *Social psychophysiology.* New York: Guilford Press.

Zillmann, D., & Cantor, J. R. (1976). Effects of timing of information about mitigating circumstances on emotional responses to provocation and retaliatory behavior. *Journal of Experimental Social Psychology, 12,* 38–55.

Zillmann, D., Katcher, A. H., & Milavsky, B. (1972). Excitation transfer from physical exercise to subsequent aggressive behavior. *Journal of Experimental Social Psychology, 8,* 247–259.

Zimbardo, P. G. (1969). The human choice: Individuation, reason and order versus de-individuation, impulse and chaos. In W. J. Arnold & D. Levine (Eds.), *Nebraska Symposium on Motivation, 1969.* Lincoln: University of Nebraska Press.

Zimmerman, D. H., & West, C. H. (1975). Sex roles, interruptions and silences in conversations. In B. Thorne & N. Henley (Eds.), *Language and sex: Difference and dominance.* Rowley, MA: Newbury.

Zimmerman, R., & DeLamater, J. (1983). Threat of topic, social desirability, self-awareness and accuracy of self-report. Unpublished manuscript.

Zipf, S. G. (1960). Resistance and conformity under reward and punishment. *Journal of Abnormal and Social Psychology, 61,* 102–109.

Zollar, A., & Williams, J. (1987). The contribution of marriage to the life satisfaction of black adults. *Journal of Marriage and the Family, 49,* 87–92.

Zuckerman, D. M. (1981). Family background, sex-role attitudes, and life goals of technical college and university students. *Sex Roles, 7,* 1109–1126.

Zuckerman, M. (1978). Actions and occurrences in Kelly's cube. *Journal of Personality and Social Psychology, 36,* 647–656.

Zuckerman, M., DePaulo, B. M., & Rosenthal, R. (1981). Verbal and nonverbal communication of deception. In L. Berkowitz (Ed.), *Advances in experimental social psychology* (Vol. 14). New York: Academic Press.

Zuckerman, M., & Driver, R. E. (1984). Telling lies: Verbal and nonverbal communication in deception. In A. W. Seigman & S. Feinstein (Eds.), *Nonverbal communication: An integrated perspective.* Hillsdale, NJ: Erlbaum.

Zuckerman, M., Koestner, R., & Alton, A. O. (1984). Learning to detect deception. *Journal of Personality and Social Psychology, 46,* 519–528.

Zurcher, L. A., & Snow, D. A. (1981). Collective behavior and social movements. In M. Rosenberg & R. H. Turner (Eds.), *Social psychology: Sociological perspectives* (pp. 447–482). New York: Basic Books.

# Credits

## Photographs

**Chapter 1   Page xvi** ©Karl Kummels/Super Stock; **9** Paul Conklin/Monkmeyer Press Photo Service; **12** ©Bob Daemmrich/Stock, Boston; **14** ©Tony Freeman/Photoedit; **17** ©Jean-Marie Simon/Taurus Photos.

**Chapter 2   Page 26** ©Alan Oddie/Photoedit; **38** Stock, Boston; **43** Lightwave.

**Chapter 3   Page 50** ©Alvis Upitis/Super Stock; **56** ©Erika Stone/Peter Arnold, Inc.; **62** ©James Holland/Stock, Boston; **65** ©James L. Shaffer; **71** ©Peter Menzel/Stock, Boston; **78** Paul Conklin.

**Chapter 4   Page 82** ©Loren Santow/Click Chicago; **88** Elizabeth Crews; **90** ©Robert Kalman/The Image Works, Inc.; **99** Black Star; **107** Wide World Photos; **111** ©Loren Santow/Click Chicago.

**Chapter 5   Page 114** ©Alan Carey/The Image Works, P.O. Box 443, Woodstock, NY 12498. All rights reserved; **119** ©Michael Siluk/EKM-Nepenthe; **129** Don Smetzer ©Click Chicago; **134** Steve Allan ©Peter Arnold, Inc.; **139** Michael Sullivan; **143** ©Robert Brenner/Photoedit.

**Chapter 6   Page 146** ©Bill Leissner/TexaStock; **152** ©Barbara Alper/Stock, Boston; **163** (left) Courtesy of the American Cancer Society; (right) Courtesy of R.J. Reynolds Tobacco Co.; **170** ©Sylvia Johnson 1989/Woodfin Camp & Associates; **177** ©Peter Menzel/Stock, Boston.

**Chapter 7   Page 180** ©Bob Daemmrich/Stock, Boston; **185** ©Van Nostrand/From the National Audubon Society/Photo Researchers, Inc.; **195** Stock, Boston; **198** Courtesy of Dr. Paul Ekman; **207** ©Will Rhyns/Woodfin Camp & Associates.

**Chapter 8   Page 214** ©Sepp Seitz/Woodfin Camp & Associates; **221** ©Rhoda Sidney/Monkmeyer Press Photo Service; **229** Courtesy of Revlon, Inc.; **235** Wide World Photos; **239** ©Ray Young/Photo Genesis.

**Chapter 9   Page 246** ©Ron Cooper/EKM-Nepenthe; **251** ©Sybil Shelton/Monkmeyer Press Photo Service; **252** Palmquist/Lightwave; **256** Richard Younker ©Click Chicago; **266** Union Tribune Publishing Co.

**Chapter 10   Page 274** ©John Coletti/Stock, Boston; **278** Brad Markel/TIME Magazine; **283** ©Oscar Palmquist/Lightwave; **296** ©Jack Prelutsky/Stock, Boston; **299** ©Bob Pacheco/EKM-Nepenthe.

**Chapter 11   Page 308** ©Andrew Gillespie/Click Chicago; **312** ©Michael Kagon/Monkmeyer Press Photo Service; **316** ©Eric Kroll/Taurus Photos; **332** ©Richard Bilick/Photoedit.

**Chapter 12   Page 340** ©Mike L. Wannemacher/Taurus Photos; **345** ©Ellis Herwig/Stock, Boston; **349** Monkmeyer Press Photo Service; **353** ©Peter Vandermark/Stock, Boston.

**Chapter 13   Page 366** ©Donald L. Miller/Monkmeyer Press Photo Service; **375** ©1989 Tony Freeman/Photoedit; **378** Comstock; **385** ©Mark Antman/The Image Works; **387** ©John Lawlor/Click Chicago.

**Chapter 14   Page 396** ©Hugh Rogers/Monkmeyer Press Photo Service; **401** AP/Wide World; **414** ©Phillis Graber Jensen/Stock, Boston; **415** ©R. Eckert/EKM-Nepenthe; **422** ©Bill Leissner/TexaStock.

**Chapter 15   Page 426** William Campbell/TIME Magazine; **430** AP/Wide World; **436** ©Michael Siluk/EKM-Nepenthe; **442** ©Mark Richards/Picture Group; **449** ©Larry Downing/Woodfin Camp & Associates.

**Chapter 16   452** ©Robert Brenner/Photoedit; **460** ©Myrleen Ferguson/Photoedit; **464** ©Frank Siteman/EKM-Nepenthe; **465** Photoedit; **471** ©Peter Freed/Newsweek; **482** ©James Schnepf/Wheeler Pictures.

**Chapter 17   Page 490** ©Ben Lyon/Monterey Peninsula Herald; **495** ©Ruth Block/Monkmeyer Press Photo Service; **502** Picture Group; **507** ©Bill Leissner/TexaStock; ©**516** Hugh Rogers/Monkmeyer Press Photo Service.

**Chapter 18   Page 520** ©Margaret Thompson/The Picture Cube; **524** ©Robert Eckert/Stock, Boston; **535** ©Bohdan Hrynwice/Southern Light; **537** Wide World Photos; **544** Applewhite-AP Photos; **548** ©Bill Gillette/Stock, Boston.

**Chapter 19   552** ©Klaus D. Franke/Peter Arnold, Inc. **557** ©Allsport/Dave Cannon; **559** Wide World Photos; **563** Wide World Photos; **571** ©Barbara Alper/Stock, Boston; **573** Wide World Photos.

## Tables

**3-1** From F. Caplan. *The First Twelve Months of Life.* Copyright © 1973 by Grosset and Dunlap; **3-2** Adapted from Social structure during the school years: Onset of the degrouping

# Author Index

# Subject Index

(Page numbers in *italics* refer to illustrations)